Liquid Salts for Energy and Materials

The University of Nottingham, Ningbo, China

11–13 May 2016

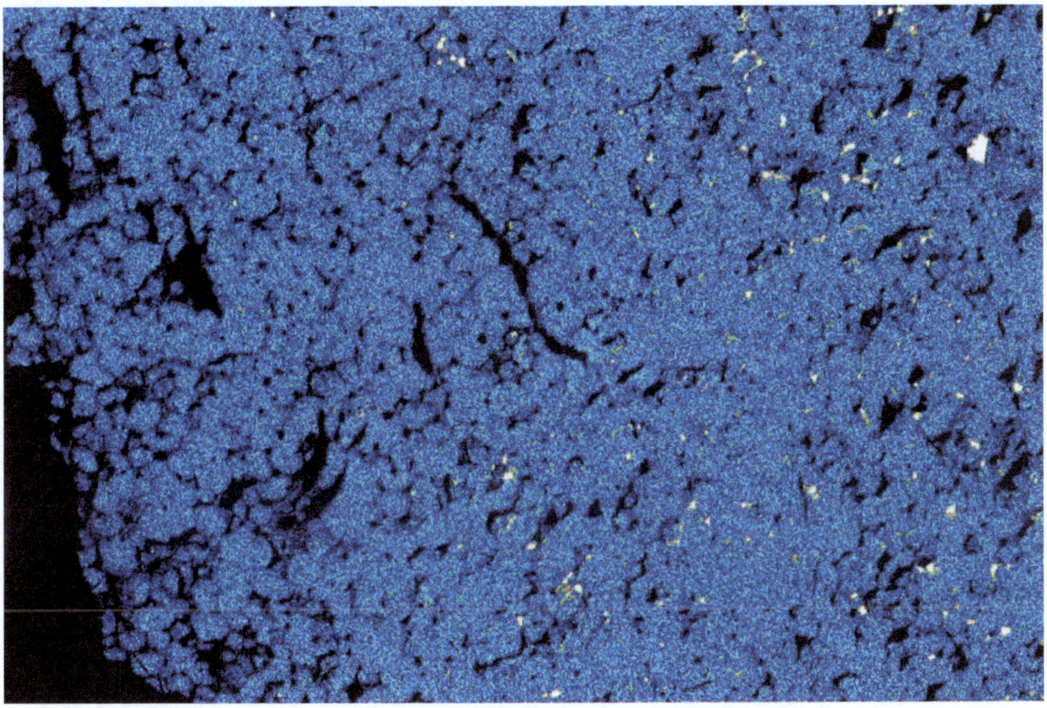

FARADAY DISCUSSIONS

Volume 190, 2016

ROYAL SOCIETY OF CHEMISTRY

The Faraday Division of the Royal Society of Chemistry, previously the Faraday Society, founded in 1903 to promote the study of sciences lying between Chemistry, Physics and Biology.

Faraday Discussions are unique international discussion meetings that focus on rapidly developing areas of chemistry and its interfaces with other scientific disciplines.

Faraday Discussions (Print ISSN 1359-6640, Electronic ISSN 1364-5498) is published 10 times a year by the Royal Society of Chemistry, Thomas Graham House, Science Park, Milton Road, Cambridge, UK CB4 0WF.

Volume 190 ISBN 13: 978-1-78262-766-1

2016 annual subscription price: print+electronic £929 US $1635; electronic only £883, US $1553. Customers in Canada will be subject to a surcharge to cover GST. Customers in the EU subscribing to the electronic version only will be charged VAT. All orders, with cheques made payable to the Royal Society of Chemistry, should be sent to RSC Order Department, Royal Society of Chemistry, Thomas Graham House, Science Park, Milton Road, Cambridge, CB4 0WF, UK.

Tel +44 (0) 1223 432398; E-mail **orders@rsc.org**

If you take an institutional subscription to any RSC journal you are entitled to free, site-wide web access to that journal. You can arrange access via Internet Protocol (IP) address at **www.rsc.org/ip**. Customers should make payments by cheque in sterling payable on a UK clearing bank or in US dollars payable on a US clearing bank.

Printed in the UK

Scientific Committee volume 190

Chair
George Zheng Chen (University of Nottingham, UK)

Anna Croft (University of Nottingham, UK)
Rasmus Fehrmann (Technical University of Denmark, Denmark)
Xianbo Jin (Wuhan University, China)
Stuart Mucklejohn (Ceravision, UK)
John Slattery (University of York, UK)

Liquid Salts for Energy and Materials

Faraday Discussions

www.rsc.org/faraday_d

A General Discussion on Liquid Salts for Energy and Materials was held in Ningbo, China on the 11th, 12th and 13th of May 2016.

CONTENTS

ISSN 1359-6640; ISBN: 978-1-78262-766-1

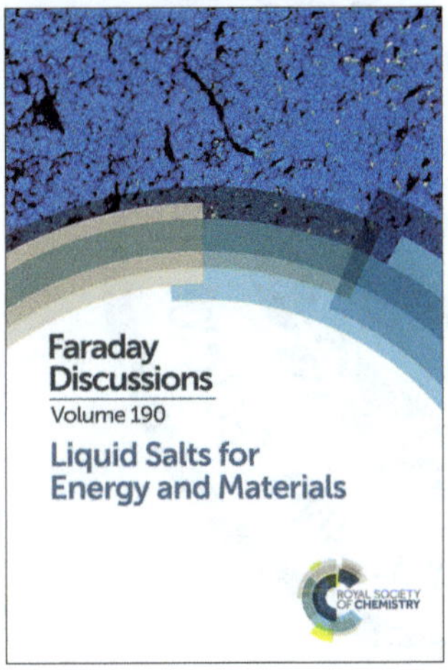

Cover
See McGregor *et al.*, *Faraday Discuss.*, 2016, **190**, 35–52.

Electron probe microanalysis map showing the distribution of niobium in a Nb-doped inert anode of composition $Nb_{0.06}Ti_{0.94}O_2$.

Front cover image courtesy of Kathie McGregor, CSIRO, Australia.

INTRODUCTORY LECTURE

PAPERS AND DISCUSSIONS

SUPPORTING ORGANISATIONS

SPONSOR

CONCLUDING REMARKS

ADDITIONAL INFORMATION

Faraday Discussions

PAPER

Molten salts and energy related materials

Derek Fray

Received 24th April 2016, Accepted 25th May 2016
DOI: 10.1039/c6fd00090h

Molten salts have been known for centuries and have been used for the extraction of aluminium for over one hundred years and as high temperature fluxes in metal processing. This and other molten salt routes have gradually become more energy efficient and less polluting, but there have been few major breakthroughs. This paper will explore some recent innovations that could lead to substantial reductions in the energy consumed in metal production and in carbon dioxide production. Another way that molten salts can contribute to an energy efficient world is by creating better high temperature fuel cells and novel high temperature batteries, or by acting as the medium that can create novel materials that can find applications in high energy batteries and other energy saving devices, such as capacitors. Carbonate melts can be used to absorb carbon dioxide, which can be converted into C, CO and carbon nanoparticles. Molten salts can also be used to create black silicon that can absorb more sunlight over a wider range of wavelengths. Overall, there are many opportunities to explore for molten salts to play in an efficient, low carbon world.

1. Introduction

The high ionic conductivities and diffusivities, high thermal and chemical stability coupled with high mutual solubilities have not changed since molten salts were first used by Davy in 1808,[1] so, what is new? There have been many developments in specialised equipment[2–4] which has enabled detailed knowledge of the speciation in melts to be elucidated.[5,6] However, most recent developments have been in creating new applications. There are three ways in which molten salts can contribute to a reduction in energy costs and use and these are improving existing processes for the extraction of energy intensive metals such as aluminium, lithium and titanium, developing completely new methods for the production of metals and, lastly, creating novel materials which may have a profound effect on the generation and storage of energy. Most existing methods for the electrochemical production of metals, using molten salts, have undergone incremental improvements which have gradually increased current and energy efficiencies such as in the Hall–Heroult process for the production of aluminium[7]

Department of Materials Science and Metallurgy, University of Cambridge, Cambridge CB3 0FS, UK. E-mail: djf25@cam.ac.uk and the electrolysis of lithium chloride for the production of lithium[8] which has increased from a few thousands of tonnes in 1960 to 35 000 tonnes today, without any major breakthroughs in the technology. However, new routes have been investigated which involve radically different technologies, such as electro-deoxidation,[9] reduction coupled with electrorefining,[10] molten salt centrifugal electrolysis,[11] the use of solid oxide membranes[12] and inert anodes[13] which will be discussed as these may lead to less polluting and more energy efficient processes.

In the fields of energy storage and carbon dioxide capture significant progress has been made using molten salt batteries,[14] combination of molten salts and solid electrolytes[15] and melts based on alkali oxides dissolved in molten salts.[16]

Many exciting new materials have been discovered such as carbon nanotubes, graphene and other novel nanoparticles which, conventionally, have been prepared by chemical vapour deposition which has an extremely poor space time yield.[17] Fortunately, many of these materials can be produced in the condensed phase using molten salts at a much higher rate and of higher quality.[17] Examples will be given for the production of carbon nanoparticles,[18] nanodiamonds[19] and graphene.[20] These materials have many promising applications in electronics due to the enhanced electrical and thermal properties and, also, in photovoltaics. Examples will be given for the use of these materials in lithium ion batteries[21] to significantly increase the capacity, converting the surface of PV silicon discs to black silicon[22] which will decrease the amount of light that is reflected as well as increasing the spread of wavelengths that can be absorbed. So far many of these studies have been at the laboratory scale but several are now being developed at a much larger scale.

2. Possible improvements in metal extraction

2.1 Aluminium

Molten salts have been used in the metallurgical industry for the more electropositive elements for over one hundred years and the changes in the energy consumption in the aluminium industry have been profound, with the current density increasing and the cell voltage falling with the overall effect of reducing the energy consumption from 30 kW h kg^{-1} to about 13 kW h kg^{-1}.[23] However, the original design of the cell has not changed significantly, it has just been made much more efficient with the cathode still being a molten pool on the bottom of the cell and with the consumable carbon anodes positioned above the cathode in a molten bath of cryolite (Na_3AlF_6) containing dissolved alumina.[7] The molten aluminium cathode sits on a graphite/carbon bed which does not wet carbon. As a result, the pool has to be quite deep in order to ensure complete coverage, otherwise the graphite/carbon bed will be exposed to the electrolyte and sodium will be deposited in preference to aluminium.[7] The very high current densities that are applied, together with the intense magnetic fields that are generated, make the aluminium pool unstable. Droplets of aluminium can be swept off into the electrolyte and reoxidised at the anode, thereby, reducing the current efficiency.[24] It would be vastly preferable if a wetted cathode could be used so that instead of an unstable pool, the molten aluminium could be restrained as a thin film which could be drained off.[25,26] Titanium diboride is wetted by aluminium but, unfortunately, no one has devised a way of using titanium boride plates in the Hall–Heroult cell which prevents the plates from moving under the operating conditions in the cell.[27] The other possible dramatic improvement is to use an inert anode which would allow oxygen to be liberated rather than the discharged oxygen ions reacting with the carbon anodes to form carbon dioxide and carbon monoxide. This leads to both a reduction in the consumption of the anodes and in the amount carbon dioxide released to the environment. The ideal arrangement would be inert anodes, with a wetted cathode, which would create a cell that was stable with a small anode to cathode distance creating aluminium and oxygen using less energy.[7] Although inert anodes have not been successfully developed for the aluminium industry, other approaches, discussed later in this paper, have been more successful.

2.2 Lithium

The demand for lithium is increasing as the use of lithium ion batteries continues to expand from mobile electronic devices to cars and, even, the domestic storage of electricity created by solar panels. The technology of the cell for the extraction of lithium has not changed for decades – essentially a steel cathode which does not react with molten lithium, a lithium chloride based electrolyte and a carbon anode which does not react with the evolved chlorine.[8] The main difficulty arises from the fact that both lithium and chlorine are less dense than the electrolyte and need to be separated by a barrier which means there is a large inter-electrode space which leads to a high IR drop, of about 4 V, in the electrolyte which is more than the decomposition potential of about 3 V. This can be compared to the IR drop in the Hall–Heroult cell for aluminium of about 1.5 V. One way of reducing the anode to cathode spacing, without increasing the back reaction between the lithium and chlorine, is to apply a centrifugal force to encourage the separation of the two phases.[28,29] This can be achieved by spinning the electrodes at modest velocities as the lithium wets the steel cathode and is expelled tangentially and horizontally whilst the chlorine collects on the carbon anode and leaves from the centre of the cell (Fig. 1). At 40 rpm, using this approach, it was shown possible to reduce the anode to cathode spacing to about 10 mm from about 100 mm with a reduction in energy consumption to 22 kW h kg^{-1} compared to the value of 40 kW h kg^{-1} found in conventional cells.[11,29]

Although this would be a welcome reduction in the energy consumption of the production of lithium, unfortunately, it does little to affect the price of lithium ion batteries as the cost of lithium is only 3% of the total cost of the battery. However, this approach has also been considered for the electrowinning of magnesium which also suffers the same problem as both anodic and cathodic products are less dense than the electrolyte.[30]

2.3 Titanium

The vast majority of titanium is produced by the Kroll process[31] where relatively pure rutile is carbochlorinated, in a fluidised bed, to form titanium tetrachloride which is purified by fractional distillation.

$$TiO_2 + 2Cl_2 + 2C = TiCl_4 + 2CO \tag{1}$$

The titanium tetrachloride is then reduced by molten magnesium in a Kroll reactor to form titanium

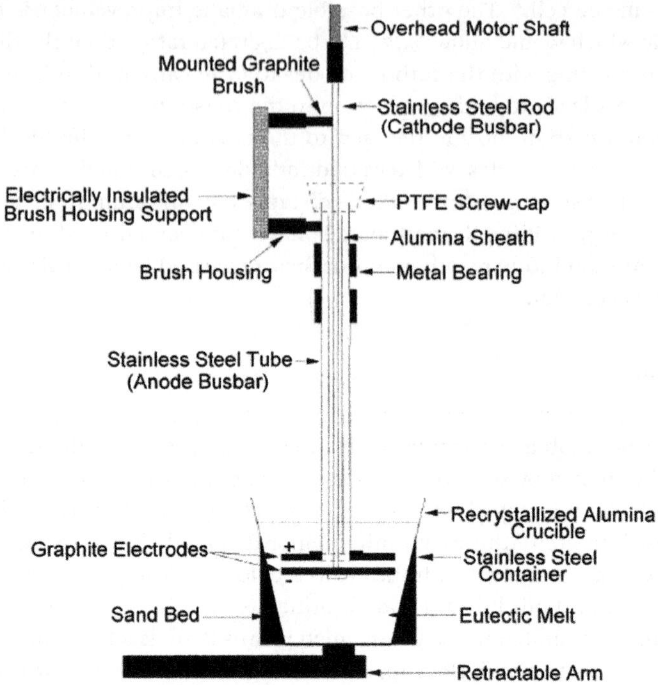

Fig. 1 Experimental arrangement for centrifugal electrolysis.[28]

$$TiCl_4 + 2Mg = Ti + 2MgCl_2 \tag{2}$$

and magnesium chloride which is normally recycled within the plant to give chlorine for carbo-chlorination and magnesium for the reduction of the titanium tetrachloride. The size of a Kroll reactor is about 10 tonnes and the time taken for the whole process from feed to product is about 21 days.[8] It is also an expensive, labour intensive process which contributes to the high cost of the titanium product compared to other metals, so titanium is usually only used where there is no alternative. This is unfortunate as titanium alloys are strong, corrosion and oxidation resistant and light which means the number of possible applications is virtually endless.

In the mid-nineteen nineties, it was discovered that it was possible to reduce TiO_2[9] and many other metal oxides including iron oxide,[32] tantalum oxide,[33] zirconium oxide,[34] niobium oxide,[35] chromium oxide[36] and silica[37] by simply making the oxide cathodic in a bath of a molten chloride of a strongly electropositive element, such as calcium or lithium, containing about 1 mol% of the respective oxide in order to maintain a concentration of oxygen ions in the melt. This electro-deoxidation process is known as the FFC Cambridge process and is shown in Fig. 2. As well as pellets of the oxide, powders[38] and fluidised particles have been used.[39] The overall cathodic reaction for titanium dioxide is

$$TiO_2 + 4e^- = Ti + 2O^{2-} \tag{3}$$

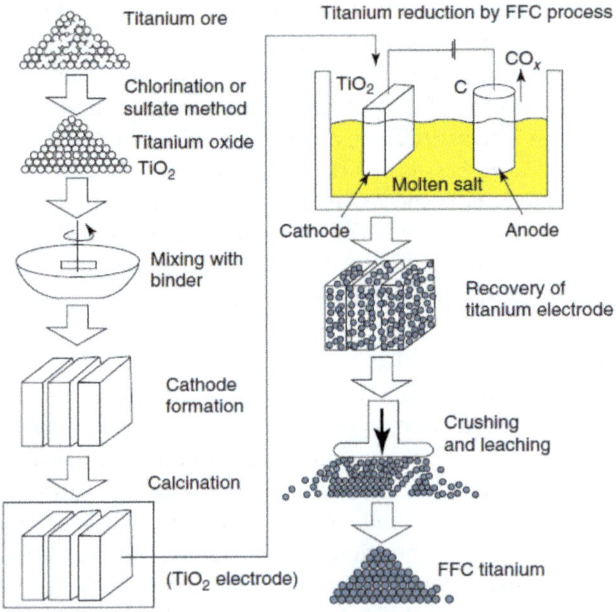

Fig. 2 Flowchart for FFC Cambridge process.

But, as was found out later, the reaction path proceeds by the formation of titanates and reduced titanium oxides, further complicated by many changes in the microstructure.[40]

The oxygen ions migrated to the carbon anode where a mixture of carbon dioxide and carbon monoxide formed, a similar reaction to that observed in the Hall–Heroult cell for the extraction of aluminium.

$$2O^{2-} + C = CO_2 + 4e^- \tag{4}$$

However, unlike cryolite which has a low solubility for carbon dioxide, the carbon dioxide can combine with the O^{2-} ions in the melt to form carbonate ions.[41]

$$CO_2 + O^{2-} = CO_3{}^{2-} \tag{5}$$

The carbonate ions can diffuse back to the cathode to form nanosized carbon particles, perhaps contaminating the metal product, and liberating oxygen ions which can migrate back to the anode continuing the reaction, reducing the current efficiency and increasing the energy consumption. Although this reaction is a nuisance in the cell, others have used these reactions to capture carbon dioxide and to then to electrochemically liberate oxygen and C as nanoparticles which can be used in lithium ion batteries,[16,42] as reported later in this paper. These effects can be overcome by using an inert anode which is fully discussed in the section entitled inert anodes.

$$CO_3{}^{2-} + 4e^- = C + 3O^{2-} \tag{6}$$

Another disadvantage of the FFC Cambridge process is that, depending on the potential at the cathode, an activity of calcium is generated in the electrolyte which raises the electronic conductivity in the melt and lowers the overall current efficiency.[40] Careful control of the potential is therefore required to maximise the current efficiency. As well as reducing oxides, sulphides can also be reduced.[43,44] A further property of the process is that there is no refining of the cationic species in the oxide as only the anodic species are removed. This can be a disadvantage if refining is required but can be a considerable advantage if alloys are desired especially those that are difficult to make when the melting point of one element is greater than the boiling point of some of the other alloying elements. A very large number of alloys and intermetallic compounds have been made including TiNi,[45–47] TiMo,[48] TiW,[49] Al_3Sc,[50] NbAl,[51] NiTiHf,[52] TaNb,[53] NbHf,[54] HfC,[55] Ti_2CO,[56] TiSi,[57] LaNiCo,[58] TaC[59] and TiC/SiC composite.[60]

At the present time, the process is being developed by Metalysis Ltd (http://www.metalysis.com) in the UK concentrating on tantalum powder for capacitors and titanium powders for near net shape processing. A further benefit of the process is that it is able to reduce oxide powders, which are readily available, to pure metal or alloy powders without going through liquid or solid phase transformations, which in the case of alloys, usually causes segregation. As is well known, there are additional benefits to be able to perform 3D printing and make near net shape components which, for the aircraft industry, greatly increases the fly to buy ratio that is extremely low for titanium components. However, this is a long term vision as it takes a long time to get regulatory approval for materials from new processes to be used in critical parts on aircraft although the automotive industry is likely to be much more receptive to these innovations.

Okabe, Hamanaka and Taninouchi, as reported in this discussion, have used electrochemical deoxidation for recycling titanium using molten magnesium chloride.[61]

In order to overcome the problem of the lack of refining of the cations in the FFC Cambridge process, another method was discovered that could perform reduction and refining at the same time. High grade titanium ores are very expensive whilst those containing about 20% iron are cheaper by an order of magnitude. The Chinuka process,[10] shown in Fig. 3, uses an oxycarbide anode (Ti_2CO) and an electrolyte of molten NaCl–KCl which has negligible solubility for oxygen species and is a variation on the approach adopted by Wainer,[62] Withers[63–66] Zhu and Jiao,[67–69] Kjos, Haarberg and Martinez.[70–72] The basic anodic reaction is the ionisation of the metallic elements and the order in which the elements ionise depends upon the ionisation potential and in a typical low grade titanium ore it would be the following elements: calcium, magnesium and titanium are likely to end up as ions in the salt whilst the others will remain as a sludge beneath the anode. Looking at the cathode, there will also be potential sequence that will encourage titanium to deposit first whilst the other elements will remain in the electrolyte. A typical current efficiency is around 70% with an energy consumption of about 7 $kW\,h\,kg^{-1}$ of titanium. Fig. 3 shows the flowchart of the Chinuka process.

The results, shown in Fig. 4, were obtained demonstrating that from a relatively impure ore, a high purity metal is obtained. As well as titanium, the process has successfully been applied to niobium and there is considerable interest from

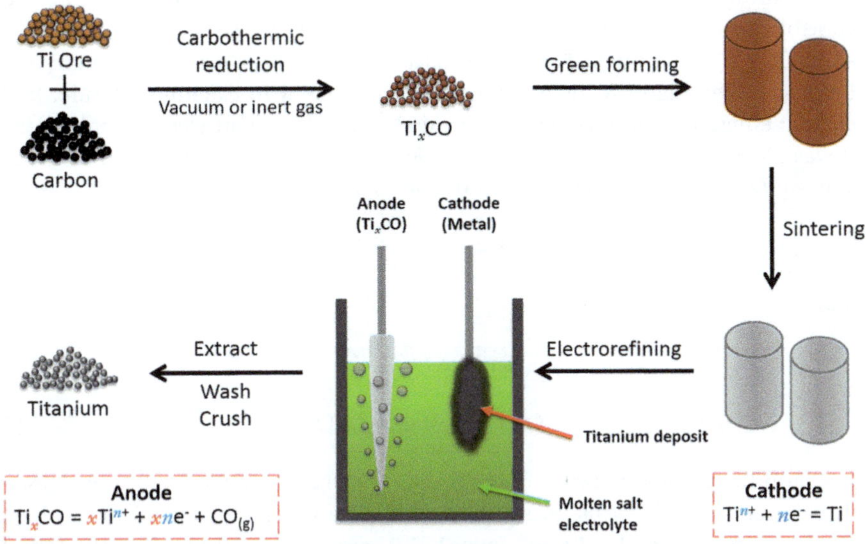

Fig. 3 Flowchart for Chinuka process.

the nuclear industry where to carry out reduction and refining of used oxide fuel rods in a single step is a considerable advantage over the existing processes where at least two steps are required.

2.4 Uranium

In the nuclear industry, there is much interest in reducing used metal oxide fuels with the amount needed to be recycled increasing to 100 000 tonne by the end of this century.[73,74] The conventional method of treating this material is to reduce it with lithium metal to form a mixture of uranium and transuranic elements and lithium oxide. The former are electrorefined and the lithium oxide sent for disposal. Electro-deoxidation has significant advantages as handling of the chemically active lithium and the disposal of radioactive lithium oxide is avoided.

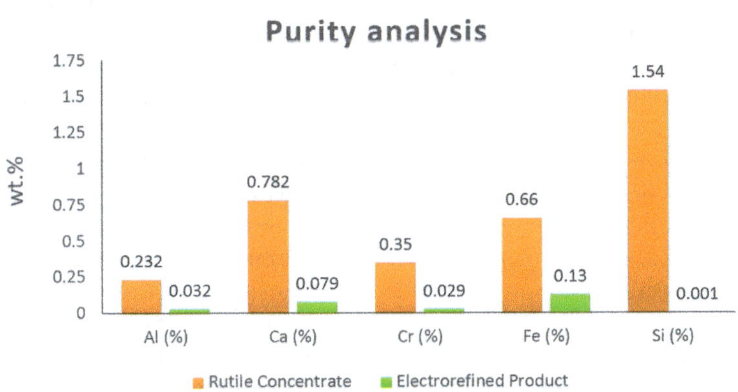

Fig. 4 Concentration of impurities before and after electroreduction and refining.

Other advantages of this approach over the existing methods is the ability to control the oxygen ion content of the melt and the overall process is much simpler.[75–77] In the nuclear industry, LiCl has preference over calcium chloride because of its relatively low melting point compared to calcium chloride, high current efficiency and its compatibility with the subsequent electrorefining step. It was concluded that U_3O_8 was electro-deoxidised to $LiUO_3$, U_4O_9 and UO_2 and, subsequently, the UO_2 was reduced by the following reactions.[78–80]

$$UO_2 + 4e^- = U + 2O^{2-} \tag{7}$$

and

$$Li^+ + e^- = Li \tag{8}$$

$$UO_2 + 2Li = U + Li_2O \tag{9}$$

Platinum anodes are used although Li_2O has to be added to prevent the anodic dissolution of the inert platinum anode. This is an international effort involving KAEIRI (South Korea), CREIPI (Japan), INL and ANL (US) and IGCAR (India). Based on these fundamental studies, work now centres on the development of an engineering sized cell encompassing both electro-deoxidation and electrorefining processes. At KAERI, a 10 tonne per year capacity facilitator is being operated to help solve existing technology issues.[73]

In this Faraday discussion, Andrew Mount describes the development of microelectrodes for sensing elements in the reprocessing of pyrochemical spent nuclear fuel.[81]

3. Inert anodes

In many high temperature molten salt processes the carbon anodes are consumed by the anodic evolution of oxygen to form carbon dioxide, a greenhouse gas, and obviously it would be preferable if an inert anode could be used.

The search for inert anode materials in the aluminium industry goes back more than 100 years[82] in an attempt to have dimensionally stable anodes to make the operation of the cell easier and, more recently, to curtail the evolution of carbon dioxide. The ideal properties that are required are: resistant to attack by electrolyte and oxygen, an electronic conductor, mechanically strong and resistant to thermal shock.[13,83] Unfortunately all materials fail in one or other of the properties. Metals are likely to oxidise, oxides are likely to dissolve in the electrolyte and their fabrication into large blocks may be difficult. There has been much interest in SnO_2 doped with Sb_2O_3 to increase the conductivity and CuO to aid sintering[84,85] but it always dissolves, to a certain extent, in the electrolyte, and the tin and copper being less electropositive than aluminium will be deposited first, contaminating the aluminium. Cermets, mixtures of metals and ceramics, were thought to be the answer with the concept that the best properties of both constituents would be the result but, unfortunately, it was the worst properties that frequently pertained. Alcoa had some success with a cermet consisting of nickel ferrite with nickel oxide and copper oxide.[86] One way to reduce the corrosion rate is to reduce the temperature but, at the same time, this reduces the solubility of alumina so the cells operate with a suspension of alumina.[87] Another approach is to add cerium fluoride to the cryolite based electrolyte which leads to the formation cerium oxyfluoride on the anode, slowing down the attack on the tin oxide anode.[88] Overall, there are several 100 patents in the literature describing inert anodes in the Hall–Heroult cell but not one is totally satisfactory.

In this Faraday Discussion, Geir Martin Haarberg discusses his group's interesting work on the use of methane to reduce both the anodic potential on the anode and also the rate of consumption, thereby reducing carbon dioxide emissions.[89]

In the FFC Cambridge process it would also be an advantage to have an inert anode, as mentioned earlier, the carbon dioxide that is created from a carbon anode can dissolve in the salt as carbonate ions with the oxygen ions which diffuse to the cathode and discharge to form carbon nanoparticles and oxygen ions. Tin oxide anodes are used in the glass industry in calcium oxide containing melts at high temperature so it was assumed that they would function satisfactorily in calcium chloride melts. However, although tin oxide electrodes were initially satisfactorily, after about 48 hours an insulating layer of calcium stannate formed.[90] One solution to this problem was found in low temperature solid state physics where it had been reported that a solid solution of calcium titanate and calcium ruthenate is electronically conducting.[91,92] According to thermodynamics, the material should be stable in both oxygen and chlorine containing environments and this was found to be the case,[93] as shown in Fig. 5. The successful development of an inert anode could lead to the use of electro-deoxidation being used to generate pure oxygen and useful materials for use in space.[94]

Other groups have reported the use of calcium titanate that has been doped with metallic elements and this is described by Kathie McGregor in this Faraday discussion.[95]

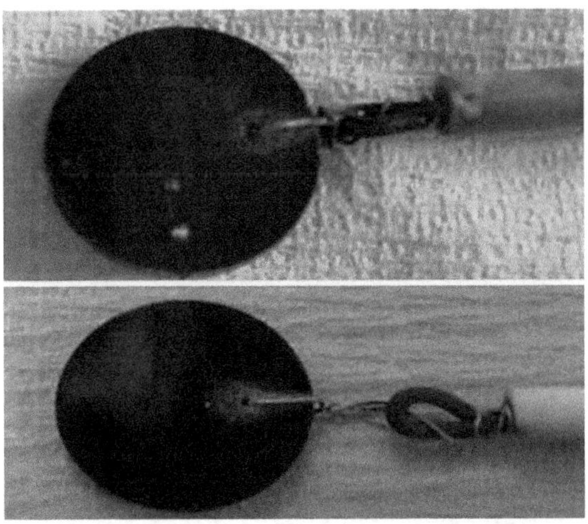

Fig. 5 Photograph of $Ca_{(1-x)}Ru_xO_3$ anodes after electrolysis $x = 0.28$ (top) and $x = 019$ (bottom).[11]

George Chen's group adopted a different approach using an electrolyte containing magnesium chloride. The overall reaction is

$$MO_x + MgCl_2 = M + xMgO + xCl_2 \quad (10)$$

This very interesting approach is somewhat more complicated than the standard electro-deoxidation reaction as $MgCl_2$ is consumed and MgO and chlorine are produced. Its take up may depend upon the cost of $MgCl_2$ and treatment of the by-products. This work is presented in this Faraday Discussion.[168]

In the US, a process for the direct electrowinning of iron from iron oxide dissolved in a molten flux of CaO, MgO, Al_2O_3 and Fe_3O_4 to produce pure iron free of carbon is being explored by Sadoway's group.[96,97] This can be compared with the conventional process where coke is used to reduce the iron oxide in a blast furnace to produce iron, CO and CO_2. The metallic product is saturated with carbon which has to be removed by oxidation to create steel and even more gaseous carbon products with the result that the steel industry generates about 10% of CO_2 that ends up in the environment. The inert oxygen evolving anode that is being used is Cr–Fe which forms a stable protective layer in use. Originally, an iridium anode was used but this would be too expensive for commercial use.[97] The applied voltage is 3.8 V which delivers a current density of 2 A cm^{-2} with a current efficiency of 34%.[97] This is being commercialised by the Boston Electrometallurgical Corporation.

4. Solid oxide membranes

A different approach, shown in Fig. 6, is also being developed in the US where an inert anode of molten silver, contained in a solid oxide membrane (SOM), is being used above the temperature of 800 °C where silver oxide is unstable. The overall reaction for the production of metals from their oxides which can be represented by

$$MeO_x = Me + x/2O_2(g) \quad (11)$$

has been successfully applied to magnesium, aluminium, titanium, tantalum, ytterbium and silicon oxides.[98–105] It can also produce alloys and intermetallics such as TiFe alloys, Ti–Si intermetallics and CeNi.[106–109] The metal oxide or oxides

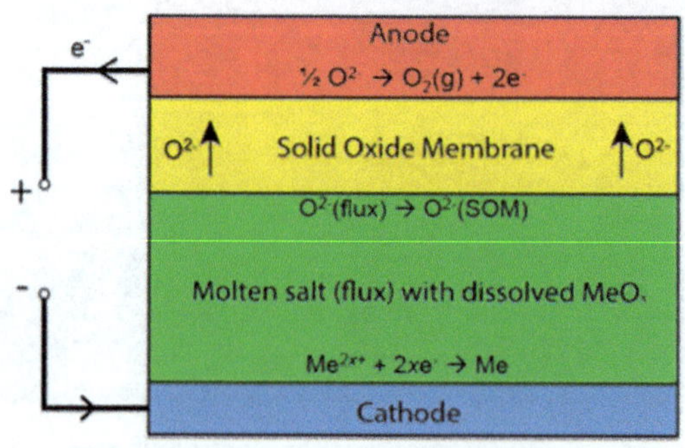

Fig. 6 Schematic of SOM cell.[104]

are dissolved in a molten salt and the cathode is separated from an inert anode by a solid oxide membrane as shown in Fig. 6. The metals are deposited on the cathode and the oxygen ions migrate through the solid oxide membrane and are discharged on the inert silver anode. The molten salt electrolyte is magnesium fluoride based and the SOM is yttria stabilised zirconia. In some case the metal is deposited on the cathode above the boiling point and is removed in a flow of argon and condensed. The main problem initially was the attack of the SOM by the fluoride salt but it is claimed that this has been overcome by adding yttria to the electrolyte.[110] If methane is fed to the liquid anode this can react in a fuel cell mode to further reduce the energy consumption of the cell. The process is being commercialised in Massachusetts by Infinium llc (http://www.infiniumc.com) and it is hoped to be able to produce rare earth metals in the near future.

A further interesting variant would be to combine the FFC cathode with the molten silver anode and this approach is described by Qian Xu for the production of metallic carbides in this Faraday Discussion.[111]

5. Carbonate melts

It has been known for more than fifty years that electrolysis of sodium and potassium carbonate melts[112–114] gives the alkali metals as the cathodic product whilst for lithium carbonate, the cathodic reaction was

$$CO_3^{2-} + 4e^- = C + 3O^{2-} \tag{12}$$

as the potential for this reaction was less than that for the liberation of lithium.

$$Li^+ + e^- = Li \tag{13}$$

Lubomirsky and co-workers[115,116] found that carbon dioxide could be converted to carbon monoxide and oxygen by electrolysis using titanium and carbon electrodes using a lithium carbonate based electrolyte through the reaction

$$CO_2(g) = CO(g) + 1/2O_2 \tag{14}$$

at around 900 °C. Below, this temperature nanosized carbon particles were the product.

$$CO_2(g) = C + O_2(g) \tag{15}$$

Both the groups of Chen[16] and Sadoway[117] found that the product was carbon nanoparticles which could find application as the anode in a lithium ion battery. Wang[118,119] and Jiao[120] also discuss the preparation of active carbons in this Faraday Discussion. Licht[121–123] obtained a similar product from CO_2 dissolved in Li_2O–Li_2O melts and the generation of carbon and hydrogen from a hydroxide/carbonate electrolyte.

Ito's group are also doing interesting work on the formation of carbon films by either the anodic oxidation of carbide ions (C_2^{2-}) or the cathodic reduction of CO_3^{2-} ions in a LiCl–KCl–$CaCl_2$ plus either CaC_2 or $CaCO_3$. The group are also working on the development of industrial processes such as the creation of ammonia where nitrogen is reduced to the nitride ion and oxygen is oxidised to oxygen ions which then combine to form ammonia. The key reaction in the molten salt is

$$N^{3-} + 3/2H_2O = NH_3 + 3/2O^{2-} \quad (16)$$

Plasma-induced discharge electrolysis is also being used to create nanoparticles of many elements and intermetallic compounds with a design that could create 7 kg day^{-1} of product.

Further details are given this Faraday Discussion by Yasuhiko Ito.[124]

Very innovative work has been done recently by Banerjee[125] who was able to use molten caesium carbonate to activate organic substrates to react with CO_2. The major finding is that CO_2 can be reacted with 2-furancarboxylate (FC) to form furan – 2,5-dicarboxylicacid (FCDA) which can be derived from biomass waste materials such as maize or sawdust. FCDA is one of the monomers used to generate polyethylene furancarboxylate (PEF) which can compete with polyethylene terephthalate (PET) which is obtained from petrochemicals. In this innovative process, the caesium carbonate can be recycled and the products can easily be extracted from the salt. Furthermore, benzene can be reacted with CO_2 to form benzoic acid which can be changed to terephthalic acid, a precursor to PET.

6. Molten salt fuel cells and batteries

6.1 Molten salt carbonate fuel cells

Molten carbonate fuel cells are regarded as being clean, with an electrical efficiency of 55%, and are suitable for stationary power generation with more than 50 installations around the world producing over 300 MW of clean electric power with a 100 MW plant being commissioned in South Korea. Overall, it is the most successful major fuel cell with more installed plants than any other fuel cell system.[126] The electrolyte in the fuel cell is a normally a eutectic mixture of potassium and lithium carbonates, operating at 650 °C. The anode is Ni, alloyed with chromium or aluminium, and the cathode is lithiated nickel oxide.

At the anode the fuel, which may be hydrogen or methane, reduces the carbonate ion to carbon dioxide, releasing electrons and generating electrical power. On the cathode side, new carbonate ions are formed by combining the carbon dioxide from the anode or from another source with oxygen from the air. A recent innovation is that the overall fuel cell can be configured to capture carbon dioxide and regenerate it in a more concentrated form.[126]

The cells operate for up to 40 000 hours with the electrolyte performing well; the main problem is the degradation of the cell components and not the electrolyte.[126]

Irvine and his group have also utilised a hybrid of solid electrolytes and molten salt fuel cells in order to maximise the properties of both phases.[15,127–130] A carbon based anode is placed in a molten carbonate electrolyte which is in contact with a stabilised zirconia oxygen ion conducting electrolyte. At the cathode, the oxygen ionises to form oxygen ions which pass through the solid oxide membrane to the carbonate melt where they react with a carbon based cathode. The great advantage of this approach is that the considerable problems of interfacing two solids together are avoided. Also, compared to solid state carbon fuel cells, there is no need to recirculate the gases. Power densities of around 300 mW cm^{-2} were obtained generating a potential difference of about 1 V. It was shown possible to upscale the arrangement to multi cells with the possibility of using domestic solid oxide waste as the fuel. This work is presented in this Faraday Discussion.[130]

There are also several other interesting batteries being developed in the US, based upon molten salts. Sadoway is working on a novel concept of a molten salt battery where one electrode is a reactive molten metal which floats on a layer of salt which in turn floats on another metal that can react with the upper metal,[131] shown in Fig. 7. The overall structure is very similar to the Hoopes cell that is used to electrorefine aluminium.[132] Many metallic systems have been investigated including Li[14,133] Mg,[135] Na[136] and Ca[134,137] as the reactive elements and many low temperature metal such as Sn, Sb, Pb and Sb as the cathode.[131] These batteries are primarily intended for load levelling and will be stationary so that the density of the molten electrodes is not of concern. However, it is worth noting that the potential developed by these batteries is modest compared to lithium-ion batteries and in order to get a constant voltage the cathode needs to be in a two phase region. However, as all the active components, anode, electrolyte and cathode are liquid there are no concerns about dendrites and interfaces suffering stress. It is being developed by AMBRI (http://www.ambri.com) in Massachusetts with extensive trials on Cape Cod and Hawaii.

6.2 Molten air batteries

Stuart Licht's group is also developing molten air batteries[138–140] in molten carbonates with Li_2O with Fe going to Fe^{3+} the overall reaction is

$$3/4O_2 + 3e^- = 3/2O^{2-} \tag{17}$$

which was made up of

$$FeO_2{}^- = 1/2Fe_2O_3 + 1/2O^{2-} \tag{18}$$

$$Fe + 2O^{2-} = FeO_2{}^- + 3e^- \tag{19}$$

$$FeO^{2-} = 1/2Fe_2O_3 + 1/2O^{2-} \tag{20}$$

The volumetric capacity was 10 000 W h l^{-1} which can be compared with 6200 W h l^{-1} for lithium ion batteries.

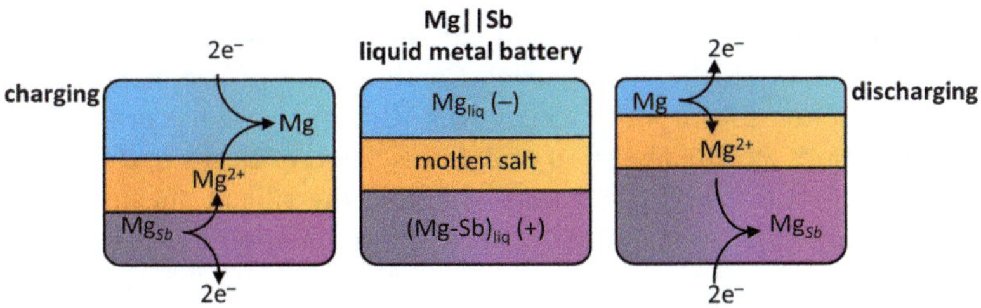

Fig. 7 Diagram showing operation of Mg–Sb molten salt battery.[133]

7. Novel energy related materials produced using molten salts

7.1 Carbon nanotubes

Lithium ion batteries have transformed portable electronics but it is still frustrating that the batteries need to be charged on a daily basis. The only way this situation can change is to increase the capacity of the batteries. Most lithium ion batteries use a graphite anode in which lithium is able to intercalate into the structure to form LiC_6. Lithium can also intercalate with a metal oxide cathode at a much lower lithium activity than the anode. When the charge is inserted into the battery, the lithium moves from the cathode to the anode and most batteries can undergo several thousand charge–discharge cycles without degradation of the battery. However, in order to increase the capacity of the anode, it is necessary to replace the graphite with something that has a significantly higher capacity for lithium. There are several metals that can form intermetallic compounds with lithium and can replace the graphite anodes. However, this higher performance comes at a cost, and this is the enormous volume changes, up to 400%, that occur on the formation of the lithium intermetallic compounds and their subsequent discharge in the battery which causes the intermetallic compound to break up into finer and finer particles which eventually lose electrical contact, greatly increasing the resistance of the anode and its eventual failure – this can occur after a few tens of cycles.[141] What is needed is a method of containing the alloy so it can expand and contract due to lithium insertion and de-insertion without losing electrical contact. This can be achieved by forming tin and other metal filled carbon nanotubes by inserting lithium from a lithium chloride–tin chloride melt at around 800 °C.[142,143] What is assumed to happen is that the lithium ions discharge on the graphite electrode, and then intercalate into the structure which causes sheets of graphite to be extruded into the melt, where to reduce the surface area exposed to the melt roll up into a scroll. If the melt contains tin ions, these ions will be preferentially discharged on the surface of the graphite to form small droplets which can be encapsulated into by carbon nanosheets to form tin filled carbon nanotubes.[18,144–146] In the lithium ion battery, the lithium can pass around the scroll to reach the tin, where it alloys, and expands the tin which causes the carbon scroll to partially unroll.[147] On discharge, the tin shrinks as the lithium is removed but still remains in contact with the conducting carbon nanotubes, thereby overcoming the normal failure route of lithium tin anodes.[147]

Silicon has a much higher capacity than tin but the melting point of silicon is higher than the temperature of the electrolyte that is used to create the carbon nanotubes, so it is harder to fill carbon nanotubes with silicon. An alternative approach is to mix silicon nanoparticles with sheets of graphene (see next section) in a molten salt or water which cause the graphene to encapsulate the silicon particles.[148] On the insertion and de-insertion of lithium, the graphene sheets simply slide over each other ensuring that the silicon does not decrepitate and continues to be in intimate electrical contact with the other coated particles.[149]

7.2 Graphene

Graphene was discovered about two decades ago[150] with single layer graphene being isolated about one decade ago using sticky tape to remove one layer at a time from a graphite block.[151] Although its remarkable properties have been investigated in many laboratories the number of commercial applications are limited. The reason for this is that high quality graphene cannot be made easily and reproducibly and, furthermore, there are no accepted standards for graphene which can consist of a single layer to multilayers with everything in between. It is also very difficult to scale up most methods of production which can only make about 1 mg cm^{-2} h^{-1}.

Based on the experience of synthesising carbon nanotubes and nanoparticles by the intercalation of lithium atoms which have a diameter that is close to the interlamellar spacing in graphite, it was thought it would be interesting to attempt to insert hydrogen atoms into graphite. This can be achieved by either exposing the LiCl melt to humid argon or, if lithium oxide is in the melt, to a hydrogen containing environment. The net result is the introduction of water into the salt which when electrolysed produces hydrogen atoms, as the decomposition of water is less than that of LiCl, which, being much smaller than the interlamellar spacing can easily intercalate to produce one or two layers of graphene, as shown in Fig. 8.[152–154] From the Raman spectrogram it was also apparent that the graphene is of very high quality. The rate of production of graphene from graphite is around 200 mg cm^{-2} h^{-1} or 50 kg m^{-2} day^{-1} which is much faster than all other methods. In our laboratory we have explored two applications, making a silicon graphene anode for lithium ion batteries,[149] mentioned in the previous section, and strengthening alumina, presented by Ali Kamali in this Faraday

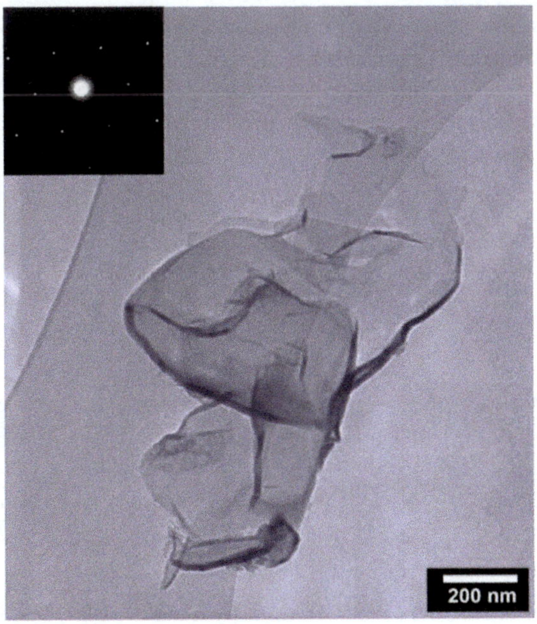

Fig. 8 TEM of graphene.[154]

Discussion.[155] In the case of the anode in the lithium ion battery, the lithium ions are discharged on the surface of the graphene and then diffuse between the graphene sheets to reach the silicon with the reverse occurring on discharge. The fracture toughness of the alumina was increased from 4.5 MPa $m^{1/2}$ to 6.98 MPa $m^{1/2}$, due to the graphene acting as a barrier to crack growth.[155] If carbon nanotubes are combined with graphene sheets, a composite structure is formed which results in very high capacity supercapacitor electrodes.[156]

7.3 Nanodiamonds

A great deal of energy is expended in drilling for oil and minerals using nanodiamonds that are created by attrition grinding from larger diamonds, made at very high temperatures and pressures, to form diamonds with many facets which bare no relation to the conventional shape of jewellery diamonds. It had been predicted that sp^3 diamond nucleation from sp^2 diamond is possible inside carbon nanotubes or nanoparticles due to surface tension effects[157] and there have been many efforts to convert carbon nanotubes to diamonds using laser radiation, plasma sintering, and shock waves, but success was only achieved at high temperatures and pressures in the presence of catalysts.[158–160] Alkali carbonates can also be used although at much higher temperatures and pressures.[161] Rather surprising, it was found that by taking the product from lithium intercalation into graphite and oxidising it in air at around 500 °C, the structure transformed to nanodiamonds of near perfect orthorhombic structure,[19] as shown in Fig. 9. The formation of nanodiamonds was proved by the presence of the diffraction pattern of the diamond 111 plane and the Raman spectra of the material which, again, showed the presence of diamonds. Due to the sharpness of the crystals, they are likely to have a much better grinding performance than conventional nanodiamonds and this may also apply to other applications of nanodiamonds, such as medical applications. The reason for this transformation was perhaps due to the presence of nano-sized lithium carbonate in the carbon nanoparticles, which could act as a catalyst, created from the carbon dioxide, produced at the anode, reacting with the lithium oxide in the melt which has already been discussed in this paper.

$$2O^{2-} + C = CO_2 + 4e^- \quad (21)$$

$$CO_2 + Li_2O = Li_2CO_3 \quad (22)$$

Fig. 9 SEM of nanodiamonds.[19]

7.4 Black silicon

At present, silicon is the preferred material to make solar cells due to its abundance and cheapness relative to other PV materials.[162] One downside is that about 40% of the incident light is reflected which means that about a quarter of the theoretical energy is lost.[22] This situation can be rectified by making the surface black by creating cylindrical holes in the surface in order to entrap more of the light. However, this not an easy task as the reagent is usually HF (a very corrosive acid), in conjunction with a silver catalyst, and the leaching process is very slow.[163,164] In the laboratory, it has been found that if the thin intrinsic layer of silica that is on the surface of silicon is electro-deoxidised in molten calcium chloride, a very robust porous silicon layer is left which absorbs much more of the light and, furthermore, unlike the leaching method, where only light that is incident at 90 degrees to the surface is absorbed, light from a much wider range of light is absorbed.[165,166] In addition, if a layer of TiO_2 is put on the surface of the porous silicon, the reflectance is reduced to below 0.1%.[167] A company, BlackSilicon Ltd., has been formed to develop this technology (Fig. 10 and 11).

8. Possible scale up

All the developments discussed in this paper are very interesting but will never be of any use unless they are scaled up. Fortunately, in the past, there are many molten salt processes that have gone from laboratory scale experiments to full scale production. Aluminium and lithium are exclusively produced by the electrolysis of molten salts so it is possible to maintain baths of molten salts at elevated temperatures so it is just a question of arranging the electrodes. At the present time, Metalysis is producing industrial quantities of tantalum for the supercapacitor market and titanium powder for 3D printing. Molten salt batteries and the SOM process are also being scaled up. For the preparation of nanoparticles, nanotubes, graphene and nanodiamonds which are being developed by two companies in the UK, all that is needed is a series of vertical plates of about 1 m^2, perhaps as bipolar electrodes, and a means of harvesting the salt from separation of the carbon based product. Overall, the technology already appears to exist so it may be just a question of applying it.

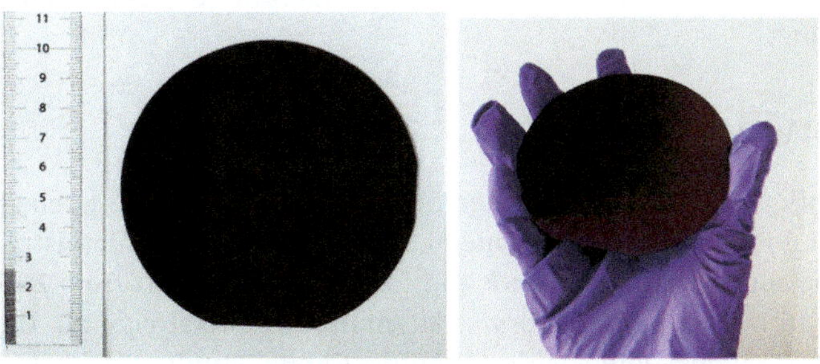

Fig. 10 Black silicon disks produced by electro-deoxidation of silica layer.[167]

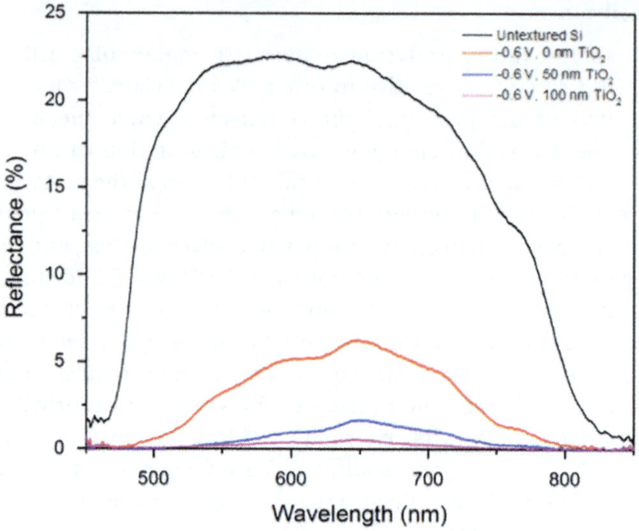

Fig. 11 Reflectance of black silicon with TiO_2 layer.[167]

9. Conclusions

The properties of molten salts such as their high melting points and corrosive nature tend to discourage innovation due to the perceived experimental and practical difficulties, but for some systems there are no alternative methods. The Hall–Heroult cell in aluminium industry and molten carbonate fuel cells can be regarded as successful examples of how these problems can be overcome. Moreover, in the past two decades, many novel applications of molten salts have been investigated. These include possible improvements in the extraction of metals, capture of carbon dioxide, molten salt batteries, the use of molten salts in making nanoparticles to increase the capacity of lithium ion batteries and the output of solar cells. In contrast to previous decades, many of these innovations from academic laboratories have been able to attract venture capital to create several spin out companies in the UK and US which, hopefully, will lead to the large successful industries of the future. Indisputably, molten salts will continue to contribute to a world that uses less energy and, at the same time, produce materials with a significantly better performance which can find innumerable useful and exciting applications.

Acknowledgements

For the work that has been presented in this paper from my research group, I am very grateful to past and present members of the Materials Chemistry Group for performing the experimental work and for financial support from EPSRC, Innovate UK, EU, DARPA, EAORD, British titanium plc., Metalysis Ltd., University of Cambridge Challenge Fund, various Venture Capital Funds and, lastly, those who work in the spin-out companies.

References

1 J. A. Paris, *The Life of Sir Humphry Davy*, Palala Press, 2015.
2 E. Robert, V. Lacassagne, C. Bessada, D. Massiot, B. Gilbert and J. P. Coutures, *Inorg. Chem.*, 1999, **38**, 214.
3 V. Lacassagne, C. Bessada, P. Florian, S. Bouvet, B. Ollivier, J. P. Coutures and D. Massiot, *J. Phys. Chem. B*, 2002, **106**, 1862.
4 I. Nuta, E. Veron, G. Matzen and C. Bessada, *Inorg. Chem.*, 2011, **50**, 3304.
5 P. A. Madden, M. Salanne and D. Corradini, *Faraday Discuss.*, 2016, DOI: 10.1039/c5fd00223k.
6 S. Cikit, Z. Akdeniz and P. A. Madden, *J. Phys. Chem. B*, 2014, **118**, 1064.
7 J. Thonstad, P. Fellner, G. M. Haarberg, J. Hives, H. Kvande and A. Sterten, *Aluminium Electrolysis: Fundamentals of the Hall–Heroult Process*, Aluminium-Verlag Marketing & Kommunikation GmbH, 2001.
8 *Handbook of Extractive Metallurgy*, ed. F. Habashi, Wiley-VCH, Weinheim, 1997.
9 G. Z. Chen, D. J. Fray and T. W. Farthing, *Nature*, 2000, **407**, 361.
10 S. Jiao and D. J. Fray, *US Patent*, 20160010232 A1.
11 A. Cox and D. J. Fray, *Trans. Inst. Min. Metall., Sect. C*, 1997, **106**, C123.
12 U. B. Pal and A. C. Powell, *JOM*, 2007, **59**, 44.
13 D. R. Sadoway, *JOM*, 2001, **53**, 34.
14 K. Wang, K. Jiang, B. Chung, T. Ouchi, P. J. Burke, D. A. Boysen, D. J. Bradwell, H. Kim, U. Muecke and D. R. Sadoway, *Nature*, 2014, **514**, 348.
15 Y. Nabae, K. D. Pointon and J. T. S. Irvine, *Energy Environ. Sci.*, 2008, **1**, 148.
16 H. V. Ijije, C. Sun and G. Z. Chen, *Carbon*, 2014, **73**, 163–174.
17 A. R. Kamali and D. J. Fray, *J. Mater. Sci.*, 2016, **51**, 569.
18 C. Schwandt, A. Dimitrov and D. J. Fray, *Carbon*, 2012, **50**, 1311.
19 A. R. Kamali and D. J. Fray, *Chem. Commun.*, 2015, **51**, 5594.
20 A. R. Kamali and D. J. Fray, *Nanoscale*, 2015, **7**, 11310.
21 A. R. Kamali and D. J. Fray, *J. New Mater. Electrochem. Syst.*, 2010, **13**, 147.
22 X. Liu, P. R. Coxon, M. Peters, B. Hoex, J. M. Cole and D. J. Fray, *Energy Environ. Sci.*, 2014, **7**, 3223.
23 G. P. Darcy, H. Kvande and A. Tabereaux, *JOM*, 2011, **63**, 104.
24 P. V. Polyakov, V. Y. Buzunov, Y. G. Mikhaeev and V. G. Putinysev, *Tsvetn. Met.*, 1993, **34**, 29.
25 C. E. Ransley, *J. Met.*, 1962, **14**, 129.
26 L. G. Boxall, A. V. Cooke and H. W. Hayden, *Light Met.*, 1984, 573.
27 R. C. Dorward and J. R. Payne, *Proceedings of the International Symposium CIM*, ed. M. Bouchard, Pergamon, 1990.
28 A. Cox and D. J. Fray, *Metall. Mater. Trans.B*, 1996, **27**, 889.
29 A. Cox, J. W. A. Morris and D. J. Fray, *Light Met.*, 1998, 1295.
30 O. Germer, Private Communication, 2016.
31 W. J. Kroll, *Trans. Am. Inst. Min. Metall. Eng.*, 1959, **215**, 546.
32 A. Cox and D. J. Fray, *Ironmaking Steelmaking*, 2008, **35**, 561.
33 R. J. Barnett and D. J. Fray, *J. Mater. Sci.*, 2014, **49**, 4148.
34 K. S. Mohandas and D. J. Fray, *Metall. Mater. Trans. B*, 2009, **40**, 685–699.
35 X. Y. Yan and D. J. Fray, *Metall. Mater. Trans. B*, 2002, **33**, 685.
36 G. Z. Chen, E. Gordo and D. J. Fray, *Metall. Mater. Trans. B*, 2004, **35**, 223.

37 J. Zhao, S. Lua, L. Hua and C. Li, *J. Energy Chem.*, 2013, **22**, 819.
38 I. Margerison and R. K. Rasheed, WO 2010092358 Patent.
39 R. Abdulaziz, L. D. Brown and D. Inman, *Electrochem. Commun.*, 2014, **26**, 2565.
40 C. Schwandt and D. J. Fray, *Electrochim. Acta*, 2005, **51**, 66.
41 A. Maeda and A. McLean, *J. Met.*, 1983, **35**, 75.
42 J. Ren, F. F. Li, J. Lau, L. González-Urbina and S. Licht, *Nano Lett.*, 2015, **15**, 6142.
43 X. Ge, X. Wang and S. Seetharaman, *Electrochim. Acta*, 2009, **54**, 4397.
44 H. Gao, M. Tan, L. Rong, *et al.*, *Phys. Chem. Chem. Phys.*, 2024, **16**, 19514.
45 Y. Zhu, M. Ma, D. Wang, J. Kai, H. Xiaohong, J. Xianbo and G. Z. Chen, *Chin. Sci. Bull.*, 2006, **51**, 2535.
46 B. Jackson, M. Jackson, D. Dye, D. Inman and R. Dashwood, *J. Electrochem. Soc.*, 2008, **155**, E171.
47 B. Jackson, D. Dye, D. Inman, R. Bhagat, R. J. Talling, S. L. Raghunathan, M. Jackson and R. J. Dashwood, *J. Electrochem. Soc.*, 2010, **157**, E57.
48 R. Bhagat, M. Jackson, D. Inman and R. Dashwood, *J. Electrochem. Soc.*, 2008, **155**, E63.
49 R. Bhagat, M. Jackson, D. Inman and R. Dashwood, *J. Electrochem. Soc.*, 2009, **156**, E1.
50 X. Liao, H. Xie, Y. Zhai, *et al.*, *J. Mater. Sci. Technol.*, 2009, **25**, 717.
51 X. Y. Yan and D. J. Fray, *J. Alloys Compd.*, 2009, **486**, 154.
52 B. X. Wang, R. Bhagat, X. Z. Lan and R. J. Dashwood, *J. Electrochem. Soc.*, 2011, **158**, D595.
53 Q. Xu, C. Schwandt and D. J. Fray, *Adv. Mater. Res.*, 2011, **160–162**, 1131.
54 M. Abdelkader and D. J. Fray, *Electrochim. Acta*, 2012, **64**, 10.
55 M. Abdelkader and D. J. Fray, *J. Eur. Ceram. Soc.*, 2012, **32**, 4481.
56 D. Tang, W. Xiao, L. Tian and D. Wang, *J. Electrochem. Soc.*, 2013, **160**, F1192.
57 X. Zou, X. Lu, Z. Zhou, W. Xiao, Q. Zhong, C. Li and W. Ding, *J. Mater. Chem.*, 2014, **2**, 7421.
58 M. Anik, A. B. Aybar, N. B. Hatirmaz and D. Özdemir, *J. Electrochem. Soc.*, 2015, **162**, A1080.
59 C. Osarinmwian, E. P. L. Roberts and I. M. Mellor, *Chem. Phys. Lett.*, 2015, **621**, 184.
60 K. Chen, Y. Hua, C. Xu, Q. Zhang, C. Qi and Y. Jie, *Ceram. Int.*, 2015, **41**, 11428.
61 T. H. Okabe, Y. Hamanaka and Y.-K. Taninouchi, 2016DOI: 10.1039/c5fd00229.
62 E. Wainer, *US Patent*, 2722509, 1955.
63 J. Withers, J. Laughlin, Y. Elkadi, J. DeSilva and R. O. Loutfy, *Key Eng. Mater.*, 2010, **436**, 55.
64 J. Withers, J. Laughlin, Y. Elkadi, J. DeSilva and R. O. Loutfy, *Key Eng. Mater.*, 2010, **436**, 61.
65 J. C. Withers and R. O. Loutfy, *US Patent*, 2005/0166706 A1.
66 J. C. Withers, R. O. Loutfy and J. P. Laughlin, *Materials Technology: Advanced Performance Materials*, 2007, **22**, 66.
67 S. Jiao, X. Ning, K. Huang and H. Zhu, *Pure Appl. Chem.*, 2010, **82**, 1691.
68 S. Jiao and H. Zhu, *J. Mater. Res.*, 2006, **21**, 2172.
69 S. Jiao and H. Zhu, *J. Alloys Compd.*, 2007, **438**, 243.
70 O. S. Kjos, G. M. Haarberg and A. M. Martinez, *ECS Trans.*, 2009, **16**, 229.

71 A. M. Martinez, K. S. Osen, E. Skybakmoen, O. S. Kjos, G. M. Haarberg and K. Dring, *Key Eng. Mater.*, 2010, **436**, 41.
72 O. S. Kjos, G. M. Haarberg and A. M. Martinez, *Key Eng. Mater.*, 2010, **436**, 93.
73 E.-Y. Choi and S. M. Jeong, *Prog. Nat. Sci.*, 2015, **25**, 572.
74 T. Inoue and L. Koch, *Nucl. Eng. Technol.*, 2008, **40**, 183.
75 M. Iizuka, Y. Sakamura and T. Inoue, *J. Nucl. Mater.*, 2006, **359**, 102.
76 B. H. Park, I. W. Lee and C. S. Seo, *Chem. Eng. Sci.*, 2008, **63**, 3485.
77 S. M. Jeong, J. M. Hur, S. S. Hong, D. S. Kang, M. S. Choung, C. S. Seo, J. S. Yoon and S. W. Park, *Nucl. Technol.*, 2008, **162**, 184.
78 S. D. Herrmann, S. X. Li, M. F. Simpson and S. Phongikaroon, *Sep. Sci. Technol.*, 2006, **41**, 1965.
79 S. D. Herrmann, S. Li and M. Simpson, *J. Nucl. Sci. Technol.*, 2007, **44**, 361.
80 K. V. Gourishanker, L. Redey and M. Williamson, in *Light Metals*, ed. W. Schneider, The Minerals, Metals and Materials Society, 2002, pp. 1075–1082.
81 D. K. Corrigan, J. P. Elliott, E. O. Blair, S. J. Reeves, I. Schmueser, A. Walton and A. R. Mount, *Faraday Discuss.*, 2016, DOI: 10.1039/c6fd00002a.
82 C. M. Hall, *US Patent*, 400, 664, 1889.
83 A. D. Mcleod, J. S. Haggerty and D. R. Sadoway, *CIM Bull.*, 1985, **79**, A38.
84 H. Alder, *US Patent*, 3,930,967, 1976.
85 H. Alder, *US Patent*, 3960678 A, 1977.
86 E. Olsen and J. Thonstad, *J. Appl. Electrochem.*, 1999, **29**, 301.
87 T. R. Beck, *Light Met.*, 1995, 355.
88 J. K. Walker, J. Kinkoph and C. K. Saha, *J. Appl. Electrochem.*, 1989, **19**, 225.
89 G. M. Haarberg, B. Khalaghi and T. Mokkelbost, *Faraday Discuss.*, 2016, DOI: 10.1039/c5fd00239g.
90 K. T. Kilby, S. Jiao and D. J. Fray, *Electrochim. Acta*, 2010, **55**, 7126.
91 S. L. Cuffini, V. A. Macagno, R. E. Carbonio, A. Melo, E. Trollundand and J. L. Gautier, *J. Solid State Chem.*, 1993, **105**, 161.
92 T. He and R. J. Cava, *Phys. Rev. B: Condens. Matter Mater. Phys.*, 2001, **63**, 172403.
93 S. Jiao and D. J. Fray, *Metall. Mater. Trans. B*, 2010, **41**, 74.
94 C. Schwandt, J. A. Hamilton, D. J. Fray and I. A. Crawford, *Planet. Space Sci.*, 2012, **74**, 49.
95 G. A. Snook, K. MacGregor, A. J. Urban, M. R. Lanyon, R. Donelson and M. I. Pownceby, 2016DOI: 10.1039/c5fd00235d.
96 H. Kim, J. Paramore, A. Allanore and D. R. Sadoway, *J. Electrochem. Soc.*, 2011, **158**, E101.
97 A. Allanore, L. Yin and D. R. Sadoway, *Nature*, 2013, **497**, 353.
98 U. B. Pal, D. E. Woolley and G. B. Kenney, *JOM*, 2001, **53**, 32.
99 A. Krishnan, U. B. Pal and X. G. Liu, *Metall. Mater. Trans. B*, 2005, **36**, 463.
100 X. Guan, U. B. Pal, S. Gopalan and A. C. Powell, *J. Electrochem. Soc.*, 2013, **160**, F1179.
101 E. S. Gratz, X. Guan, J. Milshtein, U. B. Pal and A. C. Powell, *Metall. Mater. Trans. B*, 2014, **45**, 1325.
102 X. Guan, U. B. Pal and A. C. Powell, *Metall. Mater. Trans. E*, 2014, **1**, 132.
103 M. Suput, R. DeLucas, S. Pati, G. Ye, U. B. Pal and A. C. Powell, *Inst. Min. Metall., Trans., Sect. C*, 2008, **117**, 118.
104 A. Krishnan, X. G. Lu and U. B. Powell, *Scand. J. Metall.*, 2005, **34**, 293.
105 J. Xu, B. Lo, Y. Jiang, U. Pal and S. Basu, *J. Europ. Cer. Soc.*, 2014, **34**, 3887.

106 X. Lu, X. Zou, C. Li, Q. Zhong, W. Ding and Z. Zhou, *Metall. Mater. Trans. B*, 2012, **43**, 503.

107 B. Zhao, X. Lu, Q. Zhong, C. Li and S. Chen, *Electrochim. Acta*, 2010, **55**, 2996.

108 X. Zhou, X. Lu, C. Li and Z. Zhou, *Electrochim. Acta*, 2010, **55**, 5173.

109 X. Zou, X. Lu, Z. Zhou, C. Li and W. Ding, *Electrochim. Acta*, 2011, **56**, 8430.

110 J. Milshtein, E. Gratz, S. Pati, A. C. Powell and U. B. Pal, *J. Min. Metall., Sect. B*, 2013, **49**, 183.

111 X. Zou, K. Zheng, X. Lu, Q. Xu and Z. Zhou, *Faraday Discuss.*, 2016, DOI: 10.1039/c5fd00221d.

112 P. Drossbach and D. Sauermann, *Electrochim. Acta*, 1964, **9**, 1373.

113 M. D. Ingram, B. Baron and G. J. Janz, *Electrochim. Acta*, 1966, **11**, 1629.

114 H. E. Bartlett and K. E. Johnson, *J. Electrochem. Soc.*, 1967, **114**, 4457.

115 V. Kaplan, E. Wachtel, K. Gartsmann, Y. Feldman and I. Lubomirsky, *J. Electrochem. Soc.*, 2010, **157**, B552.

116 V. Kaplan, E. Wachtel and I. Lubomirsky, *J. Electrochem. Soc.*, 2014, **161**, F54.

117 H. Yin, X. Mao, D. Tang, W. Xiao, L. Xing, H. Zhu, D. Wang and D. R. Sadoway, *Energy Environ. Sci.*, 2013, **6**, 1538.

118 W. Xiao, B. Lu, J. Zhou, H. Wang, Y. Song and D. Wang, *Faraday Discuss.*, 2016, DOI: 10.1039/c5fd00215j.

119 D. Wang, B. Deng, Z. Chen, M. Gao, Y. Song, K. Zheng, J. Tang, W. Xiao and X. Mao, *Faraday Discuss.*, 2016, DOI: 10.1039/c5fd00234f.

120 S. Jiao, J. Ge, L. Hu and Y. Song, *Faraday Discuss.*, 2016, DOI: 10.1039/c5fd00217f.

121 J. Ren, J. Lau, M. Lefler and S. Licht, *J. Phys. Chem.*, 2015, **119**, 23342.

122 J. Ren, F.-F. Li, J. Lau, L. Gonzalez-Urbina and S. Licht, *Nano Lett.*, 2015, **15**, 6142.

123 F.-F. Li, S. Liu, B. Cui, J. Lau, J. Stuart, B. Wang and S. Licht, *Adv. Energy Mater.*, 2015, **5**, 1401791.

124 Y. Ito, T. Nishikiori and H. Tsujimura, 2016DOI: 10.1039/c5fd000237k.

125 A. Banerjee, G. R. Dick, T. Yoshino and M. W. Kanan, *Nature*, 2016, **531**, 215.

126 A. Mehmeti, F. Sanyoni, M. D. Pietra and S. J. Mcphail, *J. Power Sources*, 2016, **308**, 97.

127 Y. Nabae, K. D. Pointon and J. T. S. Irvine, *Energy Environ. Sci.*, 2008, **1**, 1417.

128 D. A. Bonaccorso and J. T. S. Irvine, *Int. J. Hydrogen Energy*, 2012, **37**, 19337.

129 A. Chien and J. T. S. Irvine, *Molten Salts Chemistry: From Laboratory to Applications*, 2013, p. 403.

130 X. Yue, A. Arenillas and J. T. S. Irvine, *Faraday Discuss.*, 2016, DOI: 10.1039/c6fd00001k.

131 H. Kim, D. A. Boyson, J. M. Newhouse, B. L. Spatocco, B. Chung, P. J. Burke, D. J. Bradwell, K. Jiang, A. Tomaszowska, K. Wang, W. Wei, L. A. Ortiz, S. A. Barriga, S. P. Poizeau and D. R. Sadoway, *Chem. Rev.*, 2013, **113**, 2075.

132 T. G. Pearson, *The Chemical Background of the Aluminium Industry*, The Royal Institute of Chemistry, London, 1955.

133 X. Ning, S. Phadke, B. Chung, H. Yin, P. Burke and D. R. Sadoway, *J. Power Sources*, 2013, **241**, 239.

134 H. Kim, D. A. Boyson, T. Ouchi and D. R. Sadoway, *J. Power Sources*, 2015, **275**, 370.

135 D. J. Bradwell, H. Kim, A. H. C. Sirk and D. R. Sadoway, *J. Am. Chem. Soc.*, 2012, **134**, 1895.

136 B. L. Spatocco, T. Ouchi, G. Lambotte, P. J. Burke and D. R. Sadoway, *J. Electrochem. Soc.*, 2015, **162**, A2729.
137 T. Ouchi, H. Kim, X. Ning and D. R. Sadoway, *J. Electrochem. Soc.*, 2014, **161**, A1898.
138 J. Ren, F. Li, J. Law, L. Gonzalez-Urbina and S. Licht, *Nano Lett.*, 2015, **15**, 6142.
139 S. Licht, B. Cui, J. Stuart, B. Wanga and J. Lau, *Energy Environ. Sci.*, 2013, **6**, 3646.
140 F. F. Li, J. Law and S. Licht, *Adv. Sci.*, 2015, **2**, 1500260.
141 M. Winter and J. O. Besenhard, *Electrochim. Acta*, 1999, **145**, 31.
142 W. K. Hsu, M. Terrones, J. P. Hare, H. Terrones, H. W. Kroto and D. R. M. Walton, *Chem. Phys. Lett.*, 1996, **262**, 161.
143 W. K. Hsu, J. Li, H. Terrones, M. Terrones, N. Grobert, Y. Q. Zhu, S. Trasobares, J. P. Hare, C. J. Pickett, H. W. Kroto and D. R. M. Walton, *Chem. Phys. Lett.*, 1999, **301**, 159.
144 R. Das Gupta, C. Schwandt and D. J. Fray, *Carbon*, 2014, **70**, 142.
145 C. Schwandt, A. T. Dimitrov and D. J. Fray, *Carbon*, 2012, **50**, 1311.
146 I. A. Kinloch, G. Z. Chen, J. Howes, C. Boothroyd, C. Singh, D. J. Fray and A. H. Windle, *Carbon*, 2003, **41**, 1127.
147 R. Das Gupta, Ph.D. thesis, University of Cambridge, 2009.
148 A. R. Kamali and D. J. Fray, *Materials Today Communications*, 2015, **2**, 38.
149 A. R. Kamali and D. J. Fray, Unpublished results.
150 M. S. Dresselhaus, G. Dresselhaus and P. G. Ekland, *J. Mater. Sci.*, 1993, **8**, 2054.
151 K. S. Novoselov, A. K. Geim, S. V. Morozov, D. Jiang, M. I. Katsnelson, I. V. Grigorieva, S. V. Dubonos and A. A. Firsov, *Nature*, 2005, **438**, 197.
152 A. R. Kamali and D. J. Fray, *Carbon*, 2013, **56**, 121.
153 A. R. Kamali and D. J. Fray, *Nanoscale*, 2015, **7**, 11310.
154 A. R. Kamali and D. J. Fray, *J. Mater. Sci.*, 2016, **51**, 569.
155 A. R. Kamali, J. Feighan and D. J. Fray, 2016DOI: 10.1039/c6fd00005c.
156 H.-K. KimA. R. KamaliK. C. RohK.-B. Kim and D. J. Fray, *Energy Environ. Sci.*, 2016DOI: 10.1039/c6EE00815A.
157 Q. Liu, C. X. Wang, S. W. Li, J. X. Zhang and G. W. Yang, *Carbon*, 2004, **42**, 629.
158 Y. Zhu, T. Sekine, T. Kobayashi, E. Takazawa, M. Terrones and H. Terrones, *Chem. Phys. Lett.*, 1998, **287**, 689.
159 F. Zhang, F. Ahmed, G. Holzhuter and E. Burkel, *J. Cryst. Growth*, 2012, **340**, 1.
160 H. Yusa, *Diamond Relat. Mater.*, 2002, **11**, 87.
161 Y. N. Pal'Yanov, A. G. Sokol, Y. M. Borzdov, A. F. Khokhryakov, A. F. Shatsky and N. V. Sobolev, *Diamond Relat. Mater.*, 1999, **8**, 1118.
162 Technology Roadmap: Solar Photovoltaic Energy, IEA Energy Technology Perspectives, 2014, https://www.iea.org/publications/freepublications/publication/technology-roadmap-solar-photovoltaic-energy-2014-edition.html.
163 H.-P. Wang, D.-H. Lien, M.-L. Tsai, C.-A. Lin, H.-C. Chang, K.-Y. Laib and J.-H. He, *J. Mater. Chem. C*, 2014, **2**, 3144.
164 F. Priolo, T. Gregorkiewicz, M. Galli and T. F. Krauss, *Nat. Nanotechnol.*, 2014, **9**, 19.
165 E. Juzeliunas, A. Cox and D. J. Fray, *Electrochem. Commun.*, 2010, **12**, 1270.

166 E. Juzeliunas, A. Cox and D. J. Fray, *Electrochim. Acta*, 2012, **68**, 123–127.
167 P. R. Coxon, M. Coto, E. Juzeliunas and D. J. Fray, *Prog. Nat. Sci.*, 2015, **25**, 583.
168 Y. Yuan, W. Lei, H. Chen, Z. Wang, X. Jin and G. Z. Chen, DOI: 10.1039/c5fd00231a.

Faraday Discussions

PAPER

Development of a niobium-doped titania inert anode for titanium electrowinning in molten chloride salts†

Graeme A. Snook, Katherine McGregor,* Andrew J. Urban, Marshall R. Lanyon, R. Donelson and Mark I. Pownceby

Received 23rd December 2015, Accepted 11th March 2016
DOI: 10.1039/c5fd00235d

The direct electrochemical reduction of solid titanium dioxide in a chloride melt is an attractive method for the production of titanium metal. It has been estimated that this type of electrolytic approach may reduce the costs of producing titanium sponge by more than half, with the additional benefit of a smaller environmental footprint. The process utilises a consumable carbon anode which releases a mixture of CO_2 and CO gas during electrolysis, but suffers from low current efficiency due to the occurrence of parasitic side reactions involving carbon. The replacement of the carbon anode with a cheap, robust inert anode offers numerous benefits that include: elimination of carbon dioxide emissions, more efficient cell operation, opportunity for three-dimensional electrode configurations and reduced electrode costs. This paper reports a study of Nb-doped titania anode materials for inert anodes in a titanium electrolytic reduction cell. The study examines the effect of niobium content and sintering conditions on the performance of Nb-doped TiO_2 anodes in laboratory-scale electrolysis tests. Experimental findings, including performance in a 100 h laboratory electrolysis test, are described.

1. Introduction

Most titanium metal is still produced *via* the pyrometallurgical process first developed by William J. Kroll in 1940.[1] The process is based on the magnesiothermic reduction of titanium tetrachloride, with magnesium chloride generated as a by-product. It can produce high purity sponge titanium, but is slow and costly, due to its batch-type nature. Kroll's prediction that his process would be rapidly replaced by a less expensive electrolytic route has been challenging to fulfil. The highly reactive nature of titanium and its potential for the formation of dissolved titanium species with multiple valencies has contributed to the

CSIRO Mineral Resources, Private Bag 10, Clayton South, Vic., 3169, Australia. E-mail: Kathie.Mcgregor@csiro.au

† Electronic supplementary information (ESI) available. See DOI: 10.1039/c5fd00235d

difficult nature of this approach. As well, the energy intensive and multi-step nature of both the traditional extraction and the component manufacturing processes has meant that less than 5% of titanium ore is made into metal. The emergence of additive manufacturing or 3D printing methods that use metal powder feedstocks has drastically reduced the manufacturing steps to the final metal component, cutting down on waste and reducing costs. The development of lower cost extraction methods for high purity titanium and titanium alloy powders is vital for the growth of new markets for additively manufactured titanium components.

Electrolytic technologies such as the Fray–Farthing–Chen (FFC) Cambridge[2] and Ono-Suzuki (OS) processes[3] have been suggested as alternatives to the Kroll process[1] for producing titanium metal/alloy powder and sponge. The FFC process in particular has generated extensive studies and attracted much interest and involves the direct electrochemical deoxygenation of solid titanium dioxide (TiO_2) pellets to titanium metal in a $CaCl_2$ bath at temperatures up to 1000 °C.[2] It has been estimated that an FFC-type approach may reduce the costs of producing titanium sponge by more than half[4–6] with the additional benefit of a smaller environmental footprint. The FFC process utilises a consumable carbon anode which releases a mixture of CO_2 and CO gas during electrolysis. However, the process suffers from low current efficiency due to the occurrence of parasitic side reactions involving the consumable carbon anode. As previously noted,[7] the disadvantages of carbon anodes are as follows:

- loss/wear and cracking of anode carbon;
- regular lowering of anodes required to maintain optimum anode–cathode distance (ACD) and steady-state thermal conditions;
- regular replacement;
- disturbances in cell operation due to anode wear;
- carbon contamination of the metal product;
- cell shorting by particulate carbon;
- poor sealing of cell;
- reduced current efficiency;
- increased energy consumption;
- greenhouse gas emissions.

Replacement of the carbon anode with a cheap, robust inert anode could address most of these disadvantages. An inert anode evolves pure oxygen (rather than carbon dioxide and carbon monoxide) and, in principle, is not consumed during the electrolysis process. The overall cathode and anode reactions, when utilising an inert anode, are as follows:

$$\text{Cathode: } TiO_2(s) + 4e^- \rightarrow Ti(s) + 2O^{2-} \tag{1}$$

$$\text{Inert anode: } 2O^{2-} \rightarrow O_2(g) + 4e^- \tag{2}$$

$$\text{Overall reaction: } TiO_2(s) \rightarrow Ti(s) + O_2(g) \tag{3}$$

Inert anodes usually function *via* interaction of the anode with the electrolyte and the anode product (O_2). Upon immersion of the inert anode in the bath, there is dynamic formation of an insulating oxide film at the anode surface. Such films must be thin enough to allow electronic conduction, but thick enough to protect the surface from further reaction. One disadvantage is that the anodic potential for oxygen evolution is larger than that for carbon dioxide, but this is partly offset by the higher overpotential for carbon dioxide evolution.[8]

Whilst there have been extensive (unsuccessful) efforts in the aluminium industry to identify stable inert anode materials,[8–11] and some effort for magnesium,[12–14] few reports have appeared specifically for titanium.[7,15–18] Some success at the laboratory scale has been reported for calcium ruthenate[17] (nominally $CaRuO_3$ although different stoichiometries may apply in practice), but the high cost of ruthenium compounds is a barrier to implementation. An opportunity therefore exists for the development of alternative inert anode materials that can be prepared using compounds and process methods that are:

- low cost (both in raw materials cost and also fabrication);
- scalable to commercial quantities;
- able to produce large anodes.

We also acknowledge that although an inert anode will eliminate formation of gaseous CO_2 and CO, the environmental impact of the process will depend on the way the O_2 generated is treated. Handling large quantities of O_2 in an industrial application will require significant engineering to comply with environmental, health and safety regulations.

In earlier work, we surveyed a range of metals, alloys and conductive ceramics with limited success.[7] In this paper, we investigate one of the types of conductive ceramic anodes that Dow Chemical Company patented in 1982 (ref. 19) for possible use as an inert anode in the magnesium or aluminium electrolysis processes. Specifically, this patent included titanium dioxide doped with elements such as niobium and tantalum. Both Nb- and Ta-doped titanium dioxide materials were investigated as inert anodes for magnesium production in electrolytic reduction cells (magnesium chloride bath), and significant progress was achieved with the development of a dimensionally stable material. In a joint Dow Chemical Company–US Department of Energy project, an anode doped with 3 mol% of Nb_2O_5 met a 30 day lifetime goal with an acceptable wear rate, and was the most promising in terms of lifetime and conductivity. Unfortunately the work was not continued. This paper reports a study of Nb-doped titania anode materials for inert anodes in a titanium electrolytic reduction cell. Ta-doped titania materials were also investigated, but the preparation of conductive material proved difficult and the preliminary results were poor. For interest, the results are included as ESI.†

2. Experimental

2.1. Materials preparation

Nb-doped TiO_2 powder and sintered bar samples in the form $Nb_xTi_{1-x}O_2$ were produced *via* a solid-state reaction method at two doping levels corresponding to $Nb_xTi_{1-x}O_2$ ($x = 0.054$ and 0.060). Nb_2O_5 (99.5% purity) and rutile TiO_2 (Ferro Ticon HR, Niagara Falls, NY) were wet milled together in deionised water for 5 hours and then dried in an oven overnight at 40 °C. The dried powders were calcined in a slightly reducing atmosphere at 950 °C for 5 hours.

Sample bars of nominal dimension 5 × 5 × 50 mm were produced from the Nb-doped titania powders by milling the powders with a polyvinyl alcohol–polyethylene glycol–water binder solution and pressing uniaxially in a die press. The bars were sintered by heating at 300 °C hour^{-1} to either 1350 °C, 1400 °C or 1500 °C and held for 1 hour. This was performed to assess the effect of sintering temperature on physical characteristics such as porosity and electrical conductivity. The Nb-doped titania bars sintered at different temperatures will be referred to as NB-1350, NB-1400, and NB-1500. Typical density for the pressed bars after firing was 4.1 g cm^{-3} (Archimedean method, 97% of theoretical density).

2.2. Materials characterization

Starting materials were characterized using a combination of methods including powder X-ray diffraction (PXRD), scanning electron microscopy (SEM) and electron probe microanalysis (EPMA).

The powder specimens were examined by PXRD using a Philips diffractometer equipped with a graphite monochromator on the diffracted beam and employing Cu Kα radiation. The X-ray tube was operated at 40 kV and 40 mA with 1° divergence and scatter slits, a 0.3 mm receiving slit, and Soller slits in the incident and diffracted beams. The PXRD patterns were collected in the 10–40° 2θ range with 0.02° step size, integrating 10 s per step. The recorded patterns were analyzed with the Rietveld method using the Topas3 software package (©Bruker AXS) to determine the cell parameters. Powder X-ray diffraction (PXRD) and electron probe microanalysis (EPMA) studies confirm that the samples are comprised of a single phase material (Nb-doped rutile) and are homogeneous in composition (*i.e.* homogeneous in Ti : Nb ratio).

The composition of each starting mixture was determined by wavelength dispersive EPMA on sectioned, polished surfaces using a JEOL Superprobe microanalyzer, model JXA-8900R. Approximately 40 individual analyses were collected on each sample for Ti or Nb, whilst O was calculated by difference. Operating conditions for the quantitative analyses were: 20 kV accelerating voltage, 18 nA beam current, and an analysis volume of approximately 1–2 μm. Standards used were niobium metal and rutile (TiO_2). Counting times for all elements were 20 s on the peak (Kα for Ti; Lα for Nb) and 10 s on the background. For samples doped with 6 mol% Nb, quantitative EPMA studies of the Ti : Nb ratio from 40 sample points in multiple grains confirmed the stoichiometry as $Ti_{0.94}Nb_{0.06}O_2$.

After electrolysis, parts of selected anodes were sectioned using a diamond-tipped saw and then mounted in epoxy resin. The anode was sectioned and mounted so as to expose a cross-section through the sample from core to rim. The samples were polished flat using successively finer diamond pastes up to a final polishing size of 1 μm and then coated with a 15 nm thick carbon film prior to analysis. For high-resolution EPMA mapping, a series of maps showing the distribution of the elements O, Ti, Ca, Nb, Ta and Cl were collected for each sample. Standards used were rutile (TiO_2), wollastonite ($CaSiO_3$), niobium metal (Nb), tantalum metal (Ta) and halite (NaCl). The area covered was 900 × 1536 μm for each electrode. Operating conditions for the electron probe in mapping mode were: accelerating voltage 15 kV, current 105 nA, step size 1.5 μm, and counting times of 20 ms per step.

2.3. Instrumentation and anode testing

Electrochemical measurements were performed using an EG&G PAR Model 362 scanning potentiostat and an EG&G PAR Current Booster Model 365 (10 A). An in-house built resistometer[20] was connected between the potentiostat and the cell to supply an interrupted current pulse signal (every 60 ms) of 1 A (70 μs duration) followed by a rest period (70 μs duration) and a reverse (−)1 A pulse of the same duration. From the voltage output this instrument is able to measure the *in situ* resistance value or the ohmic voltage drop for the cell.[21–25] The measurement is so rapid that the slower Faradaic processes do not contribute to the voltage at this time scale. For the purposes of discussion, the cell voltage with the ohmic component (*IR*) subtracted will be termed the *IR* free voltage.

Experimental details for the furnace, cell and calcium chloride bath preparation are given in ref. 7 and 24.

All measurements were performed in two electrode mode (without a reference electrode) using galvanostatic (constant current) control (~0.2 A cm^{-2}) and a $CaCl_2$/0.2 wt% CaO bath at 950 °C. This current density was deemed to be sufficiently high to test the durability of the anodes and to be utilised effectively in titanium production. Evaluation of the anode samples for periods of up to 5 days continuous operation was carried out using the cell design in Fig. 1. Operating the laboratory scale cell for extended periods of time requires an oversized TiO_2 cathode that can supply sufficient amounts of oxide ions and allow formation of oxygen at the anode. This ensures that supply of oxide ions from the cathode is not the limiting factor for the solid-state deoxygenation reaction, and the lifetime of the anode can be assessed. The *IR* free voltage was monitored during the

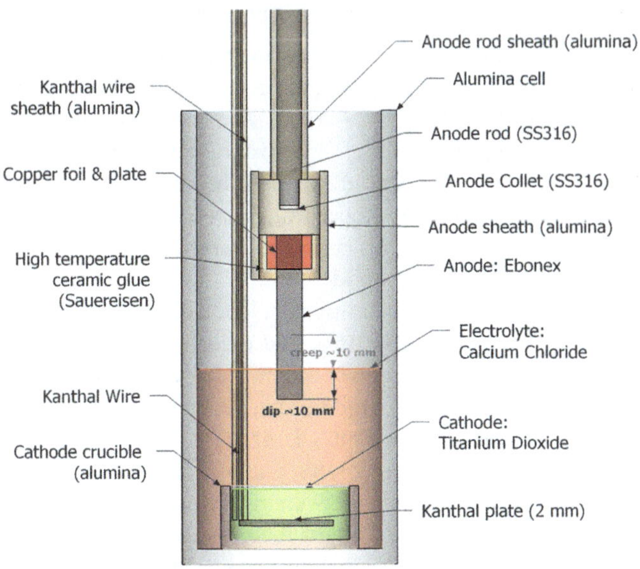

Fig. 1 Cell design for testing inert anodes in the FFC process.

reduction reactions, and remained below 3.0 V at all times. Accordingly, it is unlikely the oversized titania cathode was depleted of oxygen to the point where calcium metal was produced from the bath rather than the desired cathode reduction reaction. The off-gases from selected tests were monitored for chlorine evolution (Drager tube) and found to be negligible. The oversized TiO_2 cathode was prepared by spot welding a small piece of Kanthal strip to a length of Kanthal wire (18 gauge), and pressing the tight fitting strip into a sintered alumina dish (60 mm diameter). A slurry of titania paste, 40 g, prepared from TiO_2 (99.5%, minimum purity rutile, Alfa Aesar) and de-ionised water, was added to the dish, covering the Kanthal strip, and oven-dried for 24 h. The Kanthal wire was sheathed with a sintered hollow alumina rod to prevent corrosion and attack by the calcium chloride bath. The cathode assembly was placed at the bottom of an alumina crucible.

The inert anodes were prepared as rectangular ceramic bars. The ceramic inert anode was attached to an alumina sheathed stainless steel (SS 316) rod. The standard sized bar was around 5.4 mm × 3.8 mm; while the thicker anode bars were around 8.8 mm × 4.4 mm. Absolute measurement of when the anode was touching the top of the molten salt bath was done by measuring the cell resistance (*via* the CSIRO resistometer) while lowering the anode with the cell at open circuit.[24] This resulted in the cell resistance changing from well in excess of 10 Ω to below 10 Ω (within the measuring range of the resistometer) once the bath had made contact with the anode.

3. Results and discussion

This study investigates doped titania materials for inert anodes in an electrolytic reduction cell that converts solid TiO_2 to Ti metal in a chloride bath. Anode material developments focused on Nb-doped TiO_2.

3.1. Optimisation of Nb-doped titania inert anodes

At low concentrations and reducing conditions, niobium is most likely incorporated into titania as Nb^{5+} (from Nb_2O_5) and produces a highly conductive material, probably in the form of $Nb_x{}^{5+}Ti_x{}^{3+}Ti_{1-2x}{}^{4+}O_2$. The electronic conductivity of the material is due to the presence of Ti^{3+} sites in the rutile structure. According to literature reports,[11,12] maximum conductivity is achieved for ~6 mol% Nb dopant (3 mol% Nb_2O_5) *i.e.*, $Nb_{0.06}Ti_{0.94}O_2$. Above this concentration, the dopant tends to accumulate at the grain boundaries as a separate, unreacted phase, and not further aid the conductivity of the anode bar.

Stability to oxidation is also an important consideration.[12] Under oxidizing conditions, Ti^{3+} may be oxidised to Ti^{4+}, and oxide ions incorporated into the lattice. The lattice structure rearranges to accommodate the oxide ions, and leaves some cation vacancies to achieve charge balance. Prior to *in situ* tests, the stability of selected formulations of Nb-doped TiO_2 in air at 700 °C was investigated. *Ex situ* oxidation tests were conducted over 24, 48, 120 and 240 h, and SEM and EPMA analyses demonstrated that whilst there were no detectable changes in chemical composition, there appeared to be a slight increase in porosity with time. There was no observation of phase or grain boundary segregation of Nb_2O_5, which might be expected to form upon oxidation of Ti^{3+} to Ti^{4+}. It is possible that minor phase segregation occurs, but is below the detection limits of the characterization techniques. This may give reduced electrical conductivity of the material, depending on the extent of the segregation.

A concerted effort was undertaken to optimise the structure of the Nb-doped titania anode, and fabricate a dense, homogenous, conductive material. Experiments investigated the effect of varying calcining temperature and atmosphere, and also concentrated on the development of a suitable mechanical mixing method to enhance the desired properties. Promising results were achieved by calcining in a reducing atmosphere and using lower dopant levels.

The first generation of materials (generation 1 – $Nb_{0.06}Ti_{0.94}O_2$) were optimized for conductivity, but utilised technical grade starting materials. Some grain boundary phases were observed in SEM studies (Fig. 2a). The second generation of optimized materials (generation 2 – $Nb_{0.054}Ti_{0.946}O_2$), used a better grade of raw materials (for details see the Experimental section) and a lower Nb content of

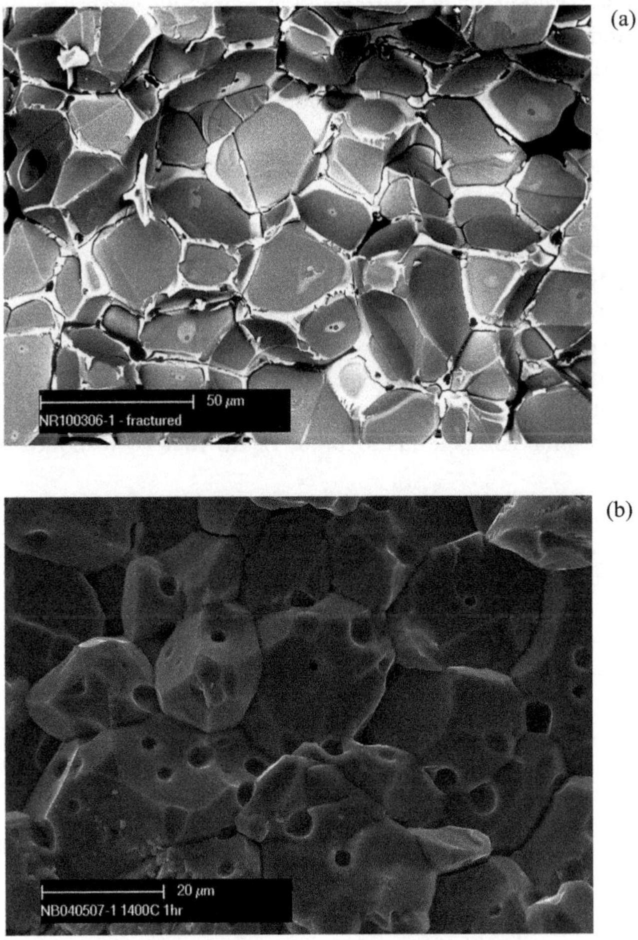

Fig. 2 SEM images of: (a) generation 1 material ($Nb_{0.06}Ti_{0.94}O_2$). Bright phase at the edges is a high niobium phase; (b) generation 2 material ($Nb_{0.054}Ti_{0.946}O_2$).

5.4 mol%. SEM studies demonstrate that this leads to the elimination of the grain boundary phase found in generation 1 samples (Fig. 2b).

Short-term (<24 h) electrolysis tests were carried out on several samples of generation 1 and 2 Nb-doped titania. The anodes were prepared in bar form and attached to a stainless steel rod and sheathed with alumina as described in the Experimental section. The cell voltage as a function of electrolysis time is shown in Fig. 3, and compares the behaviour of the generation 1 ($Nb_{0.06}Ti_{0.94}O_2$) and generation 2 ($Nb_{0.054}Ti_{0.946}O_2$) anode materials. Although not shown on the plots, the *IR* free voltage was monitored during the reduction reactions, and remained below 3.0 V at all times. The plots demonstrate that the cell voltage increases at a constant rate for the run that uses a generation 1 anode, but shows considerably more stability when utilising the generation 2 material. Optimisation of grain size, homogeneity, porosity, calcining temperature and niobium levels successfully minimised the variable performance of the anode. The minor increase in cell voltage at ~7–8 hours of electrolysis is attributed to cathode processes, not to any change in the anode material, and the evidence for this is discussed later. Once a reasonably stable cell voltage was maintained over 24 h, further development work was performed to more fully explore the capability of the inert anodes.

Initial testing of the generation 2 materials were conducted with lower absolute currents (0.5 A). Later tests were conducted at higher currents (1 A) and required a larger area of the electrode to be immersed in the molten salt to maintain the same current density (0.2 A cm^{-2}). Consequently, longer and thicker samples were manufactured for these tests.

3.2. 24 hour tests: effect of sintering temperature

Three types of generation 2 samples (NB-1350, NB-1400, NB-1500) were sintered at temperatures of 1350 °C, 1400 °C and 1500 °C and tested using constant current electrolysis (0.5 A or ~0.2 A cm^{-2}) in a $CaCl_2$/0.2 wt% CaO bath at 950 °C. Fig. 4a shows the voltage–time plots for the three anode samples (NB-1350 corresponds to 1350 °C; NB-1400 corresponds to 1400 °C; NB-1500 corresponds to 1500 °C).

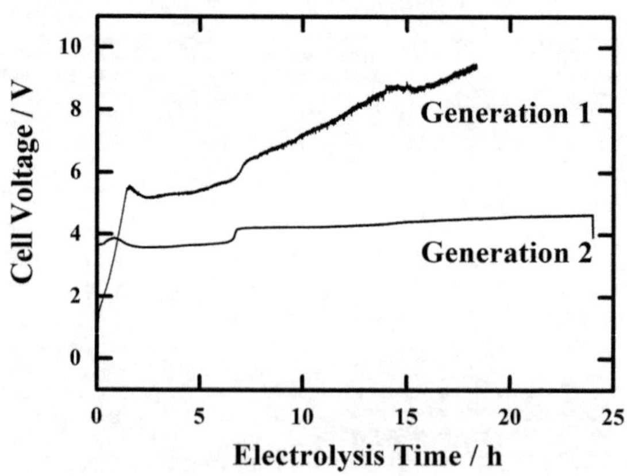

Fig. 3 Constant current (0.2 A cm^{-2}) electrolysis utilising a generation 1 ($Nb_{0.06}Ti_{0.94}O_2$) and generation 2 anode ($Nb_{0.054}Ti_{0.946}O_2$) with an oversized pasted TiO_2 cathode.

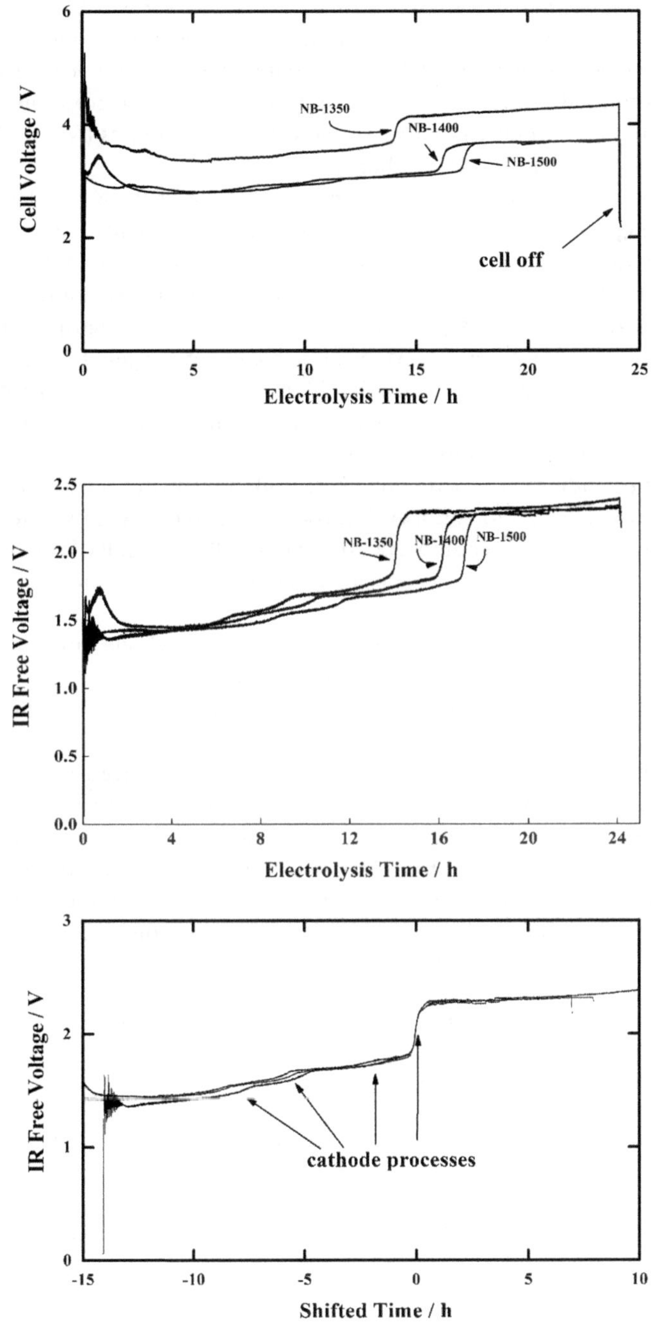

Fig. 4 Constant current (0.2 A cm^{-2} or 0.5 A) electrolysis utilising a series of $Nb_{0.054}Ti_{0.946}O_2$ electrodes sintered at different temperatures. (a) Raw data, (b) *IR* free voltage, (c) shifted in time.

The electrolysis of samples NB-1400 and NB-1500 exhibits a rising cell voltage to ~2.6 V during the initial current ramp from 0 A to 0.5 A and very gradually increased over the next hour. After around 11–12 hours, the cell voltage climbed rapidly by ~0.5 V over a 30 minute period, before reaching a plateau at ~3.1 V. The final cell voltage for NB-1400 and NB-1500 after 24 hours of electrolysis was ~3.3 V. Sample NB-1350 ran at a higher voltage due to ohmic drop considerations (confirmed below) when compared to the other two samples (NB-1400 and NB-1500), but shows similar voltage changes.

The *IR* free voltage for the three samples (NB-1350, NB-1400 and NB-1500) is shown in Fig. 4b, and was obtained from an in-house instrument (see Experimental section for details). Once the ohmic component is removed, the same voltage plateaus are reached at similar times in the electrolysis for the three runs. At all times, the *IR* free voltage remains below 3.0 V, indicating that it is unlikely the oversized titania cathode has been depleted of oxygen to a point where calcium metal is produced from the bath rather than the desired cathode reduction reaction. Post-electrolysis examination of the cell, once cooled, confirmed no obvious calcium deposits.

A series of additional experiments were undertaken with a carbon anode and a reference electrode, and confirm that the rise in voltage is linked to processes occurring at the oversized cathode, not the anode.[26] Those results demonstrate that:

• the voltage changes still occur at the same elapsed time for both a carbon anode and inert anode;

• the changes occur at twice the rate when the cathode current density is doubled and the anode current density is kept constant;

• experiments conducted with a reference electrode indicate these voltage changes are on the negative electrode (the cathode).

The processes occurring at the cathode are a series of electrochemical reduction processes resulting in the formation of mixtures of different titanium suboxides as well as calcium titanate.[27] Confirmation of the species formed was established by *ex situ* X-ray diffraction of smaller (~1 g) titanium dioxide pellet cathodes removed at fixed electrolysis times.[26] For example, the first identified products at the equivalent of ~5–6 hours electrolysis (for the results in Fig. 4) were ascribed to the following reactions:

$$7TiO_2 + Ca^{2+} + 2e^- \rightarrow Ti_6O_{11} + CaTiO_3$$

$$5TiO_2 + Ca^{2+} + 2e^- \rightarrow Ti_4O_7 + CaTiO_3$$

$$6TiO_2 + Ca^{2+} + 2e^- \rightarrow Ti_5O_9 + CaTiO_3$$

The next process at around 10–12 hours electrolysis, resulting in mostly Ti_2O_3 detected, is attributed to the two reactions:

$$4Ti_4O_7 + Ca^{2+} + 2e^- \rightarrow 5Ti_3O_5 + CaTiO_3$$

$$3Ti_3O_5 + Ca^{2+} + 2e^- \rightarrow 4Ti_2O_3 + CaTiO_3$$

The major voltage transition from ~1.6 V to 2.3 V (*IR* free voltage), which occurs between 14 and 17 hours in Fig. 4b is attributed to:

$$2Ti_2O_3 + Ca^{2+} + 2e^- \rightarrow 3TiO + CaTiO_3$$

Following more prolonged electrolysis times, the reaction progresses towards near full loss of oxygen and formation of titanium metal. For the oversized cathodes used in this study, the electrolysis does not proceed to completion.

The shapes of the voltage–time plots obtained for three types of generation 2 samples (NB-1350, NB-1400, NB-1500) are similar, but the voltage transitions occur at slightly different times. If the voltage plots are shifted slightly in time (Fig. 4c), the voltage transitions in the different plots coincide with one another. The voltage transitions are attributed to the cathode processes described above, with titania being reduced to more conductive phases. Some of our earlier work, utilising the oversized TiO_2 cathode with a carbon anode,[24] has shown that slight differences in the cathode fabrication (constructed by hand) are responsible for the variation in time of the cathodic voltage processes.

The cell resistance (measured by the in-house resistometer) for runs NB-1350, NB-1400 and NB-1500 is shown in Fig. 5. The position of the major cell voltage transition observed in Fig. 4 at around 14–16 h (3rd process) is marked with arrows on the plot in Fig. 5. As expected, there is no obvious resistance change at this transition. The sharp voltage transitions observed at constant current electrolysis are due to the logarithmic relationship between the voltage and the

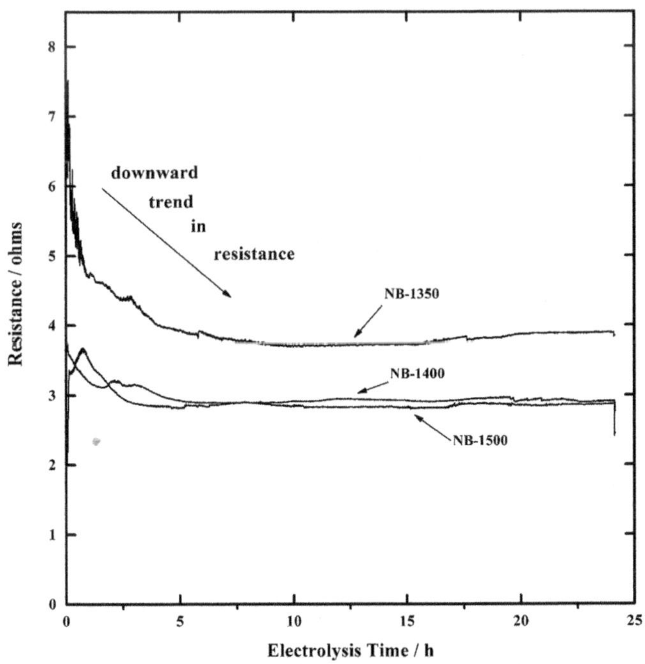

Fig. 5 Resistance plot for constant current (0.2 A cm^{-2} or 0.5 A) electrolysis utilising a series of Nb-doped TiO_2 electrodes sintered at different temperatures.

concentrations of redox species involved (according to the Nernst equation). However, resistance should be linearly proportional to the concentrations of redox species present. This resistance has a general downward trend due to increasing conductivity of the cathode with increased degree of reduction. Consequently, a gradual change in resistance was observed.

These experiments illustrate that no clear trend has emerged in the behavior of the anodes sintered at different temperatures. Any differences in voltage behavior appear linked to cathode processes.

3.3. 24 hour tests: effect of increasing current

Subsequent tests on the inert anode material (sample NB-1500 W, which was sintered at 1500 °C and was a wider sample) were carried out at 1 A current (~0.2 A cm^{-2}). These results are compared with the test at 0.5 A (~0.2 A cm^{-2}) shown in Fig. 6. Both anodes were run with cathodes fabricated in the same batch. Increasing the current with respect to the cathode shifts the time at which the major voltage transition occurs. The current density for the two tests with respect to the anode area is the same but is half with respect to the cathode area for the second test. This further confirms that the cathode process is responsible for the voltage transitions. It is likely, in the 24 hour test, the voltage at the anode (disregarding ohmic drop) has changed by only a small amount. Doubling the current with respect to the cathode has the effect of roughly halving the time the cathode takes to be consumed.

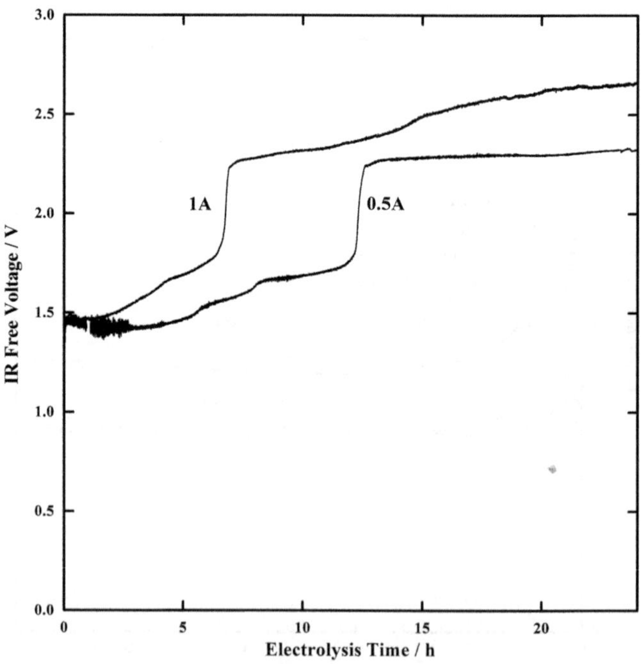

Fig. 6 Effect of absolute current (keeping current density constant at 0.2 A cm^{-2}) when carrying out constant current electrolysis utilising a $Nb_{0.054}Ti_{0.946}O_2$ electrode sintered at 1500 °C.

3.4. Extended (100 hour) electrolysis run

Comparison of the first generation Nb-doped titania anodes with the second generation material shows a much better performance from the latter over 24 hours. This gave confidence that a 100 hour run could be attempted utilising the optimised generation 2 material ($Nb_{0.054}Ti_{0.946}O_2$). A summary of all the long term measurements on the inert anode material is given in Table 1. A final modification to the cathode design for the 100 h tests was to make the cathode dish as deep as the previous designs but with a larger width. The total mass of titania pasted into this larger dish was 76 g rather than the 40 g used in the previous cathodes. This means more titania is available for the cathode reaction.

The 100 hour run for the inert anode was performed in two steps: 0 to 50 h and 50 to 100 h. Two different baths were used, and the cathode was replaced with a fresh cathode after 50 hours, so that the performance due to depletion of the cathode was not limited. The experiments were a good test of the durability of the inert anode, as the cell was cooled, the anode removed and then placed into a new cell and fresh bath in a second furnace. The voltage–time plots for the entire 100 hour run are shown in Fig. 7.

By the end of the first 50 h run, the cell voltage rose to 6 V. A rise in *IR* free voltage (2.8 V) and cell resistance is observed. At the beginning of the second 50 h run (new cathode and bath), the *IR* free voltage returned to its initial value and the resistance also returned to a lower value. This confirms that the cell voltage changes are mainly due to changes in the cathode material (*i.e.* reduction of the cathode material and increased resistivity of the cathode material). Towards the end of the second stage (50–100 h) of the electrolysis run there was a logging connection problem, as indicated on the graph, which affected the passive logging of resistance but did not affect the actual electrolysis. The resistance is roughly interpolated in this region. There was a large degree of noise in this region as well. This was most likely due to corrosion beginning to occur on the anode holder. Chemical analysis of the bath at this point (shown in Table 2)

Table 1 Constant current electrolysis runs with oversized TiO_2 cathodes to test capabilities of the inert anode

Expt. no.	Initial V/V	Final V/V	*R* (initial)/ ohm	*R* (final)/ ohm	*IR* free (initial)/ V	*IR* free (final)/ V	*I*/A	Current density/ A cm^{-2}	Dip depth/ mm	S.A./ cm^2	Electrolysis time/h
1	4.05	2.97	2.59	3.08	1.42	1.47	0.5	0.14	18	3.56	5.5
2	2.73	3.61	2.70	2.55	1.43	2.27	0.5	0.17	15.6	2.92	24
3	2.47	3.38	2.21	2.12	1.43	2.33	0.5	0.12	19.6	4.03	24
4	3.55	4.06	4.33	3.53	1.43	2.32	0.5	0.16	16	3.18	24
5	3.64	4.64	2.25	2.03	1.46	2.66	1.0	0.19	15.1	5.37	24
6	3.07	3.72	3.12	2.91	1.39	2.33	0.5	0.16	17	3.20	24
7	4.55	4.35	4.73	3.89	1.36	2.37	0.5	0.19	12.5	2.59	24
8	3.18	3.72	2.97	2.87	1.45	2.32	0.5	0.19	13	2.60	24
9	3.26	4.06	3.43	3.24	1.38	2.34	0.5	0.17	15.5	2.94	24
10	4.18	6.43	2.77	3.74	1.40	2.74	1.0	0.18	19	5.41	50
11	3.65	6.13	2.32	3.62	1.45	2.59	1.0	0.22	15.5	4.48	50 (2nd)

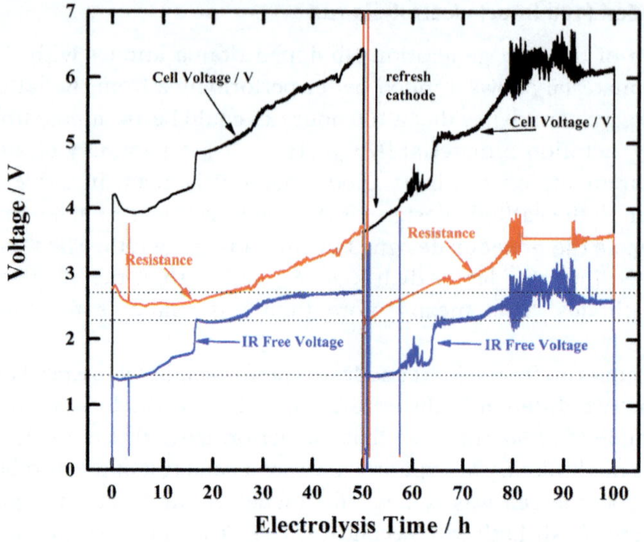

Fig. 7 100 hour electrolysis test (two 50 hour electrolysis runs) for $Nb_{0.054}Ti_{0.946}O_2$ inert anode and oversized pasted TiO_2 cathode. Voltage, resistance and *IR* free voltage plots.

reveals the presence of metal contamination. In particular, there is a large amount of nickel. Nickel deposition can contaminate the cathode product (Ti metal). This nickel has presumably come from the Inconel furnace components from trace amounts of HCl vapour in the cell that arise from moisture in the chloride bath. The increased metal concentration may explain the 0.2 V drop in *IR* free voltage towards the end of the second fifty hour run. This is not a limitation in the anode, but rather a limitation due to the bath chemistry and furnace materials.

3.5. Post-electrolysis examination

The dimensions and weight of the $Nb_{0.054}Ti_{0.946}O_2$ electrode decreased significantly in the 100 h run. A likely explanation is that niobium leaches into the bath due to oxidation of Ti^{3+} sites in the outer layers of the Nb-doped TiO_2 anode, and oxide ions are incorporated into the lattice. Nb_2O_5 is expected to form, and may segregate to the grain boundaries. The presence of Nb_2O_5 in the grain

Table 2 Bath compositions at different stages of the 100 hour (2 part) electrolysis[a]

	Electrolysis				
Stage	Nb ppm	Ti wt%	Cr μg g^{-1}	Ni μg g^{-1}	Fe μg g^{-1}
Bath 1: after fifty hours	84	0.56	1.07	2797	0.63
Bath 2: before 2nd fifty hours	3	0.05	530	27	540
Bath 2: after 2nd fifty hours	147	0.41	3310	14 400	9310

[a] Initial composition for bath 1 not recorded.

boundaries and junctions is a point of attack and wear by the chloride bath. The electrical conductivity of the bulk anode material appeared to be unaffected over the length of the 100 h electrolysis. EPMA results confirm the presence of an outer Nb-depleted region of TiO_2 (Fig. 8). The sample shows a well-defined and porous low Nb-rutile rim, of around 250 μm in width. The bulk material still appears dense and homogenous, with little bath penetration. From the change in dimensions and the change in weight (see Fig. 9), the amount of niobium leached into the bath was calculated (181 ppm); this agrees well with the analytical results obtained post-electrolysis (in Table 2). The analysis also showed significant quantities of nickel and iron in the bath originating from the Inconel cell furnace components. There is a possibility that iron oxide can form on the oxidative anode. Indeed, XRD analysis of the anode outer layer scraped off the post-electrolysis anode (which appeared orange in colour) did in fact reveal a small trace of iron oxide. The total weight loss from the anode during the 100 h run was 0.901 g.

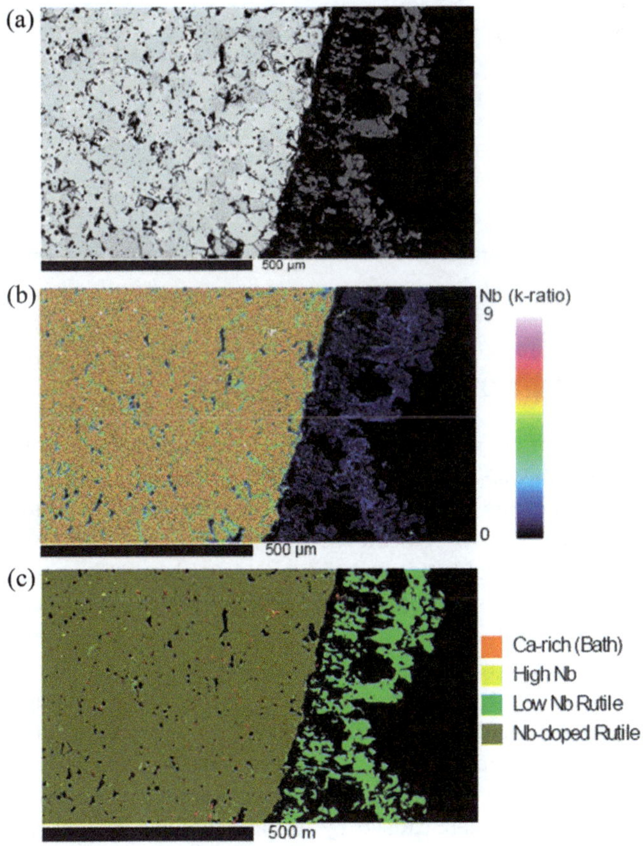

Fig. 8 Electron probe mapping images for the 24 hour electrolysis run of $Nb_{0.054}Ti_{0.946}O_2$: (a) back-scattered electron image; (b) elemental map for Nb, scale bar represents relative concentration; (c) elemental phase map for Ca and Nb.

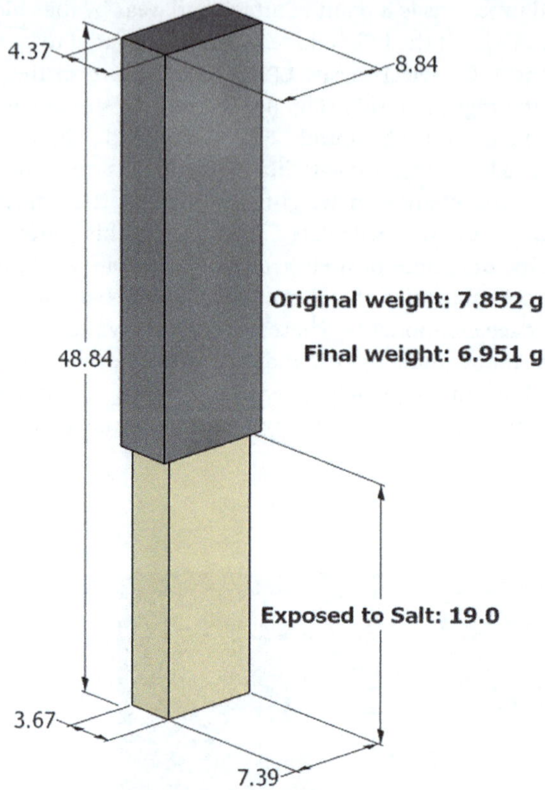

Fig. 9 Schematic representation of wear of inert anode (dimensions in mm) following the 100 h electrolysis test.

4. Concluding remarks

The performance of Nb-doped anodes as inert anodes for the solid-state electrolytic reduction of TiO_2 has been investigated. The effect of niobium content and sintering conditions on the performance of Nb-doped TiO_2 anodes was also studied. The best results were achieved by calcining in a reducing atmosphere and using dopant levels of 5.4 mol% niobium, *i.e.*, $Nb_{0.054}Ti_{0.946}O_2$. The effect of sintering temperature has little effect on the porosity or physical characteristics of the material, which performed well in 24 h electrolysis tests at a current density of 0.2 A cm^{-2}. Subsequently, $Nb_{0.054}Ti_{0.946}O_2$ anodes were subjected to an intensive 100 h electrolysis test. The anode gave an acceptable *IR* free voltage during the extended test, but post-electrolysis SEM and EPMA analyses reveal some degradation of the anode, and the presence of an outer Nb-depleted region of TiO_2 which is highly porous. Niobium was demonstrated to have leached into the bath. Anode deterioration is likely caused by oxidation of the Ti^{3+} sites in the rutile structure. Improvements in wear rate and stability to oxidation would need to be achieved for these materials to be regarded as truly 'inert', low-cost candidate anodes for the FFC process.

References

1 W. Kroll, *Trans. Am. Electrochem. Soc.*, 1940, **78**, 35–47.

2 G. Z. Chen, D. J. Fray and T. W. Farthing, *Nature*, 2000, **407**, 361–364.

3 R. O. Suzuki, M. Aizawa and K. Ono, *J. Alloys Compd.*, 1999, **288**, 173–182.

4 F. H. Froes, M. N. Gungor and M. A. Imam, *JOM*, 2007, **59**, 28–31.

5 F. H. Froes, M. N. Gungor and M. A. Imam, *Mater. Technol.*, 2006, **21**, 206–209.

6 F. H. Froes, M. A. Imam and D. Fray, in *Cost-Affordable Titanium – Symposium Dedicated to Professor Harvey Flower*, ed. F. H. Froes, M. A. Imam and D. Fray, Minerals, Metals & Materials Soc, Warrendale, 2004, pp. 3–8.

7 K. McGregor, E. J. Fraser, A. J. Urban, M. I. Pownceby and R. L. Deutscher, *ECS Trans.*, 2006, **2**, 369–380.

8 R. P. Pawlek, in *Light Metals 2008*, ed. D. H. DeYoung, Minerals, Metals & Materials Soc, Warrendale, 2008, pp. 1039–1045.

9 R. P. Pawlek, in *Light Metals 2004*, ed. A. T. Tabereaux, Minerals, Metals & Materials Soc, Warrendale, 2004, pp. 283–287.

10 R. P. Pawlek, in *Light Metals 2002*, ed. W. A. Schneider, Minerals, Metals & Materials Soc, Warrendale, 2002, pp. 449–456.

11 K. Billehaug and H. A. Ǿye, *Aluminium*, 1981, **57**, 2.

12 DOE/ID/13133, Report for US Department of Energy, June 1997, http://www.osti.gov/bridge/servlets/purl/530966-595Atx/webviewable/530966.pdf.

13 A. Mousa, X. Y. Yan, M. I. Pownceby, M. Cooksey, K. J. McDonald and M. R. Lanyon, in *Magnesium Technology 2006, Proceedings*, ed. A. A. Luo, N. R. Neelameggham and R. S. Beals, Minerals, Metals & Materials Soc, Warrendale, 2006, pp. 13–18.

14 J. F. Moore, J. N. Hryn, M. J. Pellin, W. F. Calaway and K. Watson, in *Magnesium Technology 2000*, ed. H. L. Kaplan, J. N. Hryn and B. B. Clow, Minerals, Metals & Materials Soc, Warrendale, 2000, pp. 21–26.

15 A. Martin, D. Lambertin, J. C. Poignet, A. Allibert, G. Bourges, L. Pescayre and J. Fouletier, *JOMMER*, 2003, **55**, 52–54.

16 S. Q. Jiao and D. J. Fray, *Metall. Mater. Trans. B*, 2010, **41**, 74–79.

17 (*a*) K. T. Kilby, S. Q. Jiao and D. J. Fray, *Electrochim. Acta*, 2010, **55**, 7126–7133; (*b*) S. Jiao, L. Zhang, H. Zhu and D. J. Fray, *Electrochim. Acta*, 2010, **55**, 7016–7020; (*c*) L. Hu, Y. Song, J. Ge, S. Jiao and J. Cheng, *J. Electrochem. Soc.*, 2016, **163**, E33–E38.

18 H. Yin, L. Gao, H. Zhu, X. Mao, F. Gan and D. Wang, *Electrochim. Acta*, 2011, **56**, 3296–3302.

19 S. F. Spangenberg, A. L. Finley and D. J. Searson, *US Patent*, 4448654, 1984.

20 R. L. Deutscher, S. Fletcher and J. A. Hamilton, *Electrochim. Acta*, 1986, **31**, 585–589.

21 G. A. Snook, A. J. Urban, M. R. Lanyon and K. McGregor, *J. Electroanal. Chem.*, 2008, **622**, 225–232.

22 K. McGregor, G. A. Snook, N. V. Y. Scarlett, A. J. Urban, M. R. Lanyon and I. C. Madsen, in *Light Metals 2009*, ed. G. Bearne, Minerals, Metals & Materials Soc, Warrendale, 2009, pp. 435–441.

23 G. A. Snook, K. McGregor, A. J. Urban and M. A. Cooksey, *Electrochim. Acta*, 2009, **54**, 4925–4932.

24 G. A. Snook, K. McGregor, A. J. Urban and M. R. Lanyon, *J. Solid State Electrochem.*, 2009, **13**, 591–598.

25 K. McGregor, G. A. Snook, A. J. Urban and M. R. Lanyon, *ECS Trans.*, 2011, **28**, 349–360.

26 K. McGregor, G. A. Snook, A. J. Urban and M. R. Lanyon, 2010, unpublished work.

27 C. Schwandt and D. J. Fray, *Electrochim. Acta*, 2005, **51**, 66–76.

Faraday Discussions

PAPER

Solid oxide membrane-assisted controllable electrolytic fabrication of metal carbides in molten salt

Xingli Zou,*[ab] Kai Zheng,[ab] Xionggang Lu,*[ab] Qian Xu[ab] and Zhongfu Zhou[abc]

Received 13th December 2015, Accepted 8th January 2016

DOI: 10.1039/c5fd00221d

Silicon carbide (SiC), titanium carbide (TiC), zirconium carbide (ZrC), and tantalum carbide (TaC) have been electrochemically produced directly from their corresponding stoichiometric metal oxides/carbon (MO_x/C) precursors by electrodeoxidation in molten calcium chloride ($CaCl_2$). An assembled yttria stabilized zirconia solid oxide membrane (SOM)-based anode was employed to control the electrodeoxidation process. The SOM-assisted controllable electrochemical process was carried out in molten $CaCl_2$ at 1000 °C with a potential of 3.5 to 4.0 V. The reaction mechanism of the electrochemical production process and the characteristics of these produced metal carbides (MCs) were systematically investigated. X-ray diffraction, scanning electron microscopy, and transmission electron microscopy analyses clearly identify that SiC, TiC, ZrC, and TaC carbides can be facilely fabricated. SiC carbide can be controlled to form a homogeneous nanowire structure, while the morphologies of TiC, ZrC, and TaC carbides exhibit porous nodular structures with micro/nanoscale particles. The complex chemical/electrochemical reaction processes including the compounding, electrodeoxidation, dissolution–electrodeposition, and *in situ* carbonization processes in molten $CaCl_2$ are also discussed. The present results preliminarily demonstrate that the molten salt-based SOM-assisted electrodeoxidation process has the potential to be used for the facile and controllable electrodeoxidation of MO_x/C precursors to micro/nanostructured MCs, which can potentially be used for various applications.

Introduction

Metal carbides (MCs), especially transition metal carbides, have attracted a significant amount of attention over the past few years as electrocatalyst support materials, structural materials, and high-temperature materials, *etc.*[1–11] In

[a]*State Key Laboratory of Advanced Special Steel, Shanghai University, Shanghai 200072, China. E-mail: xinglizou@shu.edu.cn; luxg@shu.edu.cn*

[b]*Shanghai Key Laboratory of Advanced Ferrometallurgy, Shanghai University, Shanghai 200072, China*

[c]*Institute of Mathematics and Physics, Aberystwyth University, Aberystwyth, SY23 3BZ, UK*

particular, nanostructured MCs materials hold promise for various applications due to their interesting properties.[1–3,6–11] MCs materials are commonly quite stable, resistant to oxidation and acid etching, electrically conducting, and superconducting.[6] In addition, MCs ceramics such as tantalum carbide (TaC) and zirconium carbide (ZrC) generally possess high melting points and thus have the potential to be used for extremely high-temperature applications.[12–19] While the high melting points of MCs are useful for high-temperature applications, they also cause problems when preparing these MCs materials. Currently, MCs can be produced by various methods such as arc melting, high-temperature annealing, carbothermic reduction, sol–gel chemistry method, combustion process, and solid-state method, *etc.*[1–20] However, these methods typically involve complex/energy-intensive processes such as a high-temperature process. In particular, the high-temperature process commonly causes aggregation during the production process and greatly decreases the surface areas of the produced MCs products.[6] Therefore, searching for new alternative process for the controllable fabrication of micro/nanostructured MCs at relatively moderate conditions is required.

Molten salt electrolytic production of homogeneous metal/alloy powders from their corresponding metal oxide (MO_x) precursors at relatively moderate temperatures (approximately 800 to 1000 °C) has been intensively investigated in recent years.[21–37] In particular, the recently developed molten salt-based electro-deoxidation process known as the FFC Cambridge process has attracted tremendous research interest towards electrochemical production of metal/alloy/composite powders from solid MO_x precursors.[21,23–25] The molten salt-based electrodeoxidation process commonly uses molten calcium chloride ($CaCl_2$) as the electrolyte, and an oxygen component contained in the solid MO_x precursors can be electrochemically removed, thus the preformed solid porous MO_x precursors can be directly converted into metal/alloy powders. In the molten salt-based electrodeoxidation process, the typical anode electrode used in the experiment is a graphite-based anode.[38] During the electrodeoxidation process, oxygen ions (O^{2-} ions) were generated from the MO_x cathode *via* the general reaction, *i.e.*, $MO_x + 2xe^- \rightarrow M + 2xO^{2-}$, then the formed O^{2-} ions electromigrated to the graphite-based anode and oxidized to form carbon monoxide (CO) and/or carbon dioxide (CO_2). Therefore, the reactive graphite-based anode was commonly degraded by the formation of CO/CO_2 and erosion during electrodeoxidation.[38] In addition, the use of a graphite-based anode is also known to pollute the molten electrolyte and decrease the current efficiency of the electrodeoxidation process.[38,39] In particular, during the electrodeoxidation process, the anodically produced CO_2 can convert to CO_3^{2-} ions in the molten salt through the reaction, *i.e.*, $CO_2 + O^{2-} \rightarrow CO_3^{2-}$.[38] Then, the formed CO_3^{2-} ions could be subsequently electrolyzed to form C at the cathode, *i.e.*, $CO_3^{2-} + 4e^- \rightarrow C + 3O^{2-}$.[23,24,38] Therefore, the electrodeoxidation process and the cathodic products produced in the graphite-based anode electrolytic cell are usually influenced by the above-mentioned side reactions.[40,41] This graphite-based anode molten salt electro-deoxidation process has been used to produce MCs (such as titanium carbide (TiC), silicon carbide (SiC), hafnium carbide (HfC), niobium carbide (NbC), and ZrC, *etc.*) in the previous work.[41–49] However, in order to produce MCs, the carbon content of the MO_x/carbon (MO_x/C) precursors used in the graphite-based anode electrodeoxidation process should be generally higher than the stoichiometric ratios of their corresponding MCs.[44,48,49] Therefore, it is difficult to control the carbon content of the MCs products fabricated using the graphite-based anode electrodeoxidation process.[41] In order to produce pure MCs with pre-designed components, the graphite-based anode needs to be improved to avoid the CO/CO_2-related anodic side reactions.

The solid oxide membrane (SOM)-assisted electrodeoxidation process employs an inertly assembled yttria-stabilized zirconia SOM (YSZ-SOM) anode to replace the graphite-based anode.[50–56] During the SOM-assisted electrodeoxidation process, only O^{2-} ions could migrate through the SOM, thus the anode reaction area was separated from the molten salt system. Consequently, the influences of the anodic products on the electrodeoxidation process, and the cathodic products, could be effectively avoided. In our previous work,[52–56] the SOM-assisted electrodeoxidation process has been successfully used to prepare metals, alloys, and composites from their MO_x precursors. Importantly, the SOM-assisted electrochemical production process has been proved to be able to precisely control the carbon content of the produced carbide-containing composite products.[52] Thus, it is reasonable to expect that the SOM-assisted electrodeoxidation process has a significant potential to be used for the controllable electrochemical production of MCs with pre-designed components from their corresponding MO_x/C precursors.

In this work, MCs such as SiC nanowires (SiC NWs), TiC, ZrC, and TaC micro/nanoparticles have been electrochemically produced from their corresponding MO_x/C precursors using the SOM-assisted electrodeoxidation process. The electrodeoxidation processes were terminated at pre-determined intervals of time, and the obtained products were characterized using X-ray diffraction (XRD), scanning electron microscopy (SEM), and transmission electron microscopy (TEM), *etc.* In addition, the reaction mechanisms of the electrodeoxidation/*in situ* carbonization processes including chemical and electrochemical reactions are also discussed.

Experimental section

Fabrication of MO_x/C cathode precursors

The homogeneous MO_x/C precursors including silicon dioxide/carbon (SiO_2/C), titanium dioxide/carbon (TiO_2/C), zirconium dioxide/carbon (ZrO_2/C), and tantalum pentoxide/carbon (Ta_2O_5/C) mixtures ($SiO_2 : C = 1 : 1$, $TiO_2 : C = 1 : 1$, $ZrO_2 : C = 1 : 1$, and $Ta_2O_5 : C = 1 : 2$) were prepared from micro/nanosized SiO_2, TiO_2, ZrO_2, Ta_2O_5, and C powders. The stoichiometric MO_x/C mixtures were then pressed under appropriate pressures to form porous SiO_2/C, TiO_2/C, ZrO_2/C, and Ta_2O_5/C pellets, respectively. The pressed MO_x/C pellet was sandwiched between two porous nickel foils to form a pellet cathode. The porous nickel foil (porosity: ~95%; pore per inch: 110; area density: 380 g m^{-2}; pore size: 0.2 to 0.4 mm) possesses a uniform structure and thus can be functioned as the extended electronic conductor to provide more initial contact points to the MO_x/C pellet. It should be noted that the pressed MO_x/C pellets would be pre-sintered in argon (Ar) gas during the temperature rising process. However, on the basis of the thermodynamic calculations/evaluations of these general reactions between MO_x and C at 1000 °C, *i.e.*, $MO_x + (x + 1)C \rightarrow MC + xCO$ ($\Delta G^o_{(M=Si,\ x=2)} = 173.22$ kJ mol^{-1}, $\Delta G^o_{(M=Ti,\ x=2)} = 95.99$ kJ mol^{-1}, $\Delta G^o_{(M=Zr,\ x=2)} = 226.29$ kJ mol^{-1}, $\Delta G^o_{(M=Ta,\ x=2.5)} = 45.08$ kJ mol^{-1}), it is obvious that the initial carbon component contained in the MO_x/C precursors would not be consumed in the inert Ar gas atmosphere during the temperature rising process. Therefore, it is possible to precisely control the initial carbon contents of MO_x/C precursors to produce the desired MCs products.

Fabrication of the SOM-based anode

The assembled YSZ-SOM-based anode was composed of a one-end-closed YSZ tube filled with carbon-saturated liquid tin, and a metallic electrode wire (such as Mo wire) was inserted into the tube to conduct electricity during the electrodeoxidation process. The YSZ-SOM tube was fabricated using a slip casting method in our laboratory, and it could be fabricated to form tube/crucible shapes with different membrane thicknesses. As YSZ-SOM commonly possesses excellent molten salt corrosion resistance,[51–56] it is reasonable to believe that the SOM-assisted electrodeoxidation process can support a long-duration electrolysis process. The liquid tin contained in the SOM tube was used as a medium to transport O^{2-} ions during the electrodeoxidation process, and the carbon saturated in the liquid tin was used as the reductant to react with oxygen in this experiment. The key feature of the SOM-assisted electrodeoxidation process is that the anode reaction area was separated by the O^{2-} ion-conducting membrane from the molten salt during the electrodeoxidation process, indicating that the anodic products such as CO and/or CO_2 could not make contact with the molten $CaCl_2$. Consequently, the side reactions, *e.g.*, $O^{2-}{}_{\text{(in molten salt)}} + CO_2/{+}Ca^{2+}{}_{\text{(in molten salt)}} \rightarrow CO_3{}^{2-}/CaCO_{3\text{(in molten salt)}}$[38] could not occur during the electrodeoxidation process. Therefore, the carbon content of the cathodic products can be predesigned and controlled. Actually, the anodic products of the SOM-assisted electrodeoxidation process can be controlled to form O_2, CO/CO_2 or H_2O with further modification.[51,57]

Electrodeoxidation process

The fabricated MO_x/C pellet cathode and the SOM-based anode were assembled in an alumina crucible filled with analytically pure $CaCl_2$ to form an electrochemical

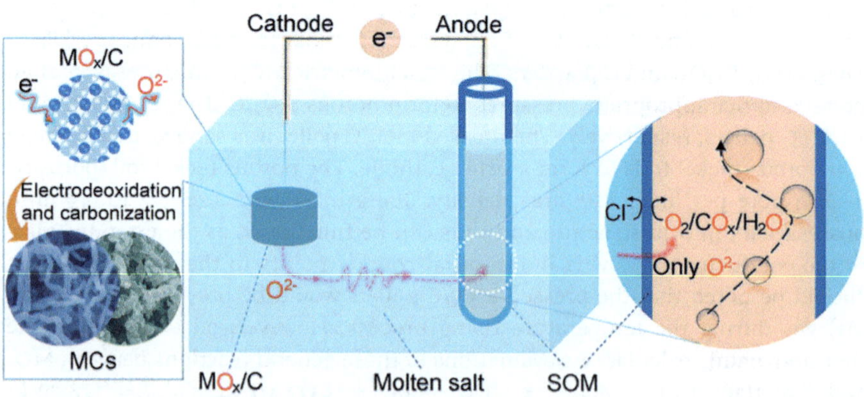

Fig. 1 Schematic illustration of the SOM-assisted controllable electrolytic cell used for the electrochemical production of MCs in molten salt.

cell, as shown in Fig. 1. $CaCl_2$ was melted to form a molten salt and served as the molten electrolyte during the electrodeoxidation process. During the whole experiment, the reaction chamber was continuously purged with ultra-high-purity Ar gas. When the temperature increased to the pre-determined electrolysis temperature (1000 °C), pre-electrolysis was carried out at 2.0 to 2.5 V between the SOM-based anode and a Fe–Cr–Al wire cathode to eliminate redox-active species from the molten $CaCl_2$. When the background/residual current of the pre-electrolysis process remained at a lower level, the pre-electrolysis process was terminated, and then the electrodeoxidation process was carried out at 3.5 to 4.0 V. A Biologic HCP-803 electrochemical workstation was used to control the electrochemical experiment and record the current–time curve. During the electrodeoxidation process, the oxygen component contained in the MO_x/C precursors was gradually removed and MO_x/C could transform into MCs through electrodeoxidation and *in situ* carbonization processes, as shown in Fig. 1. In order to produce partially electrodeoxidized MO_x/C samples, the electrodeoxidation was interrupted after different reaction times. After the experiment, the electrodeoxidized MO_x/C samples were vigorously rinsed with tap water and then dried at 100 °C. In addition, cyclic voltammetry (CV) experiments were performed in a three-electrode electrolytic cell. During the CV experiment, a metallic cavity electrode (MCE) filled with MO_x/C powders was used as the working electrode, and the SOM-based anode and a Pt wire were used as the counter electrode and reference electrode, respectively.

Characterization of products

The phase composition of the partially deoxidized MO_x/C products and the final products was determined using XRD on a Rigaku D/Max-2550 diffractometer, with Cu Kα radiation ($\lambda = 0.15406$ nm) operated at a voltage of 40 kV and 100 mA. The microstructure of the samples was investigated using SEM on a JEOL JSM-6700F microscope. TEM analysis was performed on a JEOL JEM-2010F microscope operating at 200 kV. Digital optical photographs of the samples were taken using a KEYENCE VHX-1000C digital optical microscope. N_2 adsorption and desorption isotherms were measured using a Micromeritics ASAP 2020 sorptometer at liquid nitrogen temperature (−196 °C). Before the measurement, the sample was degassed at 200 °C for 6 h. The specific surface area was calculated using the Brunauer–Emmett–Teller (BET) method. The pore size distribution curve was calculated using the desorption branch of the isotherms and the Barrett–Joyner–Halenda method. The average particle sizes of the product powders were determined using a Malvern Mastersizer 2000 laser particle size analyzer. Raman spectroscopy of the sample was performed on a Renishaw InVia Raman microspectrometer using an Ar ion laser (514.5 nm).

Results and discussion

Electrolytic fabrication of SiC NWs

Electrodeoxidation of SiO_2 to Si in molten salts has been intensively investigated in recent years.[22,32,35,36,58,59] In particular, Si NWs have been successfully produced from SiO_2 powder in molten salts and the reaction mechanism of the electrodeoxidation process was proposed.[23,35,36] Based on the similarities of the reaction conditions and the electrodeoxidation processes, the reaction mechanisms of the electrodeoxidation of SiO_2/C and SiO_2 in molten $CaCl_2$ are expected to be quite similar. Fig. 2a presents the CV curve of the electrodeoxidation of SiO_2/C in molten $CaCl_2$ at 1000 °C. Obviously, the reduction peak occurring at approximately −1.1 V is mainly responsible for the electrodeoxidation reactions. The CV curve is similar to the previous work,[23] indicating that the electrodeoxidation of SiO_2/C has a similar electrochemical reaction mechanism to the electrodeoxidation of SiO_2 in molten

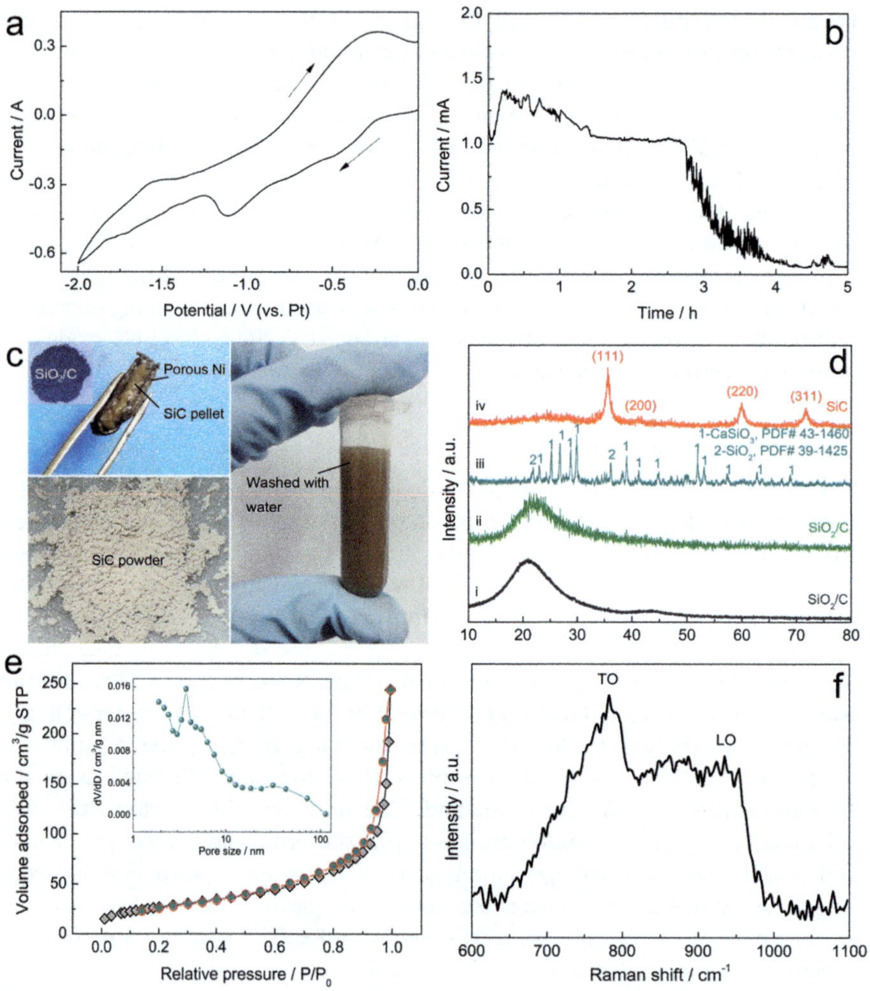

Fig. 2 (a) CV curve of the SiO_2/C precursors in molten $CaCl_2$ at 1000 °C with a scan rate of 60 mV s^{-1}. (b) Typical current–time curve of the electrodeoxidation of SiO_2/C in molten $CaCl_2$ at 1000 °C and 3.8 V. (c) Photos of SiO_2/C powder precursors, the cross section of the completely electrodeoxidized SiO_2/C pellet (*i.e.*, SiC pellet), and the obtained SiC powder product (1000 °C, 3.8 V, 5 h). (d) XRD patterns of (i) SiO_2/C precursors, (ii) the SiO_2/C precursors after being sintered at 1000 °C in Ar, (iii) the surface of the SiO_2/C pellet after being immersed/sintered in molten $CaCl_2$ at 1000 °C for 6 h, and (iv) the obtained SiC (1000 °C, 3.8 V, 5 h). (e) Nitrogen adsorption–desorption isotherms of the produced porous SiC. (f) Typical Raman spectrum of the produced SiC.

$CaCl_2$.[23–25,35,36] Therefore, the reaction mechanism of the electrodeoxidation of SiO_2/C in molten $CaCl_2$ is believed to generally involve compounding,[25] electrodeoxidation,[24] dissolution–electrodeposition,[23,35] and *in situ* carbonization processes. The typical current–time curve of the electrodeoxidation of SiO_2/C in molten $CaCl_2$ at 1000 °C and 3.8 V is shown in Fig. 2b. In comparison with the representative current–time curves obtained from the electrodeoxidation of solid MO_x/compounds in molten $CaCl_2$,[30,53,55,56] it can be seen from Fig. 2b that the current–time curve of the electrodeoxidation of SiO_2/C exhibits a relatively different variation trend. The current stays at a relatively large value (1000 to 1400 mA) during the electrodeoxidation time period of 0 to 3 h, then the current gradually declines to a small residual current level (~200 mA). The current–time curve shows a similar variation trend to our recent work,[60] in which the reaction mechanism was dominated by an *in situ* dissolution–electrodeposition mechanism. Therefore, the large current value time period of 0 to 3 h (Fig. 2b) may be mainly attributed to the dissolution–electrodeposition process occurring during electrodeoxidation.

Fig. 2c shows the photos of SiO_2/C powder precursors, the cross section of the completely electrodeoxidized SiO_2/C pellet (*i.e.*, SiC pellet), and the obtained SiC powder. Obviously, the produced SiC powder exhibits a yellow color, while the SiO_2/C precursors display a black color. During the electrodeoxidation process, the SiO_2/C mixture would be inevitably sintered in molten $CaCl_2$. Therefore, it is important to evaluate the influences of the sintering processes in Ar gas and molten $CaCl_2$ on the SiO_2/C precursors. The XRD patterns of the SiO_2/C precursors before and after being sintered in Ar gas and molten $CaCl_2$ are shown in Fig. 2d, and the XRD pattern of the obtained final SiC product is also given in Fig. 2d. Clearly, the sintering process in Ar gas had almost no influence on the SiO_2/C precursors. On the contrary, the immersion/sintering process in molten $CaCl_2$ (without being pre-electrolyzed) did influence the SiO_2/C precursors. Not surprisingly, the surfaces of the SiO_2/C pellet precursors have been partially converted into $CaSiO_3$ (PDF# 43-1460) during the immersion process in molten $CaCl_2$, this observation is consistent with the previous work.[35] Finally, after being electrodeoxidized at 1000 °C and 3.8 V for 5 h, the SiO_2/C precursors were completely transformed into SiC (PDF# 29-1129), as shown in Fig. 2d. The obtained SiC powder was further characterized using N_2 adsorption–desorption isotherms and Raman analyses, as shown in Fig. 2e and f. The BET specific surface area was calculated to be 96.64 $m^2\ g^{-1}$. The Raman spectrum of the produced SiC shows a similar characteristic to the previous work,[61] indicating that the obtained SiC is pure SiC.

In addition, the microstructures of the SiO_2/C precursors and the obtained SiC product were also examined using SEM and TEM, as shown in Fig. 3. The SiO_2/C precursors show nanosized powder structures, and the particle sizes are approximately 100 nm (Fig. 3a). Interestingly, after being electrodeoxidized at 1000 °C and 3.8 V for 5 h in molten $CaCl_2$, the powdered SiO_2/C precursors have been converted into SiC NWs structures (Fig. 3b). The SEM image of the SiC NWs obtained from the SiO_2/C mixture powders (particle sizes are about 100 nm) is shown in Fig. 3b(i), it is obvious that the SiC NWs are homogeneous with diameters of approximately 30 nm. To investigate the influence of the particle size of the initial SiO_2/C precursors on the final SiC product, the SiO_2/C precursors with particle sizes of approximately 1 μm were also used as precursors, and the SEM image of the obtained SiC NWs is shown in Fig. 3b(ii). It can be seen from Fig. 3b

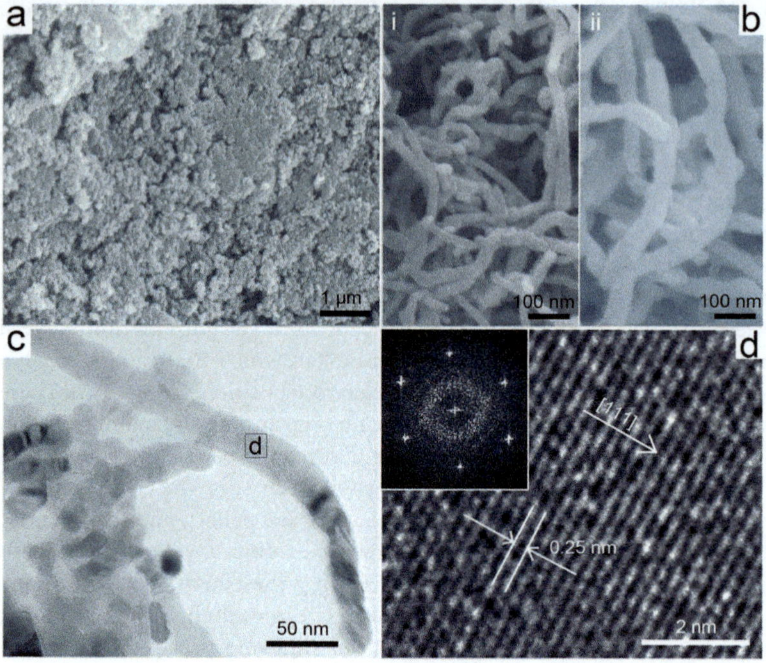

Fig. 3 Typical SEM images of (a) SiO_2/C pellet precursors and (b) the produced SiC NWs; (i) the SiC NWs produced from nanosized (~100 nm) SiO_2/C precursors (1000 °C, 3.8 V, 5 h) corresponding to (a), (ii) the SiC NWs produced from microscale (~1 μm) SiO_2/C precursors (1000 °C, 3.8 V, 5 h). (c) TEM image and (d) HRTEM image of the obtained SiC NWs produced from nanosized (~100 nm) SiO_2/C precursors, the inset in (d) is its corresponding SAED pattern.

that the diameters of the obtained SiC NWs are different, indicating that the initial particle size of the SiO_2/C precursors can, at least partially, influence the diameter of the final SiC NWs. During the electrodeoxidation process, solid SiO_2/C precursors were electrodeoxidized to form initial SiC nuclei. Simultaneously, the intermediate product $CaSiO_3$ was formed, the generated $CaSiO_3$ could immediately dissolve into molten $CaCl_2$ at the reaction front. Therefore, the dissolved $CaSiO_3$ (*i.e.*, SiO_3^{2-}) could *in situ* electrodeposit on the pre-formed SiC nuclei and further reacted with C to form a SiC nanowire structure. Consequently, the particle size of the initial SiC nuclei generated from the SiO_2/C precursors could influence the subsequent electrodeposition process. The typical TEM and high-resolution TEM (HRTEM) images of the obtained SiC NWs are shown in Fig. 3c and d, respectively. It can be seen from the figure that the lattice spacing value of the SiC NWs is 0.25 nm, which corresponds to the interplanar spacing of the [111] planes of 3C–SiC.[8,9] The selected area electro diffraction (SAED) pattern (Fig. 3d inset) further confirms that the obtained SiC product is crystalline.[9]

Electrolytic fabrication of TiC powder

Besides SiC, TiC has also been electrochemically produced from nanoscale TiO_2/C precursors (particle sizes are approximately 500 nm) using the SOM-assisted electrodeoxidation process. Fig. 4a–d shows the SEM images of the cross-section of a partially electrodeoxidized TiO_2/C pellet. It is obvious that the cross-section of the pellet exhibits different morphologies. The outside of the pellet has been electrodeoxidized and the particles show a homogeneous porous structure. The reaction area of the pellet mainly consists of large agglomerated particles, which may be attributed to the compounding process.[52] The electrodeoxidation-generated O^{2-} ions could react with Ca^{2+} ions and TiO_2 to form a $CaTiO_3$ compound in the reaction area, therefore, the initial particles grew up to form large particles during the electrodeoxidation process. The unreacted centre part of the pellet shows compact structures. These SEM image observations suggest that the reactions firstly occurred on the surfaces of the TiO_2/C pellet and then gradually extended into the pellet's interior. Fig. 4e shows the variations of the average particle size of TiO_2/C mixtures during the electrodeoxidation process. It clearly shows that the average particle size gradually increases in the first ~2 h due to the formation of $CaTiO_3$. Then the average particle size decreases with the removal of the oxygen component from the particles. When the TiO_2/C precursors have been completely electrodeoxidized to TiC within 4 h, the average particle size is about 3 to 4 μm. As the electrodeoxidation time prolongs to 6 h, the average particle size gradually increases due to the sintering effect. This observation is consistent with the results reported in our previous work.[52,53]

Fig. 5 shows the macro/microstructures and XRD patterns of the produced TiC powder. Obviously, the cross section of the completely electrodeoxidized TiO_2/C

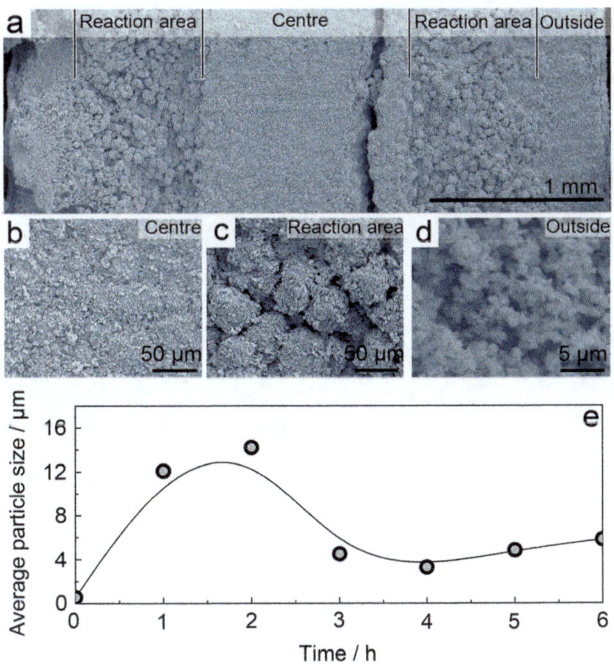

Fig. 4 SEM images of the cross-section of the partially electrodeoxidized (1000 °C, 4.0 V, 1 h) TiO_2/C pellet showing (a) the macrostructure of the pellet, and the microstructures of (b) the unreacted centre area, (c) the reaction area, and (d) the electrodeoxidized outside area. (e) Variations of the average particle size of the mixed TiO_2/C powders during the electrodeoxidation process at 1000 °C and 4.0 V.

pellet (*i.e.*, TiC pellet) exhibits a uniform structure, and the obtained TiC powder displays a gray color, as shown in Fig. 5a. Fig. 5b exhibits the XRD patterns of the TiO_2/C precursors before and after being electrodeoxidized at 1000 °C and 4.0 V at different times. The intermediate products contain Ti_2O_3 (PDF# 43-1033) and $CaTiO_3$ (PDF# 42-0423). Finally, the intermediate products have been electrodeoxidized and transformed into the TiC (PDF# 01-1222) product. It should be noted that a Ti metal phase is not observed in the XRD patterns shown in Fig. 5b. The reason may be attributed to the *in situ* carbonization reaction (Ti + C → TiC, $\Delta G^{o}{}_{1000°C} = -169.84$ kJ mol^{-1}) simultaneously occurring with the electrodeoxidation reaction during the electrodeoxidation process. The electrodeoxidation-generated Ti could immediately react with C to form TiC, indicating that only a small amount of Ti could exist in the reaction area. The typical SEM and TEM images of the produced TiC are shown in Fig. 5c and d, respectively. Microscale TiC particles with smooth surfaces are obtained (particle sizes are approximately 1 to 4 μm), and these TiC particles begin to interconnect together to form a porous nodular structure.

Electrolytic fabrication of the ZrC powder

The electrodeoxidation of nanoscale ZrO_2/C precursors (particle sizes are approximately 500 nm) to ZrC, using the SOM-assisted electrodeoxidation

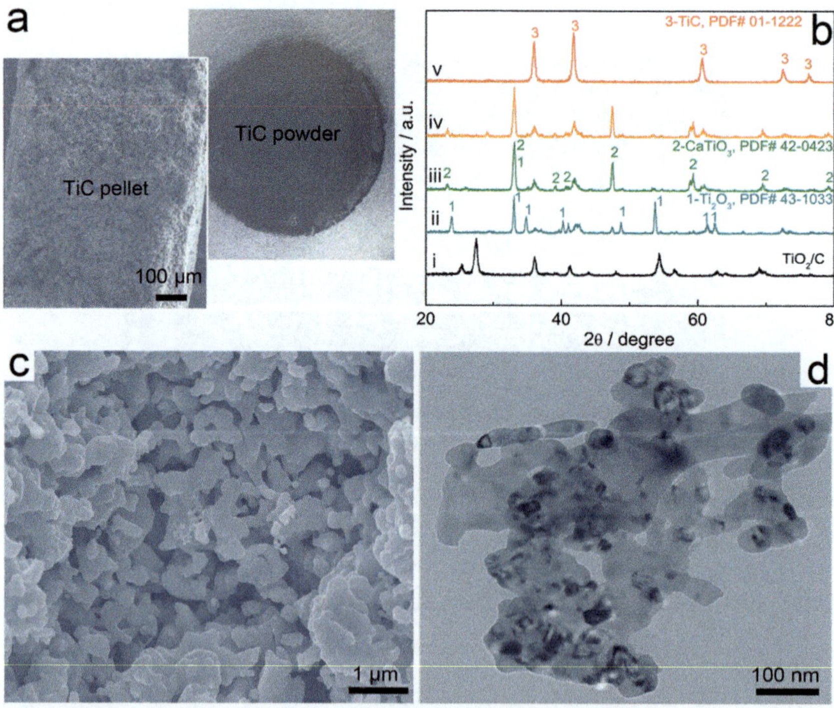

Fig. 5 (a) Photos of the completely electrodeoxidized TiO_2/C pellet (*i.e.*, TiC pellet) and its corresponding TiC powder. (b) XRD patterns of the TiO_2/C precursors (i) before and (ii)–(iv) after being electrodeoxidized in molten $CaCl_2$ at 1000 °C and 4.0 V for (ii) 1 h, (iii) 2 h, (iv) 3 h, and (v) 4 h. (c) Typical SEM image and (d) TEM image of the produced TiC powder (1000 °C, 4.0 V, 4 h).

process, was also systematically carried out in molten $CaCl_2$ at 1000 °C and 4.0 V. The XRD patterns of the ZrO_2/C precursors before and after being electrodeoxidized at different times are shown in Fig. 6a and b. XRD analysis shows that the reaction pathway generally proceeds through $ZrO_2/C \rightarrow Ca_{0.15}Zr_{0.85}O_{1.85}/C \rightarrow CaZrO_3/C/ZrC \rightarrow ZrC$. The electrodeoxidation-generated O^{2-} ions could also react with Ca^{2+} ions and ZrO_2 to form Ca–Zr–O compounds during the electrodeoxidation process. During the initial electrodeoxidation period, $Ca_{0.15}Zr_{0.85}O_{1.85}$ (PDF# 84-1827) was gradually formed (Fig. 6a and b), indicating that the reaction, *i.e.*, $ZrO_2 + 0.15Ca^{2+} + 0.6e^- + 0.15C \rightarrow Ca_{0.15}Zr_{0.85}O_{1.85} + 0.15ZrC + 0.15O^{2-}$ occurred in the reaction area. When the electrodeoxidation time increased to approximately 3 to 4 h, $CaZrO_3$ (PDF# 76-2401) was generated (Fig. 6a and b), indicating that the reaction, *i.e.*, $2ZrO_2 + Ca^{2+} + 4e^- + C \rightarrow CaZrO_3 + ZrC + O^{2-}$ presented during the electrodeoxidation time period of 3 to 4 h. The reaction pathway observed in this work is generally consistent with the reaction pathway of the electrodeoxidation of ZrO_2 in molten $CaCl_2$.[24,31] After being electrodeoxidized at 1000 °C and 4.0 V for 6 h, pure ZrC (PDF# 35-0784) has been

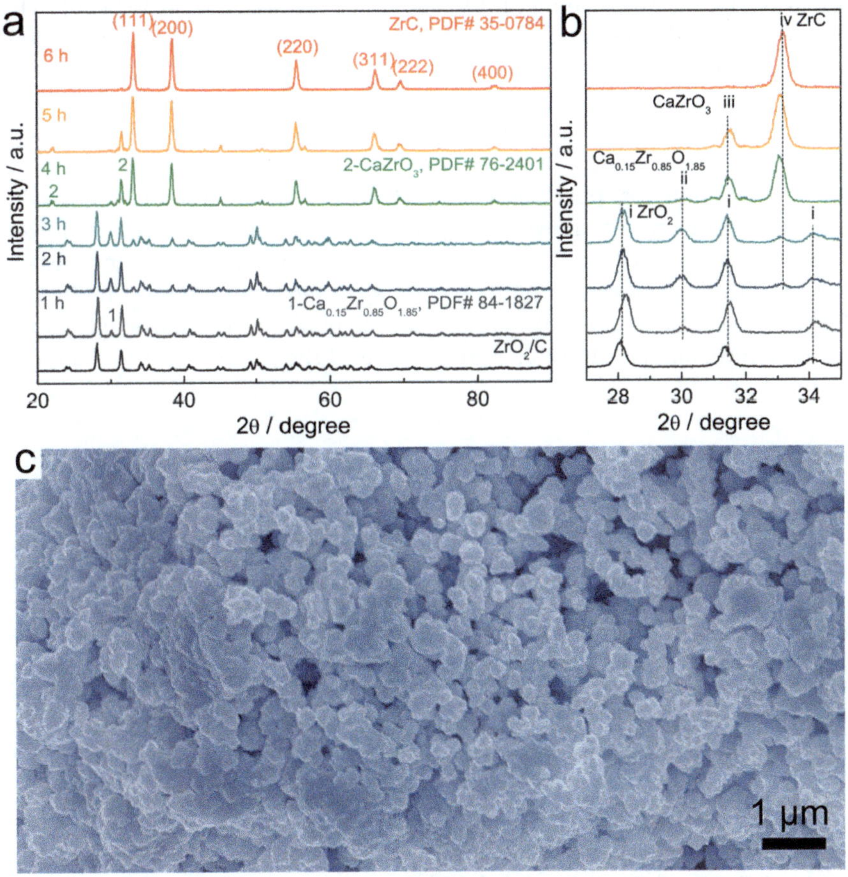

Fig. 6 (a) XRD patterns of the ZrO_2/C precursors before and after being electrodeoxidized at 1000 °C and 4.0 V at different times. (b) Partial details of the XRD patterns shown in (a). (c) Typical SEM image of the obtained ZrC powder.

produced from the ZrO_2/C precursors, as shown in Fig. 6a and b. The typical SEM image of the obtained ZrC is shown in Fig. 6c. Interestingly, the obtained nanoscale ZrC spherical particles (particle sizes are approximately 500 nm) show smooth surfaces and a porous structure, suggesting that the facile electrochemical production of homogeneous/fine ZrC micro/nanoparticles can be achieved using the SOM-assisted electrodeoxidation process at 1000 °C in molten $CaCl_2$ without a melting process.

Electrolytic fabrication of TaC powder

Micro/nanoscale TaC carbide possesses outstanding properties, and it is commonly fabricated by an energy-intensive process.[14] In this work, TaC carbide has also been electrochemically produced from Ta_2O_5/C precursors. Fig. 7a exhibits the XRD patterns of the Ta_2O_5/C precursors (particle sizes are approximately 1 μm) before and after being immersed in molten $CaCl_2$ (without being pre-electrolyzed) at 1000 °C for 5 h. In addition, the typical XRD pattern of the obtained final TaC produced at 1000 °C and 3.5 V is also shown in Fig. 7a. Similarly, the electrodeoxidation process also typically involved the formation of Ca–Ta–O compounds, such as $CaTa_2O_6$ (PDF# 39-1430). The formation of calcium-enriched oxides through combination/compounding reactions has been confirmed during the electrodeoxidation of Ta_2O_5 in $CaCl_2$-based molten salts.[23] Therefore, it is suggested that the reaction mechanism of the electrodeoxidation of Ta_2O_5/C is similar to that of the electrodeoxidation of Ta_2O_5 in $CaCl_2$-based molten salts.[23] After being electrodeoxidized at 1000 °C and 3.5 V for only 2 h, pure TaC (PDF# 77-0205) has been produced from the Ta_2O_5/C precursors, as shown in Fig. 7a. The completely electrodeoxidized Ta_2O_5/C pellet (*i.e.*, TaC pellet) shows a gray color, and the pellet can be easily separated from the wrapped porous Ni foils, as shown in the inset in Fig. 7b. The SEM observation (Fig. 7b) reveals that the obtained TaC particles possess a porous interconnected nodular structure, and the particle sizes are approximately 1 μm.

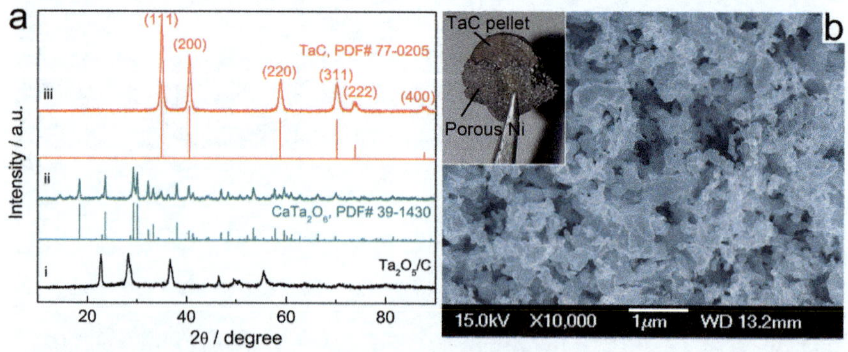

Fig. 7 (a) XRD patterns of (i) the Ta_2O_5/C precursors, (ii) the surface of the Ta_2O_5/C pellet after being immersed in molten $CaCl_2$ at 1000 °C for 5 h, and (iii) the Ta_2O_5/C precursors after being electrodeoxidized in molten $CaCl_2$ at 1000 °C, 3.5 V for 2 h. (b) Typical SEM image of the produced TaC powder (1000 °C, 3.5 V, 2 h), the inset in (b) is the photo of the completely electrodeoxidized Ta_2O_5/C pellet (*i.e.*, TaC pellet).

Reaction mechanisms of the electrodeoxidation processes

Based on the experimental results and the previous studies on the reaction mechanisms of the electrodeoxidation of SiO_2, TiO_2, ZrO_2, and Ta_2O_5 in $CaCl_2$-based molten salts,[21–29,31–37,44–46,48,52–55,62] it is reasonable to deduce that the reaction mechanisms of the electrodeoxidation of the MO_x/C precursors to MCs in molten $CaCl_2$ generally contain the chemical/electrochemical compounding, electrodeoxidation, and *in situ* carbonization processes. The chemical compounding process[25] ($MO_x + y$CaO $\rightarrow Ca_yMO_{(x+y)}$) is commonly caused by the CaO presented in the molten $CaCl_2$ (without being pre-electrolyzed) due to the hydrolysis reaction, *i.e.*, ($CaCl_2\cdot(H_2O) \rightarrow$ CaO (Ca^{2+}, O^{2-}) + 2HCl).[35] In addition, if the compounding process-generated calcium-enriched oxides (such as the intermediate compound $CaSiO_3$)[35] could partially/completely dissolve in the molten $CaCl_2$ electrolyte during electrodeoxidation, the dissolution–electrodeposition process[23,35] would thus present in the electrochemical production process.

Therefore, the reaction mechanism of the electrodeoxidation of SiO_2/C to SiC in molten $CaCl_2$ can be generally summarized as reactions (1)–(5), as shown in Fig. 8(i–iv). In our recent work,[60] the solid-to-solid electrodeoxidation mechanism combined with the dissolution–electrodeposition mechanism have been shown to be able to generate a totally different morphology of product compared to its raw material. In this work, reactions (1)–(3) can generate an initial SiC nuclei, then reactions (4) and (5) may contribute to the formation of SiC NWs through the dissolution–electrodeposition process in the reaction front, as shown in Fig. 8(iv).

$$SiO_2 + CaO\ (Ca^{2+}, O^{2-}) \rightarrow CaSiO_3\ (\Delta G^o{}_{1000^\circ C} = -90.73\ kJ\ mol^{-1}) \quad (1)$$

$$2SiO_2 + Ca^{2+} + 4e^- + C \rightarrow CaSiO_3 + SiC + O^{2-} \quad (2)$$

$$SiO_2/CaSiO_3 + 4e^- + C \rightarrow SiC + 2O^{2-}/{+}CaO\ (Ca^{2+}, O^{2-}) \quad (3)$$

$$CaSiO_3 \rightarrow Ca^{2+} + SiO_3{}^{2-} \quad (4)$$

$$SiO_3{}^{2-} + 4e^- + C \rightarrow SiC + 3O^{2-} \quad (5)$$

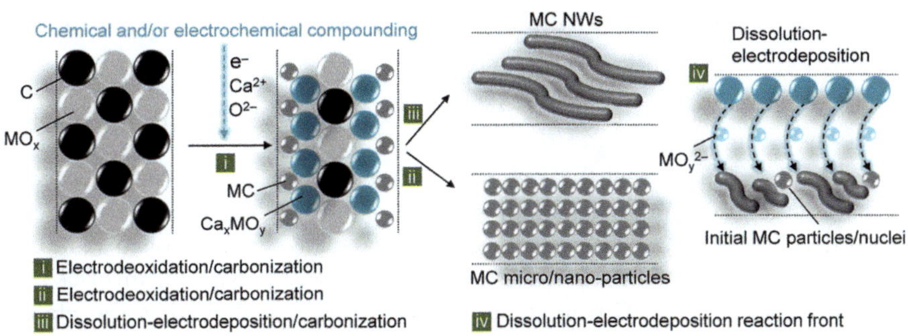

Fig. 8 Schematic illustration of the general reaction mechanisms/routes of the electrodeoxidation of MO_x/C to MCs in molten $CaCl_2$.

The reaction mechanism of the electrodeoxidation of TiO_2/C to TiC in molten $CaCl_2$ contains the chemical/electrochemical compounding process (reactions (6) and (7)) and the electrodeoxidation/*in situ* carbonization processes (reactions (8) and (9)), as illustrated in Fig. 8(i and ii).

$$TiO_2 + CaO\ (Ca^{2+}, O^{2-}) \rightarrow CaTiO_3\ (\Delta G^{o}{}_{1000°C} = -89.40\ kJ\ mol^{-1}) \tag{6}$$

$$\eta TiO_2 + Ca^{2+} + 2e^- \rightarrow CaTiO_3 + Ti_{\eta-1}O_{2\eta-3},\ (\eta = 2, 3, 4\ or\ 5) \tag{7}$$

$$TiO_2/CaTiO_3 + 4e^- + C \rightarrow TiC + 2O^{2-}/{+}CaO\ (Ca^{2+}, O^{2-}) \tag{8}$$

$$Ti_{\eta-1}O_{2\eta-3} + 2(2\eta - 3)e^- + (\eta - 1)C \rightarrow (\eta - 1)TiC + (2\eta - 3)O^{2-},\ (\eta = 2, 3, 4\ or\ 5) \tag{9}$$

Similarly, the reaction mechanism of the electrodeoxidation of ZrO_2/C to ZrC in molten $CaCl_2$ can also be generally expressed as reactions (10)–(13), as shown in Fig. 8(i and ii).

$$ZrO_2 + \delta CaO\ (Ca^{2+}, O^{2-}) \rightarrow Ca_{\delta}ZrO_{(2+\delta)} \tag{10}$$

$$Ca_{\delta}ZrO_{(2+\delta)} + 4e^- + C \rightarrow ZrC + \delta CaO\ (Ca^{2+}, O^{2-}) + 2O^{2-} \tag{11}$$

$$\sigma ZrO_2 + [2(\sigma - \varphi) - \psi]Ca^{2+} + 4(\sigma - \varphi)e^- + (\sigma - \varphi)C \rightarrow Ca_{[2(\sigma-\varphi)-\psi]}Zr_{\varphi}O_{(2\sigma-\psi)} + (\sigma - \varphi)ZrC + \psi O^{2-},\ (\sigma > \varphi, \psi \geq 0) \tag{12}$$

$$Ca_{[2(\sigma-\varphi)-\psi]}Zr_{\varphi}O_{(2\sigma-\psi)} + 4\varphi e^- + \varphi C \rightarrow \varphi ZrC + (2\sigma - \psi)O^{2-} + [2(\sigma - \varphi) - \psi]Ca^{2+},\ (\sigma > \varphi, \psi \geq 0) \tag{13}$$

In addition, the reaction mechanism of the electrodeoxidation of Ta_2O_5/C to TaC in molten $CaCl_2$ can be generally expressed by reactions (14)–(17), as illustrated in Fig. 8(i and ii).

$$Ta_2O_5 + \beta CaO\ (Ca^{2+}, O^{2-}) \rightarrow Ca_{\beta}Ta_2O_{(5+\beta)},\ (\beta = 0.5, 1, 2, 4) \tag{14}$$

$$Ca_{\beta}Ta_2O_{(5+\beta)} + 10e^- + 2C \rightarrow 2TaC + \beta CaO\ (Ca^{2+}, O^{2-}) + 5O^{2-},\ (\beta = 0.5, 1, 2, 4) \tag{15}$$

$$\gamma Ta_2O_5 + [5(\gamma - 1) - \phi]Ca^{2+} + 10(\gamma - 1)e^- + 2(\gamma - 1)C \rightarrow Ca_{[5(\gamma-1)-\phi]}Ta_2O_{(5\gamma-\phi)} + 2(\gamma - 1)TaC + \phi O^{2-},\ (\gamma > 1, \phi \geq 0) \tag{16}$$

$$Ca_{[5(\gamma-1)-\phi]}Ta_2O_{(5\gamma-\phi)} + 10e^- + 2C \rightarrow 2TaC + (5\gamma - \phi)O^{2-} + [5(\gamma - 1) - \phi]Ca^{2+},\ (\gamma > 1, \phi \geq 0) \tag{17}$$

It should be noted that these above-mentioned reactions can only be used as general guidelines for understanding the reaction mechanisms of the electrodeoxidation of MO_x/C to MCs in molten $CaCl_2$. Actually, the electrodeoxidation of MO_x (such as TiO_2) to metals (such as Ti) in molten salts typically involves individual reactions.[24,28,29] Therefore, it is reasonable to believe that the electrodeoxidation of MO_x/C in molten $CaCl_2$ is also related to individual reactions.[41–49] In comparison with the previous work,[41,44,48,49] the present work suggests that the

SOM-assisted electrodeoxidation process can be used to produce MCs with predesigned components from their corresponding stoichiometric MO_x/C precursors. Moreover, the SOM-assisted molten salt electrodeoxidation process is an environmentally friendly process, which may also have implications for the production of other MCs and MC-containing composites.[52]

Conclusions

The SOM-assisted electrodeoxidation of stoichiometric MO_x/C precursors to MCs such as SiC, TiC, ZrC, and TaC has been systematically carried out in molten $CaCl_2$ at 1000 °C and 3.5 to 4.0 V. The electrodeoxidation processes and their related reaction mechanisms were investigated, and the partially/completely electrodeoxidized MO_x/C products were characterized. The results show that homogeneous SiC NWs with diameters of 20 to 50 nm can be produced. The obtained SiC NWs possess smooth surfaces and a porous structure. The reaction mechanism of the electrodeoxidation of SiO_2/C in molten $CaCl_2$ generally contains chemical/electrochemical compounding, electrodeoxidation, dissolution–electrodeposition, and *in situ* carbonization processes. In addition, TiC, ZrC, and TaC have also been electrochemically produced from their MO_x/C precursors. The morphology analysis shows that the produced TiC, ZrC, and TaC typically exhibit porous interconnected nodular structures with micro/nanoscale particles. The reaction mechanisms of the electrodeoxidation of TiO_2/C, ZrO_2/C, and Ta_2O_5/C in molten $CaCl_2$ commonly include chemical/electrochemical compounding, electrodeoxidation, and *in situ* carbonization processes. It is suggested that the molten salt-based SOM-assisted electrodeoxidation process has the potential to be used for the facile electrochemical production of micro/nanostructured MCs materials. We are continuing our studies to investigate the detailed reaction mechanisms of the electrodeoxidation processes and the properties of the produced MCs, and further investigations will be reported in due course.

Acknowledgements

We appreciate the financial support from the National Natural Science Foundation of China (No. 51304132, 51574164), the China National Funds for Distinguished Young Scientists (No. 51225401), the National Basic Research Program of China (No. 2014CB643403), the Science and Technology Commissions of Shanghai Municipality (No. 14JC1491400) and Sichuan Province (No. 2013GZ0146), and the Young Teacher Training Program of Shanghai Municipal Education Commission. We also thank the Instrumental Analysis and Research Center of Shanghai University for materials characterization.

Notes and references

1 W. F. Chen, J. T. Muckerman and E. Fujita, *Chem. Commun.*, 2013, **49**, 8896–8909.

2 D. V. Esposito, S. T. Hunt, Y. C. Kimmel and J. G. Chen, *J. Am. Chem. Soc.*, 2012, **134**, 3025–3033.

3 Y. Liu, T. G. Kelly, J. G. Chen and W. E. Mustain, *ACS Catal.*, 2013, **3**, 1184–1194.

4 A. M. Nartowski, I. P. Parkin, A. J. Craven and M. MacKenzie, *Adv. Mater.*, 1998, **10**, 805–808.

5 A. M. Nartowski, I. P. Parkin, M. Mackenzie and A. J. Craven, *J. Mater. Chem.*, 2001, **11**, 3116–3119.

6 S. M. Schmuecker and B. M. Leonard, *Inorg. Chem.*, 2015, **54**, 3889–3895.

7 H. I. Won, N. Hayk, C. W. Won and H. H. Lee, *J. Mater. Sci.*, 2011, **46**, 6000–6006.

8 R. Wu, K. Zhou, C. Y. Yue, J. Wei and Y. Pan, *Prog. Mater. Sci.*, 2015, **72**, 1–60.

9 M. Dasog, L. F. Smith, T. K. Purkait and J. G. C. Veinot, *Chem. Commun.*, 2013, **49**, 7004–7006.

10 C. Giordano and M. Antonietti, *Nano Today*, 2011, **6**, 366–380.

11 C. Giordano, C. Erpen, W. Yao, B. Milke and M. Antonietti, *Chem. Mater.*, 2009, **21**, 5136–5144.

12 J. P. Kelly and O. A. Graeve, *Acta Mater.*, 2015, **84**, 472–483.

13 J. P. Kelly, R. Kanakala and O. A. Graeve, *J. Am. Ceram. Soc.*, 2010, **93**, 3035–3038.

14 L. K. Brar, G. Singla and O. P. Pandey, *RSC Adv.*, 2015, **5**, 1406–1416.

15 X. Tao, J. Du, Y. Li, Y. Yang, Z. Fan, Y. Gan, H. Huang, W. Zhang, L. Dong and X. Li, *Adv. Energy Mater.*, 2011, **1**, 534–539.

16 C. L. Yeh and E. W. Liu, *J. Alloys Compd.*, 2006, **415**, 66–72.

17 F. Li, X. Huang and G. J. Zhang, *Ceram. Int.*, 2015, **41**, 3335–3338.

18 D. Liu, W. F. Qiu, T. Cai, Y. N. Sun, A. J. Zhao and T. Zhao, *J. Am. Ceram. Soc.*, 2014, **97**, 1242–1247.

19 M. Dollé, D. Gosset, C. Bogicevic, F. Karolak, D. Simeone and G. Baldinozzi, *J. Eur. Ceram. Soc.*, 2007, **27**, 2061–2067.

20 S. Niksirat and S. Raygan, *Adv. Powder Technol.*, 2014, **25**, 859–864.

21 G. Z. Chen, D. J. Fray and T. W. Farthing, *Nature*, 2000, **407**, 361–364.

22 T. Nohira, K. Yasuda and Y. Ito, *Nat. Mater.*, 2003, **2**, 397–401.

23 W. Xiao and D. Wang, *Chem. Soc. Rev.*, 2014, **43**, 3215–3228.

24 A. M. Abdelkader, K. Tripuraneni Kilby, A. Cox and D. J. Fray, *Chem. Rev.*, 2013, **113**, 2863–2886.

25 D. Wang, X. Jin and G. Z. Chen, *Annu. Rep. Prog. Chem., Sect. C: Phys. Chem.*, 2008, **104**, 189–234.

26 W. Li, X. Jin, F. Huang and G. Z. Chen, *Angew. Chem., Int. Ed.*, 2010, **49**, 3203–3206.

27 K. Jiang, X. Hu, M. Ma, D. Wang, G. Qiu, X. Jin and G. Z. Chen, *Angew. Chem., Int. Ed.*, 2006, **45**, 428–432.

28 C. Schwandt, D. T. L. Alexander and D. J. Fray, *Electrochim. Acta*, 2009, **54**, 3819–3829.

29 D. T. L. Alexander, C. Schwandt and D. J. Fray, *Electrochim. Acta*, 2011, **56**, 3286–3295.

30 X. Yan and D. J. Fray, *Adv. Funct. Mater.*, 2005, **15**, 1757–1761.

31 J. Peng, Y. Zhu, D. Wang, X. Jin and G. Z. Chen, *J. Mater. Chem.*, 2009, **19**, 2803–2809.

32 X. Jin, P. Gao, D. Wang, X. Hu and G. Z. Chen, *Angew. Chem., Int. Ed.*, 2004, **43**, 733–736.

33 A. M. Abdelkader, D. J. S. Hyslop, A. Cox and D. J. Fray, *J. Mater. Chem.*, 2010, **20**, 6039–6049.

34 R. O. Suzuki, *J. Phys. Chem. Solids*, 2005, **66**, 461–465.

35 W. Xiao, X. Wang, H. Yin, H. Zhu, X. Mao and D. Wang, *RSC. Adv.*, 2012, **2**, 7588–7593.

36 W. Xiao, X. Jin and G. Z. Chen, *J. Mater. Chem. A*, 2013, **1**, 10243–10250.

37 L. Zhang, S. Wang, S. Jiao, K. Huang and H. Zhu, *Electrochim. Acta*, 2012, **75**, 357–359.

38 D. S. M. Vishnu, J. Sure and K. S. Mohandas, *Carbon*, 2015, **93**, 782–792.

39 K. S. Mohandas, *J. South. Afr. Inst. Min. Metall.*, 2013, **122**, 195–212.

40 C. Osarinmwian, E. P. L. Roberts and I. M. Mellor, *Chem. Phys. Lett.*, 2015, **621**, 184–187.

41 A. M. Abdelkader and D. J. Fray, *J. Eur. Ceram. Soc.*, 2012, **32**, 4481–4487.

42 A. M. Abdelkader, *J. Eur. Ceram. Soc.*, 2016, **36**, 33–42.

43 L. Sun, Q. Song, Q. Xu, Z. Ning, X. Lu and D. J. Fray, *New J. Chem.*, 2015, **39**, 4391–4397.

44 Q. Song, Q. Xu, J. Meng, T. Lou, Z. Ning, Y. Qi and K. Yu, *J. Alloys Compd.*, 2015, **647**, 245–251.

45 Q. Song, Q. Xu, R. Ding, J. Meng, Z. Ning, T. Lou, Y. Qi and K. Yu, *J. Electrochem. Soc.*, 2016, **163**, E49–E53.

46 X. Y. Yan, M. I. Pownceby, M. A. Cooksey and M. R. Lanyon, *Trans. Inst. Min. Metall., Sect. C*, 2009, **118**, 23–34.

47 K. Chen, Y. Hua, C. Xu, Q. Zhang, C. Qi and Y. Jie, *Ceram. Int.*, 2015, **41**, 11428–11435.

48 L. Dai, X. Wang, H. Zhou, Y. Yu, J. Zhu, Y. Li and L. Wang, *Ceram. Int.*, 2015, **41**, 4182–4188.

49 L. Dai, Y. Lu, X. Wang, J. Zhu, Y. Li and L. Wang, *Int. J. Refract. Met. Hard Mater.*, 2015, **51**, 153–159.

50 U. B. Pal, D. E. Woolley and G. B. Kenney, *JOM*, 2001, **53**, 32–35.

51 A. Krishnan, X. G. Lu and U. B. Pal, *Metall. Mater. Trans. B*, 2005, **36B**, 463–473.

52 X. Zou, X. Lu, Z. Zhou and C. Li, *Electrochem. Commun.*, 2012, **21**, 9–13.

53 X. Zou, X. Lu, Z. Zhou, W. Xiao, Q. Zhong, C. Li and W. Ding, *J. Mater. Chem. A*, 2014, **2**, 7421–7430.

54 X. Zou, X. Lu, Z. Zhou, C. Li and W. Ding, *Electrochim. Acta*, 2011, **56**, 8430–8437.

55 X. Zou, X. Lu, C. Li and Z. Zhou, *Electrochim. Acta*, 2010, **55**, 5173–5179.

56 X. Lu, X. Zou, C. Li, Q. Zhong, W. Ding and Z. Zhou, *Metall. Mater. Trans. B*, 2012, **43**, 503–512.

57 X. Guan, U. B. Pal, S. Gopalan and A. C. Powell, *J. Electrochem. Soc.*, 2013, **160**, F1179–F1186.

58 J. Zhao, J. Li, P. Ying, W. Zhang, L. Meng and C. Li, *Chem. Commun.*, 2013, **49**, 4477–4479.

59 J. Yang, S. Lu, S. Kan, X. Zhang and J. Du, *Chem. Commun.*, 2009, 3273–3275.

60 X. Zou, S. Gu, X. Lu, X. Xie, C. Lu, Z. Zhou and W. Ding, *Metall. Mater. Trans. B*, 2015, **46B**, 1262–1274.

61 J. Fréchette and C. Carraro, *J. Am. Chem. Soc.*, 2006, **128**, 14774–14775.

62 T. Wu, X. Jin, W. Xiao, X. Hu, D. Wang and G. Z. Chen, *Chem. Mater.*, 2007, **19**, 153–160.

Faraday Discussions

PAPER

Natural gas anodes for aluminium electrolysis in molten fluorides

Geir Martin Haarberg,*[a] Babak Khalaghi[a] and Tommy Mokkelbost[b]

Received 28th December 2015, Accepted 27th January 2016
DOI: 10.1039/c5fd00239g

Industrial primary production of aluminium has been developed and improved over more than 100 years. The molten salt electrolysis process is still suffering from low energy efficiency and considerable emissions of greenhouse gases (CO_2 and PFC). A new concept has been suggested where methane is supplied through the anode so that the CO_2 emissions may be reduced significantly, the PFC emissions may be eliminated and the energy consumption may decrease significantly. Porous carbon anodes made from different graphite grades were studied in controlled laboratory experiments. The anode potential, the anode carbon consumption and the level of HF gas above the electrolyte were measured during electrolysis. In some cases it was found that the methane oxidation was effectively participating in the anode process.

Introduction

Aluminium is produced by the Hall–Héroult process, which was patented independently by Hall and Héroult in 1886. The overall primary cell reaction is:

$$Al_2O_3\ (diss) + \tfrac{3}{2}C\ (s) = 2Al\ (l) + \tfrac{3}{2}CO_2\ (g) \tag{1}$$

Alumina is dissolved in a molten fluoride electrolyte based on cryolite (Na_3AlF_6) containing AlF_3 and CaF_2.[1] Modern cells are equipped with so-called prebaked carbon anodes and operate at ~955 °C in a horizontal electrode design. The theoretical anode carbon consumption is 333 g per kg Al, and consumed anodes must be replaced. Research to find inert anodes for oxygen evolution has not been successful. An alternative is to supply natural gas (methane) through the anode, which will give the following cell reaction:

$$Al_2O_3\ (diss) + \tfrac{3}{4}CH_4\ (g) = 2Al\ (l) + \tfrac{3}{4}CO_2\ (g) + \tfrac{3}{2}H_2O \tag{2}$$

[a]*Department of Materials Science and Engineering, Norwegian University of Science and Technology, NO-7491 Trondheim, Norway. E-mail: geir.martin.haarberg@ntnu.no*
[b]*SINTEF Materials and Chemistry, NO-7465 Trondheim, Norway*

The reversible cell potentials of reactions (1) and (2) are very similar; the standard potentials are ~1.2 V. The use of methane will reduce the amount of CO_2 evolved by up to 50%, and the anode process will occur at a lower anode potential due to the very low overvoltage. In principle, methane oxidation will eliminate the replacement of anodes and the so-called anode effect which leads to the formation of PFC gases (CF_4 and C_2F_6) that are strong greenhouse gases. The main challenge related to the use of methane in this process is the formation of water vapour which may lead to the presence of HF gas. Consumption of carbon anodes according to reaction (1), the yield of methane consumption in reaction (2) and possible cracking of methane are other important issues.

The anodic overvoltage during industrial aluminium electrolysis is quite high, about 0.4 V at ~0.7 A cm^{-2} for so-called prebaked anodes.[1] At lower current densities there is a linear relationship between anodic overvoltage and log current density.

Considerable research has been carried out to develop an inert oxygen evolving anode to replace the consumable carbon anode.[2] So far no inert anode material has been developed for industrial operation. The most promising inert anode candidates have been based on tin oxide and nickel ferrite. In the case of using an inert anode, the cell reaction will be the following:

$$Al_2O_3\ (diss) = 2Al\ (l) + \tfrac{3}{2}O_2\ (g) \tag{3}$$

The reversible potential of reaction (3) is ~2.2 V.

Studies of so-called gas anodes for electrowinning in molten salts have been reported to some extent. Porous carbon was mainly used as the anode and a small depolarisation effect was observed in molten cryolite for aluminium electrolysis.[3–5] Magnetite has also been studied as the anode, but it was found to be unstable in molten cryolite during electrolysis. The use of hydrogen as an anodic reactant in porous or non-porous non-carbon electrodes, including metallic, metal oxide and refractory materials, was reported in a US patent in 1972.[6] In 2000, another US patent on designing a non-consumable anode of the type used for solid oxide fuel cells (SOFC) with a solid oxide membrane supplied with reformed natural gas was suggested for aluminium electrolysis.[7] More recent papers[8–10] are related to applying the solid oxide membrane technology to produce metals from their oxides in molten salts, with the introduction of hydrogen. A similar approach was applied in magnesium production in molten chlorides, resulting in the depolarisation effect and the formation of HCl gas.[11] In addition, a thermodynamic analysis considering carbon and nickel-based hydrogen diffusion anodes in the electrolyte (Na_3AlF_6–AlF_3–Al_2O_3) was carried out in 2007 to identify optimum operating parameters in aluminium production (ref. 12), and a nickel-based hydrogen diffusion anode was used in an experimental study of aluminium electrowinning in 2011, showing measurable depolarisation of the anode potential.[13]

The concept of applying gas anodes has not been fully implemented in industrial electrolysis processes. The research has mainly been focused on the use of porous carbon anodes and the type of SOFC with a zirconia-based solid membrane. For carbon anodes, there are competing reactions between carbon and reducing gases on the anode, resulting in partial consumption of the anode. The unacceptable dissolution of zirconia-based materials in molten cryolite is an obstacle for using SOFC-type inert anodes for aluminium electrolysis.[14] Therefore, a type of stable inert anode material is needed.

In more recent activity, the depolarisation effect of hydrogen and methane was demonstrated in laboratory experiments.[15–17] The authors demonstrated the depolarisation effects of Pt and SnO_2-based gas anodes (H_2) in molten chlorides as a model system for further studies in more corrosive molten cryolite.[15–17] Fig. 1 and 2 show results obtained during electrolysis in molten cryolite–alumina based electrolytes at 850 °C.[18] The depolarisation effect of methane was demonstrated both for galvanostatic and potentiostatic electrolysis. Maintaining the depolarisation effect over a long time, more than ~30–60 minutes, was found to be challenging.

In a recent paper,[19] the production of HF was studied by gas analysis during electrolysis in molten cryolite–alumina electrolytes using a porous SnO_2 inert anode with a methane supply. High levels of HF were detected (up to 0.2 wt%)

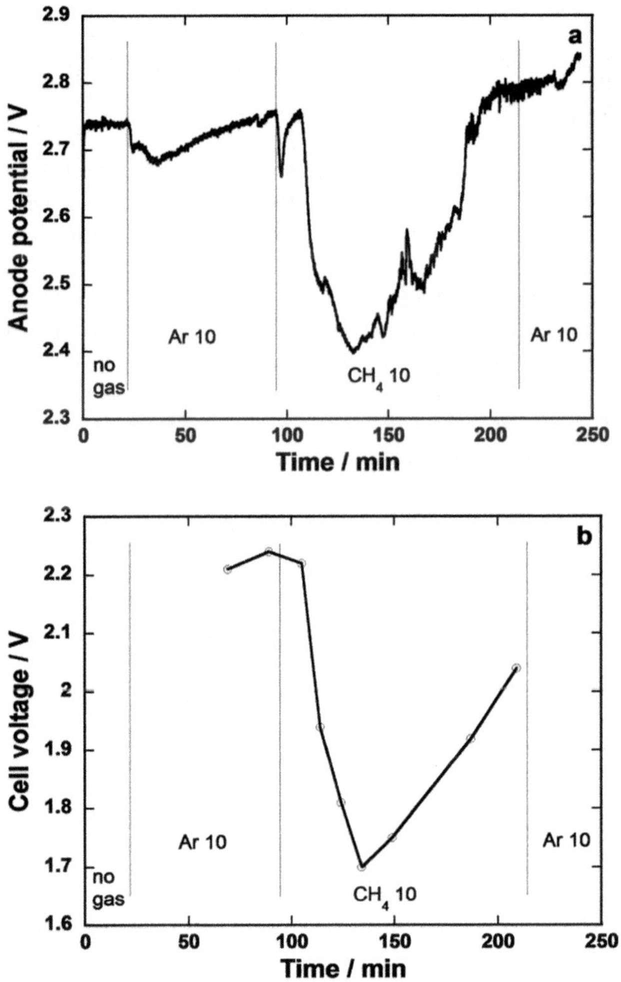

Fig. 1 Anode potential (a) and cell voltage (b) *versus* time during constant current electrolysis (0.2 A; 0.2 A cm^{-2}) using a porous SnO_2-based anode in molten Na_3AlF_6–AlF_3–Al_2O_3 (4.5 wt%) at 850 °C.[18]

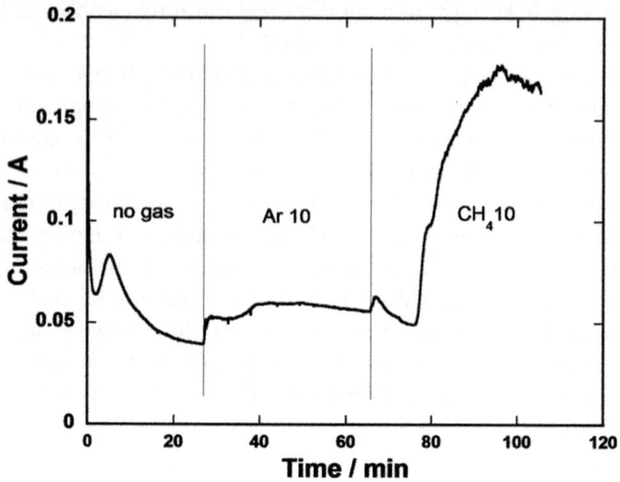

Fig. 2 Current *versus* time during electrolysis at constant anode potential (2.2 V; apparent anode area, 1 cm^2) using a porous SnO_2-based anode in molten $Na_3AlF_6-AlF_3-Al_2O_3$ (4.4 wt%) at 850 °C.[18]

during depolarisation of the anode. It was suggested that the HF problem could be resolved by capturing the produced HF and letting it react with alumina to form AlF_3, which could be fed back to the cells to make up for the loss of electrolyte in the hydrolysis reaction.[19] It was also shown that a hydrogen gas supply gave a similar depolarisation effect.

Methane is an attractive oxidisable gas due to its relative abundance, high purity and low price. In this paper porous carbon anode materials with a methane supply were used and studied as gas anodes during aluminium electrolysis in laboratory experiments. However, methane may undergo cracking, especially at higher temperatures. This would cause the formation of carbon, which may precipitate and fill up the pores of the carbon anodes. Cracking of methane takes place according to the following reaction scheme:

$$2CH_4\ (g) \rightarrow C_2H_6\ (g) + H_2\ (g) \tag{4}$$

$$C_2H_6\ (g) \rightarrow C_2H_4\ (g) + H_2\ (g) \tag{5}$$

$$C_2H_4\ (g) \rightarrow C_2H_2\ (g) + H_2\ (g) \tag{6}$$

$$C_2H_2\ (g) \rightarrow 2C\ (s) + H_2\ (g) \tag{7}$$

Decarburation may occur as follows:

$$2CH_4\ (g) \rightarrow 2C\ (s) + 4H_2\ (g) \tag{8}$$

Reaction (5) does not proceed at temperatures lower than 1350 °C without the use of a catalyst such as nickel.

The presence of CO_2 may change the methane reaction according to:

$$CH_4\ (g) + CO_2\ (g) = 2CO\ (g) + 2H_2\ (g) \tag{9}$$

Small amounts of moisture are introduced to the electrolyte by surrounding air and alumina. This will cause the formation of gaseous HF according to the following reaction:

$$AlF_3\ (diss) + \tfrac{3}{2}H_2O\ (g) = 3HF\ (g) + \tfrac{1}{2}Al_2O_3\ (diss) \qquad (10)$$

Using a methane oxidising anode will cause the formation of a significant amount of moisture according to reaction (2). This may represent a major challenge for the gas anode concept if high amounts of HF are produced. Another concern is the consumption of electrolyte in the form of AlF_3.

Experimental

The electrolyte composition was Na_3AlF_6 (natural cryolite, Greenland) with an excess of 6 wt% AlF_3 (Noralf, Boliden Odda AS) and 5 wt% CaF_2 (Merck, >97%) in addition to ~4.5 wt% anhydrous γ-Al_2O_3 (Merck, >98%). The amount of alumina was close to saturation. The electrolyte composition was close to the electrolyte commonly used in industrial aluminium electrolysis. Fig. 3 shows the schematic of the experimental set-up. The electrolyte was contained in a graphite crucible with a lining of sintered alumina, while the bottom of the crucible was the cathode. A hollow steel tube attached to the porous carbon anode was used as the current collector. The anode and cathode were positioned horizontally with respect to each other. A molybdenum wire (Norsk Spesialmetall, 99.9 pct, 1 mm Ø) was used as the current lead. The wire was passed through an alumina tube which was placed inside another alumina tube with a closed bottom. The closed tube contained aluminium at the bottom, and was fitted with a hole above the metal so that electrolyte could enter. The whole system served as an Al^{3+}/Al reference electrode. All potentials were measured against this aluminium reference electrode. The crucible containing the bath was dried in air at 120 °C overnight. The furnace and the cell with electrolyte were also dried at 200 °C in N_2 for a few hours before heating up the furnace to the working temperature of 970 °C. The furnace was continuously flushed with N_2. The inlet gas composition for the anode was controlled using mass flow controllers (Bronkhorst) and the inlet gas pressure was measured. The whole electrochemical cell was placed in a vertical tube furnace heated by resistance wires, and connected to a temperature controller. The gases were purged into the porous anode with a gas flow equal to 20 ml min^{-1}. Galvanostatic electrolysis experiments were performed using porous anodes made of three different graphite grades (Tokai Carbon Group). Some of the properties of the graphite grades used for preparing the anodes are given in Table 1. Each experiment was started by applying a constant current of 2.3 A to the cell, corresponding to an apparent current density of 0.4 A cm^{-2}. While N_2 was passed through the porous anode for the first 45 min of the electrolysis time, afterwards the gas was changed to CH_4 and the electrolysis was continued for an additional 4 h. This procedure made it possible to detect whether there was any depolarisation due to the oxidation of methane at the anode. Another series of electrolysis experiments was also carried out without using methane in order to have better insight into the process and for comparison. In this series, only nitrogen was purged through the anode; while the rest of the experimental conditions were unchanged.

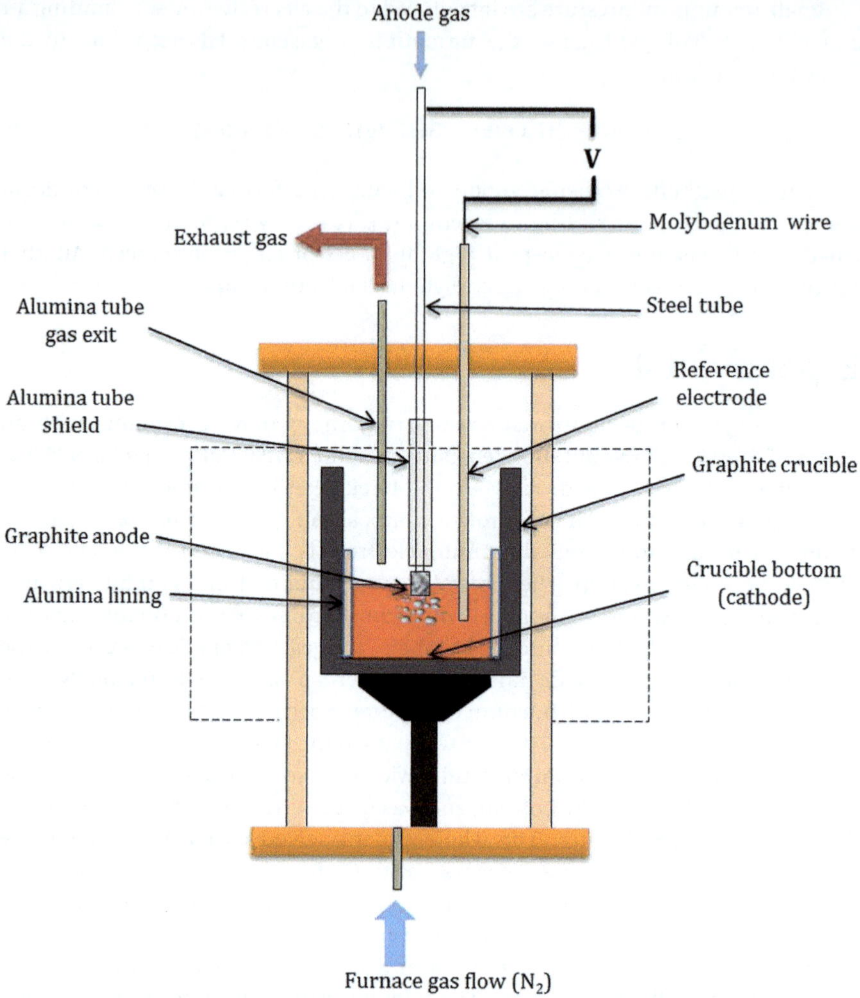

Fig. 3 Sketch of the experimental electrolysis cell.

The weights of the anodes were measured before and after each experiment to check the consumption of the anodes and they were compared with the theoretical values. The used anodes were examined by scanning electron microscopy (SEM, Hitachi S-3400 N).

Table 1 Typical properties of different graphite grades[20]

Grade name	Specific gravity (g cm^{-3})	Porosity (%)	Grain size (μm)	Grade
G347	1.85	12	11	Isotropic
G348	1.92	8	8	Isotropic
G140	1.7	20	1000	Moulded
KWPSY	1.6	20	2000[a]	Extruded

[a] Maximum grain size.

A new anode concept was designed in order to try to improve the methane oxidation reaction in more recent experiments. A denser graphite (G348, Table 1) with low permeability was used to shield the main graphite anode (G347, Table 1). The new anode is shown in Fig. 4.

Results and discussion

The desired situation is when methane oxidation, reaction (2), is the only anodic process. In this case the anode potential will be considerably lower than that during normal electrolysis due to the high overvoltage during normal electrolysis when methane isn't supplied. In reality these two oxidation reactions will take place at the same time, and the carbon anode will be depolarised based on the partial current densities of these two reactions.

Table 2 gives the consumption of the porous graphite anodes in electrolysis experiments. The theoretical carbon consumption based solely on reaction (1) is 1.2 g. The isotropic graphite (G347) was found to behave quite differently compared to the other grades. There is a significant change in weight loss of the anode during electrolysis when CH_4 was supplied through G347 compared to when N_2 was supplied through the anode. The moulded (G140) and extruded (KWPSY) grades were also consumed slightly less when CH_4 was flushed into the anode, but the difference is less significant than for G347. It seems that when isotropic grade was used, CH_4 participated in the anodic reaction remarkably, while for the other two grades the anode consumption was almost equal to the theoretical value. The weight loss difference of the extruded grade anode (KWPSY) with and without CH_4 is noticeable. But this is due to the fact that this anode showed a higher consumption when N_2 was used, compared with the two others. The reason for this is not clear at the moment.

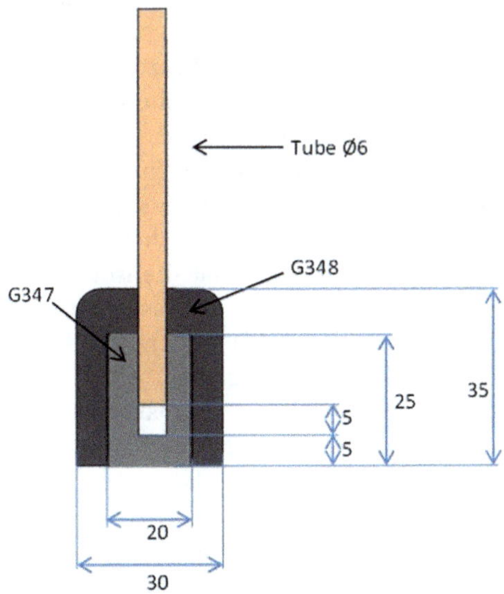

Fig. 4 Sketch of a new anode design with two different graphite grades. The values in the figure are in millimetres.

Table 2 The consumption of graphite anodes when supplied with only N_2 and when supplied with $N_2 + CH_4$ during electrolysis in cryolite-based electrolyte at 970 °C for 285 min. The theoretical carbon consumption is 1.2 g. Values including carbon deposition are shown in parentheses

Graphite type	Anode gas	Weight loss (g)	Consumption (%)
Isotropic (G347)	$N_2 + CH_4$	0.67 (0.87)	56 (73)
	N_2	1.28	107
Moulded (G140)	$N_2 + CH_4$	1.15 (1.35)	96 (113)
	N_2	1.24	103
Extruded (KWPSY)	$N_2 + CH_4$	1.16 (1.36)	97 (113)
	N_2	1.4	117

Generally the anode carbon consumption is expected to be higher than the theoretical value. This is due to the occurrence of other anode carbon consumption reactions such as air-burn, CO_2 burn (Boudouard reaction), especially when the gas penetrates into the pores of the anode, and dusting.[1] Dusting, where unreacted carbon particles detach from the anode, is not believed to be an important issue in controlled laboratory experiments.

However, it is seen here that for all of the graphite grades, when CH_4 was applied, the weight loss was less than the theoretical consumption. Hence, it is likely that CH_4 was involved in the anodic reaction – at least to some extent – for all the graphite grades. This conjecture sounds more probable when we consider the likely thermal cracking of CH_4 resulting in precipitation of carbon in the porous structure of the anode. Methane becomes unstable in terms of its elements from 530 °C. However, the kinetics is slow. The equilibrium constant for the cracking of methane is 87 at $T = 970$ °C.[21] Therefore, this reaction is most likely to happen in our experiments, at least to some extent. Precipitated carbon can add to the final weight of the anodes. An experiment was performed to check this, where all the conditions were unchanged, except that no current was passed. So, the sole factor influencing the anode weight was the amount of carbon precipitation. It was found that around 0.2 g of carbon was precipitated in 4 h (the same amount of time that CH_4 was flushed into the anodes during the electrolysis experiments). This suggests that the real carbon consumption was higher than the values reported in Table 2. By considering the possible added weight from carbon deposition, new values for the weight loss and consumption are found. These are also listed in Table 2; the values are shown in parentheses. It is clear even when considering this added value that the isotropic graphite (G347) has worked well and that introduction of methane has resulted in less consumption, which implies the participation of methane in the anodic reaction. However, it seems that for the other two graphite grades, when methane was used the consumption was similar or even higher compared to the experiments where nitrogen was supplied. The excess consumption of porous graphite anodes in the presence of reducing gases (CO and H_2) has also been reported in another study.[4] After the experiments there was some soot on the surface of the graphite anodes, which might be due to such reactions. The reaction of the anode with reducing gases such as methane might result in excess consumption of the anode. Therefore, such reactions must also be considered in the evaluation of the anode weight changes. Their occurrence depends on graphite properties such as porosity and grain size. In the case of isotropic graphite (G347), methane participation was the dominating effect.

A more recent experiment using the new anode design (see Fig. 2) gave a carbon consumption of 87% of the theoretical value during electrolysis with a mixture of nitrogen and methane supplied to the anode.

Apart from the electrochemical and cracking reactions, there exists another factor which might have caused weight loss changes. This entails the possible penetration of electrolyte into the porous structure of the graphite during electrolysis, which can add to the final weight of the anode. However, this should be negligible since the anode was pulled out of the bath and flushed with N_2 for 1 h after electrolysis. The fractured surfaces of the anodes were studied by SEM/EDS. These studies also confirmed that the graphite structure was essentially free from electrolyte after the experiments.

The gas pressure before the anode was measured while the anode was placed above the molten electrolyte. The isotropic grade showed the largest pressure increase, around 0.5 bar, the moulded grade showed a smaller increase, around 0.25 bar, while the extruded grade did not show any pressure increase. These differences can be attributed to the large differences in grain/pore sizes between these graphite grades.

Fig. 5–7 show the cell potential and pressure changes during electrolysis using an isotropic grade (G347) anode for two different experiments. The red plot in Fig. 5 shows the anode potential when only N_2 was supplied through the anode and the blue plot shows the potential when N_2 was supplied through the anode for 45 min prior to introducing CH_4 and continuing the electrolysis for 4 more hours.

As can be seen, when only nitrogen was supplied to the anode, the potential increased gradually during electrolysis (Fig. 5), but the pressure remained constant (Fig. 6). This is due to the consumption of graphite resulting in a decrease of surface area and consequently an increase in current density. This is

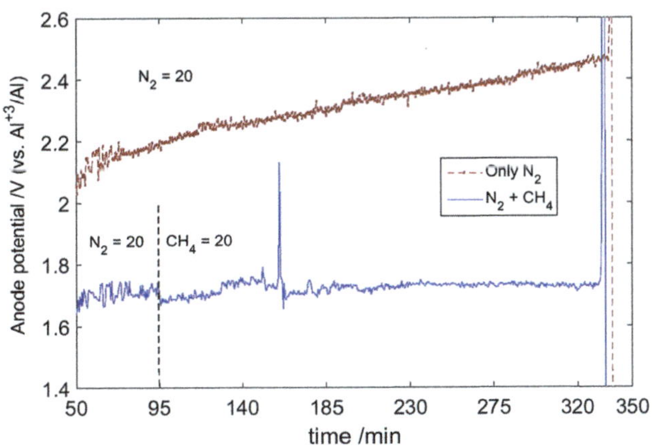

Fig. 5 Anode potential *vs.* time during aluminium electrolysis for two experiments where either only N_2 (red) or $N_2 + CH_4$ (blue) were flushed into the porous anodes. The anode was made of isotropic graphite (G347). $i = 0.4$ A cm^{-2}, $T = 970$ °C. The values in the figure are the flow of gas in ml min^{-1}.

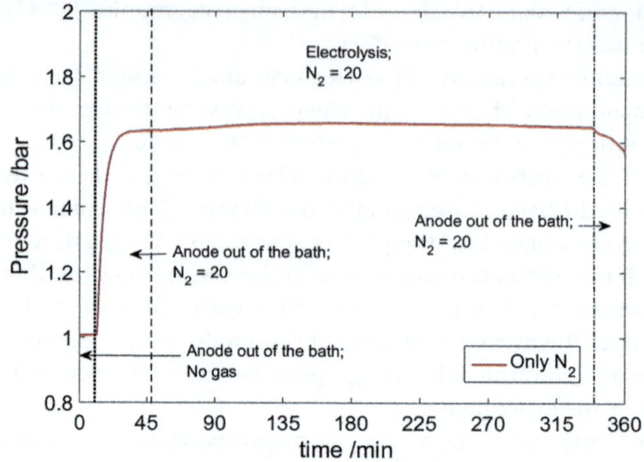

Fig. 6 Pressure changes during aluminium electrolysis where only N_2 was flushed into the porous anodes. The anode was made of isotropic graphite (G347). $i = 0.4$ A cm^{-2}, $T = 970$ °C. The values in the figure are the flow of gas in ml min^{-1}.

in agreement with previous studies demonstrating an increase in the anodic overvoltage when current density is increased.[22]

A rough calculation demonstrates that the change in the surface area was noticeable. The surface area prior to electrolysis was 6.28 cm^2 and the approximate value for the final surface area was ~5 cm^2. The final surface area was calculated based on the geometrical surface of the anode after electrolysis. Assuming the change in surface area and a constant current, the apparent current density increased from 0.37 A cm^{-2} to 0.46 A cm^{-2} during electrolysis.

However, when CH_4 was introduced to the anode the potential became stable and it remained almost constant throughout the electrolysis period. The vertical lines in Fig. 5–7 represent the time where the gas anode was shifted from nitrogen

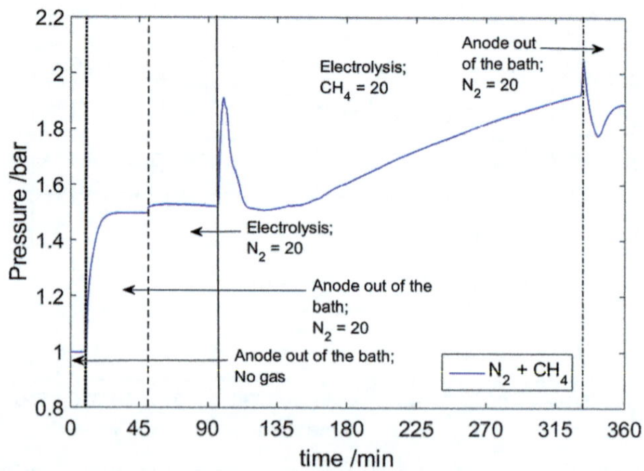

Fig. 7 Pressure changes during aluminium electrolysis where $N_2 + CH_4$ were flushed into the porous anodes. The anode was made of isotropic graphite (G347). $i = 0.4$ A cm^{-2}, $T = 970$ °C. The values in the figure are the flow of gas in ml min^{-1}.

(dashed line) to methane (solid line). There was an abrupt increase and decrease in pressure which was due to the supply of both gases during the shifting time. After that, as can be seen, introduction of methane to the anode caused a gradual pressure increase which lasted as long as methane was flushed; *i.e.* until the end of electrolysis (Fig. 7). This increase in pressure was due to carbon precipitation from the cracking reaction of methane. If the electrolysis had continued for a longer time, this might have led to clogging, as observed earlier.[4]

The change in anode potential (Fig. 5) agrees with the weight loss data (Table 2). It is clear that methane was significantly involved in the anodic reaction, so the potential was stable and did not increase. However, the problem of clogging might prevent long term electrolysis. A suitable anode design could possibly prevent the probable and undesirable clogging.

There is a small difference in the anode potential at the beginning between these two experiments, although the anodes were similar and both experiments were started by supplying N_2 to the anode. The reason for this is not clear. However, it is expected that during the supply of methane, the anode potential should be lower due to a lower overpotential.

Similar electrolysis experiments using moulded (G140) and extruded (KWPSY) graphite grades did not give significant anode potential differences when supplying nitrogen and methane. In both cases the anode potential was found to increase during the course of electrolysis due to the consumption of carbon, which will change the active anode area. These results suggest that the contribution of methane in the anodic reaction was not significant for these graphite grades, which is in agreement with the weight change results, as given in Table 2. In addition, the pressure build up was much less for these anodes.

Fig. 8–10 show results from a recent electrolysis experiment using the new anode design consisting of two graphite grades as shown in Fig. 4. Based on the carbon weight loss, the carbon consumption was ~87%, indicating that methane participated in the anode process to some extent. This is also supported by the results from measuring the HF level above the electrolyte and the anode potential

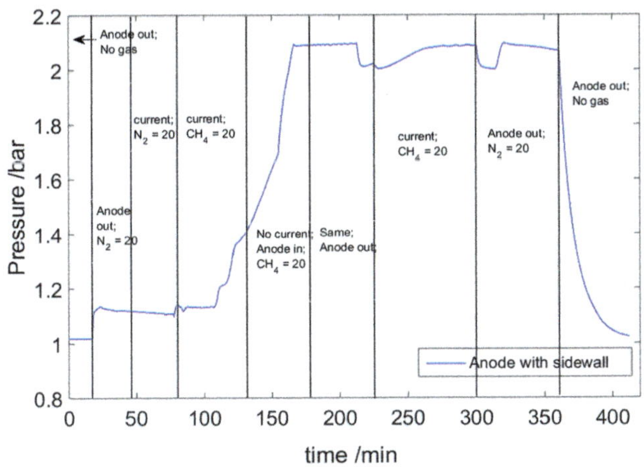

Fig. 8 Pressure changes during aluminium electrolysis when supplying N_2 + CH_4 to the porous anode. $i = 0.5$ A cm^{-2}, $T = 970$ °C. The values in the figure are the flow of gas in ml min^{-1}.

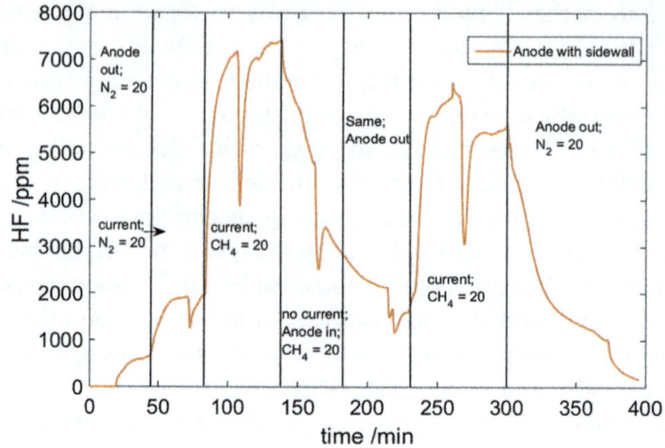

Fig. 9 Changes in HF level of the off-gas during aluminium electrolysis when supplying N_2 + CH_4 to the porous anode. $i = 0.5$ A cm^{-2}, $T = 970$ °C. The values in the figure are the flow of gas in ml min^{-1}.

variation. The anode potential (Fig. 10) was not found to increase significantly during the course of electrolysis, although short time variations were observed. This might be due to the larger surface area of the anode in this experiment. High levels of HF were measured during methane supply. These observations indicate that methane oxidation was more important than the carbon weight loss indicated, which suggests that cracking of methane must have contributed to carbon formation. The pressure build up also increased significantly upon introduction of methane, as shown in Fig. 8.

Shielding of the carbon anode using carbon is not ideal. Using an insulating shielding material is challenging due to the aggressiveness of molten cryolite. Sintered alumina may be used in alumina saturated electrolytes, but it is difficult to obtain a good seal between graphite and alumina.

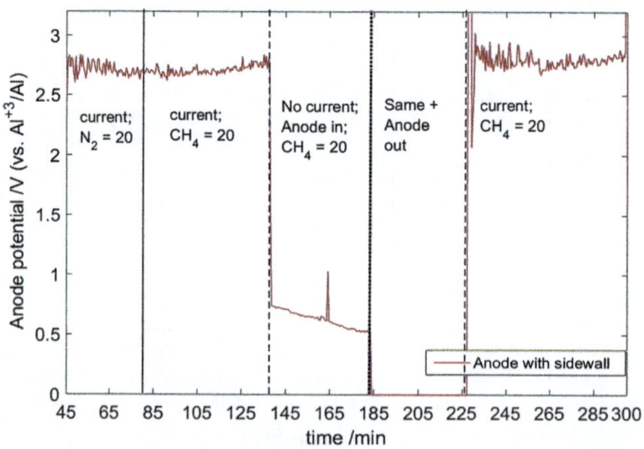

Fig. 10 Anode potential changes during aluminium electrolysis when supplying $N_2 + CH_4$ to the porous anode. $i = 0.5$ A cm^{-2}, $T = 970$ °C. The values in the figure are the flow of gas in ml min^{-1}.

Conclusions

Porous carbon anodes made from different graphite grades were studied in controlled laboratory experiments. The anode potential, the anode carbon consumption, the anode pressure build up and the level of HF gas above the electrolyte were measured during electrolysis. In some cases it was found that the methane oxidation was effectively participating in the anode process. The main challenges in promoting the methane oxidation reaction during electrolysis are related to optimising the anode porosity and graphite structure, and the gas flow, including the methane content to minimise the cracking of methane. The isotropic graphite grade (G347) showed a better performance as a gas anode compared to the moulded and extruded grades. The best results so far showed that the methane oxidation reduced the anode carbon consumption to 56% of the theoretical carbon consumption during electrolysis for 4 hours.

Acknowledgements

The authors gratefully acknowledge financial support from the Research Council of Norway in the project NOVAL (grant number 224985) of the GASSMAKS program.

References

1 J. Thonstad, P. Fellner, G. M. Haarberg, J. Híveš, H. Kvande and Å. Sterten, *Aluminium Electrolysis: Fundamentals of the Hall–Héroult Process*, Aluminium-Verlag Marketing & Kommunikation GmbH, Düsseldorf, 3rd edn, 2001.
2 I. Galasiu, R. Galasiu and J. Thonstad, *Inert Anodes for Aluminium Electrolysis*, Aluminium-Verlag Marketing & Kommunikation GmbH, Düsseldorf, 1st edn, 2007.
3 M. L. Ferrand, *Bull. Soc. Fr. Electr.*, 1957, **79**, 412.
4 V. V. Stender and V. V. Trofimenko, *Khim. Tekhnol.*, 1969, **12**, 41.
5 M. L. Kronenberg, *J. Electrochem. Soc.*, 1969, **116**, 1160.
6 J. Levitan and H. Park, *US Pat.*, US 3696008, 1972.
7 R. A. Rapp, *US Pat.*, US 6039862, 2000.
8 U. B. Pal, D. E. Woolley and G. B. Kenney, *JOM*, 2001, **53**, 32.
9 A. Krishnan, X. Lu and U. B. Pal, *Metall. Mater. Trans. B*, 2005, **36**, 463.
10 U. B. Pal and A. C. Powell IV, *JOM*, 2007, **59**, 44.
11 W. G. Van, *Utility Patent*, WO00/53826, 2000.
12 S. Namboothiri, M. P. Taylor, J. J. J. Chen, M. M. Hyland and M. A. Cooksey, *Light Met.*, 2007, 379.
13 S. Namboothiri, M. P. Taylor, J. J. J. Chen, M. M. Hyland and M. A. Cooksey, *Electrochim. Acta*, 2011, **56**, 3192.
14 R. A. Rapp and Y. Zhang, *Light Met.*, 2002, 463.
15 S. Xiao, T. Mokkelbost, G. M. Haarberg, A. P. Ratvik and H. Zhu, *ECS Trans.*, 2010, **28**, 361.
16 T. Mokkelbost, O. Paulsen, S. Xiao, G. M. Haarberg and A. P. Ratvik, *ECS Trans.*, 2010, **28**, 211.
17 G. M. Haarberg, E. Kvalheim, A. P. Ratvik, S. Xiao and T. Mokkelbost, *Trans. Nonferrous Met. Soc. China*, 2010, **20**, 2152.

18 S. Xiao, Depolarised Anodes for Electrowinning of Aluminium, PhD thesis, Norwegian University of Science and Technology, Trondheim, Norway, March 2011.

19 T. Mokkelbost, O. S. Kjos, O. Paulsen, B. Øye, H. Gudbrandsen, A. P. Ratvik and G. M. Haarberg, A concept for electrowinning of aluminium using depolarized gas anodes, *Light Met.*, 2014, 765.

20 http://en.tokaicarbon.eu/downloads/brochures.

21 C. Guéret, M. Daroux and F. Billaud, Methane pyrolysis: thermodynamics, *Chem. Eng. Sci.*, 1997, **52**, 815.

22 M. P. Taylor, B. J. Welch and J. T. Keniry, Effect of electrochemical changes on the heat balance in aluminium smelting cells, *J. Electroanal. Chem. Interfacial Electrochem.*, 1984, **168**, 179.

Faraday Discussions

ROYAL SOCIETY OF CHEMISTRY

PAPER

Electrolysis of metal oxides in $MgCl_2$ based molten salts with an inert graphite anode

Yating Yuan,[a] Wei Li,[a] Hualin Chen,[a] Zhiyong Wang,[a] Xianbo Jin*[a] and George Z. Chen[b]

Received 20th December 2015, Accepted 22nd January 2016
DOI: 10.1039/c5fd00231a

Electrolysis of solid metal oxides has been demonstrated in $MgCl_2$–NaCl–KCl melt at 700 °C taking the electrolysis of Ta_2O_5 as an example. Both the cathodic and anodic processes have been investigated using cyclic voltammetry, and potentiostatic and constant voltage electrolysis, with the cathodic products analysed by XRD and SEM and the anodic products by GC. Fast electrolysis of Ta_2O_5 against a graphite anode has been realized at a cell voltage of 2 V, or a total overpotential of about 400 mV. The energy consumption was about 1 kW h kg_{Ta}^{-1} with a nearly 100% Ta recovery. The cathodic product was nanometer Ta powder with sizes of about 50 nm. The main anodic product was Cl_2 gas, together with about 1 mol% O_2 gas and trace amounts of CO. The graphite anode was found to be an excellent inert anode. These results promise an environmentally-friendly and energy efficient method for metal extraction by electrolysis of metal oxides in $MgCl_2$ based molten salts.

Introduction

Metal materials are crucial to our society. Metals are mainly produced by pyrometallurgical methods such as carbothermic or metallothermic reductions. However, these processes usually involve multiple steps, and are energy intensive and environmentally unfriendly.[1,2] At the end of the twentieth century, Chen *et al.*[3] proposed the solid electrochemical reduction of TiO_2 to produce Ti in $CaCl_2$ molten salt, known as the FFC Cambridge process. Since then, this process has received much attention. Up till now, many metals, semi-metals and their alloys have been successfully prepared by this simple solid oxide reduction process.[4,5] The FFC Cambridge process can be summarized by following reactions.

Overall reaction:

$$MO_x \rightarrow M + x/2O_2 \text{ (inert anode)} \tag{1a}$$

[a]*College of Chemistry and Molecular Sciences, Hubei Key Laboratory of Electrochemical Power Sources, Wuhan University, Wuhan, 430072, P. R. China. E-mail: xbjin@whu.edu.cn; Fax: +86 27 68756319; Tel: +86 27 68756319*

[b]*Department of Chemical and Environmental Engineering, Faculty of Engineering, The University of Nottingham, Nottingham, NG7 2RD, UK*

$$nMO_x + xC \rightarrow nM + xCO_n \text{ (carbon anode, } n = 1 \text{ or } 2) \tag{1b}$$

Cathodic reaction:

$$MO_x + 2xe \rightarrow M + xO^{2-} \tag{2}$$

Anodic reaction:

$$2O^{2-} \rightarrow O_2 + 2e \text{ (inert anode)} \tag{3a}$$

$$nO^{2-} + C \rightarrow CO_n + 2ne \text{ (carbon anode)} \tag{3b}$$

In the above equations M represents the studied metal (*e.g.* Ti, Nb and Ta) or semi-metal (*e.g.* Si).

During electrolysis, the graphite anode is often used for its good electronic and mechanical properties. However, as shown by Reaction (3b), the anodic discharge of dissolved O^{2-} ions in $CaCl_2$, which come from the cathode by Reaction (2), will lead to anode consumption and consequently carbon emission. Moreover, the CO_2 produced at the anode can react with O^{2-} ions in the molten salt to form ${CO_3}^{2-}$ ions, which will move back to the cathode and discharge to carbon and O^{2-} ions,[6,7] leading to carbon contamination to the metal product, floating carbon on the surface of $CaCl_2$ melt, as well as low current efficiency. Although many inert anodes have been developed for O_2 evolution,[8–12] they are yet pending for commercial verification. At present, similar to the electrolysis of dissolved Al_2O_3 in cryolite based molten fluorides, carbon anodes are still the choice for the electrolysis of solid metal oxides in molten $CaCl_2$. To deal with the relevant problems, recently we have proposed a "disengaging strategy" by electrolysis of solid metal oxides in $MgCl_2$ based molten salts, in which the solubility of O^{2-} ions is insignificant to transport O^{2-} to the anode.[13] Alternatively, the released O^{2-} ions stay in the cathode by forming MgO precipitation, and the anode reaction is thus changed to oxidation of Cl^- ions to the Cl_2 gas. This disengaging strategy makes the carbon anode effectively non-consumable, eliminating CO_2 emission and all related problems. Our new process can be represented by Reactions (4–6).

Overall reaction:

$$MO_x + xMgCl_2 = M + xMgO + xCl_2 \tag{4}$$

Cathodic reaction:

$$MO_x + xMgCl_2 + 2xe \rightarrow M + xMgO + 2xCl^- \tag{5}$$

Anodic reaction:

$$2xCl^- \rightarrow xCl_2 + 2xe \tag{6}$$

As indicated in Table 1, this new process, thermodynamically, can be applied to reduce most transitional metal oxides to respective metals, including some rare earth metals, such as uranium.

Considering molten $MgCl_2$ by itself has a relatively high vapor pressure, it is better to be used together with other chloride salts, such as NaCl and KCl. Using

Table 1 Standard decomposition voltages of metal oxides in molten $MgCl_2$ at 700 °C, calculated from HSC Chemistry 6 (A. Roine, HSC Chemistry 6, Outotec Research Oy, 2006)

Metal oxides	Decomposition voltage (V)
$2KCl = 2K + Cl_2(g)$	3.566
$2NaCl = 2Na + Cl_2(g)$	3.329
$MgCl_2 = Mg + Cl_2(g)$	2.534
$2MgCl_2 + HfO_2 = Hf + 2MgO + 2Cl_2(g)$	2.388
$2MgCl_2 + UO_2 = U + 2MgO + 2Cl_2(g)$	2.335
$2MgCl_2 + ZrO_2 = Zr + 2MgO + 2Cl_2(g)$	2.317
$2MgCl_2 + TiO_2 = Ti + 2MgO + 2Cl_2(g)$	1.949
$2MgCl_2 + SiO_2 = Si + 2MgO + 2Cl_2(g)$	1.865
$3MgCl_2 + B_2O_3 = 2B + 3MgO + 3Cl_2(g)$	1.731
$5MgCl_2 + Ta_2O_5 = 2Ta + 5MgO + 5Cl_2(g)$	1.638
$3MgCl_2 + V_2O_3 = 2V + 3MgO + 3Cl_2(g)$	1.616
$5MgCl_2 + Nb_2O_5 = 2Nb + 5MgO + 5Cl_2(g)$	1.488
$3MgCl_2 + Cr_2O_3 = 2Cr + 3MgO + 3Cl_2(g)$	1.478
$3MgCl_2 + Ga_2O_3 = 2Ga(l) + 3MgO + 3Cl_2(g)$	1.298
$MgCl_2 + ZnO = Zn(l) + MgO + Cl_2(g)$	1.259
$3MgCl_2 + WO_3 = W + 3MgO + 3Cl_2(g)$	0.984
$2MgCl_2 + GeO_2 = Ge + 2MgO + 2Cl_2(g)$	0.977
$2MgCl_2 + SnO_2 = Sn(l) + 2MgO + 2Cl_2(g)$	0.944
$3MgCl_2 + Fe_2O_3 = 2Fe + 3MgO + 3Cl_2(g)$	0.939
$2MgCl_2 + MnO_2 = Mn + 2MgO + 2Cl_2(g)$	0.855
$3MgCl_2 + MoO_3 = Mo + 3MgO + 3Cl_2(g)$	0.824
$3MgCl_2 + Sb_2O_3 = 2Sb(l) + 3MgO + 3Cl_2(g)$	0.749
$MgCl_2 + NiO = Ni + MgO + Cl_2(g)$	0.743
$4MgCl_2 + Co_3O_4 = 3Co + 4MgO + 4Cl_2(g)$	0.655
$3MgCl_2 + Bi_2O_3 = 2Bi(l) + 3MgO + 3Cl_2(g)$	0.488
$MgCl_2 + CuO = Cu + MgO + Cl_2(g)$	0.316

a mix of molten salts can also further decrease the solubility of MgO.[14,15] The solubility of MgO in pure $MgCl_2$ at 730 °C was reported to be about 0.36 mol%, which is fairly low in contrast with $CaCl_2$ (>20 mol% at 850 °C) and LiCl (>10 mol % at 650 °C).[16,17] However, if 70 mol% NaCl was added, the solubility of MgO can be significantly decreased to about 0.01 mol%. On the other hand, the solubility of MgO in equimolar NaCl–KCl at 730 °C is also less than 0.01 mol%.[18] These data suggest that using the molten mixture of $MgCl_2$–NaCl–KCl (MNK) can be more effective for disengaging the carbon anode from O^{2-} ions.

In this work, we further demonstrate our new process by the electrolysis of Ta_2O_5 in the MNK melt at 700 °C. The mechanism of Ta_2O_5 reduction was investigated, and nanometer Ta powder (<50 nm) was produced at an energy consumption of about 1.0 kW h kg_{Ta}^{-1}. The inert graphite anode was confirmed with the anodic gaseous products analysed in detail.

Experimental

Anhydrous $MgCl_2$, NaCl, and KCl (analytical grade, Tianjin Guangfu Fine Chemical Research Institute, or Sinopharm Chemical Reagent Co., Ltd, China) were mixed in the eutectic ratio 5 : 3 : 2 (in mole ratio and 5 mol in total amount), and heated in a graphite or alumina crucible (inner diameter: 25–50 mm; height:

200–700 mm) in a programmable vertical furnace equipped with a sealable stainless steel retort (Wuhan Experimental Furnace Plant). The furnace temperature was slowly raised to and maintained at 400 °C for more than 12 h, and then to 700 °C until the salt mixture was fully molten. Pre-electrolysis of the molten salt was applied at 1.6 V to remove moisture and other redox-active impurities with the graphite crucible (or a graphite rod of 10 mm diameter in the alumina crucible) as the anode and a Mo wire (diameter: 2 mm) as the cathode. Pre-electrolysis lasted (*ca.* 10 h) until the current reached a low and stable background level.

Cyclic voltammograms (CVs) were recorded using a Mo cavity electrode (MCE) loading with Ta_2O_5 powder, with the potential reference to a quartz sealed Ag/AgCl electrode. For constant potential, constant cell voltage and galvanostatic electrolysis, Ta_2O_5 powder (analytical grade; particle sizes: 300 nm; Zhuzhou Cemented Carbide Works Imp. & Exp. Company) was individually die-pressed into cylindrical pellets (10 MPa, 13 or 20 mm in diameter, 0.7–2 g: 1.1–1.7 mm in thickness, porosity: ~50%), sintered in air at 1000 °C for 2 h, and sandwiched between two molybdenum (Mo) meshes to form an assembled cathode. After electrolysis for a designated time, the cathode was removed from the furnace, cooled in argon, washed in distilled water, or in dilute HCl (0.1 mol L^{-1}) and water again, and then dried in a vacuum at 80 °C before further analyses. The CVs were recorded using CHI660A Electrochemical System (Shanghai Chenhua, China), while the electrolysis was controlled by a multichannel four-electrode potentiostat (Neware, China).

The products were characterised by X-ray diffraction spectroscopy (X-ray 6000 with Cu Kα1 radiation at $\lambda = 1.5405$ Å, Shimadzu, Japan), and a FEI Sirion Field Emission Gun SEM system. GC (FULI 9790II, with a thermal conductivity detector and argon serving as a carrier gas) was used for the analysis of the anodic gaseous products. Ion Chromatography (IC) analysis for Cl^- ions was carried out using METROHM 881 Compact IC pro, with the ClO^- ions reduced to Cl^- ions by H_2O_2.

Results and discussion

Fig. 1 shows the cyclic voltammogram (CV) of Ta_2O_5 powders in MNK at 700 °C using the metallic cavity electrode (MCE). For comparison, a CV recorded on an empty MCE is superimposed. No significant reaction current can be observed on the blank CV before Mg deposition (C0) and dissolution, with an Mg/Mg^{2+} equilibrium potential of about −1.55 V *vs.* Ag/AgCl. The reduction current of Ta_2O_5 was found initially to be −0.4 V *vs.* Ag/AgCl. Preliminary comparison and analysis of the CVs of Ta_2O_5 in molten MNK and those previously reported in $CaCl_2$ have revealed fairly similar main reduction features, showing mainly three reduction processes (C1–C3).[19] However, the starting potential of C3 is unclear due to the overlap with C2.

According to Table 1, the reduction of Ta_2O_5 to Ta could occur at a potential about 0.9 V more positive than the deposition of Mg, thus, potentiostatic electrolysis of Ta_2O_5 was performed at −0.8 V *vs.* Ag/AgCl, a potential 0.75 V more positive than the Mg deposition potential so Ta was predicted to be produced. The current–time plot is shown in Fig. 2a. The reduction current rises initially to a peak suggesting the reduction followed the propagation of the three-phase interlines, which was generally observed during reduction of insulating oxides to metals.[20]

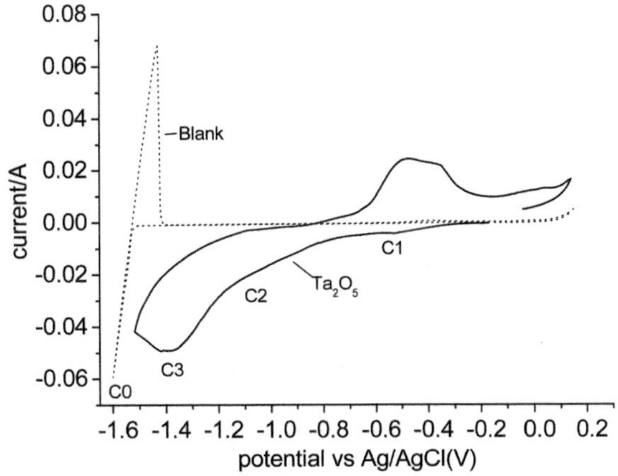

Fig. 1 Cyclic voltammograms (CVs) of the blank MCE and the MCE loading with Ta_2O_5 powder in molten $MgCl_2$–NaCl–KCl at 700 °C.

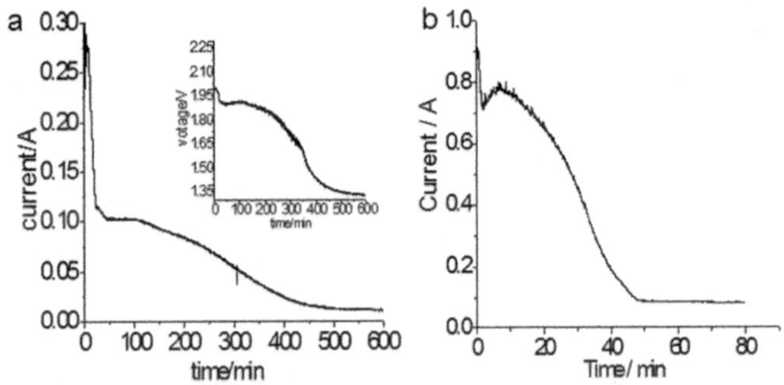

Fig. 2 Current–time plots of (a) potentiostatic electrolysis of a Ta_2O_5 pellet in MNK at −0.8 V *vs.* Ag/AgCl and 700 °C. Insert: corresponding cell voltage profile. (b) Constant voltage electrolysis of a 0.7 g Ta_2O_5 pellet at 2.0 V and 700 °C.

In line with the CV, Fig. 2a also shows several reduction processes. The first reduction displayed a high reduction current of about 0.25 A in the initial 30 minutes. The second reduction corresponded to the plateau current at about 0.1 A, which began to decrease at about 110 minutes, then a slope change at about 250 minutes could indicate a third reduction process. Finally, the current levelled off to background current after 500 minutes. To identify these reductions, XRD patterns of the electrolysis products after different times were scanned and displayed in Fig. 3. It can be seen that after 27 minutes, the Ta_2O_5 was almost completely reduced to TaO_2. The electrolysis product at 300 minutes included Ta and $Mg_4Ta_2O_9$. The appearance of $Mg_4Ta_2O_9$ after TaO_2 may suggest the possible disproportionation reaction of TaO_2, possibly like its congener VO_2, and the generation of an excess of MgO may have shifted the reaction to form the stabilized $Mg_4Ta_2O_9$. The final electrolysis product after washing with water consisted of Ta and MgO. The presence of MgO in the cathode indicates its low solubility in the MNK melts as discussed above.

The cell voltage (the inset in Fig. 2a) during the −0.8 V constant potential electrolysis was also monitored. In the main reduction stage of Ta_2O_5, the cell

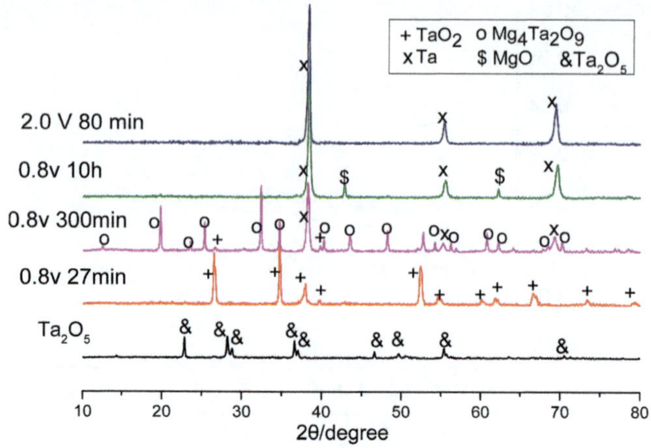

Fig. 3 XRD patterns of the products collected at the indicated times of potentiostatic electrolysis (−0.8 V *vs.* Ag/AgCl, water washed only) and constant voltage electrolysis of Ta_2O_5 (2.0 V, water and acid washed) at 700 °C in MNK.

voltage was about 1.9 V, which was reasonably higher than the decomposition voltage ~1.64 V of Ta_2O_5 in the $MgCl_2$ based melt as calculated in Table 1. In addition, it can be seen that after the cell voltage decreased to about 1.64 V, it dropped suddenly. These may indicate that the electrolysis was mainly *via* the assumed Reaction (4) or specifically Reaction (7) for Ta_2O_5:

$$5MgCl_2 + Ta_2O_5 = 2Ta + 5MgO + 5Cl_2(g) \tag{7}$$

Constant cell voltage electrolysis at 2.0 V was also carried out, and the current–time plot is shown in Fig. 2b. It can be seen that the reduction completed in about 50 minutes, and the dimensions of the cathode pellet remained almost unchanged before and after the reduction (Fig. 4a), suggesting the possibilities of both high electrolysis speed and high Ta recovery (nearly 100% in this work). Mass analysis was carefully carried out after the product was washed with water and 0.1 mol L^{-1} HCl respectively, and this confirmed that most of the MgO

Fig. 4 (a) Ta_2O_5 pellet collected after 80 min electrolysis in 700 °C MNK at 2.0 V. (b) Scanning Electron Microscopy (SEM) image of the Ta nanoparticles shown in (a) (after acid washing).

generated through Reaction (7) stayed in the cathode. It is interesting that although the porosity of the Ta_2O_5 pellet was only about 50% of that which is commonly used for electro-reduction in molten $CaCl_2$, the influence of MgO accumulation in the oxide cathode was insignificant for the electro-reduction. However, it can be derived that the porosity of the Ta_2O_5 cathode needs to be higher than 30% to prevent the ion passages from becoming blocked.

On the other hand, although the cathode pellet taken out from the molten salt was perfectly strong, it dispersed to a powder after washing in water. After being further washed with acid, pure Ta was obtained, and the XRD pattern is shown in Fig. 3. The SEM images show very fine particles (~50 nm) of the resulting Ta powder, which were smaller than those produced in molten $CaCl_2$ at 900 °C, suggesting higher specific capacitance.[19] This phenomenon may be partly associated with the relatively low molten salt temperature (700 °C) and also partly attributed to the MgO precipitated inside the pellet which prevented physical sintering of the Ta particles. The current efficiency was higher than 92%, with an energy consumption of about 1.3 kW h kg_{Ta}^{-1} for the electrolysis of 0.7 g Ta_2O_5 and 1 kW h kg_{Ta}^{-1} for the electrolysis of 2 g Ta_2O_5.

On the graphite anode, for verification of chlorine evolution, galvanostatic electrolysis of Ta_2O_5 at 1.5 A was performed in MNK. Since the MNK melt does contain some O^{2-} ions, although with a concentration in the order of 10^{-4} in molar fraction, the discharge of O^{2-} ions on graphite would occur. According to Scheme 1, the amounts of O_2, CO_2, CO and Cl_2 in the anodic gaseous product were determined.

First, about 500 mL of anodic gas was collected as shown in Fig. 5a. The colour of the anodic product was yellowish green. This is straightforward evidence for formation of Cl_2 on the anode. After being absorbed by NaOH solution, a starch iodide test indicated the existence of ClO^- in the solution, again convincing evidence for anodic Cl_2 gas, which has been reacted with NaOH. In addition, ion chromatography (IC) analysis indicated that the Cl^- and ClO^- ion concentration in the NaOH solution approximately matched the absorbed gas volume (about 460 mL). About 40 mL colourless gas remained, and GC analysis (Fig. 6) indicated the existence of O_2 and CO with concentrations in the total anodic gas of about 1.2 mol% and 0.2 mol%, respectively.

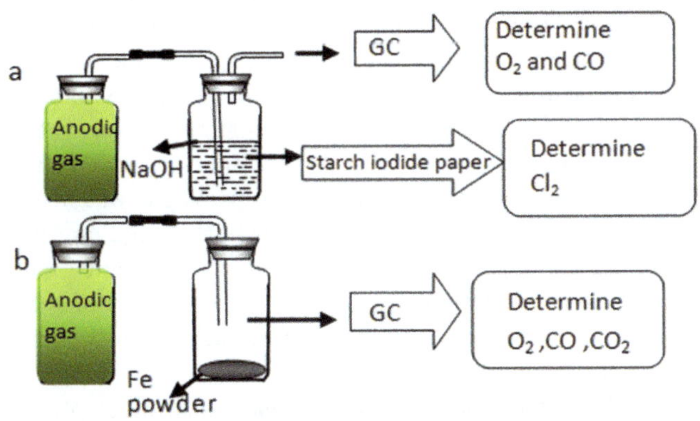

Scheme 1 Illustration of the method for identification of anodic gas.

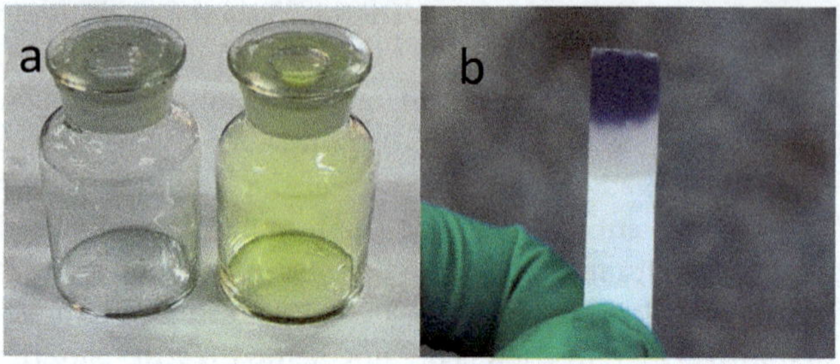

Fig. 5 (a) Bottles with 500 mL air (left) and anodic gas (right). (b) Photo of the colour change of the starch iodide paper after immersing into anodic gas bubbled through NaOH solution.

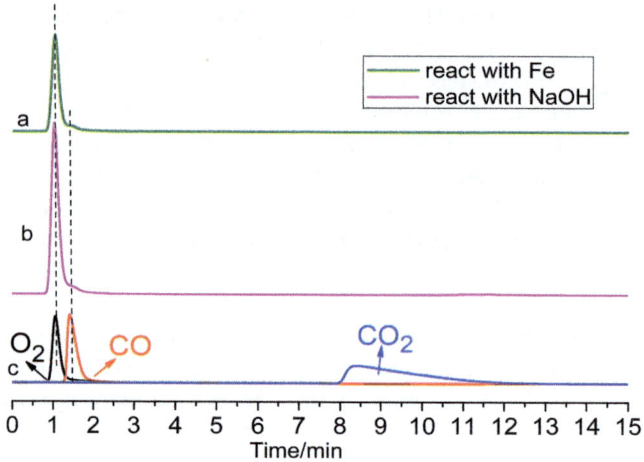

Fig. 6 GC spectrogram of anodic gas. Line (a): after reacting with Fe powder. Line (b): after being absorbed by NaOH solution. Line (c): superimposed GC spectrograms of pure O_2, CO and CO_2.

No CO_2 was detected by GC, however, even if it was formed on a graphite anode, according to the Cl_2 volume indicated by IC, the CO_2 concentration would be very low. Similarly, it was reported that no CO_2 gas was found as an anodic product during electrolysis of MgO in $MgCl_2$–$NdCl_3$ system at 700 °C.[21] However, it seems strange that there was O_2 as an anodic product without CO_2 gas on a graphite anode. For confirmation, we designed another experiment. As shown in Scheme 1b, the anodic gaseous product was allowed to react with the reducing Fe powders thoroughly, and the remaining colourless gas was also detected by GC. According to Fig. 6, CO_2 was in truth absent in the anodic product.

According to the above tests, the composition of the anodic gaseous product during galvanostatic electrolysis of Ta_2O_5 at 1.5 A consisted of mainly Cl_2 gas, about 1 mol% O_2 gas and trace amounts of CO gas, as summarized in Table 2. No CO_2 was formed on the graphite anode.

To determine the origination of anodic O_2 gas, CVs of the graphite electrode in the MNK melt were recorded as shown in Fig. 7. According to Fig. 7, the evolution

Table 2 Composition of the anodic gaseous product during galvanostatic electrolysis of Ta_2O_5 at 1.5 A

Compositions	Content (mol%)
Cl_2	98.6
O_2	1.2
CO	<0.2
CO_2	None

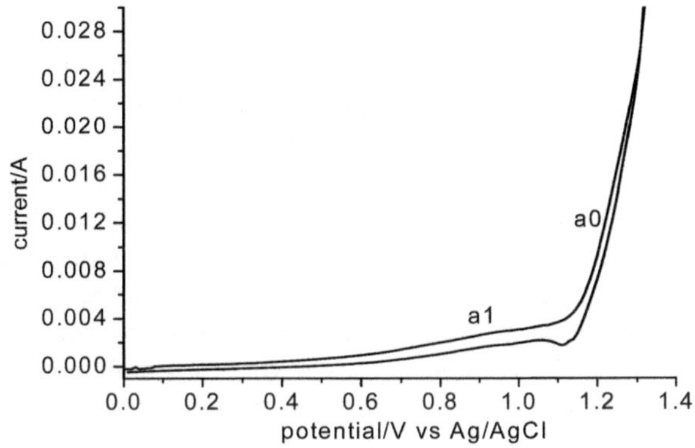

Fig. 7 Cyclic voltammogram of graphite in eutectic $MgCl_2$–NaCl–KCl (700 °C) *vs.* Ag/AgCl, scan rate 200 mV s^{-1}.

of chlorine (a0) was initiated at 1.1 V *vs.* Ag/AgCl. Coupled with the deposition potential of Mg (see Fig. 1a) on MCE, the decomposition voltage of $MgCl_2$ was determined to be 2.65 V, which is fairly close to the thermodynamic value (2.534 V) calculated from HSC Chemistry 6. The measured value was slightly higher, which is reasonable considering that the molar fraction of $MgCl_2$ in the MNK was only 0.5. The current corresponding to chlorine evolution increased rapidly with the positive scan, indicating that there is no kinetic difficulty on the discharge of Cl^- ions on a graphite anode. Since Cl^- ions are the main anions in the MNK melt, concentration polarization for the chlorine evolution is also negligible. On the contrary, in molten $CaCl_2$, the anodic discharge of O^{2-} ions on graphite is well known to be kinetically difficult both in terms of charge transfer and mass transfer.

Before a0, there is an apparent anodic current at a1, which is very stable according to the multi-cycle scans and could be attributed to the discharge of O^{2-} ions on the graphite anode. However, unlike the discharge of O^{2-} on a carbon based anode in molten $CaCl_2$, which exhibited complicated voltammograms involving formation of CO_2, CO and O_2 gases,[22–24] here the CVs of O^{2-} discharge in MNK featured a single reaction, probably the formation of O_2 gas as mentioned above. To confirm it, potentiostatic electrolysis was performed at 1.0 V *vs.* Ag/AgCl against a Ta_2O_5 pellet cathode. The anodic current remained very stable at about 15 mA during the total 3 hour electrolysis. However, no anodic gas has automatically flowed out from the anodic gas chamber. The chamber was then sealed and the anodic gas was sampled at different times for GC analysis.

Fig. 8 shows the GC analysis results. As can be seen, after 0.5 h electrolysis, there was a small peak of O_2 gas but no CO and CO_2 gas was found. After about 1 h electrolysis, the O_2 concentration increased significantly, and small peaks of CO and CO_2 became just perceptible, and the $O_2 : CO : CO_2$ ratio was about 8.2 : 1.3 : 0.5, which changed to 7.9 : 1.0 : 1.1. Considering initially there is only O_2 evolution and the CO_2 content became larger and larger, it is likely that the formation of CO_2 is a result of a reaction between the graphite anode and the accumulated O_2 in the anodic chamber. Combining the CV result and the GC analysis of the anodic gas during constant potential electrolysis at 1.0 V, it might be reasonable that the a1 current in Fig. 7 corresponds to the discharge of O^{2-} ions to O_2 gas on the graphite anode in the MNK melt at 700 °C. However, the small steady current (15 mA) at 1.0 V *vs.* Ag/AgCl was most likely to be the limited diffusion current of O^{2-} ion discharge, which is only about 1.5% of the total current considering a 1.0 A g^{-1}-Ta_2O_5 or higher current would be applied in the actual electrolysis.

The optical images of the graphite anode before and after repeated electrolysis for *ca.* 100 h in the eutectic MNK melt were compared in Fig. 9. In line with the Cl_2 and O_2 gas evolution at the anode, the fine spiral machining marks of *ca.* 100 nm in width on the side of the graphite rod remained almost intact, even though these protruding parts are usually the most vulnerable sites to chemical or electrochemical attack. These observations suggested that the graphite anode is basically inert during the electrolysis of solid oxide cathode in the $MgCl_2$ based molten salts. However, it is well-known that the anodic reaction of the solid oxide cathode electrolysis at a graphite electrode in $CaCl_2$ was the discharge of the O^{2-} to carbon oxides and carbonate ions, leading to significant erosion of the graphite anode, shuttle reactions of the oxygenated species (O^{2-} and ${CO_3}^{2-}$), and formation of a thick layer of floating carbon powder on the surface of the molten electrolyte. Since there was no corrosion to the graphite anode and consequently no carbon generated on the cathode due to the reduction of ${CO_3}^{2-}$ in a $MgCl_2$ based electrolyte, no floating carbon was supposed to form on the surface of molten MNK. This was confirmed by the clear surface of MNK after 100 h electrolysis of Ta_2O_5 as shown in Fig. 9c.

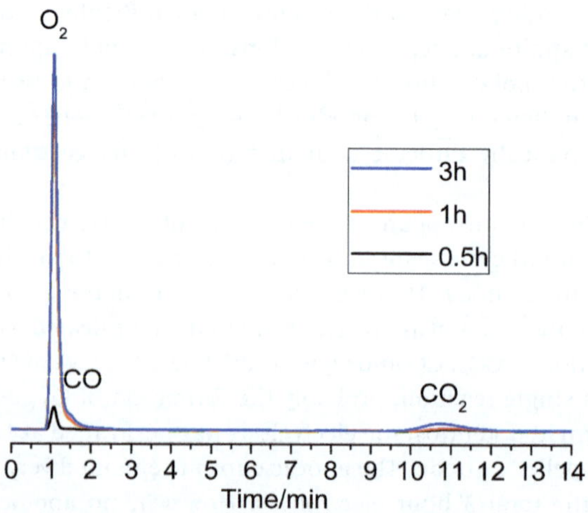

Fig. 8 GC spectrograms of anodic gas produced by potentiostatic polarization of graphite in 700 °C MNK at 1.0 V *vs.* Ag/AgCl (time indicated).

Fig. 9 Optical photos of the graphite rod anode (dia. 2.0 cm) (a) before and (b) after electrolysis experiments for *ca.* 100 h in the eutectic MNK melt. (c) Photo of cooled MNK salts after 100 h electrolysis of Ta_2O_5.

These results indicate that both the graphite anode and the MNK salt are ready for lengthy electrolysis. Of course, the consuming $MgCl_2$ needs to be supplemented, for example, along with the metal oxide pellet. In addition, the Cl_2 gas has many applications and can be collected as already commercially demonstrated in the chlor-alkali industry and the electrolysis of molten LiCl and $MgCl_2$. It is also possible to use the Cl_2 gas to regenerate $MgCl_2$, noting the recent progress in developing H_2–Cl_2 fuel cells,[25] which generate electricity and HCl while HCl is needed to convert MgO in the cathode back to $MgCl_2$.

Conclusions

We have demonstrated the high speed, low energy consumption electrolysis of Ta_2O_5 in $MgCl_2$–NaCl–KCl melt using an inert graphite anode. The reduction process was investigated by cyclic voltammetry, and potentiostatic and constant voltage electrolysis. The main intermediate products of the electro-reduction of Ta_2O_5 were identified to be TaO_2 and $Mg_4Ta_2O_9$. The final reduction product was nanometer Ta powder with sizes of about 50 nm. The anodic product included mainly Cl_2 gas, plus about 1 mol% O_2. There was only about 0.2 mol% CO in the anodic gas, and no CO_2 was detected, suggesting that the graphite anode is fairly stable in our new process. The process should be applicable for the preparation of many other metals and alloys with high energy efficiency and low carbon emission.

Acknowledgements

We thank the NSFC, Ministry of Education of China, the EPSRC, Guangdong Nat. Sci. Found., Nat. Key Fundamental R&D Program of China and Nat. Hi-Tech R&D Program of China for financial support (Grant No. 20773094, 21173161, 51204060, NCET-11-0397, EP/J000582/1, S2012040007501).

References

1 W. Kroll, *Trans. Electrochem. Soc.*, 1940, **78**, 35–47.
2 T. Okabe, K. Nikami and K. Ono, *Bulletin of the Iron and Steel Institute of Japan*, 2002, 7, 39–45.
3 G. Z. Chen, D. J. Fray and T. W. Farthing, *Nature*, 2000, **407**, 361–364.
4 D. H. Wang, X. B. Jin and G. Z. Chen, *Annu. Rep. Prog. Chem., Sect. C: Phys. Chem.*, 2008, **104**, 189–234.
5 A. M. Abdelkader, K. T. Kilby, A. Cox and D. J. Fray, *Chem. Rev.*, 2013, **113**, 2863–2886.
6 L. Cassayre, P. Palau, P. Chamelot and L. Massot, *J. Chem. Eng. Data*, 2010, **55**, 4549–4560.
7 C. Schwandt and D. Fray, *Electrochim. Acta*, 2005, **51**, 66–76.
8 D. R. Sadoway, *JOM*, 2001, **53**, 34–35.
9 K. T. Kilby, S. Jiao and D. J. Fray, *Electrochim. Acta*, 2010, **55**, 7126–7133.
10 C. Chen and X. Lu, *Acta Metall. Sin. (Chin. Ed.)*, 2008, **44**, 145.
11 S.-W. Kim, E.-Y. Choi, W. Park, H. S. Im and J.-M. Hur, *Electrochem. Commun.*, 2015, **55**, 14–17.
12 S. Jiao, K.-N. P. Kumar, K. T. Kilby and D. J. Fray, *Mater. Res. Bull.*, 2009, **44**, 1738–1742.
13 W. Li, Y. T. Yuan, H. L. Chen, X. B. Jin, Z. Y. Wang and G. Z. Chen, *Progress in Natural Science: Materials International*, 2015, **11**, 002.
14 S. Boghosian, A. Godø, H. Mediaas, W. Ravlo and T. Østvold, *Acta Chem. Scand.*, 1991, **45**, 145.
15 K. H. Sten, *Electrochim. Acta*, 1979, **24**, 509.
16 T. Usami, M. Kurata, T. Inoue, H. E. Sims, S. A. Beetham and J. A. Jenkins, *J. Nucl. Mater.*, 2002, **300**, 15.
17 D. A. Wenz, I. Johnson and R. D. Wolson, *J. Chem. Eng. Data*, 1969, **14**, 250.
18 R. Combes, F. De Andrade, A. De Barros and H. Ferreira, *Electrochim. Acta*, 1980, **25**, 371–374.
19 T. Wu, X. B. Jin, W. Xiao, X. H. Hu, D. H. Wang and G. Z. Chen, *Chem. Mat.*, 2007, **19**, 153–160.
20 G. Z. Chen, E. Gordo and D. J. Fray, *Metall. Mater. Trans. B*, 2004, **35**, 223.
21 K. Cathro, R. Deutscher and R. Sharma, *J. Appl. Electrochem.*, 1997, **27**, 404–413.
22 H. L. Chen, X. B. Jin, L. P. Yu and G. Z. Chen, *J. Solid State Electrochem.*, 2014, **18**, 3317.
23 M. Mohamedi, B. Børresen, G. M. Haarberg and R. Tunold, *J. Electrochem. Soc.*, 1999, **146**, 1472–1477.
24 R. Tunold, G. M. Haarberg, K. S. Osen, A. M. Martinez and E. Sandnes, *ECS Trans.*, 2011, **35**, 1–9.
25 B. Huskinson, J. Rugolo, S. K. Mondal and M. J. Aziz, *Energy Environ. Sci.*, 2012, **5**, 8690.

Faraday Discussions

PAPER

Template-free electrodeposition of AlFe alloy nanowires from a room-temperature ionic liquid as an anode material for Li-ion batteries

Gang Chen, Yuqi Chen, Qingjun Guo, Heng Wang and Bing Li*

Received 3rd December 2015, Accepted 29th February 2016
DOI: 10.1039/c5fd00211g

AlFe alloy nanowires were directly electrodeposited on copper substrates from trimethylamine hydrochloride (TMHC)–$AlCl_3$ ionic liquids with small amounts of $FeCl_3$ at room temperature without templates. Coin cells composed of AlFe alloy nanowire electrodes and lithium foils were assembled to characterize the alloy electrochemical properties by galvanostatic charge/discharge tests. Effects of $FeCl_3$ concentration, potential and temperature on the alloy morphology, composition and cyclic performance were examined. Addition of Fe into the alloy changed the nanowires from a 'hill-like' bulk morphology to a free-standing morphology, and increased the coverage area of the alloy on Cu substrates. As an inactive element, Fe could also buffer the alloys' large volume changes during Li intercalation and deintercalation. AlFe alloy nanowires composed of a small amount of Fe with an average diameter of 140 nm exhibited an outstanding cyclic performance and delivered a specific capacity of about 570 mA h g^{-1} after 50 cycles. This advanced template-free method for the direct preparation of high performance nanostructure AlFe alloy anode materials is quite simple and inexpensive, which presents a promising prospect for practical application in Li-ion batteries.

1. Introduction

Over the past two decades, Li-ion batteries have become the primary energy storage devices for a wide range of portable electronic devices.[1–3] Conventional graphite anodes have relatively low capacity and safety problems, which urge people to search for new anode materials with higher capacities and longer cycle lives.[4] There have been plenty of attempts to replace graphite by different types of anodes. Among many other promising candidates, aluminum seems to be an attractive anode material owing to its high specific capacity (993 mA h g^{-1} for AlLi alloy), light weight and low cost.[5,6] However, the large volume change during

East china University of Science and Technology, 130 Meilong Road, Shanghai, 200237, China. E-mail: bingli@ecust.edu.cn

lithium intercalation and deintercalation processes causes aluminum cracking and gradual degradation, resulting in rapid capacity loss and poor cycle ability.[6,7] Usually, nano-structured electrode materials,[8,9] intermetallic compounds,[10,11] and active/inactive nanocomposites[12,13] are used to overcome the electrode deterioration caused by volume expansion. For instance, Endres *et al.*[6] used polycarbonate membranes as templates to electrodeposit Al nanowires with an average diameter of 90 nm on gold films. The high-cost Al-nanowires as anodes for Li-ion batteries delivered a charge and discharge capacity of 790 and 719 mA h g^{-1} for the first cycle, and withstand 50 successive Li deposition/stripping cycles without significant disintegration. Lindsay *et al.*[14] prepared an Fe_2Al_5 intermetallic compound by arc melting the metallic Fe and Al, crushing to a suitable size and then ball milling. The highest discharge capacity observed was 485 mA h g^{-1}. In Cao's research,[15] Al–Fe–C composites with a particle size of about 5–7 μm were prepared by high-energy ball milling, which delivered an initial capacity of 436 mA h g^{-1} with a reversible capacity of 255 mA h g^{-1} after 15 cycles. Obviously, Al nanowire structures significantly improved the performance of the batteries, because of higher specific electrode/electrolyte interface area and better toleration of volume change during the cycling process. However, the Al nanowire electrode capacity and cyclability still cannot meet the critical requirements for the anode of Li-ion batteries. In view of the advantages of a combination of an active/inactive nanocomposite and nano-structured Al electrode, here we propose to electrodeposit AlFe nanowires from a much cheaper ionic liquid (IL) as the anode of Li-ion batteries.

Al nanowires and some other metal nanowires such as Ag, Zn, Ge have been electrodeposited from ILs using a polycarbonate (PC) membrane or aluminum oxide (AAO) as templates.[16–18] However, the templates have to be prepared before the electrodeposition and dissolved after the electrodeposition. On the contrary, the template-free method is a simple and time-saving process. For example, Ti, SmCo, SnSi and Al nanowires have been obtained by electrodeposition from ILs without using any templates.[10,19–21] Furthermore, the transition metal elements like Fe, Co and Ni can act as inactive elements to buffer the large volume change of Al nanowires and act as barriers against the aggregation of Al nanowires into large bulks during Li-ion intercalation and deintercalation processes.[14]

Based on the above work, we electrodeposited AlFe alloy nanowires from a much cheaper trimethylamine hydrochloride (TMHC/$(CH_3)_3N\cdot HCl$)–$AlCl_3$–$FeCl_3$ IL without templates. The as-prepared AlFe alloy nanowires can be directly used as anode materials for Li-ion batteries, thus no conductive additives and binder are needed and the anode utilization rate is improved. AlFe alloy nanowires exhibited an outstanding cyclic performance and delivered a specific capacity of about 570 mA h g^{-1} after 50 cycles.

2. Experimental

The IL was prepared by slowly adding TMHC (Alfa Aesar, 98%) into a certain mixture of $AlCl_3$ (Alfa Aesar, 99.9%) and $FeCl_3$ (Alfa Aesar, 98%) in a beaker and continuously stirred for about 24 h using a magnetic stirrer bar. Electrochemical measurements and electrodepositions were performed under nonconvection conditions at room temperature. A three-electrode cell controlled by a PARSTAT2273 (PAR-Ametek Co, Ltd) with PowerSuite software package was employed for electrochemical measurements and electrodeposition. A Cu sheet (99.9%) was used as the working electrode for electrodeposition, while a tungsten wire was used as the working electrode for voltammetric study, and the reference electrode was prepared by immersing an Al wire (99.9%) into a 40 : 60 mol% TMHC–$AlCl_3$ ionic liquid located in a fritted glass tube.[22] For stabilization of the IL composition during electrodeposition, an Al sheet (99.9%) was selected as a counter electrode. All the electrodes were firstly polished with emery papers to a mirror polish, and washed with distilled water. Then, Cu sheets were dipped in dilute sulphuric acid for 15 min and flushed with distilled water, dried in a vacuum oven, cleaned with acetone and finally dried in air. After each electrodeposition, a Cu sheet covered with AlFe alloy nanowires was quickly moved into acetonitrile and cleaned for three times. It could then be directly assembled into a coin-type cell as the anode with a lithium plate (Φ 16 mm × 1.5 mm, China Energy Lithium Co, Ltd) as the cathode in 1 M $LiPF_6$ in ethylene carbonate (EC), dimethyl carbonate (DMC) and ethyl methyl carbonate (EMC) (1 : 1 : 1 in volume) as the electrolyte with a Celgard 2400 separator. All the experiments were performed inside an argon-filled glove box (O_2 and H_2O < 1 ppm). Scanning electron microscopy (SEM, S-3400N, Hitachi, Japan), transmission electron microscopy and high-resolution transmission electron microscopy (TEM/HRTEM, JEM-2100, 200 kV) were employed to characterize the morphology. Energy dispersive X-ray spectroscopy (EDS, Falcon) was used to monitor the chemical composition of the nanowires. Selected area electron diffraction (SAED, JEM-2100, 200 kV) was employed to analyze the structure of the nanowires. A charge/discharge battery tester (NewWare TC5.X, Shenzhen) was used to test the performance of the batteries during the cut-off voltages between 0.01 and 2.9 V *vs.* Li/Li^+, and all the capacities were calculated based on the total mass of the deposits.

3. Results and discussion

3.1. Cyclic voltammetry

We prepared the IL by slowly adding TMHC to $AlCl_3$ in a beaker whilst stirring. When the mole ratio of TMHC to $AlCl_3$ is higher than 42 : 58, the IL cannot be used for electrodeposition at room temperature due to its high viscosity. When the mole ratio of TMHC to $AlCl_3$ is lower than 38 : 62, $AlCl_3$ cannot be totally dissolved in IL, and the IL is not transparent. The IL with a mole ratio of 40 : 60 for TMHC/$AlCl_3$ shows suitable viscosity and melting points. $FeCl_3$ presents the highest solubility in the TMHC–$AlCl_3$ IL in a mole ratio of 40 : 60 : 0.7. Fig. 1 compares the cyclic voltammograms (CVs) recorded on W electrodes in various TMHC–$AlCl_3$–$FeCl_3$ ILs. As the mole ratio is 40 : 60 : 0, a reduction peak A at −0.18 V and the corresponding oxidation peak B at 0.12 V in curve 1 are observed, which are related to Al^{3+} ion reduction and subsequent oxidation. The reduction process of Al^{3+} ions is as follows: $4A1_2C1_7^- + 3e^- \rightarrow Al + 7A1C1_4^-$. As $FeCl_3$ is added to the above IL with the mole ratio 40 : 60 : 0.2 for TMHC–$AlCl_3$–$FeCl_3$, two reduction peaks and three oxidation peaks were found in curve 2 of Fig. 1. Compared with curve 1, a reduction peak C at 0.14 V corresponds to Fe^{3+} reduction, and a reduction peak D at −0.16 V is related to Al^{3+} reduction, which is a slightly more positive than peak A in curve 1. The three oxidation peaks, peak E (0.07 V), peak F (0.2 V) and peak G (0.56 V) in curve 2 should be related to the oxidation of Al, AlFe alloy and Fe. The electrochemical windows of TMHC are just a little more negative than the reduction potential of Al both in curve 1 and curve 2.

3.2. Electrodeposition of Al and AlFe nanowires

Al nanowires are obtained by electrodeposition from a TMHC–$AlCl_3$ IL (40 : 60 mole ratio) for 1200 s at 25 °C at different potentials. As the reduction potential changes from −0.12 V to −0.2 V, Al nanowires become more uniform and the average diameters of Al nanowires decrease from 330 nm to 120 nm, as shown in Fig. 2. Nevertheless, the reduction potential of Al is limited to be more positive than −0.2 V because of the quite narrow electrochemical window of TMHC. So the average diameter of the Al nanowires cannot be reduced any more.

Upon increasing the content of $AlCl_3$ in the TMHC–$AlCl_3$ IL or raising the electrodeposition temperature, the morphologies of the deposits changed as shown in Fig. 3. Deposits from the 38 : 62 mol% TMHC–$AlCl_3$ IL at −0.15 V at 25 °C are composed of larger particles, in Fig. 3(a). Meanwhile the deposits obtained from the 40 : 60 mol% TMHC–$AlCl_3$ IL at 50 °C have shown interconnected wires with larger diameters, as shown in Fig. 3(b).

The nanowire diameters and morphologies are controlled by both ion mass transport rate (IMTR) and ion consumption rate (ICR) during electrodeposition.[23] The formation mechanism of Al nanowires can be explained as shown in Fig. 4. As a reduction potential more negative than −0.1 V is imposed on the working electrode at 25 °C, the double layer is firstly charged, then Al^{3+} ions are reduced onto the Cu substrate according to the equation $4A1_2C1_7^- + 3e^- \rightarrow Al + 7A1C1_4^-$ and new Al nuclei are formed.

$Al_2Cl_7^-$ ions among Al nuclei are subject to two unequal forces, as shown in Fig. 4(a), in which the force F_1 is electric field force, and the force F_2 is caused by a concentration gradient. Under these two unequal forces, some of the $Al_2Cl_7^-$ ions diffuse to the nuclei for further electrodeposition on the nuclei, and other $Al_2Cl_7^-$ ions move to the counter electrode, which lead to the fast consumption of $Al_2Cl_7^-$ ions among the growing nuclei. Therefore, $Al_2Cl_7^-$ ions in bulk IL diffuse to the working electrode for supplement, and are reduced on the top of the nuclei as soon as they come into contact with the nuclei, which causes longitudinal growth of the nuclei. Meanwhile, the lateral growth is limited by the extremely low

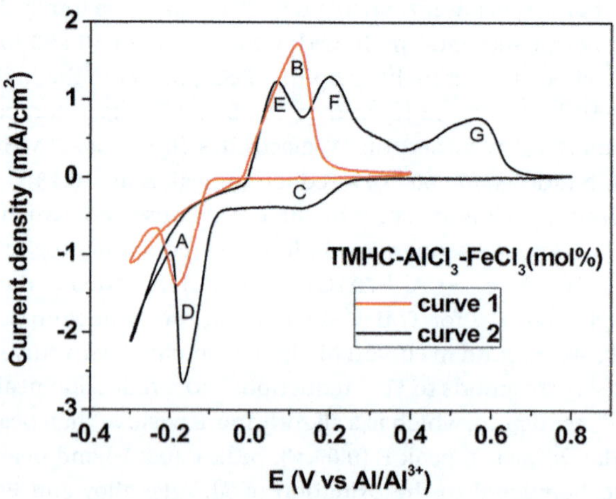

Fig. 1 Cyclic voltammograms recorded on W electrodes in a different mole ratio TMHC–$AlCl_3$–$FeCl_3$ ILs (curve 1: 40 : 60 : 0; curve 2: 40 : 60 : 0.2) at 25 °C, scan rate = 10 mV s^{-1}.

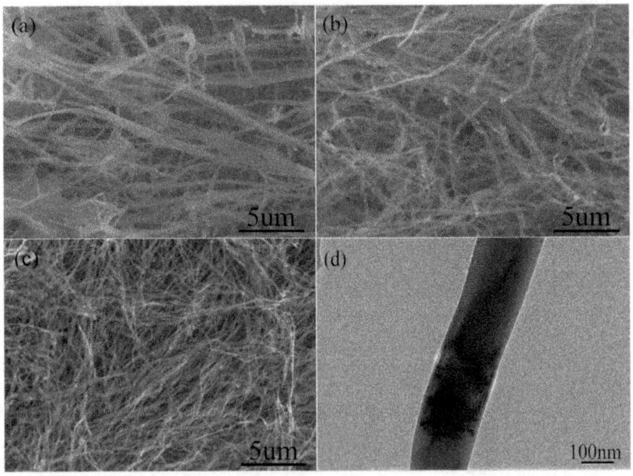

Fig. 2 SEM images of the deposits obtained from a 40 : 60 mol% TMHC–$AlCl_3$ IL for 1200 s at 25 °C. (a) −0.12 V; (b) −0.15 V; (c) −0.20 V; and (d) TEM image for (b).

concentration of $Al_2Cl_7^-$. As a result, the nuclei grow longitudinally to form nanowires as presented in Fig. 4(b). While increasing the content of $AlCl_3$ to 32 : 68 mol% for TMHC–$AlCl_3$, the fast consumption of $Al_2Cl_7^-$ among nuclei can be simultaneously supplied due to a higher concentration. As the equilibrium conditions necessary to form nanowires disappear, bigger particles are gained. Increasing the temperature can accelerate the diffusion of $Al_2Cl_7^-$ and lead to the timely addition of $Al_2Cl_7^-$ among nuclei, so overlapping rods with larger diameters are formed.

Fig. 5 gives the SEM and TEM images of deposits obtained from the 40 : 60 mol % TMHC–$AlCl_3$ IL with different additions of $FeCl_3$ at −0.15 V and 25 °C. As 0.2 mol% $FeCl_3$ is added into TMHC–$AlCl_3$ (40 : 60 mol%), the average diameter of AlFe alloy nanowires by electrodeposition is increased to 140 nm in Fig. 5(a). However, as $FeCl_3$ concentration increases to 0.5 mol% and 0.7 mol%, the average diameters of AlFe alloy nanowires are increased to 280 nm and 370 nm in Fig. 5(b) and (c).

Addition of $FeCl_3$ into the TMHC–$AlCl_3$ IL can effectively improve the Al nanowire morphology. Firstly, by comparing Fig. 5 with Fig. 2, as $FeCl_3$ concentration increases from 0 mol% to 0.7 mol%, the obtained AlFe nanowires are more

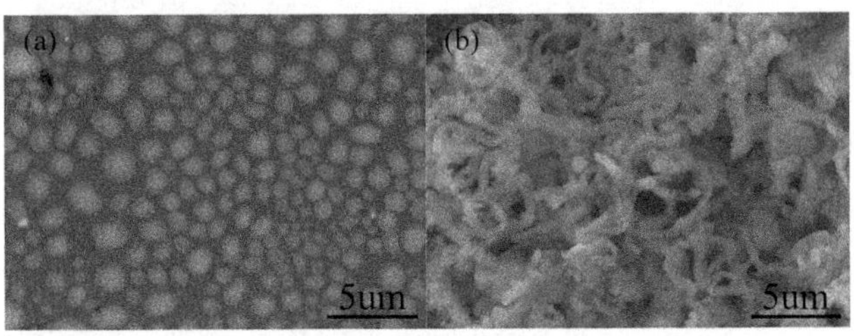

Fig. 3 SEM images of deposits obtained from a TMHC–$AlCl_3$ IL at −0.15 V 1200 s, (a) 38 : 62 mol%, 25 °C, and (b) 40 : 60 mol%, 50 °C.

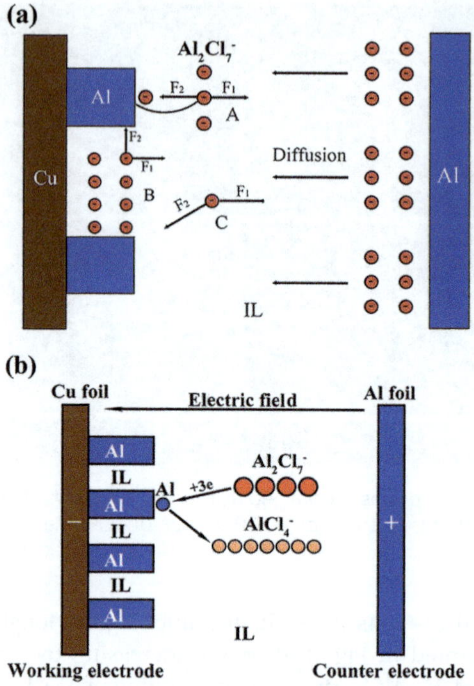

Fig. 4 The formation of Al nanowires in 40 : 60 TMHC–$AlCl_3$ IL.

ordered. But the average diameter increases from 120 nm to 370 nm, because addition of Fe^{3+} ions can accelerate the lateral growth of the nanowires. Secondly, AlFe alloy nanowires exhibit rough surface by comparison of Fig. 2(d) and

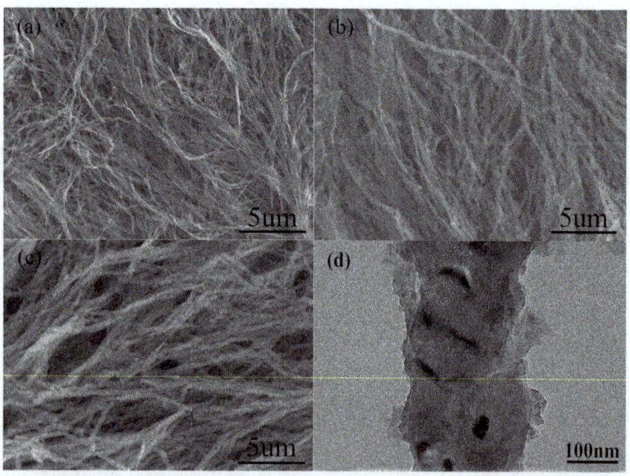

Fig. 5 SEM images of deposits obtained from the TMHC–$AlCl_3$–$FeCl_3$ IL with different addition of $FeCl_3$, −0.15 V, 1200 s, 25 °C (a) 40 : 60 : 0.2 mol%; (b) 40 : 60 : 0.5 mol%; (c) 40 : 60 : 0.7 mol%; and (d) TEM image for (a).

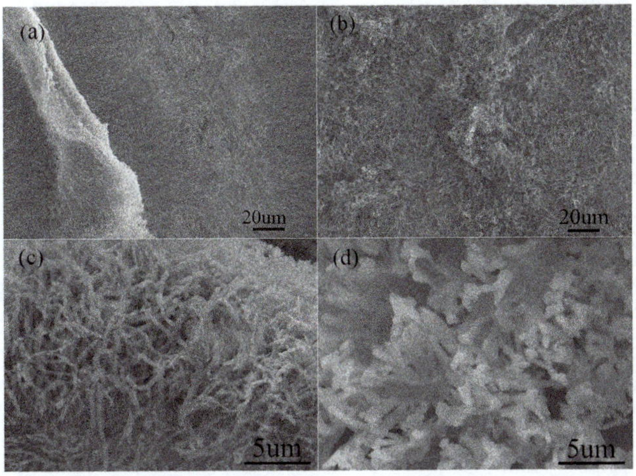

Fig. 6 SEM images of deposits obtained from TMHC–$AlCl_3$–$FeCl_3$ IL, −0.15 V, 1200 s, 25 °C. (a) 40 : 60 : 0 mol%; (b) 40 : 60 : 0.2 mol%; (c) 40 : 60 : 0.2 mol%; and (d) 40 : 60 : 0.5 mol%.

Fig. 5(d). Thirdly, AlFe deposits totally cover the Cu substrate, while Al nanowires only cover parts of the substrate, as shown in Fig. 6(a) and (b). Obviously, addition of $FeCl_3$ can improve the nucleation rate of deposits on the Cu substrate. During potentiostatic electrolysis, the 40 : 60 : 0.2 mol% TMHC–$AlCl_3$–$FeCl_3$ IL shows a higher current density for nucleation than the 40 : 60 mol% TMHC–$AlCl_3$ ILs in Fig. 7, because Fe^{3+} can increase the simultaneous nucleation rate of deposits on the Cu substrate. Finally, the obtained Al nanowires grow in random directions similar to grass in a garden, and several fallen parts of the Al nanowires may accumulate together and grow into 'hill-like' formations due to the relatively low hardness of Al, which can seriously damage the performance of the nanowires as anodes for Li-ion batteries. However, with a small addition of $FeCl_3$, free-standing nanowires take the place of 'hill-like' formations of Al deposits as shown in

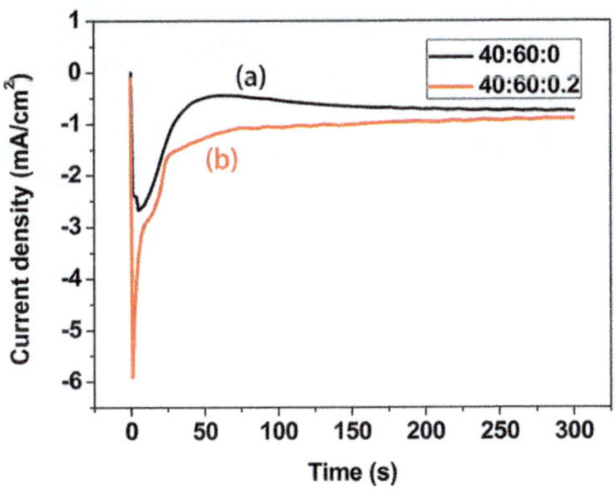

Fig. 7 Potentiostatic electrolysis recorded from TMHC–$AlCl_3$–$FeCl_3$ ILs at −0.15 V for 300 s. (a) 40 : 60 : 0 mol%, (b) 40 : 60 : 0.2 mol%.

Table 1 The EDS analysis of AlFe alloy nanowires

Composition of TMHC–$AlCl_3$–$FeCl_3$ (mol%)	Applied potential (V)	AlFe alloy nanowire composition (Al : Fe mol%)
40 : 60 : 0.2	−0.10	10.78 : 1
40 : 60 : 0.2	−0.15	15.07 : 1
40 : 60 : 0.2	−0.20	20.64 : 1
40 : 60 : 0.5	−0.15	5.24 : 1
40 : 60 : 0.7	−0.15	4.37 : 1

Fig. 6(c) and (d). These changes can effectively improve the performance of AlFe alloys as anode materials for Li-ion batteries.

3.3. Composition and structure analysis of deposits

The mole ratios of Al and Fe elements in the AlFe deposits were analysed by EDS and are listed in Table 1. We define the deposits from the 40 : 60 : 0.2 mol% TMHC–$AlCl_3$–$FeCl_3$ IL as AlFe1, and deposits from the 40 : 60 : 0.5 mol% TMHC–$AlCl_3$–$FeCl_3$ IL as AlFe2. By comparison, the Fe content in AlFe alloy nanowires is increased by applying a smaller overpotential or higher $FeCl_3$ concentration. Selected area electron diffraction (SAED) can observe the crystallinity of a very small area. SAED analysis of the Al nanowires in Fig. 8(a) obtained from the 40 : 60 : 0 mol% THMC–$AlCl_3$–$FeCl_3$ IL reveals the single crystal nature of these nanowires. Differently, AlFe1 alloy nanowires present consistent multi-crystal structures as evidenced by a few diffraction spots highlighting the rings underneath in Fig. 8(b). However, it is hard to distinguish the interatomic spacing of both Al and AlFe1 nanowires from the HRTEM images in Fig. 8(c) and (d), which may due to oxidation of the nanowires' surface.

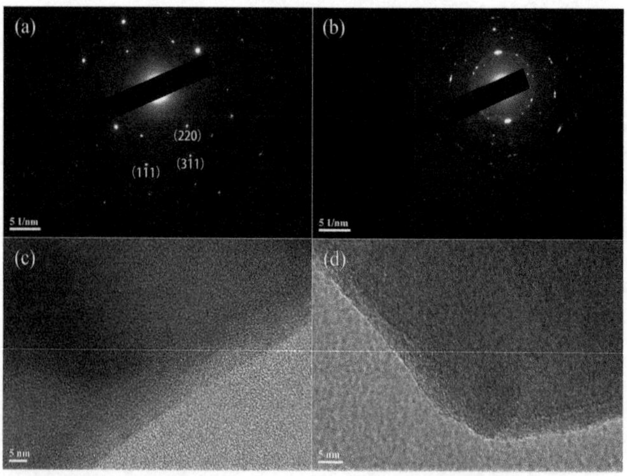

Fig. 8 SAED of deposits obtained from (a) 40 : 60 : 0 mol% and (b) 40 : 60 : 0.2 mol% TMHC–$AlCl_3$–$FeCl_3$ ILs at −0.15 V for 1200 s. (c) HRTEM for (a); and (d) HRTEM for (b).

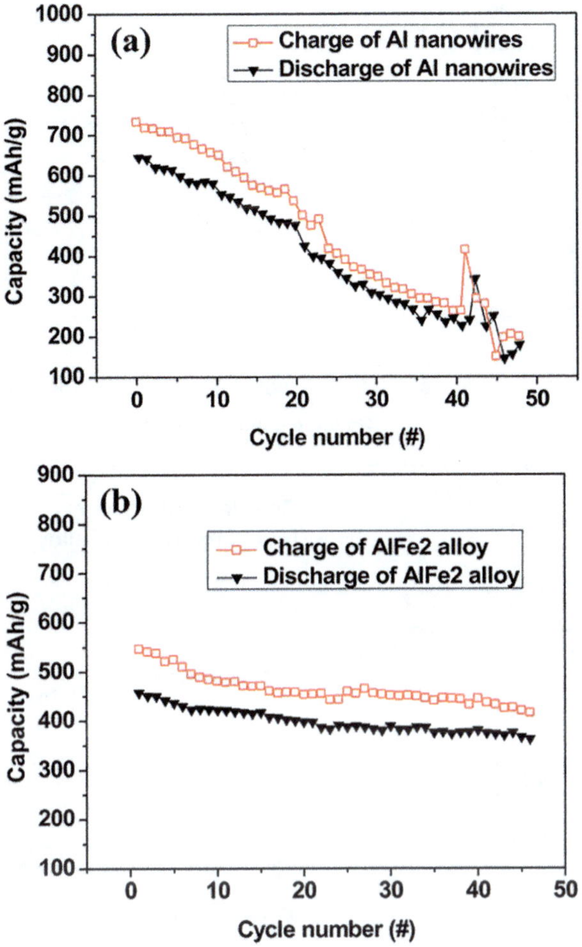

Fig. 9 Cyclic performances of the (a) Al electrode and (b) AlFe2 alloy electrode at a galvanostatic current density of 300 mA g^{-1} and the potential range between 0.01 and 2.9 V (*vs.* Li/Li^{+}).

3.4. Electrochemical performance of Al and AlFe

It has to be mentioned that the coin cells we used here are half-batteries, which means that lithium foils were used as counter electrodes. We define the intercalation of Li-ions to the alloy as the charge progress and deintercalation of Li-ions from the alloy as the discharge progress of the batteries. From the cyclic performances in Fig. 9(a), pure Al nanowires electrode electrodeposited from the 40 : 60 mol% TMHC–$AlCl_3$ IL at −0.15 V show a fast fading capacity either in charge or discharge at galvanostatic current densities of 300 mA g^{-1} in the potential range between 0.01 and 2.9 V (*vs.* Li/Li^{+}). Although the first discharge capacity is about 650 mA h g^{-1}, it decreases to lower than 200 mA h g^{-1} after 45 cycles. Fig. 9(b) shows the cyclic performance of the AlFe2 alloy electrode. Clearly, addition of Fe in the AlFe alloy enhances the cycling ability of the Al anode material but reduces the specific capacity of the AlFe alloy to 550 mA h g^{-1} for the first charge.

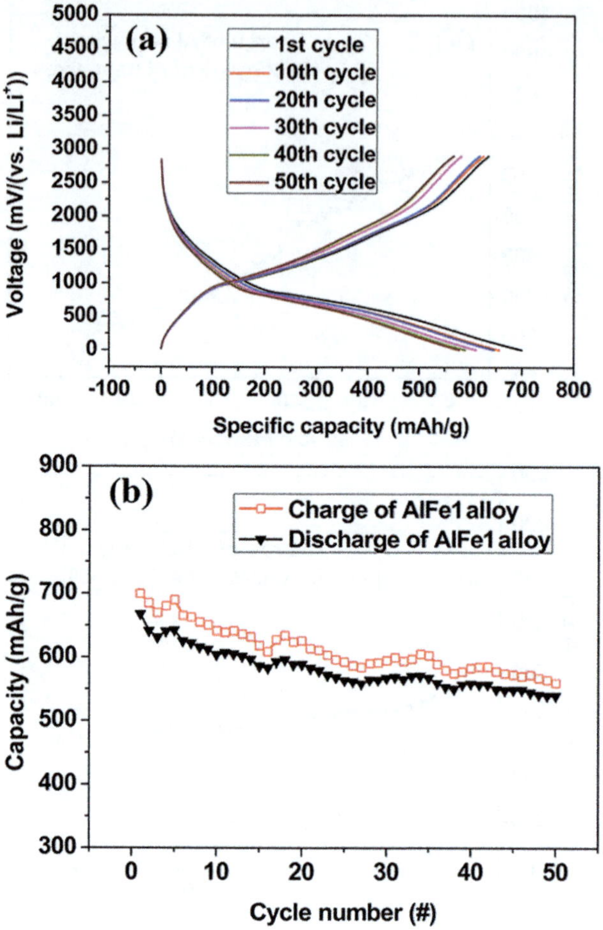

Fig. 10 (a) Charge/discharge behavior; and (b) cyclic performance at galvanostatic current densities of 300 mA g^{-1} and the potential range between 0.01 and 2.9 V (*vs.* Li/Li^{+}) of the AlFe1 alloy electrode.

As the $FeCl_3$ concentration in the IL is reduced to about 0.2 mol%, the AlFe1 alloy obtained from the 40 : 60 : 0.2 mol% TMHC–$AlCl_3$–$FeCl_3$ IL at −0.15 V and 25 °C shows a very stable and higher specific capacity than AlFe2 at the same current

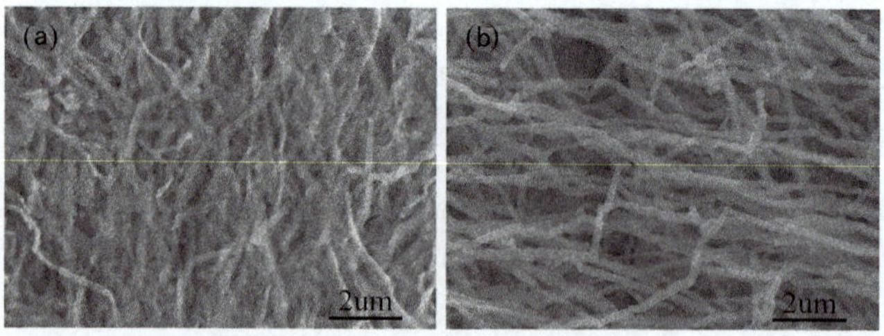

Fig. 11 SEM images of AlFe1 nanowire electrode after 50 successive charge/discharge cycles: (a) intercalation of lithium; and (b) deintercalation of lithium.

densities and potential range, as shown in Fig. 10(a). Since the progress of lithiation/delithiation reactions depends on the phase structure of the electrode,[24] the voltage profile in Fig. 10(b) maintained for 50 cycles indicates that the AlFe1 alloy undergoes stable lithiation/delithiation reactions without destroying its phase structure.

By comparing Fig. 9 with Fig. 10, the AlFe1 nanowire electrodes exhibit better cycling ability than pure Al nanowires and a higher capacity than AlFe2. Each battery was activated under a current density of 100 mA g^{-1} for one cycle by a cut-off voltage of 0.01 V and 2.9 V *vs.* Li/Li^+. Fig. 10(a) shows charge/discharge profiles of the AlFe1 alloy for the first to fiftieth cycles at the same conditions as above. The capacity of the first charge is 700 mA h g^{-1}, and the discharge capacities for the first and fiftieth cycles are 634 mA h g^{-1} and 570 mA h g^{-1} with coulombic efficiencies of 90.6% and 81.4%, respectively. Compared with the Al nanowire electrode in Fig. 9(a) and AlFe2 alloy electrode in Fig. 9(b), AlFe1 alloy shows a higher capacity and better cycle stability. SEM images of AlFe1 nanowire electrodes after 50 charge/discharge cycles are shown in Fig. 11. The nanowire structure does not collapse under such charge/discharge conditions and the initial structure is not visibly altered. Some lithium can be seen on the nanowires as a result of the deposition of lithium, as shown in Fig. 11(a). Lithium intercalation increases the AlFe1 nanowire diameters and makes them more interconnected with each other. We can conclude that the nanowire arrays can accommodate the large volume change caused by the intercalation of Li into the AlFe alloy nanowires.

4. Conclusions

AlFe alloy nanowires were successfully electrodeposited from a cheap TMHC–$AlCl_3$–$FeCl_3$ IL at constant potential without templates. The formation mechanism of Al nanowires in the 40 : 60 TMHC–$AlCl_3$ IL was explained. The average diameter of the AlFe alloy nanowires obtained in 40 : 60 : 0.2 mol% TMHC–$AlCl_3$–$FeCl_3$ at −0.15 V for 1200 s was about 140 nm. The AlFe alloy nanowire electrode exhibited an improved cyclic performance with a relatively high specific capacity. The results showed that nanowire arrays were able to accommodate the large volume change caused by the intercalation and deintercalation of Li-ions. The fabricated AlFe nanowire electrodes exhibited a capacity of 570 mA h g^{-1} after 50 cycles at a current density of 300 mA g^{-1}. Two main reasons may explain the phenomenon that a small amount of Fe can effectively improve the performance of the Al anode material. One is that addition of Fe can change the morphology of the nanowires from the 'hill-like' bulk structures to free-standing nanowires, which can greatly cover the Cu substrates and increase the electrode/electrolyte interface area. The other is that transition metal element Fe can act as an inactive element to buffer the large volume changes and act as a barrier against the aggregation of Al nanowires into large grains during the Li-ion insertion and extraction progress. On the basis of this research, other elements with higher specific capacity like Ge and Si could be introduced to the AlFe alloy to further increase the capacity. Electrochemically made Al based alloys could be used as potential anodes for Li-ion batteries.

Acknowledgements

This work was financially supported by National Natural Science Foundation of China (No. 51474107).

Notes and references

1 S. R. Gowda, V. Pushparaj, S. Herle, G. Girishkumar, J. G. Gordon, H. Gullapalli, X. B. Zhan, P. M. Ajayan and A. L. M. Reddy, *Nano Lett.*, 2012, **12**, 6060–6065.

2 T. Huang, Y. Yao, Z. Wei, Z. Liu and A. S. Yu, *Electrochim. Acta*, 2010, **56**, 476–482.

3 J. M. Tarascon and M. Armand, *Nature*, 2001, **414**, 359–367.

4 D. H. Nam, R. H. Kim, C. L. Lee and H. S. Kwon, *J. Electrochem. Soc.*, 2012, **159**, A1822–A1826.

5 O. Mao, R. A. Dunlap and J. R. Dahna, *J. Electrochem. Soc.*, 1999, **146**, 405–413.

6 S. Z. E. Abedin, A. Garsuch and F. Endres, *Aust. J. Chem.*, 2012, **65**, 1529–1533.

7 S. K. Sharmaa, M. S. Kimc, D. Y. Kimb and J. S. Yuc, *Electrochim. Acta*, 2012, **87**, 872–879.

8 C. Yang, D. W. Zhang, Y. B. Zhao, Y. H. Lu, L. Wang and J. B. Goodenough, *J. Power Sources*, 2011, **196**, 10673–10678.

9 J. S. Zhu, D. L. Wang, T. F. Liu and C. F. Guo, *Electrochim. Acta*, 2014, **125**, 347–353.

10 C. J. Su, Y. T. Hsieh, C. C. Chen and I. W. Sun, *Electrochem. Commun.*, 2013, **34**, 170–173.

11 P. Nithyadharseni, M. V. Reddy, B. Nalini, M. Kalpana and B. V. R. Chowdari, *Electrochim. Acta*, 2015, **161**, 261–268.

12 X. F. Lei, J. F. Xiang, X. L. Ma, C. W. Wang and J. T. Sun, *J. Power Sources*, 2007, **166**, 509–513.

13 X. F. Lei, C. W. Wang, Z. H. Yi, Y. G. Liang and J. T. Sun, *J. Alloys Compd.*, 2007, **429**, 311–315.

14 M. J. Lindsay, G. X. Wang and H. K. Liu, *J. Power Sources*, 2003, **119–121**, 84–87.

15 Z. X. Chen, J. F. Qian, X. P. Ai, Y. L. Cao and H. X. Yang, *Electrochim. Acta*, 2009, **54**, 4118–4122.

16 S. Z. E. Abedina and F. Endres, *Electrochim. Acta*, 2009, **54**, 5673–5677.

17 C. Fournier and F. Favier, *Electrochem. Commun.*, 2011, **13**, 1252–1255.

18 A. Willert, S. Z. E. Abedin and F. Endres, *Aust. J. Chem.*, 2014, **67**, 875–880.

19 J. Szymczak, S. Legeai, S. Diliberto, S. Migot, N. Stein, C. Boulanger, G. Chatel and M. Draye, *Electrochem. Commun.*, 2012, **24**, 57–60.

20 Y. Q. Chen, H. Wang and B. Li, *RSC Adv.*, 2015, **5**, 39620–39624.

21 A. M. Elbasiony, M. Olschewski, S. Z. E. Abedin and F. Endres, *ChemElectroChem*, 2015, **2**, 1361–1365.

22 Y. T. Hsieh, M. C. Laia, H. L. Huang and I. W. Sun, *Electrochim. Acta*, 2014, **117**, 217–223.

23 W. F. Wei, X. H. Mao, L. A. Ortiz and D. R. Sadoway, *J. Mater. Chem.*, 2011, **21**, 432–438.

24 L. Q. Mai, X. C. Tian, X. Xu, L. Chang and L. Xu, *Chem. Rev.*, 2014, **114**, 11828–11862.

Faraday Discussions

PAPER

Direct oxygen removal technique for recycling titanium using molten $MgCl_2$ salt

Toru H. Okabe,*[a] Yuki Hamanaka†[b] and Yu-ki Taninouchi[c]

Received 17th December 2015, Accepted 19th January 2016

DOI: 10.1039/c5fd00229j

Deoxidation of Ti, or direct removal of O dissolved in metallic Ti, is known to be extremely difficult when Mg is used as the deoxidizing agent. This difficulty arises because the chemical potential of O_2, p_{O_2}, under Mg/MgO equilibrium is high (approximately 10^{-41} atm at 1200 K) and is equivalent to that of Ti containing ~2 mass% O at 1200 K. Therefore, when deoxidizing Ti to the commercial level of high-grade pure Ti (below 0.05 mass% O) using an Mg reductant at 1200 K, the activity of the reaction product MgO (a_{MgO}) must be decreased to below ~0.025, which is difficult in practice. In this study, the removal of O in Ti in molten $MgCl_2$ salt using an electrochemical technique was examined at ~1173 K with the objective of obtaining Ti containing less than 0.05 mass% O. Ti samples and graphite electrodes immersed in molten $MgCl_2$ served as the cathode and anode, respectively. A constant voltage was applied between the electrodes using an external DC source. Molten $MgCl_2$ was employed to produce the deoxidizing agent Mg and to facilitate deoxidation of Ti by decreasing the activity of the reaction product MgO. By applying a voltage of approximately 3.1 V between the electrodes, the chemical potential of Mg in the molten $MgCl_2$ was increased at the surface of the Ti cathode, and the Ti samples were deoxidized. The resulting O species, mainly formed O^{2-} dissolved in the molten $MgCl_2$, was removed from the molten salt by reacting with the C anode to form CO (or CO_2) gas. Ti wires containing 0.12 mass% O were deoxidized to less than 0.02 mass% O. In some cases, the O concentration in the Ti samples was reduced to the level of 0.01 mass%, which cannot be accomplished using the conventional Kroll process. The possible application of this deoxidation technique to practical industrial recycling processes is discussed.

[a]International Research Center for Sustainable Materials, Institute of Industrial Science, The University of Tokyo, 4-6-1 Komaba, Meguro-ku, Tokyo 153-8505, Japan. E-mail: okabe@iis.u-tokyo.ac.jp

[b]Department of Materials Engineering, Graduate School of Engineering, The University of Tokyo, 7-3-1 Hongo, Bunkyo-ku, Tokyo, 113-8656, Japan

[c]Institute of Industrial Science, The University of Tokyo, 4-6-1 Komaba, Meguro-ku, Tokyo 153-8505, Japan

† Currently at Mitsubishi Materials Corporation.

Introduction

In recent years, the recycling of Ti and Ti alloys has become an important issue.[1,2] The demand for Ti and Ti alloys in the aerospace industry is increasing, and large amounts of Ti scraps are generated from the fabrication of Ti and Ti alloy products. Fig. 1 shows the material flow of Ti in the US.[1] The volume of scrap generated is greater than the volume of fabricated Ti products.

Commercial Ti and its alloys contain oxygen at approximately 0.1 mass% (1000 mass ppm) as a major impurity. Oxygen contamination of the product must be avoided, because the presence of O deteriorates the mechanical properties of the metals. Table 1 shows the ASTM grades and acceptable impurity concentrations of Ti and its alloys.[3] In producing commercially pure Ti of ASTM Gr. 1, low-O-containing Ti feed such as high-purity Ti sponge (0.03–0.1 mass% O) produced by the Kroll process[4] is used. Using large amounts of scrap metal to produce low-O Ti is hindered by the high O level in Ti scrap, which typically exceeds 0.1 mass%. Titanium and its alloys containing more than 0.4 mass% O cannot be recycled into titanium alloys, and these metals are mainly used in cascade recycling as additive elements for steel, aluminum, *etc.*

O removal directly from Ti and its alloys to levels below 0.1 mass% (1000 mass ppm) is very difficult because O dissolves in Ti to form an interstitial solid solution, and because Ti has a strong affinity with O.[5,6] Table 2 lists some effective methods for direct removal of O from Ti and from TiAl.[2,6–48] For example, the deoxidation of solid Ti by reaction with chemically active Ca dissolved in $CaCl_2$ in the temperature range of 1273 to 1473 K has been examined. Okabe *et al.* successfully produced Ti containing less than 50 ppm by mass of O with a high residual resistivity ratio (RRR > 120).[18,19] Ca metal is one of the most effective deoxidation agents because it has an extremely strong affinity for O and because it

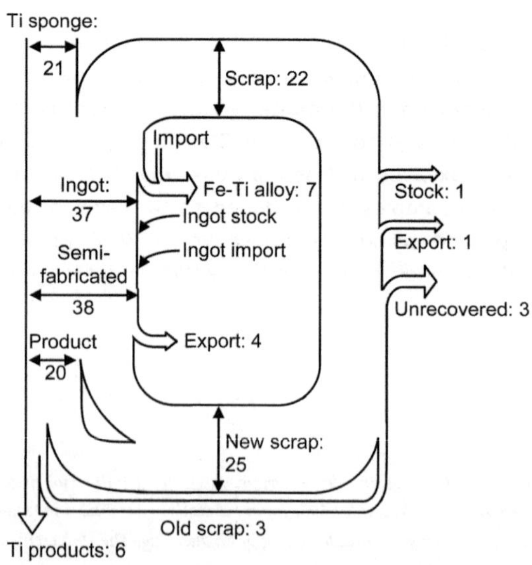

Fig. 1 Estimated material flow of titanium scrap in the U.S. in 2004.[1] Units in the figure are 10^3 ton. Values are rounded to the nearest thousand metric tons of contained titanium.

Table 1 Grades and impurity concentrations of pure titanium and titanium alloys[3]

		Concentration of element i, C_i (mass%)						
	Specification	C	H	O	N	Fe	Al	V
Pure Ti	ASTM Gr. 1	≤0.08	≤0.015	≤0.18	≤0.03	≤0.2	—	—
	ASTM Gr. 2	≤0.08	≤0.015	≤0.25	≤0.03	≤0.3	—	—
	ASTM Gr. 3	≤0.08	≤0.015	≤0.35	≤0.05	≤0.3	—	—
	ASTM Gr. 4	≤0.08	≤0.015	≤0.40	≤0.05	≤0.5	—	—
Ti–Al–V alloy	ASTM Gr. 5	≤0.08	≤0.015	≤0.20	≤0.05	≤0.40	5.5–6.75	3.5–4.5
	ASTM Gr. 23	≤0.08	≤0.0125	≤0.13	≤0.03	≤0.25	5.5–6.5	3.5–4.5

has very low solubility in solid Ti. In the Ca-halide flux deoxidation process shown in Table 2, the $CaCl_2$ flux contains the deoxidation agent Ca and facilitates the reaction by diluting the reaction product CaO, thus decreasing the activity of the by-product. However, the purity of flux and the initial O content of the Ti are subject to limits, since the ultimate limit of deoxidation depends on the amounts of both O present as an impurity in the $CaCl_2$ flux and CaO produced during deoxidation.

An effective technique for the preparation of O-free Ti by electrochemical deoxidation using $CaCl_2$ molten salt has been developed.[21] Removal of O^{2-} (present as CaO in the flux) and production of Ca deoxidant using electrochemical techniques improved the deoxidation limit of the Ca-halide flux deoxidation process. Under certain conditions, Ti with 0.001 mass% (10 mass ppm) O has been successfully produced by the electrochemical deoxidation technique.

Hashimoto *et al.* proposed a method for the deoxidation of Ti during electrowinning using molten fluoride salt baths,[45–47] but contamination of Ti by O or C is unavoidable using this method. In 2001, Chen *et al.* developed a novel process, called the FFC process, for producing Ti directly from TiO_2 immersed in $CaCl_2$ molten salt using electrochemical methods, and they succeeded in obtaining low-O Ti using this process.[26–28] Ono and Suzuki developed a new Ti reduction process that utilizes electrochemically produced Ca reductant in molten $CaCl_2$ salt.[32] To obtain low-O Ti, a highly reducing atmosphere is required. For this reason, molten $CaCl_2$ with high metallic Ca activity is employed as the reaction medium. Other recent developments in deoxidation techniques for Ti and its alloys are also listed in Table 2.

Deoxidation of Ti using Mg as the deoxidizing agent is believed to be almost impossible. This is because the deoxidation limit of a Ti–O solid solution by Mg under MgO saturation is about 2 mass% (20 000 mass ppm) at 1200 K; reducing the deoxidation limit to the level of 100 mass ppm is difficult from a practical perspective. However, if deoxidation of Ti using a $MgCl_2$ medium becomes feasible, the reduction and electrolysis facilities of the conventional Kroll process[4] could be utilized. Therefore, developing an effective deoxidation process for Ti and its alloys using molten $MgCl_2$ is expected to significantly enhance the recovery rate for Ti scraps and the production of low-O-containing Ti.

Table 2 Some examples of the upgrading process for titanium metal investigated in previous studies[2,6–48]

Method	Reference	Advantages (deox. limit)	Disadvantages
Deoxidation (Deox.)			
Deox. by M/MOx eq. (DOSS process)	Ono and Miyazaki,[6] Fisher,[7–10] Okabe *et al.*,[11–14] Oh *et al.*,[15,16] Roh *et al.*[17]	Ti powder can be deoxidized. (~500 ppm O)	Low capability of deox.
Calcium halide flux deox.	Okabe *et al.*[18–20]	High capability of deox. (~50 ppm O)	Not suitable for removing large amounts of O. Leaching of molten salt bath is needed
Electrochemical deox.	Okabe *et al.*,[21,22] Nakamura *et al.*,[23] Hirota *et al.*,[24] Taninouchi *et al.*[25]	Ultra-high capability of deox. $CaCl_2$ bath (<10 ppm O), $MgCl_2$ bath (<100 ppm O)	Long processing time
FFC process	Chen *et al.*,[26–28] Fray,[29] Mohandas and Fray,[30] Tripathy *et al.*[31]	Production of low-oxygen titanium directly from oxides	Long processing time
OS process	Ono and Suzuki,[32] Suzuki *et al.*[33–37]	Production of low-oxygen titanium directly from oxides	Long processing time
Deox. during EB remelting with Al	Yahata *et al.*[38]	Short processing time (~100 ppm O)	Excess Al is contained in Ti
VIM with $CaAl_2$	Rotmann *et al.*[2]	Short processing time (~1500 ppm O)	Capability of deox. is insufficient
PESR with Ca and CaF_2 flux	Reitz *et al.*[39]	Short processing time (~250 ppm O)	Subsequent VAR is required to remove residual Ca. Applicability to deoxidation of pure Ti is uncertain
Electric arc remelting under H_2–Ar atmosphere	Su *et al.*[40]	Short processing time (~300 ppm O)	Capability of deox. is insufficient. Applicability to deoxidation of pure Ti is uncertain
Deox. of Ti alloy by HDH	Oh *et al.*,[16] Roh *et al.*[17]	Simple process	Capability of deox. is insufficient
Others			
Electrorefining in molten salt	Nettle *et al.*,[41] Suchkov *et al.*,[42] Takeuchi *et al.*,[43,44] Hashimoto *et al.*,[45–47] Miyazaki *et al.*[48]	Almost all impurities can be removed	Contamination from molten salt. Difficulty of morphology control of Ti deposition

Table 3 Some representative results for titanium deoxidation using various solid state deoxidation techniques

	Oxygen (mass ppm)			Nitrogen[c] (mass ppm)			
Method[a]	Init.		After deox.	Init.		After deox.	Note and references (Exp. no.)
Ca-CaO	430	→	460	20	→	340	1273 K, 86 ks, ref. 11
	1430	→	470	100	→	470	Eqn (3)
$Y\text{-}Y_2O_3$	200	→	120		N.A.[f]		1231 K, 260 ks, ref. 12
	1100	→	150, 170		N.A.[f]		(#2U)
$Y\text{-}Y_2O_3$	210	→	90	10	→	(30)	1213 K, 551 ks, ref. 14
	950	→	100, 220	20	→	(50)	(#U)
	670	→	80, 100	30	→	(60)	
$Y\text{-}Y_2O_3$	210	→	40, 80	10	→	(50)	1173 K, 518 ks, ref. 14
	950	→	160, 160	20	→	(120)	(#T)
	670	→	30, 240	30	→	(80)	
$Ca\text{-}Y\text{-}Y_2O_3$	210	→	130	10	→	(40)	1173 K, 604 ks, ref. 14
	950	→	60, 90	20	→	(50)	(#W)
	670	→	60, 90	30	→	(120)	
$Ca\text{-}CaCl_2$	130	→	16	10	→	41	1273 K, 90 ks, ref. 19
	130	→	17	10	→	34	(#17B1)
							$RRR_{4.2}$[d] = 98, Hv[e] = 85
$Ca\text{-}CaCl_2$	130	→	35	10	→	49	1373 K, 40 ks, ref. 19
	130	→	22	10	→	45	(#19B1)
							$RRR_{4.2}$[d] = 95, Hv[e] = 92
$Ca\text{-}CaCl_2$	150	→	62	10	→	86	1273 K, 86 ks, ref. 19
							(#3C)
							$RRR_{4.2}$[d] = 122, Hv[e] = 82
Electrochem.	900	→	<10	40	→	20	1223 K, 22 ks, ref. 21
$(CaCl_2)$[b]	1400	→	<10	20	→	20	(#49, contaminated by carbon)
Electrochem.	110	→	10	<10	→	<10	1223 K, 27 ks, ref. 23
$(CaCl_2)$[b]	900	→	20	30	→	50	(#69, carbon: 20 → 30, 50 → 80 ppm)
							$RRR_{4.2}$[d] = 155, Hv[e] = 64
Electrochem.	210	→	40, 50	10	→	10	1189 K, 36 ks, ref. 24
$(CaCl_2)$[b]							(#12)
	(O in Gd: 2100 → 10, 50 ppm, O in Tb: 2600 → 20, 20 ppm, O in Dy: 2300 → 20, 30 ppm O in Er: 2800 → 40, 40 ppm, O in Y: 11200 → 340, 380 ppm)						
Electrochem.	110	→	100	<10	→	40	1173 K, 86 ks, ref. 23
$(MgCl_2)$[b]	900	→	60	30	→	70	(#75, carbon: 20 → 50, 50 → 120 ppm)
	840	→	90	20	→	440	1173 K, 86 ks, ref. 25 (D13_150121[g])
	1200	→	270	40	→	260	1173 K, 86 ks, ref. 25 (D15_150123[h])

[a] "M-MOx": deoxidation by metal/metal oxide, "$Ca\text{-}CaCl_2$": calcium-halide flux deoxidation, "Electrochem.": electrochemical deoxidation. [b] Salt in parentheses shows molten salt used for electrochemical deoxidation. [c] Values in parenthesis include uncertainties. [d] Residual resistivity ratio, $\rho_{298}/\rho_{4.2}$, determined by measurement at 298 and 4.2 K. [e] Vickers micro hardness (kg f mm^{-2}), measured using 500 g load at room temperature. [f] "N.A.": not analyzed or not reported. [g] Ti wire (ϕ1.0 mm) was used as sample. Oxygen and nitrogen contents were analyzed by KOBELCO Research Institute, Inc. [h] Ti rod (ϕ3.0 mm) was used as sample. Oxygen and nitrogen contents were analyzed by KOBELCO Research Institute, Inc.

Principle

Because the principle of electrochemical deoxidation of Ti has been reported in detail previously,[5,21] the deoxidation process is only briefly outlined here. A solid Ti–O solution is deoxidized by Mg in the following reaction:

$$\text{O (in Ti)} + \text{Mg (in flux)} = \text{MgO (in flux)} \tag{1}$$

The equilibrium O concentration in Ti depends thermodynamically on the temperature T according to the following equation,

$$[\text{mass\% O}] = (a_{\text{MgO}}/a_{\text{Mg}})(1/f_{\text{O}})\exp(\Delta G^{\circ}/RT) \tag{2}$$

where ΔG° is the standard Gibbs free energy change associated with the reaction described by eqn (1), a_{MgO} and a_{Mg} are the activities of MgO and Mg, respectively, and f_{O} is the activity coefficient of O in solid Ti.

Fig. 2 shows the binary phase diagram for the Ti–O system.[49] The equilibrium O concentration of Ti under Mg/MgO equilibrium at 1273 K is plotted as an open circle. For reference, the equilibrium O concentration of Ti under Ca/CaO equilibrium at 1273 K is plotted as a filled circle (about 0.05 mass% O). The equilibrium O concentration under Mg/MgO equilibrium is approximately 2.6 mass% (7.4 mol%) at 1273 K. This shows that it is thermodynamically difficult to remove O directly from Ti to levels below 1 mass% by Mg deoxidant when the activity of MgO (a_{MgO}) is high.

The O potentials of β-Ti–O solid solution at temperatures between 1173 and 1473 K have been determined,[11] and the feasibility of preparing low-O Ti by decreasing the activity of the by-product CaO (a_{CaO}) has been examined using various Ca-halide fluxes in the presence of Ca metal (Ca-halide flux deox. shown in

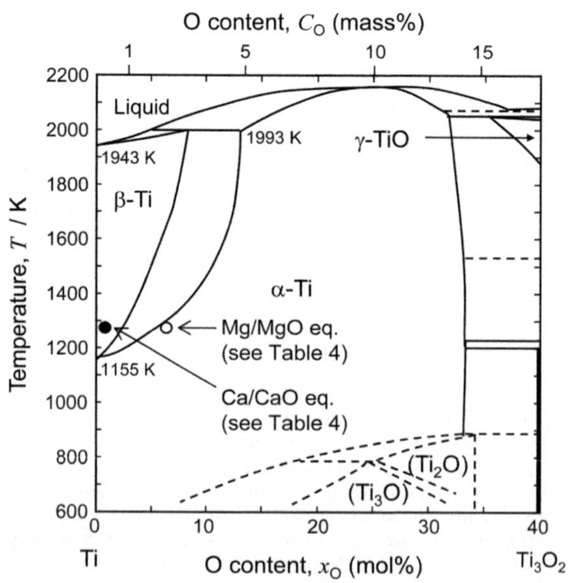

Fig. 2 Phase diagram of the Ti–O binary system.[49] Compositions of Ti–O solution under Ca/CaO and Mg/MgO equilibrium at 1273 K are plotted in the figure.

Table 2).[18,19] Several factors have been considered in selecting a flux for deoxidation. $CaCl_2$ has been found to be the most desirable flux at temperatures of ~1300 K because the deoxidation product CaO dissolves easily in $CaCl_2$ flux, which decreases its activity and thus lowers the limit of deoxidation. By submerging Ti wires and small pieces in Ca-saturated $CaCl_2$, Ti samples has been deoxidized to a level of 50 mass ppm (see also Table 3).[5,19]

In the Ca-halide flux deoxidation method, the ultimate limit of deoxidation depends on the final O concentration in the flux, which means that the purity of the flux and the initial O content of the Ti will rule the deoxidation limit. To decrease the deoxidation limit set by a_{CaO}, an electrochemical method was developed (electrochemical deox. shown in Table 2).[21,22] The modified Ca-halide flux deoxidation method is characterized by both the production of Ca from the $CaCl_2$ flux and the effective removal of O^{2-} (present as CaO) dissolved in the flux by electrochemical means. In this method, as expressed in eqn (3), a Ti cathode is deoxidized by Ca, which forms electrochemically on the surface of the cathode according to eqn (4). O^{2-} species in the flux are carried continuously to the C anode, and O in the flux is removed from the system as CO (or CO_2) gas by the anodic reaction expressed in eqn (5).

$$\text{O (in Ti)} + \text{Ca (in flux)} = \text{Ca}^{2+}\text{ (in flux)} + \text{O}^{2-}\text{ (in flux)} \tag{3}$$

$$\text{Ca}^{2+}\text{ (in flux)} + 2e^- = \text{Ca (on Ti cathode: in flux)} \tag{4}$$

$$\text{O}^{2-}\text{ (in flux)} + \text{C (C anode)} = \text{CO (gas)} + 2e^- \tag{5}$$

Unlike the Ca-halide flux deoxidation, the electrochemical method does not require the addition of metallic Ca as a deoxidant, because the activity of Ca near the cathode can be increased by controlling the applied voltage between the Ti cathode and the C anode. In some cases, Ca is precipitated on the cathode using an applied voltage exceeding the theoretical decomposition voltage of $CaCl_2$ (3.25 V at 1223 K). Furthermore, $CaCl_2$ flux can be purified through pre-electrolysis. This electrochemical method therefore has the advantage that contamination-free Ca deoxidant and $CaCl_2$ flux can be used in the deoxidation process. By applying the electrochemical technique to molten $CaCl_2$, Ti containing O below 10 mass ppm has been successfully produced under certain conditions.[5,21]

In this study, the feasibility of the electrochemical deoxidation technique using $MgCl_2$ as flux was examined. A schematic of the electrochemical deoxidation of Ti in molten $MgCl_2$ is shown in Fig. 3.

$$\text{O (in Ti)} + \text{Mg (in flux)} = \text{Mg}^{2+}\text{ (in flux)} + \text{O}^{2-}\text{ (in flux)} \tag{6}$$

Based on the thermodynamic calculation shown in Fig. 4,[5,11,50] when a_{MgO} is decreased to the level of 10^{-2} in the presence of metallic Mg at 1200 K, the chemical potential of O_2 (p_{O_2}) in the system is decreased to ~10^{-45} atm, and Ti containing 0.018 mass% (180 mass ppm) O can be obtained. In other words, when Ti is deoxidized down to commercial O levels for high-grade Ti (0.05 mass% O, or 500 mass ppm) using Mg as reductant, a_{MgO} at 1200 K must be decreased to the level of 0.025.

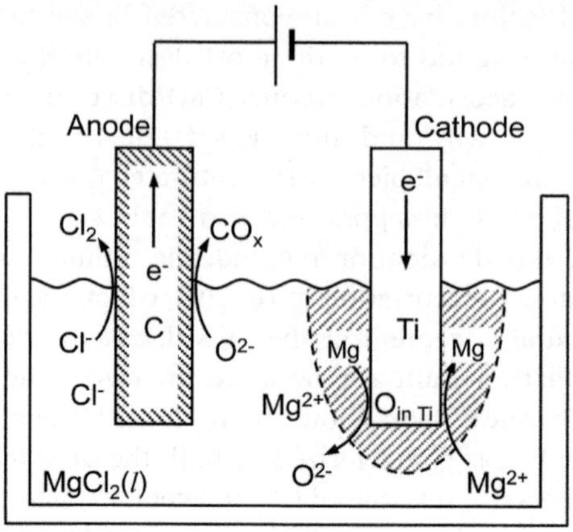

Fig. 3 Schematic illustration of electrochemical deoxidation in $MgCl_2$ molten salt. Electrolysis is achieved using titanium as the cathode and carbon as the anode. At the cathode, a deoxidant of magnesium is produced by electrolysis, and the deoxidation product of magnesium oxide is dissolved in molten salt. At the anode, an oxide anion generated by the cathode is removed as carbon monoxide or dioxide gas. The anode reaction occurs in parallel with chlorine evolution. During this process, the activity of MgO can be kept at a low level on the cathode surface.

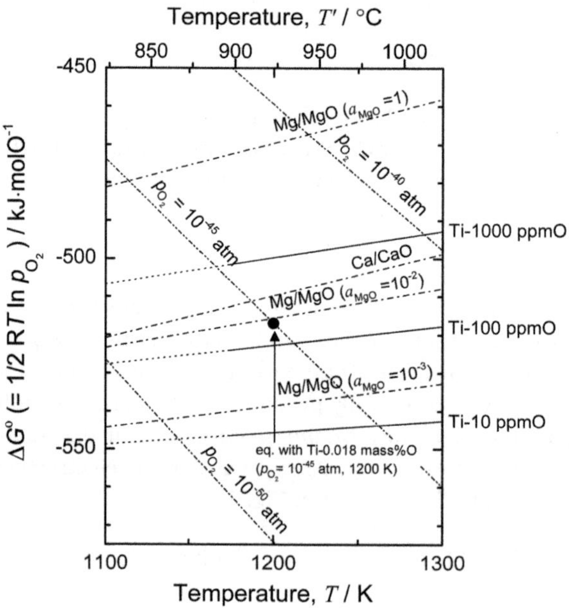

Fig. 4 Ellingham diagram of selected oxides[50] and solid solutions of Ti–O.[5,11] Oxygen chemical potential for Ti-0.018 mass% O ($p_{O_2} = 10^{-45}$ atm) at 1200 K is plotted for reference.

Experimental procedure

Fig. 5 shows a flowchart of the experimental procedure for the electrochemical deoxidation of Ti in $MgCl_2$. A schematic of the experimental apparatus used for

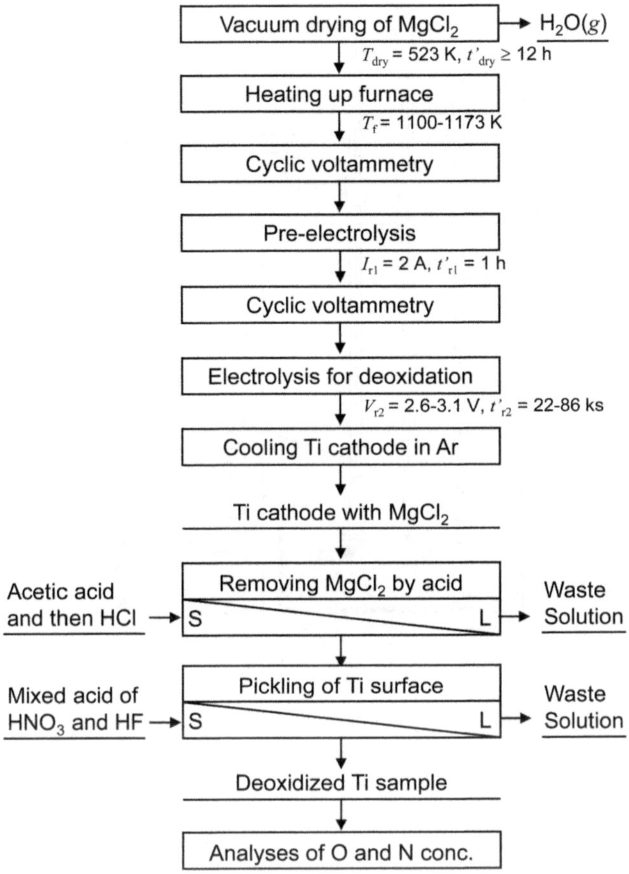

Fig. 5 Flowchart of experimental procedure for electrochemical deoxidation of titanium in molten $MgCl_2$.

the electrochemical deoxidation of Ti in $MgCl_2$ is shown in Fig. 6. Ti in the form of small pieces and wires with different O concentrations and configurations was used as the starting material. The flux was reagent-grade anhydrous $MgCl_2$ (>97.0%) in flake form, dried at 523 K for more than 43 ks (12 h).

The dehydrated $MgCl_2$ flux was contained in a Ti or mild steel crucible (89 mm in diameter, 200 mm in height) and set in a gas-tight stainless steel chamber, as shown in Fig. 6. The Ti cathode consisted of several strands of Ti wire ~50 mm in length and 1 mm in diameter. The total electrode surface area was approximately 1000 mm^2. In the same experiment, a Ti rod (approximately 30 mm in length, 3 mm in diameter) was used as the cathode. High-purity (99.9%) graphite was used as the C anode (approximately 35 mm in length, 3 or 6 mm in diameter). These electrodes, as well as thermocouple tubing, were inserted into the reaction tube through a gas-tight water-cooled stainless steel head.

After the assembled cell was evacuated at ~673 K to ensure a completely gas-tight and moisture-free system, Ar was introduced to the reaction tube. The furnace was then heated to a constant temperature between 1100 and 1173 K. After the $MgCl_2$ was melted, the graphite anode was inserted into the molten salt,

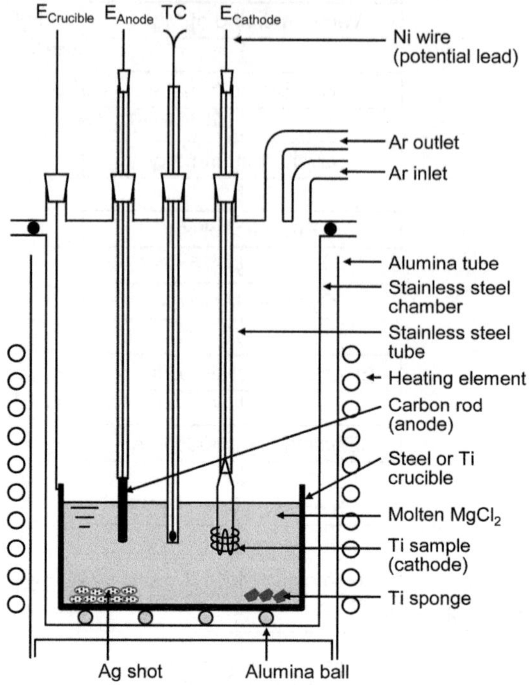

Fig. 6 Schematic illustration of experimental apparatus for electrochemical deoxidation of titanium in $MgCl_2$.

and pre-electrolysis was conducted. In the pre-electrolysis, a 2 A current was passed for ~3.6 ks (1 h) between the Ti crucible (or Ni electrode immersed in molten salt) and the C anode to eliminate residual metal and gaseous impurities in the molten salt. Approximately 50 g of silver was placed at the bottom of the crucible to absorb the magnesium metal deposited during electrolysis. The temperature of the molten salt was measured directly using a chromel–alumel thermocouple protected by a stainless steel sheath. After pre-electrolysis, the Ti cathode containing the Ti samples was inserted into the molten $MgCl_2$, and deoxidation experiments were performed by applying voltages between 0 and 3.1 V between the electrodes, which were separated by a distance of ~30 mm. In most cases, the deoxidation of the Ti cathode occurred with the application of voltages exceeding 2.6 V at 1100–1173 K for longer than 22 ks (6 h).

Before and after the pre-electrolysis of $MgCl_2$ and the deoxidation experiments, the electrochemical properties of the flux were measured by cyclic voltammetry (CV). An electrochemical interface (Solartron SI1287) was used for the CV measurements.

After the deoxidation experiment, the electrodes were removed from the molten $MgCl_2$, and the cell was cooled in a stream of Ar gas. In some cases, only the Ti cathode was replaced after cooling for the next experimental run. The fused salt that adhered to the surface of the Ti electrode was removed by leaching with (1 + 1) acetic acid, after which the Ti samples of the electrode were carefully cleaned in (1 + 10) aqueous HCl solution, followed by water, alcohol, and acetone, and then dried.

The Ti samples obtained were subjected to O and N analyses using LECO analyzers. Prior to the analyses, the samples were surface-etched with a 1 : 4 : 10 mixture of HF–HNO_3–H_2O. To increase the accuracy of the O and N analyses, 1.1 g of nickel capsule containing 8.8 ± 1.4 μg O and 0.2 ± 0.2 μg N was used as an extraction bath for each 0.05–0.1 g Ti sample. The instrument for the O analysis was calibrated using Ti standard samples containing 470 ± 30 and 980 ± 40 mass ppm O.

Results and discussion

Fig. 7(a) shows the CV of the $MgCl_2$ salt at 1173 K using a Ni quasi-reference before pre-electrolysis. The deposition of Mg occurs at approximately −0.7 V with respect to the Ni quasi-reference, and gas evolution occurs at approximately +1.6 V. The

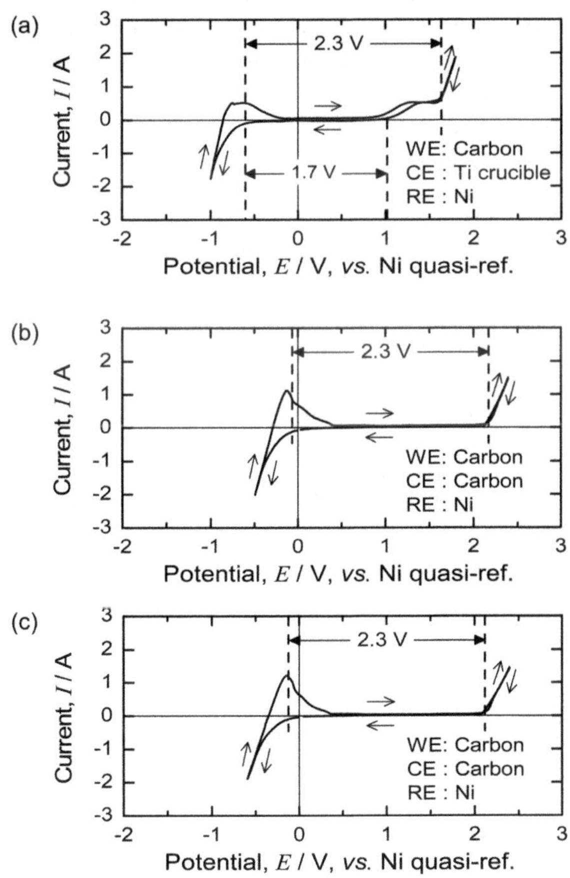

Fig. 7 Cyclic voltammograms in $MgCl_2$ molten salt at 1173 K. Surface area of electrode: 2 cm^2 (C rod, Ni rod). Surface area of electrode: 260 cm^2 (Ti crucible). Scan rate: 100 mV s^{-1}. (a) Before pre-electrolysis. (b) After pre-electrolysis. (c) After electrolysis for Ti deoxidation.

oxidation wave between 1 and 1.6 V probably results from oxygen impurities in the molten salt, and the gas evolution peak at ~1.6 V corresponds to the evolution of Cl_2 gas.

During pre-electrolysis using a Ni electrode as a cathode, Mg metal was electrochemically deposited on the Ni electrode. After pre-electrolysis, the Mg deposition and gas evolution peaks shifted to 0 and 2.3 V, respectively. Fig. 7(b) and (c) show the CV of the molten $MgCl_2$ before and after the deoxidation experiments. As the theoretical decomposition voltage of $MgCl_2$ at 1173 K is 2.3 V,[50] the reference electrode worked correctly, and the Ni quasi-reference electrode became a reference for Mg/Mg^{2+} after pre-electrolysis.

The results shown in Fig. 7 indicate that the electrochemical properties of $MgCl_2$ are unchanged before and after the deoxidation experiments. After pre-electrolysis, only the Mg deposition wave and Cl_2 evolution are observed, and therefore, the reaction system in the molten salt can be regarded as contamination-free $MgCl_2$.

By applying a voltage of ~3.1 V between the electrodes, the chemical potential of metallic Mg (a_{Mg}) in $MgCl_2$ was increased on the surface of the Ti cathode to a value approaching 1, and the Ti samples were deoxidized by the electrochemically produced Mg. The resulting O species, present mainly as O^{2-} dissolved in the molten $MgCl_2$, reacted at the C anode to form CO (or CO_2) gas, which was removed from the molten salt system. For example, after electrochemical deoxidation at 1173 K for 86 ks, Ti wires 1.0 mm in diameter and containing 0.084 mass % (840 mass ppm) O were deoxidized to 0.009 mass% (90 mass ppm) (Exp. no. D13_150121 shown in Table 3).[25] When a Ti rod of 3.0 mm diameter containing 0.12 mass% (1200 mass ppm) O was deoxidized at 1173 K for 86 ks, Ti with 0.027 mass% (270 mass ppm) O was obtained (Exp. no. D15_150123 shown in Table 3).[25] The N concentration in the sample was increased from ~10–40 to 40–440 mass ppm after deoxidation.

In Table 3, some representative analytical results for the O and N levels in Ti before and after deoxidation are listed, as well as results from previous studies.[11,12,14,19,21,23–25] The analytical values below 0.01 mass% (100 mass ppm) listed in the table include some uncertainty because gas analysis of Ti samples below 0.01 mass% (100 mass ppm) is difficult. However, notably, Ti with an oxygen level of about 0.01 mass% (100 mass ppm) was obtained by using Mg as the deoxidant when using the electrochemical technique.

Table 4 summarizes the deoxidation limits for the Mg/MgO, Ca/CaO, and Y/Y_2O_3 systems. The activity of the metal oxides for obtaining Ti with 100 mass ppm O when using the respective metal deoxidants was calculated using the available thermodynamic data[11,50] and are listed in the table. In this study, Ti wires containing 0.12 mass% (1200 mass ppm) O were deoxidized to less than 0.02 mass% (200 mass ppm) O. In some cases, the O concentration in the Ti samples was reduced to below 0.01 mass% (100 mass ppm) O, which cannot be accomplished using the current Kroll process. These experimental results show that the activity of MgO, a_{MgO}, was decreased to below 0.01.

When electrolysis in $CaCl_2$ is utilized to remove O directly from Ti, the deoxidation limit can be easily decreased to below 0.01 mass% (100 mass ppm). However, after deoxidation, $CaCl_2$ flux attached on the surface of Ti products must be removed by dissolving in aqueous solution, which causes O pickup. It is

Table 4 Equilibrium oxygen concentrations calculated under M/MOx equilibrium at 1273 K using available thermodynamic data.[11,50] Calculated activities of metal oxides (a_{MO_x}) used to obtain low-oxygen-containing titanium (100 and 1000 mass ppm O) are also shown for reference

System	Equilibrium O conc. in Ti at 1273 K, $C_{O,eq}$ (mass ppm)	Required a_{MO_x} to obtain Ti with 1000 ppm O at 1273 K, a_{MO_x}	Required a_{MO_x} to obtain Ti with 100 ppm O at 1273 K, a_{MO_x}
Mg/MgO	26 000	0.039	0.0039
Ca/CaO	500	2.0	0.20
$Y/YO_{1.5}$	180	13	0.41

impractical to remove $CaCl_2$ by evaporation, because the vapor pressure of $CaCl_2$ is low even at elevated temperatures. Fig. 8 shows the temperature dependence of the vapor pressures of $CaCl_2$ and $MgCl_2$.[50] The vapor pressure of $MgCl_2$ is two or three orders of magnitude higher than that of $CaCl_2$. The high vapor pressure of $MgCl_2$ might enhance operational challenges during electrolysis.[51] However, $MgCl_2$ can be removed directly from titanium products by vacuum treatment; a technique for $MgCl_2$ removal *via* vacuum distillation is already established in the Kroll process. This $MgCl_2$ removal technique can easily be applied to the deoxidation process investigated in this study.

Fig. 9 shows the new recycling process for Ti scrap using electrochemical deoxidation in molten $MgCl_2$ followed by the removal of $MgCl_2$ by vacuum distillation. The reduction chamber used in the conventional Kroll process can be utilized in the electrochemical deoxidation process, as schematically shown in

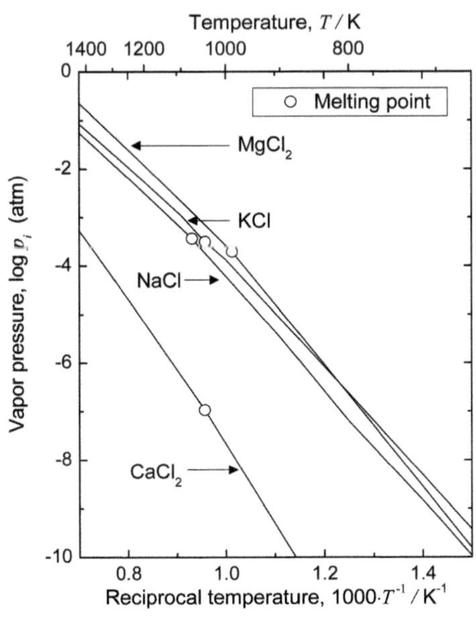

Fig. 8 Vapor pressures of metal chlorides as functions of reciprocal temperature.[50]

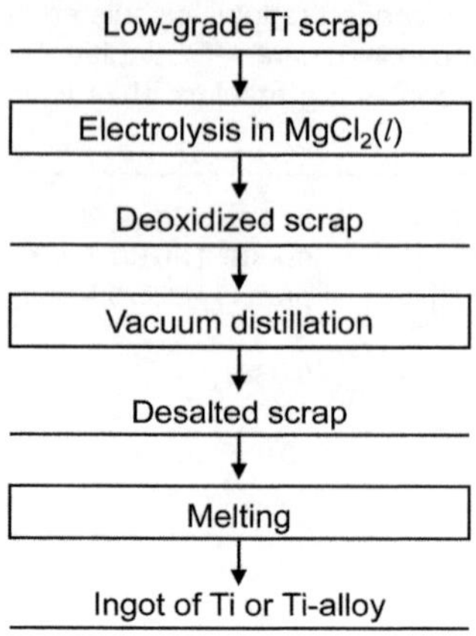

Fig. 9 Recycling process proposed in this study. Oxygen dissolved in Ti scrap is directly removed by molten salt electrolysis in $MgCl_2$.

Fig. 10. As Ti production is increased, scrap treatment becomes increasingly important. In the future, the direct O removal technique proposed in this study may be applied to recycling Ti scrap.

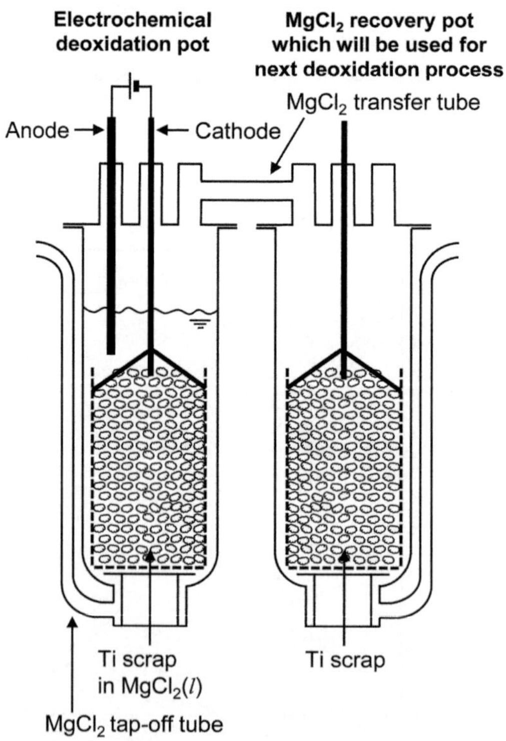

Fig. 10 Industrial application of the recycling process based on electrolysis in $MgCl_2$. Process is feasible in existing Ti smelters based on the Kroll process.

Conclusions

The removal of O from Ti using an electrochemical technique in molten $MgCl_2$ was examined at temperatures of ~1173 K. The activity of Mg was increased near the Ti cathode surface and the activity of MgO was decreased by applying voltages between the Ti cathode and C anode immersed in molten $MgCl_2$. These activity changes facilitated the deoxidation of a Ti cathode consisting of the Ti samples. A Ti sample containing approximately 1000 mass ppm O were deoxidized to less than 200 mass ppm by applying 3.1 V between the Ti and C electrodes for 86 ks. In some cases, the O concentration in Ti was decreased to below 100 mass ppm, which cannot be accomplished using the conventional Kroll process. The application of this deoxidation technique to industrial recycling processes may become practical in the future.

Acknowledgements

The authors are grateful to Professor Masafumi Maeda and Professor Kazuki Morita of The University of Tokyo and to Dr Tetsushi Deura, and Messrs. Daisuke Matsuwaka and Fumiaki Kudo of Kobe Steel, Ltd. for their sample analysis. The authors thank Dr Katsuhiro Nose and Mr Hisao Kimura for their valuable suggestions and technical assistance. This research was partly funded by a Grant-in-Aid for the Next Generation of World-Leading Researchers (NEXT Program) for the Research Project for Development of Environmentally Sound Recycling Technology of Rare Metals (NEXT Program #GR019), and a Grant-in-Aid for Scientific Research (S) (KAKENHI Grant #26220910) by JSPS.

References

1 T. G. Goonan, *Titanium Recycling in the United States in 2004, Flow studies for recycling metal commodities in the United States*, ed. S. F. Sibley, U. S. Geological Survey Circular, 1196, 2010, ch. Y, pp. Y1–Y16.

2 B. Rotmann, C. Lochbichler and B. Friedrich, Challenges in Titanium Recycling-Do We Need a New Specification for Secondary Alloys?, *Proceedings of EMC 2011*, 2011.

3 American Society for Testing and Materials, *Standard Specification for Titanium and Titanium Alloy Strip, Sheet, and Plate, B265–06b*, ASTM International, West Conshohocken, PA, 2006.

4 W. Kroll, The Production of Ductile Titanium, *Trans. Electrochem. Soc.*, 1940, **78**, 35–47.

5 T. H. Okabe, K. T. Jacob and Y. Waseda, Removal of Oxygen in Reactive Metals, in *Purification Process and Characterization of Ultra High Purity Metals*, ed. Y. Waseda and M. Isshiki, Springer, Berlin, 2001, pp. 3–37.

6 K. Ono and S. Miyazaki, Study on the Limit of Deoxidation of Titanium and the Reduction of Titanium Dioxide by Saturated Calcium Vapors, *J. Jpn. Inst. Met.*, 1985, **49**(10), 871–875, (in Japanese).

7 R. L. Fisher, Deoxidation of Titanium and Similar Metals Using a Deoxidant in a Molten Metal Carrier, *US Pat.* No. 4923531A, 1990, (UK Patent No. GB 2224749A, 1989).

8 R. L. Fisher, Deoxidation of a Refractory Metal, *US Pat.* No. 5022935, 1991.

9 R. L. Fisher and S. R. Seagle, Removal of Oxide Layers from Titanium Castings Using an Alkaline Earth Deoxidizing Agent, US Patent No. 5211775 A, 1993.

10 R. L. Fisher and S. R. Seagle, DOSS, An Industrial Process for Removing Oxygen From Titanium Turnings Scrap, *Titanium Science and Technology*, ed. F. H. Froes and I. Caplan, The Minerals, Metals & Materials Society (Proceedings of the 7th World Conference on Titanium, 1992), 1993, vol. 3, pp. 2265–2272.

11 T. H. Okabe, R. O. Suzuki, T. Oishi and K. Ono, Thermodynamic Properties of Dilute Titanium-Oxygen Solid Solution in Beta Phase, *Mater. Trans. JIM*, 1991, **32**(5), 485–488.

12 T. H. Okabe, T. Deura, T. Oishi, K. Ono and D. R. Sadoway, Thermodynamic Properties of Oxygen in Yttrium Solid Solution, *Metall. Trans. B*, 1996, **27B**, 841–847.

13 T. H. Okabe, K. Hirota, Y. Waseda and K. T. Jacob, Thermodynamic Properties of Oxygen in Ln–O (Ln = La, Pr, Nd) Solid Solutions and Their Deoxidation by Molten Salt Electrolysis, *J. MMIJ*, 1998, **114**(11), 813–818.

14 T. H. Okabe, K. Hirota, E. Kasai, F. Saito, Y. Waseda and K. T. Jacob, Thermodynamic Properties of Oxygen in RE–O (RE = Gd, Tb, Dy, Er) Solid Solutions, *J. Alloys Compd.*, 1998, **279**, 184–191.

15 J.-M. Oh, B.-K. Lee, C.-Y. Suh, S.-W. Cho and J.-W. Lim, Preparation Method of Ti Powder with Oxygen Concentration of < 1000 ppm Using Ca, *Powder Metall.*, 2012, **55**(5), 402–404.

16 J.-M. Oh, K.-M. Roh, B.-K. Lee, C.-Y. Suh, W. Kim, H. Kwon and J.-W. Lim, Preparation of Low Oxygen Content Alloy Powder from Ti Binary Alloy Scrap by Hydrogen-Dehydrogenation and Deoxidation Process, *J. Alloys Compd.*, 2014, **593**, 61–66.

17 K.-M. Roh, C.-Y. Suh, J.-M. Oh, W. Kim, H. Kwon and J.-W. Lim, Comparison of Deoxidation Capability for Preparation of Low Oxygen Content Powder from TiNi Alloy Scraps, *Powder Technol.*, 2014, **253**, 266–269.

18 T. H. Okabe, R. O. Suzuki, T. Oishi and K. Ono, Preparation of Extra Low Oxygen Titanium by Calcium-Halide Flux Deoxidation, *J. Iron Steel Inst. Jpn.*, 1991, **77**(1), 93–99, (in Japanese).

19 T. H. Okabe, T. Oishi and K. Ono, Preparation and Characterization of Extra-Low-Oxygen Titanium, *J. Alloys Compd.*, 1992, **184**, 43–56.

20 T. H. Okabe, T. Oishi and K. Ono, Deoxidation of Titanium Aluminide by Ca–Al Alloy under Controlled Aluminum Activity, *Metall. Trans. B*, 1992, **23B**, 583–590.

21 T. H. Okabe, M. Nakamura, T. Oishi and K. Ono, Electrochemical Deoxidation of Titanium, *Metall. Trans. B*, 1993, **24B**, 449–455.

22 T. H. Okabe, T. Deura, T. Oishi, K. Ono and D. R. Sadoway, Electrochemical Deoxidation of Yttrium–Oxygen Solid Solution, *J. Alloys Compd.*, 1996, **237**, 150–154.

23 M. Nakamura, T. H. Okabe, T. Oishi and K. Ono, Electrochemical Deoxidation of Titanium, *Proc. Int. Symp. Molten Salt Chem. Technol.*, 1993, 529–540.

24 K. Hirota, T. H. Okabe, F. Saito, Y. Waseda and K. T. Jacob, Electrochemical Deoxidation of RE–O (RE = Gd, Tb, Dy, Er) Solid Solution, *J. Alloys Compd.*, 1999, **282**, 101–108.

25 Y. Taninouchi, Y. Hamanaka and T. H. Okabe, submitted.

26 G. Z. Chen, D. J. Fray and T. W. Farthing, Direct Electrochemical Reduction of Titanium Dioxide to Titanium in Molten Calcium Chloride, *Nature*, 2000, **407(6802)**, 361–364.
27 G. Z. Chen, D. J. Fray and T. W. Farthing, Cathodic Deoxidation of the Alpha Case on Titanium and Alloys in Molten Calcium Chloride, *Metall. Trans. B*, 2001, **32B**, 1041–1052.
28 G. Z. Chen, D. J. Fray and T. W. Farthing, Removal of Oxygen from Metal Oxides and Solid Solutions by Electrolysis in a Fused Salt, *US Pat.* No. 2004/0159559 A1, 2004.
29 D. J. Fray, Emerging Molten Salt Technologies for Metals Production, *JOM*, 2001, **53**(10), 26–31.
30 K. S. Mohandas and D. J. Fray, FFC Cambridge Process and Removal of Oxygen from Metal-Oxygen Systems by Molten Salt Electrolysis: an Overview, *Trans. Indian Inst. Met.*, 2004, **57**(6), 579–592.
31 P. K. Tripathy, M. Gauthier and D. J. Fray, Electrochemical Deoxidation of Titanium Foam in Molten Calcium Chloride, *Metall. Trans. B*, 2007, **38B**, 893–900.
32 K. Ono and R. O. Suzuki, A New Concept for Producing Ti Sponge: Calciothermic Reduction, *JOM*, 2002, **54**(2), 59–61.
33 R. O. Suzuki and S. Inoue, Calciothermic Reduction of Titanium Oxide in Molten $CaCl_2$, *Metall. Trans. B*, 2003, **34B**(3), 277–285.
34 R. O. Suzuki, K. Ono and K. Teranuma, Calciothermic Reduction of Titanium Oxide and In-situ Electrolysis in Molten $CaCl_2$, *Metall. Trans. B*, 2003, **34B**(3), 287–295.
35 R. O. Suzuki and S. Fukui, Reduction of TiO_2 in Molten $CaCl_2$ by Ca Deposited during CaO Electrolysis, *Mater. Trans.*, 2004, **45**(5), 1665–1671.
36 R. O. Suzuki, Calciothermic Reduction of TiO_2 and In-situ Electrolysis of CaO in Molten $CaCl_2$, *J. Phys. Chem. Solids*, 2005, **66**(2), 461–465.
37 R. O. Suzuki, Direct Reduction Process for Titanium Oxide in Molten Salt, *JOM*, 2007, **59**(1), 68–71.
38 T. Yahata, T. Ikeda and M. Maeda, Deoxidation of Molten Titanium by Electron-Beam Remelting Technique, *Metall. Trans. B*, 1993, **24B**, 599–604.
39 J. Reitz, C. Lochbichler and B. Friedrich, Recycling of Gamma Titanium Aluminide Scrap from Investment Casting Operations, *Intermetallics*, 2011, **19**, 762–768.
40 Y. Su, L. Wang, L. Luo, X. Jiang, J. Guo and H. Fu, Deoxidation of Titanium Alloy Using Hydrogen, *Int. J. Hydrogen Energy*, 2009, **34**, 8958–8963.
41 J. R. Nettle, D. H. Baxer Jr and F. S. Wartman, *Electrorefining Titanium Metal, United States Bureau of Mines, Report of Investigations 5315*, Washington, D.C., U.S., 1957.
42 A. B. Suchkov, Z. A. Tubyshki, Z. I. Sokolova and N. V. Zhukova, Electrolytic Refining of Titanium Alloy Scrap Using Solid Intermediate Electrodes, *Russ. Metall.*, 1969, **6**, 52.
43 S. Takeuchi and O. Watanabe, Studies on the Method for the Electro-Refining of Titanium from the Molten Salts (1) Effect of the Various Molten Salt Baths on the Electro-Refining, *J. Jpn. Inst. Met.*, 1964, **28**(10), 633–638, (in Japanese).
44 S. Takeuchi and O. Watanabe, The Investigation on the Electrorefining of Titanium from the Molten Salts (II) Crystal Shapes and Growth of Titanium

from the Various Molten Salt in the Electro-refining, *J. Jpn. Inst. Met.*, 1964, **28**(11), 728–734, (in Japanese).

45 Y. Hashimoto, K. Uriya and R. Kono, Electro-winning of Titanium Metal from Its Oxide by Fused Salt Electrolysis at Temperatures above the Melting Point of the Metal, *Denki Kagaku*, 1971, **39**(6), 516–522, (in Japanese).

46 Y. Hashimoto, Influences of Fluoride Salt Baths on Fused-Salt Electrodeposition of Titanium Metal from TiO_2, *Denki Kagaku*, 1971, **39**(12), 938–943, (in Japanese).

47 Y. Hashimoto, Electro-Winning of Titanium from TiO_2 or $CaTiO_3$ in CaF_2-MgF_2 Molten Salt Bath, *Denki Kagaku*, 1972, **40**(1), 39–44, (in Japanese).

48 H. Miyazaki, Y. Yamakoshi and Y. Shindo, Electrorefining of Titanium in Molten Salts, *Mater. Jpn.*, 1994, **33**(1), 51–54, (in Japanese).

49 T. B. Massalski, *Binary Alloy Phase Diagrams*, ASM international, Materials Park, OH, United States of America, 2nd edn, 1990.

50 I. Barin, *Thermochemical Data of Pure Substances*, VCH Verlagsgesellschaft mbH, Weinheim, Germany, 3rd edn, 1995.

51 W. Li, Y. Yuan, X. Jin, H. Chen and G. Z. Chen, Environmental and Energy Gains from Using Molten Magnesium-sodium-potassium Chlorides for Electro-metallisation of Refractory Metal Oxides, *Prog. Nat. Sci.*, 2015, **25**(6), 650–653.

Faraday Discussions

PAPER

Developing energy efficient lignin biomass processing – towards understanding mediator behaviour in ionic liquids†

Majd Eshtaya,*[a] Andinet Ejigu,[b] Gill Stephens,[a] Darren A. Walsh,[b] George Z. Chen*[ac] and Anna K. Croft*[a]

Received 16th December 2015, Accepted 7th January 2016
DOI: 10.1039/c5fd00226e

Environmental concerns have brought attention to the requirement for more efficient and renewable processes for chemicals production. Lignin is the second most abundant natural polymer, and might serve as a sustainable resource for manufacturing fuels and aromatic derivatives for the chemicals industry after being depolymerised. In this work, the mediator 2,2′-azino-bis(3-ethylbenthiazoline-6-sulfonic acid) diammonium salt (ABTS), commonly used with enzyme degradation systems, has been evaluated by means of cyclic voltammetry (CV) for enhancing the oxidation of the non-phenolic lignin model compound veratryl alcohol and three types of lignin (organosolv, Kraft and lignosulfonate) in the ionic liquid 1-ethyl-3-methylimidazolium ethyl sulfate, ([C_2mim][C_2SO_4]). The presence of either veratryl alcohol or organosolv lignin increased the second oxidation peak of ABTS under select conditions, indicating the ABTS-mediated oxidation of these molecules at high potentials in [C_2mim][C_2SO_4]. Furthermore, CV was applied as a quick and efficient way to explore the impact of water in the ABTS-mediated oxidation of both organosolv and lignosulfonate lignin. Higher catalytic efficiencies of ABTS were observed for lignosulfonate solutions either in sodium acetate buffer or when [C_2mim][C_2SO_4] (15 v/v%) was present in the buffer solution, whilst there was no change found in the catalytic efficiency of ABTS in [C_2mim][C_2SO_4]–lignosulfonate mixtures relative to ABTS alone. In contrast, organosolv showed an initial increase in oxidation, followed by a significant decrease on increasing the water content of a [C_2mim][C_2SO_4] solution.

[a]Department of Chemical and Environmental Engineering, Faculty of Engineering, University of Nottingham, Nottingham NG7 2RD, UK. E-mail: majdkha@yahoo.com; george.chen@nottingham.ac.uk; anna.croft@nottingham.ac.uk

[b]Department of Chemistry, University of Nottingham, Nottingham NG7 2RD, UK

[c]Department of Chemical and Environmental Engineering, Faculty of Science & Engineering, University of Nottingham Ningbo China, Ningbo 315100, P. R. China

† Electronic supplementary information (ESI) available. See DOI: 10.1039/c5fd00226e

1. Introduction

Lignin is the second most abundant naturally synthesised compound after cellulose, comprising 15–30% of total dry wood. It is a complex natural polymer composed of a diversely linked network of phenolic and non-phenolic compounds and hence could provide a potential source for the sustainable production of fuels and aromatic chemicals that are currently obtained from petroleum-based feedstocks.[1–3] Currently, lignin is produced mainly as a major by-product of both the pulp and paper industry and biomass ethanol industry, and is either discharged in black liquor or used as a low-grade fuel. Therefore, depolymerisation of lignin into aromatic products would greatly benefit both the fine chemicals industry and the further development of sustainable chemical processes.[4–6]

Over the years, several methods have been explored and documented for the depolymerisation of lignin, such as pyrolysis,[4,7] chemical oxidation, including electrochemical oxidation,[6,8,9] hydrolysis[10,11] and enzymatic reactions.[12–14] Among these methods, enzymatic reactions and electrochemical oxidation are preferable due to their environmentally friendly characteristics and lower energy demands.

A number of different types of enzymes are available for the biodegradation of lignin, such as laccase (EC 1.10.3.2, *p*-benzenediol:dioxygen oxidoreductases).[15] Laccases alone have only a very limited effect on lignin degradation due to their low redox potentials (0.45–0.80 V *vs.* NHE). This means that laccases can oxidise phenolic groups only, but not the non-phenolic aromatic structures that comprise more than 80% of lignin. Laccases can react indirectly with non-phenolic compounds and potentially degrade lignin in the presence of a redox mediator.[16] A mediator is a low-molecular weight compound that, once oxidised by the enzyme, diffuses away from the enzymatic pocket and then can oxidise other substrates, thus regenerating its reduced form. Overall, the laccase-mediator system acts as a catalyst to assist oxidation of lignin by mediation of electron transfer from the oxidisable groups in lignin to oxygen molecules. This simplified mechanism is depicted in Fig. 1a.[16–21] Generally, the ideal redox mediator should

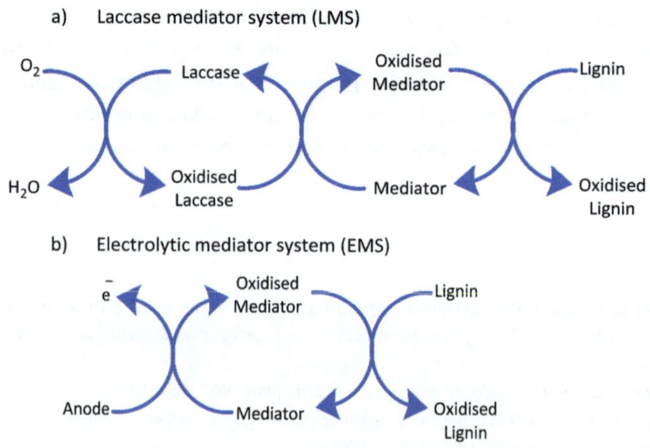

Fig. 1 Laccase mediator system (LMS) and electrolytic mediator system (EMS).[27]

be a good laccase substrate, able to generate stable radicals upon oxidation that do not inhibit the enzymatic reaction, and also should not be consumed during the redox transformation.[17,19,22]

In the past two decades, the range of compounds identified as mediators for the laccase mediator system (LMS) has increased dramatically and 2,2′-azino-bis(3-ethylbenthiazoline-6-sulfonic acid) diammonium salt **1** (ABTS) (Fig. 2) is currently regarded as the best mediator for laccase.[21–26]

Despite the obvious advantages, LMS has a number of drawbacks such as high cost of the enzymes and their sensitivity to the experimental conditions (pH, temperature and inhibitors). In addition, there are difficulties associated with the use of LMS in conventional solvents.[28–31]

In contrast to LMS, an electrolytic mediator system (EMS) (Fig. 1b) could provide a promising and effective alternative for lignin depolymerisation. Instead of using a laccase to oxidise the mediator, the mediator can be oxidised electrochemically instead. By avoiding the requirement for a sensitive enzyme, the electrochemical process can be performed under a wide range of pH and temperature conditions with mediators that have a high redox potential.[17,19,27]

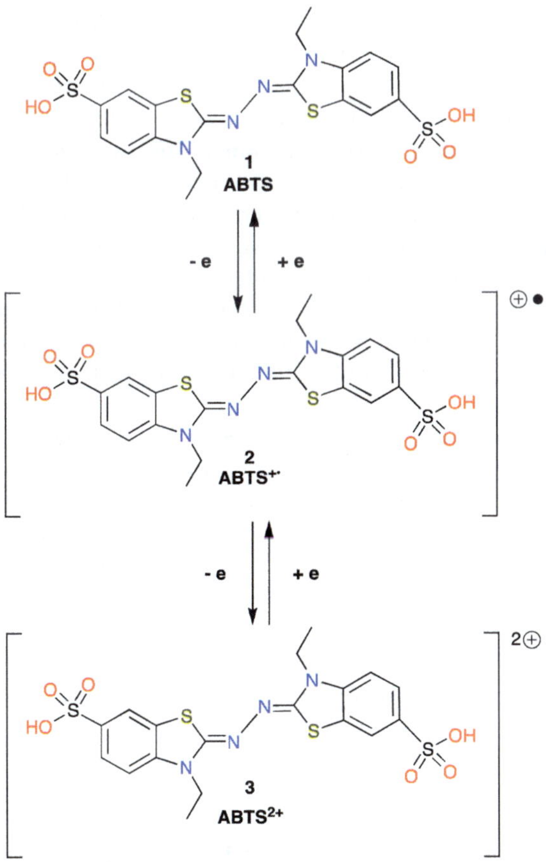

Fig. 2 Chemical structure of ABTS mediator **1** and formation of the cation radical **2** and the dication **3** by removal of one and two electrons from ABTS **1**, respectively.[24]

One of the main obstacles to the degradation of lignin is its insolubility in most commonly-used solvents. Over recent years, ionic liquids (ILs) have been used for the dissolution of biomass. Very high solubility of lignin has been achieved using 1-ethyl-3-methylimidazolium acetate ([C_2mim][C_1CO_2]), 1-butyl-3-methylimidazolium chloride ([C_4mim]Cl) and 1-ethyl-3-methylimidazolium ethyl sulfate ([C_2mim][C_2SO_4]),[32–34] and this good solubility provides new opportunities for lignin degradation.

Unlike conventional organic solvents, room temperature ILs have many favourable properties in electro-organic synthesis including low vapour pressure at room temperature, a wide liquid range, inherent conductivity, non-volatility, good dissolving power towards many substrates, and excellent thermal and chemical stability. Also, they possess a wide window of electrode potential and no addition of supporting electrolyte is required, which can assist electrochemical processes such as the oxidation of lignin.[35,36] When compared to conventional solvents, however, ILs are commonly 10–100 times more viscous, and the diffusion rates of species in these media are lowered accordingly.[36,37]

Numerous studies have been done on the electrochemical behaviour of ABTS **1** in aqueous solutions,[21–27,38] but to our knowledge, no study of the electrochemical behaviour of ABTS **1** for oxidising either lignin or a lignin model compound in ionic liquids has so far been reported. More recently, Reichert and co-workers successfully depolymerised alkali lignin through electro-catalytic oxidative cleavage without using any mediator. They performed the depolymerisation in a special protic ionic liquid, triethylammonium methanesulfonate ([HN_{222}][C_1SO_3]), using an active electrode of mixed ruthenium/vanadium/titanium oxides. A wide range of aromatic compounds, such as aldehydes and ketones, were identified as the cleavage products, as detected by GC-MS and HPLC after diethyl ether extraction.[6] Another study performed by Zhu *et al.*[9] achieved electro-oxidative cleavage of lignin in alkaline solutions (2 wt%) by anodic oxidation coupled with electro-generated H_2O_2 oxidation in a non-diaphragm electrolytic cell. Owing to the effective breakage of the C–C and C–O–C linkages among C9 units, lignin macromolecules were depolymerised into low molecular weight aromatic products with different functional groups, such as aldehydes, ketones, phenols and acids.

This work is aimed at studying the electrochemical behaviour of ABTS **1** in oxidising the non-phenolic lignin model veratryl alcohol **4** and three technical lignins: Kraft lignin, organosolv lignin and sulfonated lignin (lignosulfonate), in [C_2mim][C_2SO_4]. Due to its high capacity to dissolve lignin, [C_2mim][C_2SO_4] was chosen in this study.[39,40] Cyclic voltammetry was applied as a quick, accurate and efficient way to explore the ABTS-mediated oxidation process.

2. Materials and methods

2.1. Chemicals

Organosolv lignin was supplied by Lignol Innovations Ltd, Canada. The ionic liquid [C_2mim][C_2SO_4] (≥95% purity), sodium lignosulfonate, alkali (Kraft) lignin, 3,4-dimethoxybenzyl alcohol **4** (veratryl alcohol), silver trifluoromethane sulfonate (AgTfO) (≥98% purity), 2,2′-azino-bis-(3-ethylbenzothiazoline-6-sulfonic acid) diammonium salt **1** (ABTS) and all other chemicals and solvents used in this study were purchased from Sigma Aldrich (UK) and used without any further purification.

2.2. Viscosity measurements

The viscosities of $[C_2mim][C_2SO_4]$, 0.1 M sodium acetate buffer solution (pH 4.5), 15% (v/v) $[C_2mim][C_2SO_4]$/buffer solution and lignin/$[C_2mim][C_2SO_4]$ mixtures were measured at room temperature with a Kinexus pro$^+$ rotational rheometer (Malvern Instruments Ltd., UK) equipped with cone-and-plate geometry (40 mm in diameter; with a cone angle of 4° and a gap of 150 μm). A strain rate sweep was performed from 1.0 to 1000 s^{-1}.

2.3. Cyclic voltammetry (CV) measurements

Cyclic voltammetry and chronoamperometry were carried out using Autolab Type III potentiostat/galvanostat from EcoChemie (Metrohm, Netherlands) controlled by the Autolab GPES software, version 4.9, in an undivided three-electrode cell. A glassy carbon (GC) disc electrode (3.0 mm in diameter) was used as the working electrode, which was polished prior to use and between each measurement using aqueous alumina (0.05 μm) suspensions. After polishing, the electrode surface was rinsed with deionised water. The reference electrode was an IL-based Ag/Ag^+ electrode, which has been described previously (10 mM AgTfO, $[C_2mim][C_2SO_4]$),[41] and a graphite rod was used as the counter electrode. All the solutions were degassed with nitrogen prior to measuring and all experiments were performed at room temperature.

The electrochemical characterisation of ABTS **1** was first carried out alone in either 0.1 M sodium acetate buffer solution (pH 4.5), 15% (v/v) $[C_2mim][C_2SO_4]$/buffer solution or as-received $[C_2mim][C_2SO_4]$ at different scan rates. The voltammetric responses of ABTS mediator **1** were also recorded separately at the same potential scan rate in the presence of either the lignin model compound veratryl alcohol **4** (13.8 mM) in 15% (v/v) $[C_2mim][C_2SO_4]$/buffer solution or as-received $[C_2mim][C_2SO_4]$, or lignin (Kraft, organosolv and lignosulfonate) (5 wt%) dissolved in as-received $[C_2mim][C_2SO_4]$. The current response of ABTS-mediated organosolv lignin at different concentrations of sodium phosphate buffer (pH 12) in $[C_2mim][C_2SO_4]$ was also measured. Finally, voltammograms of ABTS **1** in the presence of lignosulfonate, either in sodium acetate buffer solution or in 15% (v/v) $[C_2mim][C_2SO_4]$/buffer solution, were recorded.

3. Results and discussion

3.1. Electrochemical behaviour of ABTS in $[C_2mim][C_2SO_4]$

The cyclic voltammogram of 10 mM ABTS **1** in $[C_2mim][C_2SO_4]$ with a scan rate of 200 mV s^{-1} is shown in Fig. 3. Two oxidation peaks ($E_{p,a}$) at potentials of 0.500 V (A) and 0.815 V (B) were observed, corresponding to the oxidation of ABTS **1** to its cation radical **2** ($ABTS^{+\cdot}$) and subsequently to its dication **3** ($ABTS^{2+}$). The reverse scan showed two reduction peaks ($E_{p,c}$) at 0.738 V (C) and 0.423 V (D) for the reduction of the dication and the cation radical, respectively. Analysis of the two redox couples at a range of scan rates (see inset in Fig. 3) revealed that the oxidation ($i_{p,a}$) and reduction ($i_{p,c}$) peak currents were proportional to the square root of the scan rate. Additionally, at each scan rate, the ratio of $i_{p,a}$ to $i_{p,c}$ was

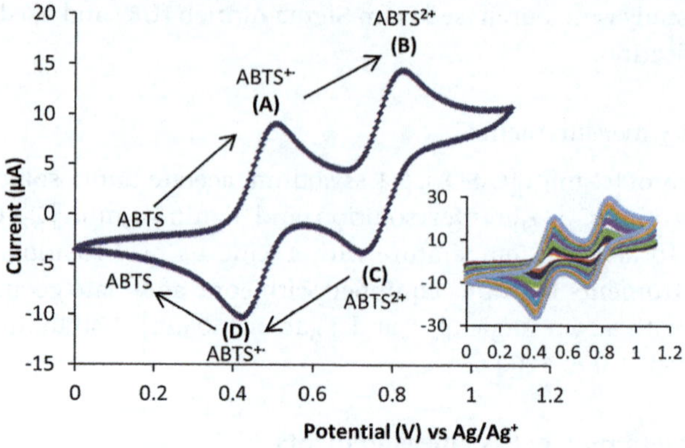

Fig. 3 Cyclic voltammogram of 10 mM ABTS **1** in $[C_2mim][C_2SO_4]$ recorded at a scan rate of 200 mV s^{-1}. The inset shows the cyclic voltammogram of 10 mM ABTS **1** in $[C_2mim][C_2SO_4]$ with varying scan rates (from inside to outside (mV s^{-1}): 50, 100, 200, 400, 600, 800 and 1000).

close to one, indicating that the mass transport of ABTS **1** to the electrode surface was diffusion controlled in $[C_2mim][C_2SO_4]$, and confirming that the radical cation $ABTS^{+\cdot}$ **2**, dication $ABTS^{2+}$ **3** and $[C_2mim][C_2SO_4]$ do not interact with one another, *e.g. via* ion pairing or proton transfer, to a significant extent. These observations are consistent with the electrochemical behaviour of ABTS in aqueous systems reported by Bourbonnais *et al.*[21]

For an electrochemically-reversible reaction with a one-electron transfer process, the peak–peak potential separation ($\Delta E_p = E_{p,a} - E_{p,c}$) for each redox couple at each scan rate should be approximately 59 mV at room temperature and independent of the scan rate. However, the ΔE_p for the ABTS radical cation **2** and dication **3** in $[C_2mim][C_2SO_4]$ were significantly greater than 59 mV and increased as the scan rate increased (77 mV at 50 mV s^{-1} and 101 mV at 1000 mV s^{-1}), indicating a quasi-reversible redox process. These results might indicate slow electron transfer kinetics, and may also be partly attributed to the effect of uncompensated ohmic resistance of the electrolyte and the measuring circuit.[42]

The CV features observed at high (200 mV s^{-1}) and low (50 mV s^{-1}) scan rates differed markedly, as shown in Fig. 4. Thus, at a low scan rate and a higher concentration of ABTS (50 mM), no cathodic signal corresponding to the reduction of the ABTS dication **3** was detected. Instead, the intensity of the $ABTS^{+\cdot}$ **2** reduction peak was about 1.4 fold that of its oxidation peak ($i_{p,a}/i_{p,c} = 1.4$). This behaviour was previously observed in aqueous solutions and attributed to the comproportionation reaction between ABTS **1** and its dication **3** with the formation of $ABTS^{+\cdot}$ **2**, which has resulted in the disappearance of a cathodic peak due to the reduction of $ABTS^{2+}$ **3**.[21] This comproportionation reaction depends on the concentration of the electroactive species close to the surface of the electrode so that, in the case when the concentration of ABTS **1** is high, the comproportionation reaction of the dication **3** will take place before its reduction at the electrode surface.

Diffusion coefficient of ABTS in $[C_2mim][C_2SO_4]$. The impact of solvent viscosity on the oxidation reaction can be related to the diffusivity of the

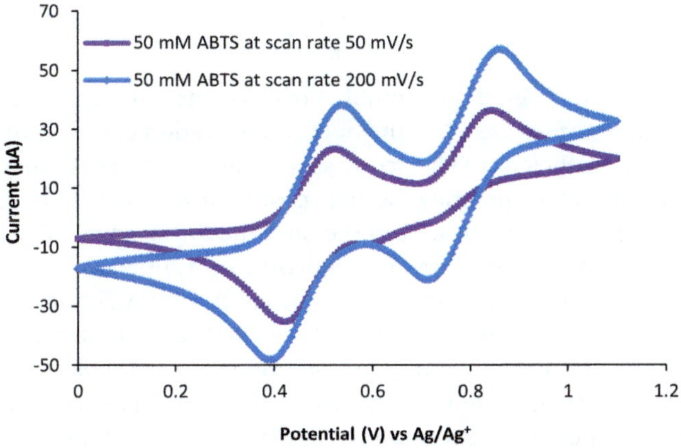

Fig. 4 Cyclic voltammograms of 50 mM ABTS **1** in $[C_2mim][C_2SO_4]$ recorded at scan rates of 50 mV s^{-1} and 200 mV s^{-1}.

substrate. Single potential step chronoamperometry was employed in order to determine the diffusion coefficient of ABTS **1** in $[C_2mim][C_2SO_4]$. Fig. 5A shows a typical chronoamperogram obtained using 10.0 mM ABTS **1** in $[C_2mim][C_2SO_4]$ after a potential step from a potential where no Faradaic reaction occurred to a potential at which the oxidation of ABTS **1** was diffusion controlled. The current, i, is proportional to $t^{-0.5}$ as shown by the Cottrell equation, eqn (1)

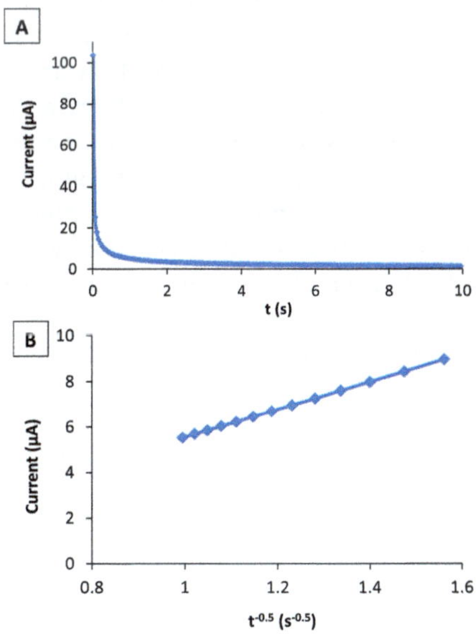

Fig. 5 (A) Potential step chronoamperogram obtained at a 3 mm diameter glassy carbon (GC) disk electrode in 10.0 mM ABTS **1** in $[C_2mim][C_2SO_4]$. The potential was stepped from 0.3 V to 0.6 V *vs.* ABTS **1**/ABTS$^{+\cdot}$ **2** and held for 10 s. (B) Cottrell plot of $i(t)$ *vs.* $t^{-0.5}$ for the oxidation process.

$$i(t) = \frac{nFACD^{0.5}}{\pi^{0.5}t^{0.5}} \quad (1)$$

where i is the peak current, n is the number of electrons transferred per molecule, A is the electrode surface area, C is the bulk concentration of the redox species, D is the diffusion coefficient of the redox species and F is the Faraday constant. Fig. 5B shows the corresponding Cottrell profile of $i(t)$ *vs.* $t^{-0.5}$ and, from the gradient of the straight line best-fit to the plot, the diffusion coefficient of ABTS **1**/ABTS$^{+\cdot}$ **2** in $[C_2mim][C_2SO_4]$ was calculated to be 1.9 (±0.3) × 10^{-8} cm^2 s^{-1}, which is typical for molecular species in viscous ILs.[42,43] A similar analysis was performed to find the diffusion coefficients of ABTS **1** in sodium acetate buffer solution and the 15% (v/v) $[C_2mim][C_2SO_4]$/buffer mixture (Table 1). Table 1 also shows the viscosities of these solvents. There was good agreement between the diffusion coefficients of ABTS **1** and the viscosities of both buffer solutions and $[C_2mim][C_2SO_4]$ with literature values.[42,44] The value of the diffusion coefficient of ABTS **1** in $[C_2mim][C_2SO_4]$ was almost three orders of magnitude lower than that determined in the buffer solution (3.5 (±0.2) × 10^{-6}), which is consistent with the high viscosity of this IL and reflects the low mass transport rate of ABTS **1** in $[C_2mim][C_2SO_4]$, compared to that in the buffer solution.

3.2. Cyclic voltammetry of veratryl alcohol (VA)

The redox reaction between the mediator ABTS **1** and the non-phenolic lignin model compound veratryl alcohol **4** (VA) was also included in this study as a benchmark, and was analysed using CV. As can be seen in Fig. 6, VA **4** alone did not show any oxidation peak below 900 mV but, in the presence of ABTS **1**, the recorded peak height increased by more than four-fold at the appropriate potential, indicating the $ABTS^{2+}$ dication **3** catalysing VA **4** oxidation. No oxidation of this lignin model **4** occurred in the potential range corresponding to the formation of ABTS$^{+\cdot}$ **2** between 0.70 and 0.95 V. This finding is consistent with those previously reported in the literature.[37] The effect of VA **4** on redox catalysis of an electrochemical reaction, as described by Bourbonnais *et al.*,[21] (Scheme 1) is for the ABTS dication **3** to oxidise the VA **4** to veratryl aldehyde **5** with concurrent regeneration of the ABTS radical **2**, resulting in an increase in the peak current compared to that seen on the CV of the ABTS solution without veratryl alcohol **4**.

The electrochemical behaviour of ABTS **1** in the presence of VA **4** in as-received $[C_2mim][C_2SO_4]$ was also explored. Fig. 7 shows the CVs of ABTS **1** (1.0 mM), veratryl alcohol **4** (13.8 mM) and the two compounds mixed together in $[C_2mim]$

Table 1 Diffusion coefficients of ABTS **1** in different solvents using potential step chronoamperometry

Solution type	Diffusion coefficient (cm^2 s^{-1})	Viscosity (mPa s)
1 mM ABTS in sodium acetate buffer	3.5 (±0.2) × 10^{-6}	0.76 ± 0.02
1 mM ABTS in 15% (v/v) $[C_2mim][C_2SO_4]$/buffer	1.74 (±0.04) × 10^{-6}	1.23 ± 0.01
1 mM ABTS in $[C_2mim][C_2SO_4]$	2.6 (±0.1) × 10^{-8}	96.6 ± 4.4
10 mM ABTS in $[C_2mim][C_2SO_4]$	1.9 (±0.3) × 10^{-8}	96.6 ± 4.4

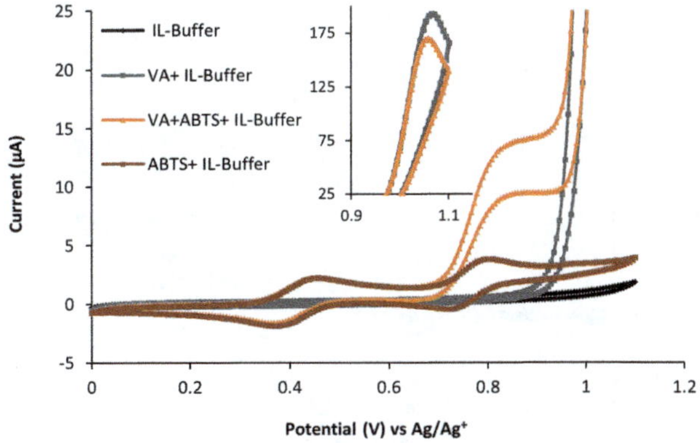

Fig. 6 Cyclic voltammograms of 1 mM ABTS **1**, 13.8 mM veratryl alcohol **4**, and 13.8 mM veratryl alcohol **4** mixed with 1 mM ABTS **1**, in 15% (v/v) $[C_2mim][C_2SO_4]$ in 0.1 M sodium acetate buffer (pH 4.5) recorded at a scan rate of 10 mV s^{-1}. The inset shows the portion of the CVs with current above 25 μA between 0.9 and 1.1 V.

$[C_2SO_4]$ under the same conditions as described above for the 15% (v/v) $[C_2mim][C_2SO_4]$/buffer mixture. The resulting CVs confirm the stability of $[C_2mim][C_2SO_4]$ in the oxidative potential range of 0 to 1.1 V *vs.* Ag/Ag^+. As with the 15% (v/v) $[C_2mim][C_2SO_4]$/buffer mixture, the simultaneous electrochemical oxidation of ABTS **1** and VA **4** resulted in an increase of the second oxidation peak of ABTS, indicating a regeneration of ABTS **1** at the electrode. However, the peak current of the second oxidation in the presence of VA **4** is reduced by 10 fold in $[C_2mim][C_2SO_4]$, compared with in the 15% (v/v) $[C_2mim][C_2SO_4]$/buffer mixture as can be

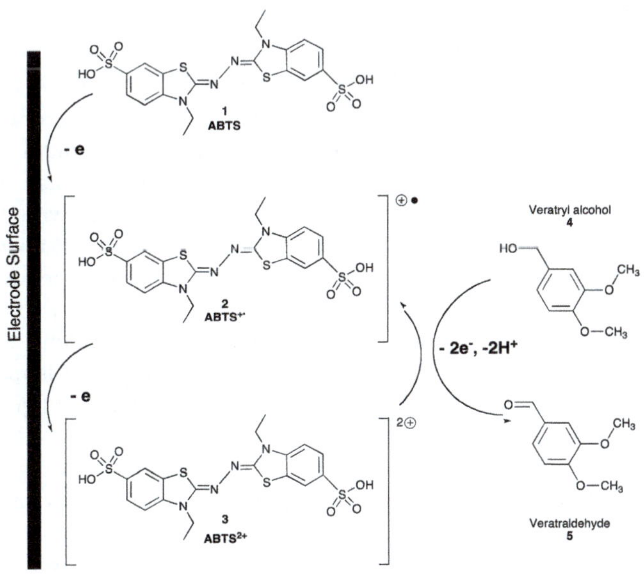

Scheme 1 Schematic representation of the reaction of ABTS **1** and VA **4** at the working electrode.

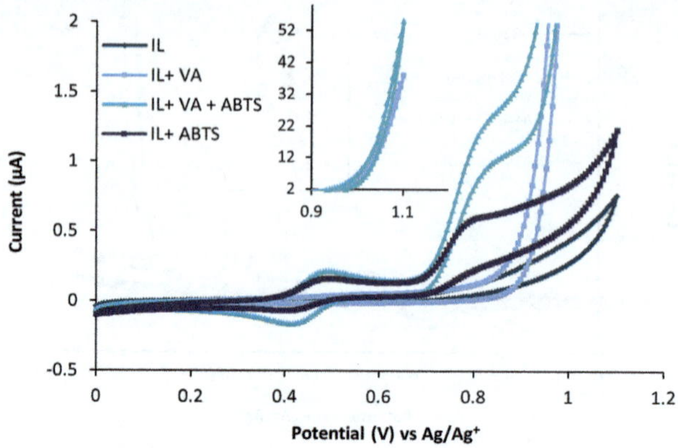

Fig. 7 Cyclic voltammograms of 1 mM ABTS **1**, 13.8 mM veratryl alcohol **4** and 13.8 mM veratryl alcohol **4** mixed with 1 mM ABTS **1**, in [C_2mim][C_2SO_4] recorded at a scan rate of 10 mV s^{-1}. The inset shows the portion of the CVs with current above 2.0 μA between 0.9 and 1.1 V.

seen by comparing Fig. 6 and 7. This is probably due to the high viscosity of [C_2mim][C_2SO_4] (96.6 mPa s at 22 °C, Table 1), and can be explained by VA **4** diffusion through the higher viscosity solvent becoming slower and hindering the transport of analytes towards the working electrode. Therefore, lower peak currents were measured. Fig. 6 and 7 also indicate that the electrochemical behaviour of ABTS **1** in [C_2mim][C_2SO_4] was different to that found in 15% (v/v) [C_2mim][C_2SO_4]/buffer mixtures. Whereas ABTS **1** showed reversible electrochemical behaviour in 15% (v/v) [C_2mim][C_2SO_4]/buffer mixtures, the CVs of ABTS **1** in [C_2mim][C_2SO_4] did not show the cathodic peak corresponding to the reduction of the dication **3** ($ABTS^{2+}$) at this particular ABTS **1** concentration and at the scan rate used.

3.3. Cyclic voltammetry of different lignins in [C_2mim][C_2SO_4]

Mainly because of the structural complexity of lignin (Fig. 8), most of the studies reported to date have been performed using lignin model compounds. However, the challenge is to valorise the widely available authentic lignins that are by-products of cellulose isolation. Therefore, the reactions of ABTS **1** with three types of technical lignin (organosolv, Kraft and sulfonated lignin) were studied by cyclic voltammetry in [C_2mim][C_2SO_4] to determine the reactivities of the two oxidised states of ABTS, **2** and **3**, towards lignin (Fig. 9–12). Initially the voltammetric response of ABTS **1** (10 mM) in [C_2mim][C_2SO_4] was assayed alone at different scan rates and then the voltammetric responses of each type of lignin (5 wt%) were also recorded separately at the same potential scan rate. Each lignin (5 wt%) was dissolved in [C_2mim][C_2SO_4] and the mixture was stirred at 60 °C for 2 h in order to ensure complete dissolution. Subsequently, nitrogen was purged through the resultant solutions before the CVs were measured in order to reduce potential interference from oxygen. Finally, the CVs of ABTS **1** (10 mM) in the presence of lignin (5 wt%) in [C_2mim][C_2SO_4] were recorded as indicated above.

As can be seen in Fig. 9, an increase in the oxidation current was detected at potentials around 1.0 V for all types of lignin in the absence of ABTS **1**. This is

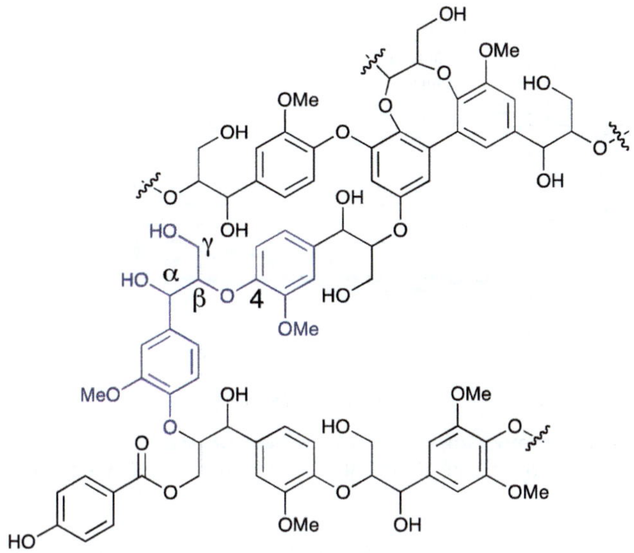

Fig. 8 Representative structure of a fragment of lignin.[45]

possibly related to the oxidation of the hydroxyl (OH) group on either the α-carbon or γ-carbon (see Fig. 8).[36] ABTS **1** behaved differently in response to addition of different lignins, as shown in Fig. 10–12. Its behaviour was fairly electrochemically reversible in the presence of lignosulfonate and Kraft lignin in $[C_2mim][C_2SO_4]$ (Fig. 11 and 12). However, organosolv lignin caused a significant increase in the second oxidation peak of ABTS (Fig. 10).

In Fig. 10, the anodic current corresponding to the formation of the radical cation **2** decreased slightly and shifted towards more positive potentials. These features seem to indicate direct interactions between ABTS and organosolv lignin, which may be due to the high polydispersity of organosolv lignin and the presence of low molecular weight fractions that could activate cascade reactions between

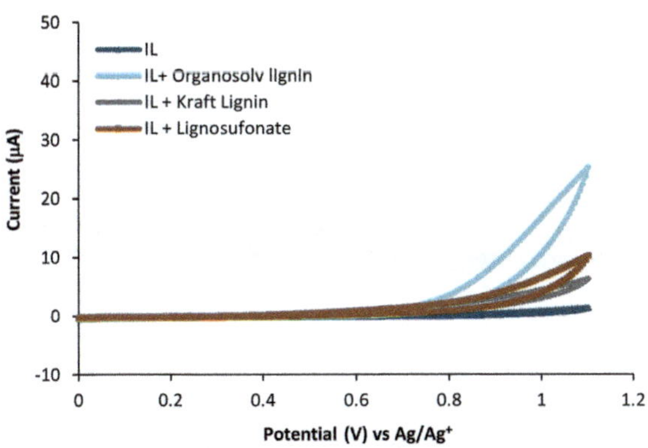

Fig. 9 Cyclic voltammograms of organosolv, Kraft and lignosulfonate lignin (5 wt%) in $[C_2mim][C_2SO_4]$ recorded at a scan rate of 50 mV s^{-1}.

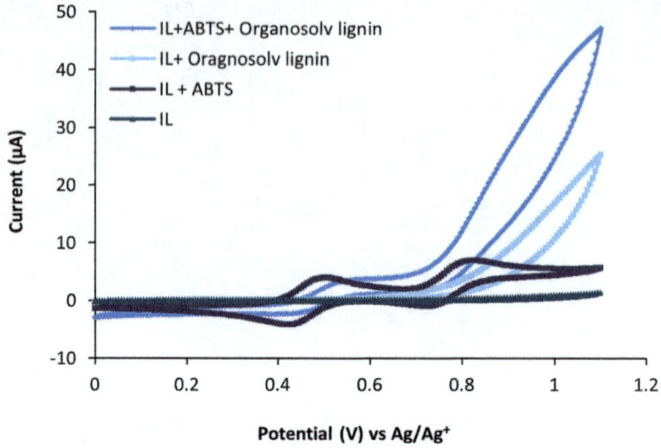

Fig. 10 Cyclic voltammograms of ABTS **1** (10 mM), organosolv lignin (5 wt%) and organosolv lignin (5 wt%) mixed with 10 mM ABTS **1**, in $[C_2mim][C_2SO_4]$ recorded at a scan rate of 50 mV s^{-1}.

their phenolic and non-phenolic structures.[19,38,46] Generally, however, organosolv lignin contains less contaminants than lignin obtained from other processes, such as Kraft lignin or lignosulfonate,[47] and is also closest in structure to native lignin.

It can also be seen in Fig. 10 that the dication formation current peak seems to have merged into and also enhanced the current from oxidation of organosolv lignin. This cyclic voltammetric response was highly likely caused by the peak potential for formation of the ABTS dication **3** being very close to the current onset potential for the oxidation of organosolv lignin. Nonetheless, the significantly greater currents between 0.7 and 1.1 V in the CV of the mixed solution of ABTS and organosolv lignin compared to those in the CVs individually recorded is still evidence for the ABTS dication being responsible for the mediated or catalytic

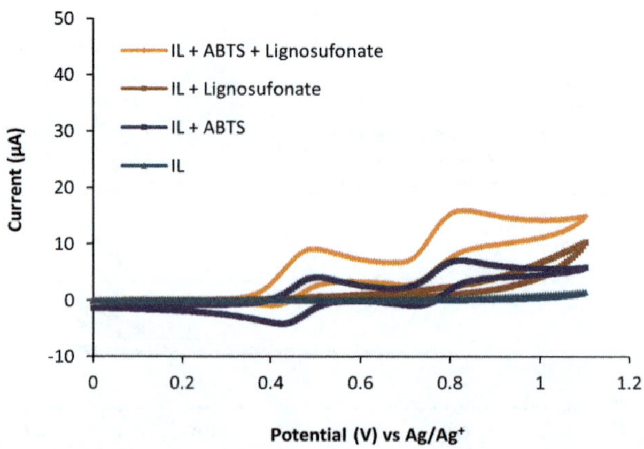

Fig. 11 Cyclic voltammograms of ABTS **1** (10 mM), lignosulfonate (5 wt%) and lignosulfonate (5 wt%) mixed with 10 mM ABTS **1**, in $[C_2mim][C_2SO_4]$ recorded at a scan rate of 50 mV s^{-1}.

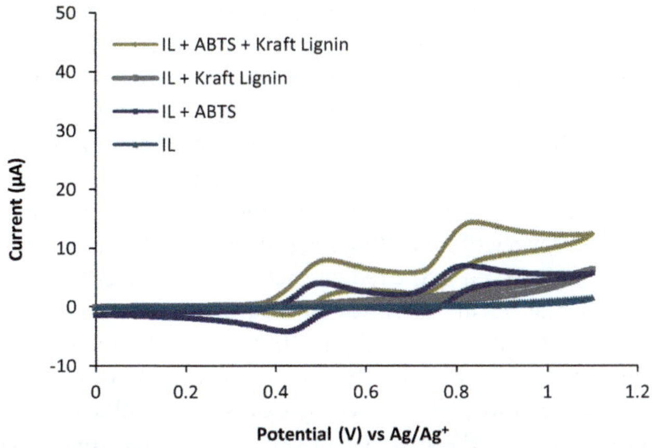

Fig. 12 Cyclic voltammograms of ABTS **1** (10 mM), Kraft lignin (5 wt%) and Kraft lignin (5 wt %) mixed with 10 mM ABTS **1**, in $[C_2mim][C_2SO_4]$ recorded at a scan rate of 50 mV s^{-1}.

oxidation of organosolv lignin. A possible factor of influence on the differing oxidation behaviour of organosolv lignin with ABTS **1** is that, because organosolv lignin comprises relatively low molecular weight lignin units, relative to the other lignins examined herein, it is also the most amenable to IL treatment, affording a more deconstructed lignin, thus making it more accessible to ABTS **1**.[48]

The CVs of ABTS **1** in the presence of lignosulfonate and Kraft lignin were quite similar, as can be seen in Fig. 11 and 12. In these two cases, both the peak currents for the formation of the radical cation **2** and dication **3** of ABTS increased noticeably after addition of either lignosulfonate or Kraft lignin. These results indicate possible interactions between the lignin and both **2** and **3**, leading to the increase of the oxidation currents. However, in comparison with Fig. 6, 7 and 10, whilst these interactions might have caused depolymerisation of the lignin, the effect, if any, was insignificant and controlled by diffusion. These differences could possibly be due to a solubility issue. Organosolv lignin is soluble in $[C_2mim][C_2SO_4]$ at room temperature, but lignosulfonate and Kraft lignin exhibit low solubility, or the dissolution is kinetically very slow in $[C_2mim][C_2SO_4]$ at room temperature. A mixing process at 60 °C for about 2 h was needed to make sure these two lignins completely dissolved. This heating step could potentially initiate adverse coupling reactions of the phenols in the lignin structure. Additionally, in contrast to organosolv lignin, Kraft lignin and lignosulfonate contain sulfur groups and therefore, a strong interaction between $[C_2mim][C_2SO_4]$ and sulfur species on the lignin could have possibly occurred in such a way that it limits the accessibility of ABTS **1** to the lignin macromolecules. Thus, further studies are required to understand the intermolecular interactions between the lignin and ABTS mediator **1** with the IL ions to explain the differences between the reactivity of ABTS **1** with each of these types of lignin.

3.4. Cyclic voltammetry of lignosulfonate (LS)

In order to elucidate the effect of viscosity on the ABTS **1** catalytic efficiency (as defined later), CVs of lignosulfonate in the sodium acetate buffer (pH 4.5) and the 15% (v/v) [C_2mim][C_2SO_4]/buffer mixtures were studied. Most of the lignins available are practically insoluble in water. However, sodium lignosulfonate contains hydrophilic functional groups, which make this type of lignin water-soluble. Therefore, sodium lignosulfonate was used in this part of the study. Fig. 13 shows the CV of lignosulfonate (1 wt%) at a low potential scan rate (10 mV s^{-1}) in the 15% (v/v) [C_2mim][C_2SO_4]/buffer solution. The increase in the anodic peak currents of ABTS **1** in the presence of lignosulfonate, compared with those measured in the presence of ABTS **1** alone, indicate that both the cation radical **2** and dication **3** were involved in the redox reactions of lignosulfonate. This finding is consistent with the results obtained by Bourbonnais and co-workers for the electrochemical oxidation of Kraft lignin (Indulin) in buffer solutions.[17,21] The increase of the peak current for the formation of the ABTS cation radical **2** has been related to the quantity of more easily oxidisable residues in lignin such as phenolic groups, whereas the increase of the peak current for formation of the ABTS dication **3** corresponds more likely to (non-phenolic) groups in lignin with higher redox potentials.[49] The reaction mechanism is already known for the oxidation of phenols by ABTS cation radical **2** and consists of at least two homogeneous chemical reactions: fast and reversible electron transfer (Scheme 2), consistent with the increase in the cation radical peak in Fig. 13, followed by the reaction of the formed phenoxy radicals with another ABTS-derived radical cation.[44,50]

Additionally, in the presence of lignosulfonate, the CVs present well-shaped oxidation peaks (Fig. 13), but only one small reduction peak was observed, indicating the absence of ABTS **1** regeneration. A likely explanation is that a portion of the ABTS dication **3** and almost all of the radical cation **2** formed coupled with phenoxy radicals and other reactive intermediates produced during the reaction to form covalently-linked derivatives.[51]

The catalytic efficiency of ABTS **1** can be expressed by the ratio i_k/i_c, where i_k (catalytic current) is the anodic peak current of ABTS **1** in the presence of lignin, and i_c (diffusion current) is the anodic peak current of ABTS **1** alone.[21,27,38] According to this relationship, the catalytic efficiencies of the two oxidised species

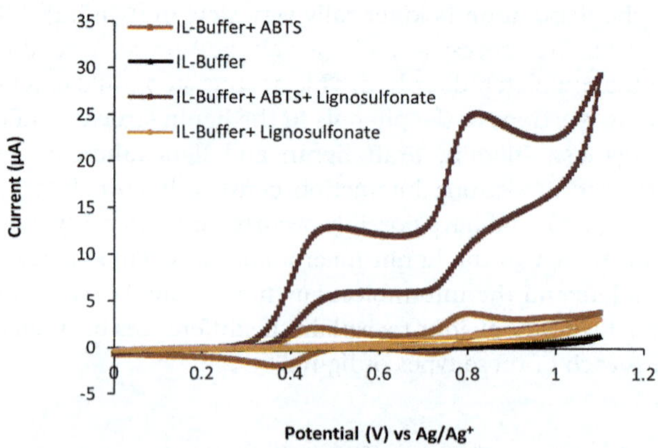

Fig. 13 Cyclic voltammograms of 1 mM ABTS **1**, 1 wt% lignosulfonate and 1 wt% lignosulfonate mixed with 1 mM ABTS **1**, in 15% (v/v) [C_2mim][C_2SO_4] in 0.1 M sodium acetate buffer (pH 4.5) recorded at a scan rate of 10 mV s^{-1}.

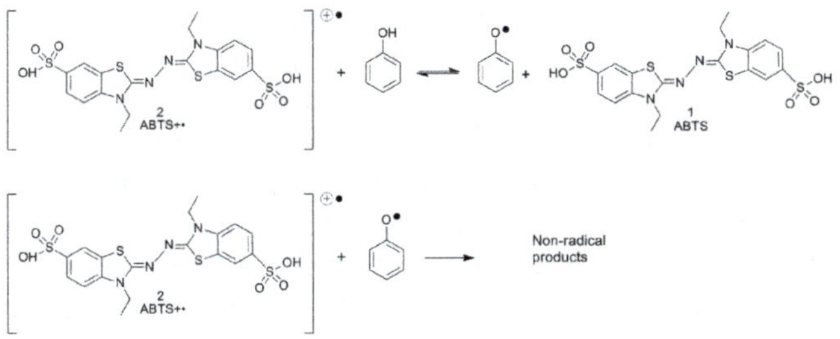

Scheme 2 Proposed reaction scheme for the oxidation of phenol with ABTS cation radical **2**.

of ABTS, **2** and **3**, in the presence of lignosulfonate in 15% (v/v) IL/buffer solution were calculated. On this basis, the anodic current corresponding to the ABTS radical cation **2** increased by 5.9 fold and the ABTS dication **3** by 6.7 fold, reflecting that the oxidation capability of ABTS dication **3** is higher than that of the ABTS cation radical **2**.

The absence of the cation radical peak as well as the presence of only one small reduction peak were observed in the CVs of lignosulfonate oxidation (1 wt%) by ABTS **1** in the buffer solution (Fig. 14). Noticeably, the dication peak rose strongly in the buffer solution, its catalytic efficiency being intensified by 17.6 fold. The strong current increase at 809 mV indicates that the reaction between $ABTS^{2+}$ **3** and lignosulfonate is relatively fast compared to the reaction with ABTS radical cation **2**.

Electrochemical oxidation of ABTS **1** at a concentration of 1.0 mM was reversible in the buffer solution at the scan rate of 10 mV s^{-1} (Fig. 13). However, when the scan rate was increased to 400 mV s^{-1}, ABTS **1** (1.0 mM) exhibited a more reversible oxidation process (data not shown). This can be explained by the comproportionation reaction between ABTS **1** and its dication **3** to give two equivalents of the radical cation **2** at the slower scan rate. This ABTS **1** behaviour

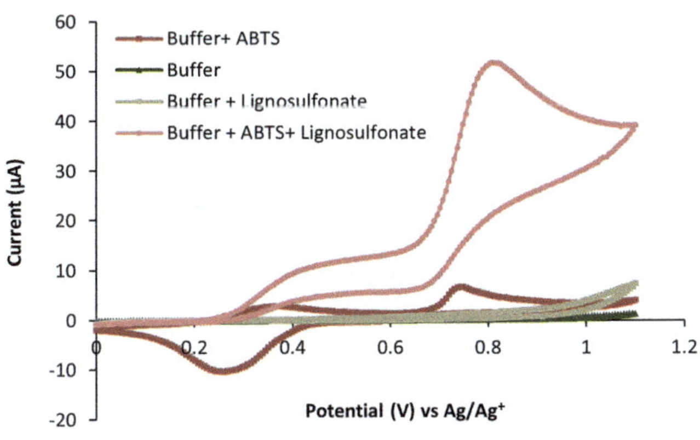

Fig. 14 Cyclic voltammograms of 1 mM ABTS **1**, 1 wt% lignosulfonate and 1 wt% lignosulfonate mixed with 1 mM ABTS **1**, in 0.1 M sodium acetate buffer (pH 4.5) recorded at a scan rate of 10 mV s^{-1}.

was also found in ionic liquid and discussed previously in the context of the electrochemical oxidation behaviour of ABTS **1** in $[C_2mim][C_2SO_4]$.

It is likely that the viscosity effect could be the reason behind the high increases in the anodic peak currents of ABTS **1** in the presence of lignosulfonate in the buffer solution, compared with 15% (v/v) $[C_2mim][C_2SO_4]$/buffer solutions, where the viscosity of the buffer is around half that of the mixture containing 15% (v/v) IL (Table 1). Through this decrease in viscosity, a significant increase, about two-fold, in the diffusion coefficient of ABTS **1** in the buffer solution would promote mass transfer at the electrode surface–bulk solution interface and thus also increase the catalytic efficiencies.

3.5. Cyclic voltammograms of organosolv lignin in buffer/$[C_2mim][C_2SO_4]$ solutions of different concentrations

The reaction between ABTS **1** and organosolv lignin at different concentrations of sodium phosphate buffer (pH 12) in ionic liquid was also explored. As shown in Fig. 15, the addition of buffer to the IL resulted in an increase in the second oxidation peak of ABTS **1**. The likely reason behind the initial increase in the dication **3** peak current is a dilution effect, whereby the viscosity of the IL is reduced, resulting in improved mass transfer. This is consistent with the similar electrochemical behaviour seen in solutions of 5% (v/v) buffer/IL, 7% (v/v) buffer/IL and 10% (v/v) buffer/IL, since small quantities of water are known to dramatically affect viscosity. For example, addition of only 5% (v/v) of water to $[C_2mim][C_2SO_4]$ could lead to a 30% reduction in the IL viscosity.[37,52] Increasing the concentration of phosphate buffer by more than 10% (v/v) lowered the peak current measured for ABTS **1**. The reduction in peak current was most likely caused by a solubility effect, since organosolv lignin is insoluble in the sodium phosphate buffer (pH 12). A layer of organosolv lignin could have then formed at the working electrode surface, hence either slowing or preventing ABTS **1** from accessing the electrode. These results, taken together, emphasise that viscosity

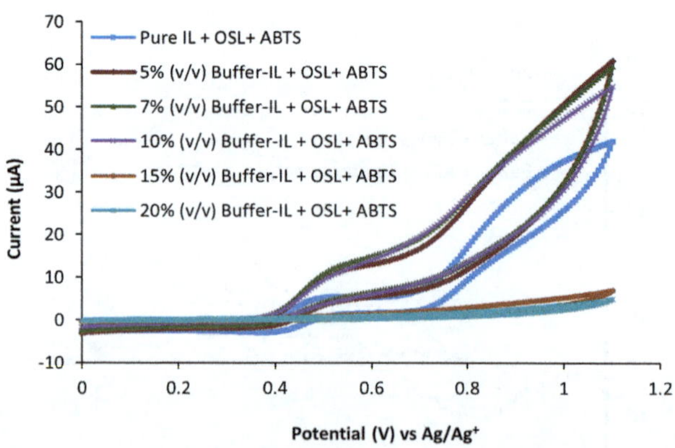

Fig. 15 Cyclic voltammograms of organosolv lignin (OSL) (5 wt%) mixed with 10 mM ABTS **1** in pure $[C_2mim][C_2SO_4]$, 5% (v/v) buffer (pH 12)/$[C_2mim][C_2SO_4]$, 7% (v/v) buffer/$[C_2mim][C_2SO_4]$, 10% (v/v) buffer/$[C_2mim][C_2SO_4]$, 15% (v/v) buffer/$[C_2mim][C_2SO_4]$ and 20% (v/v) buffer/$[C_2mim][C_2SO_4]$, recorded at a scan rate of 50 mV s^{-1}.

and solubility are important driving factors to consider in lignin-based electrochemical oxidation reactions.

4. Conclusions

Electrochemical oxidation has been explored for the potential degradation of three types of lignin (lignosulfonate, Kraft and organosolv) in $[C_2mim][C_2SO_4]$, utilising the mediator ABTS **1**. The electrochemical behaviour of ABTS **1** alone in $[C_2mim][C_2SO_4]$ was examined and is quite similar to that previously seen in buffered aqueous solutions, in terms of proportionality of peak currents to the square root of the scan rate and the ratio of anodic to cathodic peak currents.

The oxidation currents in the CVs for the formation of either or both the ABTS radical cation **2** and the ABTS dication **3** were observed to increase after addition of all types of lignin, proving ABTS **1** to be an effective mediator for the electrochemical oxidation of lignin. However, the use of ABTS **1** was found to affect more significantly the oxidation of organosolv lignin, which is consistent with the greater solubility of this lignin in this IL over the other types of lignin examined in this study. Similarly, solubility of this lignin decreased dramatically at around 10% (v/v) buffer concentration and resulted in a concomitant reduction in the oxidation peak current.

Solubility of lignin and viscosity of the solvent or solution are likely to play a vital role in the electrochemical oxidation reaction of lignin. The viscosities and interactions of ionic liquid–lignin solutions will influence the mass transfer of ABTS **1** from the electrode surface into the bulk solution, and IL–solute interactions have the potential to limit the electron transfer kinetics, as has been shown for other reaction types.[53–55] Overall, the rates of heterogeneous and homogeneous electron-transfer reactions are lower in $[C_2mim][C_2SO_4]$ compared to those measured in the buffer solution. The CV studies of ABTS mediator **1** with lignin demonstrate a high level of catalytic efficiency toward lignosulfonate in buffer (dication 17.6 fold) and 15% (v/v) $[C_2mim][C_2SO_4]$/buffer (cation 5.9 fold and dication 6.7 fold) but there was no increase of catalytic efficiency associated with the presence of $[C_2mim][C_2SO_4]$. A future study investigating the molecular interactions between ABTS **1**/ionic liquid/lignin and lignin/ionic liquid would help to clarify the influence of ionic liquids, in terms of their solubilisation of lignin components, specific solute–solvent interactions, and viscosity, and delineate their impact on reactivity.

Finally, one of the more significant findings emerging from this study is that electrochemical techniques can be quickly and easily applied to study ionic liquid/lignin/mediator systems for the prediction and selection of an appropriate ionic liquid and mediator combination for the lignin degradation process. In the future this may lead to optimised electrochemical processes for improved utilisation of lignolytic biomass in an energy efficient manner.

Acknowledgements

Majd Eshtaya is grateful to the Islamic Development Bank (IDB) and the University of Nottingham for a doctoral fellowship grant and An-Najah National university for support. AKC acknowledges the support of COST programme CM1206 EXIL. GZC thanks the EPSRC (EP/J000582/1) for funding of the electrochemical equipment used in this work.

Notes and references

1 U. Bornscheuer, K. Buchholz and J. Seibel, *Angew. Chem., Int. Ed.*, 2014, **53**, 10876–10893.

2 W. Boerjan, J. Ralph and M. Baucher, *Annu. Rev. Plant Biol.*, 2003, **54**, 519–546.

3 G. Chatel and R. D. Rogers, *ACS Sustainable Chem. Eng.*, 2014, **2**, 322–339.

4 P. Azadi, O. R. Inderwildi, R. Farnood and D. A. King, *Renewable Sustainable Energy Rev.*, 2013, **21**, 506–523.

5 T.-Q. Yuan, F. Xu and R.-C. Sun, *J. Chem. Technol. Biotechnol.*, 2013, **88**, 346–352.

6 E. Reichert, R. Wintringer, D. A. Volmer and R. Hempelmann, *Phys. Chem. Chem. Phys.*, 2012, **14**, 5214–5221.

7 G. W. Huber, S. Iborra and A. Corma, *Chem. Rev.*, 2006, **106**, 4044–4098.

8 K. Barta, G. R. Warner, E. S. Beach and P. T. Anastas, *Green Chem.*, 2014, **16**, 191–196.

9 H. Zhu, L. Wang, Y. Chen, G. Li, H. Li, Y. Tang and P. Wan, *RSC Adv.*, 2014, **4**, 29917–29924.

10 A. Toledano, L. Serrano and J. Labidi, *J. Chem. Technol. Biotechnol.*, 2012, **87**, 1593–1599.

11 H. Wang, M. Tucker and Y. Ji, *J. Appl. Biochem.*, 2013, **2013**, 9.

12 A. Hatakka, in *Biopolymers Online*, Wiley-VCH Verlag GmbH & Co. KGaA, 2005, DOI: 10.1002/3527600035.bpol1005.

13 F. J. Ruiz-Duenas and A. T. Martinez, *J. Microb. Biotechnol.*, 2009, **2**, 164–177.

14 T. D. H. Bugg, M. Ahmad, E. M. Hardiman and R. Rahmanpour, *Nat. Prod. Rep.*, 2011, **28**, 1883–1896.

15 M. Lahtinen, L. Viikari, P. Karhunen, J. Asikkala, K. Kruus and I. Kilpeläinen, *J. Mol. Catal. B: Enzym.*, 2013, **85–86**, 169–177.

16 R. Bourbonnais, M. G. Paice, I. D. Reid, P. Lanthier and M. Yaguchi, *Appl. Environ. Microbiol.*, 1995, **61**, 1876–1880.

17 D. Rochefort, D. Leech and R. Bourbonnais, *Green Chem.*, 2004, **6**, 14–24.

18 R. Bourbonnais and M. G. Paice, *FEBS Lett.*, 1990, **267**, 99–102.

19 E. Aracri, T. Tzanov and T. Vidal, *Ind. Eng. Chem. Res.*, 2013, **52**, 1455–1463.

20 L. P. Christopher, B. Yao and Y. Ji, *Frontiers in Energy Research*, 2014, **2**, 1–13.

21 R. Bourbonnais, D. Leech and M. G. Paice, *Biochim. Biophys. Acta, Gen. Subj.*, 1998, **1379**, 381–390.

22 O. V. Morozova, G. P. Shumakovich, S. V. Shleev and Y. I. Yaropolov, *Appl. Biochem. Microbiol.*, 2007, **43**, 523–535.

23 D. Rochefort, R. Bourbonnais, D. Leech and M. G. Paice, *Chem. Commun.*, 2002, 1182–1183, DOI: 10.1039/b202621j.

24 M. Fabbrini, C. Galli and P. Gentili, *J. Mol. Catal. B: Enzym.*, 2002, **16**, 231–240.

25 G. P. Shumakovich, S. V. Shleev, O. V. Morozova, P. S. Khohlov, I. G. Gazaryan and A. I. Yaropolov, *Bioelectrochemistry*, 2006, **69**, 16–24.

26 M. Díaz-González, T. Vidal and T. Tzanov, *Appl. Microbiol. Biotechnol.*, 2011, **89**, 1693–1700.

27 T. Shiraishi, Y. Sannami, H. Kamitakahara and T. Takano, *Electrochim. Acta*, 2013, **106**, 440–446.

28 A. Kunamneni, F. J. Plou, A. Ballesteros and M. Alcalde, *Recent Pat. Biotechnol.*, 2008, **2**, 10–24.

29 M. Lorenzo, D. Moldes, S. Rodríguez Couto and M. Sanromán, *Chemosphere*, 2005, **60**, 1124–1128.

30 S. Shipovskov, H. Q. N. Gunaratne, K. R. Seddon and G. Stephens, *Green Chem.*, 2008, **10**, 806–810.

31 A. P. M. Tavares, O. Rodriguez and E. A. Macedo, *Biotechnol. Bioeng.*, 2008, **101**, 201–207.

32 S. H. Lee, T. V. Doherty, R. J. Linhardt and J. S. Dordick, *Biotechnol. Bioeng.*, 2009, **102**, 1368–1376.

33 N. Sun, M. Rahman, Y. Qin, M. L. Maxim, H. Rodriguez and R. D. Rogers, *Green Chem.*, 2009, **11**, 646–655.

34 P. Mäki-Arvela, I. Anugwom, P. Virtanen, R. Sjöholm and J. P. Mikkola, *Ind. Crops Prod.*, 2010, **32**, 175–201.

35 H. Zhao, *J. Chem. Technol. Biotechnol.*, 2010, **85**, 891–907.

36 A. Chen, E. I. Rogers and R. G. Compton, *Electroanalysis*, 2010, **22**, 1037–1044.

37 N. Harwardt, N. Stripling, S. Roth, H. Liu, U. Schwaneberg and A. C. Spiess, *RSC Adv.*, 2014, **4**, 17097–17104.

38 K. González Arzola, M. C. Arévalo and M. A. Falcón, *Electrochim. Acta*, 2009, **54**, 2621–2629.

39 H. Liu, L. Zhu, M. Bocola, N. Chen, A. C. Spiess and U. Schwaneberg, *Green Chem.*, 2013, **15**, 1348–1355.

40 K. Stärk, N. Taccardi, A. Bösmann and P. Wasserscheid, *ChemSusChem*, 2010, **3**, 719–723.

41 G. A. Snook, A. S. Best, A. G. Pandolfo and A. F. Hollenkamp, *Electrochem. Commun.*, 2006, **8**, 1405–1411.

42 K. R. J. Lovelock, A. Ejigu, S. F. Loh, S. Men, P. Licence and D. A. Walsh, *Phys. Chem. Chem. Phys.*, 2011, **13**, 10155–10164.

43 S. O'Toole, S. Pentlavalli and A. P. Doherty, *J. Phys. Chem. B*, 2007, **111**, 9281–9287.

44 I. Schroder, E. Steckhan and A. Liese, *J. Electroanal. Chem.*, 2003, **541**, 109–115.

45 A. Rahimi, A. Ulbrich, J. J. Coon and S. S. Stahl, *Nature*, 2014, **515**, 249–252.

46 S. Nanayakkara, A. F. Patti and K. Saito, *Green Chem.*, 2014, **16**, 1897–1903.

47 I. Cybulska, G. Brudecki, K. Rosentrater, J. L. Julson and H. Lei, *Bioresour. Technol.*, 2012, **118**, 30–36.

48 A. George, K. Tran, T. J. Morgan, P. I. Benke, C. Berrueco, E. Lorente, B. C. Wu, J. D. Keasling, B. A. Simmons and B. M. Holmes, *Green Chem.*, 2011, **13**, 3375–3385.

49 R. Bourbonnais and M. Paice, *J. Electrochem. Soc.*, 2004, **151**, E246–E249.

50 A. M. Campos and E. A. Lissi, *Int. J. Chem. Kinet.*, 1997, **29**, 219–224.

51 E. Matsumura, E. Yamamoto, A. Numata, T. Kawano, T. Shin and S. Murao, *Agric. Biol. Chem.*, 1986, **50**, 1355–1357.

52 H. Rodriguez and J. F. Brennecke, *J. Chem. Eng. Data*, 2006, **51**, 2145–2155.

53 H. M. Yau, S. J. Chan, S. R. D. George, J. M. Hook, A. K. Croft and J. B. Harper, *Molecules*, 2009, **14**, 2521–2534.

54 H. M. Yau, S. T. Keaveney, B. J. Butler, E. E. L. Tanner, M. S. Guerry, S. R. D. George, M. H. Dunn, A. K. Croft and J. B. Harper, *Pure Appl. Chem.*, 2013, **85**, 1979–1990.

55 H. M. Yau, S. A. Barnes, J. M. Hook, T. G. A. Youngs, A. K. Croft and J. B. Harper, *Chem. Commun.*, 2008, 3576–3578, DOI: 10.1039/b805255g.

Faraday Discussions

PAPER

Molten-salt treatment of waste biomass for preparation of carbon with enhanced capacitive properties and electrocatalytic activity towards oxygen reduction†

Beihu Lu,‡[a] Jing Zhou,‡[a] Yuqiao Song,[a] Hailong Wang,[b] Wei Xiao*[a] and Dihua Wang*[a]

Received 9th December 2015, Accepted 6th January 2016

DOI: 10.1039/c5fd00215j

Carbon powders are building blocks for electrochemical energy storage/conversion devices. Green, cost-affordable and facile preparation of carbon with applicable electrochemical properties is therefore essential for effective utilization of fluctuating renewable energy. Herein, the preparation of carbon nanoflakes *via* impregnation of waste biomass *i.e.* boiled coffee beans in molten Na_2CO_3–K_2CO_3 (with equal mass) at 800 °C and molten $CaCl_2$ at 850 °C is reported. The microstructure and surface chemistry of the obtained carbons are specified. The correlations between synthetic conditions and microstructure/surface chemistry of the obtained carbons are rationalized. The derived carbon nanosheets are tested and compared as active materials for supercapacitors in a configuration of symmetric full cells in 1 M $MeEt_3NBF_4$ in acetonitrile and electrocatalysts towards the oxygen reduction reaction (ORR) in O_2-saturated 0.1 M aqueous KOH. Despite the lower surface area, the carbon nanosheets derived in molten Na_2CO_3–K_2CO_3 exhibit enhanced capacitive properties and electrocatalytic ORR activity. The present study highlights the importance of thermal media on the microstructure, surface chemistry and electrochemistry of carbon from biomass.

Introduction

Inorganic high-temperature molten salts (MSs) are highly regarded in terms of energy, environmental and resource sustainability.[1] Due to enhanced reaction kinetics at elevated temperatures, MS is capable of sustaining high space-time

[a]*School of Resource and Environmental Sciences, Hubei International Scientific and Technological Cooperation Base of Sustainable Resource and Energy, Wuhan University, Wuhan 430072, PR China. E-mail: gabrielxiao@whu.edu.cn; wangdh@whu.edu.cn; Fax: +86 27 68775799; Tel: +86 27 68775799*

[b]*Advanced Energy Storage Materials and Devices Lab, School of Physics Electrical Information Engineering, Ningxia University, Yinchuan 750021, China*

† Electronic supplementary information (ESI) available. See DOI: 10.1039/c5fd00215j

‡ Equal contribution.

yields for materials preparation/processing and triggering reactions at appreciable rates without the assistance of costly and sophisticated catalysts.[2] MSs generally possess strong capability for dissolving and/or absorbing reactants, facilitating desired reactions and removal of inconvenient impurities.[2,3] Their large electrochemical windows render MSs as versatile electrolytes for electrochemical metallurgy and electrosynthesis.[4,5] As ionic liquids with less solvation than aqueous solutions, species tend to occur in less solvated states in MSs. This hence provides new reaction characteristics. The present cations and/or anions can also be used as reactants for materials preparation and modifications.[6,7] Their high polarity is also beneficial for the formation of nanostructured materials with unusual physicochemical properties.[6,8] As thermal media with high heat capacity, MS processes can be easily integrated into solar-thermal grids to attain high energy efficiency.[9,10] Last but not least, their low cost, low toxicity and technological maturity (*e.g.* Al extraction industry) ensure good affordability and environmental benignity for MS processes.

Carbon powders are building blocks for electrochemical energy conversion and storage devices such as supercapacitors, batteries, fuel cells and metal–air batteries. Preparation of carbon powders in MSs have recently become a research hot spot. Fray *et al.* pioneered electrochemical preparation of carbon nanotubes by electrochemical intercalation of cations from molten chlorides for exfoliation of graphite electrodes.[7,11–14] Up-scalable preparation of graphene was recently reported by the same group by electrolysis of graphite in molten LiCl, in which intercalation of *in situ* generated hydrogen facilitates the exfoliation of layered graphite and hence the formation graphene nanosheets.[15] A more recent innovation combining carbon preparation and CO_2 capture in MS electrolysis was reported, in which CO_2 is firstly absorbed in molten carbonates and/or molten chlorides and thereafter effectively electro-reduced to carbon in the same MSs without the presence of any costly electrode materials.[2,9,16–20] Carbon powders with different microstructures including nanoparticles,[2] sheets,[16] fibres,[16,19] and hollow structured carbon[18] were successfully generated in such a MS electrolysis.

Besides MS electrolysis, MS can also be used as a reaction media for carbon preparation and/or modifications. Using MS as a thermal medium for carbonization/pyrolysis of carbon-rich biomass in molten carbonates and molten chlorides was recently reported.[3] MSs can also function as structure directing agents/modifiers for tailoring the microstructure and therefore physicochemical properties of carbon materials. Molten carbonates were reported to be an effective re-activation agent for altering the surface chemistry of commercial activated carbon and significantly enhanced capacitance of the treated carbon was reported and rationalized.[21] Controlled preparation of exfoliated carbon sheets, pitted particles and carbon nanorods was successful *via* molten chloride corrosion of graphite.[22] Preparation of nanodiamonds in a relatively moderate manner was fulfilled by simply heating Li_2CO_3-containing carbon nanostructures in molten chlorides at atmospheric pressure.[23]

Generally, carbon powder is quite stable in MSs, which can be regarded as an inert liquid-phase thermal media. Compared with combustion in gas-phase media, carbonization of biomass in MSs tends to cause less carbon emission. Such inert thermal media also have strong capability for dissolving electrochemically inert impurities (*e.g.* silica in biomass), ensuring high purity of MS-derived carbon. The agglomeration of ultrafine carbon into large-sized carbon is retarded in high-temperature MSs due to insufficient delocalized electrons in metalloid carbon. MSs containing an abundance of ions can also restrain the aggregation/coalescence ultrafine carbon. These help to retain the nanostructure of carbon in MSs. Importantly, the presence of ions in MSs is helpful in engineering surface chemistry of carbon and accelerating the process of breaking chemical bonds between precursors.[24]

The above successes for the MS preparation of carbon highlight the advantages of MSs, which also provokes investigation on the applicability of MS-generated carbon for electrochemical energy storage/conversion devices. In this sense, specifying correlations between parameters of MS treatment and microstructure of MS-derived carbon is of importance. Herein, the preparation of carbon nanoflakes *via* the impregnation of biomass, *i.e.* boiled coffee beans, in two kinds of typical inorganic molten salts *viz.* Na_2CO_3–K_2CO_3 with equal mass at 800 °C and $CaCl_2$ at 850 °C is investigated. The microstructure and surface chemistry of the obtained carbons are carefully characterized and specified. After rationalization of the correlations between synthetic conditions and microstructure/surface chemistry of the obtained carbons, the derived carbon nanosheets were tested as active materials for supercapacitors in a symmetric full-cell configuration in 1 M $MeEt_3NBF_4$ in acetonitrile and electrocatalysts towards the oxygen reduction reaction (ORR) in 0.1 M aqueous KOH. Despite the lower specific surface area, the carbon nanosheets derived in molten Na_2CO_3–K_2CO_3 exhibited enhanced capacitive properties and electrocatalytic ORR activity than those derived in molten chlorides. The present study highlights the importance of thermal media on the microstructure, surface chemistry and electrochemistry of carbon from biomass.

Experimental

Commercial coffee beans (Arabica coffee beans, LIM's drop coffee roasters, blue mountain flavoring) were purchased from supermarkets. $CaCl_2$, Na_2CO_3 and K_2CO_3 with purities higher than 99.9% purity were purchased from Sinopharm Chemical Regent Co. Ltd., China. The purchased coffee beans were then boiled in deionized water for 1 hour. Then the boiled coffee beans were thoroughly rinsed in deionized water and then dried in air at 120 °C overnight. The derived powder was directly used in the following MS treatment. 500 g anhydrous $CaCl_2$ or Na_2CO_3–K_2CO_3 (equal in mass, eutectic point at 709 °C) contained in an alumina crucible were firstly heated at 250 °C for 24 h. The alumina crucibles were then placed and sealed in a stainless steel reactor, in which argon gas (>99.999%) was continuously purged to guarantee an anaerobic environment. The MS reactor was then heated to the target temperatures (800 °C for Na_2CO_3–K_2CO_3 and 850 °C for $CaCl_2$). After the MS melted, 20 g boiled coffee beans wrapped with a nickel foam sheet were immersed in the MS. After MS treatment for 1 h, the MS-treated samples were washed with 0.1 M HCl, hot deionized water and deionized water until a neutral pH value was reached. The samples were then dried in air at 120 °C. The as-obtained carbons were then characterized, with the experimental details being provided in the ESI.†

Results and discussion

Fig. 1a shows the XRD patterns of the samples derived upon treatment in the molten carbonates (800 °C Na_2CO_3–K_2CO_3) and molten chlorides (850 °C for

$CaCl_2$), respectively. In both patterns, only two broad and weak peaks occur, indicating poor crystallinity of the two samples. The peaks located at around 22 degrees are assigned to a (002) reflection of a graphitic-type lattice, whilst the shoulders occurring at 43 degrees are indexed to disorder stacking of micro-graphite.[21] The discrepancy on peak intensity and width suggests a higher crystallinity and more ordered structure of the sample treated in molten carbonates.

Fig. 1b shows the Raman spectra of the two samples. Two characteristic peaks ascribed to D and G bands of carbon appear in both spectra. The two spectra can be well indexed to disordered carbon. The D band located at 1360 cm^{-1} only becomes active in the presence of disorder. The G band reflecting the relative motion of sp^2 carbon atoms is ascribed to ordered carbon. In the spectrum for the sample treated in molten chlorides, the intensity of the D band is higher than that of the G band, whilst a more intense G band than D band appears in the spectra for the molten carbonated treated sample. This suggests that the sample treated in molten carbonates has a more ordered carbon structure, in a good agreement with the XRD patterns.

Surface functional groups over the MS-treated samples were examined by FT-IR. Fig. 1c displays the FT-IR spectra of the precursor (boiled coffee beans) and MS-treated samples. A typical O–H stretching vibration peak at 3400 cm^{-1} corresponding to hydroxyl groups appears in the three samples. The peak at 2920 cm^{-1} corresponds to the –CH–, $-CH_2-$ or $-CH_3$ stretching. The peak at 1620 cm^{-1} may be ascribed to the presence of coupled carbonyl groups and aromatic ring stretching. The peak located at 1420 cm^{-1} corresponds to the C–C stretching vibration. The peak centered at 1060 cm^{-1} is associated with the C–O vibration of peroxide and/or heterocyclic C=O rings. After MS treatment, all the peaks belonging to organic species become much smaller, indicating occurrence of carbonization of the precursory biomass. The peak at 613 cm^{-1} only appears in the molten chloride treated sample, probably attributed to the C–Cl vibration due

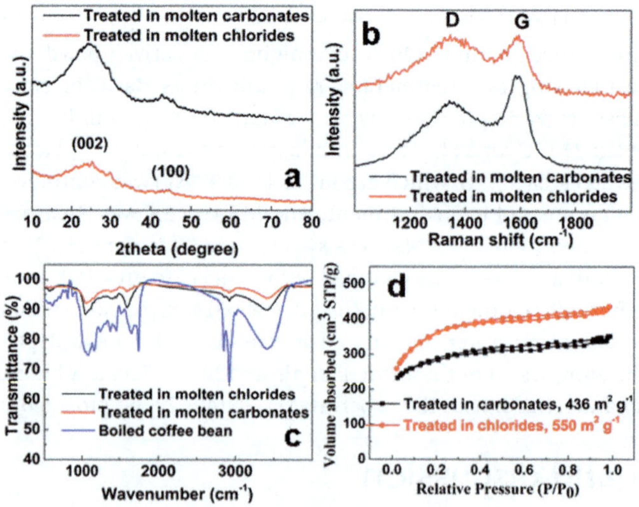

Fig. 1 Powder XRD patterns (a), Raman spectra (b), FT-IR spectra (c) and N_2 adsorption/desorption isotherms (d) of the precursory boiled coffee beans and derived carbons.

to the presence of trace residual chlorides.[3] In comparison with the molten chlorides treated sample, the spectrum of the molten carbonate treated sample shows lower intensities of the surface functional groups. This suggests fewer surface functional groups in the molten carbonates treated sample. The presence of surface functional groups generally prompts formation of less ordered carbon, agreeing well with the Raman spectra.

N_2 adsorption/desorption isotherms at 77 K of the two MS treated samples are shown in Fig. 1d. The two isotherms are of typical type I isotherms with clear H4-type hysteresis loops according to the IUPAC classification.[21] The hysteresis loop for the MS carbonate treated sample terminates at a lower relative pressure than that of the sample treated in molten chlorides, indicating more micropores in the former. The BET surface areas are determined to be 436 and 550 $m^2\ g^{-1}$ for the molten carbonate and chloride treated samples, respectively. A high specific surface area occurs in the sample treated in molten chloride.

Surface chemistry of the MS treated samples is further examined by XPS. As shown in Fig. 2, all the XPS survey spectra (Fig. 2a1, b1 and c1) only show the peaks of C 1s and O 1s. After MS treatment, the O 1s peaks become significantly weakened, indicating carbonization upon MS treatment.

Table 1 summarizes all the XPS data. The survey spectra also indicate a much smaller oxygen content for the sample treated in molten carbonates, compared to

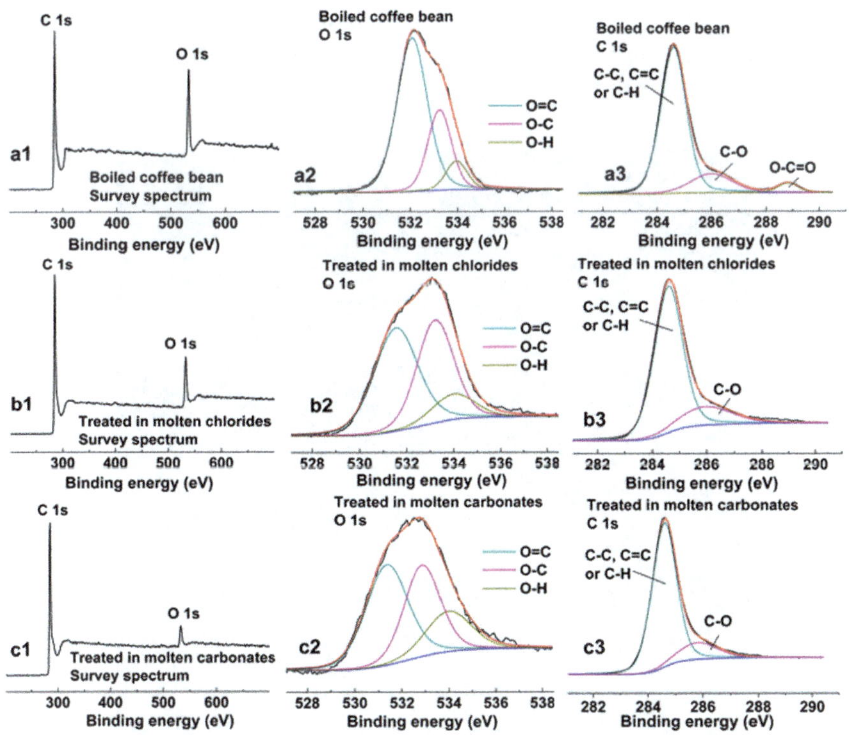

Fig. 2 Survey XPS spectra (a1, b1 and c1), high-resolution O 1s spectra (a2, b2 and c2) and high-resolution C 1s spectra (a3, b3 and c3) of the precursory boiled coffee beans (a1, a2 and a3) and the samples after being treated in molten chlorides (b1, b2 and b3) and molten carbonates (c1, c2 and c3).

Table 1 Summary on the XPS data (all percentages are exhibited as atomic ratios)

	C 1s spectra			O 1s spectra			Survey spectra
	C–C, C=C or C–H	C–O	O–C=O	O=C	O–C	O–H	O/C
Boiled coffee beans	79.14%	16.61%	4.25%	65.08%	26.31%	8.61%	29%
Treated in molten chlorides	81.20%	18.80%	N.A.	48.47%	40.01%	11.52%	17%
Treated in molten carbonates	84.20%	15.80%	N.A.	46.34%	33.76%	19.90%	8%

that treated in molten chlorides. For high-resolution C 1s spectra (Fig. 2a3, b3 and c3), the peaks located at 284.6, 286.0 and 288.8 eV are ascribed to C–C/C=C/C–H, C–O and O–C=O bonds. The O–C=O peak only exists in the precursor, due to the presence of bulk organic species. Smaller C–O signals appear in the carbonate-treated sample, suggesting fewer oxygen-containing surface functional groups in the sample. All the high-resolution O 1s spectra can be de-convolved into three peaks appearing at 531.6, 533.2 and 534.0 eV, which are assigned to O=C, O–C and O–H bonds. Smaller O=C signals exist in the carbonated treated sample, indicating its higher purity. FI-IR and XPS confirm the carbonization of biomass upon MS treatment. It is also verified that oxygen-containing groups including O=C, O–C and O–H exist at the surface of MS-derived carbon and fewer oxygen-containing groups appear in the sample treated in molten carbonates.

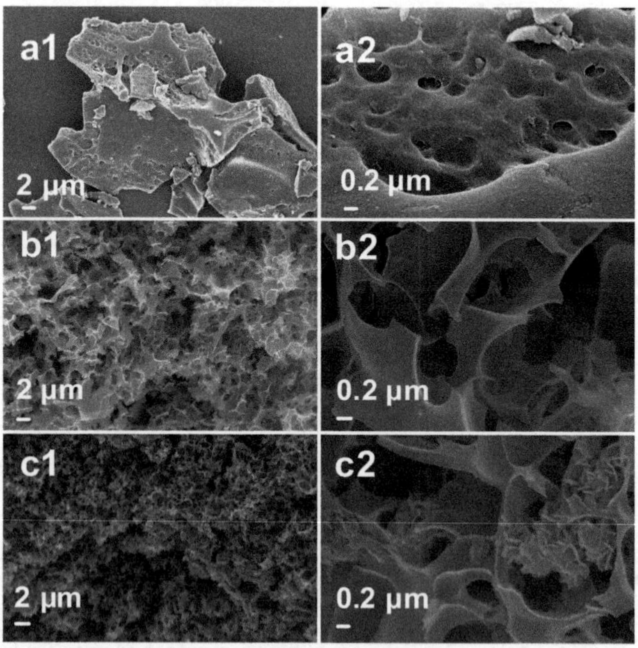

Fig. 3 SEM images of the precursory boiled coffee beans (a), the samples after being treated in molten carbonates (b) and in molten chlorides (c).

SEM images of the samples are shown in Fig. 3. It is exhibited in Fig. 3a that the precursory boiled coffee beans consist of less porous granules with sizes ranging from 5 to 30 μm. After MS treatment, the samples become much more porous and nanoflakes becomes the predominant morphology (Fig. 3b and c). The generated nanoflakes are interconnected, with pores existing in the interstitial sites. Decomposition of biomass results in such interstitial sites, with hydrates and other organic species becoming released from the solid. The remaining carbonized species then restack in the MSs, generating the obtained nanoflakes. By comparing Fig. 3b1 and c1 (in the same magnitude of magnification), it is clearly shown that the chloride-treated sample has a smaller particle size. Therefore the chloride-treated sample shows a higher BET surface area. Fig. 3b2 shows the formation of uniform nanoflakes after molten carbonate treatment. The same nanoflakes also appear in the molten chloride treated sample (Fig. 3c2), with co-existence of more tiny nanoflakes. The formation of smaller nanoflakes also contributes to the higher specific surface area of the sample.

The molten carbonate treated sample was further examined by AFM. As shown in Fig. 4 and S1 (see ESI†), the generated nanoflakes are of thicknesses ranging from 30 to 80 nm. A graphite-like lamellar structure is clearly shown in the surface of the nanoflakes. The corresponding line-scan plots disclose the fine microstructures of the nanoflakes. As shown in the AFM images, the region near the line 1 has a larger thickness and the surface of such thicker nanoflakes consists of an ordered layered microstructure. The thickness of an individual layer is determined to be 2–3 nm. This indicates the generated amorphous carbon tends to restack into a more ordered structure in MSs. Such restacking is more manifest on the surface of the nanoflakes, with inner parts retaining the disordered structure. Line 2 represents such a disordered structure. It is found that the line 2 part exhibits a consistent thickness of 25 nm and the layered microstructure is absent in the disordered region.

The above results confirm the successful MS carbonization of biomass. The corresponding characterizations show that the chloride-treated carbon has a higher specific surface area, more surface functional groups and less ordered microstructure. Such difference is believed to be relevant to the nature of the MSs. It is well known that $CaCl_2$ is a strong desiccant due to its intrinsic affinity to moisture. Even when anhydrous $CaCl_2$ is employed, hydrolysis of $CaCl_2$ is still significant and the hydrolysis products, *i.e.* $Ca(OH)_2$ and HCl tend to dissolve (calcium hydroxide) or become released (HCl) in/from the melts. Therefore, the hydrolysis of molten $CaCl_2$ is inevitable in practical applications. In addition, the generated $Ca(OH)_2$ is readily decomposed to CaO and H_2O. The former has high solubility in the melts,[5,8,15,25–27] and the latter could bring about further hydrolysis. Therefore, molten $CaCl_2$ always contains appreciable amounts of oxygen anions (O^{2-}). In contrast, molten carbonates only show very minor hydrolysis effects, if any. Therefore, much less O^{2-} occurs in molten carbonates. The presence of O^{2-} increases the corrosive capability of the melts, therefore many more tiny nanoflakes coexist in the carbon derived in molten $CaCl_2$ and a higher specific surface area of the sample appears. The O^{2-} could also modify the surface of the derived carbon, with the formation of surface functional groups. Therefore, more oxygen-containing surface functional groups occur over the surface of the chlorides-treated sample. The oxygen-containing surface functional groups tend to retard

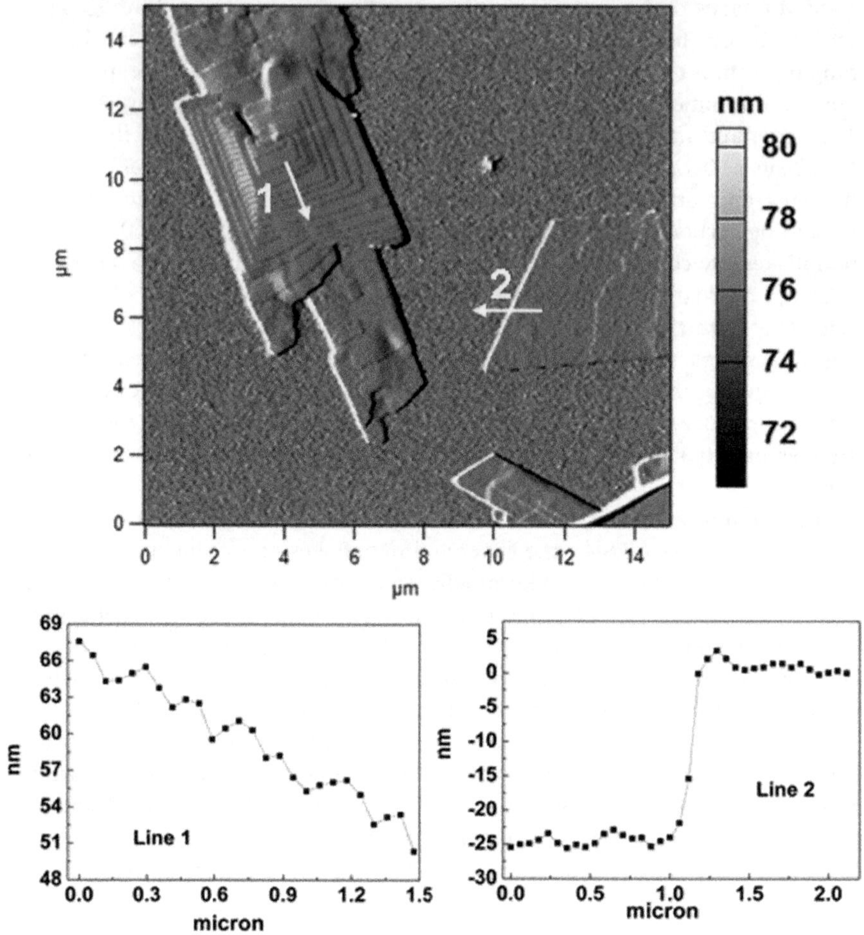

Fig. 4 AFM images and the corresponding line-scanning plots of the sample derived in molten carbonates.

the restacking of the amorphous carbon, resulting in the formation of less ordered carbon in the chlorides-treated sample.

Carbon possesses a predominant share in commercial supercapacitor markets. The capacitive properties of the as-obtained carbon are thus tested in a symmetric full-cell configuration in 1 M $MeEt_3NBF_4$ in acetonitrile ($MeEt_3NBF_4$/AN). Cyclic voltammetric (CV) and galvanostatic charge–discharge (GCD) scans at different rates are recorded at voltages ranging from 0 to 2 V. Fig. 5a and b show the CV plots based on the two MS-derived carbon powders. For the molten carbonate-derived carbon capacitor, all CVs are of ideal rectangular shapes, even under a high scan rate of 200 mV s^{-1}. This suggests an excellent high-rate capability of the sample. The specific capacitances derived from CVs are 155, 154, 152, 142 and 132 F g^{-1} for scans at 10, 20, 50, 100 and 200 mV s^{-1} (Fig. 5e). With increasing scan rates from 10 to 200 mV s^{-1}, capacitance retention of 85.16% is achieved. Compared to the carbonate-derived carbon capacitor, the CVs of the chloride-derived carbon capacitor (Fig. 5b) exhibit much lower current density

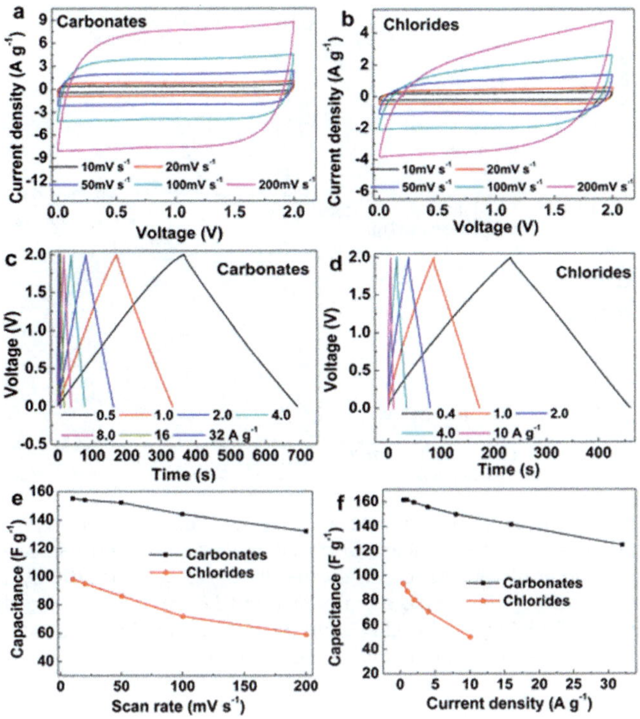

Fig. 5 Cyclic voltammetric (a and b) and galvanostatic charge–discharge plots (c and d) for the symmetric full-cell capacitors in 1 M $MeEt_3NBF_4$/AN based on carbon powders derived in molten carbonates (a and c) and molten chlorides (b and d). The panels (e) and (f) show the comparison on specific capacitance according to the CV and GCD plots. The loadings of active materials in each single electrode are 2.30 and 2.48 mg for the carbonate and chloride samples, respectively.

and more distorted CV shapes, indicating lower specific capacitance and inferior high-rate capability. As shown in Fig. 5e, the specific capacitances derived from CVs of the chloride-derived carbon are 98, 95, 86, 72 and 59 F g^{-1} for scans at 10, 20, 50, 100 and 200 mV s^{-1} respectively (Fig. 5e). Capacitance retention as low as 60.20% occurs, much lower than that of the carbonate-derived sample (85.16%).

Fig. 5c, d and f compare the GCD results of the two samples, which agree well with the CV data. At 0.5 A g^{-1}, the carbonate-derived carbon delivers a specific capacitance as high as 161 F g^{-1}. Even under a rate as high as 32 A g^{-1}, a capacitance of 125 F g^{-1} remains, highlighting its excellent high-rate capability. Much lower specific capacitance and inferior high-rate performance appear in the chloride-derived carbon. Under 0.4 A g^{-1}, a capacitance as low as 93.4 F g^{-1} occurs in the chloride-derived sample. This value sharply decreases to 50 F g^{-1} with increasing charge–discharge current density to 10 A g^{-1}.

Energy and power densities of the assembled symmetric full cell in 1 M $MeEt_3NBF_4$/AN are also demonstrated in the Ragone plots shown in Fig. 6a. The Ragone plots are derived from the GCD plots shown in Fig. 5c and d (see ESI† for details). It is clearly shown that much higher energy and power density occurs in the molten carbonate derived carbon capacitor. The maximal energy density for

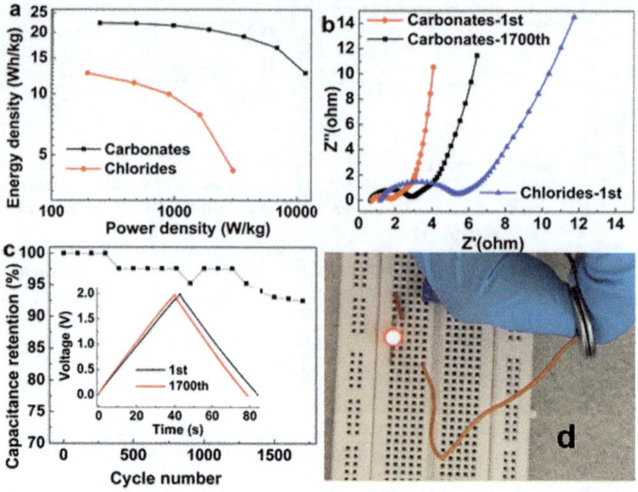

Fig. 6 Ragone plots (a) and EIS (b) of the symmetric full-cell capacitors in 1 M $MeEt_3NBF_4$/AN. The panel c shows the cycling performance of the molten-carbonates full-cell capacitor recorded at galvanostatic current density of 4 A g^{-1} for 1700 cycles, with the galvanostatic charge–discharge plots at the first and 1700th cycles being shown in the inset. The panel (d) highlights the LED lamp which got illuminated driven by the two connected-in-serial molten-carbonates full-cell capacitors.

the capacitor is as high as 22.27 W h kg^{-1}. Only a maximal energy density of 12.77 W h kg^{-1} appears for the molten chloride derived carbon capacitor, with a power density occurring at 197 W kg^{-1}. At a similar energy density (12.67 W h kg^{-1}), the molten carbonate derived carbon capacitor delivers a power density as high as 11 696.4 W kg^{-1}.

Fig. 6b shows the corresponding electrochemical impedance spectra (EIS) of the full cells. The loadings of active materials in each single electrode are 2.30 and 2.48 mg for the carbonate and chloride samples, respectively. Due to the different loadings for each capacitor, the resistance of the chloride-derived capacitor was normalized by a factor of 2.48/2.30. The intercept with the x axis at high frequency ranges indicates the equivalent serial resistance (ESR). The ESRs for the two capacitors in the 1st cycle are determined to be 0.65 and 1.17 ohm for the carbonate and chloride derived carbon capacitors, respectively. A lower ESR for the carbonate-derived capacitor suggests a higher electrical conductivity of the carbon. The semi-cycles in the EIS represent charge-transfer resistance at electrochemical interfaces, with larger charge-transfer resistance occurring at larger semi-cycles. It is found that the carbonate-derived capacitor exhibits a much smaller charge-transfer resistance.

The above electrochemical measurements confirm the enhanced capacitive properties of the carbonate-derived carbon to the chloride-derived sample, despite larger specific surface area of the latter. This can be rationalized according to the difference in surface chemistry. The carbon derived from molten chloride contains more oxygen-containing surface functional groups, which tend to deteriorate the electrical conductivity. In addition, surface functional groups incur formation of more disordered carbon, which possesses lower electrical conductivity. Therefore, the carbon derived from molten chloride exhibits a lower electrical conductivity than that derived from molten carbonates, as evidenced by the much higher ESR of the chloride-derived capacitor. For electrochemical double layer capacitors, the capacitance is proportional to the reciprocal of the distance between of the outer Helmholtz plane (OHP) and electrode surface. Surface functional groups tend to increase the distance of OHP. Therefore, a lower specific capacitance and a higher charge-transfer resistance are observed for the chloride-derived carbon capacitor. In this regard, molten carbonates are more promising media for carbonization of biomass for capacitive carbon, agreeing well with recent literature regarding enhanced capacitive properties of commercial carbon by molten carbonates treatment.[21]

The cycling performance of the molten carbonate derived carbon capacitor was tested upon GCD at 4 A g^{-1} for 1700 cycles. As can be seen in Fig. 6c, a capacitance of 92.5% is retained after the consecutive cycling, suggesting excellent cycle performance of the carbon. As shown in the EIS, the ESR slightly increases from 0.65 to 0.71 ohm after cycling for 1700 times, indicating excellent stability and durability of the molten carbonate derived carbon. An appreciable increase in charge-transfer resistance appears after cycling, mainly due to irreversible oxidation of the partial carbon surface.

We attempted to test the practical applicability of the symmetric full-cell capacitor based on the molten carbonate derived carbon. The results are shown in Fig. 6d and the video enclosed in the ESI.† A LED lamp, which requires voltages higher than 3.6 V to become illuminated, was initially connected into an electrical circuit. Two full cells were used to illuminate the LED. Each individual coin full cell was initially charged to 2.0 V by CV scanning at a scan rate of 20 mV s^{-1}. After being fully charged to 2.0 V, the two charged cells were quickly connected in serial and connected to the electric circuit. Upon connection, the LED lamp became illuminated and the duration of illumination state lasted for more than 1 minute.

Carbons are also building blocks for electrocatalysts applicable in fuel cells and metal–air batteries. The two MS derived carbons were then tested as

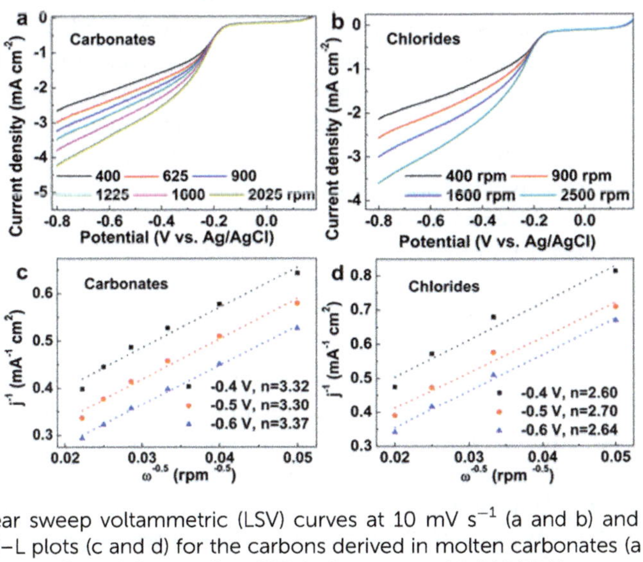

Fig. 7 Linear sweep voltammetric (LSV) curves at 10 mV s^{-1} (a and b) and the corresponding K–L plots (c and d) for the carbons derived in molten carbonates (a and c) and molten chlorides (b and d) regarding ORR in O_2 saturated 0.1 M KOH.

electrocatalysts towards the oxygen reduction reaction (ORR) in O_2-saturated 0.1 M aqueous KOH under hydrodynamic conditions (see ESI† for details).

As can be seen in the LSVs shown in Fig. 7a and b, a higher ORR current density occurs in the carbon derived from molten carbonate treatment. The K–L analysis of the LSVs of the two carbons at different rotating rates is shown in Fig. 7c and d, which was used to quantitatively understand the ORR activity (see ESI† for details).[28] All the K–L plots exhibited good linearity. According to the K–L equations, electron transfer numbers per oxygen molecule (n) could be estimated from the slopes. An average electron transfer number of 3.33 occurred in the molten carbonate derived carbon, suggesting a predominance of the direct four-electron path for ORR over the sample. A smaller electron transfer number, *i.e.* 2.65 appears for the molten chloride derived carbon, indicating higher ORR kinetic barriers in the sample. Such a number also indicates the predominance of the two-electron ORR pathway in the sample. The enhanced electrocatalytic ORR activities of the molten carbonate-derived carbon can be addressed as follows. As mentioned above, the sample exhibits higher electrical conductivity than that of carbon derived from molten chlorides. Importantly, the molten chloride derived carbon possesses more oxygen-containing surface functional groups, which tend to retard the adsorption of O_2 over the surface of carbon. An enhanced electrocatalytic activity towards the ORR hence appears in the molten chloride derived carbon.

Conclusions

In summary, carbonization of a typical waste biomass, *i.e.* boiled coffee beans, was both achieved in molten Na_2CO_3–K_2CO_3 (with equal mass) at 800 °C and molten $CaCl_2$ at 850 °C. Characterizations verified that carbon powder derived from molten chlorides was of higher specific surface area, and possessed more oxygen-containing surface functional groups and a less ordered microstructure. This can be ascribed to the hydrolysis nature of $CaCl_2$, leading to more O^{2-} in molten $CaCl_2$. O^{2-} present in the melts resulted in more oxygen-containing surface groups and a less ordered microstructure for the MS derived carbon, which in turn caused deteriorated electrical conductivity, larger outer Helmholtz plane distances at electrochemical interfaces, and higher kinetic barriers for O_2 adsorption over carbon. The above synthetic parameter–microstructure design rationales for MS carbonization of biomass towards carbon preparation for advanced electrochemical energy storage/conversion devices were evidenced by the enhanced capacitive properties and electrocatalytic activities towards ORR of the carbon derived from molten carbonate treatment.

Acknowledgements

This work was funded by the National Natural Science Foundation of China (51325102), the International Science & Technology Cooperation Program of China (ISTCPC, 2015DFA90750), the Program for Creation Team of Hubei Province (2015CFA017), Hubei Biomass-Resource Chemistry and Environmental Biotechnology Key Laboratory and the Young-talent Chenguang Project of Wuhan City. We thank the West Light Foundation of The Chinese Academy of Sciences for support.

References

1 F. Lantelme and H. Groult, in *Molten Salts Chemistry: From Lab to Applications*, ed. F. L. Groult, Elsevier, Oxford, 2013, ISBN: 978-0-12-398538-5.
2 H. Yin, X. Mao, D. Tang, W. Xiao, L. Xing, H. Zhu, D. Wang and D. R. Sadoway, *Energy Environ. Sci.*, 2013, **6**, 1538–1545.
3 H. Y. Yin, B. H. Lu, Y. Xu, D. Y. Tang, X. H. Mao, W. Xiao, D. H. Wang and A. N. Alshawabkeh, *Environ. Sci. Technol.*, 2014, **48**, 8101–8108.
4 A. M. Abdelkader, K. T. Kilby, A. Cox and D. J. Fray, *Chem. Rev.*, 2013, **113**, 2863–2886.
5 W. Xiao and D. Wang, *Chem. Soc. Rev.*, 2014, **43**, 3215–3228.
6 W. Xiao, W. Liu, X. Mao, H. Zhu and D. Wang, *Electrochim. Acta*, 2013, **88**, 756–765.
7 G. Z. Chen, X. D. Fan, A. Luget, M. S. P. Shaffer, D. J. Fray and A. H. Windle, *J. Electroanal. Chem.*, 1998, **446**, 1–6.
8 W. Xiao, X. Jin and G. Z. Chen, *J. Mater. Chem. A*, 2013, **1**, 10243–10250.
9 S. Licht, *Adv. Mater.*, 2011, **23**, 5592–5612.
10 S. Licht, B. H. Wang, S. Ghosh, H. Ayub, D. L. Jiang and J. Ganley, *J. Phys. Chem. Lett.*, 2010, **1**, 2363–2368.
11 R. Das Gupta, C. Schwandt and D. J. Fray, *Carbon*, 2014, **70**, 142–148.
12 A. T. Dimitrov, G. Z. Chen, I. A. Kinloch and D. J. Fray, *Electrochim. Acta*, 2002, **48**, 91–102.
13 I. A. Kinloch, G. Z. Chen, J. Howes, C. Boothroyd, C. Singh, D. J. Fray and A. H. Windle, *Carbon*, 2003, **41**, 1127–1141.
14 C. Schwandt, A. T. Dimitrov and D. J. Fray, *J. Electroanal. Chem.*, 2010, **647**, 150–158.
15 A. R. Kamali and D. J. Fray, *Nanoscale*, 2015, **7**, 11310–11320.
16 H. V. Ijije, R. C. Lawrence and G. Z. Chen, *RSC Adv.*, 2014, **4**, 35808–35817.
17 H. V. Ijije, R. C. Lawrence, N. J. Siambun, S. M. Jeong, D. A. Jewell, D. Hu and G. Z. Chen, *Faraday Discuss.*, 2014, **172**, 105–116.
18 H. V. Ijije, C. G. Sun and G. Z. Chen, *Carbon*, 2014, **73**, 163–174.
19 J. W. Ren, F. F. Li, J. Lau, L. Gonzalez-Urbina and S. Licht, *Nano Lett.*, 2015, **15**, 6142–6148.
20 J. W. Ren, J. Lau, M. Lefler and S. Licht, *J. Phys. Chem. C*, 2015, **119**, 23342–23349.
21 B. Lu, Z. Xiao, H. Zhu, W. Xiao, W. Wu and D. Wang, *J. Power Sources*, 2015, **298**, 74–82.
22 A. R. Kamali and D. J. Fray, *Carbon*, 2013, **56**, 121–131.
23 A. R. Kamali and D. J. Fray, *Chem. Commun.*, 2015, **51**, 5594–5597.
24 B. J. Hathaway, J. H. Davidson and D. B. Kittelson, *J. Sol. Energy Eng.*, 2011, **133**, 021011.
25 W. Xiao, X. B. Jin, Y. Deng, D. H. Wang and G. Z. Chen, *J. Electroanal. Chem.*, 2010, **639**, 130–140.
26 W. Xiao, X. B. Jin, Y. Deng, D. H. Wang, X. H. Hu and G. Z. Chen, *ChemPhysChem*, 2006, **7**, 1750–1758.
27 W. Xiao, X. Wang, H. Y. Yin, H. Zhu, X. H. Mao and D. H. Wang, *RSC Adv.*, 2012, **2**, 7588–7593.
28 J. Zhou, H. Xiao, B. Zhou, F. Huang, S. Zhou, W. Xiao and D. Wang, *Appl. Surf. Sci.*, 2015, **358**, 152–158.

References

Faraday Discussions

DISCUSSIONS

Benefits to energy efficiency and environmental impact: general discussion

Toru H. Okabe, Olga Kuzmina, John M. Slattery, Cairong Jiang, Tim Sudmeier, Linpo Yu, Han Wang, Wei Xiao, Liang Xu, Xiangling Yue, Yiyang Kong, Andrew Doherty, Geir Martin Haarberg, Shuqiang Jiao, Qian Xu, Hongmin Zhu, Dihua Wang, Paul Madden, Daniel Cooper, Kathie McGregor, Chaohui Wei, Binjie Hu, Andrew Mount, John Irvine, Ali Kamali, Babak Khalaghi, Xianbo Jin, Yingjun Liu, George Zheng Chen, Xingli Zou, Gang Chen, Ye Liu, Majd Eshtaya, Derek Fray and Yating Yuan

DOI: 10.1039/c6fd90030e

Andrew Mount opened a discussion of the paper by Derek Fray†: To what extent in the processes you outline can you tolerate water, as ideally you want precise control of the environment but this is often not achievable in industrial processes?

Derek Fray replied: There are several industrial processes using molten salts on a large scale. In the case of aluminium smelting no special precautions are taken to exclude water vapour from the air. In the electrolysis of lithium chloride, the cells are sealed as liquid lithium is highly reactive and chlorine is toxic.

George Zheng Chen responded: In the FFC Cambridge Process, air is prevented from entering the cell. However, water contained in the molten salt is a problem but can be removed by pre-electrolysis to a very low level, say below 10 ppm.

Andrew Mount addressed George Zheng Chen and Derek Fray: To what extent does the generation of highly reactive species in these molten salt industrial processes lead to corrosion of the containers? How important is this in limiting the lifetime and efficiency of the processes?

Derek Fray responded: In the case of aluminium smelting, a solidified layer of salt is used to contain the electrolyte. For the lithium cell, a refractory lining which does not react with the salt is chosen. Generally, I believe an alumina based refractory is used.

Qian Xu asked: Sixteen years have passed since the FFC Cambridge process was published in *Nature* in 2000. Would you please talk about its future on the road to the commercial-scale production?

† Derek Fray's paper was presented by George Zheng Chen, *University of Nottingham Ningbo, Ningbo, China*. All responses were communicated in reply.

George Zheng Chen answered: In terms of commercial development of the FFC Cambridge Process, you can check the webpage of the UK company, Metalysis (http://metalysis.com/). In fact, the paper presented by Prof. Juanyu Yang later in this Faraday Discussion shows another successful example of the development of the FFC Cambridge Process.

John Irvine commented: Another important technology involving molten salts is high temperature primary batteries. This is an established technology offering high power and long shelf life.

Andrew Mount said: What are the challenges in the field of liquid salt batteries/charge storage systems? What are the step changes in technological development that are required?

John Irvine answered: Molten salt batteries offer high power density in particular. Capacity is good but could be improved. Safety can be a concern with hot, corrosive liquids. There is one area of particular challenge and promise, the combination of heat and electrical storage in the same system. Molten salts can be utilised for heat storage, there are good opportunities to store electrical and thermal energy simultaneously.

George Zheng Chen asked: Some traditional batteries are very cheap and work well, such as the nickel-iron battery that has a theoretical energy capacity of 267 W h kg^{-1} and was first demonstrated more than 100 years ago.[1–3] Can we do something more with such classic batteries using liquid salts? On the other hand, the recently proposed direct carbon fuel cell (DCFC)[4–6] is a promising development. It can be in principle combined with electrolytic carbon dioxide reduction to carbon in molten salts.[7–9] If this electrolytic process is applied at large scale using renewable energy or low cost valley electricity (considering the huge amount of CO_2 emitted from power generation and other metallurgical and materials production processes), the electrolytic carbon can be partly used for materials purposes, and partly as the fuel in the DCFC to regenerate electricity. It can of course be used in combustion for heat or propelling power. In the latter case, the electrolytic carbon can be seen as an energy store of high energy density that can release the energy stored thermally or electrochemically for seasonal application, for instance. A consequence of such a combination is a rechargeable DCFC. When charging, it takes away CO_2 from the atmosphere, and when discharging, it gives the flexibility to release the energy stored in different seasons and locations. The key material for such a prospect is liquid salts, particularly at elevated temperatures that help overcome the kinetic barriers of the processes.

1 T. A. Edison., Br Patent, 1900, 20960.
2 C. Chakkaravarthy, P. Periasamy, S. Jegannathan and K. I. Vasu, *J. Power Sources*, 1991, **35**, 21–35.
3 H. Wang, Y. Y. Liang, M. Gong, Y. G. Li, W. Chang, T. Mefford, J. G. Zhou, J. Wang, T. Regier, F. Wei and H. J. Dai, *Nature Commun.*, 2012, **3**, 917.
4 N. Nakagawa and M. Ishida, *Ind. Eng. Chem. Res.*, 1988, **27**, 1181–1185.
5 K. Pointon, B. Lakeman, J. Irvine, J. Bradley and S. Jain, *J. Power Sources*, 2006, **162**, 750–756.
6 Y. Nabae, K. D. Pointon and J. T. S. Irvine, *Energy Environ. Sci.*, 2008, **1**, 148–155.

7 S. Licht, B. Wang, S. Ghosh, H. Ayub, D. Jiang and J. Ganley, *J. Phys. Chem. Lett.*, 2010, **1**, 2363–2368.
8 N. J. Siambun, H. Mohamed, D. Hu, D. Jewell, Y. K. Beng and G. Z. Chen, *J. Electrochem. Soc.*, 2011, 158, H1117–H1124.
9 H. Y. Yin, X. H. Mao, D. Y. Tang, W. Xiao, L. R. Xing, H. Zhu, D. H. Wang and D. R. Sadoway, *Energy Environ. Sci.*, 2013, **6**, 1538–1545.

Dihua Wang replied: I totally agree with this idea. This is a good way for harvesting renewable energy and is carbon neutral.

John Irvine said: It is worth noting that battery and fuel cell technologies are basically the same thing – fuel cells have external fuel stores, whereas in batteries fuel is internal. Regenerative fuel cells are equivalent to secondary batteries.

George Zheng Chen answered: I fully agree.

Paul Madden said: To what extent does the dissolution of metals in molten salts and the consequential electronic conduction through the melt limit possible technologies? Are the physical/chemical principles which determine the solubility of metals in particular salts understood?

George Zheng Chen answered: It is known that alkali and alkaline earth metals can dissolve into their salts, particularly halide salts to a significant concentration, and cause electronic conduction through the molten salts. In the FFC Cambridge Process for titanium extraction, the electronic conduction is the main cause for the current efficiency being lower than 50%. Some early papers published by Haarberg have offered explanation for the dissolution of alkali and alkaline earth metals in their own chloride salts. Prof. Haarberg may offer a better answer to this question.

Paul Madden addressed George Zheng Chen and Geir Martin Haarberg: To illustrate the previous point about technologies being limited by metal dissolution – the liquid metal battery developed by Sadoway involves a liquid Mg electrode in contact with $MgCl_2$ indicating that the dissolution of Mg in $MgCl_2$ is sufficiently low for the cell not to be electronically shorted. My understanding is that this would not work in a Ca/$CaCl_2$ couple – do we understand why there is this difference between (apparently similar) chemical systems?

Geir Martin Haarberg replied: The solubility of Mg in $MgCl_2$ is much lower than the solubility of Ca in $CaCl_2$. There is no electronic conduction associated with dissolved Mg. Heavier elements of alkali and alkaline earth systems show higher solubilities and more electronic conduction.

Andrew Mount asked: Has the shorting mechanism of the Mg–$MgCl_2$ battery been predicted/modelled and how can this be addressed?

Geir Martin Haarberg answered: The solubility of Mg in MgCl is very low, and there is no electronic conduction associated with dissolution of Mg, so shorting should not be an important issue.

Hongmin Zhu said: The solubility of metal in molten salts takes an important role in the electrochemical process. When we want to make metal by electrolysis of molten salts, we need low solubility of the metal in the molten salts so that we can lower the metallic conductivity and have enough current efficiency. That is the same case as for a battery. An example is the electrolysis of $CaCl_2$ to make Ca. Because of the high solubility of metallic Ca in molten $CaCl_2$, the current efficiency of the electrolysis is always very low, while in the case of $MgCl_2$ electrolysis, the current efficiency is much higher, because the solubility of Mg in $MgCl_2$–KCl melt is low. The dissolution of metal would be helpful for the reduction in the molten salts. We can easily reduce most oxide in molten $CaCl_2$ by Ca because Ca dissolves in the melt and reaches to the surface of the oxide. In this case the solubility will promote the reaction.

Dihua Wang remarked: Using oxides as feeding materials, we now have two major processes for electrolytic production of metals in molten salts. One is electrodeposition/electrowinning such as aluminium production by dissolving alumina into the cryolite melt in which the high solubility of alumina in the melt is the key point. Another way is electroreduction of oxides in the solid state in $CaCl_2$ based molten salts in which the solubility of oxides is quite low. There also exists a situation where the oxides have a limited but remarkable solubility in the melt, for example, silica in CaO containing $CaCl_2$. A mixture of solid-reduction and dissolution–deposition process will take place under this situation. Do you have any comments on the relative contribution of the two processes?

George Zheng Chen answered: The mixed reduction mechanism could be feasible for extraction of metals with low affinity to oxygen, or low solubility for oxygen, such as Si, but not for more reactive metals such as Ti and Zr. Nevertheless, I cannot see any technical advantage of purposely adding CaO in molten $CaCl_2$. Also, when the CaO concentration is high in molten $CaCl_2$, the current efficiency could be significantly reduced because of two connected factors. Firstly, the electrolysis current will increase resulting from the decomposition of CaO. Secondly, the decomposition produces Ca metal which dissolves in the melt, leading to electronic conduction.

Dihua Wang remarked: When we talk about a high or low solubility of oxides in molten salts, it might depend on the specific application or process. For electrowinning such as aluminium production, the solubility might be higher than 5 wt%; for the solid state reduction process, it might be lower than 0.05 wt%. For corrosion prevention of materials in molten salts, it might be expected to be even lower.

Andrew Mount asked: Would the study of nanostructured/nanoporous materials enable faster reduction (through increased solubility and shorter pore diffusion lengths)?

George Zheng Chen responded: The particle size of metal oxides should be reasonably small. Past experience shows sub-micrometres to be sufficiently good. Further reducing the oxide particle size may not lead to a significant gain because the reduction speed is more controlled by ion transport through the pores of the oxide cathode.

Kathie McGregor commented: The real advantage of the FFC process is that by reducing the metal compound in the solid-state, parasitic side reactions for highly reactive metals such titanium are avoided. Reactions such as the disproportionation of Ti(III) or formation of carbides, nitrides *etc.* occur widely for species in solution, setting up cyclic redox reactions and diminishing current efficiency.

George Zheng Chen answered: I fully appreciate and agree with this comment from Prof. McGregor.

Xianbo Jin commented: For electrolysis of solid oxide in molten salt, the porous structure of the oxide cathode is very important, and different metal oxides prefer different porosities. For example, to produce Ti, if TiO_2 is sintered to a dense solid, only the uppermost surface layer can be reduced to Ti. Things become different if you reduce silica: dense quartz can be easily reduced to Si. The reasons include the molar volume ratio between the metal and the oxide, and the sintering of metal. Dense silicon oxide to Si gives a very big porosity, but dense TiO_2 to Ti gives a very low or almost zero porosity, so after the initial reduction, a dense Ti layer will cover the TiO_2 pellet, preventing the diffusion of oxygen ions through the electrolyte in the pores. We have developed a PRS (porosity, metal to oxide molar volume ratio, sintering of metal) model to predict the optimized porosity for the electrolysis of every metal oxide.

Hongmin Zhu remarked: When you are talking about the diffusion for the rate determining step, what is the diffusion species? It seems to me that there are two possible species, one is the oxide ion (O^{2-}), the other would be dissolved Ca. Since O^{2-} is the product of the reaction, its diffusion is not necessarily the "rate determining step".

Xianbo Jin answered: This is a good comment. The reduction speed for solid oxide electrolysis in molten $CaCl_2$ finally will be determined by the O^{2-} ion diffusion. Regarding whether the possible dissolved Ca needs to be considered, first, I would like to say that the assumption of oxygen ion diffusion can well explain the observed experimental results. Second, in our model electrolysis of SiO_2, the applied potential is much more positive (> 0.7 V) than the deposition potential of metallic Ca, suggesting an activity value less than 10^{-6} mol L^{-1}, which is too low to support the large reduction current we observed during the electrolysis. However, no matter whether Ca takes part in the reduction of metal oxide or not, the complete reduction at a relatively high speed will be dependent on the O^{2-} ion diffusion in the solid oxide cathode. And thirdly, below I will discuss the influence of the diffusion of the reaction product (here O^{2-} ions) on the reduction of metal oxide to metal. Generally, an electrode reaction will have its speed determined by the current overpotential equation:

$$i = i^0\left[\frac{C_O^S}{C_O^*}\exp\left(\frac{\alpha nF(E_{eq}-E)}{RT}\right) - \frac{C_R^S}{C_R^*}\exp\left(\frac{(\alpha-1)nF(E_{eq}-E)}{RT}\right)\right]. \quad (1)$$

Where C_O^S and C_R^S are concentrations of oxidant and reductant at the electrode surface respectively, and C_O^* and C_R^* are reference concentrations of oxidant and reductant respectively, which determine an equilibrium potential E_{eq} through the Nernst equation. In case metal oxide is reduced to metal, there is:

$$i = i^0\left[\exp\left(\frac{\alpha nF(E_{eq}-E)}{RT}\right) - \frac{C_R^S}{C_R^*}\exp\left(\frac{(\alpha-1)nF\eta(E_{eq}-E)}{RT}\right)\right]. \tag{2}$$

Where C_R^S is the concentration of O^{2-} ions at the three phase electrochemical interlines (or reaction zone), and C_R^* is a reference concentration associated with an E_{eq} value. Eqn (2) indicates the influence of C_R^S (or concentration of reaction product) on the total reduction current, and the solution of eqn (2) suggests that when the charge transfer rate is sufficiently large, after the initial reduction of the surface layer of an oxide pellet, the reduction speed will be controlled by the diffusion of O^{2-} ions. For details, please refer to our earlier publications.[1,2] It seems to be true that i^0 is sufficiently large, because we observe a reduction current of about 80 A cm^{-2} for the reduction of surface silica where the concentration polarization is insignificant,[3] the quasi-equilibrium conditions can be always established, thus $C_R^S \approx C_R^*$ and

$$i = nFk^0 C_R^{*\,\alpha-1}(C_R^* - C_R^S). \tag{3}$$

Eqn (3) suggests that a net reduction current can only be observed when C_R^S is smaller than C_R^* that is determined by the equilibrium potential. However, since it is C_R^S which determines the applied potential under equilibrium conditions, the net reduction can only be realized by removal of the generated O^{2-} ions away from the reaction zone, most probably, through diffusion.

On the other hand, net reduction can also happen under non-equilibrium conditions if the applied potential becomes more negative than the equilibrium potential which determines a C_R^S equal to the concentration of O^{2-} in the SiO_2 pellet (~40 mol L^{-1} for dense silica or ~20 mol L^{-1} for the pellet with 50% porosity if all the ions can dissolve into the electrolyte), or actually the saturation concentration of O^{2-} ions in $CaCl_2$ (~4.2 mol L^{-1} at 850 °C), because precipitation of CaO would occur.[1,4] Unfortunately, the CaO precipitation will block the electrolyte passage, and the following reduction will depend on the diffusion out of O^{2-} ions with a saturation concentration in the inner part of the solid pellet.

On the contrary, if i^0 is so low that at a relatively high overpotential the second term of eqn (2) becomes negligible, there could be no need to consider the diffusion of O^{2-} ions. However, in such a case, the reduction current will be significantly lower than the above diffusion current, which will not be acceptable for electrolysis. Thus, no matter whether i^0 is high or low, the highest electrolysis speed that we pursue will be determined by the diffusion of O^{2-} ions. We have developed a PRS model based on O^{2-} ion diffusion for prediction the maximum electrolysis speeds of various metal oxides, for details, please see H. L. Chen, *et. al.*[5]

1 W. Xiao, X. Jin, Y. Deng, D. Wang, X. Hu and G. Z. Chen, *Chem. Phys. Chem.*, 2006, **7**, 1750–1758.
2 W. Xiao, X. Jin, Y. Deng, D. Wang and G. Z. Chen, *Chem. -Eur . J.*, 2007, **13**, 604–612.

3 X. B. Jin, P. Gao, D. Wang, X. Hu and G. Z. Chen, *Angew. Chem. Int. Ed.*, 2004, **43**, 733 –736.
4 R. Bhagat, D. Dye, S. L. Raghunathan, R. J. Talling, D. Inman, B. K. Jackson, K. K. Rao and R. J. Dashwood, *Acta Materialia*, 2010, **58**, 5057–5062.
5 H. L. Chen, Y. Zeng, W. Li, J. Peng, X. Jin and G. Z. Chen, *Electrochem. Commun.*, 2013, **26**, 33–36.

George Zheng Chen opened a discussion of the paper by Katherine McGregor: If Nb in your reported inert anode material is replaced by Ta, what happens?

Kathie McGregor replied: The Dow Chemical Company investigated both Nb- and Ta-doped TiO_2 inert anodes for production of magnesium *via* an electrolytic route, and reported some success with both materials. We also prepared a number of pressed bars from high purity, crystalline Ta-doped TiO_2, but the bars were not very conductive and the preliminary electrolysis results were poor. For interest, the results are included in the paper as ESI. As tantalum is far more costly than niobium, and the Nb-doped bars were producing very encouraging preliminary results, we decided not to continue with the tantalum work at that point.

George Zheng Chen remarked: The conductivity of your inert material results from oxygen deficiency or vacancies as in some semiconductor materials, which are related to the available valences of the element. Nb can be stable at the V, IV, II and 0 valences, but Ta is only stable at V and 0. Therefore, it would be difficult to use Ta to create an oxygen vacancy in the material.

On the other hand, you mentioned that removing oxygen without CO_2 emission can only be achieved *via* discharging the oxygen ion at an inert anode in molten salt electrolysis. I would like to draw your attention to our paper that will be presented here later (DOI: 10.1039/c5fd00231a). In the paper, we report a way to keep the oxygen in the cathode as MgO, instead of discharging it at the anode.

Kathie McGregor replied: Dow Chemical Company demonstrated that at low concentrations and in a reducing atmosphere, Ta can also be used to dope TiO_2 and form a conductive material. The electronic conductivity arises from the formation of Ti^{3+} sites in the rutile structure. The paper presented during the Faraday Discussion by Prof. Chen and co-workers is an excellent contribution and represents an innovative new approach to the challenge.

John Slattery said: The SEM image in Fig. 2b seems to show some pores in the generation 2 material. Can you comment on whether or not there is some porosity in this material and if so, what the significance of this might be?

Kathie McGregor replied: Typical density for the pressed Nb-doped titania bars after firing was 4.1 g cm^{-3}, which is 97% of the theoretical density (Archimedean method). It is clear from the SEM images that there are some pores present, and these are presumed to be mostly closed pores. This is not ideal, and further optimization of the preparation and sintering methods would presumably improve the material's performance.

John Irvine asked: In the reduced Nb-doped titania can you be certain that Ti^{3+} is produced rather than Nb^{4+}, or both?

Kathie McGregor replied: We have analysed the Nb-doped titania anode materials *via* electron spin paramagnetic spectroscopy, and have established it is very likely that the paramagnetic species present is Ti^{3+}, not Nb^{4+}. The results are unpublished, but we expect to publish them in the near future.

John Irvine asked: Did you consider pre-reducing the niobium titanates to form rutile Nb, TiO_2, Ebonex, *i.e.* titana reduced to Ti_4O_7 is well known as a stable electrode materials in these types of applications.

Kathie McGregor replied: We have performed detailed studies with Magneli phase electrode materials such as Ti_4O_7, using both in-house and commercially obtained materials. The best-performed electrode had a lifetime of no more than 8 hours. Detailed *ex-situ* electron microprobe and *in-situ* XRD studies demonstrated that the failure mechanism was oxidation to TiO_2. This is a poorly conducting as an anode material even at elevated temperatures. For more details see the references listed below.[1–4]

1 G. A. Snook, M. R. Rowles, M. J. Styles, K. McGregor, I. M. Madsen, A. J. Urban, N. V. Y. Scarlett and D. P. Riley, *Electrochem. Soc. Trans.*, 2013, **50**, 45–56.
2 M. R. Rowles, M. J. Styles, I. C. Madsen, N. V. Y. Scarlett, K. McGregor K, D. P. Riley, G. A. Snook, A. J. Urban, T. Connolley and C. Reinhard, *J. Appl. Cryst.*, 2012, **45**, 28–37.
3 M. R. Rowles, N. V. Y. Scarlett, I. C. Madsen and K. McGregor, *J. Appl. Cryst.*, 2010, **44**, 853–857.
4 K. McGregor, E. J. Frazer, A. J. Urban, M. I. Pownceby and R. L. Deutscher, *Electrochem. Soc. Trans. 2*, 2006, **3**, 369–380.

Andrew Mount asked: How was the current magnitude of 1A and −1A and the pulse time selected for the IR drop anode measurement and resistance compensation? Given that rapid pulses are used to control surface finish (*e.g.* in plating) and that local migration of ions in the anode is highly possible in the fields produced by such currents, how can you be sure that such pulses do not affect the reaction?

Kathie McGregor replied: The choice of pulse current magnitude depends on the expected cell resistance. For a 1 A current pulse, with a precision of approximately 1 mV over a 10 V limit, resistances within 1 mohm and up to 10 ohm can be measured. For a 0.25 A current pulse, resistances of up to 40 ohm and a 4 mohm precision are possible. So pulses of lower magnitude are used to measure higher resistances. We use 1A for measurements with molten salts, as the cell resistances are at most up to 10 ohms, but more typically 1–2 ohms (*i.e.* 1–2V cell voltage drop). The total duration of the bipolar pulse sequence (210 microseconds) is so rapid that the slower Faradaic processes do not contribute to the voltage at this time scale. For more information, please see the paper by Deutscher *et al.*[1] We have performed numerous measurements both with and without the resistance measurements, and have demonstrated that pulses do not affect the electrolytic reactions.

1 R. L. Deutscher, S. Fletcher and J. A. Hamilton, *Electrochim. Acta*, 1986, **31**, 585–589.

Andrew Mount asked: The chosen experimental method leads to the determination of the cell resistance, which is then subtracted, to give insight into the surface electrochemical reaction(s) occurring. However, the cell resistance is also important, as minimising this increases the cell efficiency. Have you also tried to optimise this parameter?

Kathie McGregor answered: The overall cell resistance is indeed important for an industrial process. Reference electrode measurements for selected laboratory experiments confirmed that most of the voltage drop is due to the oversized TiO_2 cathode, and only a small percentage is associated with the anode. For experiments of shorter duration we have used highly porous TiO_2 pellets for the cathode material, and found that the ohmic drop (and hence overall cell resistance) was reduced substantially. At the laboratory scale, it was difficult to design a practical, long-lived cathode with multiple porous pellets. The aim of the experiments was to investigate anode performance, and no significant effort was made to optimize the cathode design at this point in time, other than to provide a constant supply of oxide ions. For an industrial process, a very different cathode design would be expected.

Andrew Mount remarked: What was the reference electrode used in page 10 line 22 of your paper, and where was it placed? When using reference electrodes, one can in principal separate out the contribution of the voltages at the working electrode and hence determine the voltage drops at individual electrodes as a proportion of the total voltage. Has this been done, and if so from these voltages which process is rate determining? Is the rate determining step always the same throughout cell operation?

Kathie McGregor responded: A Ag/AgCl reference electrode was constructed according to the method of Gao *et al.*[1] A silver rod was inserted into a (molten) salt mixture (NaCl–KCl–AgCl) contained in a quartz electrode tube. A tungsten wire contact was joined to the silver rod and the assembly sealed in thick-walled quartz tubing. The electrode potentials were reasonably stable and performed satisfactorily during experiments, with the observed potentials close to those expected from the literature. A quasi-reference electrode such as Mo wire was also used on occasions. The reference electrode was placed in close proximity to the cathode, and used to monitor the half cell potential at the cathode (negative electrode) during selected experiments. For the experiments described in this paper, an oversized TiO_2 cathode was always used. Most of the voltage drop was associated with this electrode, as its design was not optimized to minimize ohmic resistance. The stability/lifetime of the anode was the limiting factor throughout all experiments.

1 P. Gao, X. B. Jin, D. H. Wang, X. H. Hu and G. Z. Chen *J. Electroanal. Chem.*, 2005, **579**, 321–328.

John Irvine said: You mentioned you were finding niobium pentoxide impurities, in the phase diagram it is not in equilibrium, I would expect to see Nb_2TiO_7 in the X-ray diffraction pattern.

Kathie McGregor responded: I would like to clarify that point, we did indeed find Nb_2TiO_7 in the grain boundaries when greater than 6 mol% of Nb dopant was used to prepare the electrode material. This was confirmed by X-ray diffraction studies.

During the high temperature electrolysis reaction, Ti^{3+} present in the Nb-doped titania was oxidised to Ti^{4+}. Oxide ions are incorporated into the lattice, and Nb_2O_5 was formed. This may segregate to the grain boundaries and be a point of attack and wear.

George Zheng Chen remarked: Professor Irvine, are you questioning if the starting material for the synthesis of the inert anode material was Nb_2O_5?

John Irvine answered: No this would be the starting material but at low levels of Nb, Nb_2O_5 is not an equilibrium phase, $Nb_2Ti_2O_7$ would be in equilibrium with Nb doped TiO_2 under oxidising conditions as used in synthesis.

Hongmin Zhu said: The inert anode for oxygen generation is a great challenge. I appreciate your efforts on the work. I have a question on the stability of Ti^{3+}. You mentioned you did not see a change during the oxidation test even over 240 h, however we saw anode deterioration during the electrolysis test. Does this mean oxidation occurs under high electrode potentials during the oxygen generation? If this is the case, do you have any ideas for the improvement?

Kathie McGregor answered: It certainly is a great challenge, particularly to make it robust and cost-effective. Degradation of the niobium-doped titania anode is observed to a much greater extent during the electrolysis test than the air oxidation test. This is likely due to the greater ease of oxygen diffusion through any pores or cracks in the material and subsequent oxidation of Ti^{3+} to Ti^{4+}. This gas produced by the electrolysis is stored by absorption-adsorption on the anode surface, until it builds up the necessary pressure to create a bubble and subsequently detaches from the anode surface. Kiss and co-workers believe that in molten salts, a transport and storage mechanism between the place of generation and the bubbles exists, and that gas can be stored in anode pores and migrate to the bubble through the pores.[1] Areas for improvement include optimization of the material preparation to minimize any porosity (closed and open), and the formation of a thin, dense, oxide coating that protects the bulk of the electrode from oxidation.

1 L. I. Kiss and S. Poncsak, in *Light Metals*, ed. W. Schneider, TMS, Warrendale, PA, 2002, pp. 217–223.

George Zheng Chen addressed Kathie McGregor and Hongmin Zhu: I think there is a difference between Prof. Zhu's and Prof. McGregor's conditions.

Ali Kamali asked: Figure 2a shows a SEM image in which there are TiO_2 grains and a bright grain boundary phase. This image may suggest that sintering has been preceded by liquid phase sintering. The phase diagram of TiO_2–Nb_2O_5 you showed in your slides also confirms the possibility of forming a liquid phase in this system at sintering temperature. This is an interesting point which may help to understand the sintering mechanism. Could you comment on this? Have you studied the chemical composition of this grain boundary phase?

Kathie McGregor answered: I agree we cannot exclude the possibility of formation of a liquid phase at the sintering temperatures employed. Small quantities of a high Nb phase were detected in the grain boundaries of the generation 1 material (Fig. 2a), but not the generation 2 material.

Ali Kamali asked: In the SEM image showing the morphology of the generation 2 material (Fig. 2), a number of pores can be seen on the grains. How do you explain this? Can it be related to the diffusion of a phase into another during the sintering process leaving holes behind?

Kathie McGregor replied: The cause of the pores is not entirely clear, but the phase diagram for the Nb_2O_5–TiO_2 system is quite complex and characterised by a number of eutectics over the temperature range 1430–1445 °C (Babich *et al.*, 1988).[1] It is possible that the pores are related to the diffusion of a phase into another during the sintering process.

1 T. G. Babich, A. V. Zagorodnyuk, G. A. Teterin, M. Y. Khodos and A. P. Zhirnova, *Zhurnal Neorganicheskoi Khimii*, 1988, **33**, 996–999.

George Zheng Chen remarked: Holes formation occurs frequently during sintering of metal oxide powders, resulting from small particles combining into large ones, leaving holes behind. In general, there is a reduction of the overall volume of the oxide precursor, but I do not think that the host body of the small holes would expand.

Ali Kamali addressed George Zheng Chen and Kathie McGregor: The shape and location of holes on the crystalline grains observed in Fig. 2b suggests that these holes should have been formed by a more complicated mechanism than a simple sintering shrinkage. I think a deeper study on the sintering process would be worthwhile.

John Irvine said: It's isolated porosity that Kamali is talking about, he could be right, but it could be due to voids. It could be something dissolving in it, so it needs more work to confirm. It's worth looking at the phase diagram to see if there's a liquid phase

George Zheng Chen commented: Prof. John Irvine's comment just reminded me that in the laboratory in Cambridge, a postdoc put Nb_2O_5 pellets on an alumina plate in a furnace for sintering at elevated temperatures, and he later found that the alumina plate and some of the furnace ceramic melted.

John Irvine commented: We looked at the NbTiO system as a solid oxide fuel cell electrode. It's a poor ion conductor, compared to perovskite, where ionic diffusion is 3 or 4 magnitudes lower. It's terrible as a fuel cell electrode but good for stable electrodes in this case

Andrew Mount addressed Kathie McGregor and Ye Liu: Dr McGregor raised nine properties in discussion, all of which would be required to be satisfied to produce an ideal oxygen generating electrode (*i.e.* one which is completely inert during operation). Given that it seems highly unlikely that all of these properties are practically achievable in an anode material in the near future, which do you and others feel are the most important to satisfy?

Kathie McGregor responded: From a practical viewpoint, I feel that materials with low electrical resistivity (*i.e.* highly conductive), good resistance to the oxygen produced at the anode, are insoluble and low cost are the key criteria to satisfy.

Ye Liu responded: I have nothing to add.

George Zheng Chen remarked: There has been continuous research and commercial efforts in development of inert anode materials for molten salt electrolysis. The most important issue is the stability of the material in use. It is possible to make ceramic materials with good conductivity and fabricate the material into the anode. The conductivity of most oxide based materials results from structural defects, such as oxygen vacancies. When used in molten salt electrolysis, there is no significant stress on the inert anode. Therefore, the physical property of the material is not a concern. The challenge is the chemical and thermodynamic issues. There is a tendency of the material with oxygen vacancies to be further oxidised to become an insulator again. Such a problem may not be serious in experiments lasting for no longer than a week, but longer term tests can reveal problems. The nature of the molten salt is another issue challenging the development of inert anode materials. Molten fluoride salts have usually high solubility for oxide ions, and hence can attack oxide based inert materials.

Kathie McGregor opened a discussion of the paper by Xingli Zou: Did you see any evidence of attack by calcium on the yttria stabilised zirconia membrane?

Xingli Zou replied: Thanks Prof. K. McGregor. The yttria stabilized zirconia (YSZ) membrane will be commonly attacked/corroded by molten $CaCl_2$ and calcium while the potential applied between the cathode and the YSZ-based anode. The reason for the corrosion process is mainly attributed to the loss of yttrium from the zirconia lattice during electrolysis. The yttrium diffusion process from YSZ to molten salt is a slow process and thus the YSZ membrane can keep relative stable during electrolysis. However, improving the molten salt corrosion resistance of the YSZ membrane is important. It is suggested that the addition of a small amount of yttria into molten $CaCl_2$ can prevent the degradation of the YSZ membrane.[1] In addition, the ceria/yttria co-stabilized membrane has been newly developed in our recent work, which exhibits excellent corrosion resistance to molten salts.

1 A. Mukherjee, N. Campagnol, J. V. Dyck, J. Fransaer and B. Blanpain, *J. Am. Ceram. Soc.*, 2015, **98**, 972–981.

Andrew Mount commented: What is origin of the enhanced stability of Ce-YSZ compared to YSZ in the molten salts?

Xingli Zou responded: Thanks Prof. A. Mount. The Ce-YSZ membrane contains two stabilizing components, *i.e.*, ceria and yttria. Element yttirum will gradually be lost from the interface of the Ce-YSZ membrane/molten salt during electrolysis. However, ceria can remain stable during the SOM electrolysis process and thus can serve as the stabilizing component for the Ce-YSZ membrane.

Andrew Mount asked: How long are these Ce-YSZ materials stable for?

Xingli Zou answered: In our recent experiment, the Ce-YSZ membrane exhibits excellent stability during molten salt electrolysis. However, it is hard to say how long the Ce-YSZ membrane can remain stable during molten salt electrolysis. It is dependent on various parameters, such as the temperature, the applied potential and the thickness of the Ce-YSZ membrane, *etc.* We are continuing our studies to investigate the new Ce-YSZ membrane, and further investigations will be reported in due course.

John Irvine remarked: For ceria doped solid oxide membranes, do you have worries about electronic conduction under these high voltages? CeO_2 will tend to short circuit and so lose efficiency. On the positive side, this may have a beneficial effect on stability or could help with oxidation catalysis?

Xingli Zou replied: Thanks Prof. John Irvine. For the ceria doped material solid oxide membranes, the electrical conductivity and ionic conductivity would be influenced to some extent.[1] However, the addition of CeO_2 has a beneficial effect on the stability of the Ce-YSZ membrane in molten salt, so appropriate CeO_2 content is very important to the Ce-YSZ membrane. In our recent work, we are trying to fabricate a Ce-YSZ membrane with reasonable ionic/electrical conductivities and good structural stability for the SOM electrolysis process applications.

1 C. H. Lee and G. M. Choi, *Solid State Ionics*, 2000, **135**, 653–661.

Toru H. Okabe said: Regarding the possibility of carbothermic reduction of Ta oxide, I have the following questions:

When injecting the sample (Ta_2O_5 + C) into the hot zone, is there any possibility of non-electrochemical reduction taking place?

Under certain conditions, carbothermic reduction of Ta oxide ($TaO_x + C \rightarrow TaC + CO_x$) may proceed, in addition to the electrochemical process you mentioned. Does this kind of side reaction (thermochemical reaction) occur? Or does only an electrochemical reaction take place?

How did you identify the reactions?

Xingli Zou replied: Thanks for these good questions. During the molten salt electroreduction process, the possible side reactions generally include the carbothermic reduction of Ta_2O_5, *i.e.*, $Ta_2O_5 + 7C \rightarrow 2TaC + 5CO$ and/or $2Ta_2O_5 + 9C \rightarrow 4TaC + 5CO_2$. However, the thermodynamic calculations show that the Gibbs free energies (ΔG^0) of the above-mentioned two reactions at 1000 °C are both much larger than zero ($\Delta G^0 > 90$ kJ mol^{-1}), which suggests that the carbothermic reactions cannot occur during the electroreduction process at 1000 °C. The thermodynamic calculations have been discussed in the paper. In addition, the XRD pattern of the Ta_2O_5/C pellet after being immersed in molten $CaCl_2$ (without an applied potential) at the electroreduction temperature (1000 °C) for 5 h has been analyzed in the paper, which further confirms that the carbothermic reactions are difficult to afford at the experimental temperature. However, the non-electrochemical reaction (the compounding reaction) between CaO and Ta_2O_5 will occur during the immersion/electroreduction processes, *i.e.*, $CaO + Ta_2O_5 \rightarrow CaTa_2O_6$, and more details about the compounding process can be found in the paper.

George Zheng Chen addressed Toru H. Okabe and Xingli Zou: I agree with your comment that tantalum carbides (*e.g.* TaC and Ta_2C) are more stable than CO. In our work of electro-reduction of Ta_2O_5 to Ta with a graphite anode, we have found that it was challenging to avoid carbon contamination which could reach a level of up to 1000 ppm,[1] suggesting a high affinity of tantalum metal to carbon. However, according to thermodynamic data, it would be difficult to achieve carbothermic reduction as expressed below. $Ta_2O_5 + 7C = 2TaC + 5CO$ $\Delta G^0 = 90.127$ kJ at 1000 °C, and $Ta_2O_5 + 4.5C = 2TaC + 2.5CO_2$ $\Delta G^0 = 220.797$ kJ at 1000 °C. Also, because of the formation of the CO and/or CO_2 gases in the above reactions, carbon loss would have occurred, making it impossible to produce the stoichiometric amount of TaC as shown by the XRD pattern in Fig. 7.

1 T. Wu, X. B. Jin, W. Xiao, X. H. Hu, D. H. Wang and G. Z. Chen, *Chem. Mater.*, 2007, **19**, 153–160.

Xingli Zou responded: Thanks Prof. George Z. Chen. I agree with you that it is difficult to avoid the carbon contamination during the electro-reduction of Ta_2O_5 to Ta with a graphite anode. However, on the basis of the thermodynamic analysis, the carbothermic reduction reactions ($Ta_2O_5 + 7C = 2TaC + 5CO$, and $Ta_2O_5 + 4.5C = 2TaC + 2.5CO_2$) are difficult to achieve during the electro-reduction process, therefore, the carbon contamination may be mainly attributed to the anodic products (such as CO_2)-related side reactions, *i.e.*, $CO_2 + O^{2-} \rightarrow CO_3^{2-}$ and $CO_3^{2-} + 4e^- \rightarrow C + 3O^{2-}$. The electro-reduction-generated Ta could react with the side reactions-generated C to form TaC, *i.e.*, Ta + C = TaC ($\Delta G^0 = -140.575$ kJ at 1000 °C). In the solid oxide membrane (SOM)-assisted electro-reduction process, the anodic reaction area was separated from molten salt by the membrane, which means the anodic products-related side reactions can be avoided. Thus, if the initial carbon component contained in the precursors (metal oxides/carbon, MO_x/C) would not be consumed (lost) by the thermodynamic reactions ($MO_x + (x+1)C = MC + xCO$), it is possible to produce the stoichiometric amount of metal carbides from their metal oxides/carbon precursors *via* the reaction $MO_x + C + 2xe^- = MC + xO^{2-}$ using inert anode (such as the SOM anode).

Andrew Mount remarked: What are the current efficiencies of the processes (*e.g.* those that form TiC and ZrC)?

Xingli Zou responded: Typically, the current efficiency of the SOM electrolysis process is approximately 80%, it can be influenced by various parameters, such as the temperature, molten salt, cathode design and the composition of the cathode precursors, *etc.* The current efficiency is expected to increase further upon optimization of electrolysis parameters.

Andrew Mount remarked: Following on from my previous question, what is/are the reaction(s) that give rise to the 20% wasted current in the processes that form TiC and ZrC?

Xingli Zou replied: The wasted/residual current efficiency is mainly attributed to the residual current during the electrolysis process. The electronic conduction properties of the molten salt and the zirconia membrane are intrinsic, and therefore their electronic conductivities inevitably create a residual current (background current) and lower the current efficiency. More detailed discussion can be found in our previous work.[1]

1 X. Zou, X. Lu, Z. Zhou, W. Xiao, Q. Zhong, C. Li and W. Ding, *J. Mater. Chem. A*, 2014, **2**, 7421–7430.

George Zheng Chen asked: Metal carbides are used for making various devices, such as heating elements. I assume the products from your electrolysis experiments were powders. It is well known that metal powders can be easily converted to a solid body *via* sintering under appropriate conditions. Is it easy to sinter your carbide powder into a solid body?

Xingli Zou responded: Thanks Prof. George Z. Chen. Yes, the metal carbides synthesized in our work are powders, and the average particle sizes of these metal carbide powders are approximately 500 nm to 5 μm, so the small metal carbide powders should be relatively easy to sinter into a solid body. However, the densification experiments for the transformation of the synthesized metal carbide powders into compact dense monoliths by sintering process are ongoing, and the results of further investigations will be reported in due course.

Daniel Cooper said: The average particle size of their binary carbide powders is small so they should be quite easy to sinter, or at least no more difficult than powders produced by other methods.

Xingli Zou answered: Thanks. Yes, the average particle sizes of the metal carbides powders synthesized in our work are small (approximately 500 nm to 5 μm), and they should be relatively easy to sinter to form a compact dense monolith. However, the sintering behaviors of the different carbides powders are different, and the related investigation is ongoing.

Xianbo Jin asked: You used solid oxide membrane but the current efficient is only 80%, which means the membrane has some electronic conductivity. This is against my impression that the electronic conductivity of the solid oxide member in SOFCs is reported to be very low so SOFCs have very high efficiencies. Have you measured the electronic conductivity of you membrane? Did you compare the value in dry and wet conditions?

Xingli Zou responded: Thanks Prof. X. Jin. The typical current efficiency in the solid oxide membrane (SOM) process for the synthesis of carbides/composites is generally around 80%. Commonly, the YSZ membrane possesses high oxygen ionic and low electronic conductivities (dependent on temperature) during the electrolysis process, which means that the low electronic conductivity may influence, at least to some extent, the current efficiency. However, we have not compared the electronic conductivities in dry and wet conditions at the experimental temperature. It should be noted that the current efficiency can be affected by various parameters, such as the cathode design, molten salt, cathode composition and temperature, *etc.* In the SOM experiments for the production of metals, alloys and composites in recent years, the current efficiencies are typically 70–95%.

Hongmin Zhu said: For the solid oxide membrane-assisted electrolysis, the biggest challenge would be the ability of oxide ions to pass through the membrane, or the stability of the membrane whilst the oxide ions pass through. Therefore I wonder how much charge (O^{2-} ion) has passed through the membrane *vs* the total volume of oxide conductor?

Xingli Zou replied: Thanks Prof. H. Zhu. Agreed! The biggest challenge of the solid oxide membrane (SOM)-assisted electrolysis process is the stability of the membrane during the molten salt electrolysis process. The yttria stabilized zirconia-based SOM will be frequently corroded during the electrolysis process in molten $CaCl_2$, and the corrosion process generally occurs at the interface of the SOM/molten $CaCl_2$. However, the corrosion process is a relatively slow process and depends on various factors, such as the applied potential and temperature, *etc.* In our experiment, the membrane can still remain stable and maintain sufficient oxygen ion conductivity even while the outside surface of the membrane has been partially degraded. Actually, it has been proven that the SOM is not the oxygen ion migration-limited step during the SOM electrolysis process, which means that almost all of the oxygen component contained in the cathode can be ionized to form oxygen ions and pass through the SOM during the electrolysis process. However, we do not have the exact number of oxygen ions that passed through the membrane *vs* the volume of the SOM during the electrolysis process. The maximum number of oxygen ions that can pass through the membrane during the electrolysis process depends on various parameters, such as the total surface area of the SOM immersed in the molten salt and the thickness of the membrane, *etc.* In our typical SOM experiment, approximately 1–5 g metal oxides/compounds are commonly used as the cathode precursors, and one SOM tube (1–2 mm in thickness, 10–15 mm in diameter and 100–150 mm in length) is used as the anode, and almost all of the oxygen components contained in the precursors can be ionized to form oxygen ions that pass through the SOM within several hours of electrolysis. The key feature of the SOM process is that the SOM can separate the anodic reaction area from the molten salt, which means that the anodic products-related side reactions can be avoided (such as the CO/CO^{2-} related anodic side reactions). Therefore, the SOM process has the potential to be used for the synthesis of various metal carbides and carbide-containing composites with pre-designed carbon contents. However, improving the corrosion resistance of the membrane is crucial to the SOM process, in our recent work, the ceria/yttria co-stabilized membrane has been developed and used as the SOM anode, which exhibits enhanced corrosion resistance to molten salt.

Daniel Cooper remarked: You mentioned in your presentation that you have synthesised MAX phases using similar techniques. One way to produce MAX phases is to start from a non-stoichiometric binary carbide precursor.[1] Have you tried using this fabrication technique to produce non-stoichiometric binary carbides, for example Ti_3C_2? If you have not tried it, based on your understanding of the fabrication method do you think it might work?

1 D. P. Riley and E. H. Kisi, *J. Am. Ceram. Soc.*, 2007, **90**, 2231–2235.

Xingli Zou responded: Thanks. Actually, the molten salt electroreduction process has been used to synthesize Cr_2AlC in a previous work,[1] which preliminarily shows that the electroreduction process has the potential to be used to produce MAX phases in molten salts. In our recent work, we have also successfully synthesized MAX phases such as Ti_3AlC_2 (unpublished work) and metal carbides using the SOM electroreduction process. However, in our experiments for the metal carbide syntheses, the precursors used for the synthesis of binary metal carbides are generally composed of stoichiometric mixtures (such as $TiO_2 : C$, 1 : 1 for the synthesis of TiC). In my opinion, the non-stoichiometric precursors may also have the potential to be used for the production of non-stoichiometric binary metal carbides using this process, however, the synthesis process may be relatively difficult to control.

1 A. M. Abdelkader, *J. Eur. Ceram. Soc.*, 2016, **36**, 33–42.

George Zheng Chen opened a discussion of the paper by Geir Martin Haarberg: It is an interesting and good idea to introduce a fuel gas, particularly hydrocarbons, to reduce CO_2 emission from the anode reactions. Anode materials should have an impact on the performance. For example, I think tin oxide and graphite will cause different kinetics and even anode reactions. Is this the case for example when methane is used as the fuel gas? Also, for electrolysis of a metal oxide, either dissolved in the molten salt, or in the solid state *via* the FFC Cambridge Process, have you observed any oxidation of the graphite anode itself to carbon monoxide or dioxide?

Geir Martin Haarberg answered: I agree that different anode materials have different kinetics. It's very difficult to achieve a complete participation of methane, so in our experiments the usual anode process (CO and CO_2 formation) takes place in parallel with methane oxidation. We always observe consumption of the graphite anode when passing methane.

Liang Xu asked: Thanks for your nice talk. It is an interesting concept, replacing the conventional graphite anode by CH_4 for the Al electrolysis. My question is, in this case, hydrogen will be introduced into the system which could result in H_2O formation. I am afraid the generated H_2O could cause serious problems in the molten salt at high temperatures. What do you think about this?

Geir Martin Haarberg responded: I agree that water vapour will be produced in the anode reaction when using methane. This will cause the formation of HF which is undesired in the process. In the current industrial process HF will be formed due to the presence of humidity, but using methane may lead to an increased level of HF. It may be challenging to reduce the extra level of HF, but it should be addressed in a possible implementation of the methane concept. It is however believed that most of the HF exiting the cells will be captured by alumina in the dry scrubbers.

Dihua Wang remarked: What is the optimized current density for this kind of gas anode? How do you optimize and control the solid/liquid/gas three phase reaction boundary?

Geir Martin Haarberg responded: The normal current density in the current process is 0.7–0.8 A cm^2. The aim is to achieve a similar current density when using methane. It's challenging to obtain an efficient action of the methane, so we are working on an improved design of the supply of the gas.

George Zheng Chen asked: Would it be possible to use the carbonate fuel cell electrode, *e.g.* porous nickel, in your gas anode?

Geir Martin Haarberg responded: The molten fluoride electrolyte based on cryolite is very corrosive, and nickel will dissolve if used as an anode. A candidate inert anode is based on nickel ferrite, which may be a possibility for the gas anode.

Shuqiang Jiao remarked: When using Sn oxide, oxygen ion is discharged on the anode to form oxygen. However, if switching to graphite anode, what is the mechanism to explain the difference. How does the electrochemical reaction happen?

Geir Martin Haarberg answered: When using graphite, methane will be oxidised to form CO_2 and H_2O. Also on tin oxide the methane oxidation should happen, but some oxygen evolution will take place as well.

Hongmin Zhu commented: It is a nice ideal to use natural gas as the anode feeding material for aluminium electrolysis. I have a question about the electrode potential during the electrolysis. In Fig. 1, before you start the bubbling of CH_4, the anode production should be oxygen, O_2, while after the CH_4 it should be H_2O, CO and CO_2. There should be more than 1 V potential difference between these electrode reactions. But, the potential drop in Fig. 1 is only 0.5 V. Do you have any ideas to explain the loss of the potential?

Geir Martin Haarberg replied: I believe that the measured potential is a mixed potential because it's likely that the normal reaction of CO_2 formation also takes place during the introduction of methane.

Paul Madden addressed Geir Martin Haarberg and Toru H. Okabe: I have given lectures to undergraduates to explain how difficult it is to separate aluminium from oxygen and shown the famous Napoleon's Helmet to illustrate. The (brighter) students have then asked, how was the aluminium for the helmet extracted as this was made before the development of the Hall–Héroult process? Can you help?

Geir Martin Haarberg answered: Prior to the Hall–Héroult process, elemental aluminium was made by reducing the ore (bauxite) by elemental sodium or potassium in a vacuum, as explained by Prof. Okabe. Carbothermal reduction of alumina is very difficult. I believe that Napoleon III is the best person for this question.

Toru H. Okabe responded: In the old days, aluminium was produced by potassium reduction of aluminium chlorides, not by electrochemical methods.

$AlCl_3$ (or complex chlorides) + 3K → Al + 3KCl (or complex chlorides).

In some cases, amalgam (Hg alloy) was utilized for facilitating reactions. Sodium can also be utilized as a reduction agent for this type of metallothermic reduction. But production of metallic K or Na was very difficult and costly in those days. For this reason, Al was a "precious metal" worthy for Napoleon's Helmet. Development of an electrochemical technique around 1900s was really a "revolution" in Al smelting history.

Andrew Mount addressed Geir Martin Haarberg and Babak Khalaghi: To be able to judge what is a significant difference between the data in Table 2, it would be useful to know what the experimental error is in the weight data in this Table.

It also seems that only the isotropic graphite may produce a significant decrease in the amount of graphite consumption when the effects of carbon deposition are taken into account. Is there an explanation for why this is the case?

Babak Khalaghi replied: Regarding the first part of the question I could say that in Table 2 the results of specific experiments are given. But for the isotropic graphite (G347) the more accurate value for the weight loss would be: 0.60 ± 0.07 (0.80 ± 0.07) (g). And for the other samples only one experiment was run. My answer to part two of the question is that it seems the important factor and property is the porosity and pore/grain size of the graphite and not the production method. Because better porosity and pore size establish the three phase boundary (electrolyte–gas–electrode) more efficiently and lead to more effective participation of gas (CH_4) in the anodic reaction. So, the fact that the graphite is isotropic or not, does not seem to have a key role in the performance of the anodes when there is considerable difference in porosity and pore/grain size of the graphite grades.

Qian Xu commented: The Al^{3+}/Al electrode was used as the reference electrode in your reseach. Can the reaction Al + Al^{3+} → Al^{+} affect the reversibility of the Al^{3+}/Al electrode?

Geir Martin Haarberg responded: It's correct that aluminium dissolves to form Al(I). But the solubility of Al is very low (~0.05 wt%), so it won't affect the electrode potential.

Daniel Cooper commented: A question was asked earlier about water being produced as a result of the reaction which leads to the production of HF, which is potentially hazardous. There are a number of processes which use methane as a source of hydrogen. Could you couple your process to one which utilises the hydrogen in order to prevent production of water and consequently HF, or would this hinder the mechanism of operation?

Geir Martin Haarberg replied: Using hydrogen instead of methane is an option. However, water vapour will still be formed in the anode process, so it will not prevent this problem.

Linpo Yu said: Steam will be generated at the anode, I believe water can react with the graphite anode at high temperature. Did you observe any broken graphite anode after your experiment?

Geir Martin Haarberg replied: I agree that water may react with the graphite anode. However, we did not observe any serious damage of the anode after experiments. The main challenge with the formation of water is believed to be due to the formation of HF.

John Irvine remarked: I was wondering about methane cracking as well. Pass the methane without driven current in batch mode, and see if you can oxidise the carbon. Then switch on the current. Utilize the carbon that's formed from the methane. You wouldn't have water forming and the hydrogen will be given off.

Geir Martin Haarberg replied: Thanks for your comment. We would like to try this procedure in future experiments.

Qian Xu opened a general discussion of the papers by Yating Yuan: You used $MgCl_2$ instead of $CaCl_2$ as the electrolyte for the electrochemical deoxidation of TiO_2. Is there any advantage for this chloride? Do you think that chlorine as the anodic product is more environmentally friendly than CO_2?

Yating Yuan responded: The biggest advantage of our process is that carbon anode is inert. So carbon-related problems such as CO_x emission, low current efficiency and carbon contamination can be all eliminated. As for the chlorine, of cause it can not be released to the atmosphere directly like CO_2. It is not a difficult thing to utilize the formed Cl_2 in industry.

Geir Martin Haarberg addressed Yating Yuan and George Zheng Chen: This paper reminds me of old Mg processes. The oxide solubility is very low when magnesium chloride is part of the electrolyte, so a lot of solid MgO is formed. This means that the molten electrolyte is always saturated with respect to oxide, and therefore the carbon anode must be slowly consumed. This also happens in the Mg electrolysis process, where the graphite anodes are not entirely inert. The high content of magnesium chloride also suggests that other oxygen containing species such as MgOHCl will be present in the electrolyte.

Geir Martin Haarberg said: When magnesium chloride is a major part of the melt, the thermodynamic stability of the electrolyte is lower, meaning that this approach will not work for more reactive metals.

George Zheng Chen replied: We agree with the thermodynamic consideration of Prof. Haarberg in the absence of an electrode potential. However, the electrode potential applied can provide an additional driving force. In our current research, reduction of ZrO_2 to Zr metal was quite successful, although we are still working on reducing TiO_2 and other more stable metal oxides. I shall also mention that the paper from Prof. Okabe in this Faraday Discussion (DOI: 10.1039/c5fd00229j) describes some good thermodynamic analyses for using $MgCl_2$ for deoxygenation of titanium.

Qian Xu asked: In the preocess of electrochemical deoxidation of TiO_2, the MgO will be produced and aggregated on the cathode. How did you seperate the byproduct from the final product Ti, and how do you manage this MgO in the line of production?

George Zheng Chen replied: In our paper, we have described the removal of the MgO from the reduced cathode in dilute HCl solutions. In this case, the MgO is also converted to $MgCl_2$ which can be fed back to the electrolysis cell. Obviously, to recover $MgCl_2$ from the aqueous solution, additional energy input is needed. Technically, it is not a problem to close the process to a cycle as discussed above. The decision on using which technique is largely determined by the market value.

Kathie McGregor asked: Chlorine evolution at graphitic electrodes can result in the formation of carbo-chloride species. Did you see any evidence for these?

George Zheng Chen answered: We have not observed formation of any carbo-chloride species, such as CCl_4 and C_2Cl_2. Thermodynamically, formation of these two compounds at our working temperature is unfavourable.

Ali Kamali addressed Yating Yuan and George Zheng Chen: The cathode has been converted to Ti in only 18 min, which is very remarkable. Have you evaluated the Ti product in terms of purity and oxygen concentration?

George Zheng Chen answered: Our paper has reported results from electro-reduction of Ta_2O_5 to Ta metal, but not Ti. In our work, short time electrolysis usually produced a mixture of Ta metal and some sub-oxides as shown in Fig. 3.

Ali Kamali addressed George Zheng Chen and Xianbo Jin: What was the cathode current density during the process?

Xianbo Jin replied: For the pellet with a diameter of about 1.3 cm, the reduction current in the main stage was about 0.8 A.

George Zheng Chen answered: The concept of current density is strictly speaking not applicable here because the electrode surface area increases continuously with the progression of metal formation on the cathode. For comparison purposes, the speed of the electro-reduction may be normalised against the mass of the oxide. In this case, the mass normalised current is about $0.8\ A\ g^{-1}$.

Han Wang addressed George Zheng Chen and Xianbo Jin: During the electrolysis, is it possible that the formation of MgO will change the porosity of the cathode, and finally lead to the blocking of the channel for liquid diffusion?

Xianbo Jin replied: Yes, the deposition of MgO will occupy some of the pores in the cathode. But it is very interesting that Mg has a molar volume larger than its oxide. This is unique compared to other elements because generally metal oxide has a larger volume than the metal. MgO has a very small molar volume of 11.2 $cm^3 mol^{-1}$, much smaller than that of CaO (16.7 $cm^3 mol^{-1}$), and the liquid channel in a Ta_2O_5 pellet will not be blocked by the MgO precipitation when the pellet porosity is higher than 30%. Normally, the metal oxide pellets for solid cathode electrolysis have a porosity of about 50%, so fast electrolysis was observed in this work.

George Zheng Chen answered: The amount of metal oxide on the cathode is fixed, so is the amount of oxygen available to form MgO. Thus, the porosity of the metal oxide cathode can be designed to avoid blockage of the pores by the precipitated MgO.

Han Wang addressed George Zheng Chen and Xianbo Jin: Does this process still follow the mechanism of three-phase interlines in the FFC Cambridge process?

Xianbo Jin responded: Yes, because the reduction still needs the electron transfer from the current collector (Ta) to the oxide, and the Mg^{2+} from the electrolyte.

George Zheng Chen responded: The three-phase interline mechanism is still followed in this process, although the difference is that the oxide ion generated from reduction of the metal oxide does not leave the cathode. Instead, it combines with the magnesium ion to form the MgO precipitate inside the cathode.

Han Wang addressed George Zheng Chen and Xianbo Jin: Since Cl_2 gas is produced in the process, I am wondering which material you have used to make the reaction vessel, and how the vessel was after the experiment?

Xianbo Jin replied: We use an alumina tube to conduct the gas out in our laboratory studies. There are many industrial examples (*e.g.*, $MgCl_2$ electrolysis) for references to deal with the anodic Cl_2 gas evolution.

George Zheng Chen replied: Thanks for your query. The materials used in our research are described in the paper, *i.e.* alumina or graphite for the crucible and stainless steel for the retort. The concern of chlorine gas reacting with steel at elevated temperature is valid, but in the experiments, the amount of chlorine gas released from electrolysis of a pellet of 2 g metal oxide was fairly small, and hence impacted insignificantly on the stainless steel retort. For commercial applications, this should not be an issue because industrial scale electrolysis of $MgCl_2$ in molten salts to produce magnesium metal and chlorine gas has been in practice for many years.

Andrew Mount asked: Given the change in porosity (driven by the change in density) on electroreduction (which should increase active electrode area), and the simultaneous surface formation of magnesium oxide (which should decrease active electrode area), there must be a theoretical relationship between particle size and change in porosity for a given material which determines whether this process remains viable throughout electroreduction. Has this been derived and has it been validated experimentally?

George Zheng Chen replied: We have done some theoretical analysis and experimental verification as published in 2013.[1]

1 H. Chen, Y. Zeng, W. Li, J. J. Peng, X. B. Jin and G. Z. Chen, *Electrochem. Commun.*, 2013, **26**, 33–36.

Andrew Mount asked: Could the small proportion of observed oxygen gas product in Table 2 be as a result of the (thermodynamically favourable from the CV) chemical reaction between the chlorine gas product and the dissolved oxide? Could active removal of the chlorine gas product therefore not decrease the proportion of this O_2 product?

George Zheng Chen answered: I agree with Prof. Mount on the possible reaction between Cl_2 and oxide ion. However, the oxide ion can also discharge directly on the anode at the potential for Cl_2 formation.

Shuqiang Jiao said: with the formation of MgO at cathode, the concentration of magnesium chloride will drop down in bulk electrolyte – how do you treat it, how do you control it in an industrial process?

George Zheng Chen replied: This is indeed an issue. Addition of $MgCl_2$ is necessary to maintain the process in continued operation. $MgCl_2$ can be produced by reacting the recovered MgO with HCl. The HCl may be produced in a H_2–Cl_2 fuel cell which can also generate electricity to feedback to the electrolysis cell. This idea will close the cycle of the process. Alternatively, $MgCl_2$ produced by other methods can be used, whilst the by products, MgO and Cl_2, can be sold on the market, if this approach may be more beneficial commercially.

Qian Xu asked: If the $MgCl_2$-based fused salt was used for electrochemical deoxidation of TiO_2, the product on the anode would be chlorine. How do you manage this kind of harmful gas?

George Zheng Chen responded: In the current industry, the Kroll process includes electrolysis of $MgCl_2$ in molten salt to regenerate the Mg metal and Cl_2. Thus, we do not anticipate the anodic generation of the Cl_2 gas in our process to be a problem. In addition to the above discussion, the Cl_2 gas can also be used to react with the recovered MgO to regenerate $MgCl_2$ and the O_2 gas. The following reaction has a negative Gibbs free energy change. $2MgO + 2Cl_2 = MgCl_2 + O_2$ At 100 °C, $\Delta G^0 = -42.484$ kJ.

Toru H. Okabe commented: Chlorine can be used for $MgCl_2$ production or else:

Many people showed concerns about Cl_2 and MgO production during the process, but for me, it is not a concern at all. Cl_2 can be used for many applications including $MgCl_2$ production, and it is easy to handle in the practical industrial chemical plants. In real industrial Ta powder production plants, Mg vapor is practically used as a reduction/deoxidation agent for removing TaO_x on the Ta powder. The majority of high purity Ta powder is produced by the following processes:

K_2TaF_7 + Na + KF + NaF → Ta powder + salt.

The obtained Ta powder is rinsed with water and dried, then reacted with Mg vapor in order to remove the oxide layer existing outside the metal particles. Then the reduced/deoxidizied Ta powder by Mg is rinsed again with water. The high quality Ta powder produced by the Mg deoxidation process is used as a material for Ta capacitors. The process proposed in the paper may applicable to Ta powder production in the future.

Dihua Wang commented: I am wondering if the electrode process is fully reversible on the Ta_2O_5 electrode in molten $MgCl_2$. A mixed reduction product of Ta/MgO formed during the electroreduction of Ta_2O_5, Ta will be oxidised back to Ta_2O_5 during the oxidation process. Is Ta fully oxidised and can it be further used to make a battery?

George Zheng Chen answered: This is a good question. We have been thinking of reversing the FFC Cambridge Process so that new and powerful molten salt based rechargeable batteries can be developed since the early days of research. I still remember that Prof. Zhaowu Tian of Xiamen University also suggested to reverse the FFC Cambridge Process for development of new rechargeable batteries in 2002 at a NSFC organised meeting in Dalian. With the MgO remaining in the metallised cathode, the chance may be higher to reverse the reduction process. It may not necessarily be the Ta/MgO couple, but other metal/MgO couples and even semi-reduced oxide/MgO couples should also be studied for this new opportunity.

Linpo Yu addressed Yating Yuan and George Zheng Chen: When MgO is formed, is it dissolved in the molten salt? If not does it block the channel?

George Zheng Chen responded: The MgO is not soluble in the molten salt, and hence remains in the pores of the metal oxide cathode whose porosity is sufficiently large to accommodate the MgO without blocking the ion passages. The optimal porosity of the metal oxide cathode can be estimated as we have reported before.[1,2]

1 H. Chen, Y. Zeng, W. Li, J. J. Peng, X. B. Jin and G. Z. Chen, *Electrochem. Commun.*, 2013, **26**, 33–36.

2 W. Li, Y. T. Yuan, X. B. Jin, H. L. Chen and G. Z. Chen, *Prog. Nat. Sci. Mater. Int.*, 2015, **25**, 650–653.

Yating Yuan commented: The solubility of MgO in $MgCl_2$ is very low.

Chaohui Wei asked: What is the purity of Ta after the electrolysis?

Yating Yuan communicated in reply: The purity of Ta is very high. Even though the purity cannot reach 100%, but the carbon (0.015 wt%) and oxygen content (0.75 wt%) in Ta powder produced in MNK is much lower than in $CaCl_2$.

Chaohui Wei communicated: What is the special functionality of $MgCl_2$–NaCl–KCl molten salts in this process?

Yating Yuan communicated in reply: To disengage oxygen ions from transporting to, and discharging at a graphite anode.

Babak Khalaghi communicated: In the paper it is mentioned that the potentiostatic electrolysis of Ta_2O_5 was performed at −0.8 V *vs* Ag/AgCl. According to the CV (Fig. 1) and also considering the finding in ref. 19, it is not likely that reduction proceeds completely; *i.e.* metallic Ta. It was expected that the Ta_2O_5 was reduced to some intermediate such as TaO_2. However, metallic Ta was produced after electrolysis at this potential. Can you give some explanation about it?

George Zheng Chen communicated in reply: According to the thermodynamic data in Table 1 of our paper, reduction of Ta_2O_5 to Ta metal should occur at a potential of about 0.9 V more positive than the potential for Mg metal deposition. The applied potential of −0.8 V *vs* Ag/AgCl was 0.75 V more positive than the Mg deposition potential according to the CV in Fig. 1, and hence should be sufficient to produce Ta metal.

Babak Khalaghi communicated: In the paper it was shown that the products of anodic reaction are Cl_2 (mainly) and a small amount of O_2. Could you give some explanation as to why CO_2 or CO did not evolve. Does the evolution of CO or CO_2 seem to be more favourable thermodynamically?

George Zheng Chen communicated in reply: The discharge of oxide anions on a carbon anode to produce CO and CO_2 is indeed more favourable than the discharge of chloride anions in terms of thermodynamics. However, the discharge of oxide anions is kinetically very difficult and can cause polarisation close to 1.0 V on a graphite anode.[1]

1 H. L. Chen, X. B. Jin, L. P. Yu and G. Z. Chen, *J. Solid State Electrochem.*, 2014, **18**, 3317–3325.

George Zheng Chen opened a discussion of the paper by Gang Chen: Molten salts are useful for making new materials and your work demonstrates this is the case – congratulations. For electrodeposition, the current efficiency is of course important. Can you explain how this was related to your product? How much of your product were indeed nanowires? What was the current efficiency for making the nanowires, not just the deposit?

George Zheng Chen addressed Qixin Shen and Gang Chen: Your AlFe nanowires can be used for Li batteries which is very good news. The nanowires are metallic conducting and therefore you should not need to add a conductive reagent, is this right? How did you attach the nanowires to the electrode substrate to measure the electrochemical properties? Did you need to mix the nanowires with a binder into a paste? If so, what was the binder? Did you use any other additives to make the electrode?

Gang Chen responded: Thank you for your question. AlFe nanowires are deposited on a copper foil with enough binding force, and they can be directly assembled into coin cells, so no binder is need here.

John Irvine remarked: What is the electrochemistry involved here? You're at very high capacity but it's not obvious what is happening. For Li-Al I would expect a voltage of a few hundred millivolts *vs* Li, you're calculating a capacity involving over 3 V – what is the electrochemistry or is it a supercapacitor you're looking at? Are you forming LiAl? Please explain your potential curve, you should have a plateau or similar for an insertion type reaction.

John Irvine asked: Can you rationalise why you measure capacity over 3 V range? You'd be able to rationalise 0.5 V. A useful capacity should be in the range of 0–1 V for a negative electrode. A lot of people publish Li data that measure capacity over a very wide range, which means nothing, as it overlaps both positive and negative electrode potential values.

George Zheng Chen addressed Qixin Shen and Gang Chen: On your charging–discharging plots in Fig. 10a, there is not a clear plateau, indicating Li intercalation. Can you explain how Li intercalation occurs in your nanowire electrode? It seems to be a common practice in laboratory, but when you were discharging, why was the cut off potential set to 2.9 V *vs* Li/Li^+? Similarly, why did you start charging from 2.9 V? Apparently, in a real lithium ion battery, your nanowires would be used as the negative electrode which will change its potential approximately between 0 and 1.0 V *vs* Li/Li^+. If possible, please replace voltage by potential in your Fig. 10a. Also, in relation with the unit of mA g^{-1}, please replace current density by mass normalised current or gravimetric current. In electrochemistry, current density is the current normalised against electrode area, and is related to the unit of mA cm^{-2}.

Gang Chen responded: Thank you for your questions and suggestions. The reason for there being no clear plateau may be due to addition of Fe or even a little amount of Cl in the nanowires. For the other question, in the research of M.J. Lindsay *et al.*,[1] ball-milling prepared AlFe alloy didn't show the linear charge–voltage dependence as well, and the cut voltage is 3 V.

1 M. J. Lindsay, G. X. Wang and H. K. Liu, *J. Power Sources*, 2003, **119**, 84–87.

John Irvine asked: Are you testing the positive or negative electrode? Everybody tests batteries against Li, you're quoting a potential across 3 V.

George Zheng Chen addressed Qixin Shen and Gang Chen: All electrochemical tests reported in this paper were carried out in a coin type cell with a lithium metal foil as the reference and counter electrodes, and the nanowire electrode as the working electrode. The related test can be regarded as a single electrode test in a pseudo-three-electrode cell. In other words, in this coin cell, the nanowire electrode is the working electrode whose behaviour is tested and studied by using it as the negative electrode in a future and real lithium ion battery in which a different positive electrode will be used. In terminology, it is not necessary to claim the nanowire electrode to be the negative electrode in the coin cell, whilst many may think the lithium metal foil is the negative electrode. The fact is that the lithium metal foil functions as both the counter and reference electrodes in the coin cell, and has nothing to do with the name of negative electrode. If possible, please replace anode by negative electrode in your paper.

Gang Chen replied: Thank you for your suggestions very much. They are very useful for me, and I'll take them on board.

Andrew Mount addressed Gang Chen and John Irvine: As the voltage/specific capacity curves (Fig. 10a) show voltages up to 3 V, building on Professor Irvine's comment that LiAl formation in batteries only typically gives 1 V, and given that the curve shows a near linear charge–voltage dependence, it seems likely that there is significant additional supercapacitance. In this case, given that most systems require a minimum threshold voltage to power the electronics, and that excess voltage may be detrimental to circuit performance, also following on from Professor Irvine's comments, would it not be better for the community to define standard criteria for assessing charge capacity for both batteries and supercapacitors, in terms of the useable charge stored and released (*i.e.* the specific capacity between maximum and maximum (useable) voltage limits)?

Gang Chen answered: Thank you for your question. A similar question was asked earlier. The reason for no linear charge–voltage dependence should be due to addition of Fe or even a little amount of Cl in the nanowires. In the research of M. J. Lindsay *et al.*,[1] ball-milling prepared AlFe alloy didn't show the linear charge–voltage dependence as well, and the cut voltage is 3 V. This is a meaningful attempt. For the useable charge stored and released, voltage limits are needed here.

1 M. J. Lindsay, G. X. Wang and H. K. Liu, *J. Power Sources*, 2003, **119**, 84–87.

John Irvine asked: What is the electrochemistry happening? Below 1 V I can see reasonably good capacity with sensible chemistry. Above 1 V I think interfacial reactions are dominating.

Xianbo Jin replied: However, I am not sure what happened in their case at potentials higher than 1 V. I agree with you that the Li alloy and dealloy processes on pure Al should take place at potentials below 1 V.

Yiyang Kong asked: Why was trimethylamine hydrochloride chosen as the target ionic liquid? From the paper explained, its low cost was a main reason. Are there any other reasons, for example, the physical and chemical properties leading to it being superior to other ionic liquids? Have other ionic liquids been tried?

Gang Chen answered: Suitable viscosity and conductivity at room temperature are other reasons for choosing trimethylamine hydrochloride. We can only get Al films from other ionic liquids with lower viscosity and higher conductivity at room temperature. And, maybe we can get nanowires from these ionic liquids with lower temperature or by adding some other solvent which could increase its viscosity. However it will increase the time and cost of the process. For the other question, we also get Al nanowires from another ionic liquid with smaller diameter without a template, but it's much more expensive.

Yiyang Kong asked: From the report of Yong Zhou *et al.*, 1-butyl-3-methylimidazolium tetrafluoroborate (C_4mimPF_6), a kind of ionic liquid, could be employed as a template to synthesize monolithic mesoporous silica.[1] This was achieved by carrying out sol-gel reaction of silica in the mixture of C_4mimPF_6 and tetramethylorthosilicate (TMOS) in the presence of HCl as an acid catalyst. After removing the formed methanol during hydrolysis of TMOS and C_4mimPF_6, a mesporous silica with a wormhole framework could be achieved. Due to the formation of the hydrogen bonds between BF_4 and the silano group of the silica gel and the π–π stacking interaction between neighbouring imidazolium rings of C_4mim^+, C_4mimPF_6 may self-assemble as template and silica would be formed on the periphery of the template, which played a key role in the formation of the wormhole framework.

Specific to this study, I am wondering whether trimethylamine hydrochloride (TMHC), the ionic liquid used in this study, could be a kind of template that made a contribution to the formation of AlFe alloy nanowires. To say the least, TMHC may not act as C_4mimPF_6 that can assemble due to the lack of imidazolium rings, but whether it is possible that TMHC may still enhance the formation of nanowires due to the interaction between TMHC and the AlFe alloy.

1 Y. Zhou, J. H. Schattka and M. Antonietti, *Nano. Letters*, 2004, **4**, 477–481.

Gang Chen replied: Thank you for your question. You mentioned that there are not imidazolium rings in this ionic liquid, and it's apparent that there aren't any other ring structures in TMHC. It may suggest no ring structure in the ionic liquid that amorphous nanowires can be achieved by electrodeposition from a similar ionic liquid.[1] The formation of nanowires is controlled by both ion mass transport rate and the ion consumption rate during electrodeposition. The ion mass transport rate is controlled by viscosity and temperature, and the ion consummation rate is controlled by conductivity, potential and temperature. For a given ionic liquid, viscosity and conductivity are controlled by mixture ratio. So, mixture ratio, potential and temperature provide a suitable condition for forming nanowires. If any condition is greatly changed, nanowires cannot be obtained.

1 Y. T. Hsieh, M. C. Laia, H. L. Huang and I. W. Sun, *Electrochim. Acta*, 2014, **117**, 217–223.

Andrew Mount remarked: The IL is stated as not acting as a template, but it seems likely that it specifically adsorbs on the surface to confer nanowire growth.

Could this be considered as templating growth on the electrode surface? Is there evidence for this, and do other similar ILs also confer this structured surface growth?

Gang Chen answered: Thank you for your question. I am not quite sure how IL can adsorb on the surface to confer nanowire growth. For another question, we also get Al and other metals or alloys nanowires from other ionic liquids even with smaller diameter without template. We think the growth of this structure is controlled by both ion mass transport rate and ion consumption rate during electrodeposition. Ion mass transport rate is controlled by viscosity and temperature, and ion consumption rate is controlled by conductivity, potential and temperature. If any condition is changed a little, nanowires can't be achieved.

George Zheng Chen said: Prof. Irvine and I had a good discussion about the charging and discharging data of the nanowires negative electrode. Charging started at 2.9 V *vs* Li/Li^+, and similarly discharging ended also at 2.9 V. It is easy to set a very high positive potential for discharging and charging in experiment, but it does not make sense because in a real lithium ion battery, the potential of the negative electrode would by no means reach such a high positive potential. It is more sensible to set the start potential for discharging and the cut-off potential for charging to a much less positive value, *e.g.* about 1.5 V. Note that the label of the vertical axis is voltage in Fig. 10a of the paper, but it is in fact electrode potential. When reporting research on electrode materials, the general principle of electrochemistry should be applied. Voltage measures the potential difference between two electrodes through which the same amount of current flows, whilst potential measures the polarisation power of a single electrode. In practice, electrode potential is measured against that of the reference electrode, but there is no current (or a negligible current) flowing through the reference electrode.

Gang Chen answered: Thank you for your pertinent suggestions very much. After many times of experiments, we found that this kind of anode material didn't show a clear plateau. By referring to several papers, there were also other alloys that showed similar results by a cut-off potential from 0.1 to 3 V. So we increased the potential as well. Your suggestions are very meaningful. Definitely, for usage, the cut-off potential should be controlled under less than 1.5 V. And thank you for your note on the difference between potential and voltage as well, we'll change it and pay attention to the difference next time.

John Irvine said: The initial anode becomes the cathode during recharge, so basic terminology should be used more carefully. Here we also have capacity in decomposition materials, characterised by much broader discharge curves. The useful range of discharge can be only a fraction of total discharge. If you are thinking of making a battery you need 2 electrodes and electronic devices will not work if the voltage of your battery changes by 20%. When you have a supercapacitor more energy is lost to heat so when you go from a battery to a supercapacitor you have energy loss, but a supercapacitor will give you more power which can be more important than storage capacity.

George Zheng Chen commented: Thermal loss can be a big issue for supercapacitors because they are designed to work at high power and hence high current. Joule heat resulting from the passage of a high current is inevitable in the supercapacitor. The incorrect use of terminology in rechargeable battery research is unfortunately commonplace. Interestingly, Chinese textbooks will never refer to cathode and anode in battery. It is always positive electrode and negative electrode. Most Chinese authors will also write positive electrode and negative electrode in their papers in Chinese, but strangely they use cathode and anode in their papers in English. It is frustrating to see incorrect use of terminology appearing in RSC journals. It can mislead the next generation. As the Chair of this Faraday Discussion, I call all RSC journals to stop publishing papers that use such incorrect terminologies, and RSC journal editors have a great responsibility.

Shuqiang Jiao opened a discussion of the paper by Toru H. Okabe: Why is the nitrogen content going up in the product?

Toru H. Okabe responded: Titanium has a very strong affinity with nitrogen. Therefore, all nitrogen in the system will react with titanium and accumulates in titanium when processing at elevated temperature. When running deoxidation experiments, nitrogen gas or compounds have to be eliminated from the reaction system, otherwise nitrogen content in the deoxidized titanium will increase after deoxidation. The data shown in the paper is simply due to the result of luck of experimental skills: *i.e.* control of the inert atmosphere was insufficient. If we do the experiment under a nitrogen-free atmosphere, the nitrogen content in titanium after deoxidation does not increase. Or, if we place a nitrogen-getter (sacrificing titanium sponge or similar) in the reaction system, nitrogen pickup into the deoxidized titanium sample can be substantially reduced.

George Zheng Chen remarked: Not many people know that your paper published in 1996 on removing oxygen from Ti metal stimulated our research for the FFC Cambridge Process. Taking this opportunity, I would like to thank you for your great work. In your early papers, you used molten $CaCl_2$ and demonstrated thermodynamically and experimentally that molten $CaCl_2$ can dissolve and transport oxide ions. In the current paper, you report the use of $MgCl_2$ for removal of oxygen from titanium. In molten $CaCl_2$, you have decreased the oxygen level in titanium to below 10 ppm which is a significant achievement. I do not think the same can be demonstrated in molten $MgCl_2$ thermodynamically and experimentally. Is this understanding correct?

Toru H. Okabe answered: Your understanding is correct. In order to obtain titanium with 10 ppm O levels, the activity of MgO has to be decreased down to a level of 0.00039 under $Mg/MgCl_2$ equiliblium at 1273 K. This is practically very difficult. We could demonstrate that titanium can be deoxidized below 100 ppm O in molten $MgCl_2$, but these experimental conditions are very difficult to establish.

On the other hand, under $Ca/CaCl_2$ equilibrium, titanium with a level of 10 ppm O can be obtained when the activity of CaO is decreased to a level of 0.02, which can be achieved by electrochemical methods or otherwise. Roughly speaking, deoxidation by Mg is over 50 times more difficult than deoxidation by

Ca. Furthermore, the solubility of oxide in $MgCl_2$ is small compared to that in $CaCl_2$. This is also a problem when considering practical applications.

George Zheng Chen commented: Following your talk and that by Yating Yuan of Wuhan University, would you please comment on the technological feasibility of using $MgCl_2$ for oxygen removal from metals and metal oxides, particularly for Ti extraction from TiO_2? What could be the theoretical and practical challenges?

Toru H. Okabe answered: If the activity of MgO in molten $MgCl_2$ is high, around 1 (unity), the deoxidition limit of titanium by Mg is about 26 000 ppm, *i.e.* 2.6 mass%, at 1273 K. When extracting titanium directly from oxide (TiO_2) in molten $MgCl_2$, the concentration of oxide ion (O^{2-}) in the molten salt is very high and the activity of MgO increases to 1. Under these conditions, titanium containing a few percent oxygen can be obtained, which has to be deoxidized by other methods for practical applications.

If more effective methods for decreasing the activity of MgO in $MgCl_2$ system are invented, Ti extraction directly from TiO_2 in molten $MgCl_2$ may become feasible. Currently we are working on this subject but have not yet succeeded.

Toru H. Okabe commented: This is a comment for the young people in the audience, especially PhD candidates:

When I was working on electrochemical deoxidation about 25 years ago, no-one appreciated my work. At that time no-one cared about this kind of electrochemical techniques. In the 1990s, there were very few people showing interest in titanium smelting/refining (Prof. Fray and Dr Chen were the exceptions in the European academic community).

I have to say thank you for the chance I was given to concentrate on this subject. And I am also grateful I encountered wonderful researchers appreciating my efforts. As a result, I am now here having an opportunity to give a talk. Please try to work on the subject with clear vision and lofty dream. When working on the subject with rigid vision and faith, you will succeed in the future. Confront the challenge, and never give-up !

Wei Xiao commented: Although molten salt technologies are an old topic, it now becomes a new research hotspot due to the importance and versatile applications towards resource and energy sustainability. By combining the classic and fundamental concepts of molten salts with emerging molten salt technologies such as molten salt nuclear reactors, spent fuel reprocessing, liquid batteries, advanced materials extraction, the continuous innovation of molten salt technologies can be expected. I have no doubt that more and more young people will join in the research and development of molten salt technologies.

Xingli Zou responded: Molten salt electrolytic technologies have long history, and it is reasonable to believe that the molten/liquid salt technologies will become more and more important for structural/energy materials processing in the future. The development of the molten/liquid salt technologies needs more researchers to focus on the subject with persistence and faith!

Xianbo Jin replied: I agree with Professor Okabe, and thank you very much for the encouragement!

Ye Liu replied: I support this.

Hongmin Zhu asked: I appreciate your nice work. I have a question about the oxygen content in titanium after the electrolysis in molten $MaCl_2$. As you mentioned in Eq. 2, the thermodynamic equilibrium limit of oxygen content in Ti with Mg/MgO is about 2.3 wt% (23000 ppm). However you have even reached 100 ppm oxygen content by the electrolysis in $MgCl_2$. Does this mean the oxide content in your $MgCl_2$ was extremely low (< 1/230 of the MgO solubility in $MgCl_2$). Did you do any special treatment before use?

Toru H. Okabe replied: Yes. The MgO content in $MgCl_2$ was very low in the experimental conditions. If MgO content is high in $MgCl_2$, titanium containing about 2.3 wt% (23 000 ppm) will be obtained after experiment at 1000 °C. When activity of MgO is decreased by the electrochemical method down to a level of 0.042 (~1/200), 100 ppm O Ti can be produced.

As for the special treatment, dehydration treatment of $MgCl_2$, and pre-electrolysis before the deoxidation experiment are very important. Before performing the experiment, $MgCl_2$ was vacuum dried at 250 °C for more than 12 hours in order to eliminate water adhered on the surface of the $MgCl_2$ flakes. After melting $MgCl_2$ in the cell, pre-electrolysis to remove O^{2-} ions (or MgO in $MgCl_2$) was carried out. During the pre-electrolysis, the O^{2-} ion in the salt was eliminated out of the molten salt as CO_x gas from the anode, and excess Mg produced during pre-electrolysis was adsorbed into Ag shot placed at the bottom of the crucible.

Yingjun Liu asked: Table 4 shows the required activity of different metal oxides for reduced oxygen to certain levels. For Mg and Ca, a 10-fold difference in oxygen level lead to a 10-fold difference in the required activity of MO. For Y, why did a 10-fold difference in oxygen level lead to 30-fold difference in required activity of Y_2O_3?

Toru H. Okabe responded: The governing reactions, especially the stoichiometric amount of oxygen per oxide, are different between the MgO sytem and Y_2O_3 system. This results in a 30-fold difference in the required activity of Y_2O_3 or $YO_{1.5}$.

In the case of the Mg/MgO system (under high Mg activity ($a_{Mg} = 1$)), the equilibrium oxygen content in Ti ([O in Ti]) and the activity of MgO (a_{MgO}) have a proportional relationship. It is expressed by the following reaction:

$$O\ (\text{in Ti}) + Mg = MgO$$

$$K = a_{MgO}/([\text{O in Ti}] \times a_{Mg})$$

In the case of the Y/Y_2O_3 system, the equilibrium oxygen content in Ti ([O in Ti]) is NOT proportional to the activity of $YO_{1.5}$ ($a_{YO_{1.5}}$). [O in Ti] is proportional to the value of $(a_{YO_{1.5}})^{(1/1.5)}$. This is because the [O in Ti] is governed by the following reaction:

$$O\ (\text{in Ti}) + Y = (1/1.5)YO_{1.5}$$

$$K = (a_{YO_{1.5}})^{(1/1.5)}/([\text{O in Ti}] \times a_{Mg})$$

When comparing the [O in Ti] based on the activity of the oxides, the value of the $a_{YO_{1.5}}$ has to be to the power of (1/1.5) or 2/3. In this case, a 10-fold difference in $(a_{YO_{1.5}})^{(1/1.5)}$ can be obtained in the data listed in Table 4.

Xianbo Jin asked: I like your metal–oxygen diagrams very much, they are very useful. How did you get the free Gibbs energies of Ti(O)? Did you try to calculate them?

Toru H. Okabe responded: As for the ΔG value for oxygen in solid titanium at elevated temperature, we used data I measured myself when I was a graduate student. For further details, please take a look at the following book chapter, 'Removal of Oxygen in Reactive Metals'.[1] You can find a lot of data, and also diagrams which you may find of interest.

T. H. Okabe, K. T. Jacob and Y. Waseda, in *Purification Process and Characterization of Ultra High Purity Metals*, ed. Y. Waseda and M. Isshiki, Springer, Berlin, 2001, pp. 3–37.

John Irvine asked: What's the most desirable oxygen content in Ti, is it the lowest possible? Do certain oxygen contents give you better performance in some applications?

Toru H. Okabe replied: It depends on the application of titanium and its alloy. In general, the most desirable oxygen content in titanium is 500 to 2000 ppm. Oxygen in titanium solid solution is an interstitial element, and it increases the mechanical strength of titanium alloys. When the oxygen content exceeds 4000 ppm, titanium alloys become brittle and cannot be used as structural materials. The effect of the oxygen in titanium alloys is similar to the effect of carbon in steel.

Oxygen content in titanium scrap is in general very high, and oxygen content of titanium alloys increases when utilizing titanium scrap as a feed material. In some cases, oxygen content in titanium exceeds 4000 ppm when re-melting titanium scrap. In this case, titanium cannot be recycled.

In the future, amount of scrap generation of titanium and its alloys will increase. Therefore, development of a deoxidation technique below 500 ppm will be important in the future, when considering recycling.

Ali Kamali said: Which method did you use for oxygen content measurement? What is the accuracy of this method?

Toru H. Okabe answered: Oxygen content in titanium samples were determined by inert-gas fusion/infrared absorption spectroscopy, so called LECO analysis. The detection limit of the analysis is 10–20 ppm O, but O content below 100 ppm have some uncertainty as mentioned in the paper. The accuracy of the analysis is estimated to be within 10% when the oxygen content is above 200 ppm. However there is a greater variation in oxygen content between the deoxidized samples than analytical error.

George Zheng Chen addressed Toru H. Okabe, Shuqiang Jiao, Hongmin Zhu and Geir Martin Haarberg: Recently, some studies have been reported on using the principle of titanium refining in molten salts for production of pure titanium powder. In doing so, electro-deposition of titanium is the key step in which titanium dendrites form. Electro-deposition of titanium has been extensively studied between 1960's and 1970's, but none of these efforts achieved commercial success. In your opinion, can the current effort in making titanium powder by electro-refining reach commercial success? Do you think that using pure titanium may make the process easier?

Geir Martin Haarberg responded: I believe that the electrorefining of titanium in molten salts to produce pure titanium powder has a good chance of being scaled up to industrial levels. But this product will probably not compete with Kroll titanium, it will rather be a speciality product. Prof. Okabe's new proposed process seems very promising.

Shuqiang Jiao answered: Despite so many attempts and failures, there is still a possibility of making titanium powder by electro-refining reach commercial success. Compared with other process, this process has a lower-costs, higher yield and simplified procedure. The titanium dendrites may be controlled by the proper electroanalysis parameters during the electrolysis. And using pure titanium as anode is more preferable in the process due to its lower impurities.

Hongmin Zhu answered: Please refer to my answer for the following question from George Zheng Chen.

George Zheng Chen asked: I have similar concerns over similar reports on using electro-refining and electro-deposition for making titanium. Prof Zhu's group reported similar results (see my reply and comment to the previous question). In addition, I had private communications with senior scientists in China who in the 1970's tested the electro-deposition process to a scale of about 250 kg. However, their project was eventually abandoned. The question is whether the newly reported electro-refining process can overcome the problems previously encountered. Researchers in the 1970's realised that the oxygen content in their product was still too high in comparison with that from the Kroll process. However, this seems not to be the case in your work. Why could the earlier efforts achieve the same?

Hongmin Zhu replied: The oxygen content of titanium by electrode-deposition can be very low, in principle. We obtained Ti metal with oxygen content as low as 20 ppm by electrode-deposition, however sometimes it could be as high as 1%. There are two sources of oxygen. One is the oxide ion (O^{2-}) dissolved in the solvent salts. The other is oxygen on the particle surface after exposure in air and during washing in water. For the first part, the oxide ion content can be controlled by choosing the solvent salt and can also be reduced by drying, HCl treatment, pre-electrolysis and so on. This is the same case as Professor Okabe talked about for Mg reduction. For the latter case, the oxygen comes from the air or water after electrolysis. This part largely depends on the surface area of the titanium product. When you make powder of particle size down to the micrometre scale, the oxygen content is always very high. If you make dendrites or particles on the millimetre scale, the oxygen content could be kept fairly low – much lower even than the product of the Kroll process. However, it is important to note that the oxygen content of the fine powders is from the air after we expose it, not from the process of electrolysis itself. As long as we can keep the powder in a closed atmosphere, the oxygen content can also be kept at a low level. Also, the oxygen content in the product of cathode deposition does not have any relation to the anode material. It is the same whether you use spongy titanium or titanium oxycarbide as the anode material.

Chaohui Wei communicated: What percentage of oxygen can you remove from the system?

Toru H. Okabe communicated in reply: It is difficult to estimate the oxygen content in $MgCl_2$ because the accuracy of the oxygen analysis of titanium after the deoxidation experiment is not high, and also because the activity coefficient of MgO in $MgCl_2$ is not known. According to the literature data, solubility of MgO in $MgCl_2$ at around 1200 K is 1–2 mol%. From the results that show 100 ppm O Ti was successfully produced by the electrochemical deoxidation method using $MgCl_2$ at 1173 K, the activity of MgO was decreased down to a level of 0.042 (~1/200). From these values, one can estimate that the MgO content in $MgCl_2$ after the electrochemical deoxidation was kept at a level of about 0.04 mol%. The initial oxygen content of $MgCl_2$ is also difficult to evaluate. But one can see from the CV curves of O^{2-} oxidation (Fig. 7 (a)) that a fairly large amount of O^{2-} existed at the beginning of the pre-electrolysis of $MgCl_2$ molten salt.

Chaohui Wei communicated: Why do you use carbon as an anode electrode?

Toru H. Okabe communicated in reply: In order to decrease the activity of MgO in $MgCl_2$, oxide ion (O^{2-} ion) in molten $MgCl_2$ has to be removed out of the reaction system. During the electrochemical deoxidation experiments, the following reactions take place on the graphite anode:

$1/x$ C (anode) + O^{2-} (in $MgCl_2$) $\rightarrow$ $1/x$ CO_x (g) + $2e^-$

When using metal anodes, most of the metal dissolves into molten salt during electrolysis. Even platinum (Pt) dissolves into molten salt as Pt ions coordinated by Cl^- ion because the activity of the Cl^- ion in the molten salt system is high. When using Pt anode, it is difficult to remove O^{2-} in the molten salt by the following reaction:

O^{2-} (on Pt anode in $MgCl_2$) $\rightarrow$ 1/2 O_2 (g) + 2 e^-

Olga Kuzmina opened a discussion of the paper by Majd Eshtaya: What was the reason for the choice of 1-ethyl-3-methylimidazolium ethylsulfate IL and not a hydrogen sulfate, for example? Have you found a correlation between the increase of the CV peak area and degree of delignification?

Majd Eshtaya responded: In this study, 1-ethyl-3-methylimidazolium ethyl sulfate ($[C_2mim][C_2SO_4]$) was chosen due to its high capacity to dissolve lignin. A low inhibitory effect of the IL anion and cation on laccase activity has been found, rendering it suitable for combination with bioprocessing approaches. Moreover, this IL provides many desirable properties such as the fact it is a liquid at room temperature and the fact that it is water miscible.

The oxidation current or the peak height increased with the degree of lignin oxidation or with the concentration of the lignin in the solution. This correlation is also reflected by the CV peak area which is proportional to the total amount of charge passed for oxidation of lignin molecules in the electrolyte. However, on the CV time scale, there is a continuous supply of lignin molecules to the electrode surface. As a result, the number of lignin molecules being oxidised depends on the mass transfer in the electrolyte. In other words, the lower the potential scan rate (*i.e.* the longer the time for oxidation), the greater the degree of reaction with lignin (*i.e.* the number of lignin molecules being oxidised).

George Zheng Chen commented: In electrochemical experiments, the current is always proportional to the concentration of the redox active species either in the liquid electrolyte or confined to the electrode surface. Furthermore, for investigation in a liquid electrolyte, the variation of current recorded in experiments of different time scales or potential scan rates can often be an indicator if a charge transfer reaction on the electrode is diffusion controlled. Of course, the diffusion rate is relevant to the size of the active species and viscosity of the liquid electrolyte. For quantitative analytical purposes, it is feasible to use a working curve that can be obtained from the analysis of several known solutions.

John Slattery asked: The lignin that you have used in your experiments is quite different to lignin as it is found naturally in biomass. Could you use your approach to break down lignin directly within samples of biomass? This might enable you to access phenolic compounds and also to make the cellulose within the biomass more accessible to saccharification.

Majd Eshtaya responded: Biomass pretreatment is an essential step for lignin valorisation. There are several different treatments where lignin is produced as a product or by-product.

Yes, the electrochemical oxidation can be used in principle for delignification of lignocellulose matter using ionic liquids and redox mediators. It has also been reported that electrochemical delignification can be achieved in aqueous solution (Kim *et al.*, 2001) and the delignification of paper pulp with the laccase mediator system (LMS) has also been performed.[1]

1 H. C. Kim, M. Mickel, S. Bartling and N. Hampp, *Electrochim. Acta*, 2001, **47**, 799–805.

Tim Sudmeier asked: Thank you for this great talk. I was wondering if you had thoughts about covalently attaching ABTS to your electrode. This might help with diffusion limitation problems and lower the amount of ABTS needed for efficient biomass processing. People have covalently attached NADH to carbon nanoparticles and used them in enzyme electrochemistry.

Majd Eshtaya replied: Recently, Liu *et al.* demonstrated that ABTS covalently attached to silica nanoparticles (SNPs) effectively mediated dye decolourisation catalysed by laccase in a manner similar to free ABTS. However, the decolourisation rate was slower for ABTS-SNPs than for the free ABTS mediator.[1] Sadowska *et al.* also attached ABTS to the end or side wall of a carbon nanotube to mediate electron transfer.[2] Therefore, technically this method might be applied for the electrode and hence reduce the effect of the diffusion limitation, but on the other hand this may be affecting the efficiency of the process. It should also be mentioned that fixing ABTS molecules by a covalent bond onto the electrode would restrict their mobility and ability to access the redox centres that may be hidden inside the large lignin molecules in solution. Thus, we believe the ABTS molecule would be a more effective redox mediator when it interacts with the lignin molecule in the solution phase.

1 Y. Liu, M. Yan, Y. Geng and J. Huang, *J. Chem.*, 2015, 7.
2 K. Sadowska, K. Stolarczyk, J. F. Biernat, K. P. Roberts, J. Rogalski and R. Bilewicz, *Bioelectrochem.*, 2010, **80**, 73–80.

George Zheng Chen asked: It may be a good idea to have the catalyst (or mediator) affixed onto the electrode surface in some cases, *e.g.* making a sensor. However, such surface confined catalyst molecules may not always be capable of finding the reaction sites that may be hidden inside a large molecule in the solution. However, if the catalyst molecules are in solution, after they have undergone electron transfer reactions at the electrode surface, they come back to solution where they exchange the electrons (or charges) with the substrate in solution. This is why they are called mediators. Particularly, because the mediator molecules are freely mobile in solution, they can access and exchange electrons with the reaction sites inside the large substrate molecules. This is the advantage of having the mediator dissolved in solution, in comparison with affixing the mediator on the electrode surface.

Xianbo Jin said: Since ABTS is dissolved in the ionic liquid, it may cause some shuttle currents. Did you measure the current efficiency? Have you ever tried to detect the cathodic product?

Majd Eshtaya responded: We shall emphasise that the ABTS is used as a redox mediator in our study. In other words, the ABTS molecule obtains a charge from the electrode and then transfers immediately the charge to a neighbouring lignin molecule long before it reaches the counter electrode. Therefore, we do not expect any shuttle current in principle.

The CVs of ABTS alone are highly reversible, indicating a high current efficiency of ABTS oxidation. However, we have not yet fully identified the product from the ABTS-mediated oxidation of lignin, and hence are unable to report accurately the current efficiency for lignin oxidation. Nevertheless, we did observe that the CV peak current of the ABTS-mediated oxidation of lignin increased in proportion to the concentration of lignin in the explored concentration range. This is an indication of high current efficiency.

Andrew Doherty addressed George Zheng Chen and Majd Eshtaya: The heterogeneous electrochemistry of the ABTS mediator at low concentrations and fast potential sweep rates, as depicted in Fig. 3, is clearly simple facile outer-sphere electron transfer. At higher concentrations and lower potential sweep-rates voltammetric behaviour is indicative of the comproportionation reaction between ABTA and its dianion yielding the radical cation. Both the dianion and radical cation should exhibit outer-sphere redox catalysis in the pretence of an oxidisable substrate when appropriate thermodynamic conditions are met. Fig. 9 shows the onset of direct (at electrode) oxidation of all types of lignin at *ca.* 1.0 V which is attributed, possibly, to the oxidation of the "hydroxyl (OH) group on either the alpha-carbon or gamma-carbon" which requires deprotonation and an inner-sphere mechanism. In Fig. 10 the enhanced current in the potential region for the dication formation, and OH oxidation, in the presence of organsolv lignin and ABTS, is attributed to the ABTS dication mediating the oxidation of organosolv lignin. I don't understand how a simple outer-sphere electron-transfer species can simply oxidise a hydroxyl group without the formation of some type of activated complex and therefore proceed *via* an inner-sphere mechanism. Can the authors explain?

Majd Eshtaya responded: We agree with Dr. Doherty that Fig. 3 suggests the electron transfer reaction of ABTS to be outer-sphere in nature. Comproportionation between ABTS and its dication (not dianion) to form the radical monocation at more positive potentials may also occur at low scan rates and high concentrations. However, we are unable to assume that the ABTS radical cation and dication would only be capable of outer-sphere oxidation of an oxidisable substrate. It will depend on the molecular structure and chemical nature of the substrate. In Fig. 9, the CVs indicate irreversible oxidation of organosolv and two lignins, starting at potentials more positive than 0.7 V, likely related to the oxidation of the hydroxyl group in the lignin. However, we cannot ascertain, based on these CVs, if these irreversible electron transfer reactions follow the inner- or outer-sphere mechanism. On the other hand, the CVs in Fig. 10 show clearly that in the presence of organosolv, the distortion of the reversible CV of ABTS is more significant on the dication peak and the respective reduction peak disappears completely. This may be an indication of the ABTS dication being stabilised, at least on the CV time scale, *via* formation of a complex between the dication and the organosolv molecule. The absence of reduction peaks on the ABTS-organosolv CVs in Fig. 10 may also be explained by the completeness of the mediated oxidation of organosolv, consuming all ABTS dications generated electrochemically.

Andrew Mount addressed Majd Eshtaya and George Zheng Chen: In Scheme 1, it is not clear whether the electron transfer between ABTS and compound 4 involves just an electron transfer, or a combined electron, proton transfer. In principle consideration of the relative acidity of the functional groups in ABTS and compound 4 in their various redox forms, or measurement of the variation of CV peaks as a function of pH (*e.g.* in aqueous solution) could be used to derive information concerning the relative pK_as to inform the likelihood of simultaneous protonation. Has this been done by yourself and/or others in the literature, and if so what information has been obtained?

Majd Eshtaya answered: It was reported that the electrochemical response of ABTS was independent of pH on a bare GC electrode.[1] Also it has been reported that the formation of ABTS radical cation and dication are not pH-dependent within the range 2–11.[2] Veratryl alcohol oxidation activity was higher at pH 3.5 than pH 5 with ABTS using laccase, but this could be attributed to the enzyme component.[3] To the best of our knowledge no study has been done on veratryl alcohol oxidation by ABTS using CV with different values of pH. In our work, as it can be seen by comparing Fig. 11 and Fig. 13, the addition of an acetate buffer in the lignosulfonate solution led to a significant increase of the CV currents. This is good qualitative evidence of simultaneous transfer of a proton and electron, in line with Prof. Mount's comment.

1 E. Kibena, U. Maeeorg, L. Matisen and K. Tammeveski, *J. Electroanal. Chem.*, 2011, **661**, 343–350.
2 A. Majcherczyk, C. Johannes and A. Hüttermann, *Applied Microbiol. Biotechnol.*, 1999, **51**, 267–276.
3 R. Bourbonnais, M. G. Paice, B. Freiermuth, E. Bodie and S. Borneman, *Applied Environ. Microbiol.*, 1997, **63**, 4627–4632.

Binjie Hu remarked: On page 8, line 20, the mass transport rate of ABTS1 in $[C_2mim][C_2SO_4]$ is lower – although the viscosity of the ionic liquids you choose is higher compared to aquous solution – they are still much lower than those with longer alkyl carbon chains – clearly, it is not easy to improve mass transfer by replacing the type of ionic liquid, are there any other methods that can be employed to improve mass transfer, *i.e.* diffusion of ABTS in $[C_2mim][C_2SO_4]$?

Secondly, can you illustrate the type of interactions between ionic liquid and lignin solutions – physical or chemical? Will a new species form?

Majd Eshtaya answered: For the first question the mass transfer can be increased by increasing the temperature, using a co-solvent (*e.g.* acetonitrile) to decrease the viscosity of the solution, and also forced convection (*e.g.* stirring). In this study, all the electrochemical experiments were performed at room temperature in stationary solutions. The viscosity is a very important parameter in electrochemical studies due to its strong effect on the rate of mass transport within the solution. It significantly influences the diffusion of species, which are dissolved or dispersed in an ionic liquid. The viscosity can be influenced by various parameters such temperature. As the temperature increases, the viscosity of the ionic liquids decreases due to the higher mobility of ions and the diffusion coefficient will increase.

Another method can be used to improve the mass transfer is to reduce the solution viscosity by using mixtures of IL and solvents. In general, the diffusivity increases with increasing mole fraction of organic solvents (or water) to IL.

For the second question, the fact that lignin can be dispersed or dissolved in selected ILs but not in water or a common organic solvent is evidence of the presence both physical and chemical interactions.

Xiangling Yue opened a discussion of the paper by Wei Xiao: You proposed in the paper that hydrolysis of $CaCl_2$ would affect the microstructure of the carbon product, which was not happening to the same extent in molten carbonate as in $CaCl_2$. Have you considered the gasification in the case of molten carbonate treatment affecting the final product? At a high temperature, such as 800 °C in your case, the thermal decomposition of carbonate would be significant, and the evolved CO_2 can react with carbon to form CO gas. With Ar flowing through, gasification would happen significantly. To me, this might be the reason for the microstructure and performance difference in the resulting carbon in molten carbonate treatment compared to that in $CaCl_2$. Have you checked the XRD of the molten carbon and carbonate mixture before washing with acid? This would give you direct evidence of carbonate decomposition.

Wei Xiao replied: In the present study, molten Na_2CO_3–K_2CO_3 at 800 °C is used. The following is the thermodynamics on the decomposition of the melts:

Na_2CO_3 (l) = Na_2O (l) + CO_2 (g), $\Delta G = 178.9$ kJ mol^{-1}, $K = 1.955 \times 10^{-9}$; K_2CO_3 (l) = K_2O (l) + CO_2 (g), $\Delta G = 267.2$ kJ mol^{-1}, $K = 9.857 \times 10^{-14}$

Therefore, thermodynamically, the melt is stable at 800 °C, and the equilibrium constant of the above two reactions is extremely small. The inveitable trace CO_2 in the sealed cell is adequate to restrain the decomposition of molten Na_2CO_3–K_2CO_3. Hence, the utilized molten Na_2CO_3–K_2CO_3 can hardly decompose. We also consider another typical carbonate, *i.e.* Li_2CO_3:

Li_2CO_3 (l) = Li_2O (l) + CO_2 (g), $\Delta G = 81.4$ kJ mol^{-1}, $K = 1.089 \times 10^{-4}$

The possibility of decomposition of Li_2CO_3 is higher than the above two, and Li_2CO_3 is expensive. Therefore, Li_2CO_3 is excluded in the present study.

George Zheng Chen asked: Why did you not use acid to dissolve the carbonate residue? Practically, dilute acid is sufficient to remove the carbonate residue, but you can surely use more concentrated acid to deal with a more difficult sample.

Wei Xiao replied: Thanks for the concern. In the present work 0.1 M aqueous HCl was used during wash process of the molten salt treated samples. We agree with you on utilization of more concentrated acid (*e.g.* 1 M aqueous HCl) for a more effective wash.

Xiangling Yue asked: You compare the XPS of the original coffee bean carbon to the molten salt treated carbon – have you studied the coffee bean after treating at 800 °C without any molten salt? Indeed, having molten salt and carbon pyrolysed together may impact the structure of carbon, have you compared the pyrolysed carbon itself at 800 °C to that with molten carbonate?

Fig. 1 SEM of the products obtained after annealing the boiled coffee bean at 800 °C in Ar for 3 h.

Wei Xiao responded: The pyrolysis of the boiled coffee beans was investigated. The SEM of the products obtained after annealing the boiled coffee bean at 800 °C in Ar for 3 h is shown in Fig. 1, below. As can be seen, the product is much less porous than those treated in molten salts. The pyrolysis of boiled coffee beans in Ar at the same temperature is much more sluggish than those treated in molten salts. This difference highlights the important role of molten salt on enhancing carbonization of biomass for carbon preparation. Molten salts can catalyze pyrolysis of biomass, dissolve inconvenient impurities of biomass and also engineer microstructure/surface chemistry of the derived carbon. Hence, molten salt carbonization of biomass is a promising method for carbon production.

Cairong Jiang addressed Wei Xiao and Dihua Wang: What is the functionality of carbonate in your research? You have used Na_2CO_3–K_2CO_3 and $CaCl_2$, have you thought about alternative carbonates, such as other binary or ternary carbonates? What did you consider when choosing the testing temperature? Why did you carry out your test in Na_2CO_3–K_2CO_3 at 800 °C and in $CaCl_2$ 850 °C? Is this related to the melting temperature of the carbonate? If it is the case, other binary or ternary carbonate might be a good choice as they have lower melting points.

Wei Xiao responded: As to the selection of molten salt for carbonization of biomass, the reaction and stability of the melts is considered. For carbonates, lithium carbonates is less stable at high temperature and more expensive. Therefore, sodium and potassium is selected. For the temperature, the melting point for Na_2CO_3–K_2CO_3 is around 710 °C. At higher temperatures, the melts might become unstable. Therefore, molten Na_2CO_3–K_2CO_3 at 800 °C is selected. The same criteria is also considered to molten $CaCl_2$ which has a melting point of 792 °C. Therefore, 850 °C is selected for molten $CaCl_2$.

In summary, high temperature can accelerate the carbonization process, which may decrease the yield and change surface chemistry of the derived carbon. The stability and reaction capability of the melts should be balanced. We acknowledge that many other melts may also be effective for the carbonization of biomass, which deserves further study.

Dihua Wang replied: Thanks for your questions. For molten salt carbonization (MCC), the composition of molten salt does play a crucial role. We have previously investigated the effect of molten salts and temperature on the MCC,[1] which provided some reference form for the selection of molten salts in this work. Generally speaking, there are some advantages for MCC, including its high heat capacity, solubility for inorganic impurities in biomass and activation effect by the direct reaction between molten salt and biomass. These functions would be affected by the composition of molten salts such as carbonate or chloride. For the testing temperature, it depends on the melting point of the molten salts as well as the required temperature for pyrolysis. Although ternary carbonates have a lower melting point, the pyrolysis reaction under 700 °C is not good for carburization.

1 H. Y. Yin, B. H. Lu, Y. Xu, D. Y. Tang, X. H. Mao, W. Xiao, H. Zhu, D. H. Wang and A. N. Alshawabkeh, *Environ. Sci. Technol.*, 2014, **48**, 8101–8108.

Cairong Jiang asked: How much carbon have you got from your 20 g of coffee beans? In other words, what is the production yield? Is there a way to estimate it?

Wei Xiao replied: In the present case, the production yield for carbon is around 30%. Generally, the carbon content in the precursory biomass was estimated by elemental analysis. The mass of final product was directly weighed. Then the production yield for carbon preparation can be calculated.

It should be noted that minor carbon loss during the course of rinsing for removal of residual salts is inevitable.

George Zheng Chen addressed Wei Xiao and Dihua Wang: I do not understand why you have made it so difficult for carbon analysis. There are simple ways to estimate, at least. One simple way you can do this is to weigh the mass of the salt before you heat and then after you've done the process you measure the weight again. The difference between the two measures will be a good estimate of the percentage of carbon in the solid and in the gas.

Wei Xiao answered: Thanks for the concern. We acknowledge that the yield can be deduced with the suggested procedure.

Dihua Wang answered: The pyrolysis product of waste biomass can be in the solid state (carbon), liquid (oil) or gas, which is dependent of temperature, composition of molten salts as well as composition of the biomass. In our experiments, the molten salts were used repeatedly and the biomass was put in a nickel basket, by which we can run the experiments continuously. Therefore, it is difficult to follow your suggestion. On the other hand, biomass always contains some inorganic impurities which will be dissolved into the melt.

Xiangling Yue asked: In your experiments you put molten salt into the crucible, seal and heat the crucible, and then put carbon in the molten salt. I am curious as to how exactly you do this experiment. Do you have the carbon suspended in the crucible and just added it in the liquid at high temperature? My point is obviously carbon can't be outside the crucible, otherwise you have to break the sealant to add carbon.

Wei Xiao answered: The experimental details are as follows. The dried boiled coffee beans were wrapped with a porous nickel foam sheet. The nickel foam acts as a basket to hold the precursor and the product, allowing penetration of the molten salt. The nickel foam containing the precursor was then attached to the end of a Mo wire (2 mm in diameter, 50 cm in length). A hole was then drilled (less than 2mm) in a ruber stopper and the Mo wire was inserted (with the attachment) into the hole. The reactor was then sealed with the rubber. After melting the salt, the whole thing was immersed into the molten salts. The immersion or non-immersion of the precursor can be controlled by adjusting the insertion depth of the Mo wire in the hole. During the treatment, the system is sealed.

George Zheng Chen asked: In your experiments for the treatment of the coffee beans in molten salts, what did you see when you opened the furnace to remove the treated nickel foam wrapped sample? If you saw a blue flame, it indicates formation of some kind of fuel gases, such as small hydrocarbon molecules, which can result in carbon loss *via* the gas phase. How much did the temperature of the removed sample drop by before it was lifted into air? If the temperature was too high, there could also be a big loss of carbon due combustion or oxidation in air. What else did you observe in the process that may cause carbon loss?

Wei Xiao responded: Thanks for the comment. After termination of molten salt treatment, the sample wrapped by Ni foam was lifted out of the molten salt (but still in the reactor purged with Ar). If some air was introduced into the reactor, flame over the sample could be observed. Therefore, we agree with you that some gaseous and liquid products are also generated. The generation of such liquid and gaseous products account for carbon loss during the molten salt treatment.

Ali Kamali addressed Wei Xiao and Xianbo Jin: Interesting research! I would recommend that you wash the residual salt off the product as much as possible so that you can carefully measure the amount of carbon you recover and much more. Without a proper washing it is extremely difficult to analyse the quality of carbon nanomaterials produced.

Xianbo Jin remarked: We have similar experience that it is very difficult to wash away salts in the activated carbon.

Wei Xiao asked: For the impurities in the derived carbon, there are two kinds of mechanisms. Some residual salt tends to be adsorbed over the derived carbon if there is inadequate rinsing in water. Such kind of physically adsorbed impurities can be further removed by a further washing process. Another kind of impurity is chemically bonded species, which, however, are difficult to remove by washing. In such a scenario, some metal species can enter into the lattice of products.

Dihua Wang remarked: One of the significant advantages of molten carbonisation of waste biomass is its unique property for dissolving inorganic impurities in the original biomass, for example, silica in dusk shell, so that carbon material with a very high purity can be obtained by the method.

Xianbo Jin said: It seems best to covert biomass to fuel – can you make a comment on this?

Wei Xiao responded: Molten salts could be a promising reaction media for the conversion of biomass into fuel. In such a scenario, catalytic behavior of molten salts play an important role in the selective formation of fuels instead of carbon. It is believed that elaborate design of the molten salt reactors to facilitate collection of liquid and/or gaseous fuel is important. In addition, extra catalysts capable of directing the formation of fuels is also a prerequisite.

Chaohui Wei communicated: I feel that coffee beans are quite a complex biomass source – what is the purity of their obtained carbon?

Wei Xiao communicated in reply: We believe that molten salt treatment of biomass is a versatile method for carbon preparation. In our previous report in ref. 3, more than 14 kinds of waste biomass were studied and found to be successful for carbon preparation. In the present study, the exact purity of the final product remains unexplored. Some organic species and very minor inorganic impurities remain. The purity of the derived and the evolution of elements during molten salt treatment deserve further study.

Chaohui Wei communicated: Why use a symmetric cell instead of an asymmetric cell?

Wei Xiao communicated in reply: A symmetric configuration is employed here to exactly compare the difference in capacitive properties of different carbons. It is acknowledged that more elaborate design of the full cell configuration of the capacitor is effective for further enhancing the capacitive capability.

Faraday Discussions

PAPER

Temperature dependence of the electrode potential of a cobalt-based redox couple in ionic liquid electrolytes for thermal energy harvesting

Jiangjing He,[a] Danah Al-Masri,[a] Douglas R. MacFarlane[b] and Jennifer M. Pringle*[a]

Received 27th December 2015, Accepted 8th February 2016
DOI: 10.1039/c5fd00238a

Increasing the application of technologies for harvesting waste heat could make a significant contribution to sustainable energy production. Thermoelectrochemical cells are one such emerging technology, where the thermal response of a redox couple in an electrolyte is used to generate a potential difference across a cell when a temperature gradient exists. The unique physical properties of ionic liquids make them ideal for application as electrolytes in these devices. One of the keys to utilizing these media in efficient thermoelectrochemical cells is achieving high Seebeck coefficients, S_e: the thermodynamic quantity that determines the magnitude of the voltage achieved per unit temperature difference. Here, we report the S_e and cell performance of a cobalt-based redox couple in a range of different ionic liquids, to investigate the influence of the nature of the IL on the thermodynamics and cell performance of the redox system. The results reported include the highest S_e to-date for an IL-based electrolyte. The effect of diluting the different ILs with propylene carbonate is also reported, which results in a significant increase in the output powers and current densities of the device.

Introduction

While ionic liquids and molten salts have been widely used as electrolytes for electrodeposition, in batteries, capacitors *etc.*,[1] their use for thermal energy harvesting is still relatively unexplored. There are myriad sources of low level waste heat, either natural (*e.g.* geothermal activity) or man-made (industrial, automotive, domestic *etc.*) that could potentially be converted to electrical energy, to offset local power demands and thus decrease our environmental impact. Thermoelectrochemical cells (also termed thermogalvanic cells or thermocells) are

[a]*ARC Centre of Excellence for Electromaterials Science, Institute for Frontier Materials, Deakin University, 221 Burwood Highway, Burwood, VIC 3125, Australia. E-mail: jenny.pringle@deakin.edu.au*
[b]*School of Chemistry, Monash University, Wellington Road, Clayton, VIC 3800, Australia*

being increasingly investigated as an alternative to semi-conductor based thermoelectric devices, particularly for harvesting low grade (<140 °C) waste heat. Part of the drive behind this research is that the Seebeck coefficient, S_e, can be of the order of mV K^{-1} in the room temperature region, for good thermoelectrochemical systems, which is at least an order of magnitude larger than for semiconductor materials. The performance of traditional thermoelectric devices, which have been much more widely researched and developed, is normally compared using a figure of merit, which incorporates S_e.[2] However, despite significant recent improvements in the figure of merit, these devices are still not widely commercially viable, particularly for harvesting low temperature waste heat.

The use of molten salts in thermoelectrochemical cells, TECs, was also first introduced in the 1960s,[3] but without significant further development or uptake of the technology. Subsequently the majority of research into these devices has focussed on the use of aqueous electrolytes.[4–6]

In a TEC, two electrodes, in contact with an electrolyte, are held at different temperatures and it is the temperature dependence of the electrode potential that produces a potential difference across the device when a temperature gradient exists.[7–10] In the TECs discussed here, a redox couple is contained in the electrolyte. For example, in a redox reaction between species A and B:

$$\mathrm{A} + n\mathrm{e}^- \leftrightarrows \mathrm{B} \quad (1)$$

the equilibrium position is different at the hot and cold electrodes and hence a potential difference is produced. The temperature dependence of the open circuit potential of the cell at steady state, which gives S_e (alternatively termed α), can be related to the entropy change in the reaction by:[9]

$$nF\left(\frac{\partial E}{\partial T}\right)_{t=\infty} = (S_\mathrm{B} + \widehat{S}_\mathrm{B}) - (S_\mathrm{A} + \widehat{S}_\mathrm{A}) - n\overline{\overline{S}}_\mathrm{e} \quad (2)$$

where n is the number of electrons involved in the electrode reaction (eqn (1)), F is the Faraday constant, S_A and S_B are the partial molar entropies of species A and B (eqn (1), which includes the solvent reorganisation entropy), $\hat{S}_\mathrm{A}$ and $\hat{S}_\mathrm{B}$ are their respective Eastman entropies of transport (the entropy associated with movement of the ions), and $\overline{\overline{S}}_\mathrm{e}$ is the transported entropy from the electrons in the metallic electrodes, leads *etc.* which are at different temperatures. However, this equation is for a steady state open circuit potential, which can take many hours to establish. Therefore, as the Eastman entropies and $\overline{\overline{S}}_\mathrm{e}$ are generally small and slow to reach equilibrium, the practical S_e value, obtained from the initial steady open circuit voltage (here, obtained typically after 10–20 minutes thermal equilibration) can be more simply represented by the difference between the partial molar entropies of the two halves of the redox couple:

$$nF\left(\frac{\partial E}{\partial T}\right)_{t=0} = S_\mathrm{B} - S_\mathrm{A} \quad (3)$$

For the development of novel redox electrolytes for thermoelectrochemical devices, we can separate out two different aspects for investigation. The first is the thermodynamics of the system, primarily given by the Seebeck coefficient of the redox couple in the electrolyte, which can be measured using a number of different electrochemical techniques.[8,11] Here, we measure $\frac{\partial E}{\partial T}$ in a non-isothermal cell, where measurement of the open circuit voltage, V_{oc}, with increasing temperature gradient allows direct determination of S_e. Subsequently, the performance of a full TEC, containing the redox electrolyte, is analysed. In the full device, when current is drawn, parameters such as mass transport, charge transfer resistance and ohmic resistance become important influences on the performance characteristics of the cell, in addition to the thermodynamic factors that dictate V_{oc}. Thus, here we discuss the Seebeck coefficient of a range of novel ionic liquid-based redox electrolytes first, followed by assessment of the full TEC performance.

The Seebeck coefficient

The Seebeck coefficient of a redox electrolyte depends on the nature of both the redox couple and the solvent(s). The most widely used redox electrolyte is aqueous $Fe(CN)_6^{3-/4-}$, which has $S_e \approx -1.4$ mV K^{-1},[4,9,12] depending on the concentration and use of a supporting electrolyte. This represents a relatively large change in entropy between the solvated oxidised and reduced forms of the redox couple. The $Cu/CuSO_4$ couple in water, which has Seebeck coefficients up to 0.84 mV K^{-1}, has also been investigated.[13] The archetypal aqueous $Fe(CN)_6^{3-/4-}$ system has also been used for recent developments of TECs with high surface area electrodes, in particular those based on nano-dimensional carbon materials, which show significant increases in power.[4,5,12]

Although aqueous electrolyte systems presently give the highest TEC power outputs, there are clear benefits to using ionic liquid electrolytes in these devices. The use of water limits the operating temperature of the device (and hence the temperature of waste heat that can be harvested), while a non-volatile electrolyte would be beneficial for long-term stability. Thus, some high boiling molecular solvents such DMSO, γ-butyrolactone and DMF have been explored for TEC applications.[14] As an alternative, the good thermal stability of many ionic liquids could allow harvesting of higher temperatures of waste heat, *e.g.* in the 100–200 °C range, without the volatility problems of molecular solvents. Further, the ionic solvent environment afforded by ILs will be different from those of aqueous or molecular solvent systems, which will impact S_e in ways not yet well explored, and could also allow dissolution of redox couples that are insoluble in other solvents. IL properties such as conductivity and viscosity will also influence the current delivery characteristics of the TEC.

The temperature dependence of the electrode potential of a wide range of different redox couples, in different solvents, has been investigated.[8,15–17] However, significantly fewer of these couples have been trialled in full TECs.[5,13,18] The molecular interpretation of S_e is also still largely unresolved; a semi-empirical correlation between S_e and the acceptor number of different molecular solvents has been demonstrated for a variety of redox couples, but could not be fully explained.[15]

Empirically, we consider that when the electron transfer reaction is accompanied by a large change or rearrangement of the solvation shell (and possibly a change in the extended solvent structuring), this is reflected in a large ΔS_{rc},[16,17,19] where $S_e \approx \Delta S_{rc}/nF$. Our first investigation of redox couples in ILs for TEC applications focussed on the I^-/I_3^- redox couple.[20] This revealed a significant dependence of S_e on the nature of the IL, with values ranging from 0.03 to 0.26 mV K^{-1} (for $[P_{2,2,2,(101)}][NTf_2]$ and $[C_2mim][BF_4]$, respectively). Recently, Aldous and co-workers reported the benefit of using a combination of ferrocene and iodine to give S_e values up to 1.67 mV K^{-1} in $[C_2mim][NTf_2]$.[21]

We have investigated the use of a $Co^{II/III}$ couple in ILs, which allowed significant improvements to S_e and power output compared to the I^-/I_3^- couple.[22] The $Co^{II/III}(bpy)_3(NTf_2)_{2/3}$ complex, (where bpy = bipyridyl and NTf_2 = bis(trifluoromethanesulfonyl)imide) gives S_e values of between 1.4 and 1.88 mV K^{-1}, depending on the nature of the IL and the redox couple concentration. It is hypothesised, supported by experimental studies in molecular solvents and theoretical investigations,[17,19,23] that the high/low spin transition of the cobalt centres upon electron transfer contributes to a bigger difference between the partial molar entropies of the two species and thus a larger S_e. However, how the nature of the solvation environment created by the IL influences this significant ΔS_{rc} is not yet understood.

Investigations into ΔS_{rc} for a range of transition-metal based redox couples in different ILs have shown a dependence $((Z_{ox}{}^2 - Z_{red}{}^2)/r)$ on the charge, Z, and radius, r, of the electroactive species, suggesting a strong influence of electrostatic forces on the magnitude of ΔS_{rc}.[24,25] In early studies on molecular solvents, empirical relationships between ΔS_{rc} and solvent acceptor number,[15] or donor number,[17] for a variety of different redox couples were tentatively proposed. Drawing on this, Migita *et al.*[24] showed that the S_e value for an iron redox couple in one IL also fitted the linear trend between acceptor number and S_e. However, establishing a qualitative relationship between ΔS_{rc} and a specific "solvent parameter" for all ILs requires considerable further investigation. It is now well recognised that there are a complex range of interactions within ionic liquids, including coulombic, Van der Waals, hydrogen bonding and dispersion forces, all of which could conceivably influence the difference in partial molar entropies between two halves of a redox couple. For reasons that are not yet clear, with our inorganic redox couples we most commonly observe S_e values that are lower in ILs than in molecular solvents.[20,26] Similarly, S_e for molten salt-based systems are reported to be lower than for analogous aqueous systems.[9] However, an increase in S_e with increasing concentration of IL in an IL : acetonitrile system was observed by Zinovyeva *et al.*[27] for an organic redox couple.

Thermocell performance

When the redox electrolyte is used in a full TEC, the value of S_e determines V_{oc}. However, when current is drawn, the output parameters of the device also depend on the internal resistances within the cell: mass-transport and activation polarization plus ohmic resistance.[7] Therefore, when comparing the TEC performance of different redox electrolytes, it is important to note that the diffusion rate of the redox couple through the electrolyte, the thermal conductivity of the electrolyte (which also dictates the ΔT that can be maintained), the solution conductivity and the charge transfer resistance can all impact the maximum power density. The first stage in the development of a theoretical model of the device operation based on these fundamental properties has been reported.[28]

One of the disadvantages of ILs as electrolytes is the relatively high viscosity compared to molecular solvents, which can result in significant mass transport limitations. Our previous study of the $Co^{II/III}(bpy)_3$ redox couple in neat ILs has shown that when Pt electrodes are used (as they are in this study) the mass transport limitation is much more significant than the charge transfer resistance.[28] The addition of molecular solvent to the IL can help address this limitation; we have recently demonstrated the benefit of using a mixture of IL and either DMSO or methoxypropionitrile to improve both S_e and TEC power output.[26] The significant increase in current and power density that occurs with addition of molecular solvent is a result of decreased electrolyte viscosity, increased conductivity and a small increase in S_e. Here we further investigate this phenomenon, using propylene carbonate (PC) as the molecular solvent and also a wide range of different ILs designed to explore the effect of changing the cation or anion on S_e and TEC performance.

In these mixed IL/solvent systems the composition will also influence the thermal and ionic conductivity, with the optimum composition likely dependant on the T_{hot} and T_{cold} used and the separation between the two electrodes. Therefore, in this study, to investigate the influence of different ILs, we use a consistent ΔT (30 °C) and electrode spacing and compare the performance of a range of neat IL and IL : PC systems. The concentration of the redox couple can also change the mass transport characteristics and V_{oc} (as S_e is normally concentration dependant), so this was kept constant at 10 mmol l^{-1}. The average operating temperature, the ΔT and the concentration of redox couple are relatively low compared to other studies, which is reflected in the power outputs;[26] while S_e normally decreases with increasing redox couple concentration,[8,14] the use of higher concentrations can, up to a point, be beneficial for TEC current and power output,[7,13,14] particularly in a mass-transport limited system. However, the aim of this investigation was comparison of different ILs, rather than achieving maximum device powers. There is also limited solubility of the $Co(bpy)_3$ couple in some of the ILs tested here, so the low concentration allowed us to include as many ILs as possible in the study.

Results and discussion

The Seebeck coefficient

The influence of the nature of the IL on the Seebeck coefficient of the $Co^{II/III}(bpy)_3(NTf_2)_{2/3}$ redox couple was investigated by systematically varying the cation and the anion. The ionic liquids studied were 1-butyl-3-methylimidazolium bis(trifluoromethanesulfonyl)imide, $[C_4mim][NTf_2]$, 1-butyl-2,3-dimethylimidazolium bis(trifluoromethanesulfonyl)imide, $[C_4dmim][NTf_2]$, 1-butyl-2,3-dimethylimidazolium trifluoromethanesulfonate, $[C_4dmim][OTf]$, 1-butyl-3-methylimidazolium tetrafluoroborate, $[C_4mim][BF_4]$, 1-butyl-3-methylimidazolium methanesulfonate, $[C_4mim][MeSO_3]$, 1-ethyl-3-methylimidazolium methanesulfonate, $[C_2mim][MeSO_3]$, 1-butyl-3-methylimidazolium tris(pentafluoroethyl)trifluorophosphate, $[C_4mim][eFAP]$, trihexyl(tetradecyl)phosphonium bis(trifluoromethanesulfonyl)imide, $[P_{6,6,6,14}][NTf_2]$, trihexyl(tetradecyl)phosphonium *p*-toluenesulfonate, $[P_{6,6,6,14}][Tos]$ and trihexyl(tetradecyl)phosphonium tetrafluoroborate, $[P_{6,6,6,14}][BF_4]$.

Table 1 Seebeck coefficients of 0.01 M $Co^{II/III}(bpy)_3(NTf_2)_{2/3}$ in different imidazolium ILs and in 1 : 1 IL : PC mixtures

	S_e (mV K^{-1}) ± 0.02	
Ionic liquid	Neat IL	1 : 1 IL : PC
$[C_4mim][MeSO_3]$	N/A[a]	2.16
$[C_2mim][MeSO_3]$	2.13	2.14
$[C_4mim][eFAP]$	1.69	1.88
$[C_4mim][NTf_2]$	1.54	1.76
$[C_4dmim][NTf_2]$	1.49	1.71
$[C_4mim][BF_4]$	1.32	1.57
$[C_4dmim][OTf]$	N/A[b]	1.64
Propylene carbonate	1.86	

[a] Solid at RT. [b] Redox couple partially or completely insoluble at room temperature.

The Seebeck coefficients in the neat imidazolium ILs (Table 1) varied from 2.13 to 1.32 ± 0.02 mV K^{-1}, clearly illustrating the influence of the IL environment on the thermodynamics of the redox system. The S_e in $[C_2mim][MeSO_3]$ is the highest we have seen for this redox couple in any IL. The $[C_4mim][MeSO_3]$ is solid at room temperature but gives similarly high S_e values in the diluted system. To probe any influence of having a relatively acidic proton on the imidazolium cation, we compared S_e of the redox couple in imidazolium based ILs with and without methylation at the 2-position of the ring ($[C_4dmim][NTf_2]$ *vs.* $[C_4mim][NTf_2]$). The impact of this more acidic proton in imidazolium ILs is a much debated subject in the IL literature, particularly with respect to its hydrogen bonding ability and the origin of the increased melting points in the methylated species.[29] Comparison of the two NTf_2 systems reveals a slight decrease in S_e with use of the methylated IL, which would be consistent with a hypothesis of decreased solvent structuring, *e.g.* less hydrogen bonding. However, more ILs need to be studied to determine if this is a consistent trend across different methylated and non-methylated species.

To further investigate the influence of different solvents on S_e, we also report the first use of IL/propylene carbonate (PC) mixtures in this field. This solvent was chosen as prior literature has shown that $Co^{II/III}(bpy)_3(ClO_4)_{2/3}$ exhibits a significant ΔS_{rc} (46 cal K^{-1} mol^{-1} ≈ 1.99 mV K^{-1}) in PC.[17] Propylene carbonate has a relatively high dielectric constant (*ca.* 64), such that dipolar structuring and reorganisation might play a role in its solutions in the way that it seems to in aqueous solutions. The physical properties of some IL : PC electrolyte mixtures, in the absence of a redox couple, have also been previously investigated for applications such as electrochemical double-layer capacitors.[30] The Seebeck coefficient of the redox couple in PC is higher than in all of the imidazolium ILs except for $[C_2mim][MeSO_3]$. This is consistent with our prior studies into other ILs compared with methoxypropionitrile or DMSO, which also gave higher S_e values in the molecular solvents.[26] For the imidazolium ILs that produce lower S_e values than PC, mixing with PC increases S_e.

Here we also report the first Seebeck coefficient data for a $Co^{II/III}$ redox couple in phosphonium IL-based electrolytes (Table 2). Unfortunately the redox couple is

Table 2 Seebeck coefficients of 0.01 M $Co^{II/III}(bpy)_3(NTf_2)_{2/3}$ in different 1 : 1 and 1 : 3 phosphonium IL : PC mixtures

	S_e (mV K^{-1}) ± 0.02	
Ionic liquid	1 : 1 IL : PC	1 : 3 IL : PC
$[P_{2,2,2,4}]$[Tos]	2.09	1.98
$[P_{6,6,6,14}][NTf_2]$	2.04	1.89
$[P_{6,6,6,14}][BF_4]$	1.92	1.93

poorly soluble in the neat ILs, but full dissolution was achieved in the 1 : 1 IL : PC systems. The results of these phosphonium-based electrolyte systems are quite promising: in both the BF_4 and NTf_2 ILs, use of the $[P_{6,6,6,14}]$ cation gives higher S_e values than in the equivalent 1 : 1 systems with the $[C_4mim]$ cation. Further, all of the phosphonium ILs give higher S_e values than PC. Across this range of imidazolium and phosphonium ILs it is clear that both the cation and the anion of the IL can influence S_e. This indicates that although the anion can be expected to be most closely associated with the positively charged Co species, both the anion and cation are involved in the reorganisation of the solvent environment around the redox couple upon electron transfer, which is the origin of S_e.

Phosphonium ILs are generally more viscous than the imidazolium analogues, particularly when long alkyl chains are used, and thus the addition of molecular solvent is valuable for improving the diffusion rate of the redox couple. Furthermore, S_e values could not be obtained in the neat IL systems because of the low solubility of the redox couple. We therefore investigated the use of higher concentrations of PC, at 1 : 3 IL : PC volume ratio, for these phosphonium systems. Consistent with the fact that neat PC gives a lower S_e than the phosphonium ILs, further dilution with PC results in a decrease in S_e (Table 2). However, the S_e of the mixed systems remain among the highest we have seen, and the PC addition is beneficial for TEC performance, as discussed further below.

Thermoelectrochemical cell

In an operational TEC, when current is drawn, there are a number of other important factors, in addition to S_e, that ultimately determine the current, voltage and power output of the device. These include the diffusion rate of the redox couple and the conductivity of the electrolyte, both of which will vary between ILs.

Fig. 1 and Table 3 show a comparison of the TEC performance of the $Co^{II/III}(bpy)_3$ redox couple in the different imidazolium IL electrolytes, with and without PC. The V_{oc} depends on S_e, and hence varies according to the trend in Table 1. The highest powers are obtained with the pure propylene carbonate, which is attributed to the lower viscosity. Of the different ILs tested, the $[C_4mim][NTf_2]$ electrolytes give the highest currents in both the neat and 1 : 1 IL : PC systems. The poorest performance is obtained with $[C_4mim][BF_4]$, in both the neat and 1 : 1 IL : PC systems, which is consistent with the lower S_e. The higher S_e for $[C_2mim][MeSO_3]$ gives a high V_{oc} but the power and currents are significantly lower than most of the other neat IL systems. This is a relatively viscous IL (1113 mPa s at 273.15 K),[31] which is clearly detrimental to cell performance.

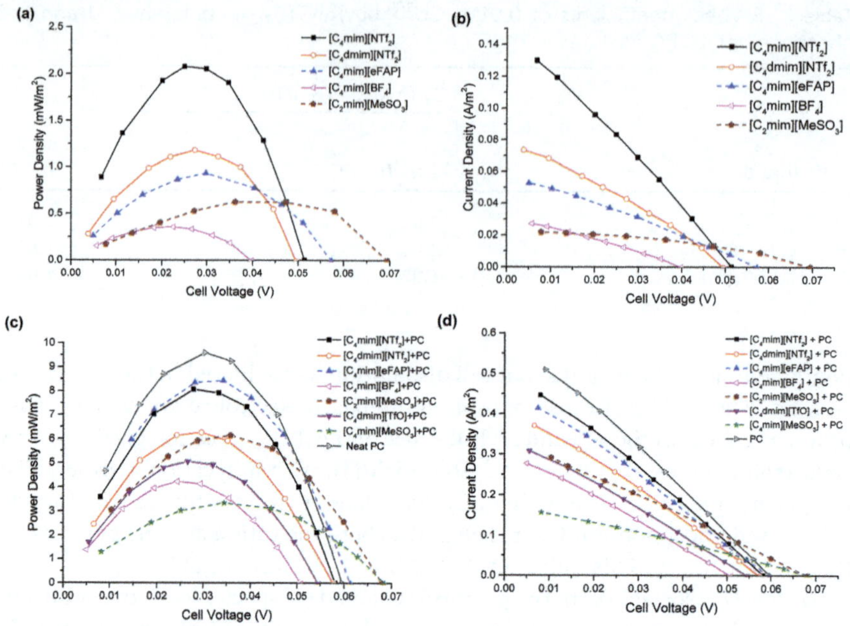

Fig. 1 (a) Power density and (b) current density *vs.* cell voltage for the neat imidazolium IL electrolytes, and (c) power density and (d) current density *vs.* cell voltage for the 1 : 1 imidazolium IL : PC and neat PC electrolytes, containing 0.01 M $Co^{II/III}(bpy)_3(NTf_2)_{2/3}$. $T_{hot} = 60$ °C, $T_{cold} = 30$ °C.

Similarly, the more viscous $[C_4dmim][NTf_2]$ gives significantly lower powers and currents than the $[C_4mim][NTf_2]$, despite the relatively similar Seebeck coefficients.

The impact of mass transport limitations was explored further by measurement of the diffusion coefficient for the redox couple in the different imidazolium IL systems (Table 4). As expected, PC allows the highest diffusion rate for the redox couple. Of the different ILs tested, $[C_4mim][NTf_2]$ supports the highest diffusion rate, which is consistent with the highest TEC currents and powers. The

Table 3 TEC performance of neat and 1 : 1 imidazolium IL : PC systems, with 0.01 M $Co^{II/III}(bpy)_3(NTf_2)_{2/3}$. $T_{hot} = 60$ °C, $T_{cold} = 30$ °C

	Power density ± 0.05/mW m^{-2}	
Ionic liquid	Neat IL	1 : 1 IL : PC
$[C_2mim][MeSO_3]$	0.67	6.02
$[C_4mim][eFAP]$	0.94	7.98
$[C_4mim][NTf_2]$	2.04	7.78
$[C_4dmim][NTf_2]$	1.15	6.44
$[C_4mim][BF_4]$	0.34	3.91
$[C_4dmim][OTf]$	N/A	4.88
$[C_4mim][MeSO_3]$	N/A	3.33
Propylene carbonate	9.59	

Table 4 Diffusion coefficients for $Co^{II}(bpy)_3$ and $Co^{III}(bpy)_3$ in the different electrolytes at room temperature

	Diffusion coefficient, $D/10^{-11}$ m^2 s^{-1} ± 5%			
	Neat IL		1 : 1 IL : PC	
Ionic liquid	$Co^{II}(bpy)_3$	$Co^{III}(bpy)_3$	$Co^{II}(bpy)_3$	$Co^{III}(bpy)_3$
$[C_4mim]$ $[NTf_2]$	1.2	1.2	7.7	7.6
$[C_4dmim]$ $[NTf_2]$	0.65	0.58	6.6	6.1
$[C_4mim]$ [eFAP]	0.45	0.40	5.9	6.0
$[C_4dmim]$ [OTf]	N/A[a]		5.8	4.6
$[C_4mim]$ $[BF_4]$	0.32	0.29	5.0	4.4
$[C_2mim]$ $[MeSO_3]$	0.22	0.27	4.0	4.3
$[C_4mim]$ $[MeSO_3]$	N/A[a]		1.8	2.4
PC	16	18		

[a] Solid at room temperature.

$[MeSO_3]$ ionic liquids produce the lowest diffusion rates, in both the neat and dilute systems. However, mixing $[C_2mim][MeSO_3]$ with PC makes a significant difference to the diffusion rate and therefore when this electrolyte is used in the TEC the mass transport limitations are not so great as they are in the neat system, and the advantage of the high V_{oc} is more evident in the power output. Similarly, the $[C_4mim]$[eFAP] systems have higher S_e values than the $[C_4mim][NTf_2]$, but in the neat ILs the diffusion coefficients in the former are significantly smaller and thus the power and current outputs are lower. In the 1 : 1 systems, which have similar diffusion coefficients, the TEC performance of these two electrolytes are similar. Thus, the benefit of PC addition is most significant for the ILs that are the most diffusion limited in their neat state.

Table 5 Ionic conductivity of the neat IL and IL:PC systems, containing 0.01 M $Co^{II/III}(bpy)_3(NTf_2)_{2/3}$. Data for neat phosphoniums not available due to poor solubility of the redox couple

	Solution conductivity (mS cm^{-1}) ± 5%	
Ionic liquid	Neat	1 : 1 IL : PC
$[C_4mim][BF_4]$	3.9	10.4
$[C_2mim][MeSO_3]$	3.7	11.0
$[C_4mim][NTf_2]$	3.7	11.1
$[C_4mim]$[eFAP]	2.8	8.1
$[C_4dmim][NTf_2]$	1.9	8.6
$[C_4mim][MeSO_3]$	N/A[a]	4.6
$[C_4dmim]$[OTf]	N/A[a]	7.2
PC	8.5	
$[P_{6,6,6,14}][NTf_2]$	N/A	2.6
$[P_{6,6,6,14}][BF_4]$	N/A	3.0
$[P_{2,2,2,4}]$[Tos]	N/A	1.9

[a] Solid at room temperature.

In addition to mass transport limitations, the different ionic liquids will impart different values of solution resistance, which also produce resistive losses in the TEC device. Measurement of the solution conductivity by EIS (Table 5) shows a significant variation across the neat IL and IL : PC solutions. It also shows that there is a valuable increase in conductivity upon addition of PC to the ILs. The most conductive system is that of PC, containing 0.01 M $Co^{II/III}(bpy)_3(NTf_2)_{2/3}$, which is attributed to the low viscosity. The least conductive system is the $[C_4mim][MeSO_3]$, which will contribute to the low power and currents produced by the TEC. However, beyond this there is no clear trend between the solution conductivity of the electrolytes and the current outputs of the TEC. Thus, while this parameter and its contribution to the total ohmic resistance in the cell will have an influence on the device performance, we conclude that it is not the most significant limitation in these systems. The trend between TEC current density and diffusion rate in the different electrolyte systems is much clearer, indicating that these are predominantly mass-transport limited systems.

The higher viscosities of phosphonium ILs is a disadvantage for their TEC performance, resulting in lower power outputs than most of the imidazolium salts (Fig. 2 and Table 6). This is also revealed in the lower solution conductivities of the phosphonium IL : PC systems compared to the imidazolium IL : PC systems at the same volume ratio (Table 5). However, the use of PC at a higher volume ratio (1 : 3) allows the TEC performance to be increased. In both the 1 : 1 and 1 : 3 IL : PC systems the highest powers and currents are achieved with the $[P_{6,6,6,14}][NTf_2]$. The $[P_{2,2,2,4}][Tos]$ shows the lowest solution conductivities of the

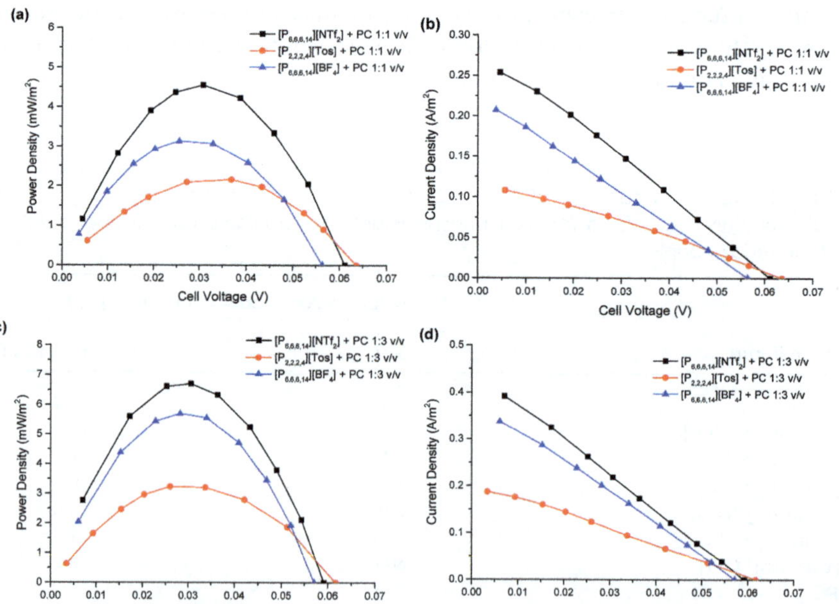

Fig. 2 (a) Power density and (b) current density *vs.* cell voltage for the 1 : 1 phosphonium IL : PC electrolytes, and (c) power density and (d) current density *vs.* cell voltage for the 1 : 3 phosphonium IL : PC electrolytes, containing 0.01 M $Co^{II/III}(bpy)_3(NTf_2)_{2/3}$. $T_{hot} = 60$ °C, $T_{cold} = 30$ °C.

Table 6 TEC performance of 1 : 1 and 1 : 3 phosphonium IL : PC systems, with 0.01 M $Co^{II/III}(bpy)_3(NTf_2)_{2/3}$

	1 : 1 IL : PC Power density $\pm$ 0.1/mW m^{-2}	1 : 3 IL : PC Power density $\pm$ 0.4/mW m^{-2}
$[P_{6,6,6,14}][NTf_2]$	4.5	6.9
$[P_{6,6,6,14}][BF_4]$	3.2	6.1
$[P_{2,2,2,4}][Tos]$	2.1	3.5

different electrolytes measured, and hence although this gives the highest Seebeck coefficients (Table 1) the TEC performance is significantly reduced.

Experimental

Materials

The Co^{II} (bipyridyl)$_3$ $(NTf_2)_2$ and Co^{III} (bipyridyl)$_3$ $(NTf_2)_3$ were synthesised following our previously reported procedures.[22] The imidazolium ionic liquids were purchased from Merck, with purity of at least >98%, except for the $[C_4mim][MeSO_3]$ that was purchased from io-li-tec (purity 99%). The phosphonium ionic liquids were obtained from Cytec or Sigma Aldrich, with a purity of at least 97%. The propylene carbonate was purchased from Sigma Aldrich (99.7% purity). All were used as received.

Measurement of the Seebeck coefficient and TEC performance

The Seebeck coefficient was measured using a non-isothermal cell as described previously.[20] For measurement of S_e and operation of the TEC, the electrolytes were prepared by dissolving equimolar amounts of each half of the redox couple in the different solvent(s). S_e values were measured using a non-isothermal cell composed of two platinum wires immersed in two sample vials containing around 1 ml of sample each. The sample vials were connected by a salt bridge containing the same redox electrolyte solution, and the temperatures of each were controlled by a heating brass block for the hot side and a water bath for the cold side. Each measurement was repeated at least twice and the average is reported. When necessary, values were rounded down rather than up to obtain the average. The error margin quoted reflects the largest deviation recorded for each solvent system in the series, quoted as half of the difference between the two measurements. Note that we also investigated $[C_4mim][PF_6]$, but the $Co^{II/III}(bpy)_3$ redox couple was not sufficiently soluble in either the neat IL or the 1 : 1 IL : PC system. The S_e and TEC performance of the redox couple in $[C_4mim][N(CN)_2]$ was also investigated, but is not reported due to instability and irreproducibility in the electrochemical behaviour.

The thermoelectrochemical cell (TEC) was composed of platinum disk electrodes (18 mm diameter) separated by an 8 mm cylindrical insert, with a 9 mm internal diameter representing the active electrode area. The whole cell was cooled and the two electrodes were heated independently to produce the desired ΔT. The power and current generated in the TEC were determined from the cell potential recorded across a sequence of known resistances. The resistances used were chosen according to the internal resistance of the cell. The values for maximum power were taken from the data point at the relevant resistance, rather than a fit to the data points.

Electrochemical impedance spectroscopy

Electrochemical impedance spectroscopy (EIS) was performed on 0.01 M solutions of the $Co^{II/III}$ couple in the different IL and IL/PC mixtures. The impedance of these systems was measured using a Biologic MTZ-35 Impedance analyzer, using a dip cell equipped with platinum electrodes. The measurement was performed at room temperature. The potential wave amplitude was 100 mV, the frequency range was 10 mHz to 2.5 MHz. The solution resistance (R_{Soln}, first touchdown) of the different electrolyte systems, with 0.01 M $[Co^{II/III}(bpy)_3](NTf_2)_2$, was determined from the Nyquist plots. The cell constant of the dip cell was determined from the Nyquist plot of a 0.01 M KCl solution of known conductivity, which was then used to calculate the solution conductivity.

Diffusion coefficient measurements

The diffusion coefficients of the Co^{II} and Co^{III} species were measured at room temperature by cyclic voltammetry in a three electrode cell using an ultramicro working electrode and a multiple-channel Biologic VMP3/Z potentiostat. A platinum electrode with 10 μm diameter, a platinum coil and a platinum wire were used as working, counter and reference electrodes respectively. The diffusion coefficients were obtained from the limiting currents at a scan rate of 20 mV s^{-1} for samples containing propylene carbonate and 3 mV s^{-1} for the neat ionic liquid samples.

Conclusions

Exploration of the Seebeck coefficient of the $[Co^{II/III}(bpy)_3](NTf_2)_{2/3}$ redox couple in a range of different ionic liquids has revealed relatively high redox reaction entropies in phosphonium based ILs, and in imidazolium sulfonate ILs. Comparison of the different S_e values for this redox couple across the range of ILs investigated shows a dependence on the nature of both the cation and the anion of the IL, indicating that the solvent restructuring that occurs upon charge transfer extends beyond the first solvation shell of anions. The addition of propylene carbonate to the ILs significantly increases the current and power densities that can be achieved in the thermoelectrochemical cell, which is consistent with a significant decrease in viscosity and increase in solution conductivity. The short circuit currents with the neat imidazolium IL electrolytes follow the same trend as the diffusion coefficients of the redox couple in these media, indicating that the performance of these devices is primarily mass transport limited. For the more viscous phosphonium IL electrolytes, the addition of propylene carbonate at a 1 : 3 IL : PC ratio is necessary to achieve power outputs closer to those of the best imidazolium ILs at a 1 : 1 IL : PC volume ratio. However, the achievement of high S_e values in the phosphonium ILs points to promising new directions in the exploration of different ILs for thermal energy harvesting; future work will investigate the use of smaller phosphonium cations, to assess the impact of different cations on S_e and also to achieve higher TEC power outputs with these less viscous ILs.

Acknowledgements

The authors gratefully acknowledge funding from the Australian Research Council through its Centre of Excellence program and fellowship support for D. R. MacFarlane (Australian Laureate Fellowship).

Notes and references

1 D. R. MacFarlane, N. Tachikawa, M. Forsyth, J. M. Pringle, P. C. Howlett, G. D. Elliott, J. H. Davis, M. Watanabe, P. Simon and C. A. Angell, *Energy Environ. Sci.*, 2014, **7**, 232; H. Ohno, *Electrochemical Aspects of Ionic Liquids*, John Wiley and Sons, Second edn, 2011; M. Smiglak, J. M. Pringle, X. Lu, L. Han, S. Zhang, H. Gao, D. R. MacFarlane and R. D. Rogers, *Chem. Commun.*, 2014, **50**, 9228; P. G. Bruce, S. A. Freunberger, L. J. Hardwick and J. M. Tarascon, *Nat. Mater.*, 2012, **11**, 19; J. Kalhoff, G. G. Eshetu, D. Bresser and S. Passerini, *ChemSusChem*, 2015, **8**, 2154.

2 J.-H. Bahk, H. Fang, K. Yazawa and A. Shakouri, *J. Mater. Chem. C*, 2015, **3**, 10362; W. Liu, Q. Jie, H. S. Kim and Z. Ren, *Acta Mater.*, 2015, **87**, 357; G. Pennelli, *Beilstein J. Nanotechnol.*, 2014, **5**, 1268.

3 K. Cornwell, *Br. J. Appl. Phys.*, 1968, **1**, 173.

4 R. C. Hu, B. A. Cola, N. Haram, J. N. Barisci, S. Lee, S. Stoughton, G. Wallace, C. Too, M. Thomas, A. Gestos, M. E. dela Cruz, J. P. Ferraris, A. A. Zakhidov and R. H. Baughman, *Nano Lett.*, 2010, **10**, 838.

5 M. S. Romano, J. M. Razal, D. Antiohos, G. Wallace and J. Chen, *J. Nanosci. Nanotechnol.*, 2015, **15**, 1.

6 T. J. Abraham, D. R. MacFarlane, R. H. Baughman, N. Li, Y. Chen and J. M. Pringle, *MRS Online Proc. Libr.*, 2013, **1575**, opl 2013 647/1.

7 B. Burrows, *J. Electrochem. Soc.*, 1976, **123**, 154.

8 A. J. De Bethune and H. O. Daley Jr, *J. Electrochem. Soc.*, 1969, **116**, 1401.

9 T. I. Quickenden and Y. Mua, *J. Electrochem. Soc.*, 1995, **142**, 3985.

10 T. I. Quickenden and C. F. Vernon, *Sol. Energy*, 1986, **36**, 63.

11 R. Koerver, D. R. MacFarlane and J. M. Pringle, *Electrochim. Acta*, 2015, **184**, 186.

12 T. J. Kang, S. Fang, M. E. Kozlov, C. S. Haines, N. Li, Y. H. Kim, Y. Chen and R. H. Baughman, *Adv. Funct. Mater.*, 2012, **22**, 477.

13 A. Gunawan, C.-H. Lin, D. A. Buttry, V. Mujica, R. A. Taylor, R. S. Prasher and P. E. Phelan, *Nanoscale Microscale Thermophys. Eng.*, 2013, **17**, 304.

14 Y. V. Kuzminskii, V. A. Zasukha and G. Y. Kuzminskaya, *J. Power Sources*, 1994, **52**, 231.

15 J. T. Hupp and M. J. Weaver, *Inorg. Chem.*, 1984, **23**, 3639.

16 S. Sahami and M. J. Weaver, *J. Electroanal. Chem. Interfacial Electrochem.*, 1981, **122**, 171.

17 S. Sahami and M. J. Weaver, *J. Electroanal. Chem. Interfacial Electrochem.*, 1981, **122**, 155.

18 H. L. Chum and R. A. Osteryoung, *Review of thermally regenerative electrochemical systems*, Solar EnergyRes. Inst., Golden, CO, USA, 1981.

19 E. L. Yee, R. J. Cave, K. L. Guyer, P. D. Tyma and M. J. Weaver, *J. Am. Chem. Soc.*, 1979, **101**, 1131.

20 T. J. Abraham, D. R. MacFarlane and J. M. Pringle, *Chem. Commun.*, 2011, **47**, 6260.

21 E. Hosseini Bab Anari, M. Romano, W. X. Teh, J. J. Black, E. Jiang, J. Chen, T. Q. To, J. Panchompoo and L. Aldous, *Chem. Commun.*, 2016, **52**, 745.

22 T. J. Abraham, D. R. MacFarlane and J. M. Pringle, *Energy Environ. Sci.*, 2013, **6**, 2639.

23 D. E. Richardson and P. Sharpe, *Inorg. Chem.*, 1991, **30**, 1412; D. E. Richardson and P. Sharpe, *Inorg. Chem.*, 1993, **32**, 1809.

24 T. Migita, N. Tachikawa, Y. Katayama and T. Miura, *Electrochemistry*, 2009, **77**, 639.

25 Y. Yamato, Y. Katayama and T. Miura, *J. Electrochem. Soc.*, 2013, **160**, H309.

26 M. A. Lazar, D. Al-Masri, D. R. MacFarlane and J. M. Pringle, *Phys. Chem. Chem. Phys.*, 2016, **18**, 1404.

27 V. Zinovyeva, S. Nakamae, M. Bonetti and M. Roger, *ChemElectroChem*, 2014, **1**, 426.

28 T. J. Abraham, N. Tachikawa, D. R. MacFarlane and J. M. Pringle, *Phys. Chem. Chem. Phys.*, 2014, **16**, 2527.

29 P. Bonhote, A.-P. Dias, N. Papageorgiou, K. Kalyanasundaram and M. Graetzel, *Inorg. Chem.*, 1996, **35**, 1168; E. I. Izgorodina and D. R. MacFarlane, *J. Phys. Chem. B*, 2011, **115**, 14659; P. A. Hunt, *J. Phys. Chem. B*, 2007, **111**, 4844; K. Fumino, T. Peppel, M. Geppert-Rybczynska, D. H. Zaitsau, J. K. Lehmann, S. P. Verevkin, M. Kockerling and R. Ludwig, *Phys. Chem. Chem. Phys.*, 2011, **13**, 14064.

30 D. J. Bozym, B. Uralcan, D. T. Limmer, M. A. Pope, N. J. Szamreta, P. G. Debenedetti and I. A. Aksay, *J. Phys. Chem. Lett.*, 2015, **6**, 2644; L. Xu, X. Cui, Y. Zhang, T. Feng, R. Lin, X. Li and H. Jie, *Electrochim. Acta*, 2015, **174**, 900; R. Zarrougui, M. Dhahbi and D. Lemordant, *J. Solution Chem.*, 2010, **39**, 921.

31 B. Hasse, J. Lehmann, D. Assenbaum, P. Wasserscheid, A. Leipertz and A. P. Fröba, *J. Chem. Eng. Data*, 2009, **54**, 2576.

Faraday Discussions

PAPER

Immobilization of a rhodium catalyst using a diphosphine-functionalized ionic liquid in RTIL for the efficient and recyclable biphasic hydroformylation of 1-octene

Yong-Qi Li, Huan Liu, Peng Wang, Da Yang, Xiao-Li Zhao and Ye Liu*

Received 26th November 2015, Accepted 11th January 2016
DOI: 10.1039/c5fd00195a

A highly efficient and stable Rh–P catalytic system in the RTIL of [PEmim]BF_4 was developed for the biphasic hydroformylation of 1-octene by using the diphosphine-functionalized ionic liquid (FIL) of **2**. While **2**-Rh(acac)$(CO)_2$ was immobilized in [PEmim] BF_4 (solvent), a typical biphasic catalysis was fulfilled with advantages of facile separation and recycling ability – 9 runs without any loss of activity. It was found that not only the acquired π-acceptor character of **2**, but also the synergetic role of the piperidyl group in [PEmim]BF_4 as an N-containing donor, cooperatively contributed to the efficient hydroformylation due to the facilitated formation and stability of the Rh-H active species (ν 2045 cm^{-1}). This was supported by the *in situ* high-pressure FT-IR spectral analysis.

1. Introduction

Hydroformylation of alkenes to give aldehydes is an atom-economic reaction which constitutes one of the most important homogeneously catalyzed processes in industry, in which the C1 resource of CO is used as the building block for carbonyl compounds.[1,2] The most efficient catalysts reported for this reaction are rhodium complexes, and a large amount of work has been devoted to the improvement of reaction rates and selectivity by ligand design.[3,4] For example, monodentate and bidentate phosphorous-based ligands,[5–11] N-heterocyclic carbene (NHC),[12–14] and the hybrid P,O- or P,N-ligands with multiple ligations to the rhodium center,[15–20] have been synthesized by finely tuning their electronic and steric properties to spur the activity and selectivity of the rhodium complexes.

However, the isolation and recycling of Rh-catalysts in homogeneous hydroformylation remains a challenge. Many efforts have been devoted to the development of catalytic systems that allow facile separation of the products from the

Shanghai Key Laboratory of Green Chemistry and Chemical Processes, School of Chemistry and Molecular Engineering, East China Normal University, 3663 North Zhongshan Road, Shanghai 200062, P. R. China. E-mail: yliu@chem.ecnu.edu.cn

catalysts and enable the catalyst to be recycled.[21] Aqueous/organic biphasic catalysis[22–24] and fluorous biphasic catalysis[25] are the most well-known methods for solving this problem. However, the low solubility of higher olefins (>C_6) in water has limited the scope of aqueous/organic biphasic hydroformylation. In recent decades, the use of functionalized ionic liquids (FILs) such as phosphine-FILs, which have long been investigated for the design of the ionic organometallates and the application to homogeneous catalysis,[26–33] in combination with room temperature ionic liquids (RTILs) as the solvents, has been developed as an efficient alternative for the immobilization of homogenous transition metal catalysts. It has advantages of improved catalytic performance and stability, facile biphasic separation and recycling of the recovered catalysts, and the tight capture of the metal catalyst in the IL phase *via* strong coordination to the phosphine-functionalized ILs.[34–41]

In this paper, for the first time, the diphosphine-FIL of **2** was applied with Rh(acac)$(CO)_2$ to catalyze the hydroformylation of 1-octene, with the RTILs ([Bmim]BF_4 and [PEmim]BF_4) as the solvents in this biphasic catalysis, in order to explore the separation and recyclability of the Rh–P catalyst (Scheme 1). Once the neutral diphosphine of **1** was incorporated into the structure of the imidazolium-based IL to afford **2**, due to the intensive electron-withdrawing effect of the adjacent imidazolium moiety, the coordinating ability of the phosphine fragments in **2** was proven to be greatly modulated, leading to the improved catalytic performance in comparison to that over **1**. Furthermore, an *in situ* high pressure spectroscopy technique was used to investigate the formation, stability, and evolution of the Rh-H (rhodium hydride) species in order to clarify the advantages over the **2**-Rh(acac)$(CO)_2$ system in the RTIL of [PEmin]BF_4, in terms of efficient hydroformylation and excellent recyclability.

2. Experimental

2.1 Reagents and analysis

The chemical reagents were purchased from Shanghai Aladdin Chemical Reagent Co. Ltd and Alfa Aesar China, and used as received. FT-IR spectra were recorded

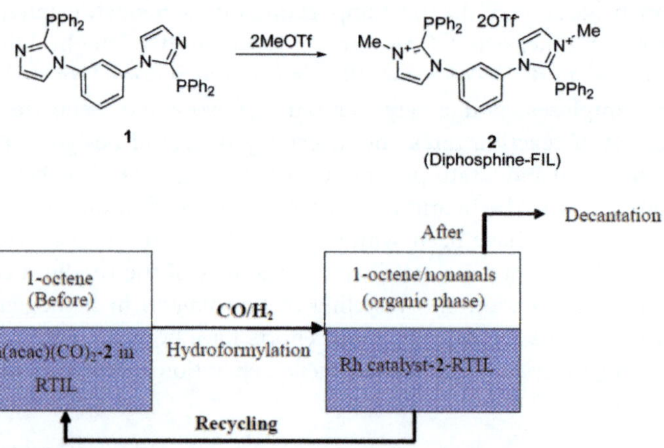

Scheme 1 Biphasic hydroformylation of 1-octene catalyzed by Rh(acac)$(CO)_2$ with the involvement of diphosphine-FIL **2**.

on a Nicolet NEXUS 670 spectrometer. The ^1H and ^{31}P NMR spectra were recorded on a Bruker Avance 400 spectrometer. The ^{31}P NMR spectra were referenced to 85% H_3PO_4 sealed in a capillary tube as an internal standard. The amount of Rh in the sample was quantified using an inductively coupled plasma atomic emission spectrometer (ICP-AES) on an IRIS Intrepid II XSP instrument (Thermo Electron Corporation). Gas chromatography (GC) was performed on a SHIMADZU-2014 equipped with a DM-1 capillary column (30 m × 0.25 mm × 0.25 μm). GC-mass spectrometric data (GC-MS) were recorded on an Agilent 6890 instrument equipped with an Agilent 5973 mass selective detector.

To measure the σ-donor ability of **1** and **2**, their corresponding phosphine–selenides were prepared by reacting elemental selenium (with 7.63% ^{77}Se) with each phosphine in deuterated solvent under the appointed conditions (see condition described in Fig. 1), which were directly analyzed without isolation by a Bruker Avance 500 spectrometer at ambient temperature.

2.2 Synthesis

L1 and **L2** were prepared according to the procedures reported by us before.[10]

2.3 General procedures for the hydroformylation of 1-octene

In a typical experiment, the ligand (**1** or **2**, 0.015 mmol), $Rh(acac)(CO)_2$ (0.005 mmol) was mixed with 1-octene (5 mmol) in the RTIL of $[Bmim]BF_4$ (1-butyl-3-methyl-imidazolium tetrafluoroborate) or $[PEmim]BF_4$ (1-piperidyl-3-methyl-imidazolium tetrafluoroborate) (3 mL). The obtained mixture was placed in a sealed

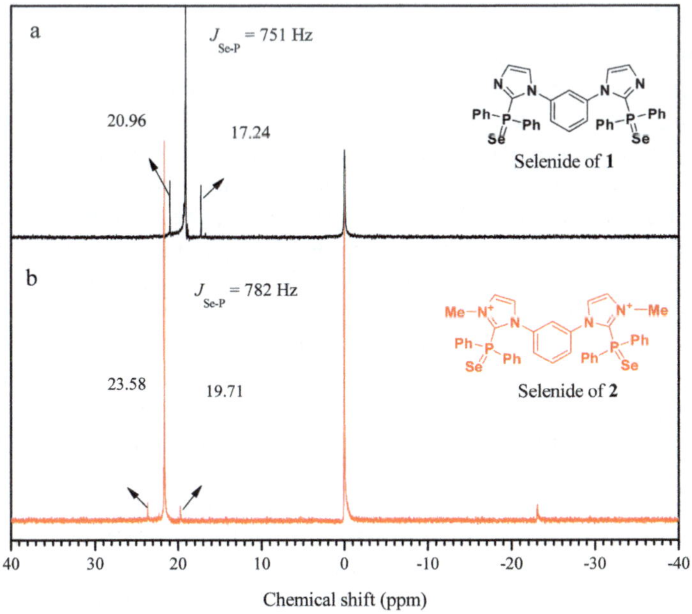

Fig. 1 ^{31}P NMR spectra of the selenides of **1** and **2** (202 MHz): (a) after reacting elemental selenium with **1** in $CDCl_3$ at 70 °C for 10 h; (b) after reacting elemental selenium with **2** in DMSO-d_6 at 70 °C for 10 h.

Teflon-lined stainless steel autoclave and purged with syngas ($CO/H_2 = 1:1$). The pressure was increased up to 4.0 MPa, and the reaction was performed under vigorous stirring at the appointed temperature. Upon completion, the autoclave was cooled down to room temperature, and the pressure was carefully released. The solution was analyzed using GC to determine the conversions (using *n*-dodecane as the internal standard) and the selectivities (*via* a normalization method).

2.4 *In situ* high-pressure FT-IR spectral analysis

All the FT-IR spectra were recorded on a Nicolet NEXUS 670 spectrometer. The spectral resolution was about 4 cm^{-1}. The mixtures of **1** (or **2**, 0.015 mmol), $Rh(acac)(CO)_2$ (0.005 mmol), $[PEmim]BF_4$ (or $[Bmim]BF_4$, *ca.* 0.1 mL), and 1-octene (0.1 mL) were added by single mixing in the specially designed high-pressure IR cell, in which cylindric CaF_2 was used as the sealing sheets. The sealed cell was then pressurised to 2.0 MPa with syngas ($CO/H_2 = 1:1$ volume ratio). Real time monitoring was performed under different temperatures.

The mixture compositions including $Rh(acac)(CO)_2$, ligand **1** (or **2**), 1-octene, the RTIL, and syngas were exactly the same as those for the real reaction in Table 1 (run 1), with the following exceptions: the much higher concentrations of $Rh(acac)(CO)_2$ and the ligand required for IR spectral detection, and the relatively lower syngas pressure (2.0 MPa instead of 4.0 MPa in the real hydroformylation) due to the limited pressure endurance capability of the designed IR cell.

3. Results and discussion

3.1 Biphasic hydroformylation over $Rh(acac)(CO)_2$ with the involvement of 1 or 2 in RTIL

2 could be regarded as a kind of diphosphine-FIL while the neutral diphosphine of **1** was incorporated into the structure of the imidazolium-based IL. Due to the good compatibility of **2** with the RTIL, the use of **2** with $Rh(acac)(CO)_2$ in combination with the RTIL solvent of $[Bmim]BF_4$ or $[PEmim]BF_4$ certainly developed a typical biphasic catalysis for the hydroformylation of 1-octene. Due to the mass transfer limitation in the biphasic system, the reaction rate had to be prolonged to 8 h to achieve the acceptable transformation. As summarized in Table 1, the RTILs of $[Bmim]BF_4$ and $[PEmim]BF_4$ were selected as the solvents. It was found that $[Bmim]BF_4$ and $[PEmim]BF_4$ could both capture **1**-$Rh(acac)(CO)_2$ or **2**-$Rh(acac)(CO)_2$ in the IL phase to afford a typical biphasic reaction, the work-up for separation and recycling was simplified greatly. Upon completion of each run, the upper organic phase was separated by decantation for GC analyses. Then, the lower IL phase containing the ligand (**1** or **2**) and Rh-complexes was reused for the next run after the extraction by diethyl ether. However, in the recycling experiments, only the catalytic system of **2**-$Rh(acac)(CO)_2$ in $[PEmim]BF_4$ could be recycled for 9 runs without loss of activity. In each run, the TON was no less than 930 and the total TON reached 8690 (Column II). The ICP analysis indicated that the leaching of **2**-$Rh(acac)(CO)_2$ into the organic phase was 3.5% for Rh and 4.7% for P. In contrast, under the same reaction conditions, when the neutral diphosphine of **1** was used, not only did the activity decrease in fresh use (TON = 590), but also the very poor 1-octene conversion of 5% was found in the second

Table 1 The recycling experiments of $Rh(acac)(CO)_2$ with the involvement of **2** (or **1**) in $[Bmim]BF_4$ and $[PEmim]BF_4$ respectively for the biphasic hydroformylation of 1-octene[a]

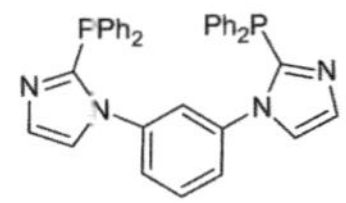

1

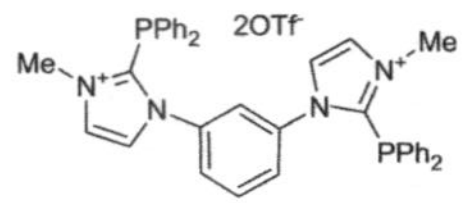

2

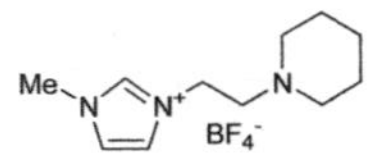

$[PEmim]BF_4$

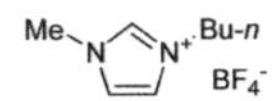

$[Bmim]BF_4$

	RTIL: $[PEmim]BF_4$								RTIL: $[Bmim]BF_4$			
	1-$Rh(acac)(CO)_2$ (Column I)				**2**-$Rh(acac)(CO)_2$ (Column II)				**2**-$Rh(acac)(CO)_2$ (Column III)			
Run	Conv.[b] (%)	Sel. to alde.[b] (%)	L/B[c]	TON	Conv.[b] (%)	Sel. to alde.[b] (%)	L/B[c]	TON	Conv.[b] (%)	Sel. to alde.[b] (%)	L/B[c]	TON
1 (fresh)	59	98	2.3	590	95	92	1.7	950	99	95	2.4	985
2	5	95	2.3	50	93	98	1.8	930	98	97	1.8	990
3	Undone	—	—	—	98	97	1.9	980	98	95	1.9	980
4	Undone	—	—	—	98	98	1.8	980	96	85	2.1	960
5	Undone	—	—	—	98	97	1.8	980	89	74	2.0	885
6	Undone	—	—	—	98	96	1.7	980	Undone	—	—	—
7	Undone	—	—	—	97	93	1.9	970	Undone	—	—	—
8	Undone	—	—	—	97	91	1.9	970	Undone	—	—	—
9	Undone	—	—	—	95	88	1.6	950	Undone	—	—	—

[a] $Rh(acac)(CO)_2$ 0.005 mmol, the ligand (1 or 2) 0.015 mmol, 1-octene 5.0 mmol, CO/H_2 (1 : 1) 4.0 MPa, P/Rh = 6 (molar ratio), RTIL 3 mL, temperature 100 °C, time 8 h. [b] Determined by GC. [c] L/B, the ratio of linear nonanal to branched ones.

run (Column I), indicating that **1**-Rh(acac)$(CO)_2$ in [PEmim]BF_4 was not able to be recycled. The ICP analysis indicated that leaching of **1**-Rh(acac)$(CO)_2$ into the organic phase in the two runs was 4.0% for Rh and 1.3% for P, which suggested that the activity drop of **1**-Rh(acac)$(CO)_2$ in [PEmim]BF_4 may be due to the rapid catalyst deactivation. These results indicated that the contribution of **2** to the catalytic performance as well as the stability of the Rh-catalyst was completely different to that of **1**.

On the other hand, when the RTIL of [Bmim]BF_4 was used instead of [PEmim]BF_4 to repeat the same recycling experiments over **2**-Rh(acac)$(CO)_2$, the deactivation of the Rh–P catalyst and the serious leaching of Rh (39%) and P (54%, ICP analysis) into the combined organic phase after 5 runs were obviously observed, leading to the gradually decreased hydroformylation efficiency (Column III).

It was noticed that while the neutral diphosphine of **1** was incorporated into the structure of the IL to generate the diphosphine-FIL of **2**, the coordinating ability of **2** should be definitely changed due to the strong electron-withdrawing effect of positive-charged imidazoliums, especially when the involved phopshine fragments were directly adjacent to the imidazolium rings. It was indeed verified that **2** was a stronger π-acceptor with $^1J_{Se\text{-}P}$ of 782 Hz in comparison to **1** with $^1J_{Se\text{-}P}$ of 751 Hz (ref. 10) as shown in Fig. 1, since an increase of the value of $^1J_{Se\text{-}P}$ indicates an increase in the character of π-acceptor ability (*i.e.*, less σ-donor ability) of a phosphine.[42] As for **2**, it not only exhibited the π-acceptor coordinating ability to mimic the behavior of the electron-deficient ligands of phosphonites or phosphites, but also had no shortcoming of liability towards hydrolysis (derived from the susceptible P–O bonds).[33b,43]

It has been widely accepted that in contrast to the electron-rich tertiary phosphine ligands, the electron-deficient π-acceptor ligands such as the typical examples of phosphonites or phosphites, are more efficient at boosting the catalytic performance of Rh-complex catalysts for the hydroformylation of olefins.[9,44–48] Hence, **2** as a good π-acceptor ligand consequently resulted in superior behaviours than that of **1** in terms of the higher conversion of 1-octene and the better chemoselectivity to aldehydes (run 1: Column I *vs.* Column II). Secondly, as an imidazolium-based ionic compound, **2** could act as a kind of phase transfer catalyst (surfactant) to facilitate the mass transfer for the biphasic reaction, which also gave rise to the reaction rate. Finally, the piperidyl group in [PEmim]BF_4 acting as an N-containing donor could work together with the phosphine of **2** to protect the Rh-catalyst in a synergetic coordination way against Rh-catalyst deactivation and leaching, which was believed to be another important factor in the more efficient hydroformylation than in [Bmim]BF_4 (Column II *vs.* Column III).

3.2 Detection and characterization of active Rh-H species using *in situ* high-pressure IR spectroscopy

The efficient and recyclable catalytic system of **2**-Rh(acac)$(CO)_2$ in [PEmim]BF_4 enhanced our understanding of the formation and stability of the active Rh-H (rhodium-hydride) species, or other rhodium complex intermediates, formed in the course of hydroformylation. The *in situ* high pressure IR spectroscopy technique is important to this research because of the valuable information on the active Rh-H species.[10,49]

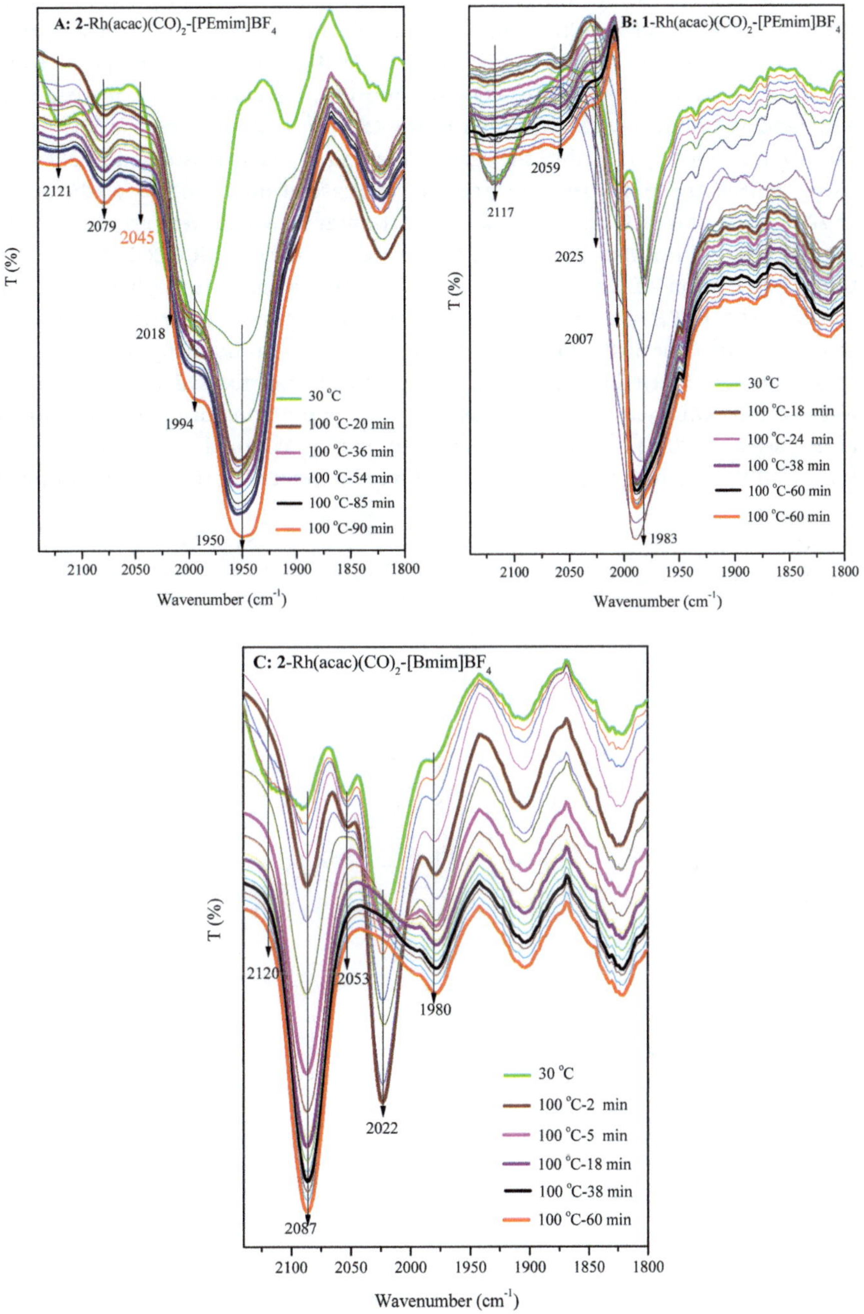

Fig. 2 *In situ* high-pressure FT-IR spectra recorded over **2**-Rh(acac)(CO)$_2$–[PEmim]BF$_4$, **1**-Rh(acac)(CO)$_2$–[PEmim]BF$_4$ and **2**-Rh(acac)(CO)$_2$–[Bmim]BF$_4$ respectively, upon mixing with 1-octene in 2.0 MPa syngas (CO/H$_2$ = 1) [the selected and stacked profiles at different temperatures were provided here for clarity].

The *in situ* spectra for the reactions over **2**-Rh(acac)(CO)$_2$ in [PEmim]BF$_4$, **1**-Rh(acac)(CO)$_2$ in [PEmim]BF$_4$, and **2**-Rh(acac)(CO)$_2$ in [Bmim]BF$_4$ in a syngas atmosphere were recorded as shown in Fig. 2. As the temperature increased from 30 to 100 °C over the **2**-Rh(acac)(CO)$_2$–[PEmim]BF$_4$ system (Fig. 2A), the peaks at 2121 and 2018 cm^{-1} were attributed to the CO vibrations in Rh(acac)(CO)$_2$ that rapidly decayed at 30 °C once the reaction mixtures were pressurised to 2.0 MPa using syngas. As the temperature gradually increased from 30 to 100 °C, the peaks at 2079, 1994 and 1950 cm^{-1} appeared concomitantly which were ascribed to the CO vibrations in non-planar Rh-complex intermediates of [Rh(CO)$_x$(**2**)$_{5-x}$] after the complexation of Rh(acac)(CO)$_2$ with **2** in the syngas atmosphere with the involvement of [PEmim]BF$_4$. While the temperature remained at 100 °C, a weak absorption peak at 2045 cm^{-1} slowly reached maximum intensity from 54 to 85 min, and then decayed to void in the monitoring afterwards, which was identified as the catalytically active Rh-H species in charge of the transformation of olefin to aldehyde, according to previously reported work.[49] When the **1**-Rh(acac)(CO)$_2$–[PEmim]BF$_4$ system (Fig. 2B) was applied for the detection, the CO vibrations at

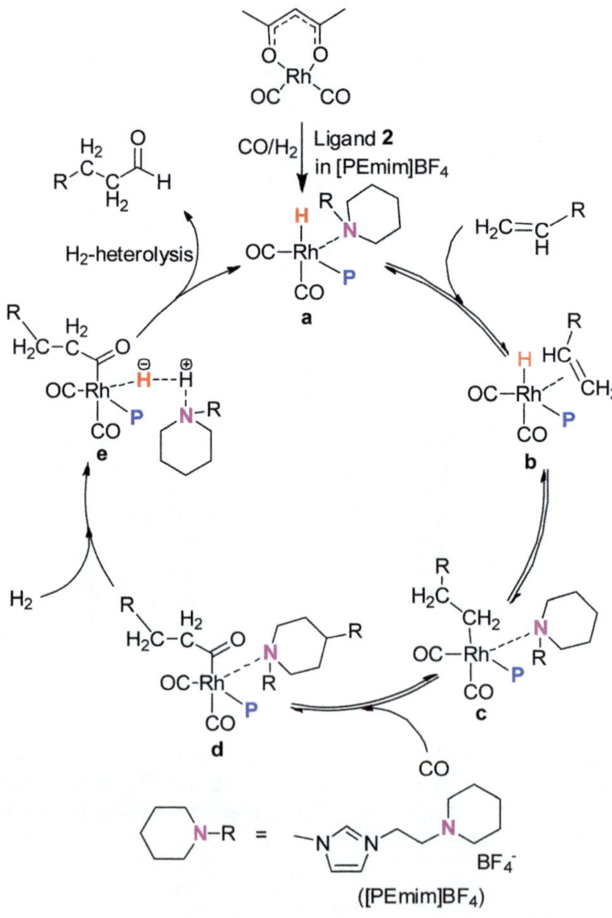

Scheme 2 Mechanism of hydroformylation catalyzed by Rh(acac)(CO)$_2$-**2** in [PEmim]BF$_4$.

2117 and 2007 cm^{-1} in Rh(acac)(CO)$_2$ rapidly decayed, and the new absorptions at 2059, 2025 and 1983 cm^{-1} appeared in a timely fashion, which was due to the formation of [Rh(CO)$_x$(**1**)$_{5-x}$]. However, the active Rh-H species with characteristic IR absorption at *ca.* 2045 cm^{-1} was not observed at all. When the RTIL of [Bmim]BF$_4$ was used in place of [PEmim]BF$_4$ for the **2**-Rh(acac)(CO)$_2$ system (Fig. 2C), the expected characteristic IR absorption at *ca.* 2045 cm^{-1} for the active Rh-H species was not observed in the overall monitoring process. This *in situ* FT-IR spectral information indicated that the different coordination abilities of the phosphine fragments in **2** and **1**, as well as the nature of the RTILs solvent, could dramatically affect the formation and stability of the active Rh-H species. The π-acceptor characteristic of **2** and the available piperidyl group in [PEmim]BF$_4$ cooperatively facilitated the formation and stability of the Rh-H species, which reasonably corresponded to the well maintained activity of **2**-Rh(acac)(CO)$_2$ in [PEmim]BF$_4$ for 9-run recycling uses (Table 1, Column II).

It is believed that the Rh–acyl complex (in Scheme 2, **d**), which is derived from the dissociation of CO and the subsequent insertion of it into the Rh–alkyl (**c**) bond, is mostly irreversibly converted into aldehydes. Therefore, all effects that support the dissociation of CO from the metal center to the formation of **d** favor hydroformylation over isomerization.[50] When **2**-Rh(acac)(CO)$_2$ was immobilized in [PEmim]BF$_4$, the surrounding piperidyl group could behave as an N-containing ligand with a relatively weak coordinating ability to Rh-center to synergistically protect the active Rh-H species (**a**) against deactivation during the catalytic cycle. It could then possibly facilitate the H$_2$-heterolysis over species (**e**) to afford Rh-H species due to the basicity of this amino group, which all contributed to the favorable formation and stability of the Rh-H species over **2**.

4. Conclusions

The catalytic system composed of the diphosphine-FIL of **2** and Rh(acac)(CO)$_2$, which was immobilized in peperidyl-containing RTIL of [PEmim]BF$_4$, was found to be highly efficient and stable for the biphasic hydroformylation of 1-octene with advantages of facile catalyst separation and excellent recyclability (without any activity loss after 9 runs), and the very low leaching of the Rh–P catalyst into the organic phase. The *in situ* high-pressure FT-IR spectral analysis verified that the formation and longevity of the active Rh-H species (2045 cm^{-1}) were facilitated not only by the involvement of **2** with a typical π-acceptor character ($^1J_{Se\text{-}P}$ of 782 Hz) in comparison to the situation over **1** with a σ-donor character ($^1J_{Se\text{-}P}$ of 751 Hz), but also by the participation of the unique RTIL of [PEmim]BF$_4$. The latter with a piperidyl tail could act as an N-containing donor to protect the Rh-catalyst cooperatively with the phosphine ligand against catalyst deactivation and leaching.

Acknowledgements

This work was financially supported by the National Natural Science Foundation of China (No. 21473058, 21273077, and 21076083), 973 Program from Ministry of Science and Technology of China (2011CB201403).

Notes and references

1 *Rhodium-Catalyzed Hydroformylation*, ed. P. W. N. M. van Leeuwen and C. Claver, Kluwer, Dordrecht, 2000.

2 C. Kubis, D. Selent, M. Sawall, R. Ludwig, K. Neymeyr, W. Baumann, R. Franke and A. Börner, *Chem.–Eur. J.*, 2012, **18**, 8780–8794.

3 R. Franke, D. Selent and A. Börner, *Chem. Rev.*, 2012, **112**, 5675–5732.

4 *Topics in Current Chemistry 342: Hydroformylation for Organic Synthesis*, ed. M. Taddei and A. Mann, Springer Verlag Berlin Heidelberg, 2013.

5 R. Lazzaroni, R. Settambolo, G. Alagona and C. Ghio, *Coord. Chem. Rev.*, 2010, **254**, 696–706.

6 F. Ungvary, *Coord. Chem. Rev.*, 2007, **251**, 2087–2102.

7 Z. Freixa and P. W. N. M. van Leeuwen, *Coord. Chem. Rev.*, 2008, **252**, 1755–1786.

8 J. Klosin and C. R. Landis, *Acc. Chem. Res.*, 2007, **40**, 1251–1259.

9 Y. Yan, X. Zhang and X. Zhang, *J. Am. Chem. Soc.*, 2006, **128**, 16058–16061.

10 Y. Q. Li, P. Wang, H. Zhang, X. L. Zhao, Y. Lu, Z. Popović and Y. Liu, *J. Mol. Catal. A: Chem.*, 2015, **402**, 37–45.

11 C. Kubis, D. Selent, M. Sawall, R. Ludwig, K. Neymeyr, W. Baumann, R. Franke and A. Börner, *Chem.–Eur. J.*, 2012, **18**, 8780–8794.

12 W. Gil and A. M. Trzeciak, *Coord. Chem. Rev.*, 2011, **255**, 473–483.

13 A. R. Almeida, A. F. Peixoto, M. J. F. Calvete, P. M. P. Gois and M. M. Pereira, *Curr. Org. Synth.*, 2011, **8**, 764–775.

14 S. D. Gonzalez, N. Marion and S. P. Nolan, *Chem. Rev.*, 2009, **109**, 3612–3676.

15 R. Maura, J. Steele, L. Vendier, D. Arquier, S. Bastin, M. Urrutigoïty, P. Kalck and A. Igau, *J. Organomet. Chem.*, 2011, **696**, 897–904.

16 G. Brancatelli, D. Drommi, G. Bruno and F. Faraone, *Inorg. Chem. Commun.*, 2010, **13**, 215–219.

17 J. Tijani and B. E. Ali, *J. Organomet. Chem.*, 2007, **692**, 3492–3497.

18 Y. S. Uh, A. Boyd, V. R. Little, P. G. Jessop, K. D. Hesp, J. C. Wechsler, M. Stradiotto and R. McDonald, *J. Organomet. Chem.*, 2010, **695**, 1869–1872.

19 J. Andrieu, J. M. Camus and P. Richard, *Eur. J. Inorg. Chem.*, 2006, 51–61.

20 I. C. Kostas, *J. Organomet. Chem.*, 2001, **634**, 90–98.

21 E. Wiebus and B. Cornils, *Catalysis by Metal Complexes: Catalyst Separation, Recovery and Recycling*, Springer, Netherlands, 2006, vol. 30, pp. 105–143.

22 F. Hapiot, L. Leclercq, N. Azaroual, S. Fourmentin, S. Tilloy and E. Monflier, *Curr. Org. Synth.*, 2008, **5**, 162–172.

23 S. K. Sharma and R. V. Jasra, *Catal. Today*, 2015, **247**, 70–81.

24 F. Joo and A. Katho, *J. Mol. Catal. A: Chem.*, 1997, **116**, 3–26.

25 X. Zhao, D. He, L. T. Mika and I. T. Horvath, *Fluorous Hydroformylation in Topics in Current Chemistry: Fluorous Chemistry*, 2012, vol. 308, pp. 275–290.

26 K. L. Luska, K. Z. Demmans, S. A. Stratton and A. Moores, *Dalton Trans.*, 2012, **41**, 13533–13540.

27 J. A. Vicente, A. Mlonka, H. Q. N. Gunaratne, M. S. Kwaśny and P. Nockemann, *Chem. Commun.*, 2012, **48**, 6115–6117.

28 R. P. J. Bronger, S. M. Silva, P. C. J. Kamer and P. W. N. M. van Leeuwen, *Dalton Trans.*, 2004, 1590–1596.

29 (*a*) J. Andrieu and M. Azouri, *Inorg. Chim. Acta*, 2007, **360**, 131–135; (*b*) J. Andrieu, M. Azouri and P. Richard, *Inorg. Chem. Commun.*, 2008, **11**, 1401–1404; (*c*) M. Azouri, J. Andrieu, M. Picquet and H. Cattey, *Inorg. Chem.*, 2009, **48**, 1236–1242.

30 D. J. Brauer, K. W. Kottsieper, C. Liek, O. Stelzer, H. Waffenschmidt and P. Wasserscheid, *J. Organomet. Chem.*, 2001, **630**, 177–184.

31 R. Wang, J. C. Xiao, B. Twamley and J. M. Shreeve, *Org. Biomol. Chem.*, 2007, **5**, 671–678.

32 B. Milde, R. Packheiser, S. Hildebrandt, D. Schaarschmidt, T. Rueffer and H. Lang, *Organometallics*, 2012, **31**, 3661–3671.

33 (*a*) N. Debono, Y. Canac, C. Duhayon and R. Chauvin, *Eur. J. Inorg. Chem.*, 2008, 2991–2999; (*b*) Y. Canac, N. Debono, L. Vendier and R. Chauvin, *Inorg. Chem.*, 2009, **48**, 5562–5568; (*c*) Y. Canac, N. Debono, C. Lepetit, C. Duhayon and R. Chauvin, *Inorg. Chem.*, 2011, **50**, 10810–10819; (*d*) I. Abdellah, C. Lepetit, Y. Canac, C. Duhayon and R. Chauvin, *Chem.–Eur. J.*, 2010, **16**, 13095–13108; (*e*) Y. Canac, C. Maaliki, I. Abdellahab and R. Chauvin, *New J. Chem.*, 2012, **36**, 17–27; (*f*) C. Barthes, C. Lepetit, Y. Canac, C. Duhayon, D. Zargarian and R. Chauvin, *Inorg. Chem.*, 2013, **52**, 48–58.

34 N. Karodia, S. Guise, C. Newlands and J. A. Andersen, *Chem. Commun.*, 1998, 2341–2342.

35 T. E. Kunene, P. B. Webb and D. J. C. Hamilton, *Green Chem.*, 2011, **13**, 1476–1481.

36 C. Kolbeck, N. Paape, T. Cremer, P. S. Schulz, F. Maier, H. Steinrueck and P. Wasserscheid, *Chem.–Eur. J.*, 2010, **16**, 12083–12087.

37 P. Wasserscheid, H. Waffenschmidt, P. Machnitzki, K. W. Kottsieper and O. Stelzer, *Chem. Commun.*, 2001, 451–452.

38 R. P. J. Bronger, S. M. Silva, P. C. J. Kamer and P. W. N. M. van Leeuwen, *Chem. Commun.*, 2002, 3044–3045.

39 F. Favre, H. O. Bourbigou, D. Commereuc and L. Saussine, *Chem. Commun.*, 2001, 1360–1361.

40 C. C. U. Englert, A. Salzer, H. Waffenschmidt and P. Wasserscheid, *Organometallics*, 2000, **19**, 3818–3823.

41 (*a*) C. Tan, P. Wang, H. Liu, X. L. Zhao, Y. Lu and Y. Liu, *Chem. Commun.*, 2015, **51**, 10871–10874; (*b*) S. J. Chen, Y. Y. Wang, Y. Q. Li, X. L. Zhao and Y. Liu, *Catal. Commun.*, 2014, **50**, 5–8; (*c*) S. J. Chen, Y. Y. Wang, W. M. Yao, X. L. Zhao, G. V. Thanh and Y. Liu, *J. Mol. Catal. A: Chem.*, 2013, **378**, 293–300; (*d*) H. X. You, Y. Y. Wang, X. L. Zhao, S. J. Chen and Y. Liu, *Organometallics*, 2013, **32**, 2698–2704; (*e*) J. Zhang, Y. Y. Wang, X. L. Zhao and Y. Liu, *Eur. J. Inorg. Chem.*, 2014, 975–985.

42 (*a*) D. W. Allen and B. F. Taylor, *Dalton Trans.*, 1982, **1**, 51–54; (*b*) A. Suárez, M. A. M. Rojas and A. Pizzano, *Organometallics*, 2002, **21**, 4611–4621; (*c*) B. Milde, D. Schaarschmidt, T. Rüffer and H. Lang, *Dalton Trans.*, 2012, **41**, 5377–5390.

43 N. Sakai, S. Mano, K. Nozaki and H. Takaya, *J. Am. Chem. Soc.*, 1993, **115**, 7033–7034.

44 C. D. Frohning and C. W. Kohlpainter, in *Applied Homogeneous Catalysis with Organometallic Compounds*, ed. B. Cornils and W. A. Herrmann, Wiley-VCH, Weinheim, 1st edn, 1996, p. 29.

45 X. F. Wu, X. Fang, L. Wu, R. Jackstell, H. Neumann and M. Beller, *Acc. Chem. Res.*, 2014, **47**, 1041–1053.
46 R. L. Pruett and J. A. Smith, *J. Org. Chem.*, 1969, **34**, 327–330.
47 S. A. Ullate, J. A. Baker, V. G. González, C. Müller, J. D. Hirst and J. J. Carbó, *Catal. Sci. Technol.*, 2014, **4**, 979–987.
48 H. Tricas, O. Diebolt and P. W. N. M. van Leeuwen, *J. Catal.*, 2013, **298**, 198–205.
49 (*a*) S. J. Chen, Y. Q. Li, Y. Y. Wang, X. L. Zhao and Y. Liu, *J. Mol. Catal. A: Chem.*, 2015, **396**, 68–76; (*b*) S. J. Chen, Y. Q. Li, P. Wang, Y. Lu, X. L. Zhao and Y. Liu, *J. Mol. Catal. A: Chem.*, 2015, **407**, 212–220; (*c*) P. C. J. Kamer, A. van Rooy, G. C. Schoemaker and P. W. N. M. van Leeuwen, *Coord. Chem. Rev.*, 2004, **248**, 2409–2424; (*d*) D. He, J. Liu, Y. Liu, T. Wang, D. Pang, Y. Chen, Y. Liang and Q. Zhu, *Catal. Lett.*, 2001, **73**, 221–224; (*e*) K. Nozaki, T. Matsuo, F. Shibahara and T. Hiyama, *Organometallics*, 2003, **2**, 594–600; (*f*) S. Deerenberg, P. C. J. Kamer and P. W. N. M. van Leeuwen, *Organometallics*, 2000, **19**, 2065–2072; (*g*) W. R. Moser, C. J. Papile and S. J. Weininger, *J. Mol. Catal.*, 1987, **41**, 271–292.
50 M. V. Herrera, L. Domke and A. Börner, *ACS Catal.*, 2014, **4**, 1706–1724.

Faraday Discussions

PAPER

High energy supercapattery with an ionic liquid solution of $LiClO_4$

Linpo Yu[ab] and George Z. Chen*[ab]

Received 21st December 2015, Accepted 5th January 2016
DOI: 10.1039/c5fd00232j

A supercapattery combining an ideally polarized capacitor-like electrode and a battery-like electrode is demonstrated theoretically and practically using an ionic liquid electrolyte containing 1-butyl-1-methylpyrrolidinium tri(pentafluoroethyl)trifluorophosphate (BMPyrrFAP), gamma-butyrolactone (γ-GBL) and $LiClO_4$. The electrochemical deposition and dissolution of lithium metal on a platinum and glass carbon electrode were investigated in this ionic liquid solution. The CVs showed that the fresh electrochemically deposited lithium metal was stable in the electrolyte, which encouraged the investigation of this ionic liquid solution in a supercapattery with a lithium battery negative electrode. The active material counted specific energy of the supercapattery based on a lithium negative electrode and an activated carbon (Act-C) positive electrode could reach 230 W h kg^{-1} under a galvanostatic charge–discharge current density of 1 mA cm^{-2}. The positive electrode material (Act-C) was also investigated by CV, AC impedance, SEM and BET. The non-uniform particle size and micropores dominated porous structure of the Act-C enabled its electric double layer capacitor (EDLC) behavior in the ionic liquid solution. The measured specific capacitance of the Act-C in this ionic liquid solution is higher than the same Act-C in aqueous solution, which indicates the Act-C can also perform well in the ionic liquid electrolyte.

Introduction

A supercapattery (= supercapacitor + battery) takes advantages of both supercapacitor (including the pseudocapacitor and electric double layer capacitor, or EDLC) and battery by combining an ideally polarized capacitor-like electrode and a battery-like electrode.[1] Although lithium ion capacitors[2,3] are also comprised of this hybrid configuration, 'supercapattery' is the more general term for this particular design.

[a]*Department of Chemical and Environmental Engineering, Energy and Sustainability Research Division, University of Nottingham, Nottingham, NG7 2RD UK. E-mail: George.Chen@nottingham.ac.uk*
[b]*Department of Chemical and Environmental Engineering, Centre for Sustainable Energy Technologies, Faculty of Science and Engineering, University of Nottingham Ningbo China, Ningbo, 315100 China*

In theory, a supercapattery can possess higher specific energy than both the battery and supercapacitor and can supply this energy at a power output almost as high as a supercapacitor. The high power output of supercapatteries is mainly a result of sharing the same electrochemically active materials with supercapacitors. Nanostructured carbons, like activated carbon (Act-C), carbon nanotubes (CNTs) and graphene are the best choice for the ideally polarized electrode because of their large surface area, porosity, stability over a wide potential window, and intrinsically low electrical resistance. In addition, pseudo-capacitive materials including MnO_2, RuO_2 and conducting polymers can also be used as capacitor-like electrode materials providing high electrode capacitance, but limiting the potential windows.

Various electrode materials from commercial battery systems can be used for the battery-like electrode, from lead–acid batteries to metal/air systems, but in practice, metal compounds[3–5] like SnO_2, MnO_2 and $LiFePO_4$ are more commonly used and commercially available. Theoretically, a hypothetical battery comprising lithium metal and fluorine gas (Li–F battery) would give a cell voltage output of about 6.1 V and offer a specific energy content of 6304 W h kg^{-1}. No battery can go beyond the specific energy content of the hypothetical Li–F battery. As for the Li-ion battery ($Li_xC_6|Li_{1-x}CoO_2$), the theoretical specific energy is 552 W h kg^{-1} at 3.5 V. A hypothetical supercapattery comprising a lithium metal negative electrode and a supercapacitor positive electrode (assumed 400 F g^{-1}) is evaluated and analyzed here. Considering the specific charge capacity of lithium is much larger than that of the supercapacitor electrode, the mass of the lithium metal is negligible. The theoretical specific energy value would be 625 W h kg^{-1} for a cell voltage varied from 3.5 V to 1.0 V. This theoretical value is even higher than the one for the Li-ion battery.

The above calculation is based on the equation for the capacitor energy, eqn (1),

$$E_{\max} = \frac{1}{2} C U_{\max}{}^2 \tag{1}$$

where C is the specific capacitance of a capacitor, and $E_{\max}$ is the maximum energy capacity of a capacitor correlated to its maximum tolerable voltage, $U_{\max}$. It should be mentioned that a hybrid cell of a battery electrode and a supercapacitor electrode does not always show typical supercapacitor behaviour, where the cell voltage (U) is always proportional to the time (t) in the plots of galvanostatic current charging and discharging. If a cell presents a battery-like feature, the term supercabattery is recommended,[6] but is not discussed in this paper. In the case of a supercapattery, the $U_{\max}$ is a key factor for the energy capacity of the device. Because the behaviour of a supercapattery is close to that of a supercapacitor,[1] several strategies that have been applied towards supercapacitors can be utilised in supercapatteries to improve the practical energy capacity, such as using the design of an asymmetric supercapacitor cell,[7] controlling the capacitance ratio of the positive and negative electrode made from the same material,[8] and serially stacking the cells through bipolar electrodes.[9] Apart from these efforts based on aqueous electrolytes, there is a strong desire for changing from aqueous to organic electrolytes to achieve a high working voltage.

The recent studies revealed both the aqueous and non-aqueous supercapatteries using lithium metal[10] and Li-ion battery material[3] as the negative electrode materials. An aqueous supercapattery consisting of a MnO_2 positive electrode and a Li/LISICON/PEO–LiTFSI/Li^+ negative electrode achieved a specific energy capacity of 114 W h kg^{-1} with a 4.3 V cell voltage.[10] The Li/LISICON/PEO–LiTFSI/Li^+ electrode is a multi-layered Li electrode, which consists of lithium metal, a LISICON-type solid glass ceramic as the water-stable solid electrolyte, and a buffer layer consisting of polyethylene oxide with $Li(CF_3SO_2)_2N$ polymer electrolyte (PEO–LiTFSI) between the lithium metal and the solid electrolyte. If MnO_2 was replaced by RuO_2 as the positive electrode material, the specific energy capacity of the device reached to 520 W h kg^{-1} with a 3.8 V cell voltage.[10] However, the current density of the aforementioned devices was only 0.255 mA cm^{-2}, which was limited by the solid/liquid interphase. This disadvantage prevents the use of these high energy capacity devices in any high power applications. Another non-aqueous supercapattery using $LiFePO_4$ as the positive electrode and Cabot carbon black as the negative electrode possessed good cycling stability at high current density in a Li-ion contained propylene carbonate (PC) electrolyte.[3] However, the potential range of the Cabot carbon black could only keep its capacitor-like behaviour from 2.80 to 1.25 V *vs.* Li/Li^+ in the electrolyte, which limited the cell voltage.

A special feature of ionic liquids is their zero or negligible volatility, whilst they are still able to offer a highly ionic environment, and wide temperature and potential windows.[11] They have brought about unique opportunities in synthesis,[12] trace analysis,[13,14] thermochromic/cryochromic materials,[15–17] and electrochemical energy storage.[18–20] Consequently, the ionic liquid solutions can be chosen as the competitive candidates for the electrolytes of a supercapattery when the battery-like electrode is a Li-ion or lithium battery negative electrode.

Here, we report our recent supercapattery work based on an activated carbon positive electrode and a Li/Li^+ negative electrode using an ionic liquid electrolyte, 1-butyl-1-methylpyrrolidinium tri(pentafluoroethyl)trifluorophosphate (BMPyrr-FAP) containing gamma-butyrolactone (γ-GBL) and $LiClO_4$. The activated carbon was characterised by cyclic voltammetry (CV), AC impedance, SEM and BET. A demonstrated supercapattery cell showed a clear capacitor-like behaviour in the CV and plots of galvanostatic charge and discharge (GCD). This particular hybrid design achieved the specific energy content up to 230 Wh kg^{-1}, and basically maintains the high power of the supercapacitor.

Experimental

In this work, a supercapacitor grade commercial product of activated carbon (Act-C, YP50F, Kuraray Chemical Co.) was used. The other chemicals, 1-butyl-1-methylpyrrolidinium tri(pentafluoroethyl)trifluorophosphate (BMPyrrFAP, Merck), 1-ethyl-3-methylimidazolium tetracyanoborate (EMIM[$B(CN)_4$], Merck), gamma-butyrolactone (γ-GBL, Sigma Aldrich) and $LiClO_4$ (Sigma Aldrich), were commercially available and used without further purification. Lithium metal (foil, Sigma Aldrich) was kept and handled in an argon filled glove box. The homemade electrochemical cells (2-electrode/3-electrode cells and sandwich type cell) were fabricated in an argon-filled glove box (O_2 < 10 ppm, H_2O < 10 ppm) and transferred outside the glove box for the electrochemical experiments using an AUTOLAB 302N potentiostat. The membrane from Celgard was used as the separator in the sandwich type cell.

For the electrochemical tests, the Act-C powder was made into pellet form and cast into electrodes with PTFE and PVDF, respectively. The details of fabricating the Act-C/PTFE pellet electrode can be found in a previous publication from this laboratory.[21] The Act-C/PVDF electrode was fabricated by the following process. 10 μL of an Act-C suspension (0.950 g Kuraray Act-C and 0.050 g PVDF powder in 20 mL DMSO suspension) was dropped on a 5 mm diameter graphite disc electrode. The electrode loaded with 0.5 mg of Act-C composite (95% wt Act-C and 5% wt PVDF) was dried in a vacuum oven at 75 °C overnight, and was then ready for the electrochemical tests.

All the experiments were undertaken at room temperature. More experimental details are specified in the following sections.

Results and discussion

The pore size and volume distribution of the Act-C materials are the most important physical characteristics. Fig. 1 presents the SEM images of an Act-C pellet containing 5% wt PTFE. The Act-C powder is non-uniform in particle size, and the PTFE binds the Act-C particles like a spider web to maintain the mechanical strength of the Act-C pellet. The surface area of the Kuraray Act-C sample is 1724 $m^2\ g^{-1}$, determined from the nitrogen absorption/desorption isotherms at 77 K (ASAP 2420) by applying density functional theory (DFT). In previous work,[21] the dominant range of the pore widths of the same Kuraray Act-C sample was between 1.5 and 2.5 nm. For an aqueous electrolyte, its wetting ability with the Act-C pellet will affect the contact between the electrolyte and Act-C particles, and vary the charge storage performance of the Act-C sample. Wetting is not a problem for the organic electrolyte, although the interaction between electrolyte and carbon materials inside the pore will play an important role. A theory was proposed for the traditional organic electrolyte that the longer the pore

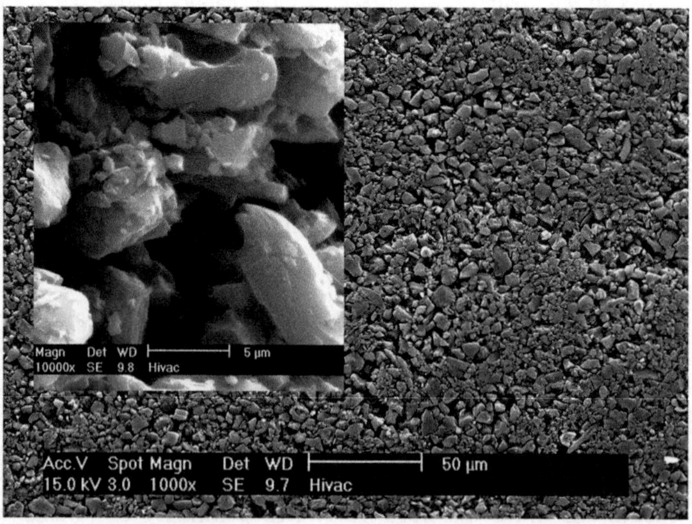

Fig. 1 SEM images of the activated carbon pellet (including 5% wt PTFE). The insert is an enlarged view of the background image.

inside the particle is, the poorer the capacitance performance of the porous carbon materials.[3] This is mainly due to the ion block of the pores by the non-ionic solvent molecule. A similar phenomenon was also observed in the case of aqueous electrolyte with organic additive.[21]

Before undertaking work on the supercapattery, we investigated the Kuraray Act-C in a commercial ionic liquid, 1-ethyl-3-methylimidazolium tetracyanoborate (EMIM[B(CN)$_4$]). Fig. 2 shows a photo of the electrochemical tube cell and CV, GCD and AC impedance test results. A small piece of Act-C pellet (about 1 mg, containing 5% wt PTFE) was loaded on a 6 mm diameter graphite disc electrode. Another graphite disc with 30 mg Act-C was used as the counter electrode. The reference electrode was a Ag/Ag$^+$ electrode filled with 0.01 mol L^{-1} $AgNO_3$ acetonitrile solution, which was connected to the electrolyte *via* a salt bridge filled with EMIM[B(CN)$_4$]. This reference can be prepared following the method described in ref. 22. It should be mentioned that the reference electrode was changed to a Li/Li$^+$ electrode in the latter discussion except in Fig. 2. Any potential mentioned in this paper will be suffixed with the reference electrode used in the experiment.

Both CVs and AC impedance results in Fig. 2 indicate that the Kuraray Act-C does not show a pure EDLC performance in EMIM[B(CN)$_4$]. The Act-C can store more charge in a potential range from −1.7 to 0 V *vs.* Ag/Ag$^+$ than the range of 0 to 1.0 V *vs.* Ag/Ag$^+$. If a symmetrical supercapacitor was fabricated using this Act-C

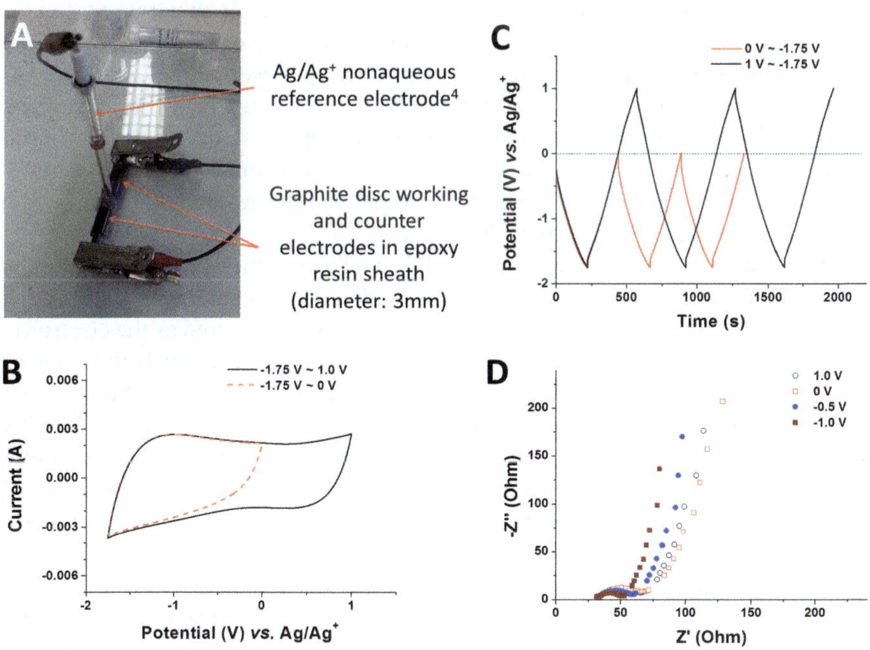

Fig. 2 (A) Photograph of the tube cell. 1.064 mg Act-C (including 5% wt PTFE) was loaded on the graphite disc as a working electrode. EMIM[B(CN)$_4$] was filled in the tube as the electrolyte. The reference electrode was a Ag/Ag$^+$ electrode in a 0.01 mol L^{-1} $AgNO_3$ acetonitrile solution connected to the electrolyte *via* a salt bridge filled only with EMIM [B(CN)$_4$]. The results in (B)–(D) were collected from this tube cell. (B) CVs of the tube cell. Scan rate: 20 mV s^{-1}. (C) Galvanostatic charge–discharge plots of the tube cell. Current: 1 mA. (D) Nyquist plots of the tube cell at potentials indicated in the figure.

and ionic liquid, the maximum voltage could not go beyond 2.7 V. The measured specific capacitance value of Kuraray Act-C in EMIM[B(CN)$_4$] was about 100 F g^{-1}, which is comparable to the value of the same sample obtained in aqueous solution.[21] The voltage of this ionic liquid based symmetrical cell does not increase dramatically, while the resistance of the cell will increase compared to the cell using an aqueous electrolyte. In this case, changing the electrolyte from an aqueous to a non-aqueous electrolyte is not economical. A new strategy should be made to improve the energy capacity of the cell.

As mentioned above, the Li-metal battery or Li-ion battery possesses very high energy capacity mainly caused by the very negative potential of Li/Li$^+$ or relevant redox couple. We chose an ionic liquid which was used as a Li-ion battery electrolyte in the following work. The ionic liquid solution containing $LiClO_4$ was used as the electrolyte for the supercapattery. The lithium salt is the Li-ion source for the electrochemical deposition and dissolution of lithium metal on the negative electrode of the supercapattery.

Fig. 3A and B shows clearly the electrochemical deposition and dissolution of lithium in the ionic liquid electrolyte containing lithium salt on a 2 mm diameter Pt disc electrode and a 3 mm diameter Glass Carbon (GC) disc electrode, respectively. Fig. 3C shows a photo of the homemade electrochemical cell, in which a working electrode (Pt or GC) and a piece of lithium metal foil as both counter and reference electrodes are sealed in a vial. The electrolyte contains 0.01 mol L^{-1} $LiClO_4$ in the mixture of BMPyrrFAP and γ-GBL (volume ratio 1 : 1). Lithium metal can be electrochemically deposited on both substrate materials, and the freshly deposited lithium is stable in the electrolyte unless a positive scan occurs showing a dissolution peak in the CVs. The over-potential on the GC electrode indicates an energy gap between the carbon and lithium surface. Another important piece of information from Fig. 3 is that the ionic liquid solution used in this work showed an electrochemical window of more than 4 V using Pt and GC electrodes. The ionic liquid solution could be a potential candidate for the supercapattery electrolyte.

In a demonstrative supercapattery cell with a lithium metal negative electrode, the redox Li/Li$^+$ reaction should be occurring during the charging or discharging process. A 0.5 mol L^{-1} $LiClO_4$ ionic liquid solution was prepared as the electrolyte to provide sufficient Li-ion in the electrolyte. The solvent content is the same as the one in the previous lithium deposition experiment, BMPyrrFAP and γ-GBL (v : v = 1 : 1). The working electrode is a piece of 0.5 mg Kuraray Act-C pellet coated on a 5 mm diameter graphite disc electrode with PVDF binder. The counter and reference electrodes are lithium metal foil. The cell was fabricated in an argon filled glove box, sealed properly and then transferred outside the box for further tests. The CV and AC impedance data of this cell is presented in Fig. 4. The blue dashed lines on the CV curve mark the potential at which the AC impedance test was run. It should be mentioned again that the real amount of lithium metal anticipating the charge storage process is negligible according to the previous calculation. The reason for placing a piece of lithium metal foil in the cell was to acquire accurate potential results during the charging and discharging processes. From the CV in Fig. 4, we found that the Kuraray Act-C could exhibit capacitor-like behaviour in the potential range from at least 1.7 to 4.3 V *vs.* Li/Li$^+$. There is no obvious turning point to the clear oxidative or reductive current in the CV. If we assumed this is capacitor behaviour, the calculated specific capacitance value is

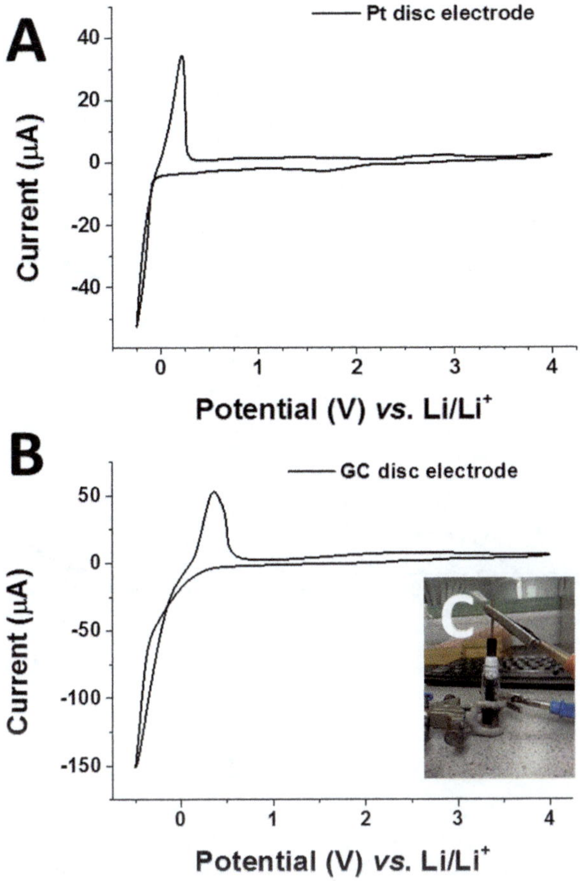

Fig. 3 CVs collected on (A) a 2 mm diameter Pt disc electrode and (B) a 3 mm diameter glass carbon disc electrode, in a mixture of BMPyrrFAP and γ-GBL (v : v = 1 : 1), containing 0.01 mol L^{-1} $LiClO_4$. Scan rate: 10 mV s^{-1}. Potential range: −0.25–4.00 V *vs.* Li/Li^+. (C) Photo of the homemade electrochemical cell.

101 F g^{-1}. The deformed rectangular CV shape should be caused by the resistance of the whole cell, which is clearly shown in the Nyquist plots of the cell at different potentials. Although the resistance of the cell shows a value of about 180 Ω, the equivalent series resistance (ESR) is about 35 Ω cm^2 (area: 0.196 cm^2), which is only 30 times larger than the value obtained from a symmetrical aqueous AC supercapacitor using the same electrode fabricating method.[21] In this case, the resistance as shown in Fig. 4 still falls in the acceptable range in practical devices. The optimisation of this hybrid design is definitely needed to increase output power.

The galvanostatic charge–discharge test was run under a current density of 1.02 mA cm^{-2} as shown in Fig. 5. Because the negative electrode is lithium foil, the voltage of the cell decays smoothly and linearly during the discharging process from 4.3 to 1.7 V. Because the charging and discharging curve is almost symmetric, we can still calculate the specific capacitance and energy capacity by the method for the supercapacitor. The Act-C specific capacitance value comes to

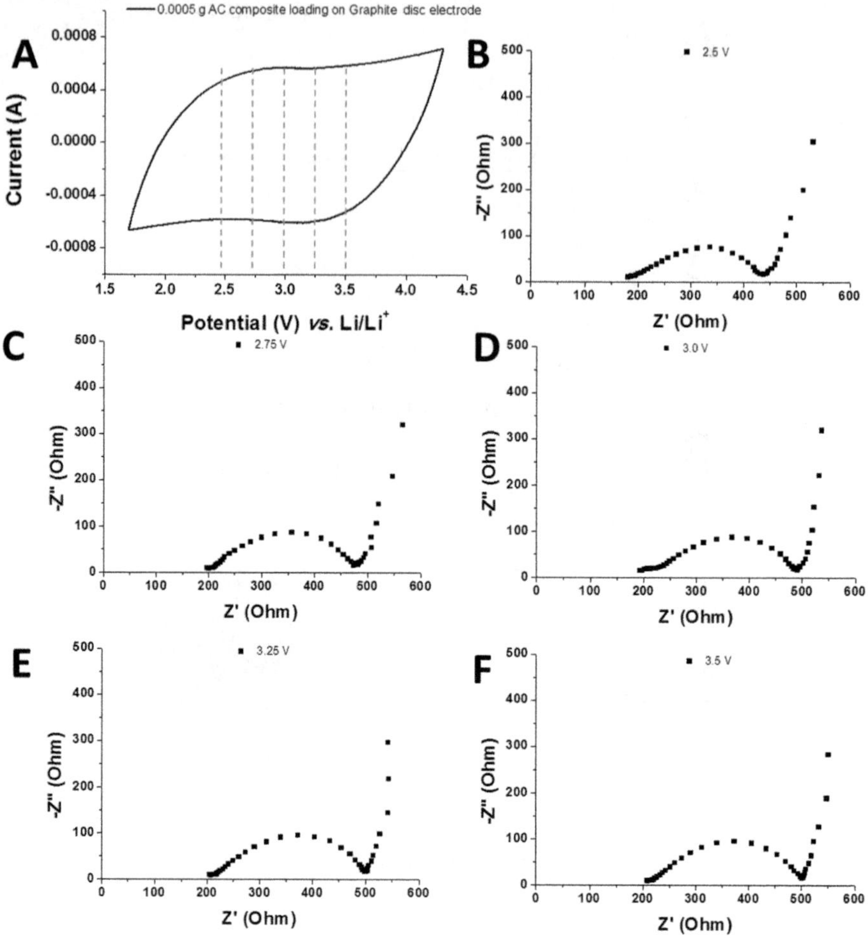

Fig. 4 (A) CV of 0.5 mg Kuraray Act-C composite (containing 5% wt PVDF) on a 5 mm diameter graphite disc electrode in mixture of BMPyrrFAP and γ-GBL (v/v = 1/1) containing 0.5 mol L^{-1} $LiClO_4$. Scan rate: 10 mV s^{-1}. Nyquist plots of the same cell at the potential of (B) 2.50 V, (C) 2.75 V, (D) 3.00 V, (E) 3.25 V, (F) 3.50 V *vs.* Li/Li^+. Lithium foil was used as the reference and counter electrodes. Frequency range: 10 kHz to 0.01 Hz.

107 F g^{-1} in the ionic liquid solution containing $LiClO_4$. We note that only a tiny amount of lithium metal takes part in the charge transfer during the charging and discharging process and the molecular weight of lithium is only 6.94 g mol^{-1}, so we can ignore the mass of the negative electrode if the active materials are counted only in the calculation of the specific energy capacity. The value is 232 W h kg^{-1}.

If we only chose an ionic liquid as the electrolyte, we can hardly use the available redox couple resources to store charge in the energy storage devices. The Li-ion source ionic liquid electrolyte can supply both a sufficient charge carrier for the charge storage on the Act-C surface and the necessary Li-ion for the chemical reaction. We do notice that the CV of the Act-C in a pure ionic liquid shows partly pseudo-capacitance behaviour. This is preliminary evidence for the charge transfer or charge isolation process on the carbon surface. The advantage of the

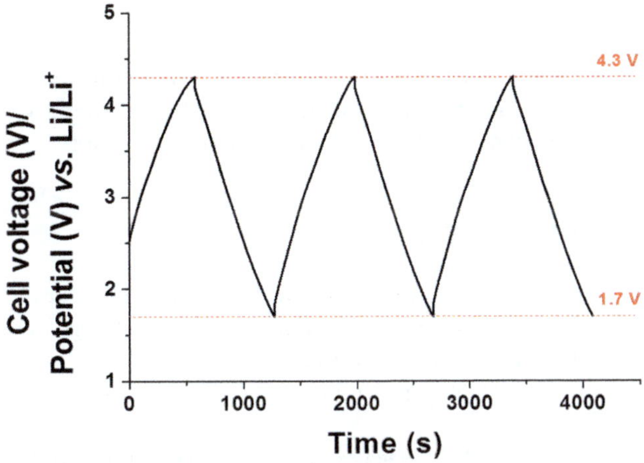

Fig. 5 Galvanostatic charge–discharge curves of a pellet of 0.5 mg Kuraray Act-C composite (containing 5% wt PVDF) on a 5 mm diameter graphite disc electrode in a mixture of BMPyrrFAP and γ-GBL (v/v = 1/1) containing 0.5 mol L^{-1} $LiClO_4$. Current density: 1 mA cm^{-2}.

organic electrolytes or ionic liquid electrolytes is that they can maintain chemical and thermal stability in a wider potential or temperature range than aqueous electrolytes. Although recent studies have revealed that some novel electrochemical energy storage devices using aqueous electrolytes can be used at higher voltage[8] and lower working temperature[23] than the traditional aqueous electrolyte based devices, the variety and particularity of ionic liquids still make themselves competitive in different applications, especially for energy storage. This paper reveals a new idea for using the Act-C materials as a positive electrode material in a supercapattery. The follow-up work to further understand the charge storage mechanism of Act-C in ionic liquids and optimization of the hybrid design of the supercapattery in this paper is ongoing.

Conclusions

A hybrid supercapattery design comprising a lithium metal negative electrode and an activated carbon positive electrode is presented and demonstrated using an experimental cell in this paper. An ionic liquid was used as one component of the electrolyte containing $LiClO_4$. The Kuraray activated carbon showed both EDLC and pseudo-capacitor behaviour in an ionic liquid electrolyte. In the ionic liquid solution, the sufficient amount of Li ions in the ionic liquid electrolyte provides additional charge carriers and enough charge storage species through the Li/Li^+ redox reaction. The energy capacity of the demonstrative supercapattery reached to 232 W h kg^{-1}, while the device can present capacitor-like behaviour when the galvanostatic charge–discharge current density is more than 1.0 mA cm^{-2}, maintaining its high power output characteristic. In addition, the part pseudo-capacitor behaviour of the activated carbon in ionic liquid revealed a possibility to study the interaction of redox species inside the micropore structure of the activated carbon materials.

Acknowledgements

This work received funding from the University of Nottingham Ningbo China and Ningbo Municipal Government (3315 Plan and IAMET Special Fund, 2014A35001-1). Responsibility for the content of this paper lies with the authors.

Notes and references

1 A. J. Stevenson, D. G. Gromadskyi, D. Hu, J. H. Chae, L. Guan, L. P. Yu and G. Z. Chen, in *Nanocarbons for Advanced Energy Storage*, ed. X. Feng, Wiley-VCH, Weinheim, 2015, vol. 1, pp. 179–210.

2 S. W. Woo, K. Dokko, H. Nakano and K. Kanamura, *Electrochemistry*, 2007, **75**, 635–640.

3 A. Krause, P. Kossyrev, M. Oljaca, S. Passerini, M. Winter and A. Balducci, *J. Power Sources*, 2011, **196**, 8836–8842.

4 K. C. Ng, S. W. Zhang, C. Peng and G. Z. Chen, *J. Electrochem. Soc.*, 2009, **156**, A846–A853.

5 S. W. Zhang, C. Peng, K. C. Ng and G. Z. Chen, *Electrochim. Acta*, 2010, **55**, 7447–7453.

6 Z. Dai, C. Peng, J. H. Chae, K. C. Ng and G. Z. Chen, *Sci. Rep.*, 2015, **5**, 09854.

7 C. Peng, S. Zhang, X. Zhou and G. Z. Chen, *Chem. Commun.*, 2010, 1499–1502.

8 J. H. Chae and G. Z. Chen, *Electrochim. Acta*, 2012, **86**, 248–254.

9 X. Zhou, C. Peng and G. Z. Chen, *AIChE J.*, 2012, **58**, 974–983.

10 S. Makino, Y. Shinohara, T. Ban, W. Shimizu, K. Takahashi, N. Imanishi and W. Sugimoto, *RSC Adv.*, 2012, **2**, 12144–12147.

11 *Ionic Liquids in Synthesis*, ed. P. Wasserscheid and T. Welton, Wiley-VCH Verlag, Weinheim, Germany, 2002.

12 M. Oromi-Farrus, J. Eras, G. Villorbina, M. Torres, V. Llopis-Mestre, T. Welton and R. Canela, *Anal. Sci.*, 2008, **24**, 1341–1345.

13 L. Chen, Y. Chi, X. Zheng, Y. Zhang and G. Chen, *Anal. Chem.*, 2009, **81**, 2394–2398.

14 T. Li, B. Li, S. Dong and E. Wang, *Chem.–Eur. J.*, 2007, **13**, 8516–8521.

15 X. J. Wei, L. P. Yu, D. H. Wang, X. B. Jin and G. Z. Chen, *Green Chem.*, 2008, **10**, 296–305.

16 X. J. Wei, L. P. Yu, X. B. Jin, D. H. Wang and G. Z. Chen, *Adv. Mater.*, 2009, **21**, 776–780.

17 L. P. Yu and G. Z. Chen, *RSC Adv.*, 2014, **4**, 40281–40285.

18 M. Anouti, E. Couadou, L. Timperman and H. Galiano, *Electrochim. Acta*, 2012, **64**, 110–117.

19 A. Orita, K. Kamijima and M. Yoshida, *J. Power Sources*, 2010, **195**, 7471–7479.

20 M. Galinski and I. Stepniak, *J. Appl. Electrochem.*, 2009, **39**, 1949–1953.

21 D. G. Gromadskyi, J. H. Chae, S. A. Norman and G. Z. Chen, *Appl. Energy*, 2015, **159**, 39–50.

22 L. P. Yu, H. J. Sun, J. He, D. H. Wang, X. B. Jin, X. H. Hu and G. Z. Chen, *Electrochem. Commun.*, 2007, **9**, 1374–1381.

23 Y. Gao, Z. Qin, L. Guan, X. Wang and G. Z. Chen, *Chem. Commun.*, 2015, **51**, 10819–10822.

Faraday Discussions

PAPER

Molten salt CO_2 capture and electro-transformation (MSCC-ET) into capacitive carbon at medium temperature: effect of the electrolyte composition

Bowen Deng, Zhigang Chen, Muxing Gao, Yuqiao Song, Kaiyuan Zheng, Juanjuan Tang, Wei Xiao, Xuhui Mao and Dihua Wang*

Received 21st December 2015, Accepted 1st March 2016
DOI: 10.1039/c5fd00234f

Electrochemical transformation of CO_2 into functional materials or fuels (*i.e.*, carbon, CO) in high temperature molten salts has been demonstrated as a promising way of carbon capture, utilisation and storage (CCUS) in recent years. In a view of continuous operation, the electrolysis process should match very well with the CO_2 absorption kinetics. At the same time, in consideration of the energy efficiency, a molten salt electrochemical cell running at lower temperature is more beneficial to a process powered by the fluctuating renewable electricity from solar/wind farms. Ternary carbonates (Li : Na : K = 43.5 : 31.5 : 25.0) and binary chlorides (Li : K = 58.5 : 41.5), two typical kinds of eutectic melt with low melting points and a wide electrochemical potential window, could be the ideal supporting electrolyte for the molten salt CO_2 capture and electro-transformation (MSCC-ET) process. In this work, the CO_2 absorption behaviour in Li_2O/CaO containing carbonates and chlorides were investigated on a home-made gas absorption testing system. The electrode processes as well as the morphology and properties of carbon obtained in different salts are compared to each other. It was found that the composition of molten salts significantly affects the absorption of CO_2, electrode processes and performance of the product. Furthermore, the relationship between the absorption and electro-transformation kinetics are discussed based on the findings.

Introduction

The sustained increase of CO_2 concentration in the atmosphere has caused great concern for its effect on global climate change. In fact, carbon dioxide is not only a greenhouse gas but also an abundant C^1 resource. Therefore, in addition to CO_2 capture and storage (CCS),[1,2] converting CO_2 into useful products such as

School of Resource and Environmental Sciences, Hubei International Scientific and Technological Cooperation Base of Sustainable Resource and Energy, Wuhan University, Wuhan 430072, PR China. E-mail: wangdh@whu.edu.cn; Fax: +86-27-6877-5799; Tel: +86-27-6877-4216

hydrocarbons and other chemicals is highly desirable for sustainable development.[3,4] Carbon capture, utilisation and storage (CCUS) is very attractive as a mid-term solution owing to the reduced burden on limited geological storage sites in CCS processes.[5,6] Electrochemical transformation of CO_2 into chemicals, fuels and functional materials, especially driven by renewable electricity is one of the most promising ways of CCUS.

The investigated electrolytes for CO_2 electro-transformation include aqueous solutions,[7] ionic liquids,[8] high temperature molten salts[9] and solid electrolytes.[10] Electrochemical reduction of CO_2 into low-carbon fuels in aqueous solution is a proven technology.[11,12] However, the reaction kinetics, current efficiency and the selectivity of products in aqueous solution is still unsatisfactory due to the poor solubility of CO_2 and limited electrochemical window of the electrolyte, which has driven research into efficient electro-catalysts for CO_2 conversion.[13,14] Ionic liquids are more satisfying candidates due to their low vapour pressure, higher CO_2 absorption capacity and electrochemical stability, but the high cost as well as the slow reaction kinetics currently prevent them from large scale application.[15,16]

High temperature molten salts are cheaper, more abundant and available with a high ionic conductivity, low vapour pressure and wide electrochemical window.[17,18] The reaction kinetics can be easily accelerated at elevated temperature without the use of complex and costly catalysts.[19–23] Besides, there are plenty of industrial experiences in primary aluminium and magnesium production to scale up the process. Although there is very limited data on the solubility of CO_2 in high temperature molten salts so far, it was recently reported that the absorption capacity of solid absorbents for CO_2 can be activated and improved by molten salts.[24–27] In fact, the absorption of CO_2 in molten salts can be enhanced by a chemical reaction between the acidic CO_2 and alkaline absorbents. Alkali metal oxides and alkali earth metal oxides have strong basicity and would be preferential absorbents. Table 1 presents the equilibrium activity of M_xO (M = Li, Na, K, Ca) under CO_2 pressure of 1 atm at 450 °C. All the data is lower than 10^{-4} M, in other words, the equilibrium partial pressure of CO_2 is in the range of 10^{-2} to 10^{-19} atm if the activity of the oxides in the melt could be kept at 0.01 M. From the data in Table 1 and as evidenced by experiments from different groups,[9,20,28,29] carbon is the preferential product when Li_2CO_3 and/or $CaCO_3$ is a component of molten salts. Therefore, CO_2 can be absorbed and decomposed in the high temperature molten salts with Li_2CO_3 and/or $CaCO_3$, which forms the foundation of the molten salt CO_2 capture and electrochemical transformation (MSCC-ET) process.

Table 1 The equilibrium activity of corresponding metal oxide and the deposition potential of metals (E_M) and carbon (E_C) in different molten carbonates at 450 °C under a CO_2 atmosphere of 1.0 atm

Molten salt	Activity of M_xO^a	E_M/V vs. CO_2–O_2/CO_3^{2-}	E_C/V vs. CO_2–O_2/CO_3^{2-}
Li_2CO_3	1.04×10^{-8}	−3.183	−1.883
Na_2CO_3	2.46×10^{-16}	−2.757	−2.681
K_2CO_3	2.05×10^{-21}	−2.829	−3.251
$CaCO_3$	2.31×10^{-5}	−3.230	−1.523

[a] The activity of metal carbonate is referred to 1.

The possible products of electrochemical transformation from CO_2 in high temperature molten salts mainly include carbon (in the form of carbon powder, film or compounds with metals) and/or carbon monoxide, depending on the temperature and applied cell voltage.[30,31] In some cases, syngas can be obtained when water vapour is present.[32–34] Table 2 briefly summarises the major molten salt systems and the transformation products of the MSCC-ET process reported in literature in recent years. The supporting electrolyte (molten salt) can be roughly divided into two categories, molten carbonates and molten chlorides. Usually the melt contains alkaline carbonate or oxide as absorbent, Chen *et al.* recently pointed out that alkaline earth metal oxides/carbonates (CaO, BaO) can also be used as absorbents for CO_2.[28]

Carbonate melts have been successfully applied in molten carbonate fuel cells (MCFCs) for a long time. It recent decades, ternary, binary or single molten alkali carbonates have been applied as the electrolyte for MSCC-ET in the temperature range of 450–950 °C as listed in Table 2. In the early 2000s, Groult *et al.* synthesised nanostructured carbon by electro-reduction of carbonate anion in fused ternary alkaline carbonates and investigated its potential application as negative material for lithium ion batteries.[20,35] Kaplan *et al.* reported the process of converting CO_2 to CO in molten Li_2CO_3 at 850–900 °C using a titanium cathode and a graphite anode,[36] and the reported thermodynamic efficiency exceeded 85%. Licht *et al.* proposed a solar thermal electrochemical photo (STEP) carbon capture process in a molten Li_2CO_3 electrolyte with a nickel cathode and a Pt anode, CO_2 was reduced to CO and C at 750–950 °C.[22] Wang *et al.* reported that CO_2 can be converted into value-added carbon and oxygen using a SnO_2 inert anode in molten Li_2CO_3–Na_2CO_3–K_2CO_3 at 500 °C.[23] The influences of temperature and cell voltage on the process have recently been investigated in detail.[37] Chen *et al.* also reported carbon powder by electro-reduction of CO_2 in Li_2CO_3–K_2CO_3 at 500–700 °C.[19] More recently, Chery *et al.* analysed and reviewed the thermodynamics of the reduction of CO_2 in molten carbonates.[30,31]

As shown in Table 2, the reduction of CO_2 can also be achieved directly or indirectly in carbonate-containing molten halides.[39] Ito and Kawamura reported that carbon film deposited on an aluminium electrode in LiCl–KCl–K_2CO_3 melts when CO_2 was fed into the cell. The current density was 100 mA cm^{-2}, much

Table 2 CO_2 reduction and obtained product in different molten salt

Molten salt	Temperature (°C)	Product	Ref.
Li_2CO_3	750–950	CO/carbon	22 and 36
Li_2CO_3–K_2CO_3	540–700	Carbon nanopowder	19
Li_2CO_3–K_2CO_3	800	Fe_3C	38
Li_2CO_3–Na_2CO_3–K_2CO_3	450–650	Carbon powder	20,23 and 37
$CaCl_2$–CaO	900	Carbon	39
LiF–NaF–Na_2CO_3	670–750	Carbon particle	40
LiCl–Li_2O	650	Carbon	39
LiCl–Li_2CO_3	650	Carbon particle	41
LiCl–KCl–K_2CO_3	450	Carbon film	21
LiCl–KCl–$CaCl_2$–$CaCO_3$	520	Carbon powder	28
NaCl–Na_2CO_3	800	Fe_3C	38

higher than that reported in aqueous solution.[21] Thermodynamically K_2CO_3 was converted into Li_2CO_3 in the melt for the Gibbs energy change of $2LiCl + K_2CO_3 = Li_2CO_3 + 2KCl$ is -105.31 kJ mol^{-1}. Similar works were reported in LiF–NaF–Na_2CO_3 melt and LiCl–Li_2CO_3 melt.[40,41] More recently Chen *et al.* reported carbon deposition in $CaCl_2$–$CaCO_3$–LiCl–KCl, confirming the possibility of CO_2 capture and reduction from $CaCO_3$-containing molten chlorides.[28] It was also reported that mild steel or titanium can be electro-carburized using CO_2 feedstock in molten carbonate salts to form a hardened surface.[38,42] These studies demonstrated that the electrochemical reduction of carbonate/CO_2 are significantly influenced by electrolysis temperature, cell voltage and melt composition, *etc.*

The mechanism for the MSCC-ET process can be briefly presented as eqn (1)–(4). In the molten salts containing M_xCO_3 (M = Li, Ca, ...), the carbonate is reduced to carbon (eqn (1) and (2)) or carbon monoxide (eqn (1a) and (2a)) with release of metal oxide and oxygen. The difference between eqn (1) and (2) is due to the difference in the anodic reaction. Depending on the experimental conditions, the formed CO can be further reduced to C (eqn (3)). The as-formed oxides can absorb CO_2 to regenerate carbonates (eqn (4)), maintaining the sustainable operation of the process. The molten salt suitable for production of carbon from CO_2 should be able to dissolve oxygen ions by which CO_2 can be absorbed and converted to carbonate ions.

$$3M_xCO_3 = 3M_xO + C + 2CO_2 + O_2 \tag{1}$$

$$4M_xCO_3 = 4M_xO + 2CO + 2CO_2 + O_2 \tag{1a}$$

$$M_xCO_3 = M_xO + C + O_2 \tag{2}$$

$$2M_xCO_3 = 2M_xO + 2CO + O_2 \tag{2a}$$

$$2CO = 2C + O_2 \tag{3}$$

$$M_xO + CO_2 = M_xCO_3 \tag{4}$$

If the MSCC-ET process was driven by electricity from fossil fuels, it will generate more CO_2. A desirable process is driven by renewable energy and/or industrial waste heat, therefore, maintaining a molten salt bath at relatively low temperature would be more preferential to prevent heat loss and corrosion of structural materials. Furthermore, carbon thermodynamically tends to be the reduction product from CO_2 at lower temperature.[30,31] It is well known that Li_2CO_3–Na_2CO_3–K_2CO_3 (43.5 : 31.5 : 25 mol%) and LiCl–KCl (58.5 : 41.5 mol%) are typical eutectics with a low melting point, 393 °C and 353 °C, respectively. Besides, they all have a wide electrochemical window and are able to dissolve alkaline oxides or carbonates. As shown in Table 2, the reduction products were significantly influenced by the composition of molten salts as well as experimental conditions such as temperature and electrode materials. In fact, due to their different physicochemical properties of molten salts such as basicity, solubility, viscosity and *etc.*, both CO_2 absorption and electrochemical reaction processes would be affected by the molten salt composition. In this work, the absorption kinetics, electrochemical behaviour, energy efficiency, and properties of the products of the two kinds of low temperature MSCC-ET electrolyte were investigated and compared in detail. Furthermore, the kinetic coupling between the CO_2 absorption and electrochemical transformation was also discussed based on the obtained results.

Experimental

CO_2 absorption

CO_2 absorption studies were carried out by a home-made gas absorption experimental set-up. The set-up consisted of two high temperature vessels (a gas storage vessel and a CO_2 absorption vessel, interconnected by a valve) with identical volume (3.5 L). An alumina crucible with 100 g salts (typically, 180 mmol kg^{-1} Li_2O-containing molten Li–Na–K carbonates or 180 mmol kg^{-1} Li_2O/CaO-containing molten Li–K chlorides) was placed on the bottom of the absorption vessel. Before absorption, the connected vessels were purged with argon and evacuated to near vacuum repeatedly at room temperature, and then they were separated from each other by turning off the valve. The set-up was then heated to 450 °C at a rate of 5 °C min. A certain amount of CO_2 gas (99% purity) was purged into the gas storage vessel after the temperature was stable. After equilibrium, CO_2 was introduced into the absorption vessel by turning on the valve quickly. The internal pressure and temperature was automatically monitored by a digital pressure indicator with a precision of ±0.001 kPa and a thermal couple with an accuracy of ±0.01 K, respectively, which were recorded and stored in a personal computer. All trials were carried out for at least 3 times, and the mean value is reported here. The absorption amount of CO_2 was calculated according to the transient pressure of the reactor:

$$n_{CO_2} = \frac{P_g V_g - P(V_g + V_e - V_m) + P_v V_m}{RT} \tag{5}$$

where n_{CO_2} (mmol) is the absorption amount of CO_2 at given time t (min), P_g (kPa) is the initial CO_2 pressure in the gas storage vessel, P (kPa) is the CO_2 pressure at time t (min), and P_v (kPa) is the saturated vapour pressure of eutectic melts. V_g (cm^3) and V_e (cm^3) are the volume of the gas storage and the absorption vessel, respectively. V_m (cm^3) and V_c (cm^3) represent the volume of molten salts and crucible, respectively.

Cyclic voltammetry and electrolysis

Cyclic voltammetry was applied to investigate the electrochemical behaviour of the eutectic melt with different compositions under CO_2 atmosphere. Molten Li–Na–K carbonates (43.5 : 31.5 : 25 mol%) or molten Li–K chlorides (58.5 : 41.5 mol%) with 2 mol% Li_2CO_3/$CaCO_3$ were used as the electrolytes. Approximately 500 g salt was dried at 250 °C under argon for up to 24 h to remove original moisture before it was heated to the desired temperatures under CO_2. CO_2 was continuously bubbled into the melt through an alumina tube during the tests. CV measurements were performed on an electrochemical workstation (CHI-1140, CH Instrument Co. Ltd., USA) using a three-electrode configuration. The reference electrode consisted of a silver wire placed in an alumina tube filled with 0.5 mol% Ag_2SO_4-containing molten Li_2CO_3–Na_2CO_3–K_2CO_3 or 0.5 mol% AgCl-containing LiCl–KCl respectively. The working and counter electrode were a nickel (Ni) wire

(1 mm in dia., 0.64 cm^2 in electrolyte) and a graphite rod (6 mm in dia., 4.00 cm^2 in electrolyte), respectively.

Galvanostatic electrolysis at a current density of 25 to 100 mA cm^{-2} was carried out in different melts using a computer-controlled power source (Shenzhen Neware Electronic Co. Ltd., China). A Ni sheet (24 cm^2 area) and graphite rod (44 cm^2 area) served as cathode and anode, respectively. CO_2 was bubbled into the melts during electrolysis. Products deposited on the cathode were collected and rinsed with 1 M HCl and distilled water to remove the adherent frozen electrolyte, and were dried in a vacuum oven at 80 °C for 12 hours.

Characterization and electrochemical performance of the obtained carbon

The obtained products were characterized by scanning electron microscopy (SEM, FEI Sirion field emission), X-ray diffraction spectroscopy (XRD, Shimadzu X-ray 6000 with Cu Kα radiation at 40 kV and 250 mA, $\lambda = 0.154$ nm), Laser Confocal Raman Microspectroscopy (Renishaw RM1000, UK) at room temperature with an excitation wavelength at 514.5 nm from a diode pumped solid-state laser and a Micromeritics ASAP 2020 automatic analyser at liquid N_2 temperature.

The capacitive performance of carbon samples was investigated in 6 M KOH with a two-electrode testing cell. The electrodes were prepared into carbon films by mixing the carbon with polytetrafluoroethylene (PTFE) and commercial acetylene black (8 : 1 : 1). Nickel foam was used as a current collector and was covered by the carbon film with a surface area of about 0.8 cm^2. The mass loading of the active material was approximately 5 mg cm^{-2}. Two electrodes with almost identical mass loading were assembled in a cell for the measurements. Their capacitive performances were evaluated by cyclic voltammetry and galvanostatic charge–discharge measurements, which were carried out on a PGSTAT 302N Autolab Electrochemical Workstation. The gravimetric capacitances C_g (F g^{-1}) were then calculated from the measured capacitance and the mass of active material.

Results and discussion

The static absorption curves of CO_2 by different molten salts at 450 °C under the CO_2 partial pressure of 50 kPa are shown in Fig. 1. As can be seen, absorption of CO_2 in blank molten LiCl–KCl is negligible, while CO_2 solubility in oxide-free molten ternary carbonates was much higher, approximately 6 mmol CO_2 was absorbed by 100 g blank Li–Na–K carbonates. The results demonstrated that the structure and composition of molten salts has a significant effect on the CO_2 absorption. Molten carbonates have a stronger interaction with CO_2 than molten chlorides. The following chemical capture mechanism[43,44] (eqn (6) and (7)) could be possible for molten carbonates.

$$CO_2 + CO_3{}^{2-} = C_2O_5{}^{2-} \tag{6}$$

$$CO_2 + nCO_3{}^{2-} = C_{n+1}O_{3n+2}{}^{2-} \tag{7}$$

Interestingly, very different absorption behaviour was present when 18 mmol Li_2O or CaO was added into 100 g salts, respectively. CO_2 can be rapidly captured

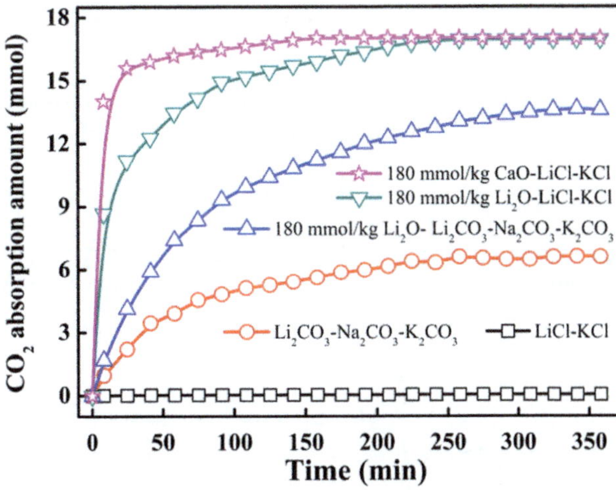

Fig. 1 CO_2 absorption curves in different salts (100 g) at 450 °C under a CO_2 partial pressure of 50 kPa.

when Li_2O or CaO was added into molten LiCl–KCl. The LiCl–KCl melt with CaO and Li_2O absorbed 17 mmol and 16.9 mmol CO_2, respectively when the equilibrium was reached. According to the absorption reaction as shown in eqn (4), the conversion efficiency was around 94% for Li_2O and CaO in the molten chlorides. Besides, it took less time to reach absorption equilibrium in molten chlorides than in molten carbonates with CaO or Li_2O, totally different from the blank molten salts. The absorption kinetics of CaO in molten chlorides is much faster and reached equilibrium within 2 hours although its absorption equilibrium constant was much lower than that of Li_2O. The conversion efficiency of Li_2O in molten carbonate (approximately 45%) is much lower than that in molten chlorides (approximately 94%), suggesting the property of molten salt affects both the absorption thermodynamics and kinetics of Li_2O. This phenomenon might be due to the lower activity of Li_2O in the molten ternary carbonates and the stronger interaction between CO_2 and carbonate slow down the diffusion rate of CO_2 in the melt. It seems that CaO was a stronger absorbent for CO_2 in molten LiCl–KCl due to its faster absorption rate and higher conversion efficiency. The experimental results demonstrated that Li_2O and CaO can serve as a good absorbent in molten salts so that it can act as a shuttle to maintain the MSCC-ET process given by eqn (1)–(4). It should be pointed out that the absorption behaviour of CO_2 is affected by its partial pressure, temperature and the concentration of Li_2O/CaO as well as the composition of molten salts. The curve shown in Fig. 1 is a boundary condition where no forced convention is involved.

To evaluate the electrochemical transformation behaviour of the captured CO_2 (in the form of carbonate), cyclic voltammetry measurements were conducted in molten Li_2CO_3–Na_2CO_3–K_2CO_3, 2 mol% Li_2CO_3–LiCl–KCl and 2 mol% $CaCO_3$–LiCl–KCl under CO_2 atmosphere at 450 °C with a Ni wire working electrode. As shown in Fig. 2, there are obvious reduction peaks before the deposition of alkali metal in the three kinds of electrolyte, suggesting the reduction of carbonate anion can take place before the decomposition of electrolyte. However, there

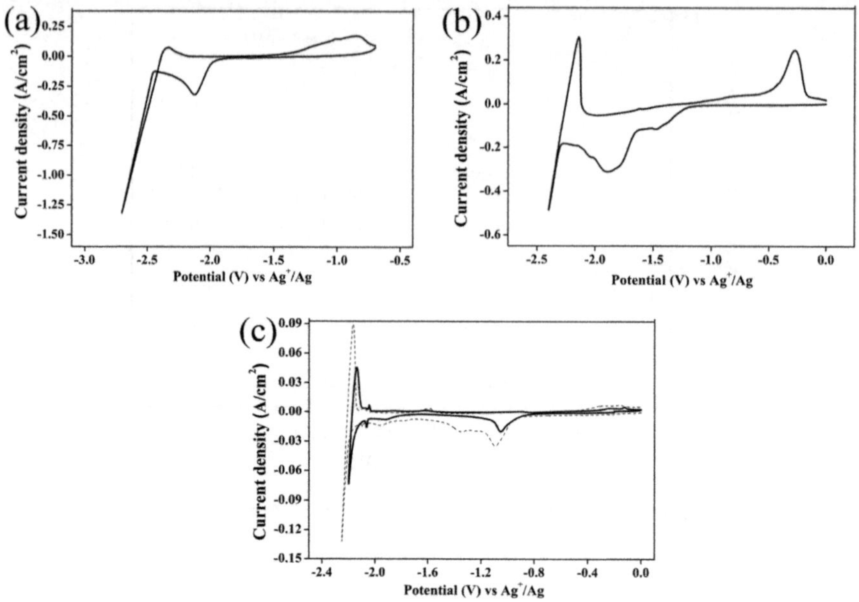

Fig. 2 Cyclic voltammograms of a nickel wire electrode in (a) Li_2CO_3–Na_2CO_3–K_2CO_3 (b) 2 mol% Li_2CO_3–LiCl–KCl (c) 2 mol% $CaCO_3$–LiCl–KCl under 1.0 atm CO_2 atmosphere at 450 °C. Solid line: cyclic voltammetry conducted in original molten mixture. Dashed line: cyclic voltammetry conducted after potentiostatic electrolysis at −2.2 V.

exists some differences in the CV curves obtained in the three molten salts. From Fig. 2a, there is only one reduction peak at −2.1 V (*vs.* Ag$^+$/Ag) before the cathodic limit in molten Li–Na–K carbonates, which could be ascribed to the reduction of CO_3^{2-} to carbon with 4-electron transfer, as presented by eqn (1) and (2). However, there are two reduction peaks before the deposition of alkaline metals on the CVs in molten LiCl–KCl containing Li_2CO_3 as shown in Fig. 2b. The reduction of carbonate/CO_2 in molten LiCl–KCl containing 2 mol% Li_2CO_3 occurred in the potential range from −1.2 to −2.2 V (*vs.* Ag$^+$/Ag). Two reduction peaks were presented at −1.5 V and −1.9 V, respectively. The two-step electrochemical reduction might be caused by the quasi-reversible reduction of CO_2 to a CO_2^{2-} radical (eqn (8)) and followed by the reduction of chemically as-formed CO (eqn (9)) to carbon (eqn (10)).[9,30] Also, it was reported that carbonate ions can be reduced in another two steps as represented by eqn (12) and (13),[9] firstly reduced to CO_2^{2-}, and CO_2^{2-} was then reduced to carbon.

$$\text{First step: quasi reversible reduction: } CO_2 + 2e^- = CO_2^{2-} \quad (8)$$

$$\text{Second step: chemical formation of CO: } CO_2^{2-} = CO + O^{2-} \quad (9)$$

$$\text{Third step: irreversible electro-reduction of CO to carbon: } CO + 2e^- = C + O^{2-} \quad (10)$$

$$\text{Overall reaction: } 3CO_2 + 4e^- = C + 2CO_3^{2-} \quad (11)$$

$$CO_3^{2-} + 2e^- = CO_2^{2-} + O^{2-} \quad (12)$$

$$CO_2^{2-} + 2e^- = C + 2O^{2-} \quad (13)$$

In the salt of 2 mol% $CaCO_3$–LiCl–KCl, a different CV profile was recorded for the first cycle as shown by the solid line in Fig. 2c. There is only one reduction peak in the potential range of −1.2 V to −0.8 V, and no carbon deposit was found by potentiostatic electrolysis using a nickel sheet electrode in the potential range. It is likely that the reduction peak is related to the reduction of the carbonate ion to CO on nickel electrode in the melt. It is different from that in Li_2CO_3 containing chloride melt, and also different from that in the $CaCO_3$–$CaCl_2$–LiCl–KCl melt with a high $CaCO_3$ concentration in which a one-step reduction to carbon was proposed.[28] However, a carbon deposit was obtained by electrolysis at a more negative potential (−2.2 V) in the melt. More interestingly, after that, two similar reduction peaks were found in 2 mol% $CaCO_3$–LiCl–KCl (dashed line in Fig. 2c), the reduction peaks at −1.1 V and −1.4 V, respectively were more positive than that for Li_2CO_3 reduction. And then carbon deposit was found under potentiostatic electrolysis of −1.8 V. The possible reason might be the partial transformation of $CaCO_3$ to Li_2CO_3 under the electrolysis condition, as the standard equilibrium constant for the reaction of $2LiCl(l) + CaCO_3(l) = CaCl_2(l) + Li_2CO_3(l)$ is only 2.97×10^{-3} at the temperature. The different CV behaviours demonstrate that the reduction of carbonate ion was very sensitive to the composition of molten salts as well as electrode materials, which remains to be investigated. Besides, it should be noted that a weak anodic peak was found at −0.3 V (*vs.* Ag^+/Ag), which can be ascribed to the oxidation of deposited carbon on Ni wire working electrode by eqn (14) and/or eqn (15). By contrast, more intensive oxidation peaks were found in Li_2CO_3-containing molten salt as presented on Fig. 2a and b. This difference might be caused by poorer adherence for the deposited carbon and stronger evolution of formed CO in $CaCO_3$-containing LiCl–KCl.

$$C + 2CO_3^{2-} = 3CO_2 + 4e^- \quad (14)$$

$$C + 2O^{2-} = CO_2 + 4e^- \quad (15)$$

Galvanostatic electrolysis was carried out in the three kinds of melts at a current density ranging from 25 mA cm^{-2} to 100 mA cm^{-2} by passing the same amount of electrical charge (2000 mA h). The cell voltage–time plots were recorded and displayed in Fig. 3. As can be seen, cell voltage in Li_2CO_3–Na_2CO_3–K_2CO_3 was much lower than that in molten chlorides at either 25 mA cm^{-2} or 100 mA cm^{-2}. The cell voltage for carbon deposition at 25 mA cm^{-2} was in the vicinity of 2.6 V for molten carbonates while it was nearly 3.2 V in carbonate-containing chlorides. It is noted that the cell voltage in $CaCO_3$ containing system is slightly lower than that in Li_2CO_3 containing chloride. When the current density was increased to 100 mA cm^{-2}, the voltage–time profile is quite different in the three salts as shown in Fig. 3b. The cell voltage in the ternary carbonates was low and gradually decreased with time, probably due to the decrease in real current density on the cathode. While the cell voltage profile in carbonate containing chlorides showed a sharp increase, followed by a gradual increase and then reached stability. These might be partially due to the concentration polarization

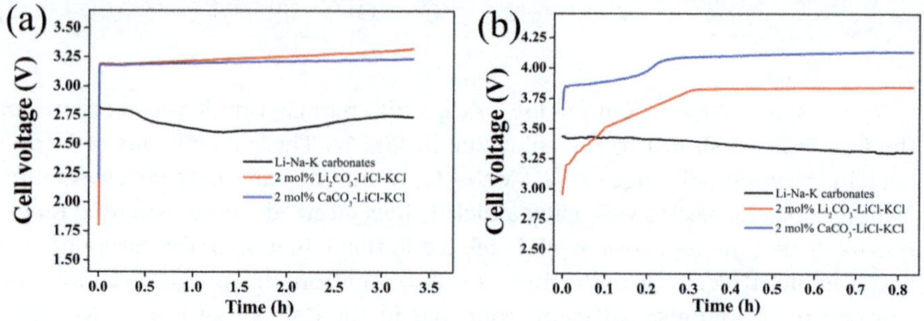

Fig. 3 The voltage–time plots for galvanostatic electrolysis in different molten salt at various current densities under CO_2 atmosphere at 450 °C with charge of 2000 mA h (a) 25 mA cm^{-2}. (b) 100 mA cm^{-2}.

of carbonate ion in the electrode/electrolyte interface. It is noted that the cell voltage in $CaCO_3$ containing salt is 200–300 mV higher than that in Li_2CO_3-containing chlorides, due to the more complicated and sluggish reduction kinetics in the $CaCO_3$ containing melt, consistent with its CV behaviour as shown in Fig. 2. The lowest cell voltage for CO_2 reduction in Li–Na–K molten carbonates is beneficial to decrease the energy consumption of the process.

Current efficiency at different current density in the melts was calculated based on the mass of the carbon product and depicted in Fig. 4. As can be seen, the highest current efficiency of approximately 90% was obtained in the ternary carbonate melt, and the current efficiency decreased with increasing current density. Current efficiency in Li_2CO_3 containing chlorides is higher than that in chlorides with $CaCO_3$. The efficiency is higher than 65% for all experimental conditions, which demonstrated that carbon is the preferential reduction product at the temperature in all electrolytes. Combing this with the cell voltage and current efficiency data, the energy consumption for electrolytic reduction of CO_2 to carbon in ternary carbonates is the lowest.

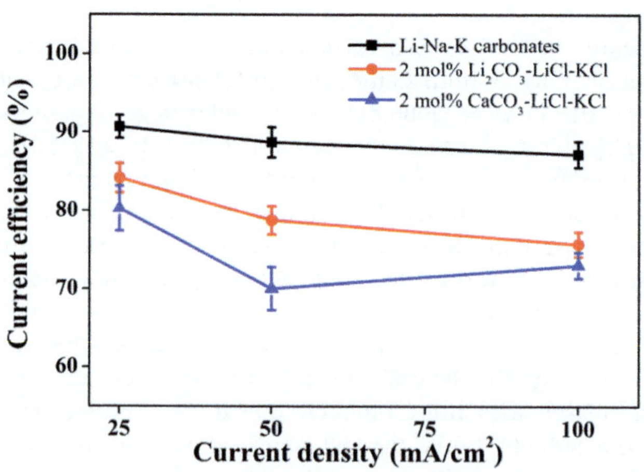

Fig. 4 Current efficiency for galvanostatic electrolysis in different molten salts under CO_2 atmosphere at 450 °C with a total charge of 2000 mA h.

The SEM images of carbon deposited in various molten salts at the same current density (50 mA cm^{-2}) are shown in Fig. 5. As can be seen, carbon obtained in molten Li–Na–K carbonates consisted of highly aggregated nanoparticles of 50 nm to 100 nm in size. Meanwhile the morphology of carbon deposited in 2 mol% Li_2CO_3–LiCl–KCl was quite complex, including nanoparticles, quasi-spherical carbon as well as micron-sized carbon sheets. Likewise, similar morphology of the deposited carbon was found in 2 mol% $CaCO_3$–LiCl–KCl, which mainly consisted of micron-sized carbon sheets and shell-like carbon, and a small quantity of nanoparticles was also observed. It is interesting that the morphology of deposited carbon is significantly influenced by the composition of the molten salts. The results inferred that the appearance of micron-sized carbon sheets is relevant to Li–K chlorides and electrolysis temperature, since no carbon sheet was found in molten LiCl–Li_2CO_3 at 650 °C.[41] However, carbon particles with much smaller size were obtained in Li_2CO_3–Na_2CO_3–K_2CO_3, which might be ascribed to a much higher concentration of Li_2CO_3 in the ternary carbonates, compared with the quasi-spherical carbon in a size of 500–800 nm obtained in Li_2CO_3-containing molten halides.[41,45] Possibly due to the template effect of chloride salts, carbon sheets were obtained in the melts at the experimental temperature. While the shell-like morphology observed in $CaCO_3$ containing chlorides might be related to its special reduction mechanism, *i.e.*, CO as an intermediate product.

The deposited carbon at 50 mA cm^{-2} in various molten salts was further studied by XRD as shown in Fig. 6a. They are amorphous as reported elsewhere[23,37] since no obvious diffraction peak was found. Fig. 6b displays the Raman spectra of the carbon obtained in different salts. Two bands located at 1395 and 1592 cm^{-1} represent the characteristic disordered (D) and graphitic (G) planes of carbon material, respectively. The lower D band peak suggests the appearance of

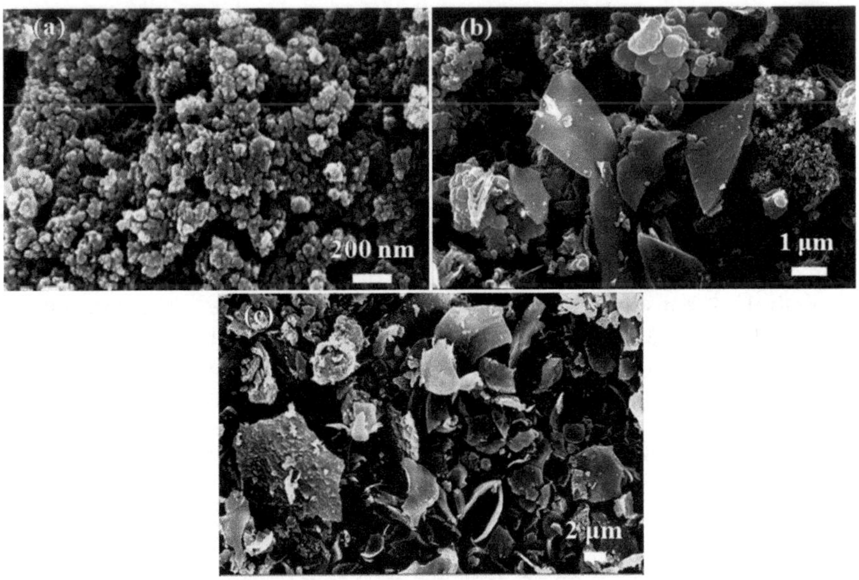

Fig. 5 The morphology of carbon deposited at 50 mA cm^{-2} in various molten salts under CO_2 atmosphere at 450 °C. (a) Li_2CO_3–Na_2CO_3–K_2CO_3. (b) 2 mol% Li_2CO_3–LiCl–KCl. (c) 2 mol% $CaCO_3$–LiCl–KCl.

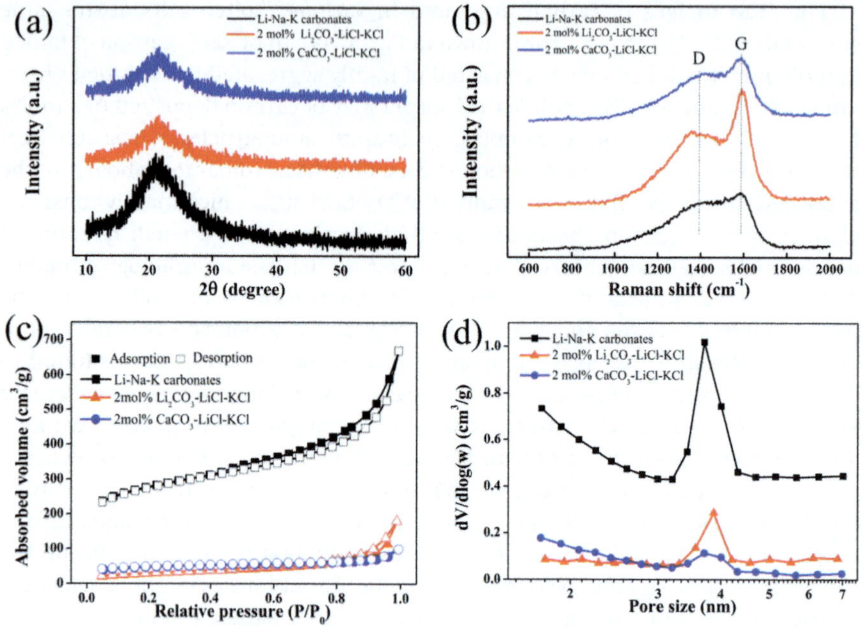

Fig. 6 Characterization of carbon obtained in various molten salts at 50 mA cm^{-2} under CO_2 atmosphere at 450 °C (a) XRD. (b) Raman spectrum. (c) Nitrogen adsorption–desorption isotherm. (d) Pore size distribution.

a low degree of structure defects in the carbon material. The ratio of intensity of D and G bands (I_D/I_G) for the carbon obtained in the three melts are shown in Table 3. It is around 0.7 for the carbon obtained from molten chlorides, but it is 0.86 for the carbon from the ternary carbonates. The low I_D/I_G indicates the relatively high degree graphitic structure in the as-prepared carbon in molten chlorides, this is related to the existence of carbon sheets as found in products obtained in carbonate-containing LiCl–KCl. Meanwhile there are more defects in the carbon obtained from molten carbonates under the experimental conditions, suggesting a highly activate surface carbon with a large amount of defects. The specific area and the porosity information of the products were further measured by N_2 adsorption–desorption automatic analysis and calculated by the Brunauer–Emmett–Teller (BET) method and the Barrett–Joyner–Halender (BJH) method, respectively. The results are shown in Fig. 6c and d and Table 3. The porosity and the specific surface area of carbon deposited in Li–Na–K carbonates is much larger than that of carbon obtained in carbonate-containing LiCl–KCl. The BET

Table 3 Characterisation summary of Raman and nitrogen adsorption–desorption tests for carbon deposited at 50 mA cm^{-2} in different molten salts

Molten salt	$I_D : I_G$	S_{BET} (m^2 g^{-1})	Pore volume (cm^3 g^{-1})	Average pore size (nm)
Li_2CO_3–Na_2CO_3–K_2CO_3	0.86	897.4	0.862	4.62
2 mol% Li_2CO_3–LiCl–KCl	0.67	113.9	0.279	9.97
2 mol% $CaCO_3$–LiCl–KCl	0.72	157.9	0.125	3.98

area of carbon from molten carbonates is 897.4 m^2 g^{-1}, much higher than that from molten chlorides (113.9 and 157.9 m^2 g^{-1}, respectively). Also, its pore volume is much larger.

Fig. 7 and 8 present the electrochemical performances of the carbon obtained in different molten salts investigated by cyclic voltammetry and galvanostatic charge–discharge tests, respectively using a carbon film electrode. As shown in Fig. 7a and b, CV plots for carbon obtained in Li–Na–K molten carbonates and $CaCO_3$-containing LiCl–KCl were more rectangular, and the corresponding current density of the plateaus were larger than those of carbon deposited in Li_2CO_3-containing LiCl–KCl. Accordingly, the results of galvanostatic charge–discharge experiments also demonstrate the promising capacitive properties of carbon obtained in Li–Na–K molten carbonates with 240 F g^{-1} at a current density of 0.2 A g^{-1}, while capacitance for carbon deposited in 2 mol% Li_2CO_3–LiCl–KCl and 2 mol% $CaCO_3$–LiCl–KCl were 102 F g^{-1} and 145 F g^{-1} at 0.2 A g^{-1}, respectively (Fig. 8a). The results of capacitance measurement at 2 A g^{-1} of charge–discharge current density were 202 F g^{-1}, 71 F g^{-1} and 110 F g^{-1} respectively as presented in Fig. 8b. This result suggests better electrochemical performance of deposited carbon can be obtained in Li–Na–K molten carbonates attributed to its higher surface area and porosity. Capacitive property of carbon obtained in $CaCO_3$-containing LiCl–KCl seems to be more attractive than carbon deposited in Li_2CO_3-containing LiCl–KCl due to its higher surface area as shown on Table 3. Different morphology, structure and electrochemical performance for carbon obtained in molten salts with different composition may suggest a different application field for the relative products.

Prospects

For the continuous operation of the MSCC-ET process, it is expected that the kinetics of CO_2 absorption and the diffusion of carbonate ion should be well matching with the electrochemical kinetics on the electrodes, just like industrial production of primary aluminium in which the dissolution of alumina should be well coupled with the electrode kinetics. Therefore, based on the absorption curve of CO_2 as shown in Fig. 1, the CO_2 transportation rate in molten salt was

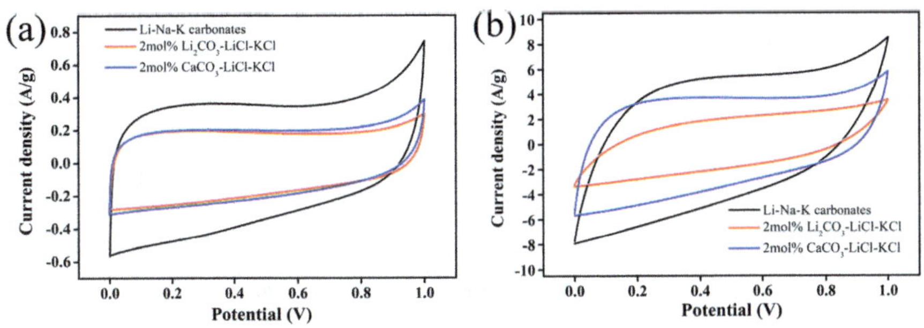

Fig. 7 Cyclic voltammograms of the carbon obtained at 50 mA cm^{-2} in various molten salts at 450 °C as working electrode in 6 M KOH aqueous solution under different scan rate. (a) 5 mV s^{-1} (b) 100 mV s^{-1}.

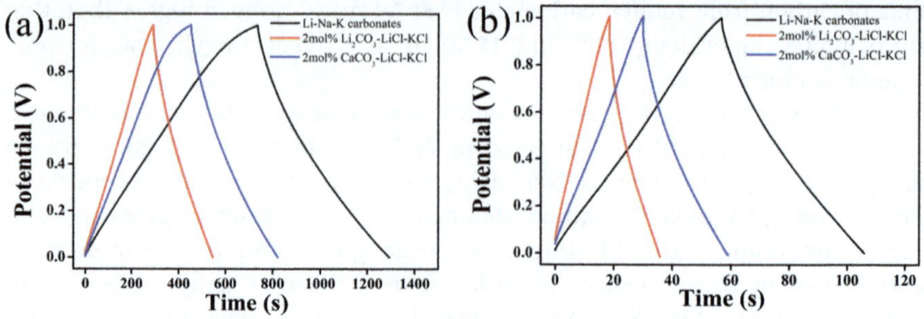

Fig. 8 Charge–discharge curves for carbon obtained at 50 mA cm^{-2} in various eutectics. (a) Charge–discharge at 0.2 A g^{-1}. (b) Charge–discharge at 2 A g^{-1}.

calculated and analysed according to damping-film theory.[46] The relationship between absorbed CO_2 and absorption time can be stated as follows:

$$\ln[(n_e - n)/n_e] = -Kt \tag{16}$$

where n_e (mmol) denotes the uptake amount at absorption equilibrium and n (mmol) is the mole of absorbed CO_2 at any time, which can be calculated by eqn (5). And K represents the reaction rate constant.

Fig. 9 shows the absorption rate of CO_2 calculated by eqn (16) based on the data in Fig. 1. The slope of each curve represents the instant absorption rate of the melt. As can be seen, absorption rate of CO_2 in CaO-containing LiCl–KCl was the fastest. Also it can be found that the curves in the three molten salts were pseudo-linear within the initial 300 seconds, suggesting a constant uptake rate during the first 5 minutes. Thus the apparent absorption rate of CO_2 capture can be calculated by eqn (17) according to the results of CO_2 absorption kinetics. The optimum coupling electrolysis is assumed when the electrochemical regeneration

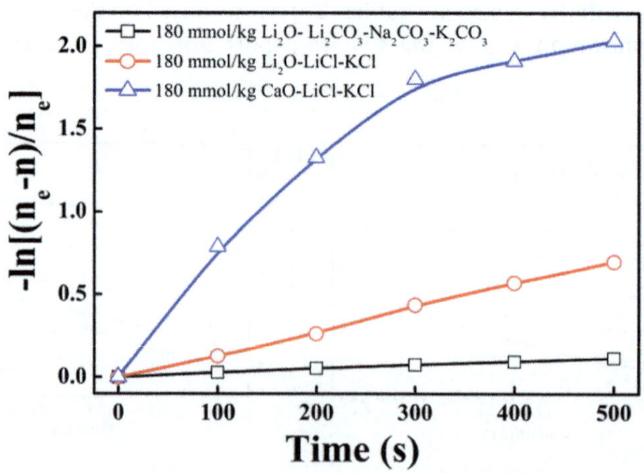

Fig. 9 CO_2 absorption rate in various molten salts under 50 kPa CO_2 partial pressure at 450 °C.

rate of oxygen ions (eqn (18)) equals the consumption rate of oxygen ions by CO_2 absorption (eqn (20)). Following this idea, the suitable current density for electrolysis with 100% efficiency can be calculated by eqn (21) according a 4-electron transfer mechanism as shown by eqn (18).

$$v_{CC} = \frac{dn}{dt} = \frac{n_{CO_2}}{\frac{m}{M}t} \tag{17}$$

n (mmol) – absorbed amount of CO_2, m (g) – the weight of molten mixture, M (g mol^{-1}) – the molar weight of molten mixture, t (h) – pseudo-linear absorption time.

$$CO_3^{2-} + 4e^- = C + O^{2-}\ \text{(regeneration)} + 2O^{2-}\ \text{(anodic-oxidation)} \tag{18}$$

$$2O^{2-}\ \text{(anodic-oxidation)} = O_2 + 4e^- \tag{19}$$

$$O^{2-}\ \text{(consumption)} + CO_2 = CO_3^{2-} \tag{20}$$

$$v_{ET} = \frac{4Fv_{CC}}{3600S} \tag{21}$$

F (C mol^{-1}) – Faraday constant, S (cm^2) – area of cathode.

As an example, Table 4 presents the calculated value of the electrolysis current density matching the static apparent absorption rate under the selected conditions based on a Ni cathode with an area of 24 cm^2. As can be seen, the process rate of MSCC-ET in $CaCO_3$-containing CaO–LiCl–KCl can be run above 239.5 mA cm^{-2} mol$_{salt}^{-1}$, followed by Li_2CO_3-containing Li_2O–LiCl–KCl and Li_2O-containing Li–Na–K molten carbonates, with 150.5 mA cm^{-2} mol$_{salt}^{-1}$ and 53.6 mA cm^{-2} mol$_{salt}^{-1}$, respectively. Although the data are not very accurate and are obtained under specific experimental conditions, it can provide some insights on the process. (1) The kinetics of CO_2 capture and reduction in $CaCO_3$-containing molten chlorides is more likely limited by electrochemical kinetics, which might be improved by applying strong electrochemical polarization within the electrochemical window; (2) the electrochemical kinetics in Li_2CO_3-containing salts is faster compared with the absorption kinetics, the process kinetics can be improved by increasing the concentration of Li_2O and partial pressure of CO_2 or temperature, and by stirring or bubbling the electrolyte. It was found in the

Table 4 The optimum coupling kinetics between a static CO_2 absorption and electrolysis under 50 kPa CO_2 partial pressure and at 450 °C in different molten salts

Molten salt composition	Molar amount of molten mixture (g mol^{-1})	CO_2 absorption rate (mmol$_{CO_2}$ h^{-1} mol$_{salt}^{-1}$)	Coupling electrolysis rate (mA cm^{-2} mol$_{salt}^{-1}$)
180 mmol kg^{-1} Li_2O Li–Na–K molten carbonates	100.1	12.0	53.6
180 mmol kg^{-1} Li_2O LiCl–KCl	56.0	33.7	150.5
180 mmol kg^{-1} CaO LiCl–KCl	56.2	53.6	239.5

authors' lab that the absorption rate can be accelerated by adding more Li_2O into the melt.

Summary

The effect of the composition of molten salts on CO_2 absorption and the electrochemical reduction process as well as the properties of the converted carbon at 450 °C was investigated. With the same mole amount of alkali/alkali earth oxide, CaO containing LiCl–KCl melt has the fastest absorption kinetics but the electrochemical kinetics in the melt is more sluggish and complicated. The electrochemical reduction in molten ternary carbonates is a one-step reduction process while it is a two-step process in Li_2CO_3 containing molten chlorides. For the constant current electrolysis in a two-electrode system, the cell voltage in the ternary carbonates is the lowest with highest current efficiency, demonstrating the best energy efficiency in molten carbonate. The morphology, density of structure defects, specific surface area as well as pore volume of the obtained carbon is also significantly influenced by electrolyte composition. Nano-carbon particles with a specific surface area of approximately 900 $m^2 g^{-1}$ and high I_D/I_G ratio (0.86) were obtained in the molten carbonates at a current density of 50 mA cm^{-2}. While carbon sheets with a specific area of approximately 150 $m^2 g^{-1}$ and low I_D/I_G ratio (around 0.7) were obtained in carbonate containing molten chlorides, the cation also has some effect on the property of the products. The carbon obtained in Li_2CO_3–Na_2CO_3–K_2CO_3 has a specific capacity of more than 200 F g^{-1} due to its higher surface area and porosity. The coupling kinetics between the absorption reaction and electrochemical transformation process was further discussed based on the measured absorption and electrode kinetics. However, the process kinetics are also affected by the concentration of absorbent, temperature, pressure as well as the convention conditions in the melt, which should be the subject of further investigation in future.

Acknowledgements

This work was funded by National Natural Science Foundation of China (51325102), the International Science & Technology Cooperation Program of China (2015DFA90750), the Program for Creation Team of Hubei Province (2015CFA017) and the Young-talent Chenguang Project of Wuhan City.

References

1 M. Mikkelsen, M. Jørgensen and F. C. Krebs, *Energy Environ. Sci.*, 2010, **3**, 43.
2 N. MacDowell, N. Florin, A. Buchard, J. Hallett, A. Galindo, G. Jackson, C. S. Adjiman, C. K. Williams, N. Shah and P. Fennell, *Energy Environ. Sci.*, 2010, **3**, 1645.
3 I. Ganesh, *Renewable Sustainable Energy Rev.*, 2014, **31**, 221–257.
4 B. Hu, C. Guild and S. L. Suib, *J. CO_2 Util.*, 2013, **1**, 18–27.
5 D. T. Whipple and P. J. A. Kenis, *J. Phys. Chem. Lett.*, 2010, **1**, 3451–3458.
6 M. Aresta and A. Dibenedetto, *Dalton Trans.*, 2007, 2975–2992, DOI: 10.1039/b700658f.
7 M. Gattrell, N. Gupta and A. Co, *J. Electroanal. Chem.*, 2006, **594**, 1–19.

8 M. Alvarez-Guerra, J. Albo, E. Alvarez-Guerra and A. Irabien, *Energy Environ. Sci.*, 2015, **8**, 2574–2599.

9 H. V. Ijije, R. C. Lawrence and G. Z. Chen, *RSC Adv.*, 2014, **4**, 35808–35817.

10 J. W. Fergus, *Sens. Actuators, B*, 2008, **134**, 1034–1041.

11 R. J. Bernot, E. E. Kennedy and G. A. Lamberti, *Environ. Toxicol. Chem.*, 2005, **24**, 1759–1765.

12 M. Bevilacqua, J. Filippi, H. A. Miller and F. Vizza, *Energy Technol.*, 2015, **3**, 197–210.

13 Y. Hori, H. Wakebe, T. Tsukamoto and O. Koga, *Electrochim. Acta*, 1994, **39**, 1833–1839.

14 R. Aydin, H. O. Dogan and F. Koleli, *Appl. Catal., B*, 2013, **140**, 478–482.

15 M. J. Muldoon, S. Aki, J. L. Anderson, J. K. Dixon and J. F. Brennecke, *J. Phys. Chem. B*, 2007, **111**, 9001–9009.

16 D. M. D'Alessandro, B. Smit and J. R. Long, *Angew. Chem., Int. Ed.*, 2010, **49**, 6058–6082.

17 T. Murakami, T. Nohira, Y. H. Ogata and Y. Ito, *Electrochem. Solid-State Lett.*, 2005, **8**, E1–E3.

18 W. Xiao and D. H. Wang, *Chem. Soc. Rev.*, 2014, **43**, 3215–3228.

19 H. V. Ijije, C. Sun and G. Z. Chen, *Carbon*, 2014, **73**, 163–174.

20 B. Kaplan, H. Groult, A. Barhoun, F. Lantelme, T. Nakajima, V. Gupta, S. Komaba and N. Kumagai, *J. Electrochem. Soc.*, 2002, **149**, D72.

21 H. Kawamura and Y. Ito, *J. Appl. Electrochem.*, 2000, **30**, 571–574.

22 S. Licht, B. H. Wang, S. Ghosh, H. Ayub, D. L. Jiang and J. Ganley, *J. Phys. Chem. Lett.*, 2010, **1**, 2363–2368.

23 H. Yin, X. Mao, D. Tang, W. Xiao, L. Xing, H. Zhu, D. Wang and D. R. Sadoway, *Energy Environ. Sci.*, 2013, **6**, 1538.

24 X. Yang, L. Zhao and Y. Xiao, *Energy Fuels*, 2013, **27**, 7645–7653.

25 D. J. Fauth, E. A. Frommell, J. S. Hoffman, R. P. Reasbeck and H. W. Pennline, *Fuel Process. Technol.*, 2005, **86**, 1503–1521.

26 V. Tomkute, A. Solheim and E. Olsen, *Energy Fuels*, 2013, **27**, 5373–5379.

27 T. Harada, F. Simeon, E. Z. Hamad and T. A. Hatton, *Chem. Mater.*, 2015, **27**, 1943–1949.

28 H. V. Ijije, R. C. Lawrence, N. J. Siambun, S. M. Jeong, D. A. Jewell, D. Hu and G. Z. Chen, *Faraday Discuss.*, 2014, **172**, 105–116.

29 K. Le Van, H. Groult, F. Lantelme, M. Dubois, D. Avignant, A. Tressaud, S. Komaba, N. Kumagai and S. Sigrist, *Electrochim. Acta*, 2009, **54**, 4566–4573.

30 D. Chery, V. Albin, V. Lair and M. Cassir, *Int. J. Hydrogen Energy*, 2014, **39**, 12330–12339.

31 D. Chery, V. Lair and M. Cassir, *Electrochim. Acta*, 2015, **160**, 74–81.

32 S. Cavallaro and S. Freni, *J. Power Sources*, 1998, **76**, 190–196.

33 P. Greppi, B. Bosio and E. Arato, *Int. J. Hydrogen Energy*, 2009, **34**, 8664–8669.

34 K. Hemmes and M. Cassir, *J. Fuel Cell Sci. Technol.*, 2011, **8**, 051005-1–051005-5.

35 H. Groult, B. Kaplan, S. Komaba, N. Kumagai, V. Gupta, T. Nakajima and B. Simon, *J. Electrochem. Soc.*, 2003, **150**, G67–G75.

36 V. Kaplan, E. Wachtel, K. Gartsman, Y. Feldman and I. Lubomirsky, *J. Electrochem. Soc.*, 2010, **157**, B552–B556.

37 D. Tang, H. Yin, X. Mao, W. Xiao and D. H. Wang, *Electrochim. Acta*, 2013, **114**, 567–573.

38 N. J. Siambun, H. Mohamed, D. Hu, D. Jewell, Y. K. Beng and G. Z. Chen, *J. Electrochem. Soc.*, 2011, **158**, H1117–H1124.
39 K. Otake, H. Kinoshita, T. Kikuchi and R. O. Suzuki, *Electrochim. Acta*, 2013, **100**, 293–299.
40 L. Massot, P. Chamelot, F. Bouyer and P. Taxil, *Electrochim. Acta*, 2003, **48**, 465–471.
41 J. B. Ge, L. W. Hu, W. Wang, H. D. Jiao and S. Q. Jiao, *ChemElectroChem*, 2015, **2**, 224–230.
42 V. Kaplan, E. Wachtel and I. Lubomirsky, *J. Electrochem. Soc.*, 2012, **159**, E159–E161.
43 P. Claes, D. Moyaux and D. Peeters, *Eur. J. Inorg. Chem.*, 1999, **4**, 583–588.
44 D. Peeters, D. Moyaux and P. Claes, *Eur. J. Inorg. Chem.*, 1999, **4**, 589–592.
45 L. X. Li, Z. N. Shi, B. L. Gao, J. L. Xu, X. W. Hu and Z. W. Wang, *Electrochemistry*, 2014, **82**, 1072–1077.
46 J. W. Ma, Z. Zhou, F. Zhang, C. G. Fang, Y. T. Wu, Z. B. Zhang and A. M. Li, *Environ. Sci. Technol.*, 2011, **45**, 10627–10633.

Faraday Discussions

PAPER

An investigation into the carbon nucleation and growth on a nickel substrate in $LiCl-Li_2CO_3$ melts

Jianbang Ge, Liwen Hu, Yang Song and Shuqiang Jiao*

Received 10th December 2015, Accepted 11th March 2016

DOI: 10.1039/c5fd00217f

The electrochemical deposition of carbon materials has been performed in $LiCl-Li_2CO_3$ melts using a Pt anode and a nickel cathode. Cyclic voltammetry and constant voltage electrolysis are conducted to investigate the electrode reactions, and the results prove that solid carbon is the only product from the cathodic reduction. Short-term electrolysis at 750 °C for 3, 10 and 20 s has been applied to study the formation and growth of the varied carbon microstructures. All of the results demonstrate that the morphologies of the deposited carbon are significantly affected by the cathode substrates, which may show different catalyzing effects on carbon nucleation. Two primary morphologies, quasi-spherical and nanofiber structures are observed at the nickel plate cathodes during the electrolysis and the formation and growth of carbon nanofibers are easily enhanced by using a high cell voltage. However, only a quasi-spherical structure is found on the molybdenum cathode substrate.

Introduction

In recent years increasing attention has been paid to the electrochemical conversion of CO_2 in molten salts to mitigate global warming because of the excellent properties (high electronic conductivity, low vapor pressure) of molten salts.[1,2] Unlike the various hydrogenated products obtained *via* the electrochemical conversion process in aqueous solutions and ionic liquids, carbonaceous materials are the main product due to the lack of elemental hydrogen in molten salts, and apparently the morphologies of the deposited carbon materials have a significant influence on their chemical and physical properties as well as their applications.[3–5] So far carbonaceous materials with different microstructures (such as spherical particles, flakes, nanosheets, nanofibers) have been obtained in molten salts under different electrochemical conditions.[6–14] Van *et al.* prepared quasi-spherical carbon particles with an average diameter of between 30 and 50 nm in molten alkali carbonates using a graphite anode and a nickel cathode.[6] The

State Key Laboratory of Advanced Metallurgy, University of Science and Technology Beijing, Beijing 100083, P. R. China. E-mail: sjiao@ustb.edu.cn; Fax: +86-10-62334204; Tel: +86-10-62334204

small and quasi-spherical carbon also has been achieved in LiCl–KCl–K_2CO_3 between an oxide-scale-coated titanium plate cathode and the graphite crucible anode.[7] CO_2 gas could be also decomposed successfully to nanometer-scale amorphous carbon and rod-like graphite crystals in molten $CaCl_2$–CaO or LiCl–Li_2O melt.[8] Tang *et al.* proved that the morphologies and particle size could be affected by the cell voltage and working temperature. The prepared carbon exhibited a variety of morphologies, such as nanoparticles, nanoflakes, nanosheets, *etc.*[9] Hu *et al.* acquired graphite sheets in $CaCl_2$-based melts.[10] Ren *et al.* prepared carbon nanofibers and the structure could be tuned by the addition of trace transition metals (Zn or Ni) on the cathode, the addition of zinc as an initiator and by control of current density.[11] Although carbon materials with different morphologies can be prepared and affected by electrolysis conditions, such as temperature, cell voltage, and cathode substrate, the mechanism of the formation and growth of different carbon structures is still not clear. Besides, the electrolysis times in these studies lasted for several hours in order to obtain enough carbon products for subsequent analysis, which may also conceal the mechanism of the carbon nucleation and its growth because the time of carbon nucleation is very short ($\sim$several milliseconds).[12]

In the present work, cyclic voltammetry has been conducted to reveal the electrochemical reduction of carbonate ions in molten LiCl–Li_2CO_3. Both long-term (2 h) and short-term constant voltage electrolysis (3, 10 and 20 s) were carried out in a two-electrode cell to investigate the mechanism of carbon nucleation and its growth in the Ni and Mo substrates using a Pt wire anode. Then the deposited carbon was characterized by X-ray diffraction analysis (XRD), scanning electron microscopy (SEM), and Raman spectrometry.

Experimental

The anhydrous LiCl with a certain amount Li_2CO_3 (2.5 mol% used for cyclic voltammetry tests and 5 mol% used for constant voltage electrolysis) were added into an alumina crucible, which was placed in a sealed vertical tubular reactor. The salt (about 180 g) was then dried at 300 °C for 12 h under vacuum to remove moisture before it was heated to a predetermined temperature under argon atmosphere. Platinum wire, nickel plates (15 mm × 5 mm × 0.2 mm), nickel rods (3 mm in diameter) and molybdenum plates were polished with abrasive papers and then washed ultrasonically in anhydrous ethanol. Finally, the electrodes were dried in a drying oven at 120 °C.

Cyclic voltammetry in a three-electrode cell was performed at 750 °C. A platinum wire (0.4 mm in diameter, $\sim$0.05 cm^2 in area) and a graphite rod (6 mm in diameter) served as a working electrode and counter electrode, respectively. A quartz Ag/AgCl electrode was selected as the reference electrode. For long-term constant cell voltage electrolysis, the processes were carried out for 2 h between a Pt wire (1 mm in diameter, $\sim$0.95 cm^2 in area) and a nickel rod (3 mm in diameter, $\sim$2.2 cm^2 in area). The outlet gas from the reactor was monitored by resolution mass spectra (Hidden, SPR-20). After electrolysis, the cathode with deposited carbon was then taken out and rinsed with 1 mol L^{-1} HCl solution and deionized water to remove the frozen electrolyte, and then dried in a drying oven. For short-term (3, 10 and 20 s) constant cell voltage electrolysis, the prepared nickel and molybdenum plates served as cathodes.

After finishing the tests, the plates were washed only by deionized water and dried in a drying oven. To ensure the accuracy and reliability of the results, each experiment was conducted in at least duplicate in short-term electrolysis experiments. All the electrochemical tests were carried out using an electrochemical workstation (CHI-660E).

The products obtained at the cathode were characterized by X-ray diffraction analysis (XRD, Rigaku-TTRIII), energy dispersive X-ray spectroscopy (EDX, Thermo scientific-Ultra dry), scanning electron microscopy (SEM, JEOL-JSM-6701F), and high resolution Raman spectrometry (Horiba-Labram HR evolution) with excitation at 532 nm. Moreover, due to the unevenness and defects of the nickel substrate, the SEM photographs taken from the middle area of the deposited carbon were chosen as the end results.

Results and discussion

Fig. 1 presents the cyclic voltammograms performed on a Pt working electrode in LiCl–Li_2CO_3 molten salt at 700 °C. When the potential scan range was between −2.0 V and 0.8 V, a reduction peak C_1 and its corresponding oxidation peak O_1 could be observed. The reduction peak C_1 apparently represents the reduction of carbonate ions to carbon according to our previous work, as noted by reaction 1 (ref. 13). The oxidation peak O_1 should correspond to the electrochemical dissolution of the deposited carbon according to reaction 2 with the aid of oxygen ions. When the Pt electrode potential is held at −1.8 V *vs.* Ag/AgCl for 20 s, the current of peak O_1 is clearly enhanced. This suggests that more carbon has been produced and re-oxidized immediately with the assistance of the oxygen ions remaining in or near the deposited carbon.

$$CO_3^{2-} + 4e^- \rightarrow C + 3O^{2-} \tag{1}$$

$$C + 2O^{2-} \rightarrow CO_2 + 4e^- \tag{2}$$

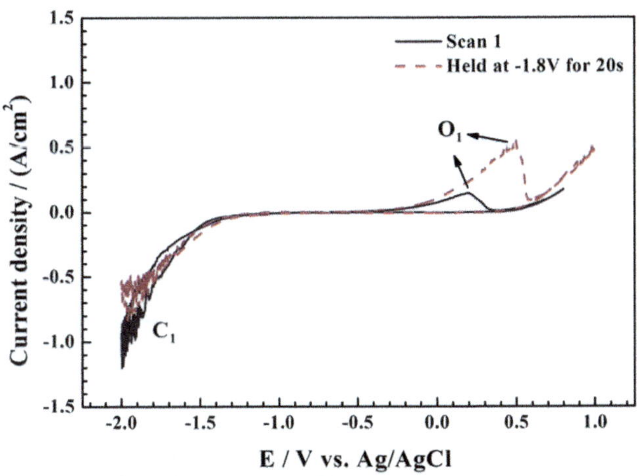

Fig. 1 Cyclic voltammograms of the LiCl–Li_2CO_3 melt at 700 °C on a Pt wire working electrode. Scan rate = 100 mV s^{-1}.

Then a long-term potentiostatic electrolysis with a cell voltage of 2.4 V was carried out in a two-electrode cell using a Pt wire anode and a nickel rod cathode at 650, 700 and 750 °C for 2 h, and the corresponding current–time plots are shown in Fig. 2a. The current becomes steady after an initial decrease. This phenomenon may be explained as the carbonate ions in the melt near the cathode are adequate to undertake the cathode reaction without involving diffusion.[14] The amount of anode gas as well as the current–time plot recorded at 750 °C and 2.4 V are shown in Fig. 2b. The amount of oxygen gas keeps rising over the course of electrolysis and falls quickly after the electrolysis, confirming the production of oxygen. However, the content of carbon monoxide remains almost unchanged, indicating no CO is generated at the cathode during the electrolysis. The variation of hydrogen gas shows that a weak peak is detected during the first several minutes with the decomposition of residual water in the melt. The results indicate that, except small amounts of hydrogen, carbon is the only product of the cathodic reduction. The Raman spectra presented in Fig. 2c shows the characteristic peaks of carbon materials, the D band at 1348 cm^{-1} and G band at 1588 cm^{-1}. Furthermore, the XRD spectra of the deposited carbon at 650, 700 and 750 °C at 2.4 V disclose that they are all amorphous carbon (Fig. 2d).

FESEM investigations were further used to characterize the morphologies of the carbon products, as shown in Fig. 3. The carbon products obtained at 650, 700 and 750 °C are presented in Fig. 3a, c and e, respectively, and the zones in the yellow dotted box in Fig. 3a, c and e are enlarged as Fig. 3b, d and f, respectively. As shown in Fig. 3a and b, at 650 °C, two different kinds of morphologies, nanofibers and quasi-spherical particles with a diameter size in the range of 1–3 μm can be observed. At 700 and 750 °C, a new microstructure, a huge carbon plate, can be seen beneath spherical carbon and nanofibers in Fig. 3c as well as in Fig. 3e. The SEM results herein are consistent with those in

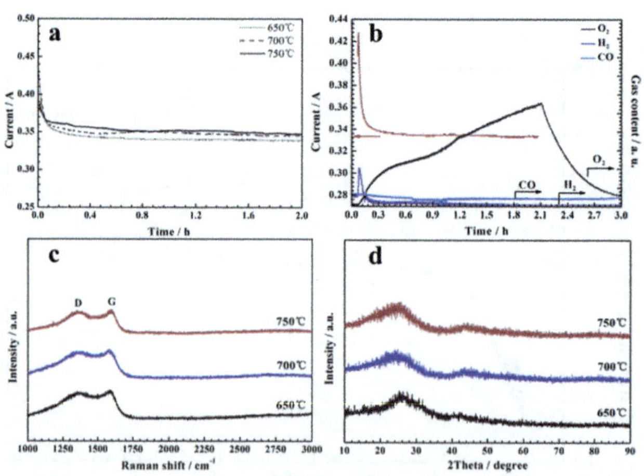

Fig. 2 (a) Current–time plots recorded at 650, 700 and 750 °C at 2.4 V for 2 h. (b) Typical variation of anode gas as well as current–time plot recorded at 750 °C and 2.4 V. Raman spectra (c) and XRD spectra (d) of the carbon materials deposited at 650, 700 and 750 °C at 2.4 V for 2 h.

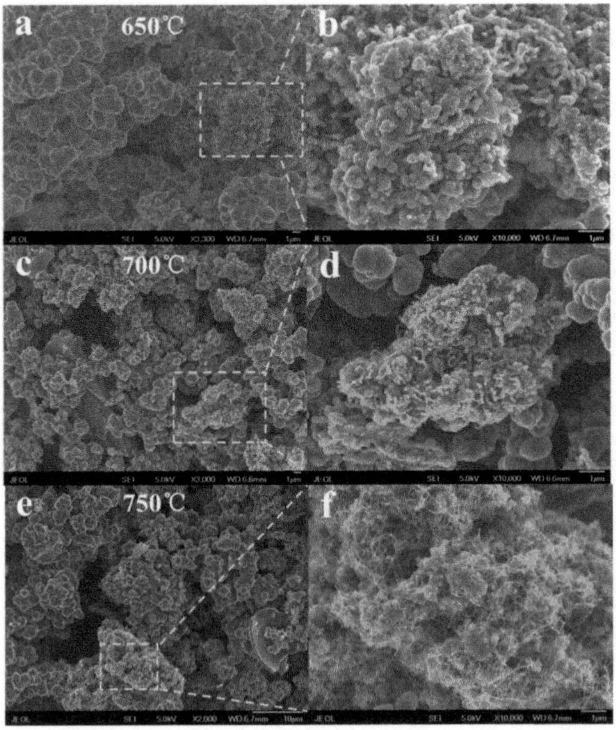

Fig. 3 FESEM images of the carbon materials deposited at 650 (a), 700 (c) and 750 °C (e) at 2.4 V for 2 h. (b, d and f) Enlargement of the zones in the yellow dotted box in (a, c and e), respectively.

the literature and also indicated that, after the long-term electrolysis, the carbon products exhibited varied microstructures.[8,9,14] It is noteworthy that the temperature variation has little influence on the microstructures of the deposited carbon materials because the morphologies of the acquired carbon do not differ significantly at different working temperatures. The reasons for the varied microstructures could be complex, including the carbon nucleation and growth during the electrolysis and the effects from the cathode substrates.[9,14] Therefore, the constant voltage electrolysis in short-times (3 s, 10 s and 20 s) using a Pt wire anode and a nickel plate cathode were conducted to dent the effect from nuclei growth and investigate the formation mechanism of various microstructures. Short-term electrolysis was performed at different cell voltages (2.0 V, 2.4 V and 2.8 V) to study its influence on carbon nucleation.

The SEM images of the carbon products obtained at different electrolysis times with 2.0 V at 750 °C are presented in Fig. 4. Before electrolysis, the bare nickel cathode has a bumpy surface (Fig. 4a). In Fig. 4b, the nickel substrate is partially covered by small carbon nanoparticles with diverse diameters of 100–200 nm after 3 s electrolysis. More carbon particles will be obtained under longer electrolysis and a few carbon nanofibers with a diameter size of ~20 nm also appear under higher magnification shown in Fig. 4c and d. The formation of carbon

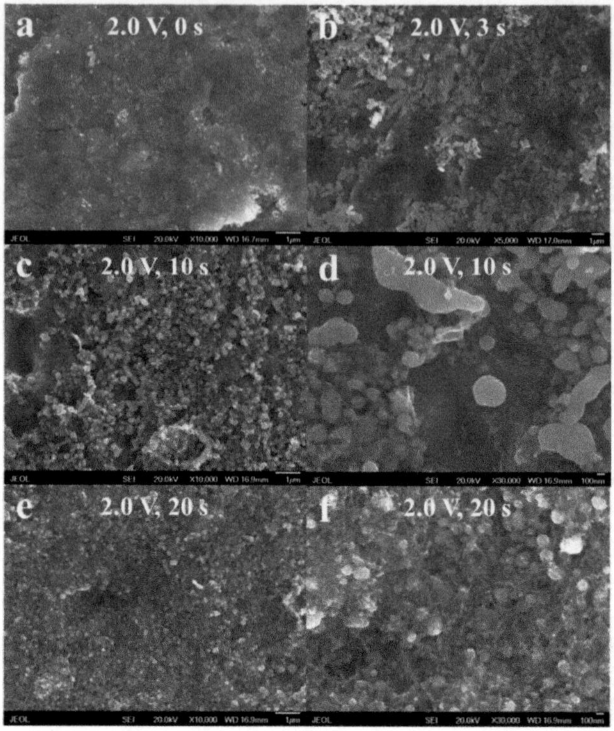

Fig. 4 FESEM images of the carbon materials deposited at nickel plate cathode under 750 °C and 2.0 V for 0 s (a), 3 s (b), 10 s (c and d), and 20 s (e and f) in molten LiCl–Li_2CO_3 melt.

nanoparticles should be related to the rapid nucleation of carbon and the size of these particles is not uniform as the nucleation of carbon during the electrolysis is progressive.[12] The appearance of carbon nanofibers can be explained by the catalysis effect of nickel. When the electrolysis time is set at 20 s, the nickel substrate was fully covered by large amounts of small carbon particles and carbon nanofibers (Fig. 4e and f). The results indicate that, when the cell is applied with a suitable voltage, the formation of carbon nanoparticles and carbon nanofibers could take place in a very short time.

Fig. 5 depicts the SEM images of the carbon products prepared at 2.4 V. Both the carbon nanoparticles and carbon nanofibers have been observed in the nickel cathode after a 3 s electrolysis, as shown in Fig. 5a and b. The nickel substrate is fully covered by the two carbon microstructures after 10 s electrolysis (Fig. 5c and d). Stepping up the cell voltage apparently improves the deposition rate during the electrolysis. However, as the electrolysis time increased to 20 s, the deposited products are mostly carbon nanoparticles (Fig. 5e), and it could be clearly seen that some of the nanofibers are hidden beneath the outermost small carbon particles at high magnification (Fig. 5f). This phenomenon could be explained as below. Under the catalysis of a nickel substrate, the carbon nanofibers are produced at the nickel surface during the initial electrolysis. When adequate carbon products are deposited to cover the nickel surface, the nucleation of carbon nanofibers stops, and they would then be covered by subsequent

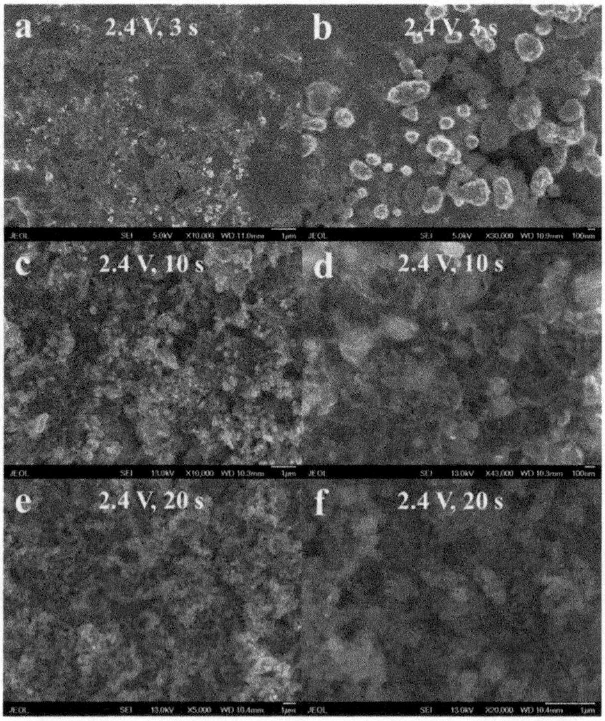

Fig. 5 FESEM images of the carbon materials deposited at the nickel plate cathode under 750 °C and 2.4 V for 3 s (a and b), 10 s (c and d), and 20 s (e and f) in molten LiCl–Li_2CO_3 melt.

deposited carbon particles. It should be noted that the size of nanoparticles and nanofibers grows over the electrolysis time, for example, the diameter size of carbon nanofibers increased from ~20 nm (10 s electrolysis) to ~50 nm (20 s electrolysis).

Unexpectedly, increasing the voltage to 2.8 V led to the enhancement of the catalysis of nickel substrate and large amounts of carbon fibers were produced during the electrolysis, as shown in Fig. 6. In Fig. 6a and b, the deposited carbon nanofibers are much more abundant at 2.8 V than at 2.0 and 2.4 V after the 3 s electrolysis. The carbon nanofibers show a rapid growth rate and are more frequently observed with a diameter size in a range of 100–150 nm during the 10 s electrolysis (Fig. 6c and d). As the electrolysis time increased to 20 s, the carbon fibers with a length of 30–50 μm were found to dominate the main microstructure of the prepared carbon products, as presented in Fig. 6e and f. This phenomenon may be explained by the splitting growth rate of carbon fibers at 2.8 V, inhibiting the formation and growth of carbon particles. Aggregation of the deposited carbon fibers is also observed in Fig. 6e, revealing the formation of large carbon plates in Fig. 3.

The above results indicate that, when the nickel plates serve as cathode substrates, two different kinds of microstructures, quasi-spherical particles and nanofibers, will be obtained during the electrolysis and the cell voltage has

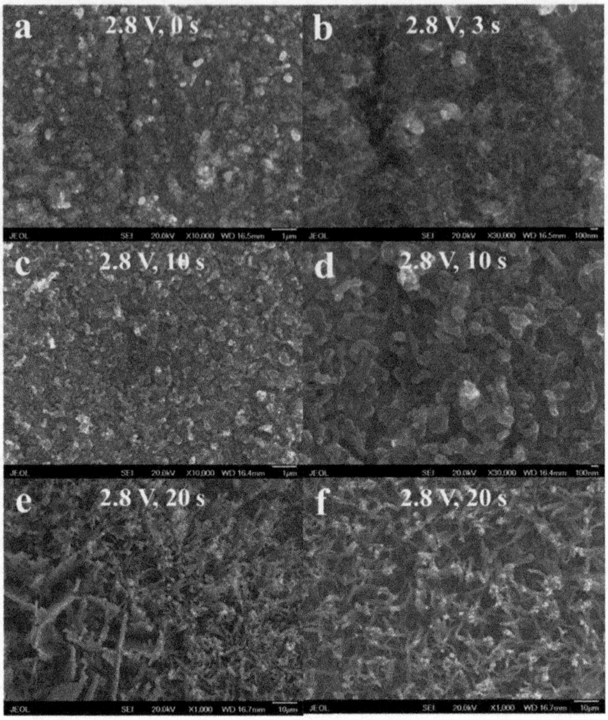

Fig. 6 FESEM images of the carbon materials deposited at nickel plate cathode under 750 °C and 2.8 V for 3 s (a and b), 10 s (c and d), and 20 s (e and f) in molten $LiCl-Li_2CO_3$ melt.

a significant influence on the latter microstructure. The higher the voltage, the more enhancement of the growth of carbon fibers. The nickel substrate may activate the carbon nucleation and the nucleation is more likely to propagate in the direction parallel to the nickel substrate, leading to the formation of carbon nanofibers. From Ren's work,[11] the formation of carbon nanofibers was controlled by the addition of trace transition metals (Zn or Ni) on the cathode to act as nucleation sites. The vast majority of nickel nanoparticles are located at nanofiber tips, which is not in accordance with our results. The difference may be caused by the different electrolysis conditions or the detecting instrument.

To confirm the catalysis effect of the nickel cathode, molybdenum plates were also used as the cathode for the short-term electrolysis at 2.4 V, with the results shown in Fig. 7. Fig. 7a shows the SEM images of the bare molybdenum plate before the electrolysis. The morphologies of carbon products obtained at 2.4 V for 3, 10 and 20 s are presented in Fig. 7b–d, respectively, and the observed cathode product conformation consists of quasi-spherical carbon structure gathered in clusters. No carbon fibers are observed in Fig. 7b–d, indicating that the deposited carbon could not be catalyzed by molybdenum to form carbon fibers. Similar to the electrolysis at the nickel cathode, the size of the carbon particles increased with the electrolysis time.

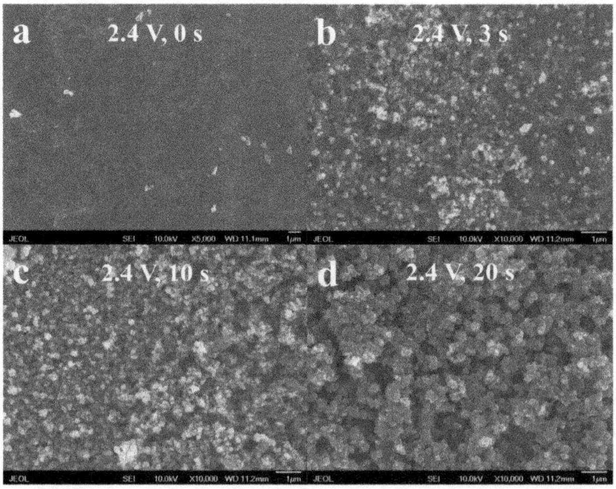

Fig. 7 FESEM images of the carbon materials deposited at molybdenum plate cathode at 750 °C and 2.4 V for 0 s (a), 3 s (b), 10 s (c), and 20 s (d) in molten LiCl–Li_2CO_3 melt.

Conclusion

Short-term electrolysis under different electrolysis conditions has been performed in LiCl–Li_2CO_3 melts, revealing that morphologies of the deposited carbon are significantly affected by cathode substrates. When selecting nickel plates as the cathode in the two-electrode cell, both quasi-spherical and nanofiber structures can be acquired at the cathodes while only quasi-spherical structures are observed at the molybdenum substrate cathodes. It is also confirmed that an increase in the cell voltage enhances the carbon nucleation rate and the catalysis effect of the nickel cathode to form carbon fibers. This result is important for making carbon fibers or carbon coating in the molten salt. The short-term electrolysis method avoids the influence from the growth of carbon materials and only primary morphologies are observed in the short-time electrolysis, which is beneficial to reveal the formation mechanism of the various carbon microstructures prepared under long-term electrolysis (several hours) in other eutectic molten salts (containing alkali carbonates).

Acknowledgements

The work was supported by the National Natural Science Foundation of China (No. 51322402) and the Fundamental Research Funds for the Central Universities (No. 230201406500005).

Notes and references

1 T. Sakakura, J. C. Choi and H. Yasuda, *Chem. Rev.*, 2007, **107**, 2365–2387.
2 M. Mikkelsen, M. Jorgensen and F. C. Krebs, *Energy Environ. Sci.*, 2010, **3**, 43–81.
3 M. Gattrell, N. Gupta and A. Co, *J. Electroanal. Chem.*, 2006, **594**, 1–19.

4 N. V. Rees and R. G. Compton, *Energy Environ. Sci.*, 2011, **4**, 403–408.
5 J. E. Bara, D. E. Camper, D. L. Gin and R. D. Noble, *Acc. Chem. Res.*, 2010, **43**, 152–159.
6 K. Le Van, H. Groult, F. Lantelme, M. Dubois, D. Avignant, A. Tressaud, S. Komaba, N. Kumagai and S. Sigrist, *Electrochim. Acta*, 2009, **54**, 4566–4573.
7 Q. S. Song, Q. Xu, Y. Wang, X. J. Shang and Z. Y. Li, *Thin Solid Films*, 2012, **520**, 6856–6863.
8 K. Otake, H. Kinoshita, T. Kikuchi and R. O. Suzuki, *Electrochim. Acta*, 2013, **100**, 293–299.
9 D. Tang, H. Yin, X. Mao, W. Xiao and D. H. Wang, *Electrochim. Acta*, 2013, **114**, 567–573.
10 L. Hu, Y. Song, J. Ge, J. Zhu and S. Jiao, *J. Mater. Chem. A*, 2015, **3**, 21211–21218.
11 J. Ren, F.-F. Li, J. Lau, L. González-Urbina and S. Licht, *Nano Lett.*, 2015, **15**, 6142–6148.
12 L. Massot, P. Chamelot, F. Bouyer and P. Taxil, *Electrochim. Acta*, 2003, **48**, 465–471.
13 J. Ge, L. Hu, W. Wang, H. Jiao and S. Jiao, *ChemElectroChem*, 2015, **2**, 224–230.
14 H. V. Ijije, C. Sun and G. Z. Chen, *Carbon*, 2014, **73**, 163–174.

Faraday Discussions

PAPER

Application of infiltrated LSCM–GDC oxide anode in direct carbon/coal fuel cells

Xiangling Yue,*[a] Ana Arenillas[b] and John T. S. Irvine[a]

Received 4th January 2016, Accepted 24th February 2016
DOI: 10.1039/c6fd00001k

Hybrid direct carbon/coal fuel cells (HDCFCs) utilise an anode based upon a molten carbonate salt with an oxide conducting solid electrolyte for direct carbon/coal conversion. They can be fuelled by a wide range of carbon sources, and offer higher potential chemical to electrical energy conversion efficiency and have the potential to decrease CO_2 emissions compared to coal-fired power plants. In this study, the application of (La, Sr)(Cr, Mn)O_3 (LSCM) and (Gd, Ce)O_2 (GDC) oxide anodes was explored in a HDCFC system running with two different carbon fuels, an organic xerogel and a raw bituminous coal. The electrochemical performance of the HDCFC based on a 1–2 mm thick 8 mol% yttria stabilised zirconia (YSZ) electrolyte and the GDC–LSCM anode fabricated by wet impregnation procedures was characterized and discussed. The infiltrated oxide anode showed a significantly higher performance than the conventional Ni–YSZ anode, without suffering from impurity formation under HDCFC operation conditions. Total polarisation resistance (R_p) reached 0.8–0.9 Ω cm^2 from DCFC with an oxide anode on xerogel and bituminous coal at 750 °C, with open circuit voltage (OCV) values in the range 1.1–1.2 V on both carbon forms. These indicated the potential application of LSCM–GDC oxide anode in HDCFCs. The chemical compatibility of LSCM/GDC with carbon/carbonate investigation revealed the emergence of an A_2BO_4 type oxide in place of an ABO_3 perovskite structure in the LSCM in a reducing environment, due to Li attack as a result of intimate contact between the LSCM and Li_2CO_3, with GDC being stable under identical conditions. Such reaction between LSCM and Li_2CO_3 was not observed on a LSCM–YSZ pellet treated with Li–K carbonate in 5% H_2/Ar at 700 °C, nor on a GDC–LSCM anode after HDCFC operation. The HDCFC durability tests of GDC–LSCM oxide on a xerogel and on raw bituminous coal were performed under potentiostatic operation at 0.7 V at 750 °C. The degradation mechanisms were addressed, especially on raw coal.

1. Introduction

With the ever increasing energy demands from human society, technologies have been developed concerning utilization of renewable resources and more efficient

[a]School of Chemistry, University of St Andrews, St Andrews, Fife KY16 9ST, UK. E-mail: xy57@st-andrews.ac.uk
[b]Instituto Nacional del Carbon (CSIC), 33080 Oviedo, Spain

use of traditional fossil fuel sources, coupled with efforts to reduce CO_2 emissions. Amongst the different fuel cell technologies, direct carbon fuel cells (DCFCs) can be fuelled by a variety of carbon forms, such as petroleum coke, coal, pyrolysed hydrocarbons and biomass *etc.*, which is quite competitive with fuel cells based on gaseous fuels. By electrochemical oxidization of solid carbon to CO_2 product, a 100% thermodynamic efficiency can be obtained theoretically as a result of a nearly zero entropy change, DCFCs thus offer a higher chemical to electrical energy conversion efficiency and lower levels of CO_2 emissions due to ease of sequestration compared to traditional coal-fired power plants.[1] Furthermore, DCFCs provide high energy density by utilising solid carbon without intermediate reforming processes, compared to solid oxide fuel cells (SOFCs) that are powered by gaseous hydrocarbons.[2] These advantages make DCFC a very attractive technology.

Different types of DCFCs are in use depending on the electrolyte, for instance, molten carbonate,[3–5] molten hydroxide,[6] and solid oxide[7] *etc.* Hybrid DCFCs (HDCFCs) have gained increasing interest due to the benefits they bring.[8] HDCFCs utilise a binary electrolyte system, *i.e.* molten carbonate and solid oxide electrolyte, and therefore combines molten carbonate fuel cell (MCFC) and solid oxide fuel cell technologies by mixing a carbonate eutectic with solid carbon, with the resulting mixture supported on a solid electrolyte. Such a design simplifies the cell assembly on the one hand, with solid electrolyte separating the cathode and anode compartments, reducing corrosion issues and eliminating CO_2 circulation/management encountered in MCFCs. It enhances the interactions between solid fuel and electrolyte, by extending the active reaction zone from a two-dimensional anode/solid electrolyte interface to three-dimensional slurry on the other hand, thus promoting the DCFC performance remarkably. Despite the prospects it offers, HDCFC faces a number of challenges due to the complexity in its system. So far, literature on this technology regarding anode materials, anode reaction mechanisms, performance stability, and proper fuels is limited.

Attempts have been made in our lab to solve some of the problems mentioned above.[9–11] A high DCFC power output that is comparable to a commercial SOFC was demonstrated.[12] Also a range of carbon forms including graphite, activated carbon, coals at different ranks and pine charcoal were tested in HDCFC, in an effort to link the carbon properties to fuel cell performance and stability.[13] Nonetheless, in all these studies, the anode material applied in for DCFC operation was restricted to the conventional Ni–YSZ cermet, which is subject to several drawbacks that have been found in the development of SOFCs, such as sintering, poor redox stability and low tolerance to coke deposition and sulphur contaminations.[14] Indeed, Ni/NiO equilibrium was found to govern the anode reaction in HDCFCs when carbon activity was low, giving a low open circuit voltage (OCV) and decreased thermodynamics.[13] The occurrence of Ni/NiO equilibration can also pose problems to mechanical integrity in the SOFC owing to the volume expansion when Ni is oxidized, resulting in failure in long term operation. Moreover, the endurance to sulphur poisoning is a significant concern when Ni-cermet is employed, as most inexpensive and readily available coals/biomass contain sulphur impurities. Alternative anode materials that can be applied in HDCFCs therefore need to be found.

(La, Sr)(Cr, Mn)O_3 (LSCM) perovskite-type materials have been reported as an alternative anode in SOFCs, originating from its good redox stability, flexibility in fuel gas options and excellent resistance to coking and sulphur poisoning.[15,16] Other options include doped ceria, *e.g.* gadolinium doped ceria (GDC), which offers excellent electrochemical activity towards a series of oxidation/reduction reactions.[17,18] Previously, a GDC–LSCM electrode derived by wet-impregnation fabrication was demonstrated to have competitive performance to a well-behaved Ni–YSZ cermet in CO_2 electrochemical reduction in a solid oxide electrolyser.[19] In the present work, GDC–LSCM oxides have been applied in HDCFCs with their electrochemical performance towards carbon oxidation evaluated and their chemical stability in carbon–molten carbonate environments investigated. The carbon fuel studied covers organic xerogel, a synthetic carbon formed from sol–gel processing of resorcinol and formaldehyde under controlled conditions, as well as a bituminous coal. Additionally, the electrochemical stability of GDC–LSCM oxide anodes with carbonate have been examined after HDCFC operation on xerogel as well as on bituminous coal, with degradation mechanisms proposed.

2. Experimental

2.1 SOFC base fabrication

SOFCs with a GDC impregnated LSCM–YSZ (abbreviated as GDC–LSCM hereafter) anode and $(La_{0.8}Sr_{0.2})_{0.95}MnO_3$ (LSM)/YSZ//LSM cathode were prepared, details of which have been described previously.[19] Briefly, these cells were based on an YSZ electrolyte (1 mm or 2 mm thick) supported structure with a two electrode geometry. YSZ powder (Pi-KEM) was pressed and fired to form a dense pellet, after which, a porous YSZ layer followed by a porous LSCM ($(La_{0.75}Sr_{0.25})_{0.97}Cr_{0.5}Mn_{0.5}O_3$, EMPA) layer was screen-printed on one side of YSZ pellet and fired at 1300 °C. Subsequently, a composite LSM/YSZ (50/50 by weight) layer followed by a pure LSM layer was introduced on the other side of the dense YSZ pellet by screen-printing and sintering at 1100 °C. Wet impregnation of gadolinium and cerium nitrate precursor solution, mixed according to the stoichiometric ratio of $Gd_{0.1}Ce_{0.9}O_2$, into the porous LSCM–YSZ backbone under a vacuum pressure of 0.15 bar and thermal treatment at 500 °C was undertaken to decompose the nitrate salts. The impregnation was repeated until a desirable loading amount of GDC, 50 wt% over LSCM–YSZ, was reached. The resulting anode was then sintered at 1100 °C.

As a comparison, an SOFC with a Ni/YSZ anode was fabricated by screen printing anode and cathode ink on each side of YSZ pellet. The anode ink had NiO/YSZ in the ratio of 65/35 and the cathode ink was the same as in preparation of the SOFC with a GDC–LSCM anode. The calcination temperatures for the Ni/YSZ anode and LSM/YSZ//LSM cathode were 1350 °C and 1100 °C respectively. A symmetric cell with LSM/YSZ//LSM introduced at both sides of YSZ pellet was prepared and measured in air at 650–800 °C to determine the cathode polarization contribution to total cell resistance. The results are shown in Table 1. Prior to electrochemical characterization, Ag wires coated with Ag paste were attached to the SOFCs as current collector and connections to electrochemical instrument.

2.2 DCFC testing and post-mortem analysis

The as-prepared SOFC was attached to the end of an alumina tube using ceramic adhesive (Aremco552), with the alumina tube acting as an anode chamber sitting

Table 1 Resistance from LSM/YSZ//LSM electrode in a symmetric cell in static air at 650–800 °C

	Operation temperature (°C)				
	650	675	700	750	800
R_p (Ω cm^2)	1.68	1.12	0.81	0.41	0.22

vertically in a temperature-programmed furnace. The solid fuel was placed inside the anode chamber at room temperature, with the fuel physically located on top of the SOFC anode. The furnace was slowly heated to ~300 °C to properly cure the ceramic adhesive before ramping up to operational temperature, 650–750 °C (some cells were tested up to 800 °C). The cell was tested in heating mode, in other words, it was measured at 650 °C and then at 700 and 750 or 800 °C respectively. At each temperature, the cell was stabilised until a steady OCV value was achieved before undertaking electrochemical testing.

Polarization curves and electrochemical impedance spectroscopy (EIS) were recorded for initial cell performance evaluation. For the former, the current density was recorded while sweeping the cell potential from the OCV value to −0.3 V in 3 min. As for EIS, the impedance spectra were recorded in a frequency range from 150 kHz (unless stated otherwise) to 1 Hz with an ac amplitude of 10 mV using a potentiostat (Solartron1470) at OCV or at a certain current density (galvanostatic mode). After initial performance assessment, stability measurements were conducted at 750 °C under potentiostatic mode (0.7 V), with current density variation recorded as a function of operation time. Over the course of electrochemical testing, the cathode was exposed to static air with the anode chamber flushed with N_2 purging gas at a flow rate of 20 ml min^{-1}. The N_2 gas was introduced by an alumina tube inserted into the anode chamber, with the end of the alumina tube a small distance from the top of the pile of solid fuel. The exhaust gas from the anode was carried by N_2 to an online gas chromatograph (GC) for composition analysis. For the symmetric cell, impedance spectra were measured under OCV at 650–800 °C with both electrodes exposed to static air.

After electrochemical characterization, the HDCFC anode was examined using a PANalytical X-ray diffractometer with Cu Kα radiation. X-ray diffraction (XRD) was recorded in a two theta range 10–90° at room temperature. The anode/cathode microstructure was inspected by scanning electron microscopy (SEM) using a JSM 5600 instrument, coupled with an Inca 2000 instrument for energy dispersive X-ray spectroscopy (EDX) measurement.

2.3 Solid fuel characterizations and preparation

The solid fuels studied here included organic xerogel and a mineral bituminous coal (denoted as B-II-M), both supplied by INCAR-CSIC, Spain. The organic xerogel was synthesized by the sol–gel condensation of resorcinol (R) with formaldehyde (F) in deionised water followed by a vacuum evaporative drying process. The pH value of the precursor solution was maintained at pH 4 using Na_2CO_3 as the basification agent, and the R/F molar ratio was equal to the stoichiometric value (0.5), with the dilution ratio, D (the total solvent/reactant molar ratio) fixed at 5.7 during gel synthesis. Detailed synthesis procedures can be found elsewhere.[20,21] After drying, the resulting xerogel was sieved under 75 μm for usage without any further treatments. With regard to the coal sample, raw B-II-M coal was simply dry-milled in a bench mortar mill and sieved under 75 microns, without any other processing. Table 2(a) shows the chemical composition of the as-supplied xerogel and B-II-M samples. The chemical composition of the carbon xerogel, obtained after pyrolysis of as-received organic xerogel at 700 °C in N_2 (described in detail in part 2.4) was also included in Table 2(a). The standards used for sample composition analysis are as follows: ASTM D-5373 in a LECO CHN 2000 for C, H, and N element, ASTM D-4239 for oxygen and sulphur in a LECO TRUSPEC Micro-O and LECO S632 respectively, ASTM 7582 for moisture and ash with ISO 562 for volatile matter in a LECO TGA-601 apparatus. Ash composition of the raw bituminous coal was also analysed using X-ray fluorescence in a Bruker SRS 3000 instrument, and the results are shown in Table 2(b).

The pore texture of the organic xerogel was measured by N_2 adsorption–desorption isotherms at −196 °C using a Sorptomatic Carlo Erba 1900 device. Analysis of the N_2 isotherms gave the BET specific surface area, S_{BET}; the micropore volume calculated by the Dubinin–Radushkevich equation, V_{micro}. The mesopore volume (V_{meso}), macropore volume (V_{macro}) and the mean pore diameter (d_{pore}) were determined by mercury porosimetry with a Carlo Erba

Table 2 Chemical composition of xerogel and raw B-II-M coal (a), the ash content of B-II-M coal (b) and texture properties of the organic xerogel (c)

(a)

	Ultimate analysis (wt%, dry ash free basis)					Proximate analysis (wt%)				
Carbon type	C	H	N	O	S	FC[a]	VM[a]	Ash[a]	Moisture	ΔH (kcal kg^{-1})
Organic xerogel	64.0	3.9	0.2	31.9	0.0	53.6	46.4	0.0	13.1	5862
Carbon xerogel	97.1	1.0	0.1	1.8	0.0	96.4	0.9	0.0	2.7	7762
B-II-M	90.4	4.6	1.5	2.8	0.7	73.9	19.4	6.7	0.7	5268

(b)

B-II-M ash composition (%)

Fe_2O_3	Na_2O	MgO	Al_2O_3	SiO_2	P_2O_5	K_2O	CaO	TiO_2
11.9	0.9	2.0	20.0	35.5	0.0	2.6	14.0	1.1

(c)

	S_{BET} (m^2 g^{-1})	V_{micro} (cm^3 g^{-1})	V_{meso} (cm^3 g^{-1})	V_{macro} (cm^3 g^{-1})	d_{pore} (nm)
Organic xerogel	98	0.04	0.33	0.73	60

[a] dry basis

Porosimeter 2000. The textural properties of the organic xerogel are provided in Table 2(c).

Regarding fuel preparation for the HDCFCs, the as-supplied xerogel/coal was pre-mixed with Li_2CO_3–K_2CO_3 carbonates (Li : K at a molar ratio of 62 : 38) by ball milling, at a weight ratio of carbon : carbonate 4 : 1, before loading into the DCFC anode chamber. This was believed to give a homogeneous dispersion of xerogel/coal in the Li_2CO_3–K_2CO_3 eutectic at operational temperature, thus improving contact and enhancing the electrochemical oxidation of carbon. 2 g of the carbon/carbonate mixture was used for each test, with the fuel cell operated in a battery mode. After fuel cell testing, the remaining fuel was collected with its crystalline structure determined by XRD.

2.4 Compatibility of solid fuel, carbonates and SOFC anode components

The chemical compatibility between LSCM/GDC and carbon/carbonate mixture was examined. This involved the treatment of pre-mixed LSCM or GDC and carbon/carbonate powders at 700–800 °C in N_2 or 5% H_2/Ar atmosphere and the identification of phases in residual powder after treatment. Organic xerogel and Li–K carbonates were mixed in a similar manner to the preparation of solid fuel for the electrochemical tests, then LSCM or GDC powder was added with a weight ratio of 1 : 2 to the carbonate and ground for ~10 min with a pestle and mortar. The above prepared powder mixtures were placed in a crucible and thermally treated using a tube furnace. The furnace was heated to 150 °C and left for 1 hour to remove moisture, and then heated to 700–800 °C for 1 hour before cooling to room temperature for XRD measurement. In this whole process, N_2 (otherwise, 5% H_2/Ar) at a flow rate of 20 ml min^{-1} was flowed through the tube furnace. The xerogel/carbonate mixture was subjected to pyrolysis during this treatment.

In addition, a LSCM–YSZ pellet was prepared and treated. LSCM and YSZ powder in a ratio of 50 : 50 were mixed, pressed into pellet form and fired at 1300 °C. The as-prepared pellet was immersed in Li_2CO_3–K_2CO_3 (at a molar ratio of 62 : 38) mixture powder in an alumina crucible. The crucible was transferred into a tube furnace and heated to 700 °C in 5% H_2/Ar flow. The sample was kept at 700 °C for 10 h before cooling. This treatment was intended to mimic the conditions in HDCFC operation, where carbonate is in contact with the SOFC anode in a reducing environment. The LSCM–YSZ pellet was retrieved from the molten salts at room temperature and measured by XRD for phase analysis, in comparison to a fresh LSCM–YSZ pellet.

3. Results and discussion

3.1 Fuel cell performance on different carbons

3.1.1 DCFC performance on organic xerogel and properties/characterizations of xerogel. Fig. 1 shows the electrochemical performance of the HDCFC with GDC–LSCM SOFC anode operating on xerogel with N_2 as the purge gas. The operation was conducted in heating mode, *i.e.* the cell was tested stepwise from 650 to 750 °C. The purpose was to reduce the loss of carbon by chemical reaction due to the Boudouard reaction (eqn (1)) and decomposition of carbonate (eqn (2)) which becomes much faster at high temperatures.

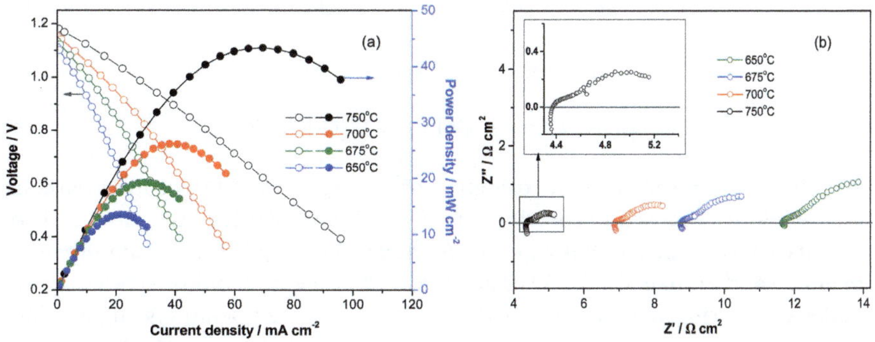

Fig. 1 Current density–voltage and current density–power density (*I*–*V* and *I*–*P*) curves (a) and impedance spectra (b) of the DCFC with the GDC–LSCM SOFC anode operating at 650–750 °C on xerogel. Note that the SOFC has a 2 mm thick YSZ electrolyte and a LSM/YSZ//LSM cathode. The inset shows the expanded impedance spectra at 750 °C under OCV conditions.

$$C + CO_2 \rightarrow 2CO \quad (1)$$

$$CO_3^{2-} \rightarrow CO_2 + O^{2-} \quad (2)$$

In Fig. 1(a), a fairly linear dependence of cell voltage on current density can be seen over the whole operation temperature range, with cell potential dropping slightly quicker at high current densities. With a 2 mm thick YSZ electrolyte support, the power output for the HDCFC running with xerogel is nearly 45 mW cm^{-2} at 750 °C. However, a total R_p value of ~0.8 Ω cm^2 and the cell area specific resistance (ASR) of ~5.3 Ω cm^2 are obtained in the impedance spectra in Fig. 1(b) at 750 °C. The ohmic resistance R_s, mainly from the thick YSZ electrolyte, takes up to 82% of the total cell resistance (ASR). Therefore, one can expect much higher current density and power output if the electrolyte was made much thinner.

In Fig. 1(b), two semi-circular arcs are observed from the impedance spectra under OCV at different temperatures, with a small one at high frequency regime (10^4 to 10^3 Hz) and a dominant one in the low frequency region with a characteristic frequency lower than 10 Hz. As temperature decreases, the cell resistance increases, with the low frequency contribution being affected more significantly. The characteristic frequency for the low frequency arc shifts to an even lower value (<1 Hz) at 650 °C, while the value for the high frequency arc stays unchanged compared to that at high temperatures. The relevant processes for the above high frequency and low frequency arc are likely to be the charge transfer at the electrode–electrolyte interface and the mass transportations, respectively. Therefore, mass transfers become controlling processes, due to lower thermodynamics and electrode kinetics as the temperature drops.

The OCV value increases when the operation temperature is raised in Fig. 1(a), and it reached 1.11, 1.16, and 1.18 V at 650, 700 and 750 °C respectively for the DCFC with an oxide anode. These values are higher than one may expect if the oxidation of carbon to carbon dioxide (eqn (3)) occurred, for which the theoretical value is 1.02 V.

$$C(s) + 2O^{2-} \rightarrow CO_2(g) + 4e^- \quad (3)$$

$$H_2(g) + O^{2-} \rightarrow H_2O(g) + 2e^- \quad (4)$$

$$H_2(g) + CO_3^{2-} \rightarrow CO_2(g) + H_2O(g) + 2e^- \quad (5)$$

Other reactions such as the oxidation of H_2 (eqn (4) and (5)) or hydrocarbons are therefore supposed to contribute to the higher OCV value. At 650 °C, a theoretical OCV value of 1.12 V is expected for a fuel cell supplied with humidified H_2 to anode and air to cathode. A GC analysis, performed to monitor HDCFC anode outlet gas during heating, revealed H_2 evolution at temperatures higher than 550 °C with H_2 concentration reaching a peak at 700 °C. However, H_2 alone cannot explain the high OCV at high temperatures such as 750 °C, as the OCV for H_2 oxidation is anticipated to decrease with temperature. The decomposition of carbonate (eqn (2)) was reported to impact OCV of HDCFC significantly, which was evidenced by adding 40 mol% of Li_2O to carbon/carbonate slurry and the demonstration of OCV values in the range of 1.2–1.3 V at 550–700 °C.[10]

Additionally, the high activity of the xerogel towards oxidation cannot be ruled out for the observed high OCV values. The synthetic xerogel has a unique structure and its textural properties can be tuned to specific applications.[21,22] It possesses a high surface area, high pore volume and a wide range of pore sizes, as shown in Table 2(c). The XRD diffraction patterns of as-supplied xerogel and as-mixed xerogel–carbonate mixture are presented in Fig. 2(a) and (b). As one can see, a very broad feature of diffraction appears in the two theta range of 10–50° for the xerogel as well as xerogel/carbonate mixture. According to the literature, the peak at $2\theta = 20$–$30°$ can be ascribed to the (002) reflection, with the one at $2\theta = 40$–$50°$ ascribed to the (100) reflection for carbon materials.[3,4] These two peaks are very broad in the XRD pattern of the xerogel sample, indicating a very disordered structure that is beneficial for carbon oxidation in the DCFC. It has been agreed by several researchers that less crystalline carbon tends to have more edges sites and

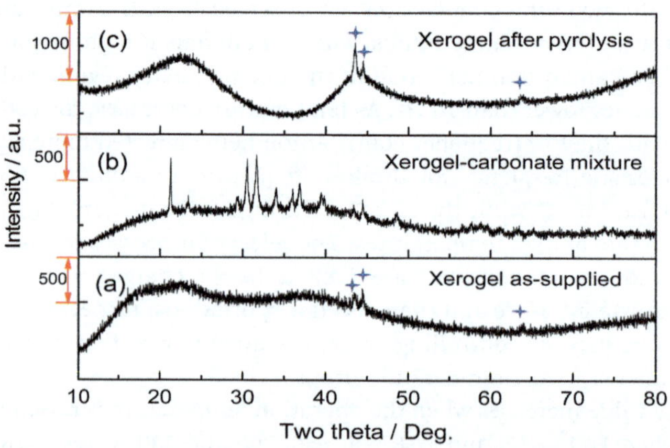

Fig. 2 XRD patterns of the xerogel and xerogel–carbonate mixture: (a) as-supplied xerogel, (b) xerogel–carbonate power as-mixed, and (c) xerogel after pyrolysed at 700 °C in N_2 without premixing with carbonate. Note that the cross labelled peaks are from the sample holder.

defects which are considered reactive sites for carbon oxidation in DCFC.[3,4,13] One may argue that xerogel has quite a low percentage of fixed carbon (53.6%) and a very high amount of volatile matter (46.4%), both of which were believed to impact DCFC performance negatively.[13] It should be pointed out that a carbonization process took place readily during heating of the fuel cell to operational temperature, which changed the carbon chemical composition considerably. This was generally neglected in previous publications. The chemical composition of xerogel after pyrolysis (carbonization) in N_2 at 700 °C is compared to fresh xerogel in Table 2(a). Apparently, the fixed carbon ratio and thus the calorific value (ΔH) is markedly promoted, with most volatile matter and moisture removed after carbonization. Despite the alteration of chemical composition, two well separated broad diffraction peaks at $2\theta = 24°$ and $2\theta = 44°$ respectively are seen in the xerogel after pyrolysis at 700 °C, as shown in Fig. 2(c), suggesting that low crystallinity and consequently high activity in xerogel remains after carbonization.

Illustrated in Fig. 3 is the electrochemical performance of the HDCFC with a Ni–YSZ anode operated with xerogel at 700–800 °C. Similar OCV values (1.1–

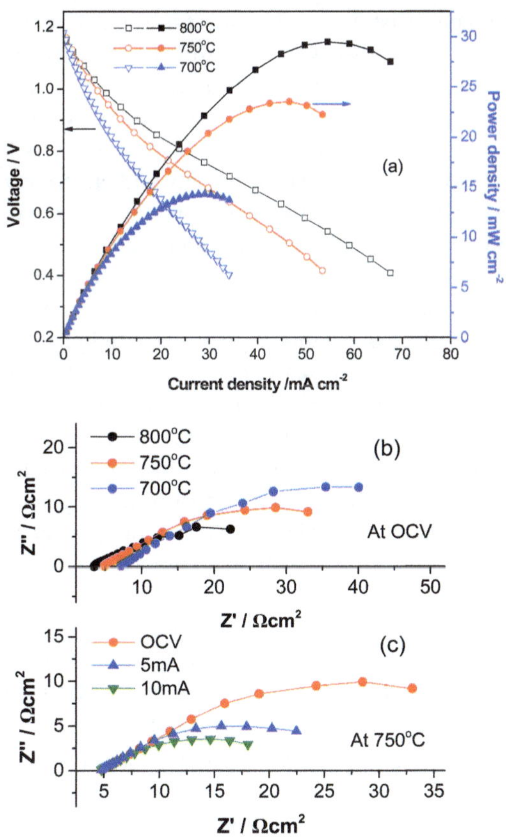

Fig. 3 I–V and I–P curves (a) and impedance spectra at OCV (b) as well as at different current density at 750 °C (c) of the HDCFC with a Ni–YSZ anode tested at 700–800 °C on xerogel. Note that the SOFC has an identical structure to that of the GDC–LSCM anode except for the anode material.

1.2 V) are achieved compared to the cell with the oxide-only anode. In Fig. 3(a), the *I–V* curves from the cell with a Ni–YSZ SOFC anode are subject to curvature in the region where current density is low, due to high activation polarization losses. As the current density increases, the cell is gradually controlled by ohmic polarization losses. In contrast, the *I–V* curves from the GDC–LSCM anode cell are almost linear in the whole measured current density range. These are supported by the declined R_p value and the depressed impedance arc in Fig. 3(c) from Ni–YSZ anode cell against increasing current density, whereas identical R_p values are attained upon drawing current up to 10 mA at 750 °C on the oxide-only anode cell (not shown here).

With the same configuration, a DCFC with a Ni–YSZ anode has a maximum power density of only 30 mW cm^{-2} at 800 °C, much lower than that of the GDC–LSCM anode cell at 750 °C. This observation is supported by the impedance comparison in Fig. 1(b) and 3(b). At 750 °C, the cell with the Ni–YSZ anode exhibits an R_s of 5 Ω cm^2 and an ASR larger than 35 Ω cm^2. The appreciably high polarization resistance from the Ni–YSZ anode cell implies a poor electro-catalytic activity from the Ni–YSZ cermet towards carbon oxidation and retarded electrochemical reaction processes taking place on the anode, in comparison to the anode composed of GDC–LSCM oxides. Furthermore, the R_s value takes up only 10–14% of the total cell resistance with respect to the cermet anode cell, a much smaller portion than was the case when the oxide anode was applied. This means that diminishing the solid electrolyte thickness is less likely to boost the cell performance tremendously in the cell with a Ni–YSZ SOFC anode.

The enhanced performance from the DCFC with a GDC–LSCM SOFC anode is probably attributed to the high ionic conductivity and electro-catalytic activity the oxide anode offers. GDC is a well-known oxide ion conductor, especially at temperatures below 800 °C. Besides, it offers mixed electronic and ionic conductivity in a reducing atmosphere, which is beneficial for facile charge transfer and surface mass diffusion compared to the electrode which has dual phases with separate electronic and ionic conductivity, for instance, Ni–YSZ cermet. It was reported by some researchers that the conductivity and/or electro-catalytic activity was further reinforced when the active phase had a particle size down to the nano scale,[23] as encountered in the oxide anode. The wet impregnation-derived anode features a microstructure having fine GDC particles highly dispersed on the surface of a LSCM support in the outer region and on the surface of a YSZ substrate on the inner side of the SOFC anode, which is believed to offer an extended active reaction zone as well as enhanced electro-catalytic properties towards electrode reactions.

3.1.2 DCFC performance on bituminous coal. It is desirable to utilise raw coal directly in DCFCs due to the abundance and low cost of this carbon source, however, it is quite challenging to run coal directly in DCFCs not only because of poor knowledge of the link between the physicochemical properties of coal and fuel cell performance, but also because of the impurities such as sulphur as well as ash content in coal which might negatively influence DCFC long-term performance. Attempts have been made to correlate the chemical composition of coals with fuel cell performance, and tune the coal properties to promote DCFC performance by demineralization.[5] However, these require extra processing and cost. Alternatively, searching for catalyst or anode materials that are robust and durable in the harsh conditions subjected to the use of raw coals is necessary.

Here the electrochemical results of a DCFC with GDC–LSCM oxides as a SOFC anode utilising bituminous raw coal will be discussed.

The as-supplied B-II-M coal was mixed with Li_2CO_3–K_2CO_3 to serve as fuel, in the same manner as in a xerogel. A 1 mm thick YSZ pellet was applied as a solid electrolyte with infiltration-derived GDC–LSCM oxide as a SOFC anode. Fig. 4(a) presents the polarization behaviour and power output of the HDCFC operated on B-II-M coal in heating mode. The linear response of cell voltage against current density is indicated in Fig. 4(a) across the whole potential range measured. A maximum power density in excess of 60 mW cm^{-2} was reached with raw coal at 750 °C. The power output is again limited by ohmic loss from the thick YSZ electrolyte, as seen from impedance results in Fig. 4(b). At 750 °C, an ASR value of ~4.2 Ω cm^2 was attained with R_s taking up 79% of the total cell resistance on HDCFC with a 1 mm thick YSZ electrolyte and oxide anode running on raw bituminous coal.

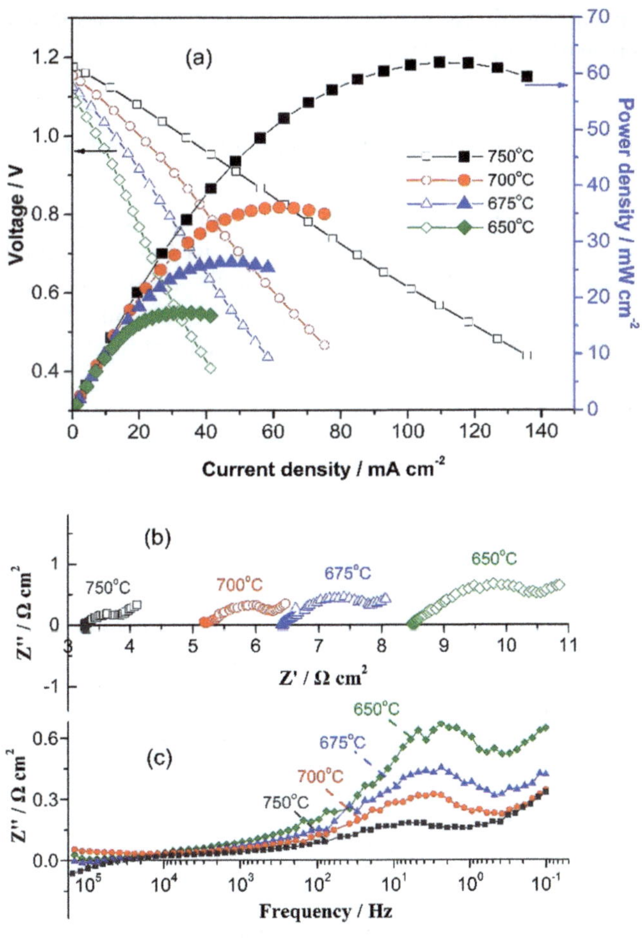

Fig. 4 (a) I–V and I–P curves and (b) complex impedance as well as (c) Bode impedance spectra of the HDCFC with a 1 mm thick YSZ solid oxide electrolyte and the GDC–LSCM oxide anode running on B-II-M coal at 650–750 °C.

Running on raw B-II-M coal, OCV values in the range 1.1–1.18 were achieved at 650–750 °C, and the higher the temperature, the higher the OCV value – a similar trend as for xerogel. As the H ratio is even higher in B-II-M coal (Table 2(a)), the electrochemical oxidation reactions involved H_2 released from coal (eqn (4) and (5)) may play a more important role in boosting OCV at the initial stage, especially at low temperatures. Besides, a low graphitic structure is also seen in the XRD pattern of B-II-M coal in Fig. 5, in which a broad peak appearing at $2\theta = 25.2°$ and another weak and broad feature at $2\theta = 35–55°$ were observed with this raw coal, apart from peaks due to mineral contents such as quartz.

In addition to the high OCV value, a total $R_p \sim 0.9\ \Omega\ cm^2$ at the lowest frequency in the impedance spectra is obtained from GDC–LSCM oxide on bituminous coal at 750 °C (Fig. 4(b)). The comparable R_p to that achieved with xerogel demonstrates that the GDC–LSCM oxide anode also performs well on raw B-II-M coal despite the existence of mineral matter and sulphur impurities. The complex impedance in Fig. 4(b) reveals a depressed arc followed by a spike at low frequencies at all temperatures measured, and a characteristic depressed arc at around 3 Hz is seen in the Bode impedance in Fig. 4(c). These suggest the limiting process for oxidation of coal on the GDC–LSCM anode is associated with the diffusion of reactant to active sites.

3.2 Compatibility study of SOFC anode component with carbon/carbonate mixture

The compatibility of the SOFC anode components with carbon/carbonates mixture is crucial for fuel cell performance and its long-term durability. The stability of LSCM and GDC with Li_2CO_3–K_2CO_3 in the atmosphere associated with the appearance of carbon was investigated under similar conditions as in DCFC operation but without the involvement of electrochemical reactions. Fig. 6 shows

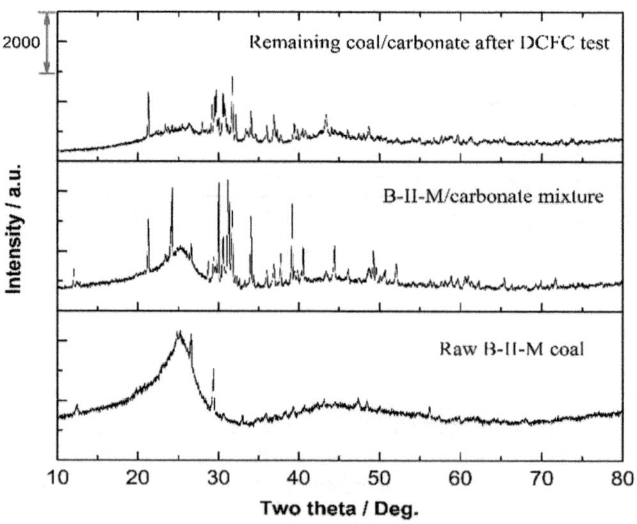

Fig. 5 XRD patterns of B-II-M raw coal and coal/carbonate mixtures before and after fuel cell testing.

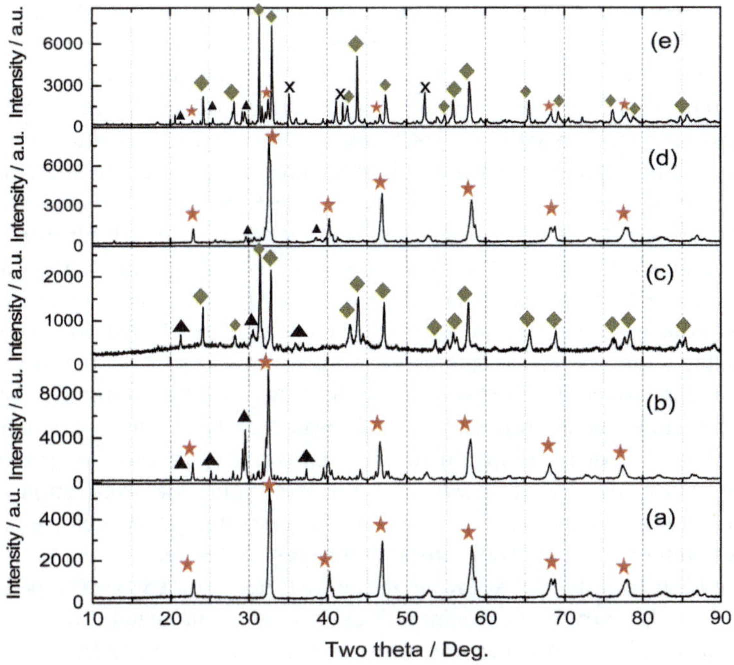

Fig. 6 XRD pattern of LSCM with xerogel and/or carbonate mixture under various treatment conditions. (a) Xerogel–LSCM at 800 °C in N_2, (b) Li_2CO_3–K_2CO_3–LSCM at 800 °C in N_2, (c) xerogel–Li_2CO_3–K_2CO_3–LSCM at 800 °C in N_2, (d) xerogel–K_2CO_3–LSCM at 800 °C in N_2, and (e) Li_2CO_3–K_2CO_3–LSCM at 800 °C in 5% H_2/Ar. The symbols interpolated stand for: ★ – ABO_3 perovskite; ◆ – A_2BO_4 type oxide; x – unknown phase, possibly MnO; ▲ – diffraction peaks from carbonate or Li/K compounds formed during pyrolysis (peaks for Li/K carbonate or compound are not differentiated here for simplicity).

the room temperature XRD patterns of the xerogel/carbonate/LSCM mixture after being treated at 700–800 °C in different environments. Apparently, a typical ABO_3 perovskite structure is observed in LSCM after pyrolysis of the xerogel–LSCM mixture at 800 °C in the absence of carbonate (Fig. 6(a)). However, diffraction peaks that can be assigned to a perovskite-related A_2BO_4 structure are identified from the xerogel–carbonate–LSCM mixture after treatment under the same conditions, besides the XRD peaks originating from Li_2CO_3, $KHCO_3$ and other K-containing compounds (Fig. 6(c)). The A_2BO_4-type oxide has a K_2NiF_4 structure which is made up of the ABO_3 perovskite layer and an AO rock salt layer arranged one upon the other. On the other hand, the ABO_3 type perovskite structure is retained in LSCM after being treated with Li–K carbonate at 800 °C in the absence of carbon (Fig. 6(b)), implying that the reduction of ABO_3-type LSCM to A_2BO_4 occurs only when carbon is present in the sample which results in reductive treatment. As confirmation, the XRD result of the Li_2CO_3–K_2CO_3–LSCM mixture that was subjected to treatment in 5% H_2/Ar atmosphere at 800 °C is shown in Fig. 6(e). Fig. 6(e) reveals the emergence of A_2BO_4 type diffraction as well as weak ABO_3 type perovskite diffraction in LSCM.

LSCM itself is usually believed to be stable under either reducing or oxidizing atmospheres, and has been frequently applied as a promising anode material in

H_2/hydrocarbon-fuelled SOFCs, though there is variation from rhombohedral/hexagonal in air to orthorhombic perovskite in H_2 atmosphere.[16] The reduction in perovskite LSCM in Fig. 6(c) and (e) can thus be accredited to the interaction/reaction between LSCM and carbonate in reducing atmospheres. It was doubted that this interaction/reaction of LSCM with Li–K carbonate was due to the well-known mobile lithium ions which diffused into the ABO_3 lattice and led to the structure transformation in LSCM. To demonstrate this, a treatment of the mixture without Li_2CO_3, *i.e.* xerogel-K_2CO_3–LSCM, was carried out in the same conditions as with Li_2CO_3. The XRD pattern in Fig. 6(d) indicates that no phase change in LSCM occurs after treatment without Li_2CO_3. Clearly, Li attack is the origin of the appearance of the A_2BO_4 type oxide in place of ABO_3 in the LSCM in a reducing atmosphere. The interaction/reaction between LSCM and Li_2CO_3 may facilitate the reduction in LSCM besides the reducing power of the carbon form. The perovskite related A_2BO_4 type oxide was reported to have a much lower electrical conductivity compared to its ABO_3 counterpart containing the same cations, especially when B = Mn, Fe.[24] Treatment and characterization of GDC with carbon/carbonate mixture powders suggests no phase change in GDC powder under the same conditions as in the case of LSCM.

An additional experiment involving the treatment of a LSCM–YSZ pellet with Li_2CO_3–K_2CO_3 slurry in a reducing atmosphere was conducted. The LSCM–YSZ (50–50 in weight ratio) pellet was immersed in Li–K carbonate (Li : K in molar ratio 62 : 38), heated to 700 °C, and left for 10 hours in a 5% H_2/Ar atmosphere.

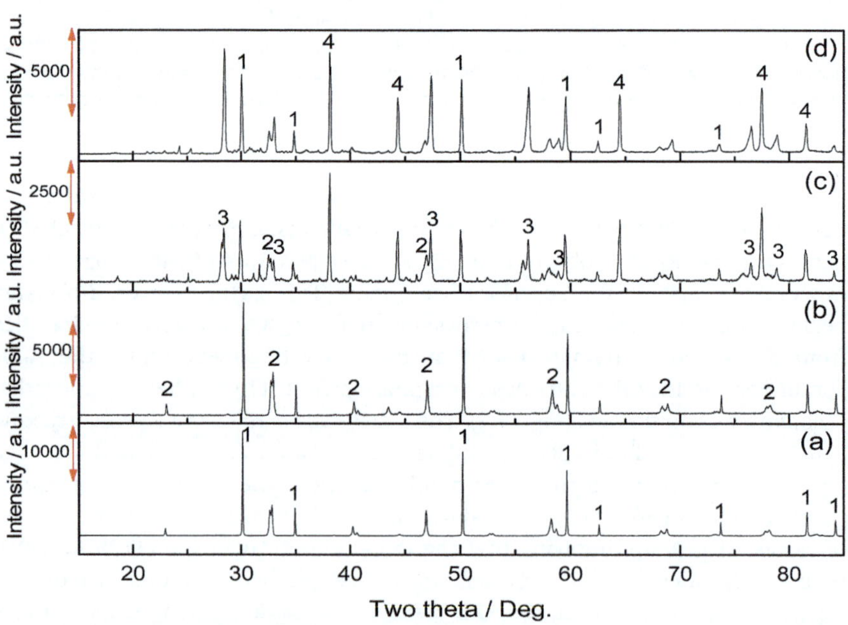

Fig. 7 XRD pattern of a (a) fresh LSCM–YSZ pellet, (b) LSCM–YSZ pellet after treatment with Li–K carbonate eutectic in 5% H_2/Ar at 700 °C for 10 h, (c) GDC–LSCM anode after HDCFC operation with xerogel (electrochemical performance shown in Fig. 1 and 8), and (d) GDC–LSCM anode after DCFC operation with coal (electrochemical performance shown in Fig. 4, 9 and 10). Note that diffraction peaks labelled 1, 2, 3 and 4 stand for YSZ, ABO_3-structured LSCM, GDC, and Ag, respectively.

Fig. 7 presents the XRD pattern of a fresh LSCM–YSZ pellet and the pellet after treatment. In the fresh pellet, only peaks from ABO_3 perovskite and YSZ are detected, as expected. In the treated LSCM–YSZ pellet, the ABO_3 type perovskite structure is maintained after being exposed to carbonate in 5% H_2/Ar for 10 h at 700 °C. These observations are obviously contrary to those involving LSCM powder. Similarly, the GDC–LSCM anode seems to be stable after HDCFC operation with xerogel and with bituminous coal at 750 °C (addressed in Section 3.3), as shown in Fig. 7(c) and (d) respectively, as no significant phase change can be found in LSCM after DCFC operation with either xerogel or raw coal. Apart from weak peaks due to Li_2CO_3 and $KHCO_3$ residuals on the surface of the GDC–LSCM anode, diffraction peaks expected from YSZ, GDC, ABO_3-type LSCM and the Ag current collector are well assigned in Fig. 7(c) and (d).

3.3 Fuel cell durability test and post-mortem characterization

3.3.1 With xerogel. Short-term HDCFC operation at 750 °C at a constant voltage of 0.7 V was carried out with xerogel after an initial performance evaluation (addressed in Section 3.1.1). The current density at 0.7 V as a function of time is exhibited in Fig. 8(a). It is seen that the current density is maintained at ~50 mA cm^{-2} for 1.5 hours at the beginning of the potentiostatic operation, probably due to the participation of H_2 and/or CO gases (liberated from xerogel or from gasification reactions) in the anode electrochemical reactions which stabilises the cell performance. After 1.5 h, the current density gradually declines and levels off to ~3 mA cm^{-2} at the end of the test which lasts for 17 h in total without interruption. A decrease in OCV value from 1.12 V to 1.06 V was noticed before and after the potentiostatic operation. This latter value is close to the equilibrium for carbon oxidation at 750 °C, consequently, carbon oxidation seems to be a residual process on the HDCFC anode, but with sluggish kinetics. The phenomena mentioned above are supported by the impedance comparison before and after the stability test, shown in Fig. 8(b). An extremely large cell resistance is observed after ~17 h discharge at 0.7 V, with both the high frequency arc characterized at 3000 Hz and the other one in the low frequency zone significantly larger than those before discharge.

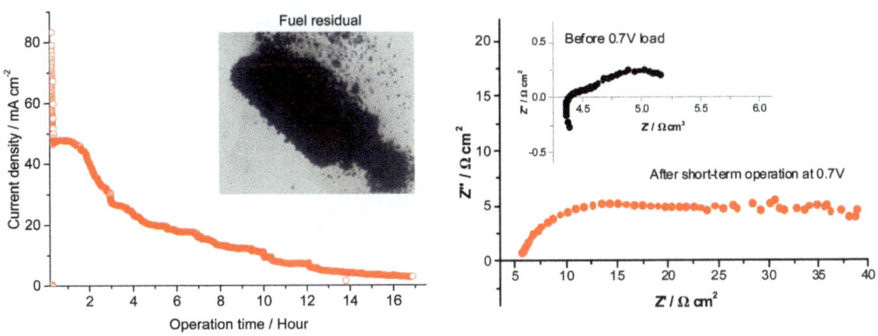

Fig. 8 (a) Current density variations in the short-term HDCFC operation at 0.7 V at 750 °C using xerogel with the GDC–LSCM oxide as the SOFC anode and a 2 mm thick YSZ pellet as solid electrolyte (the inset in (a) is a picture of the residual fuel after fuel cell operation); (b) impedance comparison before and after ~17 h HDCFC operation at 0.7 V.

Inserted in Fig. 8(a) is a picture of the remaining xerogel–carbonate mixture after DCFC operation, taken when the fuel was poured out from the anode chamber at room temperature. Interestingly, the remaining xerogel/carbonate powder ignited spontaneously once exposed to air. The vigorous ignition totally changed the composition and textural properties of the remaining fuel. This process is consistent with the presence of peroxide or superoxide which can act as an oxidant for residual organic matter. Formation of peroxidic species is anticipated in a Lux Flood basic system such as these molten carbonates.

3.3.2 With raw bituminous coal (B-II-M). The durability of a HDCFC with a GDC–LSCM oxide anode operating with B-II-M coal was also performed at 0.7 V at 750 °C for a period of ~70 h, with the initial cell performance discussed in Section 3.1.2. The discharge rate as a function of time during the stability test on raw coal is presented in Fig. 9. One can see from Fig. 9 that the current density at 0.7 V (with an initial value of 70 mA cm^{-2}) decreases at a relatively low rate in the first 9 h, it drops quickly to 15 mA cm^{-2} after 16 h. The break after 16 h brings back the discharge rate to around 32 mA cm^{-2} which maintains the cell for another 15 h at a low rate of degradation, and afterwards, the current density decays to 10 mA cm^{-2} when reaches 40 h. The cell performance was subsequently maintained until reaching 64 h, and the discharge rate declined to 5 mA cm^{-2} when the durability test was stopped at ~68 h.

In Fig. 9, the potentiostatic operation at 0.7 V was stopped at each breaking lines for impedance test, and the cell was rested at OCV for 0.5 h before impedance measurement. The impedance spectra at each break are displayed in Fig. 10. It is observed from Fig. 10 that the cell resistance degrades markedly with operation time, especially after 6 h. In the first 6 h, the increases in impedance are relatively small with slightly larger R_s and R_p values than those initially, which can be explained by the involvement of electrochemical oxidation of gaseous H_2/CO

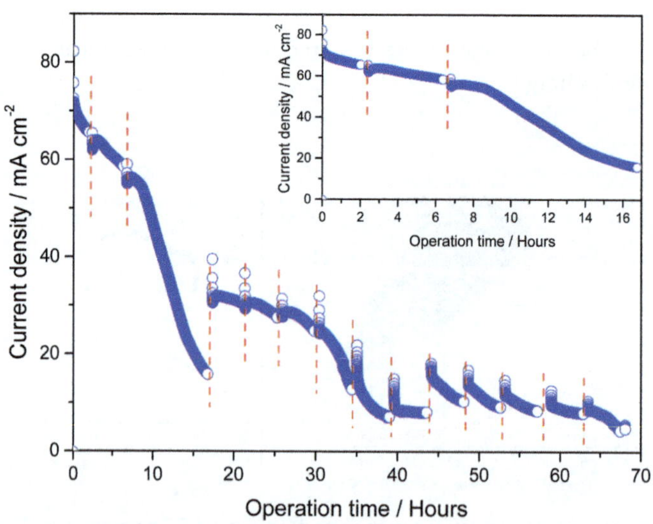

Fig. 9 Current density as a function of operation time for a HDCFC that consists of GDC–LSCM as SOFC anode and a 1 mm thick YSZ pellet as a solid oxide electrolyte working on B-II-M at 0.7 V at 750 °C for a period of ~70 h.

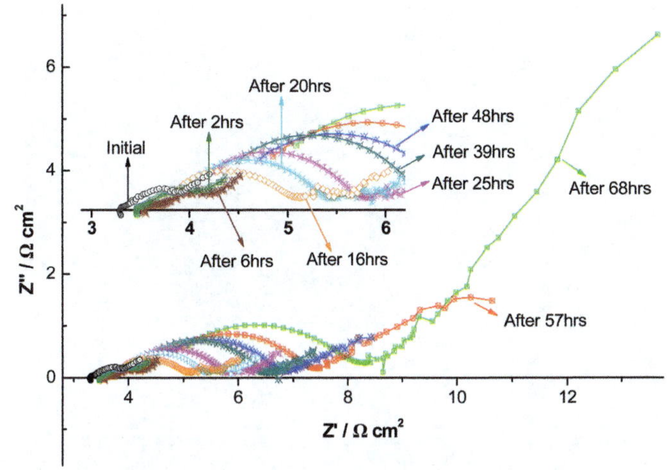

Fig. 10 Impedance evolution of the DCFC with an GDC–LSCM oxide SOFC anode and 1 mm thick YSZ solid electrolyte operating with B-II-M coal during a stability test at 750 °C at 0.7 V (the inset shows a magnified view of the impedance at different times).

released from coal or from facile gasification and oxidation kinetics as a result of low crystallinity in bituminous coal at the initial stage. After 6 h, values of both R_s and R_p, especially R_p, enlarge significantly due to inadequate production of gaseous fuel and consumption of active carbon in coal. These observations are consistent with the current density variation *vs.* operation time illustrated in Fig. 9. The impedance results also see an increasingly large high frequency arc characterised at 40 000 Hz with elongation of operation time, which is absent initially (Fig. 4(b)). Note that after 57 h, the impedance spectrum features a large high frequency arc, with an enormous spike following the high frequency arc, which suggests that the cell suffers from severe mass diffusion resistance. This might result from a lack of carbon at the active reaction zone after carbon is consumed or due to the accumulation of mineral matter/alkali metal oxides on the surface of the carbon which makes the fuel inaccessible. Correspondingly, an OCV value of 1.03 V at the end of durability test suggests the oxidation of solid carbon being the dominant reaction.

The microstructure of the oxide anode after HDCFC operation with coal was examined by SEM incorporated with EDX line-scanning, with the results shown in Fig. 11. As expected, Zr and Y were only found in the layer adjacent to the YSZ electrolyte, whereas La and Cr were found in the outer layer close to the bulk of the coal/carbonate mixture, and Ce/Gd were found in the whole depth of the SOFC anode from the EDX measurements. Apart from the above elements, K, S and C are seen mainly distributed in the outmost of region of the GDC–LSCM anode. Ag is found only in an area close to the LSCM on the outside of the anode. It is interesting that K, S and C are also seen in the bulk GDC–LSCM anode and area close to the dense electrolyte. This originates from a unique phenomenon in bituminous coal, which passed through a plastic stage during heating to DCFC operational temperature. The coal exhibited a fluid-like nature at the plastic stage. This unique characteristic of bituminous coal has been recently noticed in our

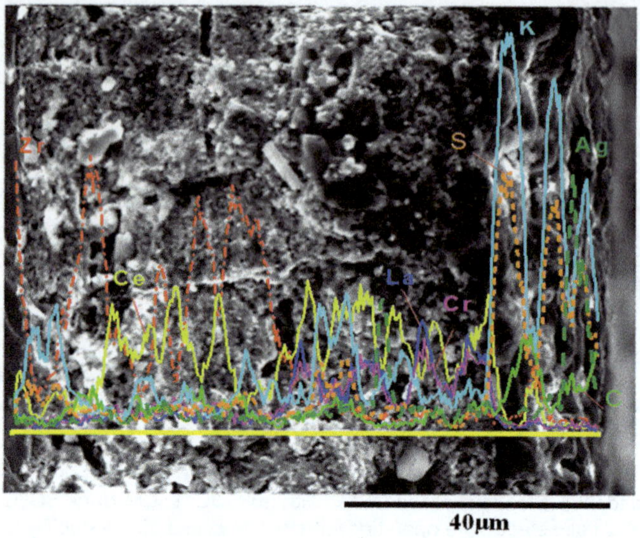

Fig. 11 SEM-EDX results for the GDC–LSCM SOFC anode after ~70 h DCFC operation with B-II-M coal at 0.7 V at 750 °C.

laboratory to impact the HDCFC electrochemical performance tremendously, compared to other raw coals and pre-treated coals.[25] The increased fluidity in raw B-II-M coal coupled with the melting of the eutectic at around 490 °C in the anode chamber bring the coal/carbonate mixture inside the bulk anode of the SOFC (as illustrated by EDX), which is beneficial for anode reactions in HDCFC as it extends the reaction region and improves further the interactions between the SOFC anode and coal once solidified at 650–750 °C after passing through its plastic stage. This also explains the reason for the superior electrochemical activity observed on raw B-II-M coal in Fig. 4.

However, the disadvantage is that impurities in raw bituminous coal, *e.g.*, sulphur, quartz and alumina (Table 2(b)), were also delivered into the bulk SOFC anode through the plastic stage, which negatively affects DCFC long-term performance, as shown in Fig. 9 and 10. Liu *et al.* exposed a series of LSCMs with different Cr/Mn ratio to 10% H_2S/H_2 at 950 °C for up to 5 days, and found MnS, La_2O_2S and MnOS impurities on $La_{0.75}Sr_{0.25}Cr_{0.5}Mn_{0.5}O_3$ decreased conductivity.[26] Although impurity phases were not identified in the GDC–LSCM anode after stability testing on raw B-II-M coal (Fig. 7(d)) in the present study, the presence of S in the bulk SOFC anode may block carbon from accessing O^{2-} for oxidation.

From the EDX line-scan data, Si and Al are not noticeable, probably due to the low ash content in B-II-M. Nonetheless, the inorganic residuals (in other words, ash) built up in the B-II-M/carbonate mixture on the top layer of the SOFC anode was visually clear when collecting the remaining fuel after finishing fuel cell testing, because of the appearance of white/light brown particulates. The precipitation of ash from coal and its accumulation with time may form a barrier layer on the anode surface, blocking diffusion of reactants to anode active sites for oxidation. Additionally, the carbonate that appears in the matrix of the SOFC anode possibly exposes the SOFC anode to harsh conditions due to the aggressive nature of the carbonate, as evidenced in Fig. 6. The decomposition of carbonate on the other hand, produces extra ash on the surface of the coal/SOFC anode. These all could contribute to the increased impedance during DCFC long-term operation, as reflected Fig. 10, resulting in a decrease in cell performance.

Last but not least, with ongoing carbon oxidation during potentiostatic operation, the coal, loaded in batch mode for initial characterization, is subjected to agglomeration and structural changes compared to fresh coal. In Fig. 5, the (002) reflection at $2\theta = 25.2°$ is hardly visible in the remaining coal after the DCFC durability test, which is reasonable as most of the H_2 releasing groups and active carbon sites were probably used up leading to an ordered structure in the remaining coal. Accompanied with the consumption of carbon inside the SOFC anode and those near the interface of the SOFC anode and coal/carbonate mixture, oxidation of coal in the bulk of the coal/carbonate mixture may become difficult due to mass diffusion problems, which may explain the large spike that appeared in the impedance spectra after 60 h in the HDCFC durability test. To develop a practical system, operation based on continuous feeding is essential to keep up high power output. Future work on HDCFC operation with a refuelling design may be conducted using a gravity based feeding tube with an auger to deliver the carbon/carbonate mixture into the anode chamber to address this issue in practical HDCFC applications.

As mentioned in the previous section, the reaction between the LSCM and carbonate leading to the formation of perovskite-related A_2BO_4 oxide was not observable after fuel cell durability tests on either the xerogel or raw coal. The ABO_3 structure was generally retained in the LSCM after HDCFC durability operation, as illustrated in the XRD pattern in Fig. 7(c) and (d). The contrast in the stability of the LSCM powder and GDC–LSCM anode with carbonate is associated with the difference in contact between the perovskite and carbonate. LSCM is susceptible to lithium carbonate attack where intimate contact occurs between the LSCM powder and carbonate in pyrolysis experiments (Fig. 6(c)–(e)), whereas in the GDC–LSCM anode, it is possible that the impregnation of GDC fine particles protect LSCM from Li attack, thus enhancing LSCM-containing anode stability. However, longer-term tests, incorporated with operation based on continuous fuel top-up, are needed to confirm or disprove this and to address as well the possibility of sulphur poisoning especially when coal is used.

4. Conclusions

In this work, an attempt at applying oxide materials such as (La, Sr)(Cr, Mn)O_3 (LSCM) and (Gd, Ce)O_2 (GDC) in hybrid direct carbon fuel cells was made, with the electrochemical performance of a HDCFC with a GDC–LSCM oxide anode (prepared by wet impregnation) evaluated for batch mode loading of two solid fuels, an organic xerogel and a raw bituminous coal. The SOFC with infiltrated oxide anode was attached and sealed to one end of an alumina tube, which was subsequently loaded with a carbon/Li–K carbonate mixture, acting as a HDCFC anode. DC polarization curves and ac-impedance spectroscopy was measured at 650–800 °C with N_2 as purging gas in the anode. The HDCFC with oxide anode showed a notably higher performance than that with the traditional Ni–YSZ anode on the xerogel as a result of high conductivity as well as high electrocatalytic properties offered by the GDC–LSCM oxide derived by wet impregnation fabrication. High OCV values, in the range of 1.1–1.2 V, and low total R_p values (0.8–0.9 Ω cm^2) at 750 °C were reached for the HDCFC with an oxide anode running on both xerogel and a bituminous coal with ash and sulphur impurities. These suggest the potential application of this GDC–LSCM oxide material in HDCFC operating with different carbon forms.

The compatibility of the LSCM/GDC with carbon/carbonate mixture was investigated at 700–800 °C under various conditions. It was found that an interaction/reaction between LSCM and lithium carbonate in reducing environment facilitated the reduction in ABO_3-type perovskite LSCM, leading to the formation of perovskite-related A_2BO_4 oxide as a result of the intimate contact between LSCM and carbonate. The GDC maintained its original structure under identical conditions. However, the above phenomenon in LSCM was not detectable on the LSCM–YSZ pellet treated with carbonate in a reducing atmosphere at 700 °C for 10 h, nor on the GDC–LSCM oxide anode after HDCFC operation at 0.7 V at 750 °C with xerogel as well as with the raw bituminous coal.

Degradation with respect to the HDCFC durability test on xerogel at 0.7 V at 750 °C over a period of ~17 h was mainly due to the loss of carbonate, which generated some active species that ignited spontaneously after exposing the remaining fuel to air at room temperature. The degradation mechanism for raw bituminous coal in a ~70 h durability test was more complicated, not only because of the existence of ash and sulphur impurities, but also the unique plastic stage raw bituminous coal passed through upon heating. Factors affecting the performance decay included sulphur poisoning and accumulation of mineral contents/oxides decomposed from carbonate, the occurrence of which was extended in the bulk GDC–LSCM anode as a result of bituminous coal passing through a plastic stage. Other factors that may account for degradation include agglomeration and structural change in coal accompanied by the using-up of active carbon in a single feed with operation time. In future work, the operation of HDCFCs with GDC–LSCM oxide anodes in the long-term, coupling with continuous fuel feeding, will be required to address the practical application of this anode material in DCFCs, especially those running on coal.

Acknowledgements

The authors would like to thank the European project 'Efficient Conversion of Coal to Electricity – Direct Coal Fuel Cells', funded by the Research Fund for Coal & Steel (RFC-PR-10007), and also thank Dr Aida Fuente Custa for her kind help on gathering the results on properties analysis of carbon and coal.

References

1 T. M. Gür, *Chem. Rev.*, 2013, **113**, 6179.
2 S. Giddey, S. P. S. Badwal, A. Kullkarni and C. Munnings, *Prog. Energy Combust. Sci.*, 2012, **38**, 360.
3 N. J. Cherepy, R. Krueger, K. J. Fiet, A. F. Jankowski and J. F. Cooper, *J. Electrochem. Soc.*, 2005, **152**, A80.
4 X. Li, Z. Zhu, R. De Marco, A. Dicks, J. Bradley, S. Liu and G. Lu, *Ind. Eng. Chem. Res.*, 2008, **47**, 9670.

5 X. Li, Z. Zhu, R. De Marco and A. Dicks, *J. Power Sources*, 2010, **195**, 4051.
6 G. A. Hackett, J. W. Zondlo and R. Svensson, *J. Power Sources*, 2007, **168**, 111.
7 X. Zhou, T. Oh, J. M. Vohs and R. J. Gorte, *J. Electrochem. Soc.*, 2015, **162**, F567.
8 F. Lantelme and H. Groult, *Molten Salts Chemistry*, Elsevier Inc., 2013, ch. 19, p. 403.
9 S. L. Jain, J. B. Lakeman, K. D. Pointion and J. T. S. Irvine, *J. Fuel Cell Sci. Technol.*, 2007, **4**, 281.
10 Y. Nabae, K. D. Pointon and J. T. S. Irvine, *Energy Environ. Sci.*, 2008, **1**, 148.
11 S. L. Jain, J. B. Lakeman, K. D. Pointion, R. Marshall and J. T. S. Irvine, *Energy Environ. Sci.*, 2009, **2**, 687.
12 C. Jiang, J. Ma, A. D. Bonaccorso and J. T. S. Irvine, *Energy Environ. Sci.*, 2012, **5**, 6973.
13 A. C. Chien, A. Arenillas, C. Jiang and J. T. S. Irvine, *J. Electrochem. Soc.*, 2014, **161**, F588.
14 S. P. Jiang and S. H. Chan, *J. Mater. Sci.*, 2004, **39**, 4405.
15 S. Tao and J. T. S. Irvine, *Nat. Mater.*, 2003, **2**, 320.
16 S. Tao and J. T. S. Irvine, *J. Electrochem. Soc.*, 2004, **151**, A252.
17 U. Hennings and R. Reimert, *Appl. Catal., A*, 2007, **325**, 41.
18 R. D. Green, C. C. Liu and S. B. Adler, *Solid State Ionics*, 2008, **179**, 647.
19 X. Yue and J. T. S. Irvine, *ECS Trans.*, 2015, **68**, 3535.
20 L. Zubizarreta, A. Arenillas, J. Piard, J. Pis and N. Job, *Microporous Mesoporous Mater.*, 2008, **115**, 480.
21 N. Job, R. Pirard, J. Marien and J. Pirard, *Carbon*, 2004, **42**, 619.
22 M. M. Rahman, *Nanomaterials*, InTech, 2011, ch. 9, p. 187.
23 K. Chen, N. Ai and S. Jiang, *J. Electrochem. Soc.*, 2010, **157**, P89.
24 M. Al Daroukh, V. V. Vashook, H. Ullmann, F. Tietz and I. A. Raj, *Solid State Ionics*, 2003, **158**, 141.
25 A. Fuente-Cuesta, C. Jiang, A. Arenillas and J. T. S. Irvine, *Energy Environ. Sci.*, 2015, submitted.
26 S. Zha, P. Tsang, Z. Cheng and M. Liu, *J. Solid State Chem.*, 2005, **178**, 1844.

Faraday Discussions

DISCUSSIONS

Improvements of energy conversion and storage: general discussion

Olga Kuzmina, John M. Slattery, Meilong Hu, Qiushi Song, Shuqiang Jiao, Geir Martin Haarberg, Qian Xu, Dihua Wang, Paul Madden, Daniel Cooper, Kathie McGregor, Tim Sudmeier, Bencan Tang, Hongmin Zhu, Ali Kamali, Linpo Yu, George Zheng Chen, John Irvine, Ye Liu, Jianbang Ge, Jennifer Pringle and Xiangling Yue

DOI: 10.1039/c6fd90029a

Kathie McGregor opened discussion of the paper by Jennifer Pringle: What value of the Seeback co-efficient was your target? I see from your paper that you show a range of values.

Jennifer Pringle answered: Yes, indeed the nature of the solvent and the redox couple can have a significant effect on the Seebeck coefficient, S_e. Of course we want this value to be as high as possible, and note that in terms of a figure of merit the S_e is squared and so a small increase in this parameter can make a big difference to the power output of the thermocell. However, we don't have a specific value of S_e that equates to a commercially viable device – in fact, we are still at the early stages of understanding why different solvents give such different S_e values and that is where our research is primarily focused at present. We are also looking at other aspects of device development, such as decreasing the mass transport resistance, which can provide additional benefits to power output.

Paul Madden commented: Did you pick the cobalt couple because of the expectation that the change in the structure of the coordination complex around this ion in different oxidation states would lead to a large entropy change? What other types of couple could have this property?

Jennifer Pringle answered: Other transition metal complexes with bipyridyl or other aromatic ligands have previously been studied in molecular solvents. However, the cobalt redox couple is particularly interesting because it undergoes a high spin/low spin transition upon electron transfer, which adds an extra electronic component to the entropy change – estimated at 80 J K^{-1} mol^{-1} – compared to, for example, the iron bipyridyl analogue.[1]

1 J. T. Hupp and M. J. Weaver, *Inorg. Chem.*, 1984, **23**, 3639–3644.

Paul Madden commented: In the semiconductor systems the Seebeck coefficient is expressed as a ratio of the electronic conductivity and the thermal conductivity which often behave in the same way and thus limit the values which can be achieved. In your type of cell are there similar competing physical factors?

Jennifer Pringle responded: Yes, although the competing factors are slightly different. We have looked at the performance of different ILs in terms of a modified figure of merit, analogous to the relationship used for semiconductor devices that you are referring to but, to make it more applicable to these liquid systems, we consider the ionic conductivity (*i.e.* by introduction of a term for the limiting diffusion rate, D, and concentration of the redox species, c) rather than the electronic conductivity. The thermal conductivity is still in the denominator of this relationship, with c and D on the numerator, but I believe that these should be easier to independently optimise than for the electronic and thermal conductivity in the semiconductor devices, *e.g.*, by increasing the redox couple concentration. The modified figure of merit is reported in Abraham *et al.*[1]

1 T. J. Abraham, D. R. MacFarlane, R. H. Baughman, L. Jin, N. Li and J. M. Pringle, *Electrochim. Acta*, 2013, **113**, 87–93.

George Zheng Chen remarked: Thank you for your interesting paper and presentation. You are looking at the electrochemical side for thermal to electrical energy conversion using ionic liquid as the electrolyte and heat absorption medium. If you combine this idea with thermal storage, would you still use ionic liquids? For heat storage, water has a far better thermal capacity. It has the highest heat capacity in liquid materials. Our recent research shows that it is possible to lower the temperature of aqueous solutions to below −60 °C,[1] and it is well known that the boiling temperature of aqueous solutions can be increased to beyond 150 °C, depending on what is added to water. Do you think you will be able to develop such a system? People think about water only when discussing the working temperature range, but it is common knowledge that addition of salt into water helps lower the freezing temperature and raise the boiling temperature of water. The heat capacity of ionic liquids does not match those of aqueous solutions. Based on these considerations, do you think it is worth studying aqueous solutions for heat storage and thermal to electrical conversion?

1 Y. Gao, Z. B. Qin, L. Guan, X. M. Wang and G. Z. Chen, *Chem. Commun.*, 2015, **51**, 10819–10822.

Jennifer Pringle replied: This is an interesting idea. Our devices are not presently intended for heat storage, only for thermal to electrical energy conversion, and the heat capacity of the electrolyte is not one of the significant parameters that influences efficiency. However, there is indeed ongoing research efforts on using water-based electrolytes for thermocells, and being able to extend the liquid range could be very advantageous. A recent report on water-based devices is Im *et al.*[1]

1 H. Im, T. Kim, H. Song, J. Choi, J. S. Park, R. Ovalle-Robles, H. D. Yang, K. D. Kihm, R. H. Baughman, H. H. Lee, T. J. Kang and Y. H. Kim, *Nature Commun.*, 2016, **7**, 10600.

Hongmin Zhu said: It is a great ideal to use redox couples for thermal energy harvesting. I have a question on the products during the energy harvesting. During the charge transfer through the electrode, the Faradaic reaction will happen on both sides of the electrodes. That means the electrode material will 'transfer' from one side to the other. How should we move them back afterwards? Can we just exchange the electrodes?

Jennifer Pringle replied: The thermocell contains a redox couple within the electrolyte, and it is this redox couple that undergoes electron transfer at the electrodes: overall, oxidation at one electrode and reduction at the other. Therefore, it is the redox couple that moves between the two sides of the cell during cell operation, not the electrode material. There should be no need to change the electrode material, and the cell should be able to run indefinitely as long as the materials are stable and there is a continuous temperature difference.

Daniel Cooper remarked: How do these devices compare with semiconductor-based thermoelectric generators in terms of both economics and required volume?

Jennifer Pringle answered: The semiconductor based devices are generally small, and can be relatively expensive as a result of the materials and fabrication techniques required. They are also a lot more efficient at high temperatures than low temperatures. In contrast, we are targeting low-grade waste heat, *e.g.* less than *ca.* 130 °C, and our ultimate aim is to develop larger-scale, cheap devices, *e.g.* flexible devices that can be made by reel-to-reel printing.

Qian Xu asked: The thermal battery can convert the heat to electricity, and the hot end (electrode), the cold end (electrode), and electrolyte are needed for the battery. Would you please tell me the working temperature range of the thermal battery when the ionic liquids serve as the electrolyte?

Jennifer Pringle answered: The upper temperature is limited by the thermal stability of the ionic liquid and/or the redox couple. We routinely run thermocell testing, with the cobalt couple in the ILs, using a hot electrode temperature of 130 °C and have not observed degradation within the timescale of the experiment. Use of a molecular solvent within the IL can lower the maximum working temperature, so we concentrate primarily on high boiling point molecular solvents.

Qian Xu commented: Is it possible to use the molten salts instead of ionic liquids for the thermal battery?

Jennifer Pringle answered: Yes, certainly that's possible, and there is published research in this area. The applicability of molten salts as electrolytes for thermoelectrochemical cells depends on the temperature of the waste heat source to be harvested and the electrochemical reaction that is going to be utilised, *e.g.* with an inert or reactive electrode. A review of this work is given by Gunawan *et al.*[1]

1 A. Gunawan, C.-H. Lin, B. A. Buttry, V. Mujica, R. A. Taylor, R. S. Prasher and P. E. Phelan, *Nanosc. Microsc. Therm.*, 2013, **17**, 304–323.

George Zheng Chen said: In comparison with inorganic molten salts, ionic liquids offer the possibility and flexibility to incorporate molecular structure variations and tune the physicochemical properties. The thermal to electrical energy conversion in ionic liquids as reported in this paper is based on the entropy change related to the electrochemical reactions. However, in comparison with energy conversion based on the enthalpy change of chemical reactions, entropy change is usually much smaller. So is the related energy conversion.

Geir Martin Haarberg remarked: Molten salts have equally high Seebeck coefficients to those reported for ionic liquids. We have reported high Seebeck coefficients in molten carbonate electrolytes using similar electrodes based on gas mixtures CO_2/O_2.

George Zheng Chen replied: As the Seebeck coefficient is associated with the entropy change of the cell reaction, it is expected that cell reactions involving a gas reactant or product can generate a high Seebeck coefficient. However, it is not easy to electrochemically reverse reactions involving gases for energy conversion and storage.

George Zheng Chen continued: For technological development of the concept as reported in this paper, the issue is how big the entropy change can be. For reactions involving gaseous reactants and/or products, entropy change may be large, but such reactions are difficult to be utilised for energy conversion, mainly because it is challenging to reverse such reactions.

John Irvine asked: A related concept is the AMTEC, alkali metal thermoelectric devic that is used in spacecraft driven by heat from radioactive decay. The potential is generated from 2 different termperatures of vapor. This delivers more power than a conventional thermoelectric device

Paul Madden addressed Jennifer Pringle and John Irvine: It has been pointed out that gas electrodes should be associated with a large entropy change for the electrode reaction but that there may be a limit to what can be achieved in a molten salt system because of the high temperatures which limit gas solubility. On the other hand the emf developed in the cell will be controlled by $T\Delta S$ and hence the high T might be an advantage.

Jennifer Pringle replied: Yes, there has been prior work on using a gaseous reaction at one electrode, in particular with chlorine gas, in molten-salt based devices. These were investigated at temperatures between *ca.* 500 and 1300 °C, and produced Seebeck coefficients of around -0.6 mV K^{-1}. The efficiencies were indeed higher at the higher temperatures.

Kathie McGregor asked: Have you been able to model any of the fundamental parameters of the ionic liquids you studied?

Jennifer Pringle replied: This is in progress. Of course, these are quite complex systems to model, as it requires accurately studying both the redox reaction and the solvent environment, but initial results suggest that it's going to be very useful.

George Zheng Chen opened discussion of the paper by Ye Liu: Your paper reports a good example for utilisation of CO in ionic liquids. In such practices, such as the metallurgical industry, CO production is commonplace, but its purity is not high. In your process, what is the requirement for CO purity? What contaminants in the CO gas are the biggest or most serious concern to your process?

Ye Liu replied: In my work, pure CO gas (>99%) was applied, since pure CO (CO/H_2 syngas) can be obtained on a large industrial scale. Many processes with CO as the feedstock have been industrialized such as hydroformylation of propene. Certainly, if CO gas is contaminated by impurities, particularly S or N-containing compounds, our reaction will definitely deteriorate due to the Rh-P catalyst poisoned by these impurities.

John Slattery remarked: If you have H_2S in your CO feedstock how stable would the rhodium catalyst be? I presume that this is not such a good thing for Rh(I).

Ye Liu responded: If H2S did exist in CO feedstck, it definitely would poison the Rh-catalyst. However, CO used in this work is very pure (>99%) which can be commercially obtained in the large industrial scale.

John Slattery said: If you did have a dirty stream of CO from an industrial source, could you use this for hydroformylation?

Ye Liu answered: No, since the impurities in the CO feedstock will impair the activity of the Rh catalyst for hydroformylation.

George Zheng Chen asked: I believe there must be an established process for purifying the CO gas. I know the price of CO_2 is reaching almost the same price as Ar. Is the price of CO similar to that of CO_2 in terms of commercial supply?

Ye Liu responded: CO can be supplied commercially.

Kathie McGregor commented: Are these types of catalysts best suited for organic synthesis, where the gases are cleaner, and less suited to metallurgical systems that may contain many contaminants? For example, in streams with sulphur-containing gases, the phosphine ligand may be quite reactive and promote instability in the catalyst?

Ye Liu replied: The phosphine-ligated Rh-catalyst is indeed not suited for metallurgical systems that contain many contaminants. However, in our organic system, the applied feedstock is very pure; and the stability of the catalysts can be guaranteed.

Bencan Tang asked: The first step in your reaction mechanism is to make Rh hydride, do you think the same conditions can be used for Cu hydride? Do you think your reaction conditions can be used for CO_2 reduction?

Ye Liu answered: No, I don't think so. Because the Cu–H is very labile and difficult to form, that is the reason why the Cu-catalyzed hydroformylation is rarely reported. Very labile Cu–H species at low concentrations are very difficult to be detected by *in situ* FT-IR spectroscopy even under high CO/H_2 pressure.

John Slattery commented: Do you know anything about the coordination chemistry of your cationic diphosphine ligand? It looks as though it might have quite a large bite angle. Is it a *cis*- or *trans*-chelating ligand, and what evidence do you have for this?

Ye Liu answered: My cationic diphosphine ligand in **2** can surely coordinate to the transition metals. Ligand **2** is still a typical bidentate chelating ligand. Its corresponding Rh(I)-complex has been firstly reported by Y. Canac and R. Chauvin (*Inorg. Chem.*, 2013, **52**, 48–58), in which the two phosphine fragments are located in *trans*-positions.

John Slattery said: What is the melting temperature of the disphosphine dication? Is it really an ionic liquid in its pure form?

Ye Liu replied: We didn't measure the melting point of the cationic diphosphine, but it is indeed in the solid state at RT. It is good news for us, since it is easy to purify the ionic solid compounds by using the recrystallization method.

John Slattery asked: Why did you choose $[BF_4]^-$ as the ionic liquid anion for these studies? Do you know anything about the stability of these ILs? Do you have any water in your reaction? These ionic liquids will decompose in the presence of water at elevated temperatures. The $[BF_4]^-$ anion can hydrolyse, with the formation of HF, under these conditions.

Ye Liu responded: The ILs with BF_4^- as the counter-anions are easy to prepare. The ILs applied in this work are quite stable with the decomposition temperature >300 °C. We also know that BF_4^- can hydrolyze to release HF under the harsh conditions. However, our reaction system is free of water and the reaction temperature is 120 °C. So we do not worry about the hydrolysis of the ILs with BF_4^- anions.

Linpo Yu commented: In your paper, you claimed the catalyst in IL cannot be recycled. Does the catalyst dissolve in the ionic liquid? If the deactivated catalyst can not be recycled and be maintained in the ionic liquid, what is the change in the catalytic efficiency?

Ye Liu replied: The catalyst used in combination with the IL with the piperidyl group as the solvent could be recycled successfully. However, if the catalyst was deactivated, the catalytic efficiency would definitely decrease no matter whether it was maintained in the IL or not.

John Slattery commented: The piperidine-functionalised ionic liquid seems to be quite effective for your catalytic reaction. Is it possible that you could form a bidentate carbene-amine ligand from the IL cation that could coordinate to Rh and influence the reaction? This might allow the possibility for the piperidine nitrogen to de-coordinate from the metal and act as an intramolecular base to facilitate H_2 activation, for example.

Ye Liu answered: It is very possible for a piperidyl-functionalized IL to act as a carbene–amine ligand since the imidazolium-based ILs have been widely regarded as carbene precursors. Hence, the piperidyl-functionalized IL as a base or carbene–amine ligand can improve the catalytic performance of the Rh–P catalyst for hydroformylation.

George Zheng Chen opened discussion of the paper by Linpo Yu: For the electrode materials used, *e.g.* activated carbon, the values of specific capacitance measured in ionic liquids are comparable to those in other electrolytes. However, the working (or cell) voltage in ionic liquids can be much higher, *e.g.* 4.5 V. In water based electrolytes, the working voltage is up to 2 V. When the working voltage increases, the energy capacity of the cell also increases.

Olga Kuzmina commented: How does your IL electrolyte performance compare to the common aqueous systems? You have compared imidazolium and pyrridinium cations. Why have you not chosen ILs with NTF_2 anions which are known to be better ion conductors?

Linpo Yu answered: Thank you for your questions. The supercapatteries with aqueous electrolytes can offer a higher power output than the ones with organic electrolytes (*e.g.* ILs), but have a relatively lower energy capacity. The conductivity of the electrolyte is an important factor, but the principal concern for the supercapattery in this paper is the involvement of the lithium metal electrode. The pyrrolidinium cation can tolerate more negative potential than the imidazolium cation. The imidazolium based ILs are available for the symmetrical supercapacitors, but are not good enough for the supercapattery in this paper. As to the anions, we will consider the choice of the NTF_2 anion in future work.

Kathie McGregor asked: You have used lithium perchlorate, have you considered substituting this with something that has less explosive hazards such as $LiPF_6$? Perchlorates are well-known to have caused nasty explosions.

Linpo Yu responded: Lithium perchlorate has been widely used as the supporting electrolyte in the fundamental research of non-aqueous electrochemistry. In this work, lithium perchlorate was mainly used as the lithium ion source in the electrolyte. The other lithium salt could be used to replace lithium perchlorate. However, the price of the salt should be considered.

George Zheng Chen commented: The main issue concerning using $LiClO_4$ in the ionic liquid is the solubility of other lithium salts being too low. This is different for carrying out electrochemical studies in conventional organic electrolytes in which salts with the PF_6^- and BF_4^- anions can be used. It should also be mentioned that the oxidation power of the perchlorate anion is not high unless it is in a highly acidic environment.

Meilong Hu remarked: In your paper you mentioned that the energy of your research can reach 230W h m^{-2}. For the present supercabattery how much energy is good for its industry use?

Linpo Yu answered: The commercial Li-ion battery can possess an energy capacity of more than 200 W h kg^{-1}, and this value is based on the weight of the whole battery, not only on the active materials. Considering the high power output of the supercapatteries, a slightly smaller energy capacity of supercapattery than the Li-ion battery is acceptable for industrial application.

George Zheng Chen commented: Energy capacity of the supercapattery depends on the electrode materials used. In this paper, the positive electrode is a commercial activated carbon whose specific capacitance was about 100 F g^{-1}. However, if we use a proper pseudocapacative material with a specific capacitance of 200 F g^{-1} to replace the activated material, the specific energy of the supercapattery should increase to 460 W h kg^{-1} from the current value of 230 W h kg^{-1}. Supercapatteries have other advantages over batteries. For example, the behaviour is more reversible in charging and discharging as shown in Fig. 5 of the paper. Consequently, energy loss is much smaller in the supercapattery than in a conventional battery.

John Slattery addressed Linpo Yu and George Zheng Chen: If storage capacity is not a challenge for this type of device, what are the challenges in this system?

Linpo Yu responded: Actually, we do believe that there is still potential for the energy capacity. The power output can also be improved if the negative electrode is optimised. The price of the electrolyte is the other concern.

George Zheng Chen commented: The work is still in its infancy and there are many challenges on the way to commercial applications. The cost of the ionic liquid is surely a concern to investors, but further increase of the ionic conductivity of, and the reversibility of lithium metal electrochemistry in ionic liquids are our next research objectives.

John Slattery addressed Linpo Yu and George Zheng Chen: The anions that you are using are expensive, in particular anions such as $[FAP]^-$. Can you make do with cheaper anions in your ionic liquids?

Linpo Yu responded: Yes, I think we can try some other ionic liquids in the future.

George Zheng Chen replied: Thank you for your suggestion. We have just started the research, and will surely test various ionic liquids with lower costs, including your suggestion.

Dihua Wang asked: In your work the supercapattery with a lithium anode and ionic liquid electrolyte has very good cyclability. I am curious about the morphology change of the lithium anode during the cycling. You know, the morphology change is quite important for chargeable lithium metal batteries.

The ionic liquid electrolyte can be any helpful for improving the cyclability of the electrode for a secondary battery according to your current research?

Linpo Yu answered: The morphology of the lithium negative electrode will definitely change after cycling. We ran the demonstrative supercapattery for 1000 cycles by CV. The CV shapes were well maintained after several cycles. We are going to test this supercapattery in a sandwich-type cell and investigate the morphology change of the lithium electrode later. As to the electrode stability enhancement by the ionic liquid electrolyte, we currently have no strong evidence.

Dihua Wang asked: The energy density you mentioned is very high, close to that of a Li-ion battery. How about the self-discharge rate for this kind of supercapattery? Another concern is during the charging process, Li cations and perchlorate anions will move to the negative and positive electrodes, respectively, which will lead to the depletion of the electrolyte, do you have any comments on its effect on the performance of the supercapattery?

Linpo Yu answered: We purposely haven't investigated the self-discharge. However, we noticed that the demonstrative cell could maintain the voltage for hours after charging. We are short for the long term self-discharge data right now. As to the concern over the charging process, the voltage loaded on the cell was smaller that the depletion voltage of lithium perchlorate. We assumed that the cation is intercalated into the carbon positive electrode when the cell is charging, leading to the contribution of the pseudo-capacitance. This reaction will not affect the cell when there is sufficient electrolyte in the cell. We are investigating the effect of the amount of electrolyte in the cell right now.

George Zheng Chen said: There are some challenging issues with supercapacitors. For example, electrical double layer capacitors built from porous activated carbon cannot retain the stored charge (energy) very long, and the cause is self-discharge. One of the proposed mechanisms is that the charge stored tends to rearrange in the porous structure of the electrode. Redox impurities can also be responsible for self-discharge. For pseudocapacitors, the situation is not much improved. When people report on their supercapacitor research, they tend to highlight the fast speed without mentioning self-discharge. The common perception is that batteries are less troubled by self-discharge. It is often true that the electrode loading of the active material is relatively high in batteries. This makes the effect of redox impurities relatively small. It is not to say that self-discharge does not occur in a battery, but it is less serious in comparison with that in supercapacitors. Another issue is that the cell voltage of a supercapacitor varies with its charging or discharging state, whilst batteries can usually output a more stable cell voltage. Therefore, when a stable voltage is needed, additional electronic circuits are needed. Regarding the use of lithium perchlorate in our work, the perchlorate anion has high electrochemical stability. I do not know what you can oxidise the perchlorate into. It is one of the most stable anions against oxidation. Ionic liquids are studied in our work because of their electrochemical and thermal stabilities. It is possible to decompose the ionic liquid electrochemically, but this is only possible when the cell is charged to a very high voltage. With added lithium ions in the electrolyte, lithium metal deposition can occur before the cathodic discharge of the ionic liquid. The lithium ion may interact with anions, which could affect the potential and kinetics of deposition of lithium metal. However, I do not expect lithium deposition would affect the anions partly because of the very negative potential of the electrode needed for lithium deposition.

George Zheng Chen opened a discussion of the paper by Dihua Wang: Your paper reports a very exciting progress in terms of CO_2 reduction in molten salts. We have done similar work and I have a strong interest in this renewed area of molten salts research. In our work (part of which was published in a previous Faraday Discussion),[1] we have noticed that the potential for reduction of carbonate ion to carbon is far more negative than the potential for re-oxidation of the deposited carbon. This is also shown clearly in Fig. 2 of your paper. Particularly, it can be seen in Fig. 2b that the far left peaks are related to Li deposition and re-oxidation (dissolution), which are fairly reversible. The previous reduction peaks at less negative potentials are related to carbonate ion reduction to carbon which are well separated by about 1.0 V from the re-oxidation peaks at the far right. Have you studied this aspect of the electrochemistry of carbonate ions? What is your explanation?

1 H. V. Ijije, R. C. Lawrence, N. J. Siambun, S. M. Jeong, D. A. Jewell, D. Hu and G. Z. Chen, *Faraday Discuss.*, 2014, **172**, 105–116.

Dihua Wang replied: Thank you for your comment. Yes – we noticed that the reduction potential for carbonate ion to carbon in Li_2CO_3 containing melt was far more negative than the oxidation potential of the deposited carbon, as evidenced in your work and in Fig.2a and b in this work. This phenomenon shall be ascribed to the high overpotential for both the cathode and anode processes. It is quite important to decrease the overpotential by optimizing the electrode material and the composition of electrolyte not only for the MSCC-ET process but also for the DCFC, and this remains something to investigate in future.

George Zheng Chen remarked: In this situation, have you obtained sufficiently good carbon deposit?

Dihua Wang answered: In Li_2CO_3 containing melt, we always got good carbon deposits at a current density lower than 100mA cm^{-2}.

Hongmin Zhu commented: Congratulations on the nice work. My understanding of the cycle for CO_2 capture is that during capture CO_2 is absorbed by the O^{2-} ion in the melt forming ${CO_3}^{2-}$, while during the electrolysis O^{2-} is enriched. Although the production of carbon on the cathode is important, I believe that the generation O_2 on the anode is more important for the whole cycle. Only with an anode material that can tolerate oxygen generation for a long time, can the whole process then become possible. So, I have question about the anode material. What kind of material have you used? How long can it be used? And have you checked the anode production during the electrolysis?

Dihua Wang answered: Yes, a stable and cost-affordable anode for O_2 generation is crucial for CO_2 transformation with oxygen gas as a by-product. In this work, we found that tin oxide is a satisfactory inert anode in molten carbonates. It has lasted for more than 600 hours without obvious damage. The anode produces a mixture of O_2 and CO_2, depending on the applied current density. Of course, it is more challenging to find an inert anode in molten chlorides that we are still working on.

Qian Xu asked: Did you investigate how the water in the molten salts affects the capture of CO_2? What about the effect of the molten salt volume on the capture of CO_2?

Dihua Wang answered: In our CO_2 capture experiment, the molten salt was pre-dried and the system was placed under vacuum several times, so the effect of water should be negligible. The volume of the molten salt has some effect on the capture of CO_2. Firstly, it will affect the capture capacity in a given experiment. Secondly, it also will affect the capture kinetics. What we measured is a static absorption process, the thickness of the molten salt will affect the diffusion of CO_2/carbonate/oxygen anion in the melt, while the thickness will be determined by the volume of molten salt in a given container with a fixed inner diameter.

Shuqiang Jiao remarked: During electrochemical reduction of CO_2, there should be two reactions happening in cyclic voltammetry. However, it is not clear to see two cathodic peaks in the CVs. Why is this?

Dihua Wang answered: In the Li_2CO_3 containing system, the reduction of carbonate is most likely a four-electron reduction to carbon so that only one reduction peak is presented before the deposition of alkaline metal. I guess what you mean is for the CV measured in a $CaCO_3$ containing system. We did not observe carbon deposition at the potential range around the reduction peak in a freshly prepared melt. A possible reason is that the reduction potential for carbonate to carbon in the system is more negative than the potential for metal deposition.

Shuqiang Jiao commented: Using pure carbonate salt, I am curious about oxygen ion solubility because the oxygen ions can form more carbonate. Have you tested the solubility of CO_2 in pure carbonate and how did you do it? Did you initially add oxide into the salt to absorb CO_2 and what is the concentration of oxide?

Dihua Wang answered: We tested the solubility of CO_2 in pure carbonate and chloride using our home-made testing system, as shown in Fig. 1. We also tested the absorption kinetics in the oxide containing carbonate and chloride, different concentrations of oxides were tested but in Fig. 1, we presented results with a concentration of 180 mmol kg^{-1}.

George Zheng Chen commented: I think Prof. Jiao was questioning which product is produced, CO or carbon in your experiments. There are two issues about producing CO gas on the cathode. The first is about the catalytic effect of the electrode material. If a Ti cathode is used, the electrolysis produces the CO gas but not carbon.[1] The second issue is that cathodic production of CO gas may be less favourable because it is a gas and, if not removed quickly, can block the electrode/electrolyte interface. Further, CO can be produced by reaction between the deposited carbon and the carbonate ions. In our paper that was published in *Faraday Discuss.* in 2014,[2] we presented a photograph of cathodes with deposited carbon on which gas bubbles formed. It seems that if CO were produced as the only product, a proper cathode material would be needed to function as a catalyst. However, if carbon deposition happens, I doubt the CO formation to be due to electrochemical reduction only. It may also be formed *via* the chemical reaction between the deposited carbon and the carbonate anions in the molten salt electrolyte.

1 V. Kaplan, E. Wachtel, K. Gartsman, Y. Feldman and I. Lubomirsky, *J. Electrochem. Soc.*, 2010, **157**, 552–556.
2 H. V. Ijije, R. C. Lawrence, N. J. Siambun, S. M. Jeong, D. A. Jewell, D. Hu and G. Z. Chen, *Faraday Discuss.*, 2014, **172**, 105–116.

Dihua Wang responded: I agree with George that the electrode material has a different catalysis activity for carbonate reduction in the melt so that it will affect the reduction product. Titanium is a good case for catalysis reduction of carbonate to CO. But in our experiment, carbon deposition still happened when the applied potential was negative enough. We also observed a bubble-shaped carbon product in $CaCO_3$ containing chlorides, suggesting CO as a possible intermediate product. The chemical reaction between the deposited carbon and carbonate is also possible, especially under the low partial pressure of CO. But at the temperature of our experiment, 450 °C, thermodynamically it is unfavourable.

Qiushi Song addressed Dihua Wang and George Zheng Chen :We have done some similar studies on electrochemical conversion of CO_2 in molten salts. It is found that carbon inevitably exists in the melt, which may cause a shortcut in the cell or a high background current. So current efficiency is obviously lower than the value of 70–80% presented in your work. How do you avoid the negative effect of carbon during the electrolysis?

Dihua Wang answered: I do not think that it is a big issue for carbon that will affect the current efficiency. Rather than liquid products such as Al, Ca and Mg on the cathode, or gas products which will transfer from the liquid phase or gas phase, carbon is in the solid state. Its negative effect can be easily avoided by using a diaphragm or a dam at the electrolyte surface. The current efficiency in our research can be more than 90%.[1–2] A high current efficiency was also reported by others.[3] On the other hand, carbon is not inevitably dispersed in the melt. Dense and adherent carbon deposits can form on the cathode if a suitable current density and electrode material are selected.

1 H. Y. Yin, X. H. Mao, D. Y. Tang, W. Xiao, L. Xing, H. Zhu, D. H. Wang, D. R. Sadoway, *Energy Environ. Sci.*, 2013, **6**, 1538–1547.
2 D. Tang, H. Yin, X. Mao, W. Xiao and D. H. Wang, *Electrochim. Acta*, 2013, **114**, 567–573
3 H. V. Ijije, R. C. Lawrence, N. J. Siambun, S. M. Jeong, D. A. Jewell, D. Hu and G. Z. Chen, *Faraday Discuss.*, 2014, **172**, 105–116

George Zheng Chen responded: I agree with the reply from Prof. Dihua Wang. In addition, carbon debris floating on the surface of molten salt may be present in a cell with a graphite anode of poor quality (*e.g.* low density) on which formation of the CO and CO_2 gases in the pores of the graphite rod may exert internal stress to break the graphite structure into carbon debris. However, in molten carbonate salts, there are more materials that are inert to oxide ion discharge, such as tin oxide.

Qiushi Song commented: Molten salt electrolysis of CO_2 is an interesting method for eliminating the effect of CO_2 emissions. But this kind of work is less significant if a suitable inert anode is not utilized. Could you give some comments on the inert anode for CO_2 molten salt electrolysis?

George Zheng Chen responded: We have used a tin oxide rod as the inert anode, and this was also used by Prof. Dihua Wang. In fact, some high nickel alloys may also behave sufficiently well when used to make the anode in molten carbonate salts.

John Irvine said: I was expecting CO to be the main product from gas phase analysis from our experience of Boudouard type processes. Have you looked at the thermodynamics of what you'd expect in these conditions. For the carbonates you're in a very different state from standard states. You have high carbonate activity and low oxygen activity and are operating at moderate temperatures so this may explain why CO is main product here.

Dihua Wang responded: Thank you for your comment. Thermodynamically carbon is the favorable product at our experimental temperature. It is possible to generate CO on a different electrode such as titanium just mentioned by George Chen and a certain electrolyte such as $CaCO_3$ containing melt as reported in this work.

Tim Sudmeier communicated: You talked about the peculiar behavior of the 2 mol% $CaCO_3$–LiCl–KCl system displaying only one reduction peak (between −1.2V and −0.8V) in the first CV scan showing no carbon deposition after prolonged electrolysis using a Ni electrode. In your paper you hypothesize that this single reduction peak corresponds to the reduction of CO_2 to CO. I was wondering now if you have any evidence for the evolution of CO during electrolysis within the potential range of −1.2V to −0.8V (without exposing it to lower potentials) or if you where able to detect higher CO concentrations on the surface of the Ni after the electrolysis.

Dihua Wang communicated in reply: Yes. We found that the composition of electrolyte did affect the reduction mechanism of the carbonate. For the system of 2 mol% $CaCO_3$–LiCl–KCl, we did a series of potentiostatic electrolysis experiments in the potential range around the reduction peak, but we did not get any carbon product. We recently detected certain amount of CO in the cathode outlet by a gas analyzer.

Ali Kamali opened discussion of the paper by Shuqiang Jiao: Interesting research! Both this study and that presented just before you have similar results in terms of producing carbon materials at the cathode. However, there are some differences between the two as well, your work is at 750 °C, whereas the previous work is at 450 °C. I'm not sure for the previous work, but carbon nanomaterials are produced in your case after just 10 seconds and nanorods after 20 seconds, which is very remarkable. There should be some catalytic activity at the cathode, but without assistance from the molten salt, for example by CVD, the same length of growth should take much longer. Do you have any idea of what the detailed mechanism would be for this catalytic activity? It might be a good idea to use cathodes with different surface roughness and conditions to see any possible effect on the nanorod growth.

Jianbang Ge replied: Thank you very much for your comments. To be honest, I am not really sure about how the nickel cathode influences the formation of carbonfibers, and as you have stated, more details are needed to investigate the mechanism. Thank you again for your constructive suggestions.

Qiushi Song asked: Is there a reference electrode in your experiments, AgCl/Ag or Pt? Is the counter electrode carbon? Is the carbon deposit uniformly dispersed on the substrate? We found that there is little carbon after 30 s of deposition. But in your work there is much more carbon. Have you calculated the current consumed for the reduction to find if it agrees with the amount of carbon deposited?

Jianbang Ge replied: There are only two electrodes in constant cell voltage electrolysis, Pt and nickel. Moreover, due to the unevenness and defects of the nickel substrate, the carbon couldn't deposit with uniform dispersion on the substrate and the SEM photographs taken from the middle area of the deposited carbon were chosen as the end results. I don't really know about your conditions for short term electrolysis (30 s). But the reasons for the little carbon may be explained below: (1) low carbonate ion concentration and cell voltage; (2) the carbon does not adhere very well to the substrate, you may lose it in the washing process; (3) other reactions take place, such as the reduction of superficial oxide on the substrate surface.

George Zheng Chen addressed Jianbang Ge and Shuqiang Jiao: I am interested in the SEM images taken from your short term experiments for the electrodeposition of carbon. Apparently, you obtained different carbon nanostructures including wires and tubes as shown in Fig. 6. It is interesting to know if these structures were formed by carbon atom addition to the lattice positions of crystalline carbon as in electrodeposition of a crystalline metal. It seems that carbons do not form crystalline structures easily, but your picture indicates that they may do, suggesting proper electrodeposition. Alternatively, the carbonate ion may be reduced to carbon by Li metal in the electrolyte that can be produced by electroreduction of Li^+ ions. Carbon has low solubility in the electrolyte, and can precipitate onto the cathode, giving rise to an amorphous deposit. Another important issue worth mentioning is the use of Li in the salt. Deposition of carbon does not happen in molten KCl or NaCl or their mixtures. My interest is to understand if the carbon formed on the cathode results from electrodeposition or precipitation of carbon. If we can confirm deposition of crystalline carbon on the cathode surface in a short time, it will prove the electrodeposition mechanism. We need more pictures like those in Fig. 6 to gain better understanding.

Jianbang Ge responded: Thank you very much for your careful logical analysis. In fact, the carbon materials obtained at 2.8 V have been characterized by TEM, and the results show that both spherical carbon and carbon fibers are mainly composed of amorphous carbon. In my opinion, considering that they both can be deposited at 2.0 V and 2.4 V, the influence of the reduction by Li metal on their formation may be low. We are not really sure about whether crystalline structures can be obtained by electrodeposition.

Kathie McGregor commented: Can you explain the mechanism of the formation and growth of the different carbon morphologies on the Ni and Mo cathode materials?

Jianbang Ge answered: Two kinds of morphologies, spherical carbon and carbon fibers can be seen on the Ni cathode. For the formation of spherical carbon, it can be obtained using many other different cathodes in many literature sources. The cathode has no effect on its formation. The mechanism of formation and growth of spherical carbon is easy to understand because the spherical structure is the most stable structure during carbon nucleation. However, the formation of carbon fibers is different. It can only be observed in the Ni cathode and, in my opinion, the Ni cathode may only have a catalytic effect on the formation of carbon fibers. The Mo substrate is just for comparison. How the Ni cathode influences the formation of carbon fibers is still not clear.

George Zheng Chen asked: According to the paper, the carbon deposited on different metal cathodes are different in morphology or structure. The deposit seemed to be more likely amorphous when Mo plate was used as the cathode, see Fig. 7.

Meilong Hu said: The raw material to obtain CO_2 is from molten salt. How do you supply the raw material to ensure the process is continuous?

Shuqiang Jiao answered: The whole process could be described as below. Firstly, the initial carbonate ions are reduced to carbon and oxide ions at the cathode, while the released oxide ions are discharged on an inert anode electrode to form oxygen during the course of the electrolysis. Then the other generated oxide ions react with carbon dioxide bubbled into the melt to regenerate carbonate ions, leading to a circulation where CO_2 could be continually absorbed and then split to carbon and oxygen. The whole reaction in this process is $CO_2 = C + O_2$.

George Zheng Chen opened a discussion of the paper by Xiangling Yue: Thank you for presenting this interesting work. I would like to propose a collaboration for us to test the carbon produced by electro-reduction of CO_2 in molten salts in your direct carbon fuel cell. As shown in your video, the almost spontaneous ignition of your carbon sample is very interesting. What was the composition of your carbon? Did it contain some impurities such as potassium? Our electrolytic carbon could ignite at about 270 °C which is much lower than the ignition temperature of normal carbon samples. We would attribute the low temperature ignition to the catalytic effect of some impurities in the carbon, such as iron and potassium. (I would think that mixing carbon with fine platinum catalyst would also lower the ignition temperature.) This feature of the electrolytic carbon, and your carbon as well, may enable them to be used as the fuel in an internal combustion engine. What was the oxygen content in your carbon samples? Did you use any method to reduce the oxygen content in your sample? My thought is that if the oxygen content is sufficiently high, the carbon sample would not ignite easily.

Xiangling Yue answered: Thank you very much for proposing the collaboration to which I am quite looking forward, it would be very interesting to see how the electrolytic carbon performs in direct carbon fuel cells, as it has quite different properties compared to other carbon forms. In our current study, the carbon we used is a synthetic carbon, it has 64% C, 32% O, 3.9% H and minor N, and no impurities from CHN analysis. Although the original carbon has quite a high oxygen content, most oxygen groups would be removed in a carbonisation process which readily happens when heating to fuel cell operational temperature. A pyrolysis experiment confirmed this, showing 97% C, 1.8% O and 1.0% H after pyrolysis. The spontaneous ignition was thus believed to be due to reactions between carbon and potassium carbonate used as a secondary electrolyte, with active potassium species formed due to the reducing power of carbon. However, this seemed to be affected by the textural properties of carbon as ignition was not happening to each carbon forms we used. Without adding carbonate, ignition did not happen. We did not use any Pt during our test. Due to the difficulty of the experiments, the exact product that caused ignition has not been identified yet.

Ali Kamali asked: Usually well graphitised carbon materials oxidise at more than 600 °C. Carbonate salts, such as lithium carbonate, can catalyse the oxidation of graphite to a much lower temperature than its normal oxidation temperature.

John Slattery addressed Xiangling Yue and George Zheng Chen: I don't see why the spontaneous ignition of the isolated carbon has to be a process that is catalysed by a potassium salt, as has been suggested. Is it possible that you are reducing some K^+ to K metal, which might give a material such as KC_8, which is pyrophoric. Only a small amount of such a species might be enough to cause the carbon to ignite on exposure to air.

Xiangling Yue answered: This is a very good comment, thank you very much! we did not think of this before. Yes, K intercalation compounds would be possible if metallic K was possible in our experimental conditions, which were not really favorable for K metal. That's why we thought more of peroxide. Ignition not only happened after electrochemical reactions, but also just after pyrolysis of the carbon and potassium carbonate mixture.

George Zheng Chen responded: This is an interesting thought, and I fully appreciate it. I do not think there was potassium metal. There might be potassium peroxide formed. If that encountered carbon, it could lead to ignition and burning.

Faraday Discussions

PAPER

Advances toward industrialization of novel molten salt electrochemical processes

Yasuhiko Ito,* Tokujiro Nishikiori and Hiroyuki Tsujimura

Received 26th December 2015, Accepted 9th March 2016
DOI: 10.1039/c5fd00237k

We have invented various novel molten salt electrochemical processes, that can be put to practical use in the fields of energy and materials. These processes are promising from both technological and commercial viewpoints, and they are currently under development for industrial application. To showcase current developments in work toward industrialization, we focus here on three of these processes: (1) electrolytic synthesis of ammonia from water and nitrogen under atmospheric pressure, (2) electrochemical formation of carbon film, and (3) plasma-induced discharge electrolysis to produce nanoparticles.

Introduction

Molten salts are, in general, chemically and physically stable, highly electrically conductive, and resistant against radiation. In many cases, they dissolve various chemical species with sufficiently high concentration. From an electrochemical aspect, some molten salts have a wide electrochemical window. These advantageous characteristics offer considerable benefits towards energy conversion and material synthesis/tailoring applications. Taking into account these advantageous characteristics of molten salt systems, traditional industrial electrochemical processes have been developed, and currently the role of molten salts in science and technology, especially in the fields of energy and materials, is becoming more and more important. Against the above background, we are now taking up the challenge of industrializing various novel molten salt electrochemical processes (MSEPs) that we have developed. Through a series of bench-scale experiments, these processes have proven to be promising from both technological and commercial viewpoints. Three of these processes are described here to showcase current research developments towards industrialization: (1) electrolytic synthesis of ammonia from water and nitrogen under atmospheric pressure, (2) electrochemical formation of carbon film, and (3) plasma-induced discharge electrolysis to produce nanoparticles.

I'MSEP Co., Ltd., Japan. E-mail: yasito@mail.doshisha.ac.jp

Electrolytic synthesis of ammonia from water and nitrogen

The production of NH_3 is crucial since it is an essential ingredient in the manufacture of fertilizer and many other important chemicals. Its distributed uses are also expected to increase: a selective catalytic reductant of NO_x emitted from vehicles, ships, and stationary facilities such as thermal power stations, it can be used as an alternative to chlorofluorocarbons (CFC), and is a safe and economical energy transport and storage media for future energy systems. Among several candidates for hydrogen storage and transport materials, ammonia is considered the most promising. However, at present, almost all NH_3 is synthesized by the Haber–Bosch process on a large scale at high temperatures and under high pressures. For the distributed use of NH_3, a new process must be created that can be operated at a lower synthesis temperature and pressure. Furthermore, it is now necessary to reduce CO_2 emissions, which requires us to create a new ammonia synthesis process that can be carried out without causing any CO_2 emission. Given the above circumstances, we propose the direct electrolytic synthesis of NH_3 from H_2O and N_2 under atmospheric pressure using molten salt electrolytes,[1–3] and we are currently developing this process for industrialization.

The principle of this process is shown in Fig. 1.

The nitride (N^{3-}) ion is produced in an electrolyte by a cathodic reduction of N_2 gas: $1/2N_2 + 3e^- \rightarrow N^{3-}$. When water vapour is supplied to the electrolyte, N^{3-} and H_2O react to produce NH_3 and O^{2-}: $N^{3-} + 3/2H_2O \rightarrow NH_3 + 3/2O^{2-}$. The produced oxide ($O^{2-}$) ion is anodically oxidized at the anode to evolve O_2 gas: $O^{2-} \rightarrow 1/2O_2 + 2e^-$. Thus, the overall reaction is expressed as $1/2N_2 + 3/2H_2O \rightarrow NH_3 + 3/4O_2$. The standard theoretical electrolysis voltage corresponding to the total reaction is calculated as 1.17 V at 600 K. Bench-scale experiments using the electrolytic cell shown in Fig. 2 revealed that the process is quite promising from a practical aspect. Some experimental results obtained with the bench-scale electrolytic cell are shown in Fig. 3.

As for the NH_3 yield data in Fig. 3, the time lag between the start of the water vapour supply and the detection of NH_3 in the outlet gas can be understood when we take into account the fact we previously found[4] that NH_3

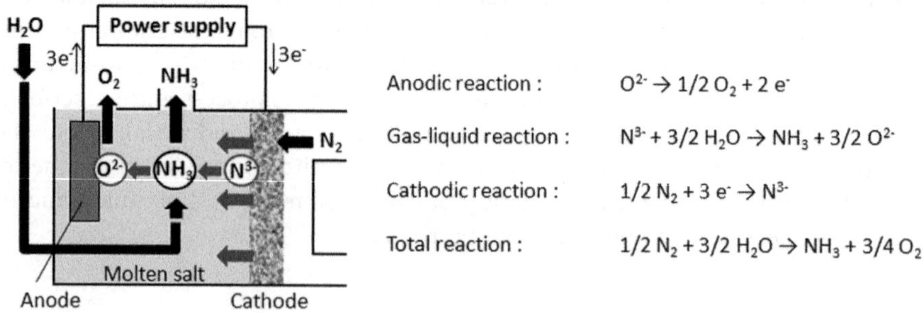

Fig. 1 Principle of NH_3 synthesis from water and nitrogen.

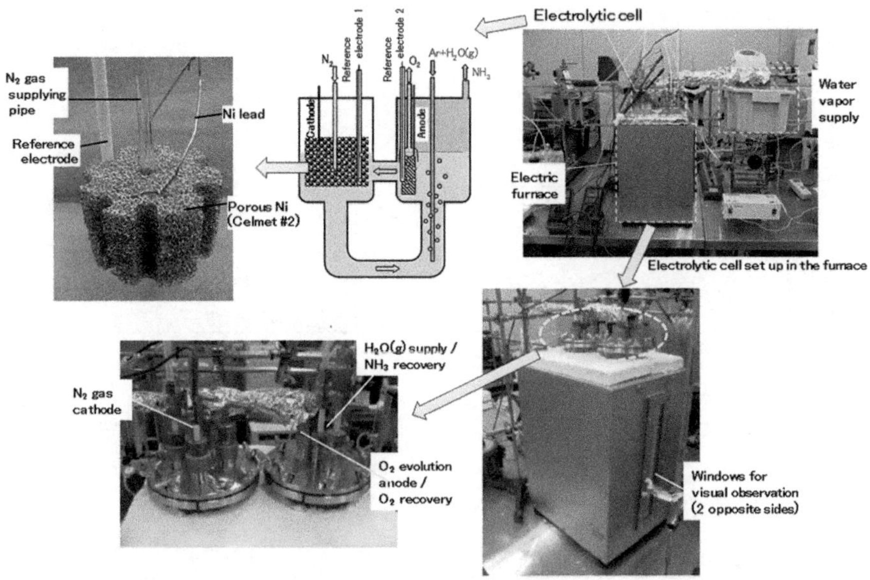

Fig. 2 Bench-scale electrolytic cell for NH_3 synthesis from water and nitrogen.

dissolves as amide (NH_2^-) and imide (NH^{2-}) ions *via* a chemical reaction: $NH_3 + N^{3-} \rightarrow NH_2^- + NH^{2-}$, and the formed NH_2^- and NH^{2-} are slowly changed back to NH_3 *via* the following reactions: $2NH_2^- \rightarrow NH_3 + NH^{2-}$ and $NH^{2-} + H_2O \rightarrow NH_3 + O^{2-}$. When utilizing the type of electrolytic cell shown in Fig. 1 (and 2), however, the required minimum electrolysis voltage is about 2.0 V, which is mainly due to the spontaneous chemical reaction inevitably included in the process: $N^{3-} + 3/2H_2O \rightarrow NH_3 + 3/2O^{2-}$. On the other hand, when we use an advanced-type electrolytic cell composed of two cells with a hydrogen permeable metal membrane between them, such as that schematically shown in Fig. 4, the required electrolysis voltage can be decreased to less than 1.5 V.

Fig. 5 shows the bench-scale advanced-type electrolytic cell currently being used in our laboratory. The experimental data obtained using this bench-scale cell will be reported in the near future.

The significant decrease in the required electrolysis voltage achieved using the advanced-type cell shown in Fig. 4 (and 5) can be explained as follows.

Fig. 6(A) shows the electrode potential for both cathode and anode corresponding to each electrode reaction. The potential difference between cathode and anode (required voltage ΔV for electrolysis) is estimated to be approximately 2.0 V. The voltage loss due to the spontaneous ammonia formation reaction ($3H_2O + 2N^{3-} \rightarrow 2NH_3 + 3O^{2-}$) can be recovered if we construct the cell shown in Fig. 6B composed of a cathode ($3H_2O + 6e^- \rightarrow 3H_2 + 3O^{2-}$) and an anode ($3H_2 + 2N^{3-} \rightarrow 2NH_3 + 6e^-$), with an emf ΔV of 0.8 V, under the assumption that we can move the O^{2-} ion from the right-hand side cell to the left-hand side cell and move the N^{3-} ion from the left-hand side cell to the

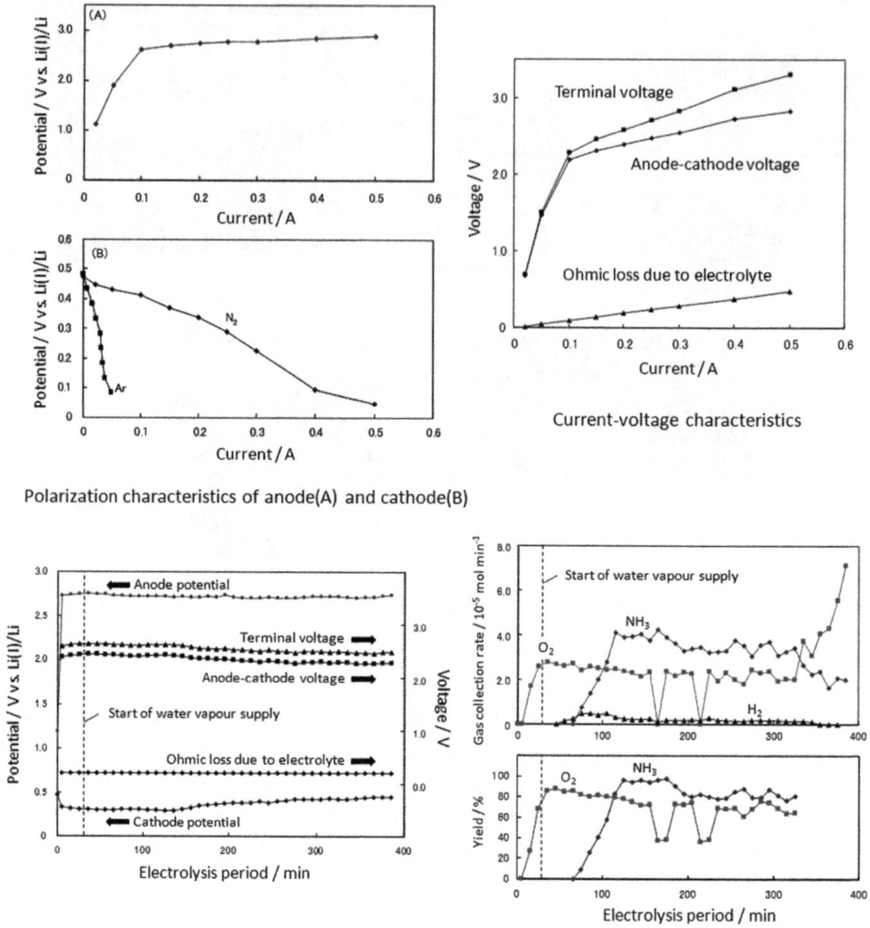

Fig. 3 Experimental results of the bench-scale electrolytic cell for NH_3 synthesis from water and nitrogen.

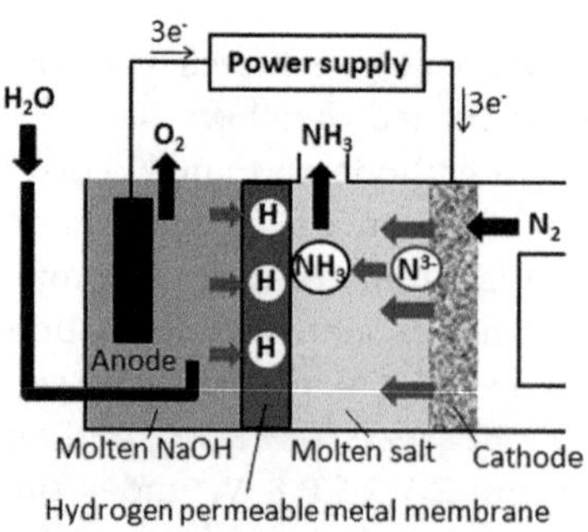

Anodic reaction : $OH^- \rightarrow 1/4\,O_2 + 1/2\,H_2O + e^-$

Cathodic reaction : $H_2O + e^- \rightarrow H_{abs} + OH^-$

Anodic reaction : $N^{3-} + 3\,H_{abs} \rightarrow NH_3 + 3\,e^-$

Cathodic reaction : $1/2\,N_2 + 3\,e^- \rightarrow N^{3-}$

Total reaction : $1/2\,N_2 + 3/2\,H_2O \rightarrow NH_3 + 3/4O_2$

Fig. 4 Schematic of an advanced-type electrolytic cell.

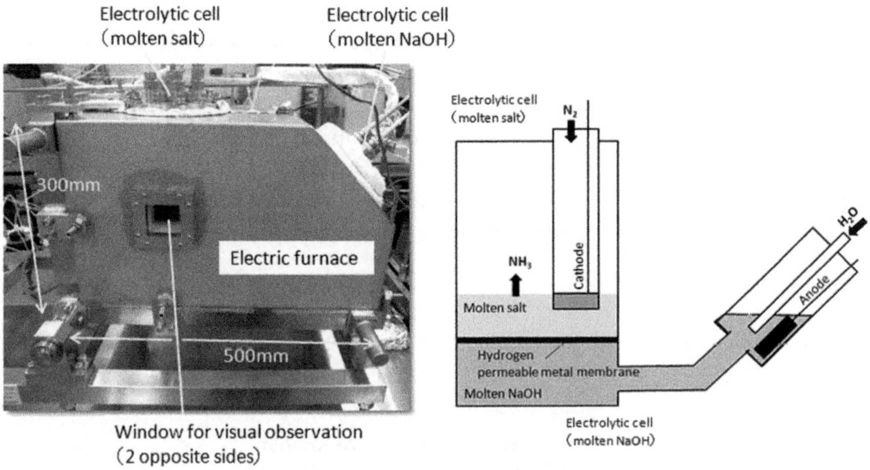

Fig. 5 Bench-scale electrolytic cell (advanced-type) for NH_3 synthesis from water and nitrogen.

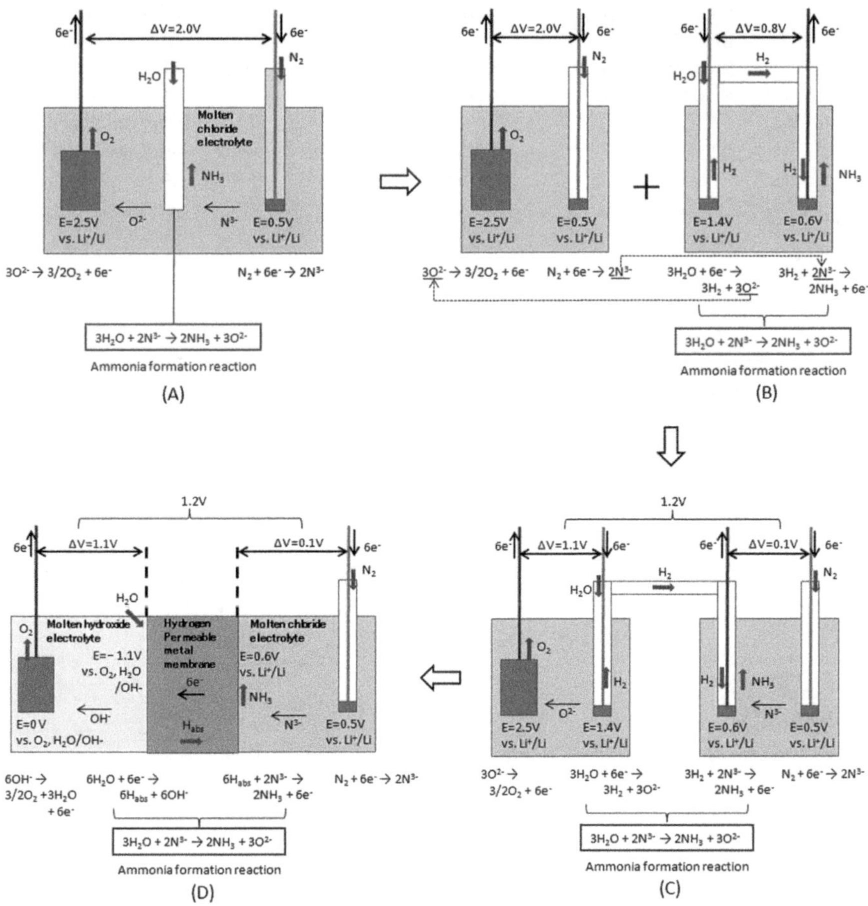

Fig. 6 Illustration explaining the difference in required electrolytic voltage between the cells of Fig. 1 and 4.

right-hand cell without any free energy loss. If we can recover 0.8 V using the right-hand side cell, then the necessary voltage for the total reaction becomes 1.2 V (=2.0 V − 0.8 V), which is conceptually possible but very difficult to achieve in practice. On the other hand, the recovery of energy lost due to the above spontaneous ammonia formation reaction can also be avoided by using two separate cells connected to each other through a hydrogen transporting pipe, as conceptually shown in Fig. 6C. An outline of the potential of each part and the approximate required voltage is also shown here. These two cells in

Table 1 Comparison of energy required for several ammonia synthesis processes

	Input energy	Energy consumption GJ/t-NH_3 Hydrogen production	Nitrogen production	Ammonia synthesis	Total	Note
Haber–Bosch process	Natural gas	→	→	→	30–35	Average
	Natural gas	→	→	→	28	Top result
	Electric power	31.2	0.46	1.48	33.2	(i) WE-NET base[5]
	Electric power	→	→	43.2	43.2	(ii) Ganley base[6]
	Electric power	31.2	→	1.43	32.7	(iii) Gosnell base[7]
Electrolytic synthesis	Electric power	—	0.46	42.6	43.1	2.5 V (based on experimental data)
	Electric power	—	0.46	34.1	34.5	2.0 V (achievable)
	Electric power	—	0.46	25.5	26.0	1.5 V[a]

[a] Advanced-type cell.

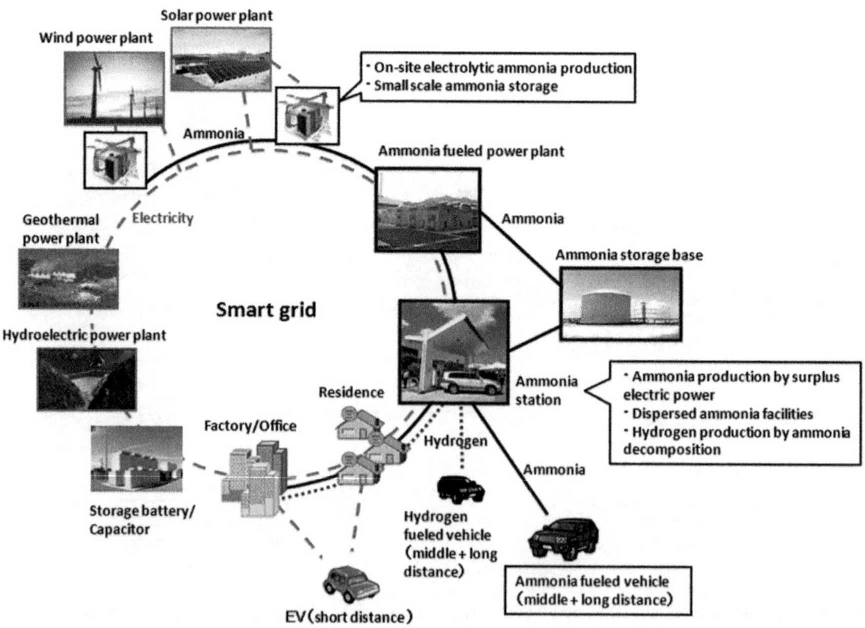

Fig. 7 Expected role of the proposed electrolytic system in an ammonia energy system.

Fig. 6C can also be connected to each other using a hydrogen permeable metal membrane as shown in Fig. 6D. In the case of Fig. 6D, we can use molten hydroxide electrolyte instead of molten chloride for the left-hand side cell, which is more advantageous than using the cell shown in Fig. 6C from an engineering perspective. The above considerations clearly explain the decrease in required voltage when the type of cell in Fig. 4 is used instead of that in Fig. 1. Note that the numerical values of potential and voltage indicated in Fig. 6 are approximate.

A comparison of the energy required for various ammonia synthesis methods, depicted in Table 1, suggests promising features of our new processes.

It should be noted here that the conventional Haber–Bosch process utilizes natural gas as a hydrogen source, and thus it emits a large amount of CO_2. The experimental data for the bench-scale experiments, together with calculated results from an economic viewpoint, encourages us to propose the ammonia energy system, conceptually shown in Fig. 7.

When we consider the role of this process from the aspect of reducing CO_2 emissions, it is promising as an effective solution to the serious and complicated interplay of two critical factors: food shortages and increased CO_2 emissions. We are now continuing our efforts to scale-up the above two processes from bench-scale to practical-sized cells under the auspices of a Japanese national project conducted by NEDO, the results of which will be reported in due course.

Electrochemical formation of carbon film

The electrochemical formation of carbon film can be achieved either by anodic oxidation of a C_2^{2-} ion or by cathodic reduction of a CO_3^{2-} ion.[8,9] The principle of the electrochemical formation of carbon film by the anodic oxidation of C_2^{2-} ion is shown in Fig. 8.

Fig. 9 shows typical cyclic voltammograms for the W electrode before and after adding CaC_2 to the LiCl–KCl–$CaCl_2$ melt at 773 K. After adding CaC_2, anodic currents arise at about 0.5 V. These anodic currents are regarded to correspond to the anodic oxidation of C_2^{2-} ion.

As shown in Fig. 10, an aluminium plate can be coated with a dense and adherent carbon film by anodic oxidation of the C_2^{2-} ion.

A cross-sectional transmission electron microscope (TEM) image of thin and dense film electrochemically formed on a stainless steel substrate is shown in Fig. 11.

The obtained film exhibited excellent adherence as shown in Fig. 12.

A simple corrosion test using an aqueous $FeCl_3$ solution (Fig. 13) as well as the anodic polarization measurement in a sulfuric acid solution (Fig. 14) demonstrated the carbon-coated plate's high corrosion resistance.

The contact electrical resistance measurement of the carbon-coated metal plate using the apparatus shown in Fig. 15 also produced very promising results as shown in Table 2. Consequently, the obtained carbon-coated metal actually has highly advantageous features for a fuel cell separator, a current collector of an Li-ion secondary battery/capacitor (Fig. 16), or a substitute for fluoric resin film used in a semi-conductor/liquid crystal production line.

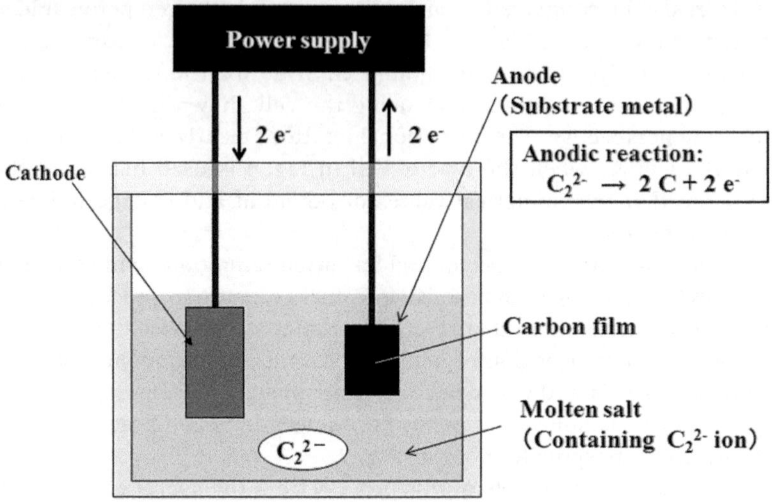

Fig. 8 Principle of carbon film formation by anodic oxidation of carbide ion.

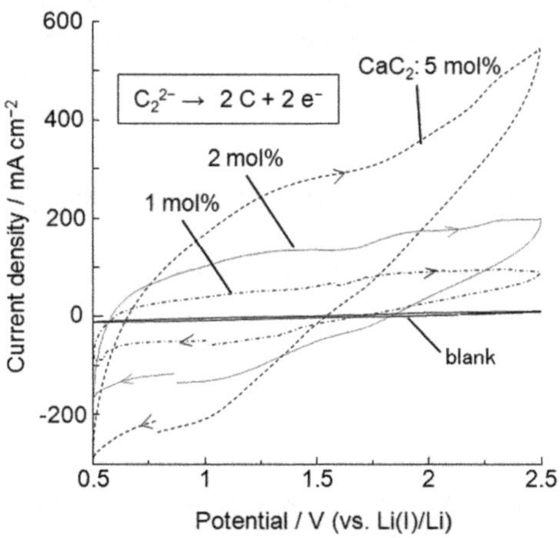

Fig. 9 Cyclic voltammograms for the W electrode in LiCl–KCl–$CaCl_2$ melt before and after adding CaC_2 (1, 2, 5 mol%) at 773 K; scan rate: 0.01 V s^{-1}.

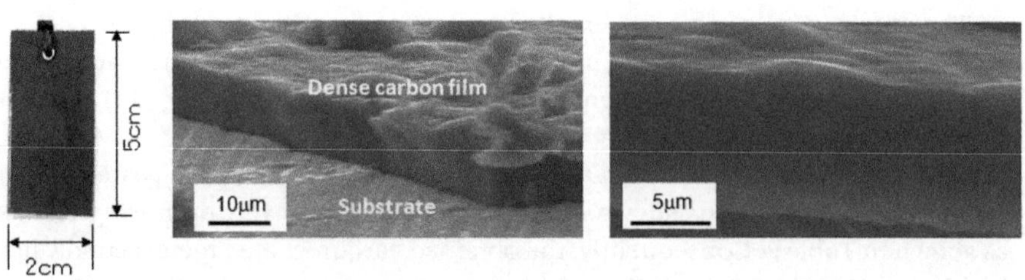

Fig. 10 Dense and adherent carbon film electrochemically formed on aluminium plate.

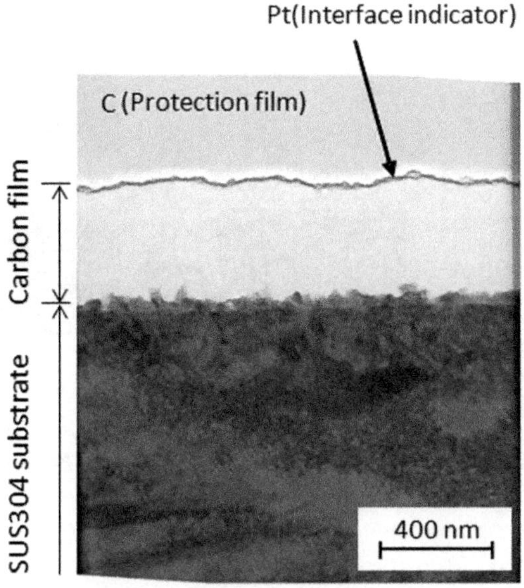

Fig. 11 Cross-sectional TEM image of carbon film electrochemically formed on stainless steel.

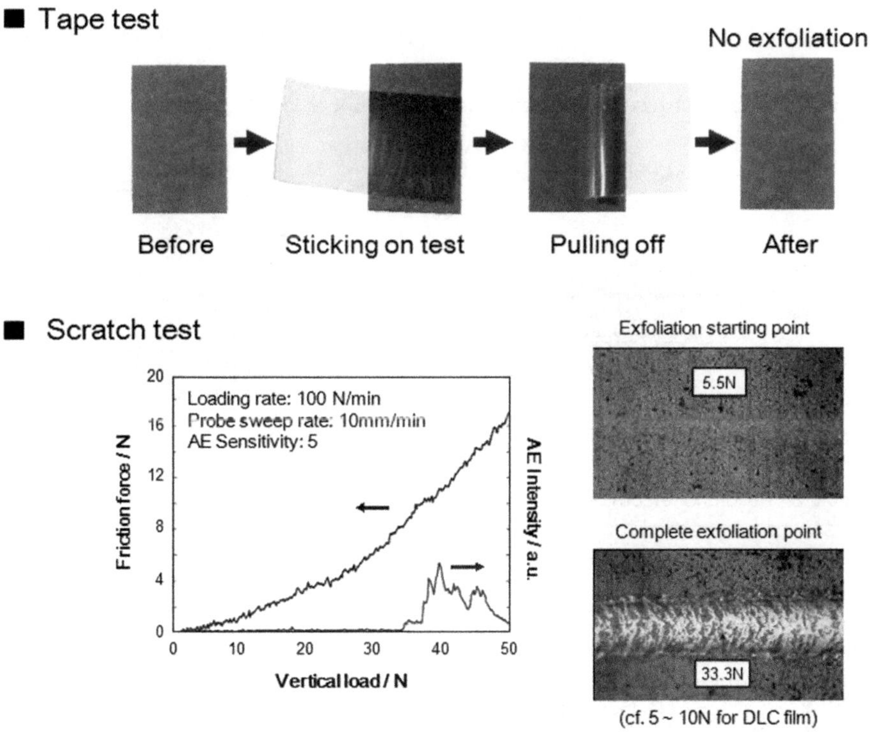

Fig. 12 Excellent adherence of an electrochemically formed carbon film.

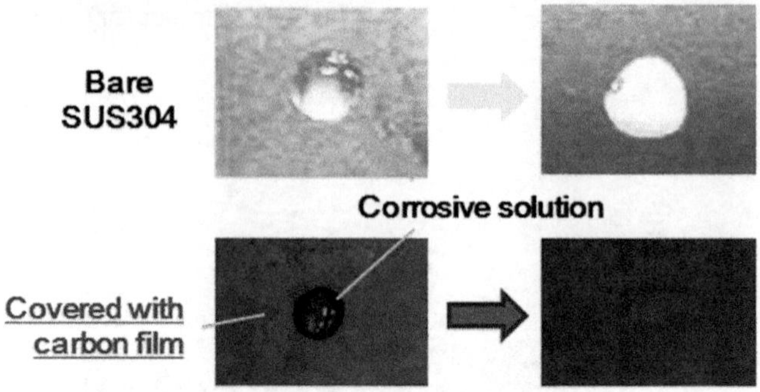

Fig. 13 Simple corrosion test ($FeCl_3$(aq.), substrate: SUS304).

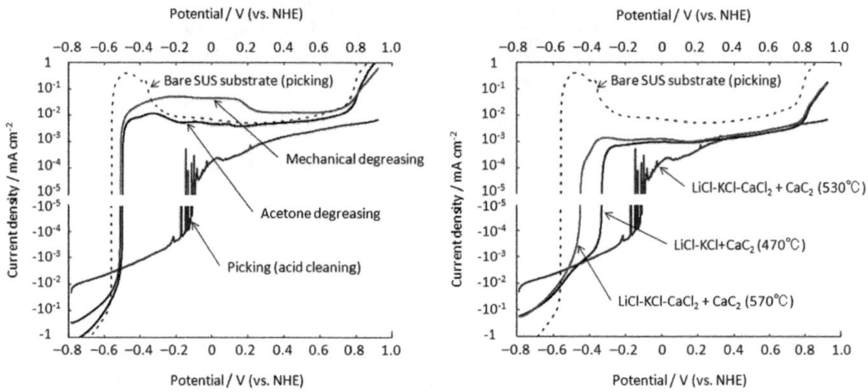

Fig. 14 Anodic polarization characteristics in a sulfuric acid solution.

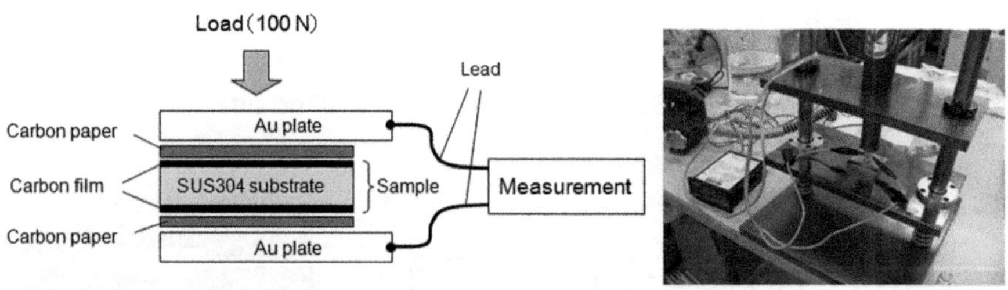

Fig. 15 Apparatus for contact electrical resistance measurement.

Table 2 Results of electrical resistance measurement

Electrolytic bath (pre-treatment)	Contact resistance before/after corrosion test mΩ cm^2	
	Before	After
SUS304 substrate (sulfuric acid)	110	220
LiCl–KCl–$CaCl_2$ + CaC_2 (530 °C) (sulfuric acid)	10	10
LiCl–KCl–$CaCl_2$ + CaC_2 (570 °C) (sulfuric acid)	7	6

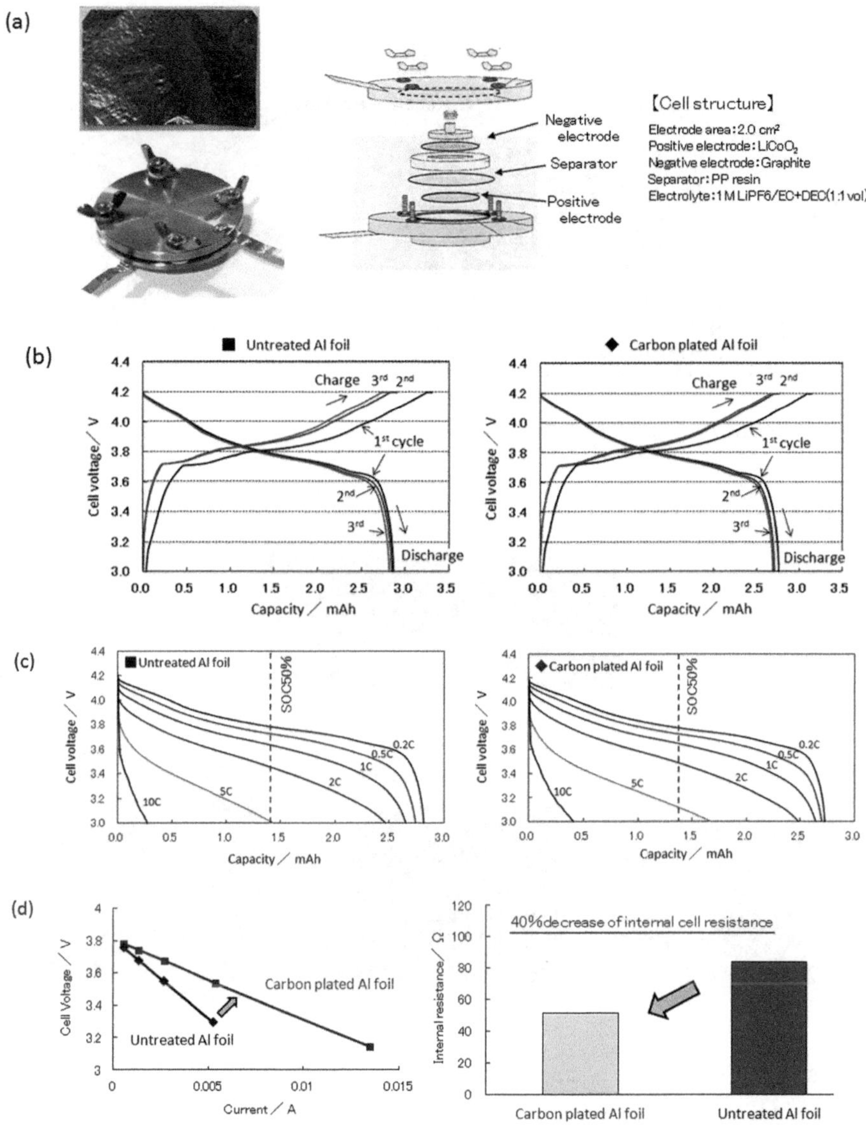

Fig. 16 Effect of carbon plating on performance of a Li-ion secondary battery: (a) photo (left) of a test cell and its structure (right); (b) charge–discharge characteristics at constant current of 0.2 C; (c) discharge characteristics at various rates; (d) current–voltage characteristics (left) and internal resistance (right) at SOC 50%.

Currently, we produce large (A4-size) carbon coated metal plates using the bench-scale carbon plating apparatus shown in Fig. 17. Carbon-plated large-sized samples are shown in Fig. 18.

Furthermore, we can even coat the surface of a screw thread with a thin and dense adherent carbon film, as well as the interior of a porous body (Fig. 19(a) and (b)). Therefore, this process can be used to produce corrosion-resistant fastening components (nuts and bolts) and metallic porous bodies.

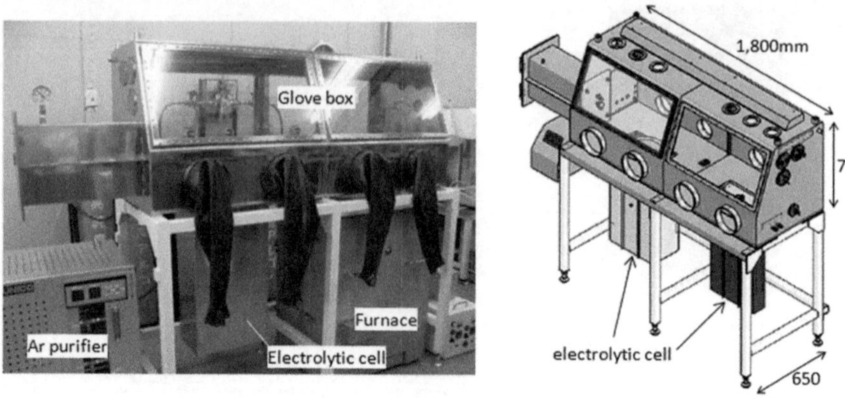

Fig. 17 Bench-scale carbon plating apparatus.

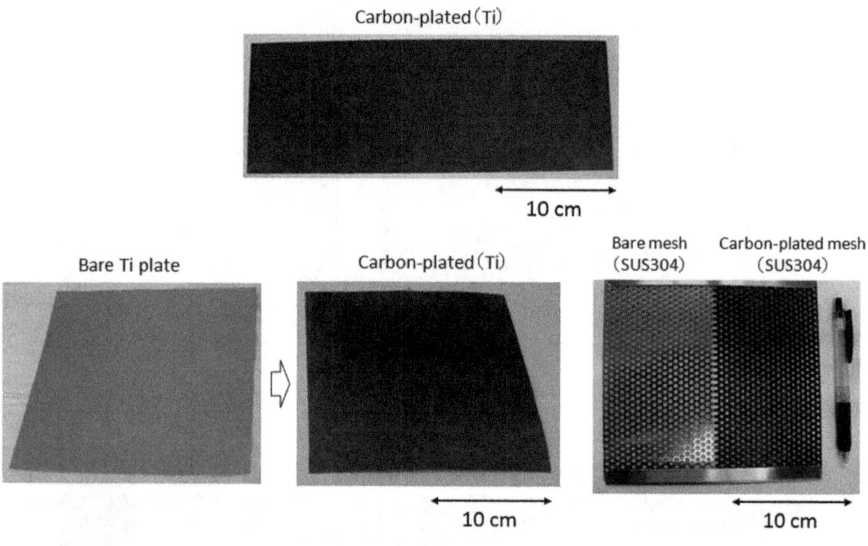

Fig. 18 Carbon-plated large-sized samples.

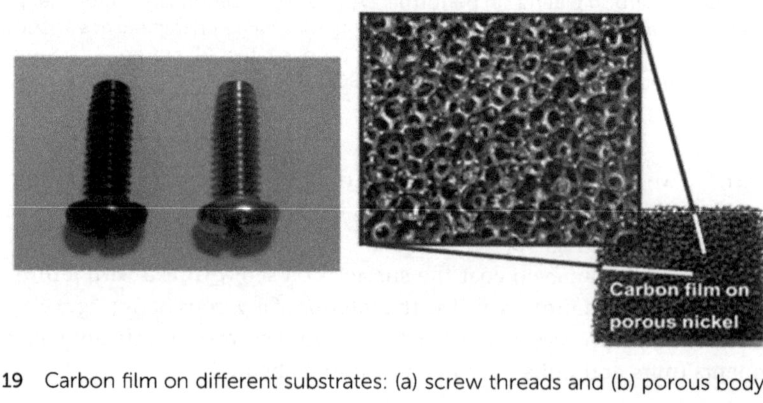

Fig. 19 Carbon film on different substrates: (a) screw threads and (b) porous body.

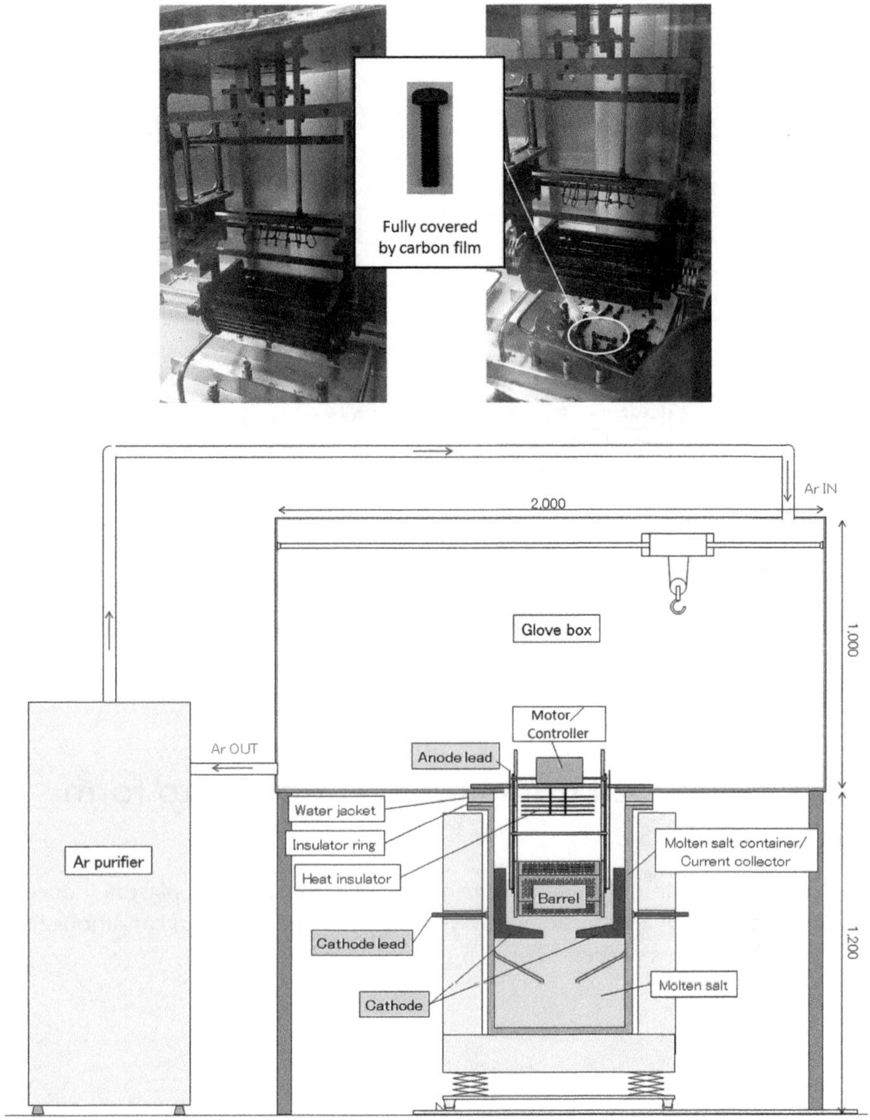

Fig. 20 Bench-scale barrel-type carbon plating apparatus.

The bench-scale barrel-type plating apparatus could successfully perform carbon plating of fastening components (Fig. 20). However, some problems remain to be solved to obtain high-quality products.

This carbon-plating technology can also be applied to the materials used in the field of mechanical engineering. As an example, the experimental results of nano-indentation are shown in Fig. 21.

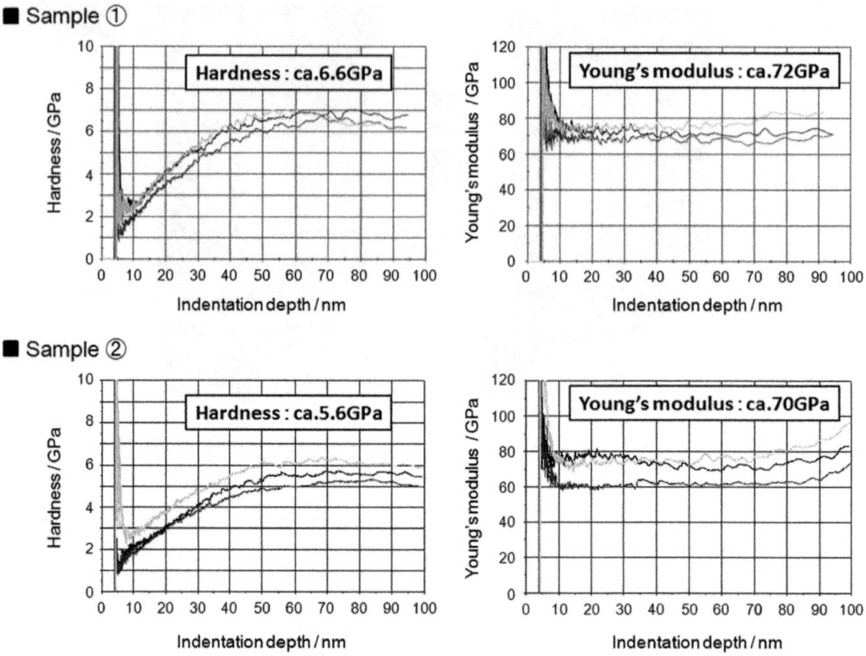

Fig. 21 Nano-indentation measurements of carbon-plated SUS304.

Plasma-induced discharge electrolysis to form nanoparticles

Even when a single electrode is outside the electrolytic bath, plasma-induced stationary discharge can make electrolysis possible under certain conditions.[10,11] The principle of this process is shown in Fig. 22.

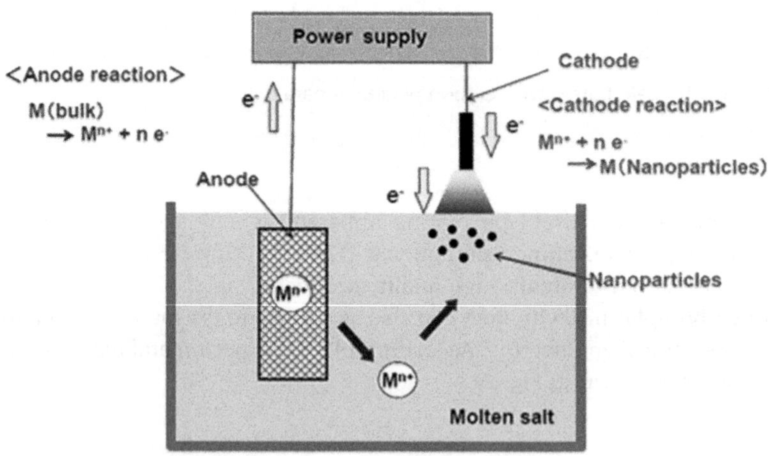

Fig. 22 Principle of plasma-induced discharge electrolysis.

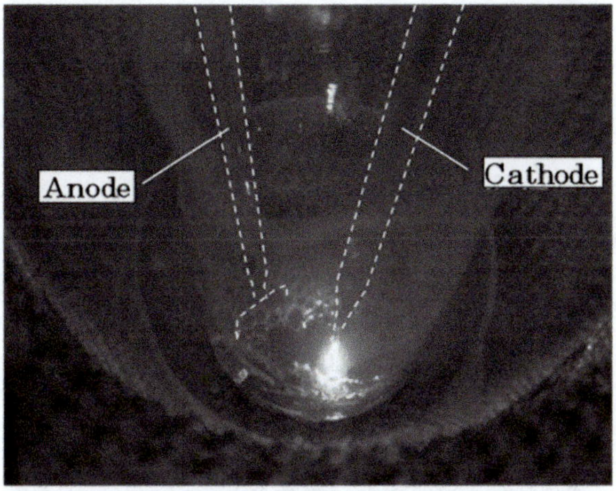

Fig. 23 Photo of cathodic plasma-induced discharge electrolysis.

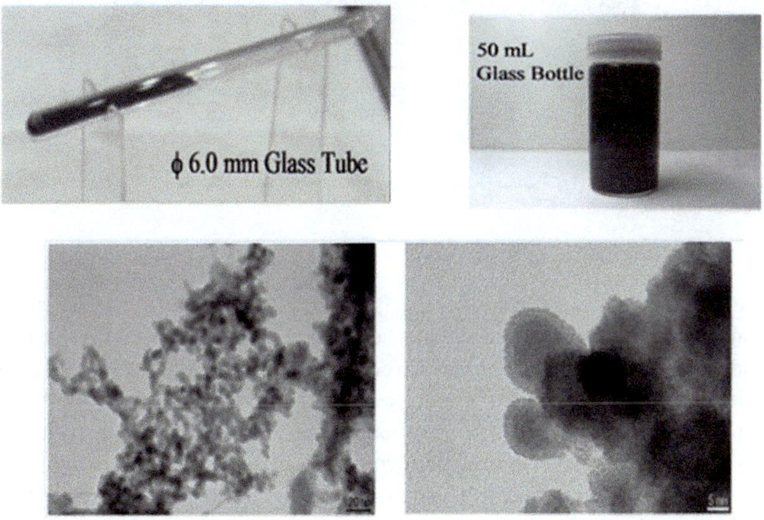

Electrolytic current 1A, 15,000C (4hours)

Fig. 24 Visual (upper) and TEM (lower) observations of Ta nanoparticle: electrolytic current 1 A, 15 000 C (4 hours).

A photograph of the interior of the electrolytic cell under stationary discharge is shown in Fig. 23.

With this discharge electrolysis method, we have so far obtained various types of metal and alloy nanoparticles: C, Al, Si, Ti, Zr, Nb, Fe, Co, Ni, Ta, Ag, Pt, NiPd, FePt, $CoPt_3$, CoPt, Co_3Pt, $SmCo_2$, $SmCo_3$, *etc.* As an example, visual and TEM observations of the obtained Ta nanoparticle are shown in Fig. 24.

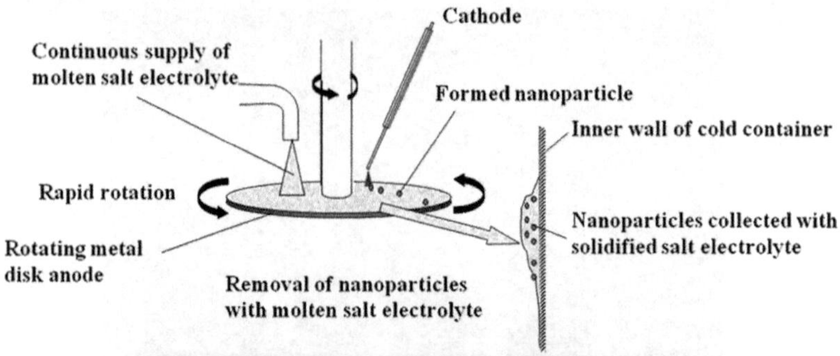

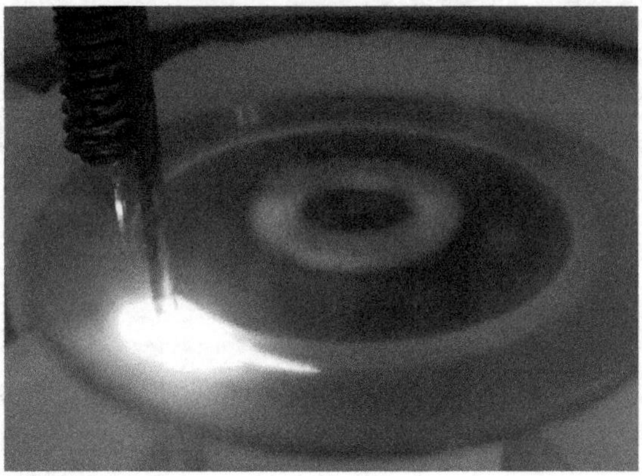

Fig. 25 Schematic and photo during operation of rotating disk anode-type plasma-induced electrolytic cell.

The primary nanoparticles sized at 5–20 nm are first formed just under the discharge level and aggregated to form secondary nanoparticles with diameters of around 100 nm. This growth process proceeds in the area just under the discharge level during the initial stage of electrolysis. Based on this observation, in order to obtain finer and more uniform nanoparticles, we constructed a rotating disk anode-type electrolytic cell and confirmed its capabilities, choosing Ni nanoparticle formation as an example task. A schematic and a photo of the cell during operation are shown in Fig. 25.

Using this type of cell, we succeeded in obtaining Ni nanoparticles with diameters of less than 10 nm at a rotation speed of 2500 rpm. Encouraged by these results, we conducted scale-up experiments choosing Ta nanoparticle production as an example task by using the bench-scale apparatus shown in Fig. 26.

Fig. 27 and Table 3 show experimental results using the bench-scale Ta particle formation apparatus.

The production of finer and uniform Ta nanoparticles is required for a high-power and high-energy-density tantalum capacitor, which is expected as a future power electronics device. Moreover, by utilizing this process, recycling and effective reuse of Ta can be achieved simultaneously. In this case, Ta scrap in the anode vessel is anodically dissolved in the molten salt to provide Ta(IV) ions,

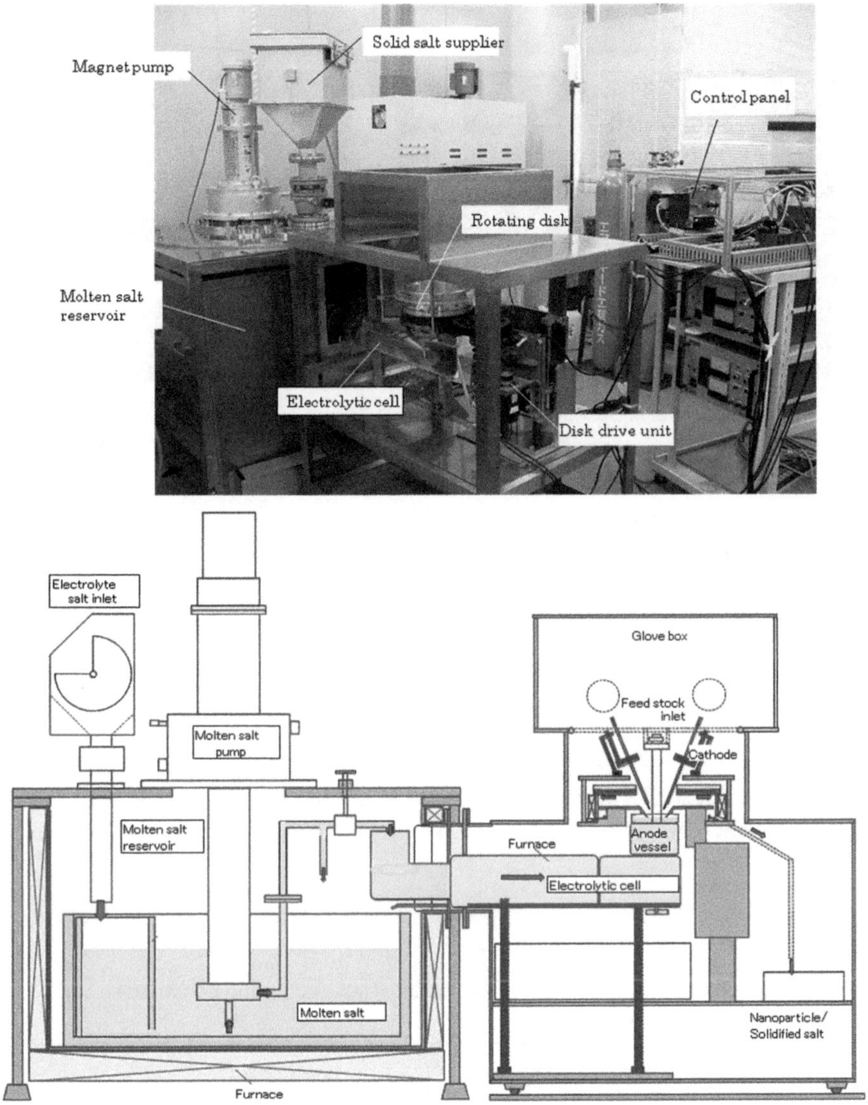

Fig. 26 Photo and schematic of bench-scale nanoparticle formation apparatus (production capacity: 200 g per day).

which are changed to Ta nanoparticles by cathodic discharge. The structure of an anode vessel for holding the Ta scrap is shown in Fig. 28.

Based on the results of the above bench-scale experiment, we have designed an apparatus capable of a further scale-up (Fig. 29), with a production capacity of 7 kg per day.

It must be emphasized that the process and the above apparatuses are also available for the recycling and effective reuse of various other crucial metals.

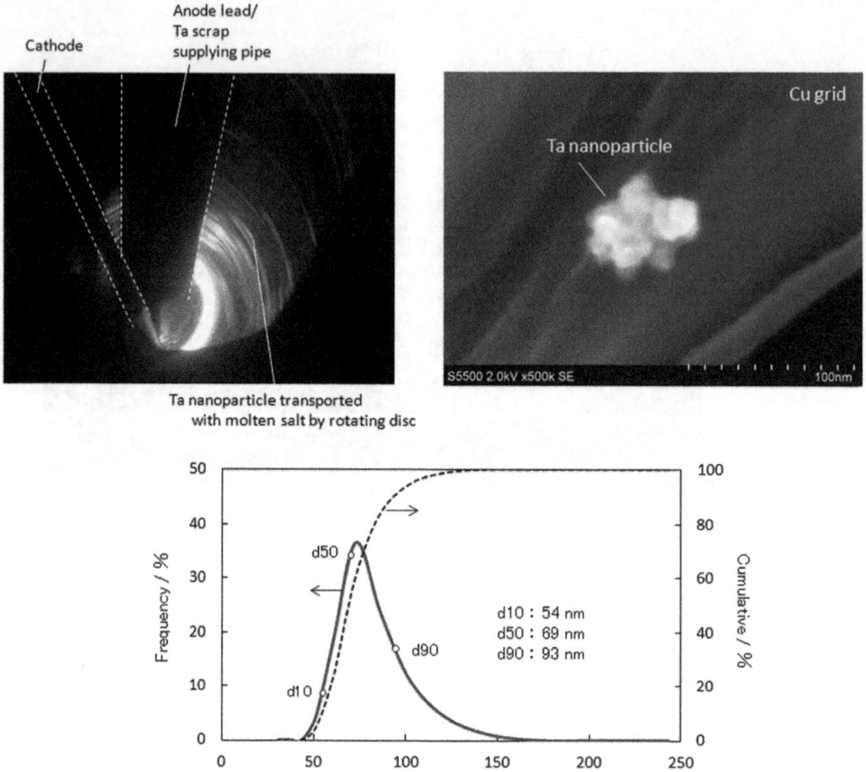

Fig. 27 Ta nanoparticles formed by plasma-induced cathodic discharge electrolysis using the bench-scale apparatus.

Table 3 Specifications and experimental results of labo-scale and bench-scale apparatus for nanoparticle formation

	Labo-scale	Bench-scale
Apparatus	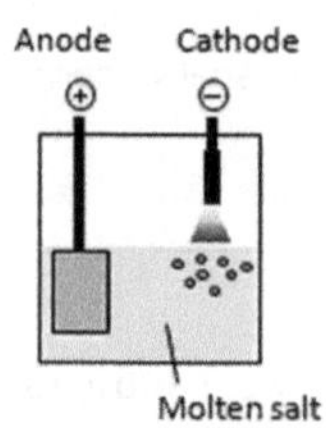	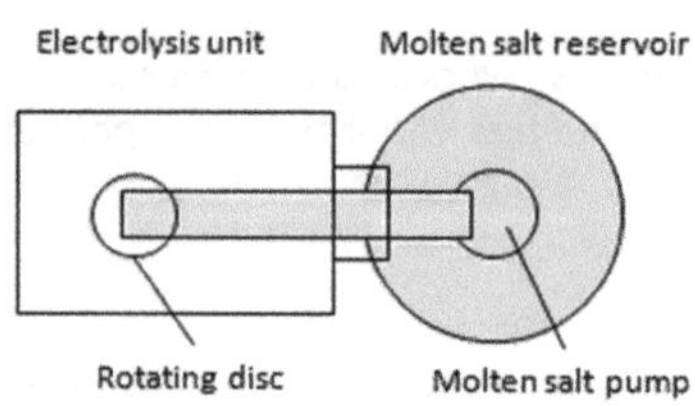
Number of cathodes	1	2 (max: 4)
Applied current	1 A	10 A (5 A × 2)
Production (Ta)	38 g per day	200 g per day
Particle size (d50)	140–200 nm	69 nm
Distribution (d90–d10)	>100 nm	39 nm

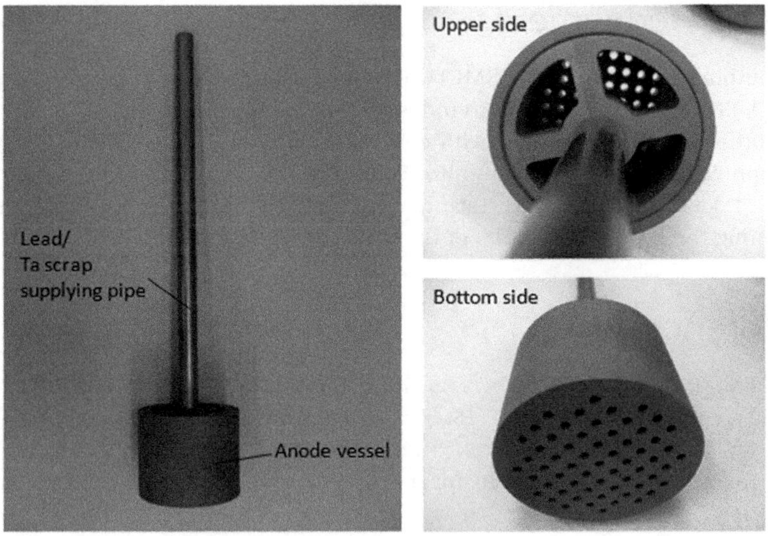

Fig. 28 Structure of anode vessel for holding Ta scrap.

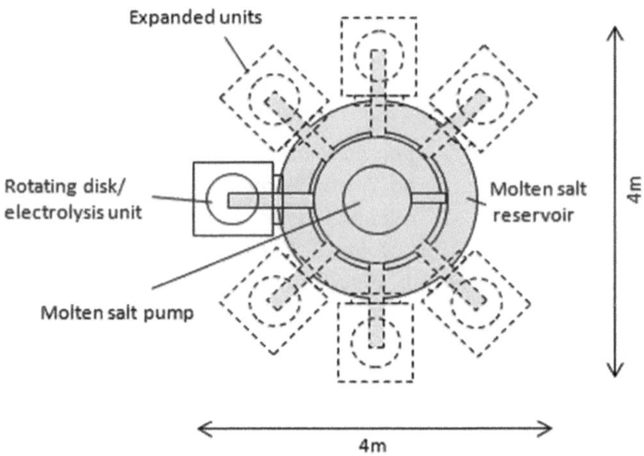

Fig. 29 Basic design of an apparatus with a 7 kg per day production capacity.

Conclusions

Although our development efforts to industrialize novel molten salt electrochemical processes are still at a bench-scale stage, we believe these processes certainly have potential to be industrialized, based on study and calculations from both technological and economic viewpoints. Furthermore, we are continuing to pursue the challenge of industrializing them in the near future. This work will contribute to the cultivation of new industrial fields of energy and materials and provide an effective solution for a future low-carbon society.

Acknowledgements

The authors greatly appreciate METI, The Small and Medium Enterprise Agency, NEDO, JST, Kyoto-city, and Kyoto Industrial Support Organization 21 for strongly supporting our R&D activities with encouragement and funds. We also express our appreciation to Takao Kunihiro, Seiji Ohata, Yoshikazu Hyodo, Ryo Nakanishi, Takaaki Hanada, and Tatsuro Sasa for their valuable contributions to obtaining the promising results of the bench-scale experiments.

References

1 Y. Ito, *Molten Salt and High Temperature Chemistry*, Yoyuen-to-Kouon-kagaku, 2003, vol. 46, no. 2, pp. 97–118.
2 Y. Ito, *Trans. Electrochem. Soc.*, 2012, **41**(46), 43–55.
3 Y. Ito, Molten Salt Electrochemical Processes Directed Toward a Low Carbon Society, *Molten Salt Chemistry and Technology*, ed. M. Gaune-Escard and G. M. Haarberg, John Wiley and Sons, 2014, ch. 7.1.
4 N. Serizawa, H. Miyashiro, K. Takei, T. Ikezumi, T. Nishikiori and Y. Ito, *J. Electrochem. Soc.*, 2010, **159**(4), E87–E91.
5 NEDO, *Report of WE-NET Subtask 3*, NEDO-WE-NET-9631, 1997.
6 J. C. Ganley, J. H. Holbrook and D. E. McKinley, *2007 Annual NH3 Fuel Conference*, San Francisco, October 15–16, 2007.
7 J. Gosnell, *2005 Annual NH3 Fuel Conference*, Argonne National Laboratory, October 13–14, 2005.
8 Y. Ito, *Tanso*, 2011, (28), 144–151.
9 H. Kawamura and Y. Ito, *J. Appl. Electrochem.*, 2000, **30**, 571–574.
10 H. Kawamura, K. Moritani and Y. Ito, *Plasmas & Ions*, 1998, **1**(1), 29–36.
11 Y. Ito, T. Nishikiori and M. Tokushige, Plasma Induced Discharge Electrolysis for Nanoparticles Production, *Molten Salts Chemistry: From Lab to Applications*, ed. F. Lantelme and H. Groult, Elsevier, 2013, pp. 269–286.

Faraday Discussions

PAPER

Determination and evaluation of the thermophysical properties of an alkali carbonate eutectic molten salt†

Xuehui An, Jinhui Cheng, Peng Zhang,* Zhongfeng Tang and Jianqiang Wang*

Received 26th December 2015, Accepted 11th February 2016
DOI: 10.1039/c5fd00236b

The thermal physical properties of Li_2CO_3–Na_2CO_3–K_2CO_3 eutectic molten salt were comprehensively investigated. It was found that the liquid salt can remain stable up to 658 °C (the onset temperature of decomposition) by thermal analysis, and so the investigations on its thermal physical parameters were undertaken from room temperature to 658 °C. The density was determined using a self-developed device, with an uncertainty of ±0.00712 g cm^{-3}. A cooling curve was obtained from the instrument, giving the liquidus temperature. For the first time, we report the obtainment of the thermal diffusivity using a laser flash method based on a special crucible design and establishment of a specific sample preparation method. Furthermore, the specific heat capacity was also obtained by use of DSC, and combined with thermal diffusivity and density, was used to calculate the thermal conductivity. We additionally built a rotating viscometer with high precision in order to determine the molten salt viscosity. All of these parameters play an important part in the energy storage and transfer calculation and safety evaluation for a system.

1. Introduction

It is an urgent task and a worldwide demand for human beings to develop renewable, sustainable and green energy. Recently, concentrating solar power (CSP) systems have aroused many researchers' interests as they present a feasible way to provide constant power without producing pollution. However, there are many practical issues that need to be resolved before their commercial promotion, such as the need for lower melting points and higher thermal stabilities. Finding the appropriate heat storage and transfer media is one of the key technological issues surrounding the application of CSP systems. Due to its low melting point, molten nitrate/nitrite salt is extensively used in CSP demonstration

Shanghai Institute of Applied Physics, Chinese Academy of Sciences, Shanghai 201800, China. E-mail: wangjianqiang@sinap.ac.cn; zhangpeng@sinap.ac.cn; Fax: +86 21 39194051; Tel: +86 21 39194051

† Electronic supplementary information (ESI) available. See DOI: 10.1039/c5fd00236b

facilities, however, its low decomposition temperature (lower than 550 °C) hinders its application. Fluoride salt can be used at high temperatures (above 700 °C), but it is so corrosive that compatible materials are rarely available. Eutectic molten carbonate salts are regarded as promising candidates for use in thermal energy storage (TES) and heat transfer fluid (HTF) materials, due to their low cost and excellent performance.[1,2]

As media for TES and HTF materials, the densities, viscosities, thermal conductivities and specific heat capacities of eutectic molten carbonate salts are key thermophysical properties for evaluating thermal storage performance, heat transfer calculations, and safety analysis. The physical properties of molten alkali carbonate binaries, such as their densities, surface tensions and electrical conductivities, have been studied extensively in order to support their applications.[3–11] However, the thermal physical properties of multi-component carbonates are sparsely reported. In particular, 32.12 wt% Li_2CO_3–33.36 wt% Na_2CO_3–34.52 wt % K_2CO_3 (denoted as $(LiNaK)_2CO_3$) eutectic molten salt has become a promising candidate for TES and HTF materials because of its low melting point and low cost. However, systematic study of $(LiNaK)_2CO_3$ is lacking and the thermal physical properties of $(LiNaK)_2CO_3$ that have been reported in the literature are inconsistent. Ward and Janz[8] reported that the density of the Li_2CO_3–Na_2CO_3–K_2CO_3 ternary system is approximately the sum of the densities of its sub-binary systems. The fusion enthalpy and entropy of the alkali carbonates were reported to be 276.1 $\pm$ 4.2 J g^{-1} and 0.414 $\pm$ 0.00418 J g^{-1} °C^{-1}, respectively.[9] Olivares *et al.*[12] concluded that the $(LiNaK)_2CO_3$ molten salt was most stable under a CO_2 atmosphere, among blanket atmospheres of CO_2, argon and air. Kojima *et al.*[13] measured its density, surface tension and electrical conductivity. Liao *et al.*[14] reported the specific heat capacities, densities, viscosities and thermal stabilities of the LiNaK eutectic carbonates, but a large deviation was presented. The densities of the $CaCO_3$–Na_2CO_3–Li_2CO_3–K_2CO_3 quaternary system and its sub-systems were measured at 555–969 °C using the double-bob Archimedean method.[15] The viscosity of the Na_2CO_3–Li_2CO_3–K_2CO_3 eutectic molten salt obtained from the oscillating crucible technique shows a large discrepancy. Recently, the rotation method has been developed for measuring the viscosities of alkali carbonates.[16]

Most of the studies on the $(LiNaK)_2CO_3$ molten salt focused on measuring density, surface tension and electrical conductivity, but viscosity, specific heat capacity and thermal conductivity, which are important properties with respect to performance in thermal storage or heat transfer systems, received less attention. In particular, limited information on the thermal conductivity of the $(LiNaK)_2CO_3$ molten salt has been available until now. In the present work, the thermal physical properties of $(LiNaK)_2CO_3$ molten salt and methods for measuring the properties of the alkali carbonate are comprehensively investigated, including with respect to specific heat capacity, thermal conductivity, viscosity, density, liquid temperature, and thermal stability.

2. Experimental details

2.1 Sample preparation

Raw materials K_2CO_3 (99.99%, from Ourchem), Li_2CO_3 and Na_2CO_3 (99.9% from Sinopharm Chemical Reagent Co. Ltd) were directly used in this work without further purification. These salts were dried separately at 200 °C under an argon atmosphere for 24 h, and then prepared in a weight ratio of 32.12% Li_2CO_3, 33.36% Na_2CO_3 and 34.52% K_2CO_3 for different thermophysical parameter determinations.

2.2 Measurements of thermal stability and specific heat capacity

Thermal stability, including simultaneous measurements for weight loss, temperature and enthalpy of phase transition, was evaluated using simultaneous thermal analysis (STA; Netzsch 449 F3) and the specific heat capacity of the $(LiNaK)_2CO_3$ molten salt was measured by differential scanning calorimetry (DSC; Netzsch 404 F3). Six certified reference materials, In, Bi, Sn, Zn, Al and Au, were used to calibrate the STA and DSC instruments at a temperature range of 100–1200 °C. Determinations of the thermal stability, melting temperature, melting enthalpy and specific heat capacity of $(LiNaK)_2CO_3$ were performed after calibration. Both the calibration and the measurements were carried out under a blanket Ar atmosphere with a heating rate of 10 K min^{-1}.

The specific heat capacity measurement was conducted through three steps.[16] Firstly, two empty crucibles with the same weight were tested; secondly, monocrystalline sapphire used as a reference material was measured; and finally, the $(LiNaK)_2CO_3$ molten salt was examined. It should be noted that the three steps were carried out under the same conditions and using the same graphite crucibles. The specific heat capacity was obtained from the ratio method[17] (eqn (1)), and data analysis was performed automatically using the Netzsch Proteus program.

$$c_p = \frac{m_R}{m_S} \times \frac{DSC_S - DSC_B}{DSC_R - DSC_B} \times c_{pR} \tag{1}$$

where DSC (μV) is the DSC signal, m (g) represents the mass, and c_p (J g^{-1} K^{-1}) denotes the specific heat capacity. The subscripts B, R and S denote the corresponding parameters of the baseline, the reference and the sample, respectively.

2.3 Measurement of thermal diffusivity

The thermal diffusivity was determined with the help of a special specimen cell and with specific sample preparation. The crucible with a lid has an inner diameter of 23 mm and a height of 8 mm. The bottom of the crucible is 0.6 ± 0.05 mm thick, which is the same thickness as the lid. The diameter of the lid is 4 mm smaller than that of the crucible, forming a space for avoiding molten salt overflow when the temperature increases. Further details on the crucible are described in ref. 18.

About 2 g of $(LiNaK)_2CO_3$ powder was pressed into a disc with a diameter of 23 mm and a thickness of 1.5–2.0 mm. The disc was put in the crucible and transferred to a vacuum furnace. Then, the molten salt was heated up to 450 °C and kept at that temperature for at least 1 hour in order to make the sample homogeneous, before finally cooling it down to room temperature. This operation was repeated at least 3 times to ensure the homogeneity of the specimen.

The crucible containing the pretreated $(LiNaK)_2CO_3$ sample disc was placed in the sample holder of the Linseis LFA 1000. The bottom of the crucible was heated by a single laser beam under a helium atmosphere, and the temperature change in the lid was detected by an infrared detector and plotted as a function of time.

The thermal diffusivity was obtained by use of a combined model, with consideration for heat loss and infinite pulse time. Details on the combined model are presented in ref. 19. Data analysis was performed automatically by the evaluation module in the LFEAuto1061 program. The reliabilities of the crucible and sample preparation have been verified in previous work.[18]

2.4 Measurement of viscosity

Determination of the viscosity was realized by an improved rotational coaxial cylinder viscometer based on Brookfield DV-III. It can be used for measuring fluids with viscosities as small as 1.6 mPa s. The $(LiNaK)_2CO_3$ salt was placed in a graphite crucible with an inner diameter of 29 mm and a depth of 220 mm, and then transferred into a furnace. An inner cylinder with a diameter of 26 mm and a height of 150 mm was placed in the graphite crucible filled with $(LiNaK)_2CO_3$ salt. It should be noted that an appropriate amount of $(LiNaK)_2CO_3$ salt is required so that the inner cylinder is completely submerged in the $(LiNaK)_2CO_3$ molten salt. The inner cylinder was rotated at a speed of 15 rpm (revolutions per minute) to keep a laminar flow. The torque was recorded at target temperatures measured by a K-type thermocouple which was placed in the molten salt. The viscosity was calculated using eqn (2).

$$\eta = \frac{\text{SMC} \times Y(\text{torque\%}) \times \varepsilon \times 100}{N(\text{rpm})} \tag{2}$$

where η is the viscosity; $\text{SMC} = \frac{1}{4\pi(h+c)}\left(\frac{1}{r^2} - \frac{1}{R^2}\right)$ is a constant related to the dimensions of the crucible and the rotator, and is calibrated using standard viscosity oil, where h and r represent the length and radius of the rotator, R is the radius of the crucible and c is a constant; Y is the torque expressed in percentage; ε is a constant and N is the rotating speed.

2.5 Measurements of density and liquidus temperature

Measurements of density and liquidus temperature were performed using the same device. The densitometer was built by ourselves and is based on the Archimedes technique. Details of the densitometer are provided in ref. 20. About 200 g of molten salt powder or ingot were weighed and melted in a furnace. Once the salt was homogeneously molten, a platinum cylinder connected to the balance with a platinum wire was completely dipped into the molten salt. The system was gradually cooled down and the balance values were recorded at target temperatures. The density of the molten carbonate salt was obtained using eqn (3).

$$\rho = \frac{m_0 - m_1 + \pi D\gamma/g}{V_0[1 + \beta(T - 25)]^3} \tag{3}$$

where ρ is the density; m_0 and m_1 are the weights of the platinum cylinder in air and in molten salt, respectively; V_0 represents the volume of the platinum cylinder in distilled water, which was calibrated before measuring the molten carbonate salt density; γ denotes the surface tension of the molten salt; D is the diameter of the platinum wire ($D = 0.2$ mm), g represents the gravitational constant; T is the temperature of the molten salt; and β is the linear expansion coefficient of platinum ($\beta = 9 \times 10^{-6}$ 1/°C).

The liquidus temperature was measured using the cooling curve method. An appropriate amount of molten salt powder or ingot was placed in the crucible and heated to 50 °C above the melting point. At least 2 hours was required in order for the temperature to be uniform throughout the sample, after which the molten salt was mixed using a graphite stirrer attached to a stainless steel rod driven by an electric motor through a hole (about 10 mm) in the crucible lid. When the molten salt reached the required temperature, its temperature began to decrease at a rate of 1 K min^{-1}. The temperature cooling curve was measured by a K-type thermocouple and recorded by a computer.

2.6 Uncertainty analysis

The uncertainty analysis of the experimental data included three steps. Firstly, the uncertainty sources (x_i, x_j) had to be determined. Secondly, the standard uncertainty needed to be calculated, including the standard deviation and other random uncertainties, such as probability distribution, statistical distribution, and other factors. Finally, the expanded uncertainty was represented as a confidence factor multiplied by the standard uncertainty, and was evaluated according to eqn (4).[21]

$$U = k\sqrt{\sum_{i=1}^{N}\left(\frac{\partial f}{\partial x_i}\right)^2 (u_{x_i})^2 + 2\sum_{1\leq i<j}^{N}\frac{\partial f}{\partial x_i}\frac{\partial f}{\partial x_j}\varepsilon_{ij}u_{x_i}u_{x_j}} \tag{4}$$

where U is the expanded uncertainty; u_{x_i} and u_{x_j} represent the standard uncertainty components caused by x_i and x_j, respectively; f represents the function of the parameters measured in this work; ε_{ij} is the correlation coefficient between x_i and x_j; and k is the confidence factor, which generally takes a value of 2. All of the devices used for the experiments, such as the balance, micrometer, DSC, viscometer, densitometer and LFA 1000, were calibrated using a special metrological instrument.

3. Results and discussion

3.1 Thermal analysis

3.1.1 Thermal stability. Melting information and the thermal stability of the $(LiNaK)_2CO_3$ molten salt were obtained using STA in an inert argon atmosphere. The results are presented in Fig. 1, which shows that two peaks are observed in the DSC curve (Fig. 1(a)). The first peak at 396.8 °C is due to melting, and the second peak at 944.3 °C is caused by decomposition. The melting peak indicated that the $(LiNaK)_2CO_3$ molten salt melts at 396.8 °C and that the melting enthalpy is −278.2 J g^{-1}. These findings are in satisfactory agreement with the literature values of 397.0 °C and −276.1 J g.[9] The second peak can be exhaustively interpreted by thermogravimetry and differential thermogravimetry curves, which are shown in Fig. 1(b) and (c), respectively. Fig. 1(b) indicates that the $(LiNaK)_2CO_3$ molten salt begins to decompose at 658 °C, and that the weight loss is less than 0.05% up to 794 °C. The decomposition rate reaches a maximum of 18.7% per minute at about 944.3 °C (see Fig. 1(c)). When the temperature reaches as high as 1010 °C, the

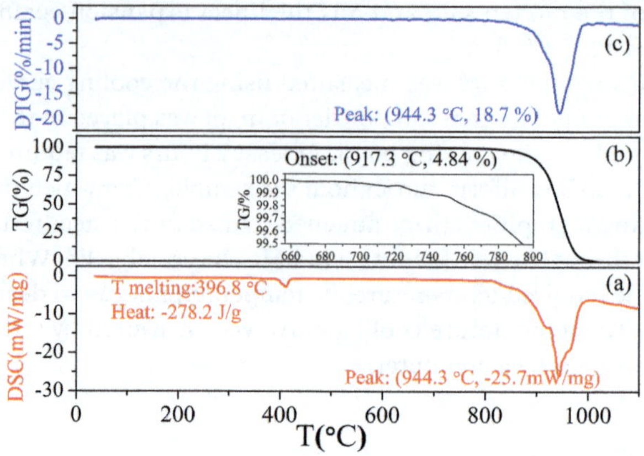

Fig. 1 Thermal analysis for the $(LiNaK)_2CO_3$ molten salt.

molten salt is completely decomposed. Therefore, it can be concluded that the $(LiNaK)_2CO_3$ molten salt remains stable below 800 °C. The thermal physical properties were measured below 650 °C to ensure a reliable result, as some parameters are seriously affected by mass loss.

3.1.2 Liquidus temperature. The reliability of the liquidus temperature measurement was verified using pure NaCl salt before the $(LiNaK)_2CO_3$ measurement was performed. The liquidus temperature of NaCl was determined to be 801.7 °C ± 0.5 °C, which is in good agreement with the literature values of 801 °C or 802 °C.[22]

The cooling curve of the $(LiNaK)_2CO_3$ molten salt is shown in Fig. 2, which presents a short plateau at 398.7 °C caused by precipitation. Inspection of Fig. 2 shows that the liquidus temperature is 398.7 °C, which is 1.9 °C higher than the melting temperature obtained from the STA measurement. The liquidus temperature was derived from a cooling process, whereas thermal analysis is a heating process. Due to the supercooling and superheating that occurs during the cooling and heating processes, the melting temperature may not be equal to the liquidus temperature. Therefore, the difference between the two methods is

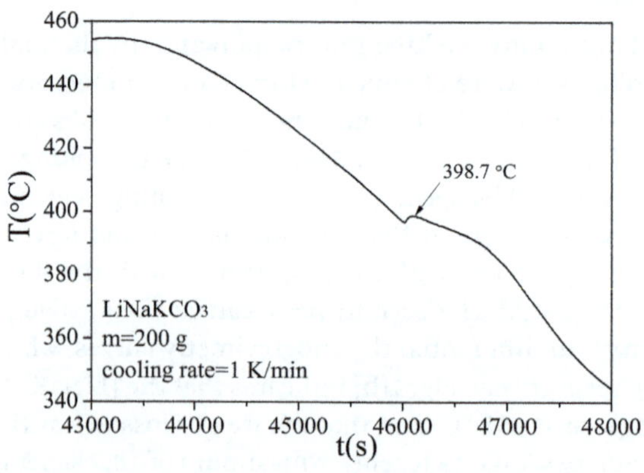

Fig. 2 The cooling curve of the $(LiNaK)_2CO_3$ molten salt.

insignificant. Additionally, only one plateau is observed in Fig. 2, indicating that the prepared sample is at its eutectic composition.

3.1.3 Specific heat capacity. The specific heat capacity obtained from the DSC technique is presented in Fig. 3, and the value can be regarded to be constant at 1.61 ± 0.08 J g^{-1} K^{-1} at the temperature range of 450 °C to 600 °C. In ref. 9, the specific heat capacity was determined by the drop calorimetry method under a CO_2 atmosphere, and it was plotted as a function of temperature, *i.e.* $c_p = 1.377 + 0.0006926T$ (T = 397–827 °C), and was found to increase slightly with increasing temperature. A much larger specific heat capacity for the $(LiNaK)_2CO_3$ molten salt was reported by Liao *et al.* using the STA technique: $c_p = -1.195 + 0.0078T$ (T = 430–750 °C).[14] Fig. 3 summarizes the results available and reveals that the specific heat capacity obtained in this work is lower than those reported in ref. 9 and 14, which might be ascribed to the different methods and gases used.

3.2 Thermal conductivity

The thermal conductivity is an important thermophysical parameter and the results available so far have been inconsistent.[23] Here, the laser flash method, which is simple, accurate and adaptable for many materials, has for the first time been adopted to measure the thermal diffusivity of the $(LiNaK)_2CO_3$ molten salt. Measurements were carried out at 50 °C intervals within the temperature range of 450 °C to 600 °C in a helium atmosphere. Each measurement was taken after the system had been kept at the respective temperature for 100 minutes, in order to ensure homogeneity. The average values of the 15 thermal diffusivity measurements (see Table S1, ESI†) are listed in Table 1, and are 0.00136, 0.00139, 0.00144 and 0.00152 cm^2 s^{-1} at 450, 500, 550 and 600 °C, respectively. The thermal conductivity was calculated for each temperature based on the obtained density, thermal diffusivity and specific heat capacity (eqn (5)), and the values are listed in Table 1.

$$\lambda = c_p \rho \alpha \tag{5}$$

where λ is the thermal conductivity in W m^{-1} K^{-1}.

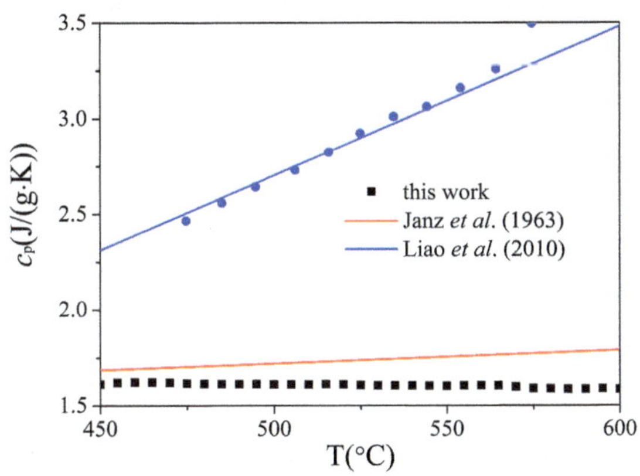

Fig. 3 The specific heat capacities obtained for the $(LiNaK)_2CO_3$ molten salt.

Table 1 Thermal conductivity of the $LiNaKCO_3$ molten salt

	450 °C	500 °C	550 °C	600 °C
α ($\times 10^{-3}$ cm^2 s^{-1})	1.36	1.39	1.44	1.52
ρ (g cm^{-3})	2.071	2.045	2.023	2.007
c_p (J g^{-1} K^{-1})	1.612	1.612	1.612	1.612
λ (W m^{-1} K^{-1})	0.454	0.458	0.470	0.492

The thermal conductivities are 0.454, 0.458, 0.470 and 0.492 W m^{-1} K^{-1} at 450, 500, 550 and 600 °C, respectively, slightly increasing with increasing temperature.

3.3 Viscosity

Standard viscosity oil (Cannon S6 oil) was used to ensure the reliability of the method, and the relative deviation was found to be less than 2%. Furthermore, the viscosity of pure KNO_3 salt was also determined and the relative deviation was less than 4%. The results for Cannon S6 oil and pure KNO_3 are listed in Table 2. The small relative deviations indicated the reliability of the viscometer.

For viscosity measurements, the SMC (spindle multiplier constant) was first calibrated to be 0.51 at 30 rpm using standard viscosity oil. About 100 g of $(LiNaK)_2CO_3$ molten salt was heated to 650 °C and then cooled down. The torque was recorded at desired temperatures and the viscosity was calculated using eqn (2). The viscosity can be fitted as a function of temperature using eqn (6). It should be noted that the unit of temperature in eqn (6) is K.

$$\eta = 0.0852 \exp\left(\frac{3.51 \times 10^4}{RT}\right) \tag{6}$$

The oscillating crucible technique was the most popular method to determine the viscosity of a molten salt, however, the precision of the viscosity severely depends on the accuracy of the density. For example, Janz *et al.*[25] measured the viscosities of the single components of Li_2CO_3, Na_2CO_3 and K_2CO_3 as well as that of the ternary eutectic molten salt, but the viscosity of the ternary eutectic mixture was more qualitative than quantitative due to the absence of precise density data. The viscosity of the ternary eutectic mixture, prepared from reagent grade and

Table 2 Viscosities of Cannon S6 oil and pure KNO_3 molten salt[24]

Cannon S6 oil				KNO_3 molten salt			
T (°C)	η_{mea} (mPa s)[a]	η_{sta} (mPa s)	Dev. (%)	T (°C)	η_{mea} (mPa s)	η_{lit} (mPa s)	Dev. (%)
25	7.76	7.897	−1.73	348	2.77	2.75	0.72
40	4.91	4.914	−0.08	380	2.35	2.32	1.28
50	3.76	3.765	−0.13	410	2.03	1.99	1.97
80	2.02	2.003	0.85	441	1.79	1.72	3.91

[a] Note: "mea" means experimental data in this work; "sta" means standard values of Cannon S6 oil; "lit" means data from literature.

commercial raw materials Li_2CO_3, Na_2CO_3 and K_2CO_3, was re-determined based on a more accurate density using the oscillating crucible technique.[26] The difference between the two samples (reagent grade and commercial raw materials) is so insignificant that the viscosity can be expressed by the same equation. The viscosity of the $(LiNaK)_2CO_3$ molten salt was measured using an oscillation viscometer,[14] and it can be expressed as a parabola against temperature. Recently, the viscosities of unary, binary and ternary eutectic alkali metal carbonate molten salts were measured by rotational methods, where the standard error was less than 5%. All of the available results are summarized and shown in Fig. 4.

3.4 Density

The reliability of the densitometer was verified using NaCl molten salt with a purity of 99.9%. The deviation between our results and the available literature values is less than 0.2% (Table 3), indicating that the device is reliable.

The $(LiNaK)_2CO_3$ molten salt was heated and kept at 610 °C for 2 hours in order to ensure homogeneity, and it was then cooled down at a rate of 1 K min^{-1}. The balance values were recorded at intervals of 5 °C, and the densities were calculated at the desired temperatures using eqn (2); the results are displayed in Fig. 5. Generally, the liquid density linearly decreases with increasing temperature, and the density of the $(LiNaK)_2CO_3$ molten salt is linearly fitted as a function of temperature, as expressed by eqn (7).

$$\rho = 2.27 - 4.34 \times 10^{-4}T \tag{7}$$

The density uncertainty was estimated using eqn (4). The major sources of experimental error for each density determination are the accuracies of the volume ($\pm 2 \times 10^{-4}$ cm^3), the balance ($\pm 2 \times 10^{-5}$ g) and the temperature ($\pm$1 K). The total uncertainty U was estimated to be $\pm$0.00712 g cm^{-3}.

Fig. 5 summarizes the density of the $(LiNaK)_2CO_3$ molten salt as measured by different techniques. Using the maximum bubble pressure method,[8] depicted as

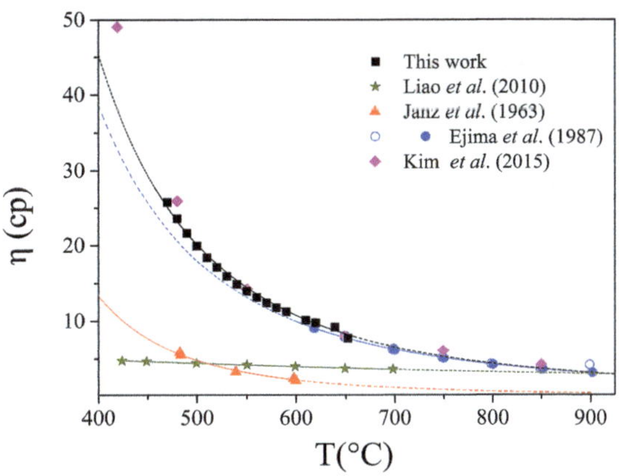

Fig. 4 The measured viscosity of the $(LiNaK)_2CO_3$ molten salt together with that available from literature results.

Table 3 Density of NaCl molten salt

	Density (g cm^{-3})		
T (°C)	This work	Ref. 27	Deviation (%)
860	1.522	1.522	0
850	1.525	1.527	0.13
840	1.530	1.531	0.06
830	1.536	1.536	0
820	1.541	1.541	0

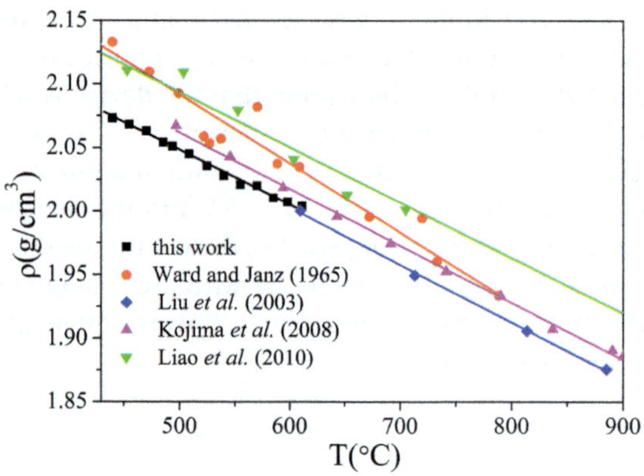

Fig. 5 The measured density of the $(LiNaK)_2CO_3$ molten salt together with that available from literature results.

the considerably scattered circular points, the error was estimated to be ±0.0151 g cm^{-3}. Scattered data points for the density of the $(LiNaK)_2CO_3$ molten salt were also reported by Liao *et al.*[14] using the Archimedean technique. The maximum bubble pressure method was also adopted by Kojima *et al.*[13] and their results are in good agreement with those obtained herein, with an estimated standard deviation of 2.8%. Liu *et al.*[3] measured the $(LiNaK)_2CO_3$ molten salt density from 610 °C to 885 °C using the double-bob Archimedean method, and their results are satisfactorily consistent with ours, with an uncertainty of about ±0.01 g cm^{-3}.

4. Conclusion

The thermal physical properties of $(LiNaK)_2CO_3$ eutectic molten salt were systematically investigated from 400–610 °C under an argon atmosphere. The density and liquidus temperature were determined using an instrument built by ourselves, which prior to use was verified using pure NaCl molten salt. The density was plotted against temperature as an expression of $\rho = 2.27\text{–}4.34 \times 10^{-4}T$ (±0.00712 g cm^{-3}), and the liquidus temperature was found to be 398.7 °C, showing good agreement with the melting point of 396.8 °C obtained by use of STA. The specific heat capacity can be regarded as a constant 1.612 J g^{-1} K^{-1}, which was calculated using DSC. Through a specially designed crucible set and with specific sample preparation, the thermal diffusivity was determined to be 0.00136–0.00152 $cm^2\ s^{-1}$ using the laser flash technique for the first time. The viscosity was also obtained using a self-developed device, and because it is related with temperature (K), it is expressed as $\eta = 0.0852\exp\left(\frac{3.51\times10^4}{RT}\right)$.

Acknowledgements

The authors are grateful for the financial support of the National Natural Science Foundation of China (21406256), the "Youth Innovation Promotion Association" and the "Strategic Priority Research Program" of the Chinese Academy of Sciences, Grant No. XD02002400.

References

1 N. Ren, Y. Wu, T. Wang and C. Ma, *J. Therm. Anal. Calorim.*, 2011, **104**, 1201–1208.
2 M. M. Kenisarin, *Renewable Sustainable Energy Rev.*, 2010, **14**, 955–970.
3 Q. Liu and R. A. Lange, *Contrib. Mineral. Petrol.*, 2003, **146**, 370–381.
4 T. Kojima, Y. Miyazaki, K. Nomura and K. Tanimoto, *J. Electrochem. Soc.*, 2007, **154**, F222–F230.
5 P. L. Spedding, *J. Electrochem. Soc.*, 1973, **120**, 1049–1052.
6 M. V. Smirnov and V. P. Stepanov, *Electrochim. Acta*, 1982, **27**, 1551–1563.
7 G. J. Janz and M. R. Lorenz, *Molten carbonate electrolytes*, 1961, **108**, 1052–1058.
8 A. T. Ward and G. J. Janz, *Electrochim. Acta*, 1965, **10**, 849–857.
9 G. J. Janz, E. Neuenschwander and F. J. Kell, *Trans. Faraday Soc.*, 1963, **59**, 841–845.
10 N. Araki, M. Matsuura, A. Makino, T. Hirata and Y. Kato, *Int. J. Thermophys.*, 1988, **9**, 1071–1080.
11 T. Kojima, Y. Miyazaki, K. Nomura and K. Tanimoto, *J. Electrochem. Soc.*, 2003, **150**, E535–E542.
12 R. I. Olivares, C. L. Chen and S. Wright, *J. Sol. Energy*, 2012, **134**, 041002.
13 T. Kojima, Y. Miyazaki, K. Nomura and K. Tanimoto, *J. Electrochem. Soc.*, 2008, **155**, F150–F156.
14 M. Liao, X. L. Wei, J. Ding, B. H. Hu and Q. Peng, *Acta Energiae Solaris Sinca*, 2010, **31**, 863–867, in Chinese.
15 Q. Liu and R. A. Lange, *Contrib. Mineral. Petrol.*, 2003, **146**, 370–381.
16 S. W. Kim, K. Uematsu, K. Toda and M. Sato, *J. Ceram. Soc. Jpn.*, 2015, **123**, 355–358.
17 ASTM E 1269-11, Standard Test Method for Determining Specific Heat Capacity by Differential Scanning Calorimetry.
18 X. H. An, J. H. Cheng, H. Q. Yin, L. D. Xie and P. Zhang, *Int. J. Heat Mass Transfer*, 2015, **90**, 872–877.
19 L. Dusza, *High Temp. - High Pressures*, 1995/1996, **27/28**, 467–473.
20 J. H. Cheng, P. Zhang, X. H. An, K. Wang, Y. Zuo, H. W. Yan and Z. Li, *Chin. Phys. Lett.*, 2013, **30**, 126501–126504.
21 D. S. Li, *Fundamental of Metrology*, China machine press, Beijing (China), 2014, in Chinese.
22 G. J. Janz and R. P. T. Tomkins, *NSRDS-NBS 61 Part IV*, 1981, 1–861

23 X. Zhang, H. Wicaksono, S. Fujiwara and M. Fujii, *High Temp. - High Pressures*, 2002, **34**, 617–625.
24 G. J. Janz, F. W. Dampier and G. R. Lakshminarayan, *Natl. Stand. Ref. Data Ser.*, 1968, **15**, 25–27.
25 G. J. Janz and F. Saegusa, *J. Electrochem. Soc.*, 1963, **110**, 452–456.
26 T. Ejima, Y. Sato, T. Yamamura, K. Tamal and M. Hasebe, *J. Chem. Eng. Data*, 1987, **32**, 180–182.
27 A. D. Kirshenbaum, J. A. Cahill, P. J. McGonigal and A. V. Grosse, *J. Inorg. Nucl. Chem.*, 1962, **24**, 1287–1296.

Faraday Discussions

PAPER

Cathodic processes of neodymium(III) in LiF–NdF_3–Nd_2O_3 melts

Chao Huang,† Xiaolong Liu,† Yuan Gao, Shizhe Liu and Bing Li*

Received 15th February 2016, Accepted 15th March 2016
DOI: 10.1039/c6fd00014b

In this paper, cyclic voltammetry and square wave voltammetry are applied to characterize the cathode processes of neodymium ions on a W electrode in LiF–NdF_3 melts with or without the metal Nd. The results indicate that neodymium ions in the LiF–NdF_3 (2 wt%) melt are reduced in two steps, *i.e.* $Nd^{3+} \rightarrow Nd^{2+}$ and $Nd^{2+} \rightarrow Nd^0$, corresponding to starting reduction potentials of 0.35 V *vs.* Li^+/Li and 0.1 V *vs.* Li^+/Li, respectively. The $Nd^{3+} \rightarrow Nd^{2+}$ process is controlled by mass transfer and the $Nd^{2+} \rightarrow Nd^0$ process is controlled by both an interfacial step and mass transfer. But in the LiF–NdF_3 melt with excess metal Nd equilibrium, the kinetics of the above two processes are controlled by mass transfer. After potentiostatic electrolysis at 0.35 V in the LiF–NdF_3–Nd_2O_3 melt NdF_2 is formed on the Mo cathode, and metallic Nd is obtained by potentiostatic electrolysis at 0.1 V in the LiF–NdF_3–Nd_2O_3–Nd melt, which validates the above electrochemical reduction results.

Introduction

As one of the rare earth elements, metallic neodymium (Nd) is an important component of the NdFeB permanent magnet. Currently, Nd is mainly produced by molten salt electrolysis from two types of electrolyte systems in the industry, one is the chloride system composed of $NdCl_3$–KCl, the other is the fluoride–oxide system consisting of LiF–NdF_3–Nd_2O_3.[1] The chloride system usually shows a lower current efficiency than the fluoride one. This is related to the two stable oxidation states in the chloride system including Nd^{3+} and Nd^{2+}, which results in a disproportionate reaction among the Nd^{3+} and Nd^{2+} ions and the Nd metal, *i.e.* $Nd^{3+} + Nd = Nd^{2+}$. A. Novoselova *et al.* and S. Vandarkuzhali *et al.*[2,3] confirmed that Nd^{3+} is reduced to Nd^0 through two consecutive steps by electrochemical methods, that is, $Nd^{3+} + e^- \rightarrow Nd^{2+}$ and $Nd^{2+} + 2e^- \rightarrow Nd$ on molybdenum electrodes in LiCl–KCl–CsCl melts and molten LiCl–KCl melts. Nd^{2+} ions in LiCl–KCl–CsCl melts are chemically stable at temperatures lower than 810–840 K.[2]

East China University of Science and Technology, 130 Meilong Road, Shanghai, 200237, P. R. China. E-mail: bingli@ecust.edu.cn

† Chao Huang and Xiaolong Liu contribute equally as first authors.

Hajimu Yamana has also demonstrated that Nd^{2+} is formed by the cathodic electrolysis of $NdCl_3$ dissolved in molten LiCl–KCl at 773 K.[4]

For the fluoride system, the current efficiency for neodymium electrowinning is generally lower than 80%, far lower than that of aluminium electrolysis. Several studies on the cathodic processes of Nd^{3+} in fluoride media have been published. C. Hamel *et al.*[5] revealed that Nd^{3+} is reduced to Nd^0 in a one-step process in LiF–CaF_2 melts exchanging three electrons by cyclic voltammetry, chronopotentiometry and square wave voltammetry. In the research led by E. Stefanidaki *et al.*,[6] the results have shown that Nd^{3+} ions are electrochemically reduced to Nd^0 *via* a three-electron reaction in LiF–NdF_3–Nd_2O_3 melts by cyclic voltammetry. But in our recent research, we found that Nd^{2+} ions can also exist stably in fluoride melts at high temperature. The relation between the current efficiency and Nd^{2+} formation in fluoride systems is under investigation. So in this paper, it is necessary to investigate cathodic processes of Nd^{3+} in LiF–NdF_3–Nd_2O_3 melts by electrochemical methods.

Experimental

In this study, the chemicals used in the experiments included LiF (99.9%), NdF_3 and Nd_2O_3 (99.9%). All the chemicals were pretreated according to the following

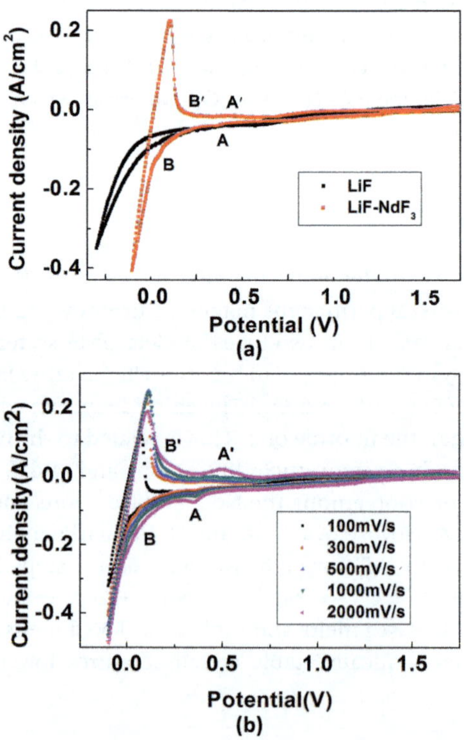

Fig. 1 Cyclic voltammograms recorded on a W electrode in (a) LiF–NdF_3 (0.26 mol L^{-1}) and LiF melts with a scan rate of 100 mV s^{-1} and (b) the LiF–NdF_3 (0.26 mol L^{-1}) melt with different scan rates at 1273 K.

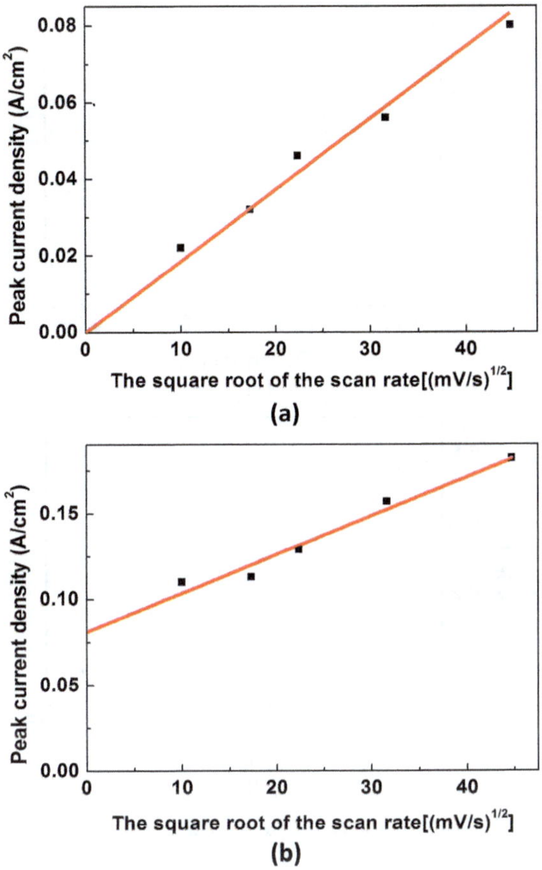

Fig. 2 (a) Peak currents of peak A and (b) peak currents of peak B *vs.* the square root of scan rates derived from Fig. 1(b).

steps: first the chemicals were placed in a graphite crucible (spectrographic purity) located in an airtight stainless steel reactor, then heated to 400 °C under vacuum and left for 8 h to remove traces of moisture, then further heated to 800 °C and left for 2 h under argon gas. Finally, the chemicals were cooled to room temperature and taken out of the reactor, and stored in an oven at 100 °C under vacuum for use. For each experiment, the chemicals were weighed and placed in a graphite crucible in the stainless steel reactor, then heated to 400 °C under vacuum for 2 h and finally heated to the required temperature under argon gas.

During the electrochemical measurements a three-electrode system was used. The working electrode and reference electrode were W wires (99.9%, diameter 1 mm), the auxiliary electrode was a graphite rod (graphite carbon for spectrography, diameter 7 mm). During each experiment, all the potentials were transferred to Li^+/Li according to the literature.[7] All the electrodes and graphite crucibles were polished with emery paper to a mirror finish, washed with distilled water, then dried in a vacuum oven prior to use. The electrode surface areas immersed into the melts were determined by measuring the immersion depth of the electrode in the melts. Cyclic voltammetry and square wave voltammetry were

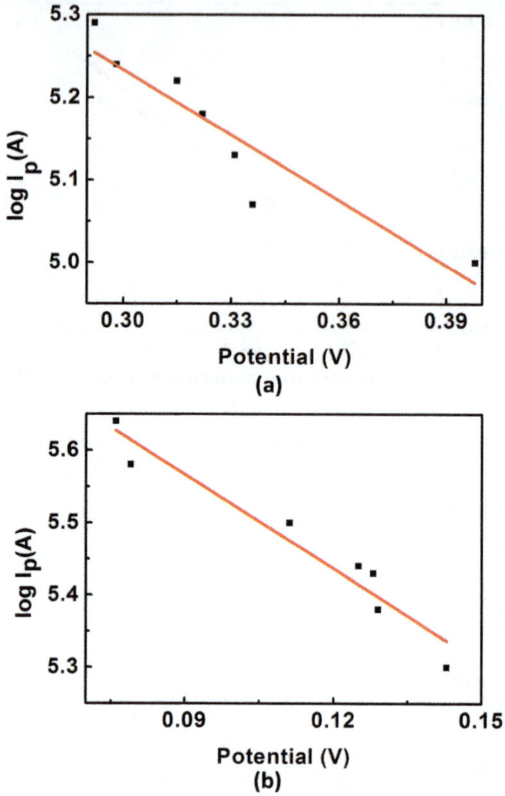

Fig. 3 (a) log I_p of peak A and (b) log I_p of peak B *vs.* the corresponding potentials derived from Fig. 1(b).

used for the investigation of the neodymium ion reduction process. The electrochemical measurements were performed using a PAR-STAT2273 (PAR-Ametek Co, Ltd.) with a PowerSuite software package.

Results and discussion

1. The electrochemical reduction of Nd^{3+} in the LiF–NdF_3 system

Fig. 1(a) shows typical cyclic voltammograms obtained on a W electrode in the LiF and LiF–NdF_3 melts at 100 mV s^{-1} at 1273 K. An obvious reduction current starting at around 0 V is attributed to the reduction of Li^+ ions in the LiF melt. As NdF_3 (0.26 mol L^{-1}) was added into the above LiF melt, two new pairs of reduction peaks A and B corresponding to oxidation peak A′ and oxidation peak B′ appeared in the cyclic voltammogram. The cathode peak A starting from about 0.35 V corresponds to the reduction of Nd^{3+} to Nd^{2+} and the anodic peak A′ is caused by the oxidation of Nd^{2+} to Nd^{3+}. The cathode peak B starting from about 0.1 V corresponds to the reduction of Nd^{2+} to Nd^0 and the anodic peak B′ is caused by the oxidation of Nd^0 to Nd^{2+}.

To investigate the reversibility of these two redox reactions in the LiF–NdF_3 system, cyclic voltammograms with different scan rates were measured and are

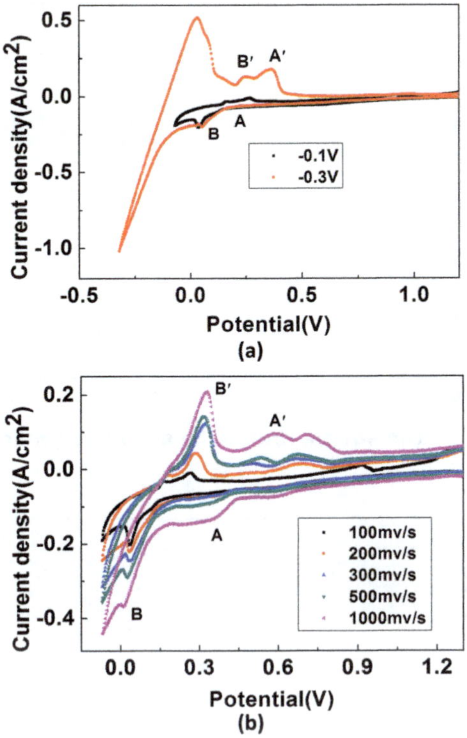

Fig. 4 Cyclic voltammograms recorded on a W electrode in a LiF–NdF_3 (6 wt%)–Nd (excess) melt with (a) different scan potential, (b) different scan rates at 1273 K.

shown in Fig. 1(b). As the scan rate increased from 100 mV s^{-1} to 2000 mV s^{-1}, the reduction current density for peak A increased and the peak potential almost did not change with the scan rate, and the reduction current for peak B and oxidation current for peak B′ also increased with the scan rate. The plot of the peak currents for peak A *versus* the square root of the scan rates is linear and passes through the origin, as shown in Fig. 2. The results indicate that the reduction process of $Nd^{3+} \rightarrow Nd^{2+}$ is controlled by diffusion of Nd^{3+}, and the reduction process of $Nd^{2+} \rightarrow Nd^0$ is controlled by diffusion of Nd^{2+} in the LiF–NdF_3–Nd (excess) system.

The plot of the peak currents for peak B *versus* the square root of the scan rates is linear and does not pass through the origin, implying that the reduction process of $Nd^{2+} \rightarrow Nd^0$ is controlled by both an interfacial step reaction and ion diffusion in the molten salts.

Both of the relationships between the logarithms of the reduction peak currents of peak A and B and the corresponding potentials are shown in Fig. 3. They are almost linear with the whole scan rates; the number of exchanged electrons can be calculated by eqn (1).[8]

$$K = \frac{\alpha n F}{2.3RT} \tag{1}$$

where K is the slope, $\alpha = 0.5$, F is the Faraday constant, R is the gas constant, T is the absolute temperature (K), and n is the number of electrons involved in the reaction.

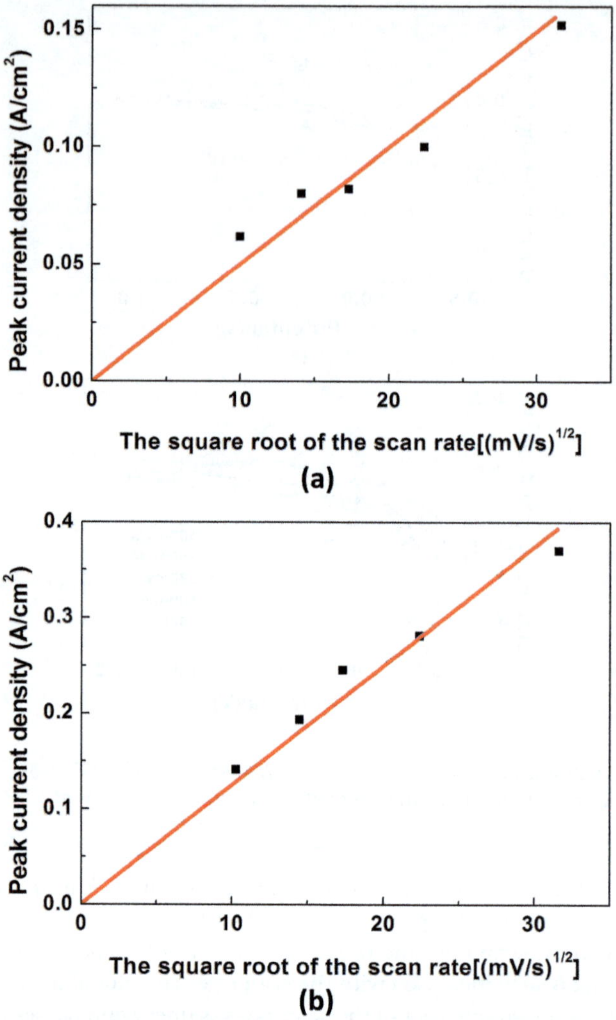

Fig. 5 (a) Peak currents of peak A and (b) peak currents of peak B *vs.* the square root of scan rates derived from Fig. 4(b).

According to the data in Fig. 3(a) and eqn (1), the slope K is 2.632 and the electron transfer number is 1.29, close to 1, implying electrochemical reduction of $Nd^{3+} \rightarrow Nd^{2+}$. According to the data in Fig. 3(b) and eqn (1), the slope K is 4.34 and the electron transfer number is 2.13, close to 2, implying electrochemical reduction of $Nd^{2+} \rightarrow Nd^0$.

2. The electrochemical reduction of Nd^{3+} in LiF–NdF_3–Nd

2.1. Cyclic voltammetry. Fig. 4 shows the cyclic voltammograms obtained on a W electrode in the LiF–NdF_3 melt with an excess amount of Nd at 1273 K. Except for Li^+ ion reduction, two pairs of oxidation/reduction peaks A/A′ and B/B′ were observed in the cyclic voltammograms in Fig. 4(a), which were supposed to be redox reactions of Nd^{3+}/Nd^{2+} and Nd^{2+}/Nd^0, respectively. Fig. 4(b) shows the cyclic

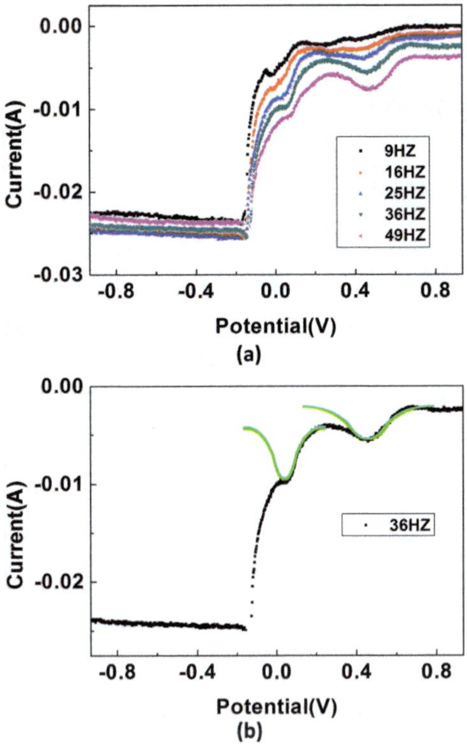

Fig. 6 Square wave voltammograms recorded on a W electrode in the LiF–NdF_3 (2 wt %)–Nd (excess) melt with (a) various frequencies and (b) fitting curves with a frequency of 36 Hz at 1273 K.

voltammograms on a W electrode with different scan rates in the LiF–NdF_3 (6 wt %)–Nd (excess) melt. As the scan rate increased from 100 mV s^{-1} to 1000 mV s^{-1}, the reduction current density for peak A increased and the peak potential almost did not change with the scan rate, and the reduction current for peak B and oxidation current for peak B′ also increased with the scan rate. Both the plots of the peak currents for peak A *versus* the square root of the scan rates, and the peak currents for peak B *versus* the square root of the scan rates in Fig. 5 show good linearity passing through the origin. These results indicate that the reduction processes of $Nd^{3+} \rightarrow Nd^{2+}$ and $Nd^{2+} \rightarrow Nd^0$ are controlled by diffusion of Nd^{3+} ions and Nd^{2+} ions, respectively, in the LiF–NdF_3–Nd (excess) system. Compared with the above results in the LiF–NdF_3 system, more Nd^{2+} ions are produced by the disproportionate reaction of $2Nd^{3+} + Nd = 3Nd^{2+}$ in the LiF–NdF_3–Nd (excess) system, which leads to a larger reduction current.

2.2. Electron transfer number. Square wave voltammetry has high sensitivity and resolution and can be used to calculate the electron transfer number.[9,10] In the case of a reversible system, a simple equation associating the half width of the peak $W_{1/2}$ and the electron transfer number can be expressed as eqn (2):[5]

$$W_{1/2} = 3.52\frac{RT}{nF} \tag{2}$$

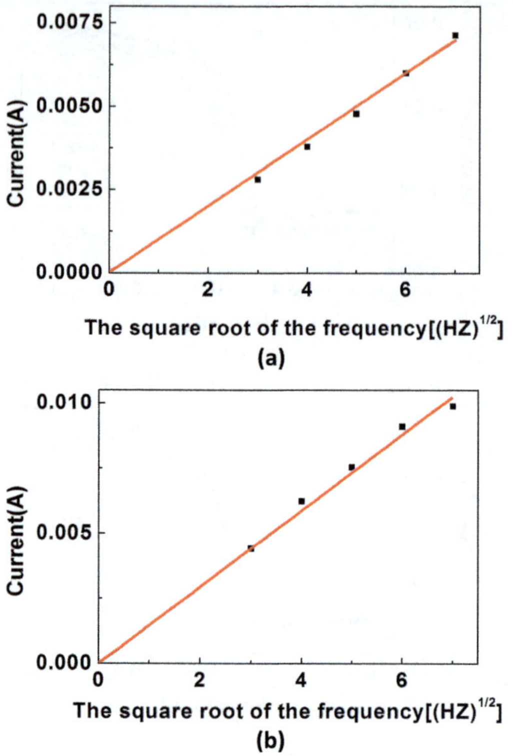

Fig. 7 Plot of the peak currents for (a) peak A and (b) peak B *versus* the square root of the frequencies ($Hz^{1/2}$) derived from Fig. 6(a).

Fig. 6 presents a typical square wave voltammogram of NdF_3 in LiF–NdF_3–Nd (excess) on the W electrode at 1273 K. Compared with the LiF melt, the square wave voltammogram in LiF–NdF_3–Nd (excess) exhibits two reduction peaks at about 0.35 V and 0.05 V, corresponding to the reduction peak A and B in the cyclic voltammograms of Fig. 4. Both the plots of the peak currents for peak A and peak B *versus* the square root of the frequencies ($Hz^{1/2}$) are linear passing through the origin (Fig. 7). The results indicate that the reduction process of $Nd^{3+} \rightarrow Nd^{2+}$ is controlled by diffusion of Nd^{3+} in the LiF–NdF_3–Nd (excess) system. According to eqn (2), the exchanged electron number for peak A is 1.28, close to 1, corresponding to the reduction of $Nd^{3+} \rightarrow Nd^{2+}$, the exchanged electron number for peak B is 2.14, close to 2, corresponding to the reduction of $Nd^{2+} \rightarrow Nd^0$ (Fig. 6(b)).

3. Constant potentiostatic electrolysis

Potentiostatic electrolysis at 0.35 V was carried out in the NdF_3 (360 g)–LiF (40 g)–Nd_2O_3 (4 g) melt at 1373 K for 11 000 s. After electrolysis, the Mo cathode with attached melt was lifted simultaneously at 1373 K from the graphite crucible. The melt in the crucible was taken out after naturally cooling to room temperature. The melt attached to the Mo electrode, as well as the melt around the Mo electrode and around the graphite anode in the graphite crucible are shown in Fig. 8. The melt attached to the Mo cathode is darker in color than that in the crucible

Fig. 8 (a) Melt attached to the Mo cathode lifted simultaneously, (b) the melt around the Mo cathode cooled to room temperature and (c) the melt around the graphite anode cooled to room temperature after potentiostatic electrolysis in the NdF_3–LiF–Nd_2O_3 melt at 0.35 V at 1373 K for 11 000 s.

and is composed of more NdF_2 than that in the crucible according to XRD analysis, as shown in Fig. 9. The results show that NdF_2 is formed during potentiostatic electrolysis at 0.35 V, so the reduction peak A is attributed to the reduction of Nd^{3+} to Nd^{2+}. After potentiostatic potential electrolysis in the NdF_3 (360 g)–LiF (40 g)–Nd_2O_3 (4 g) melt at 0.1 V at 1373 K for 6700 s, no metallic Nd was found but NdF_2 was formed after electrolysis due to the disproportionate reaction.

As 50 g of metallic Nd was added to the above NdF_3 (360 g)–LiF (40 g)–Nd_2O_3 (4 g) melt, after potentiostatic electrolysis at 0.1 V at 1373 K for 8400 s, 3.919 g of metallic Nd (larger than the theoretical value of 3.42 g) was obtained, implying the reduction peak B is attributed to the reduction of Nd^{2+} to Nd^0 and the disproportionate reaction among Nd^{3+}, Nd^{2+}, Nd^0 occurs which results in more metallic Nd formation than the theoretical value during electrolysis.

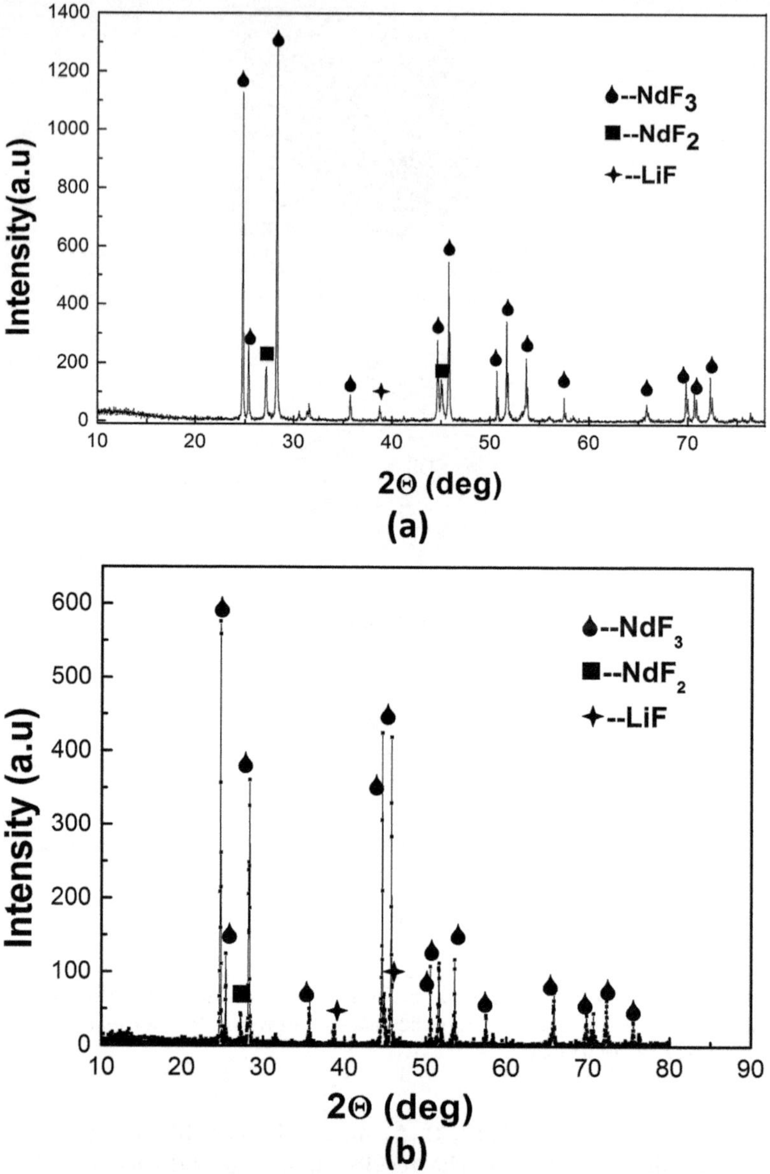

Fig. 9 XRD patterns of (a) the melt attached to the Mo cathode lifted simultaneously and (b) the melt around the Mo cathode cooled to room temperature after potentiostatic electrolysis in the NdF_3–LiF–Nd_2O_3 melt at 0.35 V at 1373 K for 11 000 s.

Conclusions

Neodymium ion electrochemical reduction processes were measured and characterized by cyclic voltammetry and square wave voltammetry on a W electrode in LiF–NdF_3 melts with or without excess metal Nd. The results indicate that neodymium ions in the LiF–NdF_3 (2 wt%) melt are reduced in two steps, *i.e.*, $Nd^{3+} \rightarrow Nd^{2+}$ and $Nd^{2+} \rightarrow Nd^0$, corresponding to the starting reduction potentials of 0.35 V *vs.* Li^+/Li and 0.1 V *vs.* Li^+/Li, respectively. The $Nd^{3+} \rightarrow Nd^{2+}$ process is controlled by mass transfer and the $Nd^{2+} \rightarrow Nd^0$ process is controlled by both an interfacial step and mass transfer. But in the $LiF–NdF_3$ melt with excess metal Nd, due to stable Nd^{2+} formation, the kinetics of the above two processes are controlled by mass transfer. After potentiostatic electrolysis at 0.35 V in the $LiF–NdF_3–Nd_2O_3$ melt NdF_2 was formed on the W cathode, and metallic Nd is obtained by potentiostatic electrolysis at 0.1 V in the $LiF–NdF_3–Nd_2O_3$–Nd (excess) melt, which validated the above electrochemical reduction results.

Acknowledgements

This work was financially supported by the National Natural Science Foundation of China (No. 51274102).

Notes and references

1 H. M. Zhu, *Encyclopedia of Applied Electrochemistry,* Springer, Berlin, 2011.
2 A. Novoselova and V. Smolenski, *Electrochim. Acta*, 2013, **87**, 657–662.
3 S. Vandarkuzhali, M. Chandra, S. Ghosh, N. Samanta, S. Nedumaran, B. Prabhakara Reddy and K. Nagarajan, *Electrochim. Acta*, 2014, **145**, 86–98.
4 H. Yamana, B. Gi Park, O. Shirai, T. Fujii, A. Uehara and H. Moriyama, *J. Alloys Compd.*, 2006, **408–412**, 66–70.
5 C. Hamel, P. Chamelot and P. Taxil, *Electrochim. Acta*, 2004, **49**, 4467–4476.
6 E. Stefanidaki, C. Hasiotis and C. Kontoyannis, *Electrochim. Acta*, 2001, **46**, 2665–2670.
7 S. Z. Liu, L. Y. Chen, B. Li, *et al.*, *Electrochim. Acta*, 2014, **147**, 82–86.
8 J.-B. Raoof, A. Omrani, R. Ojani and F. Monfared, *J. Electroanal. Chem.*, 2009, **633**, 153–158.
9 L. Massot, P. Chamelot and P. Taxil, *Electrochim. Acta*, 2009, **54**, 6361–6366.
10 C. Hamel, P. Chamelot and P. Taxil, *Electrochim. Acta*, 2004, **49**, 4467–4476.

Faraday Discussions

ROYAL SOCIETY OF CHEMISTRY

PAPER

Advances in electroanalysis, sensing and monitoring in molten salts

Damion K. Corrigan,[a] Justin P. Elliott,[a] Ewen O. Blair,[b] Simon J. Reeves,[a] Ilka Schmüser,[b] Anthony J. Walton[b] and Andrew R. Mount*[a]

Received 6th January 2016, Accepted 11th February 2016
DOI: 10.1039/c6fd00002a

Microelectrodes have a number of advantages over macroelectrodes for quantitative electroanalysis and monitoring, including reduced *iR* drop, a high signal-to-noise ratio and reduced sensitivity to convection. Their use in molten salts has been generally precluded by the combined materials challenges of stresses associated with thermal cycling and physical and corrosive chemical degradation at the relatively high temperatures involved. We have shown that microfabrication, employing high precision photolithographic patterning in combination with the controlled deposition of materials, can be used to successfully address these challenges. The resulting molten salt compatible microelectrodes (MSMs) enable prolonged quantitative microelectrode measurements in molten salts (MSs). This paper reports the fabrication of novel MSM disc electrodes, chosen because they have an established ambient analytical response. It includes a detailed set of electrochemical characterisation studies which demonstrate both their enhanced capability over macroelectrodes and over commercial glass pulled microelectrodes, and their ability to extract quantitative electroanalytical information from MS systems. MSM measurements are then used to demonstrate their potential for shedding new light on the fundamental properties of, and processes in, MSs, such as mass transport, charge transfer reaction rates and the selective plating/stripping and alloying reactions of liquid Bi and other metals; this will underpin the development of enhanced MS industrial processes, including pyrochemical spent nuclear fuel reprocessing.

1 Introduction

Molten salts (MSs) are a key constituent of a number of industrial processes, including metal refining,[1] pyrochemical reprocessing of spent nuclear fuel,[2–5] molten salt batteries[6] and catalysis.[7,8] Chloride melts are the MS of choice for a wide range of applications, due to their relatively low melting points and viscosities, their complexing ability, and their reduced corrosion compared to fluoride melts. A particularly intensively studied chloride system is the lithium

[a]*EaStCHEM, School of Chemistry, The University of Edinburgh, King's Buildings, Edinburgh, EH9 3FJ, UK. E-mail: a.mount@ed.ac.uk*
[b]*SMC, School of Engineering, The University of Edinburgh, King's Buildings, Edinburgh, EH9 3FF, UK*

chloride–potassium chloride eutectic (LKE) which has a relatively low melting point of 625 K at the eutectic composition (59 mol% LiCl : 41 mol% KCl), a relatively high ionic conductivity of 1.57 S cm^{-1} at 723 K (ref. 5) and a wide electrochemical window of 3.62 V, within which a wide range of reactions can be studied.[5]

As a result of its industrial importance, particularly in the nuclear sector, there has been a desire to develop analysis techniques suitable for MS systems, which address the challenges of withstanding the high temperatures and corrosive nature of the melts when they contain reactive species.[9–11] Electrochemical methods have typically been practically limited to macroelectrode measurements[12–20] and therefore suffer from a low signal-to-noise ratio, a time-dependent signal, *iR* distortion of the response and sensitivity to convection.[21] These effects at best complicate and often preclude quantitative analysis. Optical spectroscopies suffer from the disadvantages of the broad emissions of many of the target analytes, difficulties in identifying multiple analytes and corrosion of sampling windows.[10]

When considering electroanalysis, microelectrodes can be fabricated using a number of methods and have well-established advantages under ambient conditions of high signal-to-noise ratios, and steady-state (time-invariant) and quantitative signals which are relatively insensitive to convection when compared to macroelectrodes.[22,23] They have therefore been developed for ambient and near-ambient sensing applications[24] and for use in the monitoring of industrial processes.[25] However, there has been a significant technological barrier to the development of similar microelectrode sensors for reproducible and quantitative MS analysis and monitoring. For example, microelectrodes fabricated by drawing a wire encapsulated in glass and then exposing the end, have been employed with limited success, as this has been shown to result in initial and time-dependent variation of the response.[26] The initial variability can be attributed to differences in the wire size and cross section due to variation in the wire drawing process, which as a result requires optical and/or electrochemical characterization of electrode size and shape. The time-dependent variation in the MS can be attributed to one or more of: the tendency for chemical attack of the glass, which compromises the electrode insulation; the differential thermal expansion of the microelectrode materials, which causes materials stress and cracking; and the chemical corrosion of the electrode, which can cause it to progressively recess with time. All of these effects cause a progressive change in effective electrode area and the diffusion regime, which precludes the effective deployment of microelectrodes in MS media for physicochemical analysis.

Previously, we have reported the development and characterisation of an MSM designed to overcome the above issues.[21,27–29] In this study we have built on our previous work, employing our established photolithographic fabrication techniques to produce novel disc-shaped MSMs with precisely controlled dimensions and geometries. Disc microelectrodes have been chosen due to their well-established and quantitative ambient analytical response.[29] We first establish whether the microelectrode response of the disc MSMs under ambient aqueous conditions is similar to other microelectrodes and then compare and contrast the performance of these devices in the LKE MS[21] both with macroelectrodes and with the ambient microelectrode response. This paper has three main aims: (1) to clearly demonstrate the enhanced capabilities of the MSM in MS by benchmarking against a traditional glass pulled microelectrode, (2) to quantitate and explain the analytical response and extract physicochemical parameters in order to gain fundamental insight into the behaviour of the LKE molten salt system and (3) to investigate the potential for gaining fundamental insight into industrially relevant molten salt processes through MSM measurements. In this work we have concentrated on the characterisation of mass transport and charge transport in redox processes, along with the plating/stripping of metals and the formation of metal alloys. These are essential steps in the development of molten salt industrial processes, including pyrochemical spent nuclear fuel reprocessing, which we are developing through UK (the EPSRC REFINE programme) and EU (SACSESS programme) activities.

2 Materials and methods

2.1 Design and fabrication of the electrode devices

The disk MSM devices were produced using reliable, well-characterised silicon-based microfabrication processes (Fig. 1). First, a 500 nm thick layer of silicon dioxide insulation (Fig. 1A) was grown on the silicon substrate. Next, a 20 nm thick film of titanium nitride was deposited onto the surface as an adhesion layer for the electrode metal (which was either a 200 nm thick layer of tungsten or a 50 nm thick layer of platinum) (Fig. 1B). Patterning of the electrode metal layer was then achieved by reactive ion etching, using photolithographic masking to prevent the exposure of a band of metal around the perimeter of each device when diced (Fig. 1C). Subsequent deposition of a 500 nm thick layer of silicon nitride produced a top insulator (Fig. 1D), which was again patterned using photolithographic masking and reactive ion etching of the insulator to expose the bond pad (Fig. 1E) and the microelectrode (Fig. 1F). Fig. 1G shows a sketch of a completed device.

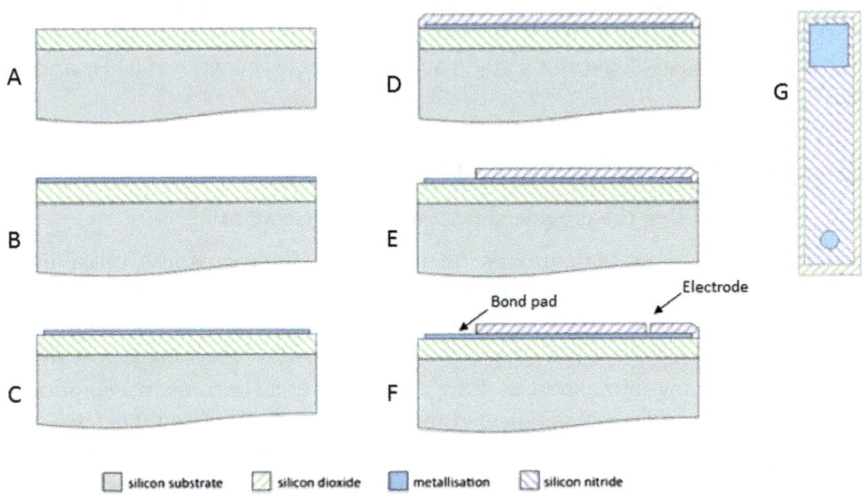

Fig. 1 Schematic showing cross-sections of the fabrication processes involved in producing the MSM (A–F) and a representative top view of a finished device with the microdisc (near the bottom) enlarged for clarity (G).

2.2 High temperature measurements in LKE

Experiments using LKE were conducted at fixed temperatures within the range 673–773 K in a glassy carbon crucible sealed inside a quartz cell within a vertical tube furnace (Carbolite, Derbyshire, UK). Tungsten wire (Goodfellow, Cambridge, UK) was employed as the counter electrode with a 1% AgCl/Ag electrode used as the reference electrode. The latter was prepared by encapsulating a silver wire in LKE containing 1% by mass silver(I) chloride (Sigma Aldrich, Poole, Dorset, UK) in a mullite tube (Multilab, Newcastle upon Tyne, UK). All potentials, E, applied to the working electrodes and quoted in this paper are with respect to this reference electrode. For each experiment, anhydrous analytical grade salts of LiCl and KCl (Sigma Aldrich, Poole, Dorset, UK) were mixed at the eutectic composition (a molar ratio of 7 : 5) in a glassy carbon crucible and then melted *in vacuo* under dry Argon to form LKE. The microelectrode (connected to a tungsten wire *via* a crocodile clip with the connector sealed in ceramic) was deployed through a port in the cell lid as in previous work.

2.3 Ambient aqueous benchmarking measurements

For ambient aqueous control experiments, the disc MSM was employed as the working electrode, with a saturated calomel reference electrode (SCE) and platinum gauze counter electrode. All aqueous solutions were prepared with deionised water (Millipore MilliQ). Electrochemical experiments were typically performed using background electrolyte solutions of aqueous potassium chloride (Fisher Scientific, lab reagent grade, >99% purity), with redox experiments using potassium ferricyanide, $K_3[Fe(CN)_6]$ (Fisons, SLR reagent grade, >99% purity) in background electrolyte. All solutions were thoroughly degassed using pure shield argon (BOC) for at least 20 minutes prior to experimentation.

All experiments in 2.2 and 2.3 were performed using a PC-controlled Autolab PGSTAT12 potentiostat (EcoChemie, Utrecht, Netherlands) and results were analysed using GPES 4.9 and NOVA 1.11 software (EcoChemie), with iterative fitting carried out using Origin 2015 (OriginLab). EIS experiments were typically carried out by applying a 10 mV rms amplitude sinusoid superimposed on the dc potential, E_{DC}, using logarithmically spaced frequencies between 100 kHz and 0.1 Hz.

3 Results and discussion

3.1 Demonstrating the enhanced fidelity of the disc MSM

In previous work, assessments have been made for the robustness, possibilities for quantitation and measurement errors associated with ambient aqueous measurements on square MSMs.[21] In terms of the measured steady-state currents, it was found that typical intra-experimental errors were relatively low, with the variation in the mass transport limiting current for one electrode in a solution of a given concentration across repeated measurements being less than 1.4%, and that for different electrodes of the same nominal edge length being less than 4.2%. In both cases these reported errors were likely to be overestimates, as the largest errors involved measurements using volatile redox couples, which results in uncertainty associated with their concentration.[21] Errors from ambient studies involving glass pulled microelectrodes can be ~10% (ref. 24) highlighting the benefit of the microfabrication approach and also the fidelity of the MSM devices in the melt. In this study, as disc MSMs have been fabricated and characterised for the first time, the response and lifetime of these MSMs in LKE is initially contrasted with a commercially available glass pulled tungsten "cone" microelectrode, which has a base diameter of the exposed cone of 26 μm (see Fig. 2A).

The quantitative response of cone microelectrodes has been established as:[30]

$$i_L = 4nFDcr(1 + qH^p) \tag{1}$$

where i_L is the mass transport limiting current, n is the number of electrons transferred, D is the diffusion coefficient, F is Faraday's constant, c is the bulk concentration, r is the cone base radius, q is 0.30611, H is the aspect ratio of the cone height to base radius and p is 1.114.

However, over the course of ten minutes in LKE these devices were observed to progressively physically deteriorate, with the etching and removal of the insulating glass seen to be the most prominent failure mechanism (see Fig. 2B–D). Thus, the exposed electrode area clearly enlarged with time. Given this, it is not

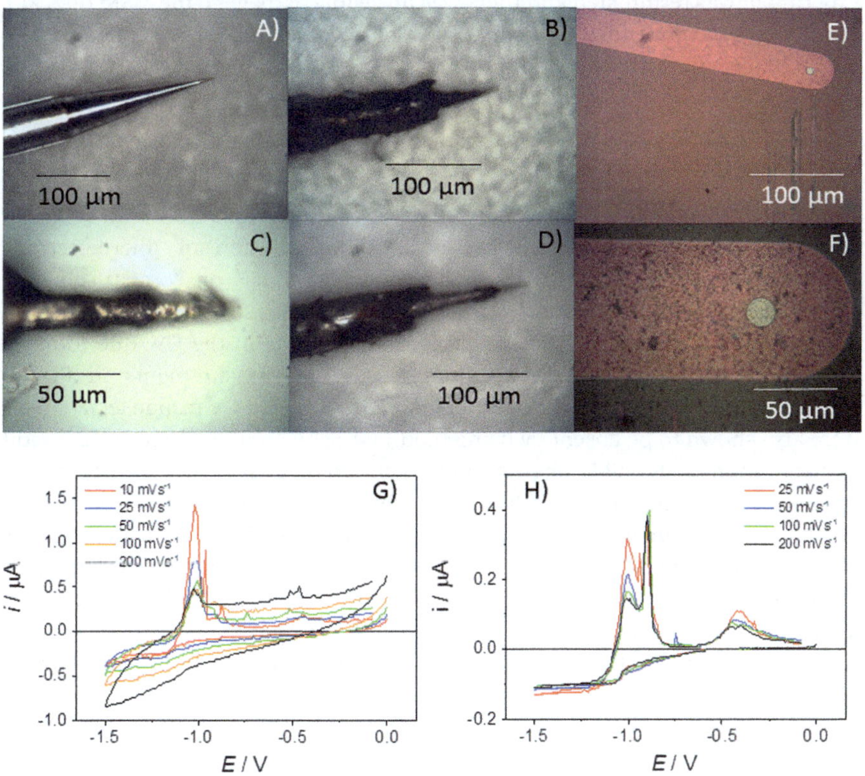

Fig. 2 Images of a commercial microelectrode (A) prior to immersion in LKE and (B–D) following progressive electrochemical cycling in LKE + 5 mM $ZnCl_2$. (E & F) Images of a Pt r = 10 μm diameter disc MSM, (E) before measurement in LKE and (F) following 30 min of electrochemical cycling in LKE (see H). (G) CVs recorded at the given range of sweep rates using the glass pulled tungsten microelectrode in LKE + 5 mM $ZnCl_2$ at 773 K. (H) CVs recorded using the r = 10 μm Pt MSM in LKE + 5 mm $ZnCl_2$ at 773 K.

surprising that the cyclic voltammograms (CVs) observed for Zn^{2+} reduction in LKE showed a change in response with scan rate and time (Fig. 2G). The observed increasing capacitance, and increasing and non-steady-state currents, together indicates the absence of characteristic microelectrode behaviour and an increasing exposed electrode area attributable to progressive insulator compromise. In contrast an $r = 10$ μm Pt MSM employed for the same measurement showed no appreciable visible deterioration and produced a stable and characteristic microelectrode response, with the zinc reduction plating currents observed at $E < -0.5$ V being invariant with time or sweep rate (Fig. 2H). The resulting diffusion coefficient for Zn^{2+} in LKE of 2.85×10^{-5} cm^2 s^{-1}, obtained from the mass transport limited current for this reduction, and the established analytical response of a disc microelectrode:

$$i_L = 4nFDcr \tag{2}$$

is in good agreement with the literature and previous measurements with MSMs.[21,31] In this case it was possible to obtain a stable electrochemical response for thirty minutes, over which the measured steady-state currents varied within the error limit previously established for square MSMs. After thirty minutes the experiment was terminated and there was no visible damage to the MSM (Fig. 2F). This therefore confirms the fidelity of the disc MSMs and their applicability to LKE MS measurement and analysis at least over these time scales.

3.2 Quantitative physicochemical parameterisation using disc MSMs and soluble redox couples in LKE MS

3.2.1 CV and chronoamperometric studies. Our previous work with solution-based reactions[21] focused on the initial demonstration of microelectrode performance in LKE solutions, with particular emphasis on comparison with macroelectrodes. The macroelectrode response was shown to suffer from convection and there were significant peak distortions in the CV, due to an *iR* drop. In contrast, consistent with the disc MSM measurements in Fig. 2, convective effects were shown to be negligible in the MSM CV response; *iR* effects were also shown to be absent by using modified Tafel analysis.[21] This gives credit to the notion that MSMs should contain inherent analytical advantages over macroelectrodes in molten salts.[21,27] In this study, the quantitative electrochemical behaviour of the Eu(III/II) system is established in LKE using the disc MSM system. The Eu(III/II) system was chosen because it is a commonly employed model redox system representative of lanthanide species.[10,12,32] Fig. 3A shows a typical CV recorded using an $r = 25$ μm Pt disc MSM.

In chloride melts, Eu(III) has been shown to react according to the following scheme:[32]

$$2EuCl_3 \rightleftharpoons 2EuCl_2 + Cl_2$$

It is clear that both oxidation and reduction waves can be seen on the CV, centred around $E = +0.4$ V, which arise from Eu(II) and Eu(III) species respectively, which we attribute to the effects of the above reactions in LKE (actually, the proportion of Eu(III) and Eu(II) is likely to vary between experiments with the

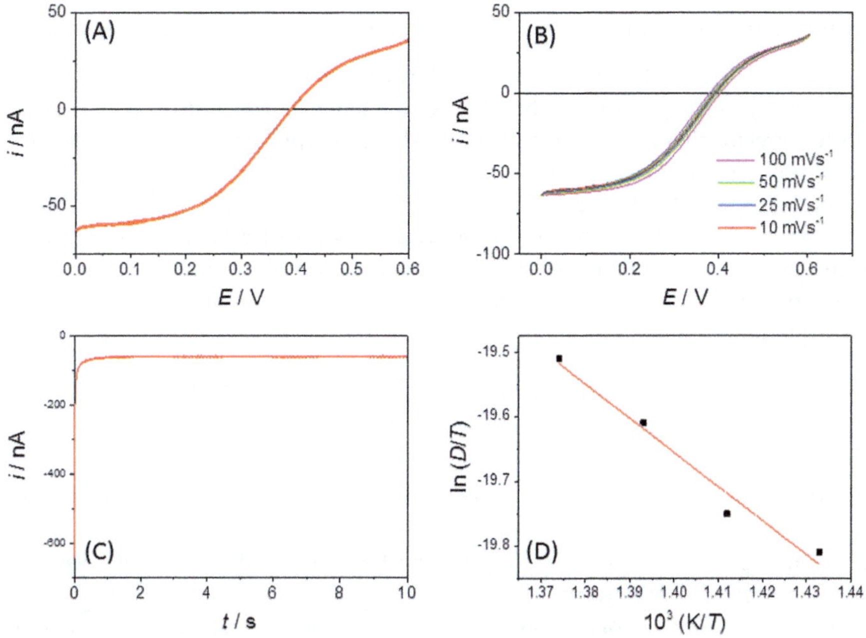

Fig. 3 (A) Typical CV obtained using a Pt disc ($r = 25$ µm) MSM in LKE + 19 mM $EuCl_3$ (initial concentration) at a sweep rate of 10 mV s^{-1} and a temperature of 703 K. (B) Corresponding CVs at the variety of given sweep rates. (C) Corresponding chronoamperometry (current, i *versus* time, t) transient recorded using an $r = 25$ µm Pt disc in LKE by stepping from $E = +0.5$ V to 0 V at $t = 0$. (D) Arrhenius plot of $\ln(D/T)$ *versus* $1/T$ used to obtain the activation energy for the diffusion of Eu^{3+} in LKE, where D is the diffusion coefficient and T is the temperature. The red line is the linear regression line of best fit.

concentration of dissolved chlorine in, and/or the overall oxidising environment of, the LKE). Reassuringly, the CV has a wave-like form characteristic of a microelectrode response, along with the characteristic microelectrode sweep rate independence (Fig. 3B) (there is only a small additional capacitive current whose importance increases as expected with sweep rate, leading to a slightly increased separation between forward and backward scans). Previous work[32] has shown the diffusion coefficients for Eu(III) and Eu(II) to be comparable. Assuming therefore a global diffusion coefficient, D, and taking i_L as the difference in oxidation and reduction mass transport limited current and $c = 19$ mM, it is possible to calculate this global diffusion coefficient from eqn (2) as $(5.0 \pm 0.2) \times 10^{-6}$ cm^2 s^{-1} in LKE at 703 K; for any CV, the combined heights of the oxidation and reduction waves (Fig. 2A and B) or the steady-state chronoamperometric currents observed at long times when carrying out both mass transport limited oxidation and reduction (*e.g.* Fig. 2C) through eqn (1) can then be used to determine the concentrations of Eu(III) and Eu(II). Fig. 3D shows an Arrhenius plot of $\ln(D/T)$ *vs.* $1/T$ derived from the global diffusion coefficient data obtained using the MSMs at various temperatures (note that this plot assumes the validity of the Stokes–Einstein equation[33] $D = \dfrac{kT}{6\pi\eta r_h}$, where k is the Boltzmann constant, r_h is the effective radius of the diffusing species and η is the viscosity of the MS, which is assumed to vary with temperature according to $\eta = \eta_0 \exp\left[\frac{E_a}{kT}\right]$, where η_0 is a constant and E_a is the activation energy). The value of E_a obtained from the slope was (43 ± 5) kJ mol^{-1} which shows good agreement (within error) with that reported previously (44 kJ mol^{-1}),[12] establishing the principle of disc MSM LKE activation energy measurement.

In fact, it should also be possible to extract values for D and c from analysing the shape of the chronoamperometric transients for disc MSMs. If one avoids fitting the currents obtained in the first few tens of milliseconds, where double layer charging dominates the current, Xiong *et al.*[34] have shown satisfactory fits to the equation developed by Shoup and Szabo:[35]

$$i(t) = 4nFDcr\left(A' + B'\frac{r}{\sqrt{Dt}} + C'\exp\left(-D'\frac{r}{\sqrt{Dt}}\right)\right) \tag{3}$$

where $A' = \frac{\pi}{4}$, $B' = \frac{\sqrt{\pi}}{4}$, $C' = 1 - \frac{\pi}{4}$ and $D' = \frac{\frac{\sqrt{\pi}}{4} - 2\pi^{-3/2}}{1 - \frac{\pi}{4}}$ for redox couples in non-aqueous solvents and ionic liquids, particularly for data recorded over the first few seconds of the transient.

Fig. 4A shows a comparable chronoamperometric transient recorded in aqueous media, when using an $r = 25$ µm Pt MSM to study 5 mM ferricyanide in 100 mM aqueous KCl. Consistent with Xiong *et al.*, it is clear that a good fit of eqn (3) to the data is observed if one limits the fit to the data recorded between 0.05 s and 1 s. The adjusted R^2 value was found to be 0.9988 and physically realistic values for ferrocyanide of $c = 5.6 \pm 0.1$ mM and $D = (4.2 \pm 0.1) \times 10^{-6}$ cm^2 s^{-1} were obtained. Also included in the plot is a fit to eqn (4), the established equation for the transient current at a hemispherical electrode.[34]

$$i = \frac{nFAD^{1/2}c}{\pi^{-1/2}t^{-1/2}} + 4nFDcr \tag{4}$$

where $A = \pi r^2$ is the area of the electrode. This simpler equation, which models the expected Cottrell behaviour at short times with the first term, and the steady-state mass transport limited current from eqn (2) at long times with the second term, would be expected to suffer from inaccuracy in the transition region due to

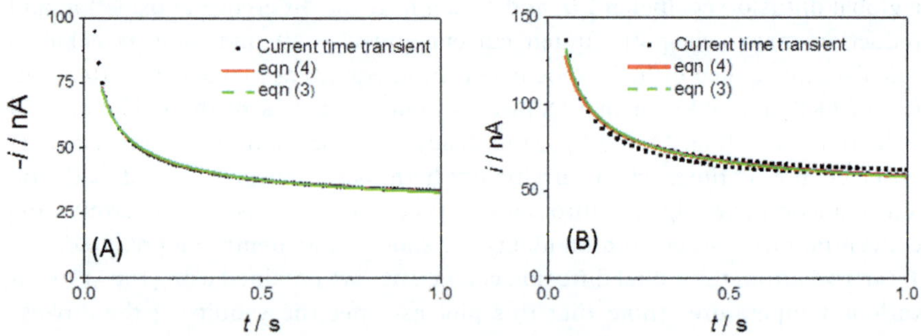

Fig. 4 Current–time transient and fits to eqn (3) and (4) for an $r = 25$ µm MSM Pt disc in (A) 100 mM aqueous KCl with 5 mM ferricyanide (E jumped from +0.34 to 0.14 V at $t = 0$) and (B) LKE with 19 mM added $EuCl_3$ (E jumped from +0.5 to +0.2 V at $t = 0$).

the switch from planar diffusional symmetry at short times to hemispherical diffusional symmetry at longer times. Interestingly it gives almost as good a fit as eqn (3), with adj. $R^2 = 0.9975$, $c = 5.2 \pm 0.1$ mM and $D = (4.3 \pm 0.9) \times 10^{-6}$ cm^2 s^{-1}.

Fig. 4B shows the typical corresponding fits to the current–time transient recorded with $EuCl_3$ in LKE. In this case the fit to the Shoup–Szabo eqn (3) was again reasonable, with $D = (4.4 \pm 0.3) \times 10^{-6}$ cm^2 s^{-1}, which is consistent with the global value of D, $c = 9.5 \pm 0.4$ mM and adj. $R^2 = 0.9694$. The fit to eqn (4) again produced comparable parameters with $c = 8.7 \pm 0.5$ mM and $D = (4.3 \pm 0.3) \times 10^{-6}$ cm^2 s^{-1} and was only marginally worse with adj. $R^2 = 0.9655$.

3.2.2 Electrochemical impedance spectroscopy (EIS). We have previously shown that accurate quantitative data can be obtained from electrochemical impedance spectroscopy (EIS) for square MSMs using solution redox couples under ambient aqueous conditions.[29]

Fig. 5A shows the typical EIS response for a r = 25 μm Pt disc MSM to an aqueous 5 mM potassium ferricyanide solution at 293 K. It is reassuring that the response fits well (with $\chi^2 = 0.002$) to the established modified Randles' equivalent circuit for the disc microelectrode[36–38] (Fig. 5C), confirming its applicability to ambient aqueous measurements with disc MSMs as with other disc microelectrodes.

Fig. 5B shows the corresponding response of an MSM to the Eu(III/II) couple at its formal potential in LKE solution at 703 K. For these and all other EIS LKE disc MSM data, the best fit was achieved by replacing the Warburg element with

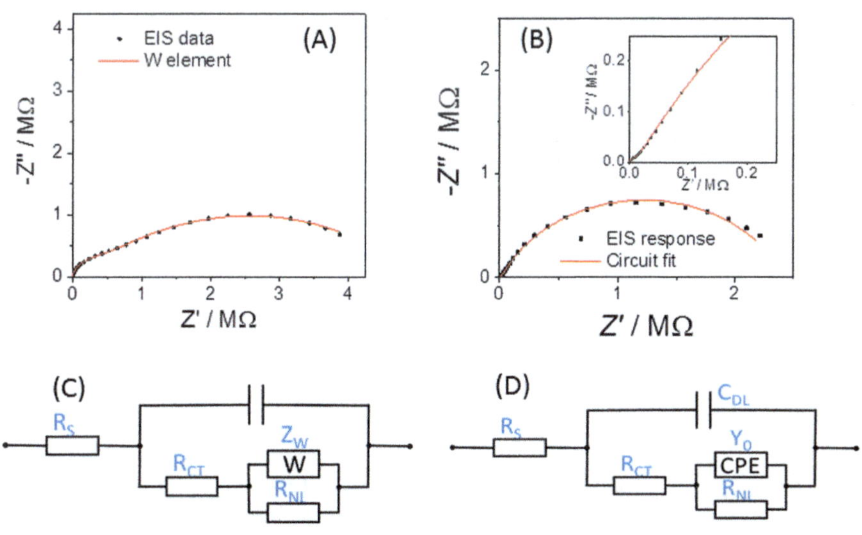

Fig. 5 (A) EIS response at $E_{DC} = +0.240$ V (points) and equivalent circuit fit (line) for an r = 25 μm Pt disc electrode in 100 mM aqueous KCl containing 5 mM potassium ferrocyanide. (B) EIS response at $E_{DC} = +0.345$ V (points) for an r = 25 μm Pt disc and equivalent circuit fits (lines) in 19 mM added $EuCl_3$ in LKE at 703 K. The inset shows an expansion of the higher frequency data, showing the greater than 45° slope consistent with a CPE. (C) The established modified Randles' circuit employed for fitting aqueous disc microelectrode EIS data. (D) The proposed modified Randles' circuit employed for fitting disc MSM EIS data in LKE MS.

a constant phase element (CPE), which is required to fit the gradient of greater than 45° observed at higher frequencies in the Nyquist plot (see inset to Fig. 5B). The fitted values for these data were: $C_{DL} = 1.66 \pm 0.02$ nF; $R_{CT} = 1590 \pm 50$ kΩ; $Q_0 = (1.12 \pm 0.03) \times 10^{-7}$ S s^n; $n = 0.695 \pm 0.005$; $R_{NL} = 2.42 \pm 0.04$ MΩ, with $\chi^2 = 0.03$. It is interesting that the equivalent data recorded in the same LKE solution using a bare tungsten macrowire electrode also requires the use of a constant phase element in place of the Warburg element in the Randles' circuit, and that the fitted value of $n = 0.69 \pm 0.03$ is the same within experimental error as that observed for the MSM in LKE. This suggests a fundamental difference in the linear diffusional properties of ambient aqueous and LKE MS solution at these frequencies. Cuervo-Reyes *et al.*[39] have recently interpreted a diffusional CPE as being due to subdiffusive transport and local trapping of charge carriers. It is tempting to infer that some of the strong electrostatic coupling and local mobile ion trapping postulated in this work in batteries and mobile glasses may also be present to some extent in LKE MS, and that this may also be the origin of the observed CPE.

Turning to the other measured EIS parameters, we have previously shown for square ambient aqueous microelectrodes that at the formal potential of a one-electron redox couple the minimum values of the two resistances R_{CT} and R_{NL} should be given by:[29]

$$R_{NL}^{min} = \frac{4RT}{nFi_L} \tag{5}$$

and

$$R_{CT}^{min} = \frac{2RT}{n^2F^2cAk^0} \tag{6}$$

with the variation of each with E_{DC} being given by:

$$R_{NL} = \frac{R_{NL}^{min}}{4}\left[2 + \exp\left(\frac{-F(E_{DC} - E^0)}{RT}\right) + \exp\left(\frac{F(E_{DC} - E^0)}{RT}\right)\right] \tag{7}$$

and

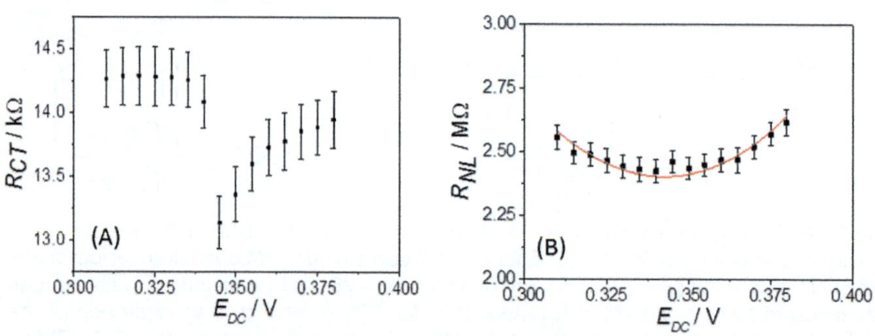

Fig. 6 Extracted values of (A) R_{CT} and (B) R_{NL} with errors from iterative fits of the EIS response of the $r = 25$ µm Pt disc electrode in 19 mM $EuCl_3$ LKE to the modified Randles' circuit (Fig. 5D). (B) also shows the fit (red line) of these values to eqn (7).

$$R_{CT} = \frac{R_{CT}^{min}}{2}\left[\exp\left(\frac{-\alpha_{ox}F(E_{DC}-E^0)}{RT}\right) + \exp\left(\frac{(1-\alpha_{ox})F(E_{DC}-E^0)}{RT}\right)\right] \quad (8)$$

where E^0 is the formal potential and α_{ox} is the transfer coefficient for the oxidation reaction.

It is clear from Fig. 6B that there is good agreement between the disc MSM data and eqn (7), and the fitted value of $R_{NL}^{min} = 2.42 \pm 0.02$ MΩ also agrees well with the value of 2.41 ± 0.03 MΩ calculated from the data in Fig. 3A and B using eqn (5). This indicates that the established EIS MSM theory (eqn (5)–(8)) can be used to extract quantitative information from disc MSM measurements in LKE MS. Using the observed value of $R_{CT}^{min} = 13.1 \pm 0.2$ kΩ a value of $k^0 = 0.26$ cm s^{-1} can be extracted from eqn (6).[29] This value is two orders of magnitude faster than that previously reported in the literature for this system (3.31×10^{-3} cm s^{-1} on a glassy carbon macroelectrode).[40] This is consistent with previous ambient aqueous microelectrode measurements, which have demonstrated that reduction of the dimensions of the electrode to the micron scale allows for markedly more efficient mass transport and the measurement of more realistic charge transfer rates.[41] Given this, it is interesting that the R_{CT} data (Fig. 6A) shows the predicted minimum at $E_{DC} = E^0$, but that these data do not fit well to eqn (8). Although the errors on these data are relatively large, this may suggest some difference between charge transfer behaviour in LKE MS and in ambient aqueous solution.

3.3 Extending the nature of disc MSM measurements – electroanalysis to characterise the formation of metal alloys

Bismuth is a metal of importance within nuclear pyrochemical processing. It melts at 271.4 °C and liquid bismuth electrodes are well known for their propensity to form alloys when electroreducing actinide and lanthanide species dissolved in LKE.[42] As a result of these properties, bismuth is one of the candidates to be used as a reactive liquid metal cathode for the selective extraction of fuel species as metal alloys during the electro-refining stage of nuclear fuel reprocessing.[43] In this study, our aim was to observe whether electrochemical Bi alloying reactions could be studied using the disc MSM.

Fig. 7A shows typical CVs of the behaviour of the bismuth(III/0) couple on Pt and W disc MSM electrodes. When cycling the W MSM electrodes, a simple plating wave (for Bi(III) reduction and deposition) and essentially single stripping peak (for Bi(0) oxidation and dissolution) were observed, centred near $E = -0.06$ V. This is consistent with the phase diagram, which indicates that W and Bi do not form alloys. Using eqn (1) and the observed mass transport limited reduction current, the diffusion coefficient for Bi^{3+} was determined to be $(6.7 \pm 0.02) \times 10^{-6}$ cm^2 s^{-1} at 698 K. In contrast, the Pt MSMs showed Bi(III)/(0) plating and associated stripping electrochemistry, centred slightly above 0 V, with three identifiable, overlapping stripping peaks. It is satisfactory that there is very close agreement for the values of i_L found for the Pt and W MSM electrodes (a 2.1% variation in i_L, which is within the error for the devices), which demonstrates that these differences are not due to changes in n or A, but rather to the effects of Pt/Bi alloying.

The Pt–Bi phase diagram[44] indicates two thermodynamically stable solid alloy phases under these conditions ($PtBi_2$ and PtBi), with a third state of intermediate composition (Pt_2Bi_3) which is thermodynamically stable above 843 K. We

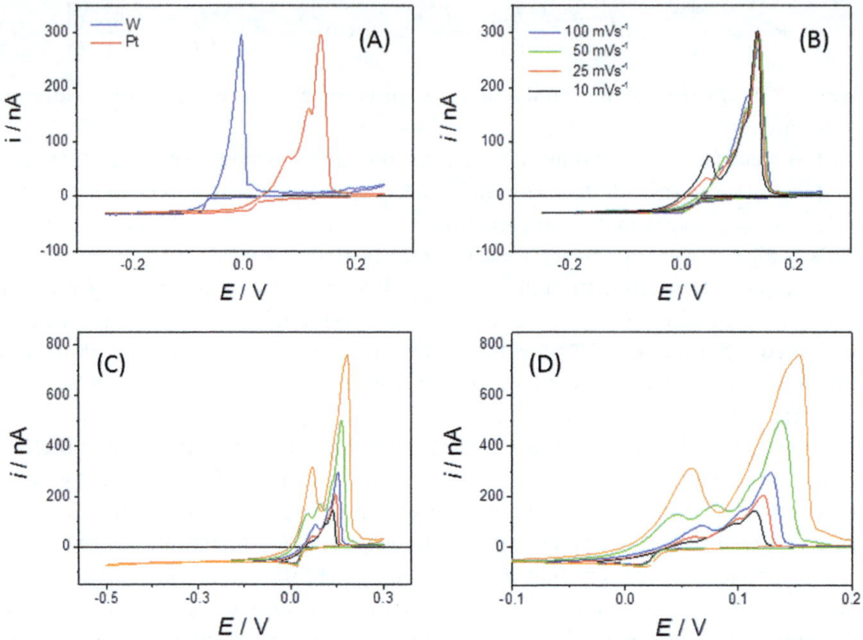

Fig. 7 (A) CVs recorded using a sweep rate of 50 mV s^{-1} with an $r = 25$ μm Pt and an $r = 25$ μm W MSM in LKE + 1.6 mM $BiCl_3$ at 668 K. (B) CVs recorded at the given sweep rates using an $r = 15$ μm diameter Pt MSM in LKE + 5 mM $BiCl_3$ at 690 K. (C) CV and (D) expanded section of the CV at 50 mV s^{-1} with a progressively increasing negative potential limit recorded using a Pt MSM ($r = 25$ μm) in LKE + 1.6 mM $BiCl_3$.

tentatively propose these three stripping peaks originate from stable $PtBi_2$, metastable Pt_2Bi_3 and stable PtBi phases respectively; further quantitative experiments are underway to establish whether this is the case. Fig. 7B also shows that when a Pt disc MSM electrode is swept over a range of scan rates a fourth stripping peak is increasingly observed at the most cathodic potential at the slowest scan rates. This we attribute to stripping of the stable liquid metal alloy (predominantly Bi) phase expected from the phase diagram at high Bi concentrations. Additionally, Fig. 7C and D show the effect of progressively increasing the negative potential limit of the CV scan (which also increases the deposition time). In this case all peaks increase in size, consistent with increased Bi deposition and alloy formation, and unlike in Fig. 7B, the peaks shift progressively to more positive potentials, indicating the contribution of stripping kinetics at these increased coverages. That significant Bi is deposited was confirmed by visual inspection following prolonged plating (Fig. 8C and D). The maximum thickness of liquid Bi plated across the $r = 25$ μm Pt disc surface, at the slowest sweep rate, was calculated to be ~9 nm. It is interesting that the reduction to bismuth metal takes place at a higher potential on the MSM than on a Pt wire macroelectrode (Fig. 8A), and distinct stripping peaks were not observed when using this macrowire, even though the calculated amount of Bi deposited per unit area was similar on each electrode. This indicates greater richness in Bi alloy characterisation information when using MSMs.

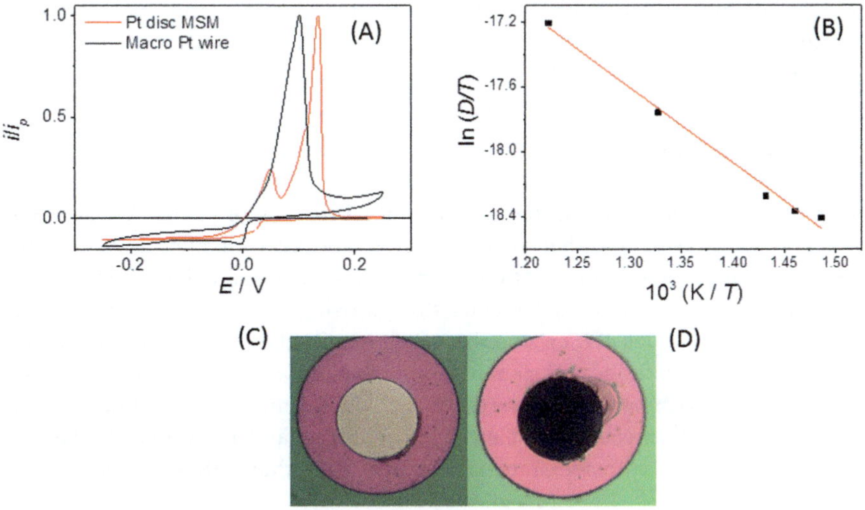

Fig. 8 (A) CVs normalised to the peak stripping current (i_p) recorded using a Pt wire macroelectrode and an $r = 25$ μm Pt disc at a sweep rate of 10 mV s^{-1}. (B) Arrhenius plot used to obtain the activation energy for the diffusion of Bi^{3+} in LKE. (C) Image of an $r = 25$ μm Pt disc before immersion in LKE + 1.5 mM $BiCl_3$. (D) Image of the same $r = 25$ μm Pt disc electrode following application of a potential of −0.2 V for 120 s in LKE + 1.5 mM $BiCl_3$ to deposit Bi.

As with Eu(III)/(II) above, the values of i_L observed for Bi(III) plating on the disc MSMs were used in eqn (1) to obtain Bi(III) diffusion coefficients, from which an Arrhenius plot was constructed (Fig. 8B). The resulting value of $E_a = 39 \pm 4$ kJ mol^{-1} was in close agreement with that found for Eu.

Bismuth binary and ternary alloys that do not involve Pt (of potential importance when carrying out coextraction using Bi electrodes) can also be studied by using W disc MSMs. Fig. 9 shows typical CVs for a mixed Bi(III), Zn(II) system in LKE, which shows Bi(III)/(0) plating/stripping (near $E = 0$ V), Zn(II)/Zn(0) plating stripping (near $E = -0.9$ V) and Li(I)/(0) plating/stripping (at and below $E = -1.8$

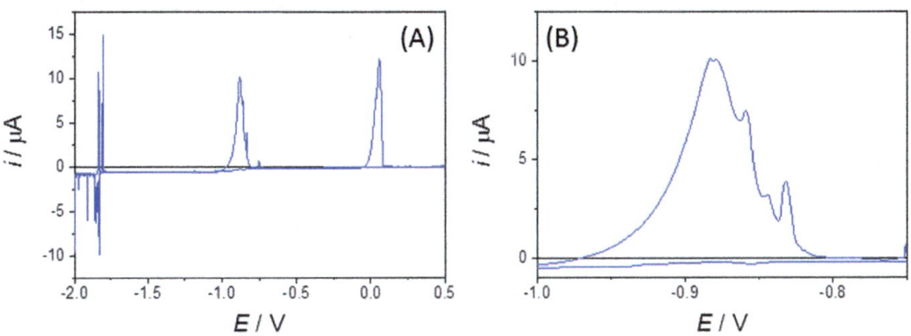

Fig. 9 (A) Bismuth, zinc and lithium codeposition studies using CV (at a sweep rate of 50 mV s^{-1}) in LKE + 5 mM $BiCl_3$ + 10 mM $ZnCl_2$ recorded using an $r = 25$ μm W MSM. (B) Expanded section of the CV showing distinct peaks in the Zn stripping region attributed to stripping from alloy phases.

V). The distinct peaks observed in Li(I)/(0) plating/stripping and Zn(0) stripping features are again attributed to alloy formation, and further indicate the richness of alloying information provided by these disc MSMs.

4 Conclusions

This work has demonstrated the feasibility of fabrication of Pt and W disc MSMs and their stability in, and their potential for, quantitative characterisation of, the LKE MS system. Much of the information that can be obtained is comparable to that obtained from ambient aqueous studies using these or other disc microelectrodes, with values of E^0, D, c and E_a being readily obtainable from established electrochemical methods using analysis established for aqueous systems. As with aqueous measurements, it appears that the enhanced mass transport rates at disc MSMs lead to the measurement of markedly larger and more realistic electrochemical charge transfer rate constants, k^0, than those observed at macroelectrodes.

In addition to these established similarities to aqueous disc MSM responses, which demonstrate the potential of disc MSMs in enhanced MS sensing, analysis and process development, there are also some interesting differences. Whilst a minimum is observed at E^0 for R_{CT} in EIS, the shape of the variation with E_{DC} is apparently not as predicted by ambient aqueous theory. There are hints from variation in the EIS equivalent circuit that there may be a contribution from subdiffusive transport in the LKE MS, as seen previously in glass melts and Li-ion battery materials. There is also the marked enhancement of characterisation information for metal alloy formation from LKE MS compared to macroelectrodes. These measurements provide real motivation for additional study to gain further physicochemical insight into the properties of LKE and other MS systems.

Acknowledgements

The authors acknowledge financial support from, and development of this work through, the UK EPSRC REFINE project (EP/J000779/1) and the EC SACSESS (Contract No. FP7-Fission-2012-323282) and SMART Microsystems (FS/01/02/10 IeMRC Flagship) programmes. We also acknowledge EPSRC and National Nuclear Laboratory (NNL) financial support for studentships. Data used within this publication can be accessed at (http://dx.doi.org/10.7488/ds/1338).

References

1 G. Z. Chen, D. J. Fray and T. W. Farthing, *Nature*, 2000, **407**, 361.

2 K. Sridharan, *Thermal Properties of LiCl–KCl Molten Salt for Nuclear Waste Separation, Project No. 09-780*, Nuclear Energy University Programs, US Department of Energy, 2012.

3 H. P. Nawada and K. Fukuda, *J. Phys. Chem. Solids*, 2005, **66**, 647.

4 D. D. Sood and S. K. Patil, *J. Radioanal. Nucl. Chem.*, 1996, **203**, 547.

5 T. Inoue and L. Koch, *Nucl. Eng. Technol.*, 2008, **40**, 183–190.

6 K. Xu, *Chem. Rev.*, 2004, **104**, 4303.

7 D. Hamon, M. Vrinat, M. Breysse, B. Durand, M. Jebrouni, M. Roubin, P. Magnoux and T. des Courieres, *Catal. Today*, 1991, **10**, 613.

8 G. W. Parshall, *J. Am. Chem. Soc.*, 1972, **94**, 8716.

9 S. Y. Oh, J. Y. Kim, S. E. Bae, Y. H. Cho, J. W. Yeon and K. Song, *J. Lumin.*, 2013, **134**, 706.

10 C. A. Schroll, S. Chatterjee, T. G. Levitskaia, W. R. Heineman and S. A. Bryan, *Anal. Chem.*, 2013, **85**, 9924.

11 A. Uehara, T. Nagai, T. Fujii, O. Shirai and H. Yamana, *J. Nucl. Mater.*, 2013, **437**, 166.

12 C. Caravaca, G. Cordoba and M. J. Tomas, *Electrochemical study of europium trichloride in molten eutectic LiCl–KCl*2005, https://oecd-nea.org/pt/docs/iem/lasvegas04/09_Poster_Session_I/PS1_01.pdf.

13 Y. Castrillejo, P. Herníndez, J. A. Rodriguez, M. Vega and E. Barrado, *Electrochim. Acta*, 2012, **71**, 166.

14 G. Cordoba and C. Caravaca, *J. Electroanal. Chem.*, 2004, **572**, 145.

15 S. Ghosh, S. Vandarkuzhali, P. Venkatesh, G. Seenivasan, T. Subramanian, B. Prabhakara Reddy and K. Nagarajan, *J. Electroanal. Chem.*, 2009, **627**, 15.

16 T. J. Kim, Y. K. Jeong, J. G. Kang, Y. Jung, D. H. Ahn and H. S. Lee, *J. Radioanal. Nucl. Chem.*, 2010, **286**, 283.

17 K. Kushkhov, M. K. Vindizheva, R. A. Karashaeva and M. R. Tlenkopachev, *Russ. J. Electrochem.*, 2010, **46**, 691.

18 H. A. Laitinen, C. H. Liu and W. S. Ferguson, *Anal. Chem.*, 1958, **30**, 1266.

19 P. Masset, D. Bottomley, R. Konings, R. Malmbeck, A. Rodrigues, J. Serp and J. P. Glatz, *J. Electrochem. Soc.*, 2005, **152**, A1109.

20 B. Prabhakara Reddy, S. Vandarkuzhali, T. Subramanian and P. Venkatesh, *Electrochim. Acta*, 2004, **49**, 2471.

21 D. K. Corrigan, E. O. Blair, J. G. Terry, A. J. Walton and A. R. Mount, *Anal. Chem.*, 2014, **86**, 11342.

22 R. J. Forster, *Chem. Soc. Rev.*, 1994, **23**, 289.

23 J. Heinze, *Angew. Chem., Int. Ed.*, 1993, **32**, 1268.

24 M. Sosna, G. Denuault, R. W. Pascal, R. D. Prien and M. Mowlem, *Sens. Actuators, B*, 2007, **123**, 344.

25 S. Legeai, K. k. Soropogui, M. Cretinon, O. Vittori, A. Heeren De Oliveira, F. d. r. Barbier and M. F. Grenier-Loustalot, *Anal. Bioanal. Chem.*, 2005, **383**, 839.

26 R. T. Carlin and R. A. Osteryoung, *J. Electrochem. Soc.*, 1989, **136**, 1249.

27 E. O. Blair, D. K. Corrigan, J. G. Terry, A. R. Mount and A. J. Walton, *J. Microelectromech. Syst.*, 2015, **24**, 1346.

28 A. Relf, D. Corrigan, C. L. Brady, J. G. Terry, A. J. Walton and A. R. Mount, *ECS Trans.*, 2013, **50**, 105.

29 H. L. Woodvine, J. G. Terry, A. J. Walton and A. R. Mount, *Analyst*, 2010, **135**, 1058.

30 C. G. Zoski and M. V. Mirkin, *Anal. Chem.*, 2002, **74**, 1986.

31 G. J. Janz and N. P. Bansal, *J. Phys. Chem. Ref. Data*, 1982, **11**, 505.

32 S. A. Kuznetsov and M. Gaune-Escard, *J. Electroanal. Chem.*, 2006, **595**, 11.

33 A. Einstein, *Ann. Phys.*, 1905, **17**, 549.

34 L. Xiong, L. Aldous, M. C. Henstridge and R. G. Compton, *Anal. Methods*, 2012, **4**, 371.

35 D. Shoup and A. Szabo, *J. Electroanal. Chem.*, 1982, **140**, 237.

36 P. Los, G. Zabinska, A. Kisza, L. Christie, A. Mount and P. G. Bruce, *Phys. Chem. Chem. Phys.*, 2002, **2**, 5449.
37 P. Los and P. G. Bruce, *Pol. J. Chem.*, 1997, **71**, 1151.
38 M. Fleischmann, S. Pons and J. Daschbach, *J. Electroanal. Chem.*, 1991, **317**, 1.
39 E. Cuervo-Reyes, C. P. Scheller, M. Held and U. Sennhauser, *J. Electrochem. Soc.*, 2015, **162**, A1585.
40 M. R. Bermejo, F. de la Rosa, E. Barrado and Y. Castrillejo, *J. Electroanal. Chem.*, 2007, **603**, 81.
41 I. Schmueser, A. J. Walton, J. G. Terry, H. L. Woodvine, N. J. Freeman and A. R. Mount, *Faraday Discuss.*, 2013, **164**, 295.
42 M. Kurata, Y. Sakamura and T. Matsui, *J. Alloys Compd.*, 1996, **234**, 83.
43 W. Dong, X. Yang, J. Zhang and J. Sheng, *J. Mater. Sci.*, 2008, **43**, 3240.
44 H. Okamoto, *J. Phase Equilib.*, 1991, **12**, 207.

Faraday Discussions

PAPER

High temperature ^{35}Cl nuclear magnetic resonance study of the LiCl–KCl system and the effect of $CeCl_3$ dissolution

H. Zhang and I. Farnan*

Received 8th January 2016, Accepted 23rd February 2016
DOI: 10.1039/c6fd00003g

This paper examines the dynamics of the LiCl–KCl system over a range of temperatures in order to understand the local structure surrounding chlorine, which is the common ion in these systems, during molten salt pyro-processing. Chlorine-35 nuclear magnetic resonance (NMR) is sensitive to the local environments of the resonant nuclei and their motion on a diffusive timescale. Thus, it is a good probe of the atomic scale processes controlling the viscosities, diffusivities and conductivities of these molten salts. The average isotropic chemical shifts ($^{35Cl}\delta$) and spin-lattice relaxation times (T_1) of ^{35}Cl in (Li,K)Cl salt mixtures have been obtained over a compositional range of 0–100 mol% KCl with an interval of 10 mol% using high temperature nuclear magnetic resonance (NMR) spectroscopy from room temperature up to 890 °C. The $^{35Cl}\delta$ in the two end member salts are consistent with the cation–anion radius ratio as previously measured on the solid halides and the average radius ratio of cation to anion, can be used to explain the variation of $^{35Cl}\delta$ with composition. The quadrupolar interaction is found to be responsible for the spin-lattice relaxation of the ^{35}Cl, and the activation energies for T_1 relaxation have been obtained for all compositions. The measured T_1 (^{35}Cl) activation energies do not vary linearly with composition and peak at 50% KCl, which also coincides with the Chemla point for this system. They also are in good agreement with the values from equivalent conductivity measurements. To investigate the response of the system to solutes, 8 wt% of $CeCl_3$ was added to the pure LiCl as a surrogate actinide. The shift induced was 120 ppm and the activation energy for the T_1 (^{35}Cl) increased by a factor of four. This is a promising preliminary result for probing the effect of actinide dissolution on the dynamics of these pyro-processing salts.

1 Introduction

Molten (Li,K)Cl binary salts with eutectic composition (40.9 mol% KCl) are used as an electrolyte in pyrochemical processing to separate actinides from spent nuclear fuel. However, in principle, the thermodynamic activity of the actinide

Department of Earth Sciences, University of Cambridge, Downing Street, Cambridge, CB2 3EQ, UK. E-mail: if203@cam.ac.uk; Tel: +44 (0)1223 333431

ions in the solvent salt can be tuned using an appropriate mixture of alkali chlorides, which could allow selective extraction of different metals from complex mixtures.[1] The composition and temperature of the electrolyte[2] can potentially be used to control the interactions of actinides with other ions in the melt in order to get this high separability. The physical consequence of tuning activities is the occurrence of ion–ion interactions, such as the formation of ionic pairs or associates, which decrease the melt conductivity.[3–5] Therefore, it is important to know the nature of the ionic species, and their interactions in the melt, in order to predict the physical and chemical properties of the melt precisely when there is a change of conditions, such as temperature, pressure or concentration. This enables the manipulation of the molten salt system with the goal of greater separation efficiency for the actinides.

This work forms part of a nuclear magnetic resonance (NMR) study of the pyrochemical process that will eventually include actinide-containing melts. Firstly, we focus on the fundamental microstructure and dynamics of the (Li,K)Cl system, as a basis for future studies of the influence of the lanthanides, actinides and other fission products on the molten salt system. The importance of studying simple molten salt solvents has been exemplified by comparing the self-diffusion rate of F^- in (Li,K)F and LiF/YF_3 salt systems, where yttrium (in YF_3) was shown to promote the formation of bridging F^- ions in the mixture.[6]

A considerable amount of research has been carried out in an attempt to understand the microscopic properties of the alkali chloride solvents and to identify the solution species and their roles in determining the bulk properties. The most relevant works involve structural investigations using neutron scattering and melt dynamics using molecular dynamics simulations.[3,7–11] In general, there is a shortening of the inter-ionic distances between metal and chloride ions and a lowering of the co-ordination number of metal ions after melting.[7] Neutron scattering studies[7] show that this effect is more pronounced in LiCl than in KCl. Partial pair distribution functions derived from molecular dynamics (MD) studies[3] of LiCl melts show that the addition of KCl also decreases the Li–Cl distance accompanied by a decrease in the $[LiCl_n]^{1-n}$ coordination number, and *vice versa* for K–Cl when mixed with LiCl. These structural details could qualitatively explain the decreasing trend of the ionic (Li^+ and K^+) diffusion rate and internal mobility (the rate of migration of one species in the melt relative to another) with increasing KCl content in the melt,[3,11] but not the different cross-over points of the two properties. Moreover, the magnitude of the calculated conductivity according to the Nernst–Einstein relation is quite similar to the true conductivity in the (Li,K)Cl system, showing that the degree of correlation between the diffusive motions of different ions is not large.[3] However, the relative change in the calculated conductivity across different compositions is not as large as for the true conductivity. This indicates that ionic association probably affects the overall charge transport in the system.[3]

It is common that the conductivity of the molten binary salts is a nonlinear function of the composition expressed as a mole fraction; usually this is a negative deviation from additivity.[12] In systems with significant deviation, long-lived complex ions are likely to exist in the mixture such as in $CdCl_2$–KCl.[12] Complexes were also proposed to exist in simple alkali halide systems, such as the (Li,K)Cl system,[5] although they show only moderate deviations from additivity in their conductivity.[5,12] In a recent study of the liquid radial distribution function using

MD simulation, an intermediate range chemical ordering of Li^+ ions was considered a general property of simple mixtures of lithium halides with other alkali halides such as the (Li,K)Cl and (Li,K)F systems.[8,11] Experimentally, Raman spectroscopy detected a new band in the (Li,K)F systems, and it was attributed to the relatively long-lived $(LiF_x)K$ configurations compared with pure alkali halides.[9,10] Both numerical simulation and Raman spectroscopy evidenced the modification of the ionic structures upon mixing in the (Li,K)F system. However, due to the difficulty of high temperature experiments caused by the corrosiveness of molten LiCl, KCl and their mixtures, there is still a lack of experimental evidence about the potential intermediate, more long-lived structures in (Li,K)Cl molten salts and their effect on dynamic properties such as conductivity.

High temperature NMR is an effective way to monitor the changes in microstructure and dynamics of salts during pyrochemical processing. For example, the identification of the local microstructure and self-diffusion coefficient of specific ions in the molten LiF/KF system has been carried out successfully despite the severe technical challenges posed by their corrosiveness.[13] In the work described here, *in situ* high temperature NMR is used to study atomic scale processes involved in molten (Li,K)Cl salt systems, and the effect of $CeCl_3$ dissolution in eutectic LiCl–KCl. Using NMR spectroscopy, we could observe the chemical shift with the compositional variation to study the ionic environment. One of the most important and useful features of NMR is its ability to probe the molecular motion by measuring the spin-lattice relaxation time, T_1, which is determined by the time varied local magnetic/electric fields created by microscopic fluctuations in the liquid. Therefore, it can be used to understand the nature of ionic motion and the rate of change of microstructure in the liquid.

2 Experimental

The hydroscopic nature of halide salts is a known problem. The chloride salts were purchased from Aldrich (KCl, 99.999% anhydrous) and Alfa Aesar (LiCl, $CeCl_3$, 99.995% ultra-dry). The samples with intermediate compositions were first weighed and sealed in glass bottles in an argon glove box (oxygen and moisture contents were both below 1 ppm, in the Department of Chemistry, Cambridge). Weighed samples were kept in desiccators (purged under argon) inside a glove bag under a pureshield argon (99.995%) atmosphere before use. All experiments were carried out within one week of sample preparation. Eleven samples with different concentrations of KCl in the range 0–100 mol% with 10 mol% intervals were produced; and seven $CeCl_3$/LiCl–KCl mixtures with a $CeCl_3$ concentration varying from 0.5% to 40%. The salts were further dried at 200 °C for two hours *in situ*, with a pre-dried pureshield argon flow in the NMR probe, and were heated to higher temperatures directly from 200 °C. For LiCl–KCl systems, the NMR sample cells were single crystal alumina (sapphire) crucibles with loose fitting lids purchased from Mettler Toledo. After the NMR runs the molten salts were diluted in deionized water, and the solution pH was measured. This showed that the solutions were slightly acidic (pH 5–6); if there had been any interaction with oxygen or moisture during the measurement the pH would have been >7. For the $CeCl_3$/LiCl–KCl ($CeCl_3$–LKE) system, the NMR sample cells were pure fused silica tubes with Swagelok fittings at the open ends.

The high temperature NMR (HT NMR) spectra were recorded on a Varian-Chemagnetics Infinity 400 MHz (9.4 T) spectrometer equipped with a 7 mm single channel home-built static probe. The probe could operate at temperatures up to 1500 °C by means of a small electric furnace surrounding the specimen coil, and the magnet and shim coils were protected by a water-cooled jacket around the furnace. The sample temperature was calibrated against the furnace temperature with an R-type thermocouple placed in the sample holder during a non-NMR run, and $CsNO_3$ and CsCl with melting points of 414 °C and 645 °C, respectively, were used as fixed points and detected by ^{133}Cs NMR. The absolute temperature error was less than ±5 °C. It took about 5 minutes for the sample temperature to become stable. The probe could be tuned over a resonant frequency range of 35–105 MHz. For more details, refer to ref. 14. The high temperature probe was operated at 39.197 MHz for ^{35}Cl, and spectra were acquired using a single pulse sequence (Bloch decay). Typically, between 8 and 64 $\pi/2$ pulses were acquired with a pulse repetition delay between 0.5 s and 1 s for ^{35}Cl, The chemical shifts were referenced to a NaCl 1 M D_2O solution at room temperature for ^{35}Cl. A small, furnace induced, shift of 1.5 ppm was detected by reversing the furnace polarity and is taken into account in all the reported data, but not in the plotted spectra. Spectra were acquired over a range of temperatures below the melting points and for all the samples at temperatures 10 °C to 20 °C above the melting point and additionally at constant temperature (above the highest melting point in the system) of 837 °C LiCl–KCl and 580 °C for the $CeCl_3$–LKE system. Each measurement was carried out 20 minutes after reaching the desired temperature. The inversion recovery pulse sequence was used to measure spin-lattice relaxation times in the melt 20 °C above the liquids for each composition, with a temperature step of 20–30 °C. The investigation of the temperature and compositional effects on the spin-lattice relaxation time was carried out within the temperature range of 636–837 °C for LiCl–KCl (except pure KCl at 837–890 °C), and 580–760 °C for the $CeCl_3$–LKE system.

3 Results

3.1 Chemical shifts

3.1.1 LiCl and KCl. Fig. 1 shows the ^{35}Cl NMR spectra of the two end member salts as a function of temperature. At low temperatures (312 °C in Fig. 1(a), below 437 °C in Fig. 1(b)), the spectra consist of a single resonance arising from all the transitions of the ^{35}Cl spin system ($I = 3/2$), with Cl located in a high symmetry site with no electric field gradient. The presence of ^{39}K and ^{6,7}Li will broaden the line through a dipolar coupling mechanism. Presumably, the greater breadth of the ^{35}Cl line for LiCl compared with KCl is a result of the greater nuclear moments of ^{6}Li and ^{7}Li compared with ^{39}K (*e.g.* at 312 °C, the maximum width at half height is 5.0 ppm and 3.7 ppm for LiCl and KCl respectively). There is a line narrowing with increasing temperature in solid LiCl as a result of cation mobility. Once the temperature exceeds T_m, as chloride ions become mobile and experience more environments during the measurement, the maximum intensities were reached. There are splittings in the peak for KCl above 768 °C and a less noticeable one in LiCl after melting, and this is due to separation of the sample from the lid and bottom of the sapphire crucible during melting. In this case, the chemical shift (δ) was determined as the centre of gravity of the resonance. The separation of salt from lid and bottom was also observed after the sample had cooled down. For both LiCl and KCl the ${}^{Cl}\delta$ is shifted toward a negative value upon melting (Fig. 1(a) and (b)), indicating that there is greater magnetic shielding of chlorine upon melting. The ${}^{Cl}\delta_{liq}-{}^{Cl}\delta_{sol}$ for LiCl and KCl at 837 °C are about −18 ppm and −12 ppm, respectively.

3.1.2 LiCl–KCl mixtures. Fig. 2(a) shows the ^{35}Cl NMR spectral changes with increasing temperature in the mixture of 70 mol% KCl/(Li,K)Cl, of which the phase evolution is indicated in the (Li,K)Cl phase diagram. The spectrum consists of one peak below 200 °C, due to the overlap of the broad single peaks of solid LiCl and KCl, and splits into two peaks at 312 °C because of the narrowing caused by the increased mobility of the cations. Above the eutectic temperature (>353 °C), there are two peaks corresponding to the eutectic liquid salt and solid KCl. The intensity of the peak due to the eutectic liquid salt increases with increasing

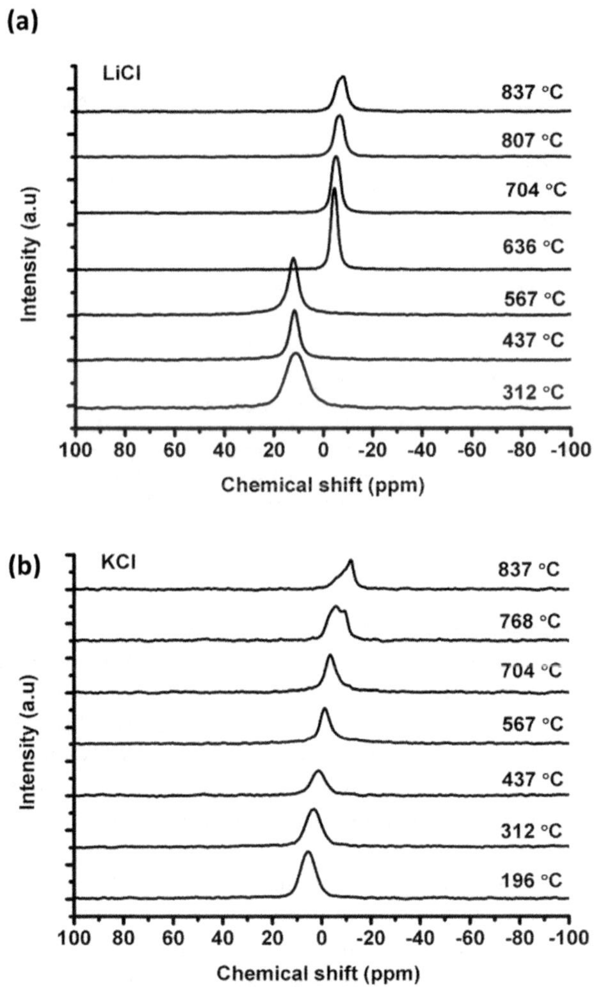

Fig. 1 Chlorine-35 NMR spectra showing change of chemical shift with temperature for (a) ^{35}Cl in LiCl (b) ^{35}Cl in KCl.

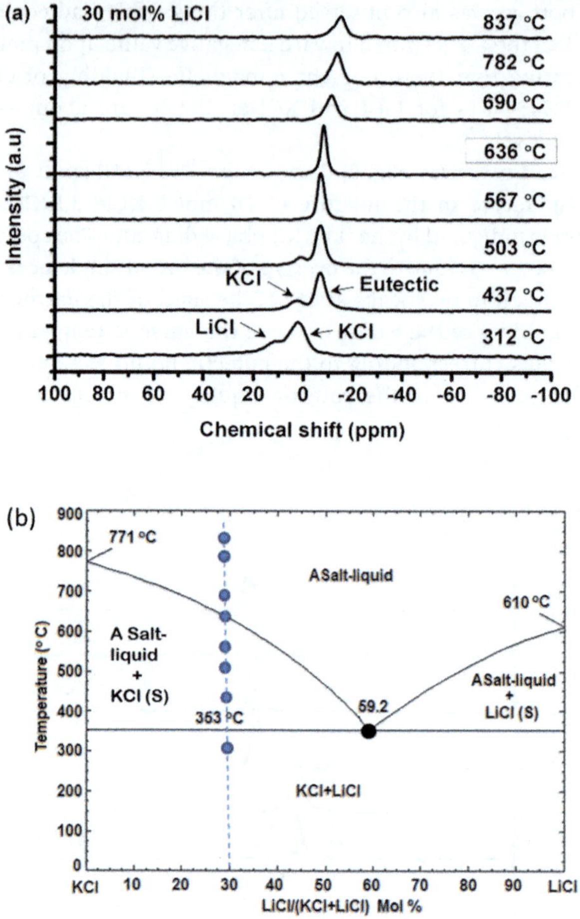

Fig. 2 (a) Chlorine-35 NMR spectra in the 70 mol% KCl/(Li,K)Cl salt system at various temperatures; (b) phase diagram of the (Li,K)Cl salt system showing the phase evolution with temperature beginning with the physical mixture of solid LiCl and KCl salts (dots on dashed line).

temperature, and is consistent with the phase diagram which indicates how the ratio of liquid to solid phase increases (Fig. 2(b)). However, the intensity ratio of the NMR spectra of the liquid to solid ratio (*e.g.* about 3.9 : 1 at 437 °C) is much higher than the value calculated from the phase diagram (*e.g.* about 1.3 : 1 at 437 °C) according to the Lever rule. In particular, the ^{35}Cl NMR signal from the solid KCl merges into the eutectic liquid at 567 °C, where the liquid to solid ratio is about 3 : 1 in the phase diagram, before reaching the melting temperature of 636 °C. This could be due to the high mobility of the chloride ions in the liquid, which appear to be exchanging with the rest of the solid KCl before melting (see Discussion).

Fig. 3(a) shows the compositional effect of adding KCl to the melt on the chemical shift of ^{35}Cl at 10–20 °C above the corresponding melting point. The $^{Cl}\delta$ is the weighted average of chlorine ions with different local environments, and it could be taken as a simple additive function of two pure salts in the (Li,K)Cl

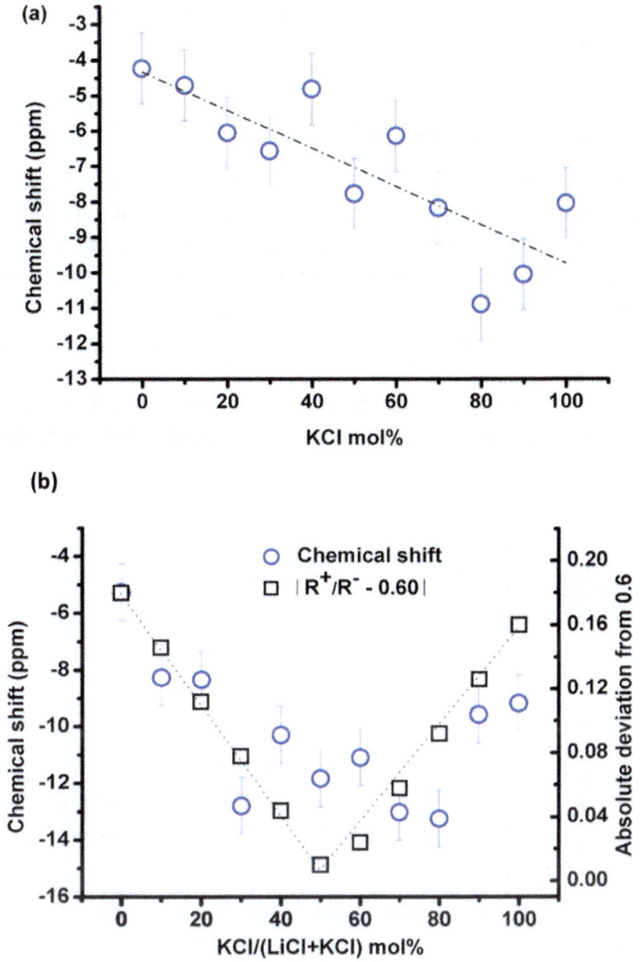

Fig. 3 (a) Variation of the ^{35}Cl chemical shifts with increasing KCl concentration in molten (Li,K)Cl at temperatures 10 °C above the melting point; (b) at a constant temperature of 837 °C (blue circles represent the chemical shift; the dotted line is a visual guide for the trend; black squares represent absolute deviation of the cation-to-anion radius ratio (R^+/R^-) from a critical value of 0.6, which will be discussed in detail in Section 4.1).

mixture if the chloride ions have no preferential association in the mixture. The $^{Cl}\delta$ decreases approximately linearly (least square fitting, the vertical deviation R^2 is 64.5%) upon mixing with KCl with a variation of ~10 ppm across the whole composition. The fitting is just for guidance, as the data become more scattered in the KCl rich zone, possibly due to the migration of the liquid salt towards the cap of the container at high temperature (*e.g.* in Fig. 1(b)).

The temperature effect on the chemical shift was similar for all compositions (−1.4 ppm/100 °C, *e.g.* in Fig. 2(a)) in the melt and was not negligible, so the compositional effect was compared at a single temperature of 837 °C, as shown in Fig. 3(b). Despite the small variation of the $^{Cl}\delta$ with composition and the scattering of the data points, it can be seen that $^{Cl}\delta$ first decreased and reached a minimum value at about 50 mol% KCl and then increased until pure KCl. This indicates that the effect of composition is nontrivial in the (Li,K)Cl mixture, which will be discussed in Section 4.1.

3.2 Spin-lattice relaxation times, T_1

3.2.1 LiCl and KCl. The temperature dependence of spin-lattice relaxation time (T_1) of ^{35}Cl in LiCl is given in Fig. 4. The relaxation is most efficient at an intermediate value of the correlation time responsible for the relaxation, which is long in a solid and short in a liquid salt. In the solid (Fig. 4(a)), T_1 decreases rapidly from 905 ms at 250 °C (523 K) to 41.0 ms at 567 °C (840 K). This is due to the increased fluctuation rate (decrease of correlation time) of the random field caused by the diffusional motion of cations in the crystal lattice with increasing temperature towards the most efficient value ($\tau_c \sim 1/\omega_0$). A T_1 minimum could not be observed. In the melt, the spin-lattice relaxation time increases as the

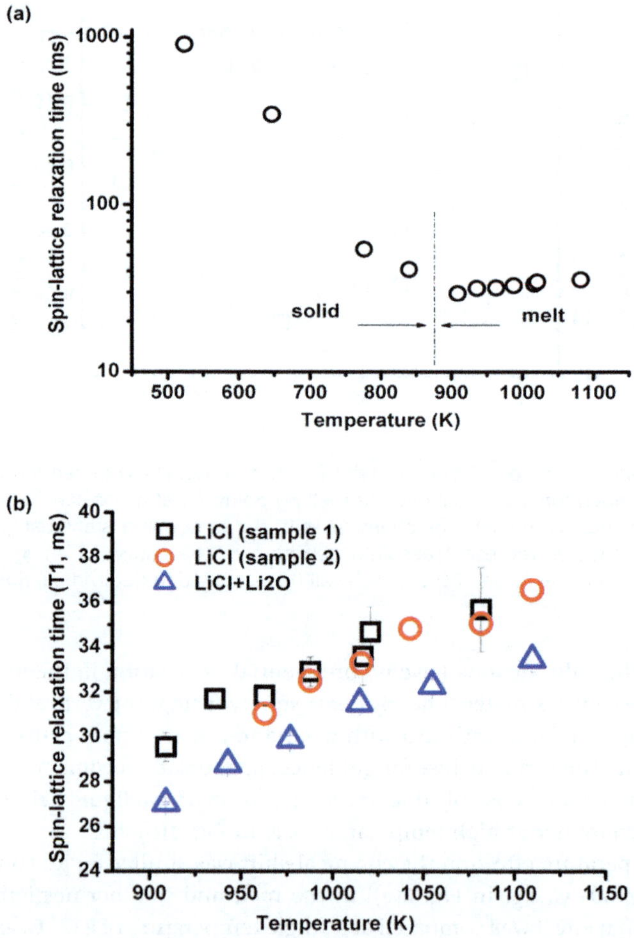

Fig. 4 Spin-lattice relaxation time of ^{35}Cl (a) with temperature in pure LiCl before and after melting, and (b) with increasing (sample 1) and decreasing (sample 2) temperature and with ~1 wt% of Li_2O.

temperature increases. The same trend of T_1 *vs.* temperature is also found for ^{6}Li, and it has much longer relaxation times, in the range of 23.5–1477 seconds in solids and about 11.3 seconds in the melt. The reliability of the ^{35}Cl T_1 measurements was ensured by testing different samples with increasing and decreasing temperatures, and the data shows good overlap (Fig. 4(b)) indicating that the salt was not oxidized in the melt during the measurements, as demonstrated by the deliberate addition of 1 wt% Li_2O in LiCl, which decreased the absolute value of the relaxation time.

In the melt, the spin-relaxation time, T_1, increases with temperature, indicating a decreasing motional correlation time with temperature and, well away from the T_1 minimum, the activation energy (E_a) of the motion responsible for the spin-lattice relaxation of Cl^- can be calculated from an Arrhenian relationship. The calculated activation energies for the spin-lattice relaxation of ^{35}Cl in pure liquid LiCl and KCl are 0.0826 ± 0.007 eV and 0.1509 ± 0.015 eV, respectively. These agree, within error, with the activation energies of 0.0746–0.0899 eV (LiCl melt) and 0.1414–0.1489 eV (KCl melt), which were obtained in high temperature equivalent conductivity measurements of LiCl and KCl.[5,12,15] Here the equivalent conductivity is the product of the specific conductivity and the equivalent volume (of 1 mol), and it refers to a state in which there is always one equivalent of the salt between electrodes at 1.0 cm distance.[5,12] Therefore, the activation energy derived from equivalent conductivity is preferred for comparisons of ionic migration. Futhermore, in LiCl and KCl melts, both cations and anions conduct, and their activation energies were taken as the same in a calculation of the conductivity,[5,12] which matched the experimental results. ^{35}Cl HT NMR detects the variation of the activation energy for molten salt equivalent conductivity. Thus, the same motional process is underlying both measurements even if their time scales are very different.

3.2.2 LiCl/KCl mixtures

3.2.2.1 Effect of temperature. The spin-lattice relaxation time of ^{35}Cl over a large temperature range of 437–837 °C (710–1110 K) was investigated in eutectic (Li,K)Cl salt. As shown in Fig. 5(a), T_1 increases monotonically with temperature. When the data are plotted as $\ln(T_1)$ against inverse temperature to obtain the activation energy, there appears to be a change in the rate of increase of T_1 with temperature above 636 °C (Fig. 5(b)).

The changing rate may be decomposed into two activation energies calculated from $\ln(T_1)$ *vs.* $1/T$ curves, which are 0.231 eV and 0. 171 eV at lower (437–636 °C) and higher (636–837 °C) temperature ranges, respectively. This indicates that the ionic mobility is enhanced at higher temperatures, which is consistent with computational results showing fewer ionic complexes in the (Li,K)Cl eutectic composition, with conductivities closer to the theoretical value of free ions with increasing temperature.[16] There is no report on the activation energy variation with temperature in the (Li,K)Cl system, but a similar phenomenon was found in the alkali and alkaline earth halide systems such as $CdCl_2$–KCl.[15] In these systems, the decrease in the activation energy for motion suggests the dissociation of the ionic complex, and this indicates there are complexes in the molten (Li,K)Cl system.

3.2.2.2 Effect of composition. Fig. 6 shows the compositional effect on the relaxation time at a constant temperature of 837 °C, and it is compared with the previous study on the variation of equivalent conductivity (Λ) at 800 °C,[5] because

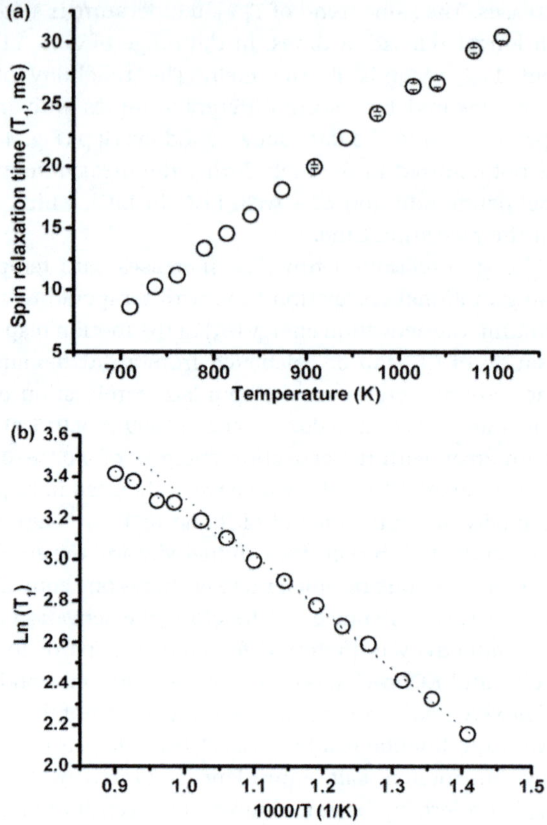

Fig. 5 (a) T_1 of chlorine-35 as a function of temperature above the eutectic temperature of $(Li_{0.59}K_{0.41})Cl$ salt; (b) ln(T_1) *vs.* 1/temperature (1/*T*), dashed lines indicate the possible existence of a change in slope, and therefore activation energy at high temperature.

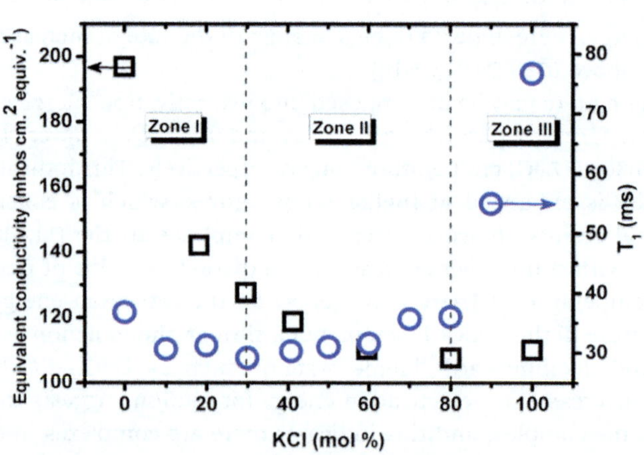

Fig. 6 The relaxation time *vs.* composition at 837 °C (1110 K, circle) measured by NMR in this work; equivalent conductivity *vs.* composition at 800 °C (1073 K, squares) from ref. 5 (classical AC technique).

the activation energy for T_1 in two end member salts is comparable with the energy derived from equivalent conductivity.

The relaxation time of Cl^- first decreases upon mixing and reaches a minimum at about 30 mol% KCl, followed by a nontrivial increase, from 29 ms to 36 ms within a wide composition range of 30–80 mol% KCl, and then increases more than two-fold from 80 mol% to pure KCl. The composition with the lowest value of T_1 is dependent on temperature and moves to lower KCl concentrations as temperature is increased. This might indicate a change in ionic mobility similar to the Chemla effect.[17] According to the relationship between T_1 and Λ, the mixture could be divided into 3 zones. In zone I (<30 mol% KCl) the two properties both decrease with increasing KCl mol%. In zone II (30–80 mol% KCl) T_1 increases but Λ decreases. In zone III (>80 mol% KCl), both increase with composition. The minimum of equivalent conductivity (80 mol% KCl) indicates a strong chance of formation of complexes or clusters in the mixture.[5] Together with the minimum value of T_1 at 30 mol% KCl, it means that the (Li,K)Cl system is not a simple mixture of two salts within a certain range of compositions.

The compositional effect on activation energy (E_a) derived from $\ln(T_1)$ *vs.* $1/T$ is shown in Fig. 7 (blue circles). The activation energy calculation is carried out within the same temperature range of 637–837 °C for samples with KCl <70 mol%; and the remainder from 20 °C above the melting point to 837 °C (pure KCl from 837–888 °C). The change of E_a corresponds well to the classification of zones in Fig. 6, and zone II has the higher activation energy than a linear combination of the activation energies of the LiCl and KCl end-members (dashed line), and it reaches a maximum at approximately 40–50 mol% KCl. Moreover, the activation energy obtained from the relaxation time is consistent with the value calculated from the equivalent conductivity (calculated based on the data presented in ref. 5) in the (Li,K)Cl mixture. The maximum activation energy indicates a great chance of forming complexes or clusters in the mixture, which can consume more energy *via* a transition from complexes to single ions with increasing temperature.

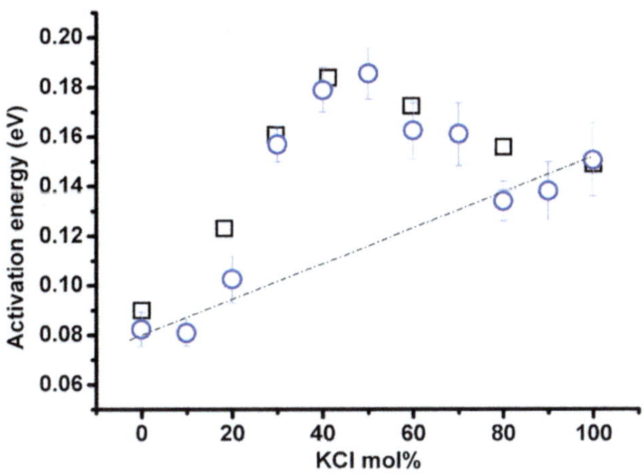

Fig. 7 Activation energy for ^{35}Cl in (Li,K)Cl molten salt mixtures calculated according to spin-lattice relaxation time *vs.* 1/temperature (blue circles); and equivalent conductivity *vs.* 1/temperature (black squares, equivalent conductivity was calculated from experimental data for specific conductivity and density presented in ref. 5).

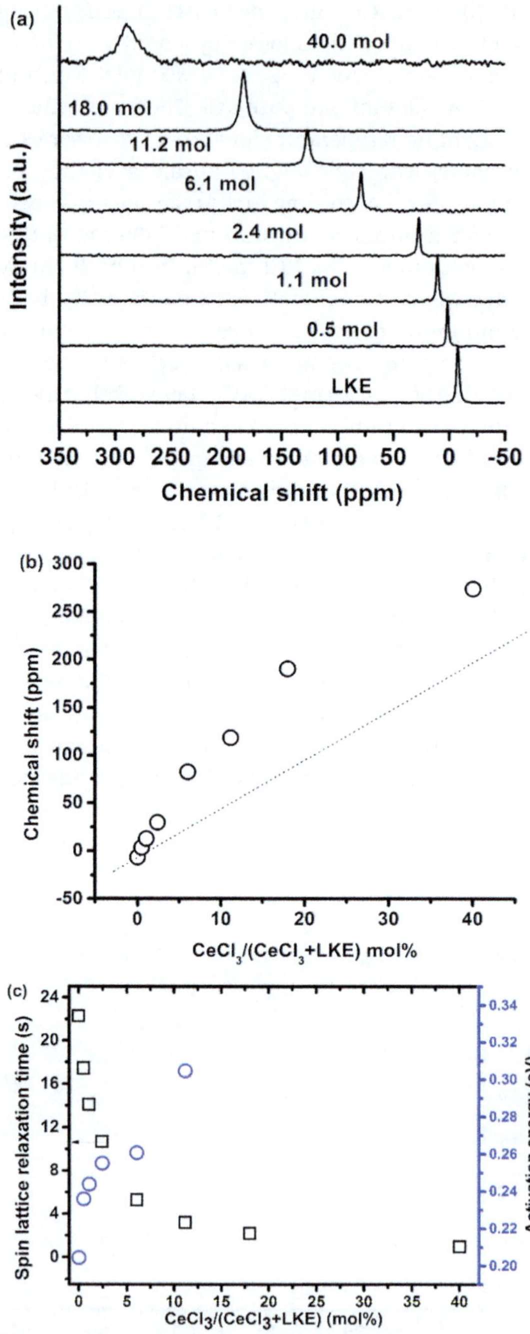

Fig. 8 (a) ^{35}Cl NMR spectra and (b) ^{35}Cl NMR chemical shift at 580 °C, and (c) spin-lattice relaxation time (squares) at 601 °C and activation energy (circle), ^{35}Cl T_1 relaxation in the $CeCl_3$–LKE system as a function of $CeCl_3$ concentration (mol%).

3.3 Effect of $CeCl_3$ dissolution

The ^{35}Cl spectra in $CeCl_3$–LKE over a wide range of $CeCl_3$ concentrations is shown in Fig. 8(a) at 580 °C, and the ^{35}Cl peak is shifted from −6.8 ppm in LKE to positive values with increasing $CeCl_3$ concentration (*e.g.* to 274 ppm in 40 mol% $CeCl_3$). Due to the paramagnetic nature of Ce^{3+} in $CeCl_3$ the ^{35}Cl resonance was adversely broadened and the chemical shift of ^{35}Cl with $CeCl_3$ higher than 40 mol%. The variation of ^{35}Cl chemical shift as a function of $CeCl_3$ mol% is presented in Fig. 8(b), the dashed line is a guide for the eye presuming the chemical shift of ^{35}Cl is only based on two local environments (Cl^- anions from LKE and $CeCl_3$). A non-linear monotonic increase is observed, and this kind of evolution has already been reported in a previous study on LaF_3–AF (A = Li, Na, K) systems.[18] The non-linearity in this case was attributed to the existence of bridging F^- before all the F^- are involved in LaF_x units in pure LaF_3.

Fig. 8(c) presents the spin-lattice relaxation time and T_1 activation energy variation with $CeCl_3$ concentration.

4 Discussion

4.1 Influences on chemical shift

The chemical shift, $^{Cl}\delta$, in KCl (−8 ppm) is more shielded than in LiCl (−4 ppm). This contradicts the empirical finding that the halide chemical shift is more negative for smaller alkali cations in the alkali halides, where the chemical shift is assumed to be controlled by the nearest neighbour interactions.[13,19] The difference in chemical shift between the two end-member chloride salts is considerably less than that in LiF and KF (*e.g.* 70 ppm difference for $^{F}\delta$).[13] This shows that the chemical shift depends on both the size of the cation and anion, and a more detailed study[19] found that the chemical shift can be related to the cation-to-anion radius ratio ($R_m = R^+/R^-$). With a decrease in this ratio, there is an increase in the shielding of the ^{35}Cl, while deshielding happens when the ratio is <0.6. This parabolic behavior of shielding and deshielding was attributed to changing interaction mechanisms from a first neighbour interaction to a second nearest neighbour induced polarisation[19] at $R_m = 0.6$. The similarity of the $^{Cl}\delta$ in LiCl and KCl in the current study is because the R_m for KCl and LiCl of 0.76 and 0.42, respectively, are located either side of the turning point at 0.6. The chemical shifts of KF and LiF decrease approximately linearly with a radius ratio of 1.04 and 0.57, respectively, resulting in a chemical shift difference of 70 ppm.

The cation-to-anion radius ratio can also explain the variation of the chemical shift with the KCl concentration at constant temperature. Given the cation-to-anion radius ratio in the mixtures is a simple addition of two-end member salts, the ratio increases with the KCl concentration, and a critical value of about 0.6 is obtained at a composition of 50% KCl. The absolute deviation of the cation-to-anion radius ratio from the critical value was plotted in Fig. 3(b), and this trend is consistent with the evolution of $^{Cl}\delta$. Therefore, the $^{Cl}\delta$ in alkali halides such as KCl with $R_m > 0.6$ is controlled by the first nearest neighbour interaction (the same as ^{23}Na, ^{87}Rb and ^{133}Cs chlorides); while for the small sized Li^+, the second nearest neighbour induced polarization will increase the paramagnetic effect resulting in an increase of $^{Cl}\delta$ in LiCl.

The coalescence of solid and eutectic liquid ^{35}Cl line shapes before reaching the melting point (as shown in Fig. 2(a)) indicates that there could be ionic exchanges between the solid and liquid phases. In a previous study on AlF_3–(Na,K)F eutectic systems,[20] the total dissolution of AlF_3 was found to be at 879 °C by thermal analysis; however, it was cloudy by visual observation, and a clear melt was only observed at approximately 900 °C. The maximum ^{27}Al NMR spectrum was observed at an even higher temperature of 935 °C. These observations show that the melting of the mixture took place over a range of temperature, and involves dissolving/melting processes. In the (Li,K)Cl system a similar process occurred, as the melting process involves the dissolving of the solid into the eutectic before reaching the melting temperature. Moreover, the temperature effect on the $^{Cl}\delta$ in molten (Li,K)Cl mixtures, where the $^{Cl}\delta$ decreases slightly (−1.4 ppm/100 °C) with increasing temperature, could be due to the change in the degree of association between the ions. The cations are more tightly bonded to the anions with an increase of temperature, and the coordination number of cations decreases according to MD simulation.[3] Therefore, this decrease in chemical shift could be interpreted with more free anions according to a previous study on the fluorine system.[13] Alternatively it could be due to thermal expansion, reducing the paramagnetic contribution in the shift causing a reduction in the Cl–Cl interaction.

It is helpful to recall that the NMR spectral life-time is in the range of milliseconds, and species that have shorter lifetimes than this are represented by an average chemical shift. Previous MD simulations[8] show that cation ordering, such as Li^+–Li^+, is the origin of the first sharp diffraction peak in Li halide mixtures. The measurement of the chemical shift of ^{35}Cl would not obtain direct evidence of intermediate range. This is because the residence time of the ions in clusters was estimated to be a few picoseconds,[3,11] and the process on this time-scale could contribute to spin-lattice relaxation. However, the increase in activation energy for spin-lattice relaxation is consistent with MD simulations.[8,11]

4.2 Spin-lattice relaxation mechanism

The NMR spin-lattice relaxation time quantifies the rate of transfer of energy from a nuclear spin system to neighbouring molecules (the lattice) by thermal motion, such as translation, rotation or internal motion of molecules. Thermal motion creates fluctuating magnetic and electric fields at the site of nuclear spins through spin interactions, which could be paramagnetic, dipole–dipole, or for ^{35}Cl, quadrupolar interactions. The purchased salt is of high purity (>99.99%), so the influence of paramagnetic elements on spin-lattice relaxation could be neglected. For spins > 1/2 ($I = 3/2$ for ^{35}Cl), an electric quadrupolar relaxation mechanism usually dominates the relaxation. The relaxation is due to interaction of Cl nuclei with a fluctuating electric field. The quadrupolar spin-lattice relaxation time under isotropic motion in a liquid is given by:[21]

$$\frac{1}{T_1} = \frac{3}{40}\frac{2I+3}{I^2(2I-1)}\left(1+\frac{\eta^2}{3}\right)\left[\frac{eQ}{\hbar}\frac{\partial^2 V}{\partial z'^2}\right]^2 \tau_c = C\tau_c \tag{1}$$

where I is the nuclear spin of nuclei, η defines the asymmetry of the electric field gradient, e is the electron charge, and Q is the nuclear quadrupole moment. The $(\partial^2 V)/(\partial z'^2)$ represents the electric field gradient at the nucleus; τ_c is the correlation time of the changing surrounding electric field due to liquid motion; and C is a constant.

T_1 relaxation is most effective (corresponding to a T_1 minimum) at a temperature where $\omega_0\tau \approx 1$. Here ω_0 is the resonant frequency of nuclei, and τ is the correlation time at this temperature. When $\omega_0\tau > 1$, the relaxation time increases with increasing correlation time (decreasing temperature), while it decreases when $\omega_0\tau < 1$. For ^{35}Cl, $\tau \sim 1/2\pi\omega_0 \sim 4.07 \times 10^{-9}$ s is expected at T_1 minimum. As T_1 of ^{35}Cl increases with increasing temperature (Fig. 4) in the melt, this means the relaxation is at the fast motion limit where $\omega_0\tau_0 \ll 1$ ($\tau_0 \ll 4.07 \times 10^{-9}$ s). This confirms the general understanding of the relatively high mobility of the ions in the molten salt reflected in the bulk transport properties, such as conductivity.

In a (Li,K)Cl melt, it is impossible to measure the T_1 minimum, so the electric field gradient could not be calculated according to eqn (1), which means that the correlation time corresponding to measurable T_1 is unknown. It is impossible to identify the nature of the thermal motion from eqn (1); therefore we calculated the correlation time from thermal motions which may contribute to relaxation, and compared the resulting T_1 (according to eqn (1)) with the measured T_1 in molten (Li,K)Cl in order to explain our observations indirectly.

We can compare our results on dynamics from nuclear magnetic resonance and molecular dynamics. The correlation time derived from translational motion (diffusion) can be estimated by $\tau_t = R_0^2/6D$, where R_0^2 is the mean square jump distance and D is the diffusion coefficient. For example in LiCl and KCl at 837 °C, taking the R_0^2 (ref. 3) and D (ref. 22 and 23) from previous work where self-diffusion was measured with radioactive tracers, the estimated translation correlation times τ_t are about 2.6 ± 0.5 ps and 5.1 ± 1.0 ps, respectively. These are considerably shorter than the τ_c correlation time at a minimumT_1. However, the activation energies derived from spin-lattice relaxation time (0.083 eV and 0.151 eV, respectively) are much less than the energies from self-diffusion of Cl^- in LiCl and KCl (0.133 eV and 0.297 eV, respectively calculated from previous measurements.[22,23] This precludes self-diffusion as the direct cause of relaxation.

The correlation times derived from the self-exchange velocity calculated from MD simulations[3] are consistent with the different T_1s in LiCl and KCl. The self-exchange velocity calculates the exchange rate of neighbouring unlike ions, $\nu = (R_2 - \langle R_2 \rangle)/\tau$, where R_2 is the distance where the radial distribution function ($g(r)$) between anions and cations crosses unity for the second time; $\langle R_2 \rangle$ is the average separation of unlike ions within R_2; and τ is correlation time of this motion ($\tau_{\text{self-exchange}}$). The $\tau_{\text{self-exchange}}$ for Li^+ and K^+ at 823 °C are 0.33 ps and 0.56 ps respectively. These values are very close to the correlation times derived from a previous study[24] in molten LiBr (0.11 ps) and LiI, which were calculated based on the similarities with the quadrupolar relaxation in liquid metal. The correlation time was interpreted as the average time that the spatial configuration of a group of near neighbours persists.

The radial distribution functions were taken to be the same for Li–Cl at 823, 727 and 627 °C, because the local structure did not change as much as ion dynamics with varying temperature.[3] The constants in LiCl calculated from eqn (1) are 8.33, 7.73 and 7.27 × 10^{13} 1/s^2 at 823, 727 and 627 °C, respectively, and 2.32 × 10^{13} 1/s^2 in KCl at 823 °C, with the correlation time calculated from self-exchange (τ) and measurable spin-lattice relaxation time (T_1). The electric field gradient in LiCl increases with increasing temperature, and the calculated Cl electric field gradient (eqn (1)) in LiCl and KCl at 823 °C were 1.34×10^{21} V m^{-2} and 0.71×10^{21} V m^{-2}, respectively. These values could potentially be compared with force field parameterized *ab initio* calculations of fluctuating electric field gradients recently published for species in aqueous solution.[25] The larger Cl electric field gradient, derived from eqn (1), in LiCl compared with KCl could be due to greater Cl core polarization due to Li, and lead to a shorter T_1 in LiCl. This is also consistent with the observation of higher chemical shifts in LiCl than KCl. Therefore, it is most likely that the relaxation is caused by motion of the Cl^- to the near neighbour unlike ion sites in the molten salts. The self-exchange velocity is the motional mechanism of the internal mobility,[3] which has similar properties to the conductivity, so this could explain the similar activation energies of the relaxation and equivalent conductivity.

4.3 Compositional effect on relaxation time

Fig. 9 compares the calculated T_1 according to eqn (1), where the correlation time ^{35}Cl is the weighted average of correlation times calculated based on self-exchange velocities of Li^+, K^+ with Cl^- in the mixture,[3] and the electric field gradient in the mixture is a weighted average of two end member salts (823 °C), with measured T_1 (837 °C). The actual lower T_1 could be due to either the longer correlation time and/or larger electric field gradient than the above weighted average of two properties assumption. The compositional effect on relaxation time and activation energies (Fig. 6–8) shows good evidence for the formation of complexes in (Li,K)Cl molten salt mixtures, which can increase both the correlation time and the electric field gradient.

Upon addition of KCl (for zone I <30 mol% KCl), both T_1 and Λ depend mainly on the LiCl concentration. Upon addition of KCl, the Li–Cl distance decreases and cation association with the anions is longer according to MD simulations.[3] These make it more difficult for ions to exchange with near neighbour unlike ions, and

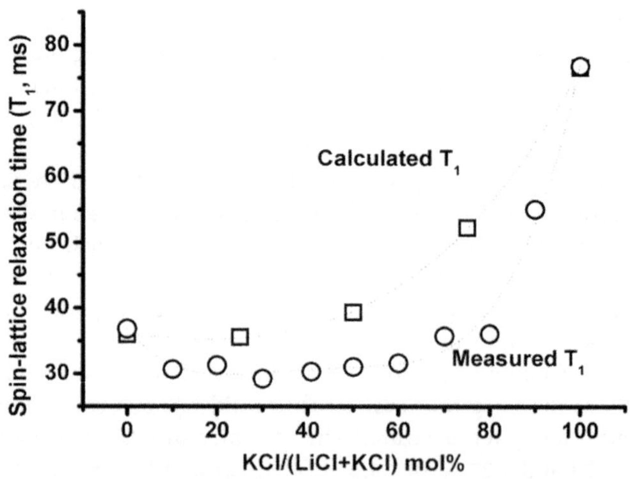

Fig. 9 ^{35}Cl NMR spin-lattice relaxation times as a function of composition calculated from eqn (1) using a linear weighted average of LiCl and KCl Cl electric field gradients and correlation times (squares) compared with measured values (circles).

increase the chlorine correlation time. Although the addition of KCl could decrease the average electric field gradient, the change is not big enough to compromise the increasing correlation time, so the T_1 decreases. The decreased mobility of Li^+ and Cl^- could lead to the negative deviation of the equivalent conductivity from the addition of the two end member salts. Furthermore, it is worth noting that the final composition of zone I is temperature dependent. Similarly as in zone III (>80 mol% KCl), KCl concentration determines the two properties. The significant increase in T_1 is due to the smaller electric field gradient in KCl (eqn (1)), and the slight increase in Λ could be due to the disappearance of clusters.

In zone II (30–80 mol% KCl), the peak value of the activation energy could be related to the formation of intermediate range chemical order on the time scale of picoseconds according to previous studies.[8,11] When there are longer-range interactions, such as clustering, in the mixture, these would be able to dissociate more readily into free ions with increasing temperature. This reduces the correlation time and electric field gradient, corresponding to a faster increase in the relaxation time, and results in greater rate of change of liquid structure with temperature and the maximum activation energy in zone II. Moreover, the concentration of $LiCl_2^-$ and K_2Cl^+ complexes in the melt calculated using density functional theory[16,26] showed a maximum in the range of 37–62 mol% KCl, in a similar plot to Fig. 7, which shows a maximum in ^{35}Cl T_1 activation energy at 40–50 mol%. However, the MD simulation showed that the intermediate range complex structure is a cluster of Li^+ ions, where the Li^+–Li^+ distance is shorter than the random ionic mixture in a mixture.[8,11] In this study, the effect of overlapping Cl^-–Cl^- pairs on the ^{35}Cl chemical shift in LiCl was observed, and the change of chemical shift in (Li,K)Cl is determined by cation-to-anion radius ratio in the whole compositional range of molten salts. The maximum activation energy for relaxation in zone II indicates that the complex could be a Li^+–Li^+ cluster, which is most significant at this compositional range calculated using radial distribution function[8,11] determined from molecular dynamics.

4.4 Effect of $CeCl_3$ dissolution

The chemical shift variation of ^{35}Cl with $CeCl_3$ dissolution shows a similar trend to a previous study on a LaF_3–ACl system,[18] though the deviation from a linear combination of end-member shifts begins at a lower concentration. It indicates the presence of at last three instantaneous chlorine environments almost throughout the whole composition.

In Glover and Madden's work,[27] the microstructure of molten mixtures of $LaCl_3$ with NaCl and CsCl has been theoretically studied by Raman spectroscopy, to examine the coordination environment around the La^{3+} ions. In pure molten $LaCl_3$ the La-centred coordination must be linked with a Cl^- belonging to the coordination of two or more La^{3+} ions, from which it is inferred that a bridging Cl^- exists in pure $LaCl_3$. Their work showed how the pure melt is broken down by the addition of alkali halides to different extents. This differs from previous studies on $LaCl_3$–AF systems,[18] where the bridging F^- exists in LaF_3–AF mixtures in the same form as in pure $LaCl_3$.

5 Conclusions

High temperature ^{35}Cl NMR measurements of chemical shift and spin-lattice relaxation time can reveal details of the microscopic structure and dynamics of (Li,K)Cl molten salts. The difference in cation-to-anion radius ratio (R_m) can explain the compositional effect on the chemical shift of chlorine at constant temperature in the melt based on second nearest neighbour effects below R_m = 0.6. Thus, the chemical shift in the mixed salt exhibits the same behavior as crystalline alkali halides. Future ^{35}Cl NMR measurements should be able to readily detect the effect of dissolved ions with R^+/R^- >0.6, such as caesium, an important fission product in spent nuclear fuel reprocessing.

It has been shown that the ^{35}Cl spin-lattice relaxation time (T_1) is a result of the interaction between the electric quadrupole moment of ^{35}Cl and the fluctuating electric field gradient caused by the rapid ionic exchange calculated by molecular dynamics simulation (self-exchange velocity).[3] The Cl motional correlation time responsible for T_1 relaxation in LiCl and KCl is on the time scale of picoseconds which is shorter than the time scale of chlorine self-diffusion of tens of ps. The self-exchange rates from MD simulations[3] and the measured T_1 relaxation times in this work at different temperatures in LiCl yield numerical values for the electric field gradients that could be directly compared with future results derived from *ab initio* calibrated force field calculations.[25] The activation energy for spin-lattice relaxation and equivalent conductivity agree very well, which means that the self-exchange rate model of internal mobility in (Li,K)Cl,[3] which has similar collective properties to conductivity is probably accurate.

Microscopic evidence for the variation of the equivalent conductivity of molten salt mixtures with composition was detected. Furthermore, in zone II (30–80 mol % KCl), the maximum activation energy occurs at the same composition for equivalent conductivities and T_1 relaxation. This suggests the presence of complexes or clusters in the melt consistent with a greater rate of change of liquid structure with temperature at intermediate compositions.

Acknowledgements

This work was carried out as part of the UK Engineering and Physical Sciences Research Council (EPSRC) funded REFINE consortium (http://www.refine.eng.ed.ac.uk/). We gratefully acknowledge this EPSRC financial support (EP/J000760/1). We also very much appreciate Prof. Brian Ralph for proof-reading the first manuscript.

References

1 Y. Okamoto, S. Suzuki, H. Shiwaku, A. Ikeda-Ohno, T. Yaita and P. A. Madden, *J. Phys. Chem. A*, 2010, **114**, 4664–4671.

2 M. Salanne, C. Simon, P. Turq and P. A. Madden, *J. Phys. Chem. B*, 2008, **112**, 1177–1183.

3 B. Morgan and P. A. Madden, *J. Chem. Phys.*, 2004, **120**, 1402–1413.

4 P. Hebant and G. S. Picard, *J. Electrochem. Soc.*, 1997, **144**, 980–983.

5 E. R. Van Artsdalen and I. S. Yaffe, *J. Phys. Chem.*, 1954, **59**, 118–127.

6 A. L. Rollet, V. Sarou-Kanian and C. Bessada, *Inorg. Chem.*, 2009, **48**, 10972–10975.

7 V. Danek, *Physico-Chemical analysis of molten electrolytes*, Elsvier, B. V., Amsterdam, 2006, ch. 1 and 8.

8 M. Salanne, C. Simon, P. Turq and P. A. Madden, *J. Phys.: Condens. Matter*, 2008, **20**, 332101–332105.

9 S. A. Kirillov, E. A. Pavlatou and G. N. Papatheodorou, *J. Chem. Phys.*, 2002, **116**, 9341–9352.

10 S. A. Kirillov, *Russ. J. Electrochem.*, 2007, **43**, 901–908.

11 M. C. Ribeiro, *J. Phys. Chem. B*, 2003, **107**, 4392–4402.

12 H. Bloom and E. Heymann, *Proc.-R. Soc. Edinburgh, Sect. A: Math. Phys. Sci.*, 1947, **188**, 392–414.

13 A. L. Rollet, S. Godier and C. Bessada, *Phys. Chem. Chem. Phys.*, 2008, **10**, 3222–3228.

14 I. Farnan and J. F. Stebbins, *J. Am. Chem. Soc.*, 1990, **112**, 32–39.

15 J. D. Edwards, C. S. Taylor, A. S. Russell and L. F. Maranville, *J. Electrochem. Soc.*, 1952, **99**, 527–535.

16 P. Hebant and G. S. Picard, *J. Electrochem. Soc.*, 1997, **144**, 980–983.

17 M. Chemla and M. Okada, *Electrochim. Acta*, 1990, **35**, 1761–1776.

18 A. L. Rollet, S. Godier and C. Bessada, *Phys. Chem. Chem. Phys.*, 2008, **10**, 3222–3228.

19 S. Hayashi and K. Hayamizu, *Bull. Chem. Soc. Jpn.*, 1990, **63**, 913–919.

20 J. H. Barner, C. Bessada and R. W. Berg, *Inorg. Chem.*, 2003, **42**, 1901–1907.

21 A. Abragam, *Principles of Nuclear Magnetism*, Oxford U. P., London, 1961, ch. 8.

22 R. Lenke, W. Uebelhack and A. Klemm, *Z. Naturforsch., A: Phys. Sci.*, 1973, **28**, 881–883.

23 J. O. 'M. Bookris, S. R. Richards and L. Nani, *J. Phys. Chem.*, 1965, **69**, 1627–1637.

24 W. W. Filho, R. L. Havill and J. M. Titman, *J. Phys. C: Solid State Phys.*, 1982, **15**, 3617–3625.

25 A. Carof, M. Salanne, T. Charpentier and B. Rotenberg, *J. Phys. Chem. B*, 2014, **118**, 13252–13257.

26 P. Hebant and G. S. Picard, *J. Mol. Struct.*, 1995, **358**, 39–50.

27 W. J. Glover and P. A. Madden, *J. Chem. Phys.*, 2004, **121**, 7293–7303.

Faraday Discussions

PAPER

Estimation of the composition of intermetallic compounds in LiCl–KCl molten salt by cyclic voltammetry

Ya L. Liu, Kui Liu, Li Y. Yuan, Zhi F. Chai and Wei Q. Shi*

Received 12th December 2015, Accepted 11th January 2016

DOI: 10.1039/c5fd00220f

In this work, the compositions of Ce–Al, Er–Al and La–Bi intermetallic compounds were estimated by the cyclic voltammetry (CV) technique. At first, CV measurements were carried out at different reverse potentials to study the co-reduction processes of Ce–Al, Er–Al and La–Bi systems. The CV curves obtained were then re-plotted with the current as a function of time, and the coulomb number of each peak was calculated. By comparing the coulomb number of the related peaks, the compositions of the Ce–Al, Er–Al and La–Bi intermetallic compounds formed in the co-reduction process could be estimated. The results showed that $Al_{11}Ce_3$, Al_3Ce, Al_2Ce and AlCe could be formed by the co-reduction of Ce(III) and Al(III). For the co-reduction of Er(III) and Al(III), Al_3Er_2, Al_2Er and AlEr were formed. In a La(III) and Bi(III) co-existing system in LiCl–KCl melts, $LaBi_2$, LaBi and Li_3Bi were the major products as a result of co-reduction.

1. Introduction

Pyrochemical reprocessing for advanced fuel cycles has been well studied by the world's nuclear power producers.[1–3] Molten salts, such as molten chlorides and molten fluorides, are chosen as mediums for the pyrochemical route.[4,5] In a typical pyrochemical reprocessing procedure, LiCl–KCl molten salt is employed as the medium for the electrorefining of spent metal fuels. Generally, actinides (Ans) in the chopped spent metal fuels are anodically dissolved into the LiCl–KCl molten salt electrolyte along with the active fission products (FPs). Meanwhile, the Ans are deposited onto the cathode, leaving the FPs in the electrolyte as ions.[6,7] In this context, various kinds of electrodes have been studied in the LiCl–KCl molten salt for the recovery of Ans and the separation of Ans from active FPs.[8] Among these studied electrodes, some active electrodes were considered to be suitable for the separation of Ans from the active FPs.[3,9–11] Liquid Cd and Bi cathodes were traditionally used for the separation of Pu and other minor actinides (MAs) from the lanthanides (Lns). Recently, solid Al electrodes have been

Laboratory of Nuclear Energy Chemistry, Institute of High Energy Physics, Chinese Academy of Sciences, Beijing 100049, China. E-mail: shiwq@ihep.ac.cn

extensively investigated as the cathode for the electrorefining process due to their excellent performance in separating Ans from Lns.[9,12,13]

In the case of using these active cathodes, intermetallic compounds might be formed due to the reaction of the deposited metal with the active cathode. A comprehensive study of these intermetallic compounds was important for the electrorefining process, because in some cases the formation of intermetallic compounds should be avoided while in some other cases it should be enhanced. However, it was difficult to obtain all the signals of the intermetallic compounds on the liquid Cd and Bi cathodes or the solid Al electrode. In our previous works, it was found that by the co-reduction process, more electrochemical signals of the intermetallic compounds were observed and the composition of the intermetallic compounds could also be estimated.[14–18] In this work, the estimation of the intermetallic compounds by the cyclic voltammetry technique was presented in detail, and it would also be of significance for further extensive study of the intermetallic compounds in other molten salt mediums.

2. Experimental

2.1 Preparation and purification of the melt

A mixture of anhydrous LiCl (44.8 g, Alfa Aesar, AR grade) and KCl (55.2 g Alfa Aesar, AR grade) was dried under a vacuum for more than 100 h at 473 K to remove the excess water. Then the salt mixture was melted in an alumina crucible located in an electric furnace. The temperature of the melt was controlled by a nickel–chromium thermocouple sheathed with an alumina tube. After the salt mixture was completely melted, the excess oxygen in the melt was removed by pure Ar gas bubbling. Then the pre-electrolysis at −2.2 V (*vs.* AgCl/Ag) was carried out for several hours (usually more than 4 hours) to remove the metal ion impurities in the melt. Lanthanum and bismuth elements were introduced into the bath in the form of dehydrated $LaCl_3$ and $BiCl_3$ powder, while cerium, erbium and aluminium elements were in the form of CeO_2, Er_2O_3, and anhydrous $AlCl_3$. Due to its volatility the exact concentration of $AlCl_3$ in the melts could not be measured, hence, the concentration of $AlCl_3$ presented in this work is the mass fraction of $AlCl_3$ initially added into the melts.

2.2 Electrochemical apparatus and electrodes

A PGSTAT302N electrochemical workstation (Autolab, Metrohm) controlled with the Nova 1.9 software package was used to obtain electrochemical data. The reference electrode was an alundum tube containing a silver wire (d = 1 mm) dipped into a solution of AgCl (1 wt%) in LiCl–KCl melts. All potentials were referred to the Ag/AgCl couple. Molybdenum wires (Alfa, 99.9+%) and tungsten wires with a diameters of 1 mm or 0.5 mm were employed for the working electrode. The lower end of the working electrode was polished with sandpaper, and then cleaned in water and ethanol using ultrasound. The active electrode surface area was determined after each experiment by measuring the immersion depth of the working electrode in the molten salts. A 6 mm diameter graphite rod (Alfa Aesar >99.99%) was used as the counter electrode.

3. Results and discussion

3.1 Estimation of Ce–Al and Er–Al intermetallic compounds

3.1.1 Ce–Al intermetallic compounds. The co-reduction behaviors of Ce(III) and Al(III) was investigated in our previous work.[15] When Ce(III) and Al(III) ions co-exist in the LiCl–KCl melt, Ce–Al intermetallic compounds could be formed by the co-reduction of Ce(III) and Al(III). From the CV result, as shown in Fig. 1, the C1/A1 and C5/A5 couples are related to the deposition and dissolution of Al and La metal, respectively.[15] The remaining couples C2(C2′)/A2, C3/A3, and C4/A4 all belong to the formation and dissolution of Ce–Al intermetallic compounds.

According to the phase diagram of Al–Ce,[19] five intermetallic compounds such as $Al_{11}Ce_3$, Al_3Ce, Al_2Ce, AlCe and $AlCe_3$ are available. However, only three couples were observed in our experiments. Hence, in this work the three couples were analyzed to determine the composition of the observed intermetallic compounds. The formation of Ce–Al intermetallic compounds, which is related to the cathodic peaks (C2′, C2, C3 and C4), should be caused by the reaction of deposited Ce and Al metal on the Mo electrode. At this experimental temperature, the formed Ce–Al intermetallic compounds were in the solid state, and could be deposited on the electrode continuously. In addition, the formation signal of Al–Li alloys was not observed in Fig. 1, indicating that the deposited Al should be almost completely reacted with the Ce metal. Furthermore, the anodic peaks (A2, A3 and A4) may be attributed to the dissolution of Ce metal into the melt as Ce(III) ions with the decomposition of the Ce–Al intermetallic compounds. Hence, by comparing the accumulated coulomb number of A2, A3 and A4 with that of A1, the Al/Ce ratios in the Ce–Al intermetallic compounds could be estimated.[20] In order to obtain the accumulated coulomb number of the anodic peaks, the CV curves obtained at different reverse potentials were then re-plotted as the current against t (i *vs.* t), and the results were shown in Fig. 2(b1)–(b4). From Fig. 2, the coulomb numbers of the anodic peaks are equal to their peak areas, which are displayed by black shadows on the curves. A small peak of A1′ is also observed, it should be due to the dissolution of Al–Mo alloys, so the released Al(III) ions at peak A1′ do not react with Ce. The results were summarized in Table 1.

For the reverse potential at −1.58 V, the ratio of Al released at peak A1 to the release of Ce at peak A2 was estimated to be 6.24. This Al/Ce value was much higher than that of any Ce–Al intermetallic compounds in the phase diagram, this result should be caused by the incomplete reaction of the deposited Al metal at an

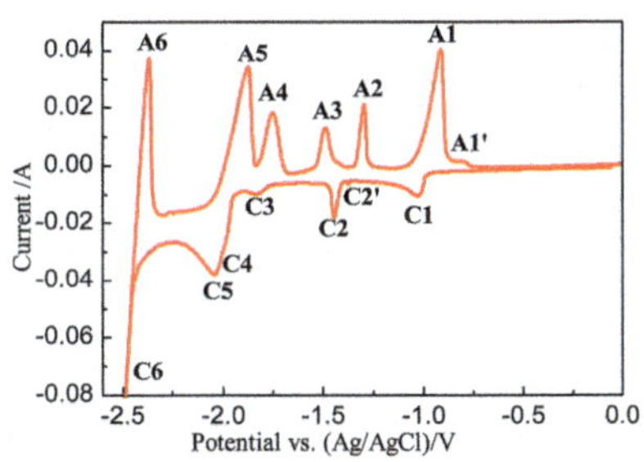

Fig. 1 The CV of the LiCl–KCl–$AlCl_3$ (~1.86 × 10^{-4} mol cm^{-3})–$CeCl_3$ (1.12 × 10^{-4} mol cm^{-3}) melts on a Mo (S = 0.13 cm^2) electrode at 773 K. Scan rate: 0.1 V s^{-1}.

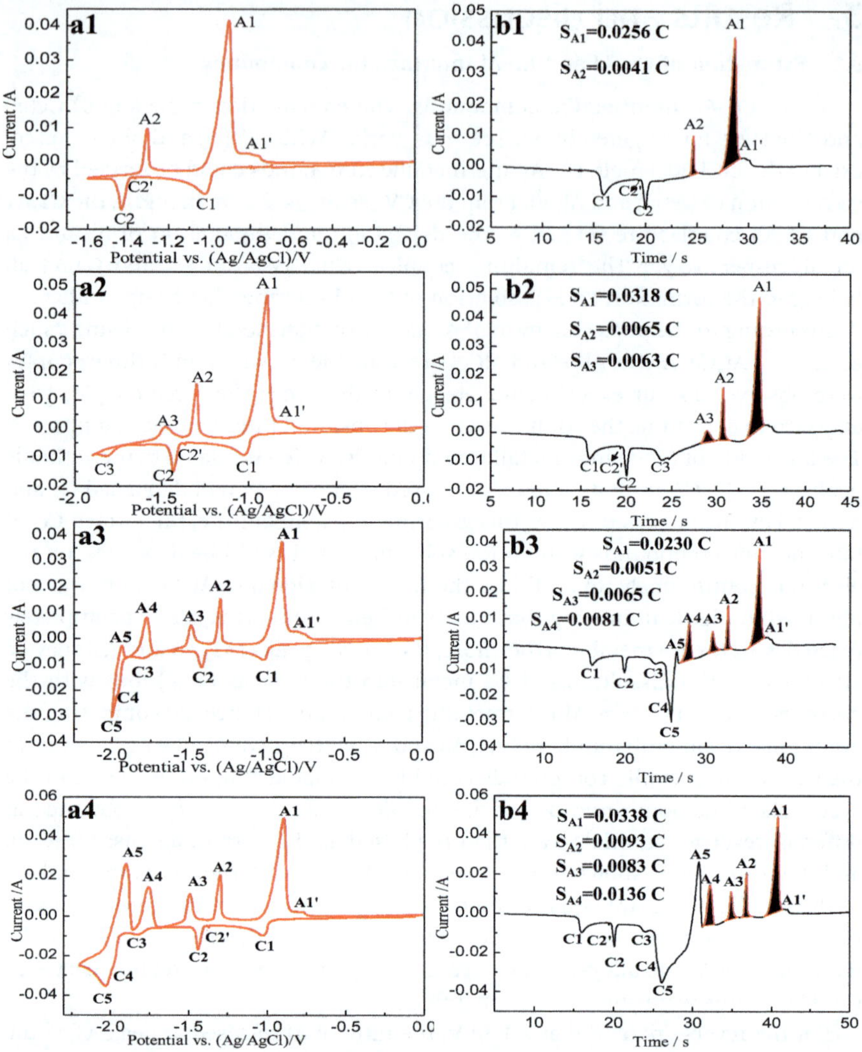

Fig. 2 CVs in the LiCl–KCl–$AlCl_3$ (~1.86 × 10^{-4} mol cm^{-3})–$CeCl_3$ (1.12 × 10^{-4} mol cm^{-3}) melts on Mo (S = 0.13 cm^2) electrode at different reverse potentials at 773 K, and the plotted curves of the CVs with *i vs. t*. Scan rate: 0.1 V s^{-1}.

Table 1 Accumulated coulomb number of the anodic peaks and the estimated values of Al/Ce ratios with different reverse potentials

Reverse potentials (V)	S_{A1}/S_{A2}	$S_{A1}/S_{(A2+A3)}$	$S_{A1}/S_{(A2+A3+A4)}$	Al_xCe_y
−1.58	6.24			
−1.90	4.89	2.48		$Al_{11}Ce_3$, Al_3Ce, Al_2Ce
−2.00	4.51	1.98	1.16	$Al_{11}Ce_3$, Al_3Ce, Al_2Ce, AlCe
−2.20	3.63	1.92	1.08	$Al_{11}Ce_3$, Al_3Ce, Al_2Ce, AlCe
−2.50	3.29	1.93	1.04	$Al_{11}Ce_3$, Al_3Ce, Al_2Ce, AlCe

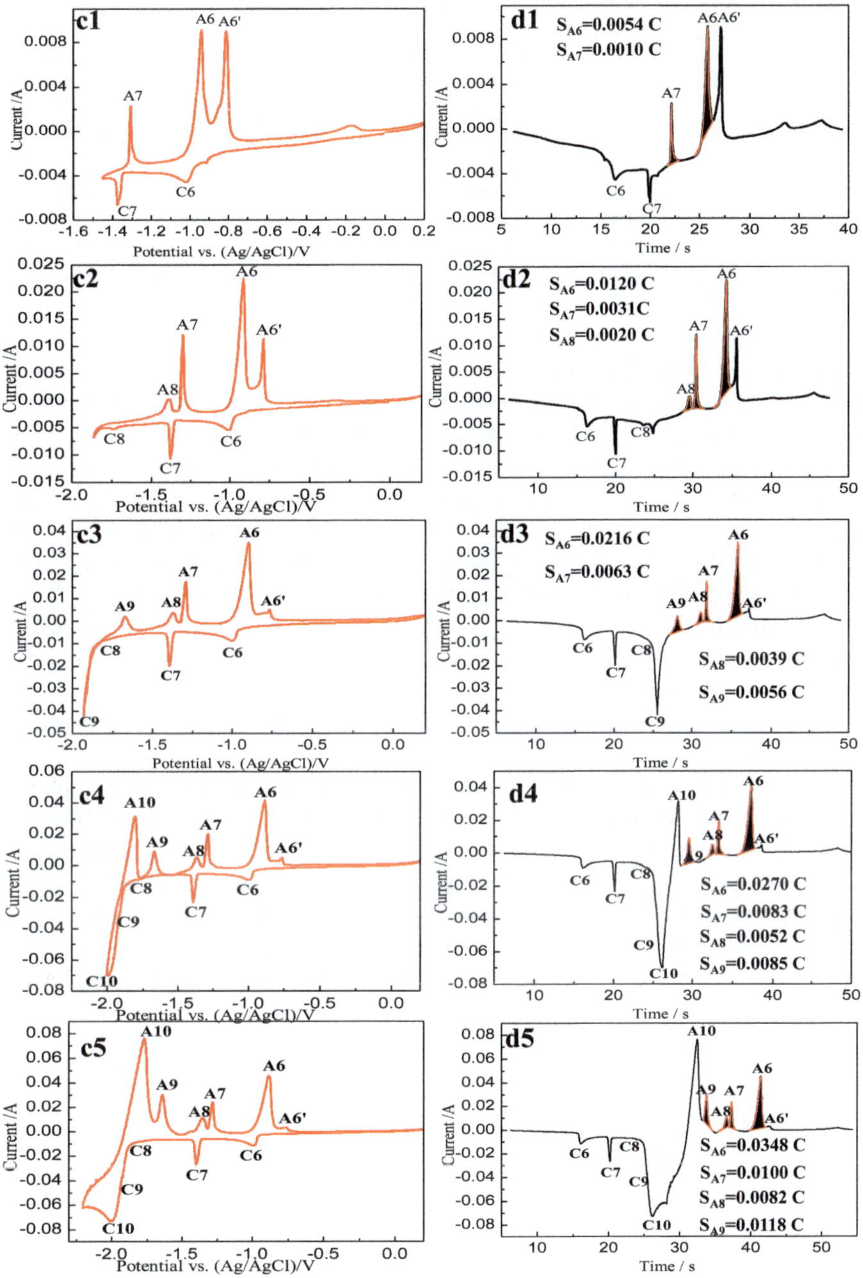

Fig. 3 CVs of the $AlCl_3$ (2 wt%)–Er_2O_3 (1.5 wt%) melts on Mo ($S = 0.39$ cm^2) electrode at different reverse potentials at 773 K, and the plotted curves of the CVs with *i vs. t*. Scan rate: 0.1 V s^{-1}.

inversion potential of −1.58 V. With the shifting of the reverse potential to a more negative value, the ratio of S_{A1}/S_{A2} became smaller and reached 3.29 when the reverse potential was −2.50 V. This result reveals that the intermetallic compound

Table 2 Accumulated coulomb number of the anodic peaks and the estimated values of Al/Er ratios with different reverse potentials

Reverse potentials (V)	S_{A6}/S_{A7}	$S_{A6}/S_{(A7+A8)}$	$S_{A6}/S_{(A7+A8+A9)}$	Al_xEr_y
−1.48	5.40			
−1.86	3.87	2.35		Al_3Er, Al_2Er
−1.93	3.42	2.12	1.37	Al_3Er, Al_2Er, AlEr
−2.00	3.26	2.00	1.23	Al_3Er, Al_2Er, AlEr
−2.20	3.48	1.91	1.16	Al_3Er, Al_2Er, AlEr

corresponding to the peak C2 might be Al_3Ce. Nevertheless, the value obtained was a little higher than 3, the reason for this might be that peak A2 corresponds to two cathodic peaks C2 and C2′. C2′ was a little anodic than C2, and generally should have a higher Al content than that of C2. From these results, C2′ might be related to the formation of $Al_{11}Ce_3$. Then by the calculated values of $S_{A1}/S_{(A2+A3)}$ at different reverse potentials, peak C3 should be attributed to the formation of Al_2Ce. When the reverse potential reaches −2.50 V, the value of $S_{A1}/S_{(A2+A3+A4)}$ is about 1, so peak C4 should be related to the formation of AlCe.

By comparing the accumulated coulomb number of A2, A3 and A4 with that of A1, four intermetallic compounds, $Al_{11}Ce_3$, Al_3Ce, Al_2Ce and AlCe were estimated. This result was in good agreement with our previous findings, in which we obtained the same four kinds of Ce–Al intermetallic compounds by potentiostatic electrolysis on an Al plate.

3.1.2 Er–Al intermetallic compounds. According to the Al–Er phase diagram,[21] five intermetallic compounds Al_3Er, Al_2Er, AlEr, Al_2Er_3 and $AlEr_3$ are available. In our previous work, three couples related to the formation and dissolution of Er–Al intermetallic compounds were also observed.[18] Here, three couples were also analyzed at different reverse potentials to determine the identity of the intermetallic compounds. The CV curves obtained at different reverse potentials are presented in Fig. 3(c1)–(c5), and the corresponding re-plotted

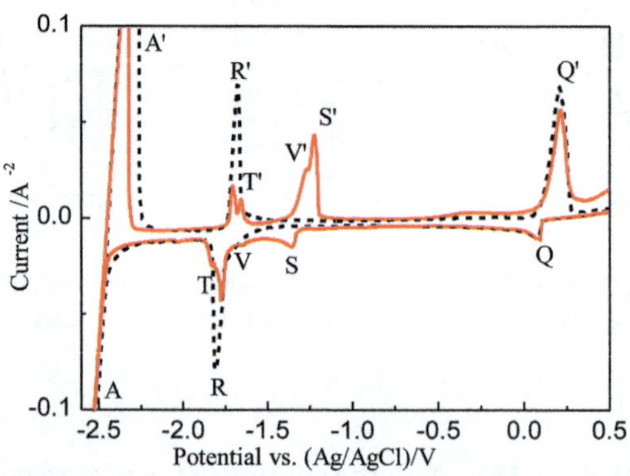

Fig. 4 Comparison of the CVs of the LiCl–KCl–$BiCl_3$ (1.03×10^{-5} mol cm^{-3}) melts (dotted curve) and the LiCl–KCl–$BiCl_3$ (1.03×10^{-5} mol cm^{-3})–$LaCl_3$ (3.33×10^{-5} mol cm^{-3}) melts (solid curve) on a W electrode ($S = 0.33$ cm^2) at 773 K. Scan rate: 0.1 V s^{-1}.

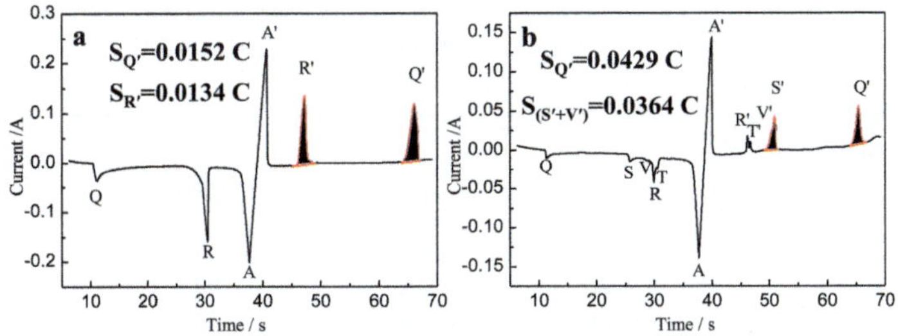

Fig. 5 The plotted curves of the CVs (in Fig. 4) with *i vs. t*: (a) CV obtained in the LiCl–KCl–$BiCl_3$ (1.03×10^{-5} mol cm^{-3}) melts; (b) CV obtained in LiCl–KCl–$BiCl_3$ (1.03×10^{-5} mol cm^{-3})–$LaCl_3$ (3.33×10^{-5} mol cm^{-3}) melts.

curves with *i vs. t* are displayed in Fig. 3(d1)–(d5). From these CVs, couples A6/C6 and A10/C10 are due to the deposition and dissolution of Al and Er metal, respectively. Peaks C7, C8 and C9 are related to the formation of various Er–Al intermetallic compounds, and the corresponding anodic peaks A7, A8 and A9 are attributed to their dissolutions. As shown in Fig. 3(a1), a noticeable anodic peak A6′ can be observed at the reverse potential of −1.48 V. It reveals that under these conditions the deposited Al can partly react with the Mo electrode and then form Al–Mo intermetallic compounds. Hence, the released Al(III) ions at peak A6′ do not react with Er but with the Mo electrode. Therefore, only the released Al(III) ions at peak A6 were counted for calculation of the formation of Al–Er intermetallic compounds through the reaction with Er.

The value of S_{A6}/S_{A7} reached nearly 3 by changing the reverse potential, which indicates that peak C7 should be related to the deposition of Al_3Er. When the reverse potential changed in the range from −1.86 to −2.20 V, the value of $S_{A6}/S_{(A7+A8)}$ remained close to 2, which reveals that peak C8 should belong to the formation of Al_2Er. Similarly, peak C9 is proposed to be related to the formation of AlEr. The formation of Al_3Er at peak C7 was confirmed in our previous work by potentiostatic electrolysis at −1.6 V.[18] In addition, the formation of Al_2Er was also confirmed by gavanoststic electrolysis at −30 mA. However, the deposition of AlEr

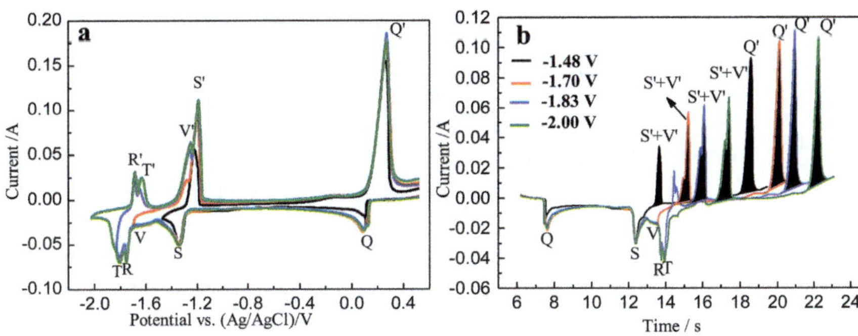

Fig. 6 (a) CVs of the LiCl–KCl–$BiCl_3$ (1.03×10^{-5} mol cm^{-3})–$LaCl_3$ (3.33×10^{-5} mol cm^{-3}) melts on W ($S = 0.33$ cm^2) electrode at different reverse potentials at 773 K; (b) the plotted curves of the CVs with *i vs. t*. Scan rate: 0.1 V s^{-1}.

Table 3 Accumulated coulomb number of the anodic peaks and the estimated values of the Bi/La ratios with different reverse potentials, in the LiCl–KCl–$BiCl_3$ (1.03×10^{-5} mol cm^{-3})–$LaCl_3$ (3.33×10^{-5} mol cm^{-3}) melts

Reverse potentials (V)	$S_{Q'}$ (C)	$S_{S'+V'}$ (C)	$S_{Q'}/S_{S'+V'}$	La_xBi_y
−1.48	0.0278	0.0083	3.35	
−1.70	0.0333	0.0160	2.08	$LaBi_2$, LaBi
−1.83	0.0326	0.0209	1.56	$LaBi_2$, LaBi
−2.00	0.0312	0.0237	1.32	$LaBi_2$, LaBi
−2.50	0.0429	0.0364	1.18	$LaBi_2$, LaBi

was not observed in our previous work, and it might be due to the use of an Al cathode for the deposition. As the deposited Er could diffuse into the Al electrode, and the amount of the deposited Er metal was relatively smaller than the solid Al electrode, and AlEr was difficult to be formed on the Al electrode. However, when an inert electrode was employed, AlEr might be formed *via* the co-reduction. A summary of the results are presented in Table 2.

Additionally, it was also found that a more negative reverse potential would be beneficial for the complete reaction of deposited Al and Ln (Ce, Er) metal, and hence made it more accurate to determine the Al/Ln ratio for each peak.

3.2 Estimation of the composition of La–Bi intermetallic compounds

A liquid Bi cathode is also one of the typical electrodes applied in the pyrochemical process. In this work, the La–Bi intermetallic compounds were also studied by the co-reduction process. The reported La–Bi phase diagram presented five La–Bi intermetallic compounds:[22] $LaBi_2$, LaBi, La_4Bi_3, La_5Bi_3 and La_2Bi.

The electrochemical behavior of Bi(III) on the W electrode was firstly studied, before studying the co-reduction properties of La(III) and Bi(III) cations. The CV result, which is given in Fig. 4 (the dotted curve), displays three couples. Peaks A and A′ are related to the deposition and dissolution of Li metal on the W electrode. Peak Q is attributed to the reduction of Bi(III) to the Bi metal, peak Q′ is its corresponding anodic peak.[17] Peaks R and R′ are related to the formation and

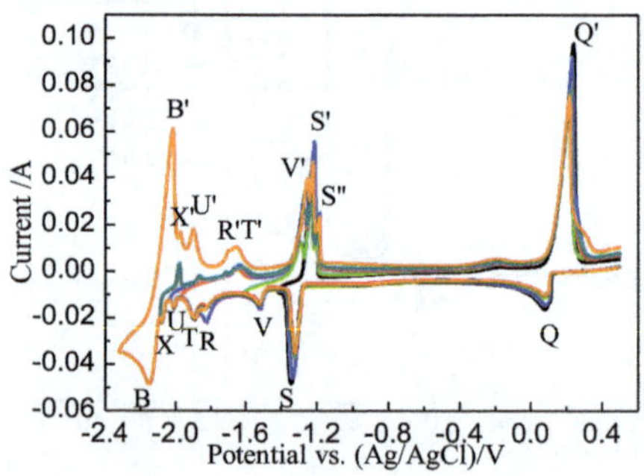

Fig. 7 CVs of the LiCl–KCl–$BiCl_3$ (1.03×10^{-5} mol cm^{-3})–$LaCl_3$ (1.33×10^{-4} mol cm^{-3}) melts on W ($S = 0.33$ cm^2) electrode at different reverse potentials at 773 K. Scan rate: 0.1 V s^{-1}.

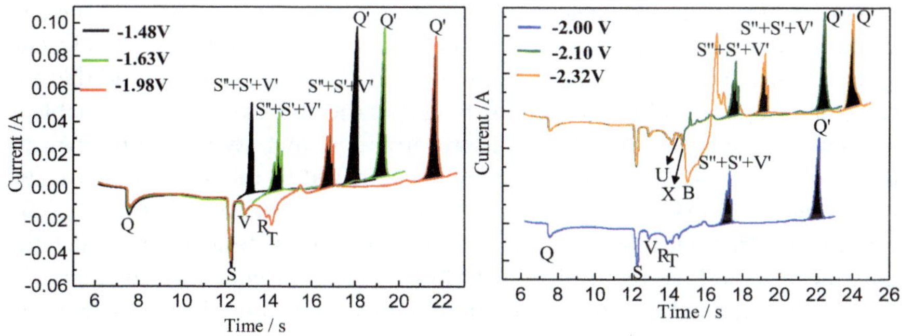

Fig. 8 The plotted curves of the CVs (in Fig. 7) with *i vs. t*.

dissolution of Li–Bi intermetallic compounds. Then the CV curve was re-plotted with *i vs. t* to estimate the composition of the Li–Bi intermetallic compounds, and the result is shown in Fig. 5(a). By comparing the value of 1/3 $S_{Q'}$ with $S_{R'}$, the ratio of Li/Bi was about 2.65, indicating that peak R should be attributed to the formation of Li_3Bi.

When La(III) ions with a concentration of 3.33×10^{-5} mol cm^{-3} co-exist in the melt, some other peaks can be observed, as shown in Fig. 4 (the solid line). The cathodic peaks S, V and T related to the formation of La–Bi intermetallic compounds are observed. The corresponding anodic peaks S′, V′ and T′ are related to the release of La(III) ions with the decomposition of the La–Bi intermetallic compounds. It can be also found that in this case peak R is still observed, indicating that Li_3Bi could also be formed beside the formation of La–Bi intermetallic compounds. The re-plotted curve with *i vs. t* is shown in Fig. 5(b). It is found that peaks R′ and T′ have some overlaps, and it is difficult to obtain the exact area of peak T′. Therefore, the intermetallic compounds formed at peak T are not estimated in this work. Peaks S′ and V′ also display large overlapping regions. Since both peaks S′ and V′ are related to the release of La(III) ions, these two peaks are calculated together. The peak areas are shown in Fig. 5(b) and the value of $S_{Q'}/S_{(S'+V')}$ is about 1.18, therefore peak V should be related to the formation of LaBi. Peak S is more anodic than V, in light of the Bi–La phase diagram, peak S should be related to the formation of the intermetallic compound $LaBi_2$.

Table 4 Accumulated coulomb number of the anodic peaks and the estimated values of Bi/La ratios with different reverse potentials, LiCl–KCl–$BiCl_3$ (1.03×10^{-5} mol cm^{-3})–$LaCl_3$ (1.33×10^{-4} mol cm^{-3}) melts

Reverse potentials (V)	$S_{Q'}$ (C)	$S_{S'+V'}$ (C)	$S_{Q'}/S_{S'+V'}$	La_xBi_y
−1.48	0.0237	0.0087	2.70	
−1.63	0.0160	0.0084	1.91	
−1.97	0.0184	0.0122	1.51	$LaBi_2$, LaBi
−2.00	0.0231	0.0159	1.45	$LaBi_2$, LaBi
−2.10	0.0180	0.0133	1.35	$LaBi_2$, LaBi
−2.32	0.0176	0.0140	1.26	$LaBi_2$, LaBi
−2.50	0.0119	0.0109	1.10	$LaBi_2$, LaBi

Furthermore, the CV measurements were carried out at different reverse potentials and were re-plotted with *i*–*t*, the obtained results are shown in Fig. 6(a) and (b). The relative peak areas are presented in Table 3. It is also found that the value of $S_{Q'}/S_{(S'+V')}$ is close to 1, confirming that the above result is reasonable.

However, the value of $S_{Q'}/S_{(S'+V')}$ is still higher than 1 even when the reverse potentials reach −2.50 V, this result could be caused by the formation of Li_3Bi. Following the formation of Li_3Bi, some of the Bi(III) ions released at peak Q′ are due to the decomposition of Li_3Bi. Nevertheless, it is difficult to deduct the amount of Bi(III) ions released from Li_3Bi in peak Q′. Hence, the obtained $S_{Q'}/S_{(S'+V')}$ value should be a little higher than the real Bi/La ratio in the Bi–La intermetallic compounds.

In order to get more information on the La–Bi intermetallic compounds, the concentration of La(III) ions was increased to 1.33×10^{-4} mol cm^{-3} in the melt. Fig. 7 and 8 present the CVs and re-plotted curves at different reverse potentials, respectively. In this case, the two other couples of U/U′ and X/X′, which can be attributed to the formation and dissolution of other La–Bi intermetallic compounds, were observed. In addition, peaks R and R′ get smaller compared with those in the above lower La(III) concentration system. It is revealed that the formation of Li_3Bi might be reduced but not completely inhibited in the higher La(III) concentration system. Hence peaks more cathodic than R are also not estimated under these conditions.

The obtained results were displayed in Table 4. It can be found that the value of $S_{Q'}/S_{(S'+V')}$ became closer to 1 in this system, giving the same conclusion that peak V should be related to the formation of LaBi. Generally, the higher the La content in La–Bi intermetallic compounds, the closer the La_xBi_y deposition potential to that of La deposition. Hence, combining the phase diagram, peaks S, T, U and X might be related to the formation of $LaBi_2$, La_4Bi_3, La_5Bi_3 and La_2Bi, respectively. However, no direct evidence for the formation of La_4Bi_3, La_5Bi_3 and La_2Bi was obtained by comparing the anodic peaks on the CV curves. To determine the composition of these intermetallic compounds some other evidence is still needed.

The previous work done by Li *et al.* showed the presence of the deposition products of $LaBi_2$, LaBi and Li_3Bi by galvanostatic electrolysis on a liquid Bi electrode.[23] The results are in good agreement with our estimations in the present work for the lower La(III) concentration system.

4. Conclusions

A method for estimating the composition of intermetallic compounds by the co-reduction process in LiCl–KCl melt was introduced in this work. In this method, CVs with different reverse potentials were re-plotted with *i vs. t*, and then the coulomb number of each anodic peak was calculated to estimate the atomic ratio in the intermetallic compounds. By this method, the composition of Ce–Al, Er–Al and La–Bi intermetallic compounds were estimated, and the obtained results were in good agreement with the previous works. It was found that a more negative reverse potential would be beneficial for a more accurate estimation. In addition, to obtain a more accurate estimation the formation of Li–M (M = Al, Li) intermetallic compounds should be avoided by adjusting the concentration of the relative metal ions in the LiCl–KCl melt.

Acknowledgements

This work was supported by the Major Research Plan "Breeding and Transmutation of Nuclear Fuel in Advanced Nuclear Fission Energy System" of the Natural Science Foundation of China (Grants 91226201, 91426302, 11275219) and the "Strategic Priority Research program" of the Chinese Academy of Sciences (Grants XDA030104).

References

1 S. A. Kuznetsov, H. Hayashi, K. Minato and M. Gaune-Escard, *Electrochim. Acta*, 2006, **51**, 2463–2470.
2 T. Kato, T. Inoue, T. Iwai and Y. Arai, *J. Nucl. Mater.*, 2006, **357**, 105–114.
3 P. Soucek, R. Malmbeck, C. Nourry and J. P. Glatz, *Energy Procedia*, 2011, **7**, 396–404.
4 K. Sridharan, T. Allen, M. Anderson and M. Simpson, *NEUP Final Report*, Project No. 09–780, 2012.
5 M. Gibilaro, L. Massot, R. Chamelot and P. Taxil, *Electrochim. Acta*, 2009, **54**, 5300–5306.
6 M. Simpson, *Idaho National Laboratory*, INL/EXT-12–25124, 2012.
7 K. Kinoshita, K. Tadafumi, I. Tadashi, M. Ougier and J. P. Glatz, *J. Phys. Chem. Solids*, 2005, **66**, 619–624.
8 P. Masset, R. J. M. Konings, R. Malmbeck, J. Serp and J. P. Glatz, *J. Nucl. Mater.*, 2005, **344**, 173–179.
9 P. Soucek, R. Malmbeck, E. Mendes, C. Nourry and J. P. Glatz, *J. Radioanal. Nucl. Chem.*, 2010, **286**, 823–828.
10 J. Serp, P. Lefebvre, R. Malmbeck, J. Rebizant, P. Vallet and J. P. Glatz, *J. Nucl. Mater.*, 2005, **340**, 266–270.
11 O. Shirai, A. Uehara, T. Fujii and H. Yamana, *J. Nucl. Mater.*, 2005, **344**, 142–145.
12 J. Serp, M. Allibert, A. Le Terrier, R. Malmbeck, M. Ougier, J. Rebizant and J. P. Glatz, *J. Electrochem. Soc.*, 2005, **152**, C167–C172.
13 P. Soucek, R. Malmbeck, E. Mendes, C. Nourry, D. Sedmidubsky and J. P. Glatz, *J. Nucl. Mater.*, 2009, **394**, 26–33.
14 K. Liu, Y. L. Liu, L. Y. Yuan, X. L. Zhao, Z. F. Chai and W. Q. Shi, *Electrochim. Acta*, 2013, **109**, 732–740.
15 L. Wang, Y. L. Liu, K. Liu, S. L. Tang, L. Y. Yuan, L. L. Su, Z. F. Chai and W. Q. Shi, *Electrochim. Acta*, 2014, **147**, 385–391.
16 L. L. Su, K. Liu, Y. L. Liu, L. Wang, L. Y. Yuan, L. Wang, Z. J. Li, X. L. Zhao, Z. F. Chai and W. Q. Shi, *Electrochim. Acta*, 2014, **147**, 87–95.
17 K. Liu, L. Y. Yuan, Y. L. Liu, X. L. Zhao, H. He, G. A. Ye, Z. F. Chai and W. Q. Shi, *Electrochim. Acta*, 2014, **130**, 650–659.
18 K. Liu, Y. L. Liu, L. Y. Yuan, X. L. Zhao, H. He, G. A. Ye, Z. F. Chai and W. Q. Shi, *Electrochim. Acta*, 2014, **116**, 434–441.
19 K. A. Gschneidner and F. W. Calderwood, *Bull. Alloy Phase Diagrams*, 1988, **9**, 669–672.
20 Y. L. Liu, G. A. Ye, K. Liu, L. Y. Yuan, Z. F. Chai and W. Q. Shi, *Electrochim. Acta*, 2015, **168**, 206–215.

21 L. Jin, Y. B. Kang, P. Chartrand and C. D. Fuerst, *CALPHAD: Comput. Coupling Phase Diagrams Thermochem.*, 2010, **34**, 456–466.

22 K. A. Gschneidner and F. W. Calderwood, *Bull. Alloy Phase Diagrams*, 1989, **10**, 440–443.

23 M. Li, Q. Q. Gu, W. Han, X. M. Zhang, Y. Sun, M. L. Zhang and Y. D. Yan, *RSC Adv.*, 2015, **5**, 82471–82480.

Faraday Discussions

DISCUSSIONS

Developments for nuclear reactors and spent fuels processing: general discussion

Toru H. Okabe, Olga Kuzmina, Tim Sudmeier, Kui Liu, Liang Xu, Xiangling Yue, Andrew Doherty, Shuqiang Jiao, Qian Xu, Paul Madden, Daniel Cooper, Chaohui Wei, Kathie McGregor, Andrew Mount, Hongmin Zhu, Ali Kamali, Dihua Wang, Xianbo Jin, Yingjun Liu, Ian Farnan, John Irvine, George Zheng Chen, Ye Liu, Yalan Liu, Peng Zhang, Yasuhiko Ito, Xiaolong Liu, Jinhui Cheng and Xuehui An

DOI: 10.1039/C6FD90031C

Tim Sudmeier opened a general discussion of the paper by Yasuhiko Ito: Firstly, I would like to thank you, Prof. Ito, for your very inspiring talk and for all of your work in this field. I have a question regarding your advanced-type ammonia synthesis cell that consists of two melts (chloride and hydroxide melt) separated by a metal membrane. The nitride ions created at the cathode will travel through the chloride melt to the metal membrane where they will lose three electrons and get adsorbed to the surface. This is very similar to what happens at the anode in your initial paper on electrochemical ammonia synthesis.[1] In that paper you observed significant amounts of dinitrogen evolving from the anode as a product of recombination of adsorbed nitrogen atoms. Do you see something similar here and if so have you thought about potential ways of inhibiting such kinds of detrimental side reactions?

1 T. Murakami, T. Nishikiori, T. Nohira and Y. Ito, *J. Am. Chem. Soc.*, 2003, **125**, 334.

Yasuhiko Ito replied: You are right to think that it is very similar to what happens at the anode in our paper on electrochemical ammonia synthesis. The most effective way of inhibiting side reactions such as nitrogen gas evolution is to decrease the over-potential of the anode by finding and developing high performance and cheap metal/alloy membranes which show both high hydrogen absorption and a high diffusion coefficient for the absorbed H atoms. It is also desirable to find an effective electro-catalyst which is robust against anodic polarization in the chloride melt. In other words, the way of inhibiting the side reaction is to keep the anodic potential as low as possible in order not to cause such a side reaction (including anodic dissolution of the anode material).

George Zheng Chen commented: The formation of H_2 on either side of the palladium membrane is a valid concern. Palladium has a high solubility and conductivity of hydrogen (atoms, not protons) and hence is often made into a membrane for hydrogen (not proton) conduction. In the cell described by Prof. Ito, water molecules are reduced to hydrogen atoms on the cathodic side (on the left side of Fig. 4 in the paper) of the palladium membrane which functions as a bipolar plate (electrode). The hydrogen atoms produced are then absorbed into, and diffuse through the palladium membrane, and they are then oxidised to protons on the anodic side of the membrane. This process enables electrons to flow continuously from the anodic side to the cathodic side of the membrane, and allows current to flow in the whole cell. Because the potential is the same on both sides of the palladium membrane (same metal and hence same potential), if H_2 gas does not form on the entering side, it will not form on the exiting side either. However, if the potential on the palladium membrane is too negative, which can result from too large a voltage or current between the anode and cathode, as labelled in Fig. 4, H_2 gas can be produced on the cathodic side of the palladium membrane.

John Irvine asked: What happens to the electrons at the membrane between the two compartments?

Yasuhiko Ito answered: You will be able to understand the function of the membrane if you carefully read the explanation given for Fig. 6 of our paper. The membrane between the two compartments serves as a bipolar plate. That is, as shown in Fig. 4, on the right-hand surface of the metal membrane, facing the molten chloride, anodic reaction $N^{3-} + 3H_{abs} \rightarrow NH_3 + 3e^-$ proceeds, and the electron generated here is used for cathodic reaction $H_2O + e^- \rightarrow H_{abs} + OH^-$ on the left-hand surface of the metal membrane, facing the molten NaOH. H_{abs} produced on the left-hand cathode surface diffuses from left to right through the membrane and is subjected to anodic reaction $N^{3-} + 3H_{abs} \rightarrow NH_3 + 3e^-$ on the right-hand surface of the metal membrane.

Andrew Mount remarked: It seems that the metal (Pd) separator in Fig. 4 acts to transport H atoms through the lattice through interstitial diffusion. At each interface there is proton transfer into/out of the metal, with the additional electron being accepted into/donated from the metal conduction band. Does this indicate that the metal separator has excess electrons, *i.e.* that it is at a highly reducing potential, and has this potential been measured?

Yasuhiko Ito replied: The membrane serves as a bipolar electrode. Therefore, we have no excess electron in the metal/alloy membrane. Please see Fig. 4 and 6 in our paper.

Shuqiang Jiao queried: Is it possible to use a hydrogen storage material as the membrane?

Yasuhiko Ito responded: In principle, yes. But practically, it is very difficult to simply use a hydrogen storage material as a form of membrane. The membrane needs to have a sufficiently high mechanical strength, together with a sufficiently high hydrogen absorption ability and high hydrogen permeability.

George Zheng Chen said: All of the materials used for making the hydrogen conducting membrane are hydrogen storage materials. However, not all hydrogen storage materials are chemically stable when in contact with the molten salt and molten sodium hydroxide, as indicated in Fig. 4.

Yasuhiko Ito responded: In order to be suitable for the cell indicated in Fig. 4, the membrane should have sufficiently high mechanical strength, hydrogen absorption ability and hydrogen permeability. In addition, it should have sufficiently high chemical stability (corrosion resistance). As for a suitable corrosion resistant material for the membrane, the left-hand surface should show high corrosion resistance against molten NaOH that contains some water and the right-hand surface should show high corrosion resistance against molten chloride. In particular, it is crucial to select a suitable material where the right-hand surface is in contact with the molten chloride and is anodically polarized. Based on the above considerations, we are now trying to select and use some suitable membrane materials.

Qian Xu asked: Could you please explain the function of the membrane in the cell for the preparation of ammonia? Does it serve as a bipolar plate?

Yasuhiko Ito answered: Yes, it serves as a bipolar plate. The function of the membrane will be clarified if you follow the explanation given for Fig. 6.

Tim Sudmeier questioned: I was wondering what the rate-limiting step is in your advanced-type ammonia synthesis set up. I suspect it is the hydrogen transport through your metal membrane. Is that the case? Also, more generally, how did you investigate the nature of the rate-limiting step?

Yasuhiko Ito replied: In our advanced-type ammonia synthesis set up shown in Fig. 4, we currently think that the most probable rate determining step is the diffusion of H absorbed in the metal membrane followed by the anodic reaction $N^{3-} + 3H_{abs} \rightarrow NH_3 + 3e^-$ that occurs on the right-hand surface of the metal membrane facing the molten chloride electrolyte. It could be investigated by separately measuring the current–potential behaviors of each of the four electrode reactions involved in the total process (see Fig. 4 in our paper).

Daniel Cooper remarked: You mentioned that you are looking at metals and alloys for the hydrogen-permeable membrane. Could you use a hydride material for this purpose in a similar manner to the use of yttria-stabilized zirconia (YSZ) for the transport of oxide anions? Furthermore, have you considered the use of a sandwich-type membrane which uses only a thin layer of the hydrogen-permeable material in order to reduce cost?

Yasuhiko Ito answered: I am not sure what your question means. However, it might be helpful for you to read my previous paper,[1] where various types of possible electrolytic cell were demonstrated. It seems rather difficult to simply use a hydride material.

1 To be confirmed.

Tim Sudmeier asked: I have a question regarding Fig. 3 of your paper. You show here the analysis of the gas stream during electrolysis, which only shows low amounts of hydrogen being produced as a by-product. Looking at the potentials I would expect to see considerable amounts of water splitting happening at the cathode. How did you prevent this from happening? Is it a continuous or a batch process?

Yasuhiko Ito replied: The data in Fig. 3 were obtained using the bench scale continuous electrolytic cell shown in Fig. 2. The low amount of hydrogen being produced as a by-product firstly shows that when the water vapour supplied into the melt reacted with N^{3-} ions to produce NH_3, almost all of the unreacted excess water vapour did not reach the cathode compartment where the cathodic reduction of water occurs, but instead went out of the cell. Secondly, it shows that very little of the NH_3 produced was thermally decomposed to produce hydrogen, even at temperatures as high as those of our experimental conditions.

Andrew Mount commented: It is interesting to see the development of a highly novel rotating disk system in the rotating disk anode-type plasma induced electrolytic cell, particularly as it is the 50th anniversary year of the original seminal rotating-ring disc papers by Albery and Bruckenstein in Trans. Faraday Soc.[1–6] (albeit a very different electrode). I suspect that you may have the ability to control the nanoparticle size through parameters such as rotation rate, molten salt flow rate, cathode placement and particle collection angle. Have you carried out a systematic study to determine the sensitivity to these parameters?

1 W. J. Albery, *Trans. Faraday Soc.*, 1966, **62**, 1915–1919.
2 W. J. Albery and S. Bruckenstein, *Trans. Faraday Soc.*, 1966, **62**, 1920–1931.
3 W. J. Albery, S. Bruckenstein and D. T. Napp, *Trans. Faraday Soc.*, 1966, **62**, 1932–1937.
4 W. J. Albery, S. Bruckenstein and D. C. Johnson, *Trans. Faraday Soc.*, 1966, **62**, 1938–1945.
5 W. J. Albery and S. Bruckenstein, *Trans. Faraday Soc.*, 1966, **62**, 1946–1954.
6 W. J. Albery and S. Bruckenstein, *Trans. Faraday Soc.*, 1966, **62**, 2584–2595.

Yasuhiko Ito responded: In order to control the nanoparticle size, we have investigated the dependence of particle size on rotation rate, however, a more detailed and systematic investigation will be necessary in the future to find out the optimal operating conditions based on theoretical consideration of the effects of other parameters such as molten salt flow rate, cathode placement and particle placement angle. We currently determine these factors rather empirically.

George Zheng Chen queried: Prof. Ito, in your paper it is shown that water vapour is introduced into your cell at elevated temperatures (Fig. 5 and 6). How do you maintain the temperature difference between the water vapour-receiving molten hydroxide phase and the molten salt phase? Or were the temperatures the same in the two molten phases?

Yasuhiko Ito answered: The temperatures are the same in the two molten salt phases. Water vapour is introduced into the molten NaOH using a carrier gas such as Ar or N_2 which is previously passed through a water bath kept at the elevated temperature, for instance 90 °C. In a large scale industrial cell, H_2O gas might be supplied into the molten NaOH without using a carrier gas. The water vapour-introducing pipe is heated in order to avoid the condensation of water. The temperatures of the introduced water vapour and the molten NaOH rapidly become equal by a gas-liquid heat exchange.

Andrew Doherty remarked: Obviously, the angular velocity of the disk is a function of r^2, where r is the radius of the disk. Moving the discharging electrode from the edge towards the axis of rotation provides a consentient method of adjusting effective velocity. Can you control the properties of the nanoparticles by controlling angular velocity?

Yasuhiko Ito replied: Yes, for instance the dependencies of particle size and particle size distribution on the rotational speed were reported in **ref. 1** below.

1 Y. Ito, T. Nishikiori and M. Tokushige, Plasma Induced Discharge Electrolysis for Nanoparticles Production, *Molten Salts Chemistry: From Lab to Applications*, ed. F. Lantelme and H. Groult, Elsevier, 2013, pp. 269–286.

Andrew Mount enquired: How many revolutions does each particle have before it spins off? Is there an angular distribution of the particle sizes when deposited?

Yasuhiko Ito replied: We can control the properties of the nanoparticles by controlling angular velocity. See for example **ref. 1 and 2** below.

1 M. Tokushige, T. Nishikiori and Y. Ito, *J. Electrochem. Soc.*, 2010, **157**, 162–166.
2 Y. Ito, T. Nishikiori and M. Tokushige, Plasma Induced Discharge Electrolysis for Nanoparticles Production, *Molten Salts Chemistry: From Lab to Applications*, ed. F. Lantelme and H. Groult, Elsevier, 2013, pp. 269–286.

Xianbo Jin said: Your work on the preparation of nano-tantalum powder is very interesting. When we reduced the Ta particle size to about 20 nm we found that the oxygen content increased due to a layer of metal oxide on the metal particles. Did this happen in your study?

Yasuhiko Ito responded: Yes, it does more or less happen during the washing of the salt that covers the produced nanoparticle with water. However, the thickness of the oxide layer on the nanoparticle surface can be reduced by properly and carefully controlling the washing conditions.

Xianbo Jin asked: Will calcium be reduced by the electrons in your process?

Yasuhiko Ito answered: When molten LiCl–KCl is used as the solvent, there is a possibility of cathodic $Li^{(I)}$ reduction followed by a reduction of dissolved aimed metallic ions, aside from the direct cathodic reduction of the aimed metallic ions. The ratio of the above two reaction pathways can be controlled by optimizing the electrolytic conditions. Where the electrolyte contains calcium ions, a similar indirect cathodic reduction pathway might be possible.

Chaohui Wei communicated: With regards to the synthesis of ammonia, do you use pure N_2 from the separation of air? Have you tried to use air as the N_2 source? Perhaps it would also work and save some cost.

Yasuhiko Ito communicated in reply: Yes, we use pure N_2 separated from air. O_2 in the air is predominantly cathodically reduced to become O^{2-}, which remains in the melt and badly affects the cathodic reduction of N_2. We dream of finding a magic electro-catalyst that can selectively reduce N_2 without reducing any O_2 in the air, so that we can use air instead of pure N_2.

Chaohui Wei communicated: Why are you using a hydrogen permeable metal membrane specifically?

Yasuhiko Ito communicated in reply: The reason why we use a hydrogen permeable metal membrane is clarified if you read the paper carefully, especially the explanations for Fig. 4 and 6.

Chaohui Wei communicated: Will the fact that you are using electricity as the power supply increase the cost of the process and weaken its competitiveness against available industrial processes on the market?

Yasuhiko Ito communicated in reply: The significance of our process is clarified in the introductory section of our paper, as well as in Table 1 and Fig. 7 of the "Electrolytic synthesis of ammonia from water and nitrogen" section.

Ali Kamali opened discussion of the paper by Jinhui Cheng: Determination of the thermophysical properties of molten salts is rather important. Why have you used a heating rate of 10 °C min^{-1} in your DSC experiments?

Jinhui Cheng answered: The melting temperature was determined using a Netzsch DSC 404 F3 instrument, which was calibrated using six certified reference materials: In, Bi, Sn, Zn, Al and Au. The calibration was completed under an argon atmosphere at the heating rate of 10 °C min^{-1} and the extrapolated onset temperature was adopted as the melting temperature, so we used a 10 °C min^{-1} heating rate for the DSC measurements.

John Irvine commented: In thermal analysis you get much better data if you use a high ramp rate — 10, 5 or 2 °C min^{-1}. A slow rate is less accurate as it gives weak, broad peaks. Most people who do thermal analysis work at 10 °C min^{-1}. To get accurate transition temperatures, take a series of rates and extrapolate back to zero rate.

Olga Kuzmina said: Heating and cooling rates affect the resolution of the obtained DSC peaks. 10 °C min^{-1} is a commonly accepted compromise between the quality of the graph (peak appearance) and the correctness of the peak position.

Ye Liu responded: Yes, I agree with this point. In TG/DTA/DSC analyses, a heating rate of 10 °C min^{-1} is usually the accepted ramp in order to guarantee the quality of the graph and measurement accuracy.

John Irvine replied: The faster you run a DTA, the more you separate the peaks. For a reversible process it's the average of the heating and cooling rate peak temperatures. As you scan faster the peak becomes easier to see, increasing in amplitude and decreasing in breadth.

Ali Kamali asked: If you use a different heating rate, the peak positions in your DSC graphs shift, leading to slightly different temperatures being recorded for the events that occur during heating. What is your comment on this?

Jinhui Cheng answered: Actually, we have tested the effect of the heating rate on the melting temperature using rates of 2, 5, 10 and 20 °C min^{-1}. Indeed, the heating rate had an influence on the peak shape — a wider peak was obtained using a smaller heating rate. However, the different heating rates had little effect on the onset temperature of the peak.

Yingjun Liu enquired: What is the procedure used to prepare the salt mixture for DSC measurements, to ensure that it remains at eutectic composition?

Peng Zhang responded: Component salts K_2CO_3, Li_2CO_3 and Na_2CO_3 were dried separately at 200 °C under an argon atmosphere for 24 h. Then about 300 g of the eutectic mixture was prepared with weight proportions of 32.12% Li_2CO_3, 33.36% Na_2CO_3 and 34.52% K_2CO_3. The mixture was kept at 500 °C for more than 48 h in order to make it homogeneous. After cooling to room temperature, the ingot was pressed into a powder, which was used for the DSC measurements.

George Zheng Chen commented: The homogeneity of the melt is important for the type of research reported in this paper. For TGA and DSC analyses, the overall mass of the sample used is in the range 20–50 mg. Therefore, weighing different salts separately and accurately and mixing these salts homogeneously is challenging laboratory practice. These authors weighed the different salts at the gram scale and then melted the mixture to achieve uniform mixing at the atomic (ionic) level. The melt was then solidified and ground into powder, from which TGA and DSC analyses could be carried out without needing to worry about the homogeneity of the sample.

Yingjun Liu said: If the mixture is not at the eutectic composition then it might separate — cooling could have an important effect.

George Zheng Chen replied: This is a good point. When the component salts do not form a solid solution, melting a mixture of the different salts and then cooling the mixture to a solid may not achieve a uniform mixture, unless the composition is at the eutectic composition. At the eutectic composition, the grains in the solidified mixture can be very fine, leading to uniform mixing. However, if the composition deviates significantly from the eutectic composition, solidification can lead to localised segregation of the components.

Jinhui Cheng answered: Yes, this is why molten salt mixtures used as heat transfer media are always at the eutectic composition. If the solid salt mixture is not uniform, there will be many problems with respect to the safety and operation of the heat transfer system.

Hongmin Zhu queried: For the measurements on the $(LiNaK)_2CO_3$ melts, you always performed the measurements at temperatures of lower than 650 °C. Is there any special reason for that? In the references the measurements were performed at temperatures as high as 900 °C.

Jinhui Cheng replied: The $(LiNaK)CO_3$ sample measured is used as the heat transfer media. For the heat transfer system, the temperature range of the operating system is generally 500–700 °C. Thermogravimetric experiments reveal that $(LiNaK)CO_3$ begins to decompose at 650 °C, so the measurements were performed below this temperature.

Xiangling Yue asked: I noticed that you used a platinum wire and cylinder in determining the density of the molten salts. To my knowledge, platinum can easily be corroded by molten carbonate, especially lithium carbonate. Would this affect the accuracy of your result?

Jinhui Cheng replied: Thank you for the question. The platinum and graphite are resistant to corrosion by most molten salts. Maybe the platinum is corroded by the carbonate and is dissolved in it, but the amount is as low as several ppm and has little impact on the measured results.

Peng Zhang added: We did observe corrosion in our experiments, but the amount of corrosion was small, so it was not taken into consideration. Changes in the volume and weight of the platinum cylinder significantly affect the accuracy of the results, but these two aspects were barely affected by the corrosion. However, the corrosion makes it more difficult to carry out experiments because the platinum wire is easily broken.

Liang Xu opened discussion of Xiaolong Liu's paper: Thanks for your very nice talk. You used CV to investigate the cathodic process of neodymium(III), and two reduction peaks were observed. My question is about the second cathodic reaction, which you mentioned is a diffusion-controlled process according to the CV curve. How did you make this evaluation? In addition, after electrolysis, very nice metallic deposits (Nd metal) were obtained on the cathode. How did you completely remove the residue salt from the metal surface (as the solubility of fluoride salt in water is very low, it seems to be impossible to clean the sample by washing)?

Xiaolong Liu responded: Thanks for your questions. Firstly, in molten salt systems, we can assume that the diffusion rate of the ions near the cathode is fast enough when we are doing the electrochemical experiments. So the plot of the peak current for peak A *vs.* the square root of the scan rate is linear, passing through the origin. But with increasing scan rates, the reaction time under the corresponding potential will be shorter, and the number of ions produced from the first reaction on the cathode will be fewer, which will affect the next reaction and lead to a relatively small current density for the next step. So the second reaction is not only affected by ion diffusion control, but also by the first step reaction's production, and the plot of the current density *vs.* the square root of the scan rate will stray from the origin.

Secondly, the density of the metal that we get from electrolysis is higher than that of the molten salts. So the neodymium metal produced on the cathode surface fell to the bottom of the crucible after formation. When the molten salts have cooled down, the spherical metal neodymium is separated from the molten salts with a very clean surface.

Andrew Mount questioned: Regarding Fig. 2(b) in your paper, which is the peak current measured, B or B'? How is the effect of any underlying Li plating/stripping current (and A and A') removed? Could this molten salt reduction be the origin of the current intercept at zero sweep rate rather than a surface step (which should be rate determining and be expected to cause plot curvature at fast sweep rates)?

Xiaolong Liu replied: Thanks for your questions. We didn't find any formation of Li–Nd from our XRD analysis. So we think that the plating/stripping of Li doesn't influence our results. With regards to your second question, when we added excess Nd to the salts, there were enough Nd^{2+} ions for the second reaction, and the kinetics changed such that they were only controlled by ion diffusion.

George Zheng Chen commented: Information from cyclic voltammetry varies, depending on the method of analysis. In the case of a CV of dissolved redox species showing two consecutive current peaks along the same potential scan direction, the second peak should include the diffusion current of the first peak. Also, in theory, the diffusion current following a current peak is a constant, disregarding the potential scan rate.

Hongmin Zhu said: The appearance of the intermediate valence ion will usually cause the low current efficiency. Your fluoride–oxide melt is almost the same as the industry neodymium electrolysis system, and the current efficiency in the industry system is pretty high. Therefore, the appearance of Nd^{2+} in your results is interesting. It would be interesting if you could do the same measurements in a chloride system as you would be able to see more clearly the appearance of Nd^{2+}.

Xiaolong Liu responded: Thanks for your suggestion, Professor Zhu, it'll be very helpful for our future work.

John Irvine opened a general discussion of the paper by Andrew Mount: In relation to the impedence analysis fits where you find a Warburg in one case and a CPE in the other, did you analyse the Warburg element with constant phase as well? CPE gives much better analysis in terms of statistical fit but it could mask information as the parameters are less clearly related to physical parameters. In comparing the two responses I would suggest using CPE for both so that the changes can be better analysed. I am concerned that the difference in fit relates to the large difference in the high frequency component and the reason that a Warburg fits better is that the responses overlap less rather than any major change in physical chemistry in the low frequency part.

Andrew Mount responded: It is generally accepted that the Warburg element is the established element in the microdisc equivalent circuit used to model the EIS redox response in aqueous solution (the "modified Randles' circuit"; see **ref. 1–3**

below). This work therefore first attempted to use this established circuit to analyse LKE EIS data. The circuit has a physically realistic form, as it models linear diffusion through the Warburg at higher frequencies when the diffusion layer is thin, and hemispherical diffusion through the non-linear resistance at lower frequencies when hemispherical diffusion is established. This is analogous to the modelling of chronoamperometry through eqn (4) in our paper, with the first term modelling linear diffusion at short times and the second term modelling hemispherical diffusion at longer times. It is worth noting that one of the reasons that a CPE can give a better fit than a Warburg is that a Warburg element has only one fit parameter (often expressed as the magnitude of the admittance at a frequency of 1 rad s^{-1}, Y_0), whereas the CPE has two (often expressed as Y_0 and n, related to the constant phase angle in the Nyquist plot). We also agree that it is generally much harder to extract physical parameters and insight from the CPE. We have therefore followed the practice of fitting to the minimum number of fit parameters required to give a satisfactory statistical fit to the real and imaginary impedance data as a function of frequency, and only using an additional parameter (ie the CPE) when both macro- and microelectrode EIS data demonstrably do not fit to the established Randles' and modified Randles' equivalent circuits respectively in LKE.

1 P. Los, G. Zabinska, A. Kisza, L. Christie, A. Mount and P. G. Bruce, *Phys. Chem. Chem. Phys.*, 2002, **2**, 5449.
2 P. Los and P. G. Bruce, *Pol. J. Chem.*, 1997, **71**, 1151.
3 M. Fleischmann, S. Pons and J. Daschbach, *J. Electroanal. Chem.*, 1991, **317**, 1.

George Zheng Chen said: It would be helpful if you could explain your simulating circuits in Fig. 5 and discuss how you built them up.

Andrew Mount replied: We have just discussed the development of the EIS equivalent circuits for macro- and microelectrodes, and the elements within. As for the Randles' circuit applied to analyse macroelectrodes, C_{DL} models and quantifies the capacitance established through the drop in potential in the liquid salt at the electrode and R_{CT}, the charge transfer resistance, models and quantifies the rate of redox reaction at the electrode surface. It should be noted that the Randles' circuit results from an exact transmission line analytical solution of coupled linear ionic transport to, and reaction at, a macroelectrode,[1–4] whilst the modified Randles' circuit for microelectrodes is a close approximation to the EIS response, developed due to the difficulties in obtaining an exact analytical solution in this case.

1 W. J. Albery and A. R. Mount, in *Electroactive Polymer Electrochemistry, Part 1: Fundamentals*, ed. M. E. G. Lyons, Plenum Press, New York, 1994
2 W. J. Albery *et al.*, *Faraday Discuss.*, 1989, **88**, 247.
3 W. J. Albery and A. R. Mount, *J. Chem. Soc., Faraday Trans.*, 1993, 89, 2487
4 W. J. Albery and A. R. Mount, *J. Chem. Soc., Faraday Trans.*, 1993, 89, 327

George Zheng Chen asked: What do you mean by non-linear resistance in the context of electrochemistry? Would you also think that it is possible to apply non-linear capacitance in the simulating circuit of an electrode or electrochemical cell? More importantly, what should these non-linear elements be correlated to in electrochemical terms?

Andrew Mount answered: The non-linear resistance in the established EIS equivalent ("modified Randles'") circuit for a microdisc is the element that relates to the observed mass transfer limited steady-state response which is observed when the sweep rate is sufficiently low in voltammetry. It gives the frequency independent response when the applied frequency is sufficiently low in EIS (such that hemispherical diffusion and a hemispherical diffusion layer have been established in the liquid salt). The element is a resistor (the applied voltage leads to an in phase current independent of the EIS frequency) and the term "non-linear" describes the fact that the diffusion to the electrode is not linear but rather hemispherical.

Kui Liu enquired: What's the durability like for your MSMs, how many times can you use them in the melt?

Andrew Mount answered: We have carried out successful repeated and quantitative MSM measurements over several hours, and we are currently undertaking a systematic study of failure mechanisms over a statistically significant number of electrodes with a view to determining their lifetime and extending this further. We have not tended to carry out measurements involving extraction, re-insertion and subsequent measurement in the melt, as this is unlikely to be their modus operandi.

George Zheng Chen commented: I believe that these microelectrodes can be designed to be disposable if the manufacturing technology is automated at low cost. Material usage and hence the material cost of each electrode are small, I think.

Kui Liu queried: Since your MSMs were fabricated using silicon materials, which have a low resistance to corrosion in the melt, the durability can't be particularly good. Recently, Argonne laboratory published a paper on the determination of *U* in the LiCl–KCl melt.[1] They employed a simple tungsten wire and a combination of an electrode cleaning and stirring process. They got a good linear relationship for the response current signal *vs.* the electrode area. If we can use a simple W wire which is very cheap and has good durability, why do we need sophisticated MSMs? In addition, what is the performance of the MSM like in a higher concentration melt or a multiple component melt, especially for the soluble–insoluble redox couple?

1 M. M. Tylka, J. L.Willit, J. Prakash and M. A. Williamson, *J. Electrochem. Soc.*, 2015, **169**, H625–H633.

Andrew Mount responded: The MSMs are indeed fabricated on a silicon substrate, as this is a relatively inexpensive and robust platform, with established processes for carrying out the controlled microfabrication of devices of high fidelity. To ensure that this silicon is protected from corrosion through exposure to the molten salt medium, the silicon substrate is entirely covered by an impermeable insulator. This also prevents electrical contact (shorting) to the MSM electrode. You are right that it has been established that a tungsten macrowire dipped into the molten salt can be used to take measurements in molten salt media. This configuration has the advantage of simplicity and durability, and like all macroelectrode measurements, if all other parameters are kept constant, it should produce a response which is linearly dependent on electrode area. Furthermore, if stirring of the molten salt can be controlled and made reproducible/constant with time (with a constant flow of solution to/across the electrode surface; typically hard to achieve), under these conditions of convective mass transfer control a reproducible current should be obtained which is proportional to species concentration. However, in practice, the extraction of characteristic and diagnostic physicochemical parameters is problematic; one would need to know the quantitative relationship between current and rate of stirring under the particular stirring regime, including its dependency on the diffusion coefficient, D and concentration c, to extract these parameters. Hence it may be that under stirred conditions, the use of a simple wire may result in the requirement for a complex, expensive or impractical stirring regime. Changes in overall response may also be hard to deconvolute into changes in D, c and/or indeed A (loss of/change in salt volume). In the absence of stirring, we have already shown that extraction of diagnostic physicochemical information (through convolution of the observed cyclic voltammetric macroelectrode response) is precluded, due to the effects of convection.[1] The MSM does not suffer from this sensitivity. Our current paper, as well as outlining the enhancement in electroanalytical performance when using MSMs, also suggests (Fig. 8a) that richer alloying information can be obtained through use of a MSM compared to a wire.

As for the detection of multiple components, the results presented in this paper and in **ref. 1** below present studies on plating/stripping reactions (soluble–insoluble redox couples) of exemplar binary and ternary systems (silver and zinc and bismuth, zinc and lithium). We are currently extending these measurements to encompass larger concentration ranges and a wider variety of couples. It is encouraging that the observed time, sweep-rate and convection rate independent currents from MSMs has enabled more ready deconvolution of the responses from individual reactions, which should also facilitate enhanced analysis of more complex mixtures.

1 D. K. Corrigan, E. O. Blair, J. G. Terry, A. J. Walton and A. R. Mount, *Anal. Chem.*, 2014, **86**, 11342.

George Zheng Chen remarked: The question on cost is a good one. It is true that people tend to choose and use the cheapest option. However, cheap may also mean missing a lot of information. The design by Prof. Mount may not be the cheapest, but when fully developed the microelectrode will help to obtain more information of better quality and accuracy.

Shuqiang Jiao asked: The limiting current of CV in Fig. 2 seems curious, should it not change with scan rate? Is there a difference?

Andrew Mount said: To clarify, Prof. Jiao is commenting on the fact that the limiting reduction current observed for mass transport limited zinc plating below $E = -1.1$ V is slightly larger at the slowest sweep rate of 25 mVs^{-1} (the red line), unlike those observed at higher sweep rates. This is most likely due to a progressive increase in electrode area arising from zinc plating at the periphery of the microdisc electrode, where the mass transport is highest. In support of this explanation, the current is seen to grow progressively with time (as the electrode is swept negatively) from near the initial mass transport limited current observed at other sweep rates. This is only seen at the slowest sweep rate, as the most zinc is deposited under these conditions.

Shuqiang Jiao questioned: In your experience, what is the largest sized microelectrode you can use and still get a microelectrode response?

Andrew Mount responded: This is an interesting question. In principle, under stagnant conditions, all electrodes, irrespective of their size, should transition into a hemispherical diffusion regime (give a microelectrode response) when the thickness of the depletion layer at the electrode surface caused by electrode reaction first becomes comparable to, and then grows greater than, the characteristic electrode dimension, L. However, given that the time to achieve this depletion layer thickness is governed by $L^2 D^{-1}$, this time becomes impractically long for macroelectrodes, and actually this regime is never achieved if there is significant additional natural convection (with the definition of "significant" being determined by the degree of convective mass transfer relative to the diffusional mass transport). As electrodes are made bigger, there is a growing sensitivity to convection, and even the relatively small amounts of natural convection typically preclude the establishment of a microelectrode response.[1] This and other published work demonstrate characteristic microelectrode responses in ambient aqueous solutions for microelectrodes with L of the order of tens of microns; it is satisfying that this is comparable to the MSM dimensions that we have studied (*e.g.* $r = 25$ μm), as the values of D in LKE are typically comparable to those found in ambient aqueous solutions. However, given that the degree of natural convection in the LKE system is likely to be higher, the upper limit of microelectrode behaviour may be lower. We are currently investigating this.

1 Gao *et al.*, *Anal. Chem.*, 1995, **67**, 1541.

Kathie McGregor asked: Are there any advantages in having arrays of microelectrodes and have you been able to fabricate them and test them in molten salts?

Andrew Mount answered: There are advantages, particularly under conditions when the diffusion fields (depletion layers) from individual microelectrodes have not overlapped, as then one obtains the characteristic microelectrode response from all of the N electrodes in the array, where N is the number of electrodes, thereby producing N times the signal. We have been able to fabricate microelectrode arrays and are in the process of testing them.

Paul Madden said: You have pointed to a different behaviour for the Eu^{II}/Eu^{III} system compared to an aqueous system. However, this should be an outer sphere charge transfer process, and hence fast and unlikely to be rate-limiting. This might suggest that atomic-scale rearrangements, not involved in the usual diffusive approach of reactants to the electrode, are involved in the molten salt system. What do you think is likely to be the distance from the electrode surface for the most important charge transfer events?

Andrew Mount replied: The power of EIS analysis using the appropriate equivalent circuit is that it enables the charge (electron) transfer resistance at the electrode surface to be extracted. We have previously established that this resistance can be measured using a microelectrode when carrying out aqueous measurements on the ferri/ferrocyanide couple (a fast outer sphere electron transfer),[1] and that the expected "bowl" shaped variation of R_{CT} with the applied dc potential, centred at the formal potential, is obtained. In this previous work the expected theoretical aqueous potential dependence for the "bowl sides" was obtained from an exponential rise in R_{CT} with a change in the dc potential away from the formal potential. On the "bowl sides", the response is dominated by the potential dependence of the smallest of the oxidation or reduction electron transfer rate constants (reduction at potentials greater than the formal potential, oxidation at potentials lower than the formal potential). We agree that the overall electron transfer rate constant will measure the slowest of the steps at/near the electrode surface required to enable electron transfer (including atomic-scale reorganisation around the redox species prior to electron transfer) and suggest that your computational modelling[2] may provide useful insight into the origin of the observed behaviour. As to the distance from the electrode surface at which the most important charge transfer events are likely to occur, we agree with the analysis in this paper that the electron transfer involves short range tunnelling on molecular length scales between the electrode and the redox species and that the rate as a function of distance is then controlled by the local potential difference between the electrode and the redox species. We note the computed differences presented in the potential profiles in this paper at the electrode surface in the molten salt medium compared to an aqueous medium containing inert ions as electrolyte, and that the modelling of this potential profile in these liquid salts is of current research interest.

1 H. L. Woodvine *et al.*, *Analyst*, 2010, **135**, 1058.
2 M. A. Pounds, M. Salanne and P. A. Madden, *Molec. Phys.*, 2015, **113**, 2451–2462.

George Zheng Chen enquired: Why do you not use the Helmholtz layer thickness?

Andrew Mount responded: Our paper does not consider the potential profile at the surface of the electrode and the molten salt, and therefore does not consider whether Helmholtz or combined Stern or Guoy-Chapman double layer are appropriate models for the potential profile at the electrode surface of the molten salt. We again note the computed differences in the potential profiles at the electrode surface in the molten salt medium compared to an aqueous medium containing inert ions as electrolyte.[1]

1 M. A. Pounds, M. Salanne and P. A. Madden, *Molec. Phys.*, 2015, **113**, 2451–2462.

Paul Madden commented: You have pointed to a difference between the behaviour of the Eu^{II}/Eu^{III} couple in the molten salt system and that typically found in aqueous systems and traced that to the kinetic steps involved in the approach of the reactants to the electrode surface — which one would normally expect to be diffusive. If electron transfer to the electrode involved a very close approach of the redox ion to the electrode surface, it could be that the desolvation of the cation in order to make contact with the electrode would become very important. Because of the strength of the attachment of the anions to these polyvalent cations and to the electrode surface, this would create a relatively high barrier to the last phase of the approach, so that this might look like an activated hopping process, rather than diffusion. We calculated such barriers in simulations of just the type of system that you have been studying.[1] In other work we looked at the slow processes which may occur in the immediate vicinity of the electrode in an aqueous system.[2] What is unclear to me is whether these very close approaches do make a substantial contribution and how they would manifest themselves in the observed electrode kinetics.

1 M. A. Pounds, M. Salanne and P. A. Madden, *Molec. Phys.*, 2015, **113**, 2451–2462.
2 D. T. Limmer, A. P. Willard, P. A. Madden and D. Chandler, *J. Phys. Chem C*, 2015, **119**, 24016–24024.

Andrew Mount replied: The charge transfer resistance determines the rate of the surface charge transfer process, independent of diffusion. We agree that it is a topic of interest as to how close the redox species have to approach the electrode surface to enable electron transfer, and whether very close approaches involving significant local reorganisation make a substantial contribution to the overall electron transfer rate.

Xianbo Jin remarked: In your microdisc electrode design, there is a hole between your electrode and the bulk electrolyte. Will the hole influence the transient response of the electrode upon, for example, chronoamperometric experiments? Have you estimated the timescale considering the depth of the hole?

Andrew Mount responded: The microdisc is indeed not inlaid (coplanar with the insulator) but rather slightly recessed by 500 nm. The effect of recessing these electrodes has been considered by Guo *et al.*[1] It has been established for microdiscs that the difference in the response of electrodes with such relatively shallow recesses compared to their area can be considered to be very close to that of inlaid electrodes.

1 Guo *et al.*, *Anal. Chem.*, 2009, **81**, 130.

Qian Xu said: The electrochemical methods can be used for analysis, sensing and monitoring of the molten salts. Are there any temperature effects on the electrochemical results? If the temperature is increasing or decreasing, can the results be stable and reproducible? For example, the positions and heights of the peaks in the CV curve might be changeable?

Andrew Mount answered: Throughout the temperature range probed in this and our previous work (typically 390–500 °C),[1] the reported results are stable and reproducible. There are indeed temperature effects that affect the electrochemical results. The height of the mass transport limited current of the CV curve (and the corresponding oxidation peak if it is a plating/stripping reaction) is proportional to, and governed mainly by, changes in diffusion coefficient D with temperature; this is governed by the activation energy (*e.g.* see Fig. 3D and Fig 5B in **ref. 1** below). A relatively minor contribution would also be due to a change in c from LKE volume expansion on heating. As to changes in peak or wave position, the position on the potential axis is determined by the formal potential, E^0, which is related to the standard reduction potential through the logarithm of the activities of the redox species. The temperature dependence of the formal potential is therefore largely determined by the change in standard reduction potential (which is itself determined by the change in entropy arising from the redox reaction) with a small contribution due to changes in activity. The width of the peaks and the shallowness of the reduction wave also change with T, determined by the parameter RT/F.

1 D. K. Corrigan, E. O. Blair, J. G. Terry, A. J. Walton and A. R. Mount, *Anal. Chem.*, 2014, **86**, 11342.

John Irvine opened discussion of Ian Farnan's paper: In Fig. 7 of your paper, is there any link between the maximum activation energy and the eutectic composition? Even though it's fully molten, there appears to be "memory" of short range order. That said, the deviation does not make sense as the activation energy is higher. Interactions should be easier for lower melting compositions.

Ian Farnan responded: We need to remember that we are measuring the activation energy for the NMR T_1 relaxation process and that we are inferring liquid mobility from this measurement. I think what we may be seeing with the longer T_1 times at close to end member compositions is a reflection of short timescale and a lower magnitude of electric field gradient fluctuations that may be less directly linked to the mobility of the liquid and more to cage fluctuation lifetimes. At intermediate compositions the electric field gradient has a much greater fluctuation, which involves a change in coordinating cation, which is much closer to the processes determining liquid mobility or diffusivity, for example. This may be why the activation energy appears higher at the intermediate composition and coincides with activation energies derived from diffusion measurements.

Paul Madden commented: You have pointed to the increase in the activation energy of the relaxation time around the equimolar Li/KCl composition and also to the fact that quadrupole coupling is the dominant relaxation mechanism for this nucleus. We imagine that the ions in the first couple of coordination shells are contributing to a fluctuating electric field gradient at the nucleus. In pure LiCl or KCl the coordination shell will, on average, be a symmetrical arrangement of Li or K ions and only the fluctuations can create a field gradient, so it is the dynamics of such fluctuations that will be reflected in the activation energy. At intermediate compositions, however, the coordination shell will normally comprise a mixture of Li and K ions which is intrinsically unsymmetrical and will cause larger field gradients (and should mean that the relaxation time itself is shorter at the intermediate composition). For this arrangement to relax, some interchange of the ionic positions is necessary (or even reorientation of the coordination shell) — a more diffusive process which is likely to have a higher activation energy than the simple in-shell fluctuations.

Ian Farnan answered: You make a very good point about the electric field gradient as a function of composition. There are two factors affecting the NMR relaxation time, T_1: the magnitude of the electric field gradient fluctuation and the correlation time itself. In the situation for these melts the T_1 will decrease with longer correlation times and, as you say, with a greater electric field gradient fluctuation. Thus, the minimum in T_1 values at intermediate compositions (at constant temperature) probably reflect this greater electric field gradient fluctuation and a longer and more diffusive timescale NMR correlation time.

Shuqiang Jiao asked: If you look at the activation energy, have you compared it with the LiCl phase diagram? Is it close to the eutectic point?

Ian Farnan replied: The maximum in the activation energy is at approximately 50% LiCl, whereas the eutectic composition is closer to 60% LiCl (59.2%). The activation energy is clearly lower than the maximum at 50%. In some way, the higher activation energy close to the eutectic point is counter-intuitive as one would expect mobility to be greater (and the activation energy lower) at the eutectic composition given that this is the lowest melting temperature in the LiCl/KCl system.

George Zheng Chen remarked: It is expected that the ion mobility should increase when the molten melt is close to the eutectic composition. Do your results agree with this expectation?

Ian Farnan responded: There are two aspects to this in our results. The activation energy and the actual value of the ^{35}Cl spin lattice relaxation time, T_1. The relaxation time T_1 is a minimum at intermediate compositions, but in the case of these intrinsically high mobility systems, this would indicate a reduction in mobility as the liquid correlation time is far above the inverse of the ^{35}Cl Larmor frequency, which is the optimum frequency for the relaxation process. Thus, to make relaxation more efficient, the motion should slow down. The activation energy also increases at intermediate compositions. All this points to the emergence of a different motional process for ^{35}Cl T_1 relaxation at intermediate compositions that is both slower and has a higher activation energy. The conclusion would be that it is this process that is increasing in frequency, reflecting the greater mobility of the liquid at these compositions.

Andrew Mount queried: Isn't it surprising that the activation energies from the conductivity measurements and from the Cl NMR spin–lattice relaxation measurements are the same in Fig. 7, given the different timescales of these processes? What does this say about the mechanisms of ion transport?

Ian Farnan answered: I think Professor Madden has made the point that the change in electric field gradient at the intermediate compositions will be greater and could require the replacement of a lithium ion by a potassium ion in the ion shell around the chloride ion, so this will be closely linked to cation transport in the liquid.

Qian Xu said: What's the difference between the NMR and Raman spectra? Can you get any information about the molten salts' structures from their NMR spectra?

Ian Farnan replied: The two techniques could be considered complimentary. The Raman spectra have a characteristic timescale of a few picoseconds and so capture a snapshot of the melt structure. The acquisition of NMR spectra takes several milliseconds, during which time the melt will have explored billions of atomic configurations. So the NMR shift will represent the average local environment of Cl. This is true for many high temperature NMR studies of molten systems. The shift tells you about local structure, usually the first coordination sphere of the resonant atom. An evolution of the structure with temperature will result in a systematic shift in the average shift. This can only be interpreted uniquely in the case of two exchanging local environments. Otherwise, only a trend towards higher or lower coordination as a function of temperature can be established.

Yingjun Liu enquired: What is the reason for the uniqueness of NaCl compared to the other alkali metal chlorides in the plot of relaxation time *vs.* ionic radii that you showed on one of your slides (but that is not presented in the paper)? In other words, what other important parameters are there to understand the relaxation process of molten alkali metal chlorides besides ionic radii?

Ian Farnan responded: I think you are referring to the plot of chemical shift *vs.* cation/anion radius ratio for alkali halides and the parabolic nature of the curve for chlorides in which sodium lies at the turning point of the parabola. There is no special significance for sodium other than its position at the turning point of the parabola. The significance is that for the other halides it is just cation radius which determines the monotonic chemical shift variation with radius ratio, whereas for the chlorides there is a parabolic behaviour, which makes the LiCl and KCl differ only by about 8 ppm. In the fluoride systems this difference is more than 70 ppm and ^{19}F NMR shows a much greater sensitivity to the melt structure than for the chlorides. We like to think of this as an advantage as any large ^{35}Cl chemical shift due to the dissolution of other elements into the melt will be entirely due to structural effects associated with the new element.

John Irvine remarked: Looking again at the maximum in activation energy it looks like a mixed alkali effect. This is in glasses for example where the conductivity decreases markedly when one ion is introduced into another's host lattice. Typically it's suggested to be linked to an entropy effect.

Ian Farnan answered: The mixed alkali effect relates to solids and the adverse affect of a mixed cation composition on ionic conductivity, *i.e.* mobility. In liquids, at least in silicate liquids, there is usually a similar but converse effect of mixed alkali or alkaline earth metals on the viscosity, reducing the viscosity, *i.e.*, in a liquid, the mobility increases with mixed alkali metals. You are right, effects like this have been directly related to entropy and the rate of structural rearrangement in the liquid. In the case of the (Li/K)Cl system this would be interpreted as a chloride ion with Li and K neighbours being less stable than a chloride with all Li or all K neighbours.

George Zheng Chen opened discussion of the paper by Yalan Liu: Thank you for your presentation with high quality cyclic voltammograms and careful analysis. My understanding is that the ultimate purpose of your work is to achieve separation of the intermetallic compounds, assuming that different metals will have different dissolution potentials. To achieve this goal, you have studied several samples, starting with the simple ones and gradually increasing the types of ions in the electrolyte. Fig. 2 and 3 have similarities and differences, and the results are for comparison between electrolytes having one type of cation and those with two types of cation. Can you explain why you would want to do that? In other words, how can your study help to achieve the goal of separation of the elements in the intermetallic compounds? Why do you want to separate these elements?

Yalan Liu answered: Thank you for the question. Actually, the different reverse potentials were chosen to determine the couples of the cathodic and anodic peaks, not for separation of the elements in the intermetallic compounds. The purpose of this work was to estimate the compositions of the intermetallic compounds formed at each cathodic peak.

George Zheng Chen said: Your CVs are quite interesting. Looking at Fig. 2 and 3, why do the CVs show very good resolution of current peaks on the oxidation branches, but on the reduction branches the resolution is not as good? For example, in Fig. 2a3 and 2a4, the reduction peaks C4 and C5 are almost merged into one, but their oxidation counterparts, A5 and A4 are well separated.

Yalan Liu replied: In the reduction branches the cathodic peaks are determined by the diffusion of the metal ions onto the electrode and the diffusion between the intermetallic compounds. The diffusion processes are relatively slow, hence, the cathodic peaks are not sharp. In the oxidation branches, the anodic peaks are related to the dissolution of a solid phase on the electrode. When the solid phase is exhausted the peak current suddenly drops, and therefore the anodic peaks are sharper than the cathodic peaks.

George Zheng Chen continued: It is obvious that the cathodic (or reduction) peaks on the CVs in both Fig. 2 and 3 are smaller than the corresponding anodic (or oxidation) peaks. Would you be able to offer an explanation? If you reverse the cathodic potential scan at a less negative potential, would the anodic peak be smaller? For example, in Fig. 2a1, if the cathodic potential scan were reversed at −1.2 V after peak C1, what might happen to peak A1?

Yalan Liu responded: Generally, when a less negative potential is applied, the anodic peak should be smaller. If the cathodic potential scan were reversed at −1.2 V after peak C1, then peak A1 should be smaller.

Hongmin Zhu remarked: The quality of your CV looks very good. Congratulations. I have some comments on the analysis of the CV. You are trying to match each anodic peak to each cathodic peak for all of the CV. This seems to be unnecessary. You almost always see more anodic (dissolving) peaks than cathodic (deposition) peaks. For example, in Fig. 1 of your paper, we see 7 anodic peaks, while we only see 5 clear peaks in the cathodic sweep. Those peaks correspond to redox couples of Ce^{3+}/Ce–Al. We will see the peak when the corresponding Ce–Al compound is on the surface of the electrode. On the anodic sweep, you usually see all of them, because dissolution happens from one Ce rich surface to the next. However, on the cathodic sweep this can't be guaranteed because diffusion of Ce into the Ce–Al compounds takes time. Because of this, the appearance of the CV will very much depend on the sweep rate. Therefore, if you really want to find each of them, you should change (lower) the sweep rate.

Yalan Liu said: Thank you very much for the comment. I agree with you and I will change the sweep rate in our future work as you suggest.

Toru H. Okabe queried: Have you tried getting the CV using a metallic aluminium electrode instead of using Mo as the working electrode? If so, how were the CV curves?

Yalan Liu replied: Yes, we have tried using an aluminium electrode instead of Mo, and only one couple related to the formation and dissolution of Al–Ln intermetallic compounds was found. Because of the abundance of Al on the Al elelctrode, the Al-rich intermetallic compound was easily formed.

Toru H. Okabe asked: How about the anodic peak when using the Al metal electrode instead of Mo? Did you observe only one peak, similar to the cathodic peak?

Yalan Liu answered: When using an Al electrode, usually only one anodic peak was found, corresponding to one cathodic peak, and the anodic peak was related to the decomposition of the Al–Ln intermetallic compounds.

George Zheng Chen enquired: In Fig. 3c2, c3, c4 and c5, there is only one reduction peak, C6, but two oxidation peaks, A6 and A6'. Your explanation for this is co-reduction, but why did the dissolution not occur at the same potential?

Yalan Liu replied: Anodic peak A6' is related to the dissolution of Al–Mo intermetallic compounds, but the corresponding cathodic peak was not observed. The released $Al^{(III)}$ ions at peak A6' do not react with Er, but instead react with the Mo electrode. However, the released $Al^{(III)}$ ions at peak A6 do react with Er, and so two peaks, A6 and A6', were observed.

Toru H. Okabe said: If you can observe the sectioned interface of the working electrode after cathodic deposition of the C6 peak, can you observe the layered Al–Ce compound? For the Ce and Al deposited electrode after peak C6, you will obtain the Ce rich Al compound. The deposit corresponding to each peak should be of different intermetallic compounds, and layer-by-layer intermetallic deposition should occur. Does this happen? Do you observe this?

Yalan Liu responded: Theoretically, the layered Al–Ce compound should be observable on the Mo electrode after cathodic deposition of the C6 peak. However, we did not find the layered Al–Ce compound; it is difficult. When different deposition potentials were employed, we obtained different Ce–Al intermetallic compounds: $Al_{11}Ce_3$, Al_3Ce, Al_2Ce and AlCe. In addition, in some cases, for other intermetallic compounds such as Al–Th and Al–Gd, layered intermetallic compounds were observed.

Dihua Wang noted: There are a lot of redox current peaks in your CVs, and some of them are quite reversible. Are there any peaks related to the reduction and oxidation of lithium? Is it possible that lithium will alloy with aluminium?

Yalan Liu answered: In the estimation of Al–Ce and Al–Er, which we presented in the paper, there are no peaks related to the reduction and oxidation of lithium. However, we did find peaks related to Li–Al intermetallic compounds when the concentration of $AlCl_3$ was high. For low $AlCl_3$ concentrations, the deposited Al metal was completely reacted with the lanthanides, hence making the formation of Li–Al intermetallic compounds difficult.

Qian Xu asked: Did you try to use the phase diagram of metals to compare with your results? Are you missing any alloys or intermetallic compounds compared to the phase diagram? Or did you obtain any more information which can be added to the phase diagram?

Yalan Liu replied: Yes, we compared our results with the relative phase diagram. Actually, the compositions of the intermetallic compounds were estimated by combining the phase diagram and the CV results. Generally, we couldn't find all of the intermetallic compounds which are presented in the phase diagram by CV. Sometimes we obtained some new information, but it needs further investigation.

Ian Farnan questioned: In your work, you studied the separation of aluminium and uranium. Do your experiments allow you to make any predictions about separating them when you have mixtures with lanthanides?

Yalan Liu responded: Yes. Actually, we have studied the Al–U, Al–Th and Al–Ln (Ln = lanthanides) intermetallic compounds and the relative deposition potentials of these intermetallic compounds were obtained. It was predicted that U and Th could be effectively separated from Ln by forming Al–U and Al–Th intermetallic compounds. In addition, we have also done some work on the separation of U from Ln.

Faraday Discussions

PAPER

The influence of fluoride ions on the equilibrium between titanium ions and titanium metal in fused alkali chloride melts

Jianxun Song,[a] Qiuyu Wang,[a] Jinyu Wu,[a] Shuqiang Jiao[b] and Hongmin Zhu*[ac]

Received 19th January 2016, Accepted 10th March 2016
DOI: 10.1039/c6fd00007j

KF is employed as a source of fluoride ions added to the melt to disclose the influence of fluoride on the disproportionation reactions of titanium ions, $3Ti^{2+} = 2Ti^{3+} + Ti$, and $4Ti^{3+} = 3Ti^{4+} + Ti$. The results reveal that the equilibrium transferred to the right direction for the first reaction and the apparent equilibrium constant increased sharply, mainly because of the formation of coordination compounds: TiF_i^{3-i}. The accurate values of the equilibrium constants referring to the formation reactions of $Ti^{3+} + iF^- = TiF_i^{3-i}$ ($i = 1-6$) in NaCl–KCl melt at 1023 K were evaluated with a best fit least squares method. It is also revealed that the stable states of the coordination compounds are TiF^{2+}, TiF_2^{+}, TiF_4^{-} and TiF_6^{3-}. Moreover, the Gibbs free energies for complex formation were estimated. Ti^{2+} was undetectable when the concentration of fluoride ion was high enough. The equilibrium constant for the formation reaction, $Ti^{4-} + 6F^- = TiF_6^{2-}$, was evaluated. The equilibrium constant, K_{c2}, for the disproportionation reaction $4Ti^{3+} = 3Ti^{4+} + Ti$, in chloride melt was determined as 0.015.

Introduction

The electrolysis processes for producing titanium are complicated due to the variety of titanium oxidation states which involve multistep processes and lead to a low current efficiency.[1,2] There are three possible valences for titanium ions in chloride melt, Ti^{2+}, Ti^{3+} and Ti^{4+}, respectively. The intermediate valence will lead to disproportionation reactions.[3,4] With the presence of metallic titanium, the equilibrium between different valence ions can be described by the following reactions:

[a]School of Metallurgical and Ecological Engineering, University of Science and Technology Beijing, Beijing, 100083, China

[b]State Key Laboratory of Advanced Metallurgy, University of Science and Technology Beijing, Beijing, 100083, China

[c]Department of Metallurgy, Graduate School of Engineering, Tohoku University, Sendai, 980-8579, Japan. E-mail: hzhu@material.tohoku.ac.jp

$$3Ti^{2+} = 2Ti^{3+} + Ti \tag{1}$$

$$4Ti^{3+} = 3Ti^{4+} + Ti \tag{2}$$

The equilibrium constants, K_{c1} and K_{c2}, for reaction (1) and reaction (2) can be expressed as:

$$K_{c1} = \frac{{x_{Ti^{3+}}}^2 x_{Ti}}{{x_{Ti^{2+}}}^3} = \frac{{x_{Ti^{3+}}}^2}{{x_{Ti^{2+}}}^3} \tag{3}$$

$$K_{c2} = \frac{{x_{Ti^{4+}}}^3 x_{Ti}}{{x_{Ti^{3+}}}^4} = \frac{{x_{Ti^{4+}}}^3}{{x_{Ti^{3+}}}^4} \tag{4}$$

where x is the concentration in molar fraction of the species, Ti^{2+}, Ti^{3+}, and Ti^{4+} respectively. The equilibrium depends on the relative chemical potentials of titanium ions of different valences. It is known that factors including temperature and solvent ions (cations and anions) have an impact on the equilibrium.

In a chloride melt, the species of titanium ions in equilibrium with metallic titanium are usually Ti^{2+} and Ti^{3+}. The relevant work has been done in detail, and especially for NaCl–KCl and LiCl–KCl systems we revalued the accurate equilibrium constant in recent reports.[5–7] It was reported that the introduction of fluoride ions to a chloride melt will lead the equilibrium reactions to the side of higher valence ions (Ti^{3+}, Ti^{4+}).[8] That is due to the relatively higher stability of the complex ions of higher valance cations with the fluoride ion (F^-), compared to lower valance cations (Ti^{2+}, Ti^{3+}). The Ti^{2+} ion will disappear when the concentration of fluoride is high enough, and then ions in equilibrium with metallic Ti will be Ti^{3+} and Ti^{4+} as in reaction (2). Investigations on the influence of fluoride have been studied for years.[8–12] However, a concrete value has not been provided, and the mechanism is still ambiguous. The present work contains new results of a study on the measurement of titanium ions in equilibrium with metallic titanium in NaCl–KCl melt with the addition of fluoride, and the influence of fluoride ion is analysed for the formation of the complexes of ${TiF_i}^{3-i}$ and ${TiF_i}^{2-i}$.

Experimental

A. The preparation of electrolytes and samples

The electrolyte used in the experiments consisted of NaCl–KCl (1 : 1) pre-dried under vacuum at 573 K for 24 h and then heated to 1023 K under an argon gas atmosphere for 12 h. HCl gas was bubbled through the melts to remove O^{2-} dissolved in the molten salt afterwards.[13] The titanium sub-chloride salts were prepared from $TiCl_4$ reacting with metallic titanium in the NaCl–KCl (1 : 1) molten salt. The prepared titanium sub-chloride salts and excess metallic titanium were held at 1023 K to reach equilibrium. High purity argon gas was used to stir the molten salts to enable adequate reaction of titanium ions and titanium metal. More details can be found in papers we published before.[5–7] Potassium fluoride KF (Sinopharm Chemical Reagent Co., Ltd. analytical grade ≥ 99.99%) was added to the chloride melts as the fluoride ions source. The KF was also pre-dried under vacuum at 573 K for 24 h and in an argon gas atmosphere up to 1173 K for 12 h. Samples were taken by extracting the melts into a piece of quartz tube.[5] After rapid-quenching, samples were transferred to the glove box to await analysis. Four parallel samples were taken from fused salt for analysis at each experiment point. The average values were considered to be the measured values of the concentration of titanium ions.

The equilibrium experiment was performed for several series with different initial concentrations of titanium ions, and for each series, KF was added to adjust the F^- concentration in the melt up to 30 times the initial concentration of titanium ions.

B. The determination of titanium ions concentration

The quantitative analysis of different valances of titanium ions consists of three steps. The concentrations of Ti^{2+} and Ti^{3+} in each sample were determined by H_2 volumetric analysis and titration, respectively. Then, the total amount of titanium ions was determined by diantipyrylmethane spectrophotometry.[5–7] Finally, the concentration of Ti^{4+} was determined.

Results and discussion

A. The equilibrium region and the color change of the samples

The analysis results showed two regions of the titanium ions in equilibrium with metallic titanium: (A) Ti^{2+} and Ti^{3+}, (B) Ti^{3+} and Ti^{4+}. In the melt before the fluoride addition and low concentration of F^-, only Ti^{2+} and Ti^{3+} were detected, Ti^{4+} was not detected. Ti^{4+} was detected in the samples with high F^- concentration, and in those samples Ti^{2+} was not detected. Ti^{3+} was detected in all the samples. Because all the experiments were carried out in the presence of titanium metal, regions (A) and (B) represent the equilibrium of reaction (1) and (2), respectively.

Fig. 1 shows a series of samples quenched from the melts with a same initial titanium salt concentration of 4.6×10^{-3} with various fluoride additions. In Fig. 1, (a)–(f) are samples in region (A), while (g)–(l) are samples in region (B). In region (A), the color of the sample changes from green to violet gradually with the addition of the fluoride to the melt. In region (B), however, the colors of the samples gradually turn lighter. This color change demonstrates the change of the dominate valence ion of titanium, $Ti^{2+} \rightarrow Ti^{3+} \rightarrow Ti^{4+}$.

B. The equilibrium between Ti, Ti^{2+} and Ti^{3+}

In the low fluoride concentration melts, in region (A), only Ti^{2+} and Ti^{3+} were detected. Therefore, it is reasonable to consider that the system is under the equilibrium of reaction (1). Fig. 2 shows the relationship between concentration of fluoride ion added $x_{F^-}^{total}$ and the apparent equilibrium constants (K'_{c1}). K'_{c1} was the same value before the addition of fluoride ions for the all samples of different initial concentration of titanium ions, which is the same as reported before ($K_{c1} = 1.55$ when $x_{F^-}^{total} = 0$).[5] It is also disclosed that the K'_{c1} increases with increasing fluoride concentration. However, K'_{c1} shows dependence on the concentration of titanium ions, even at a same concentration of fluoride.

The increase of K'_{c1} with the addition of F^- ion obviously indicates the formation of relatively stable complexes of Ti^{3+} with F^-. It is well know that in chloride melts F^- ions will be preferentially coordinated to a higher valance

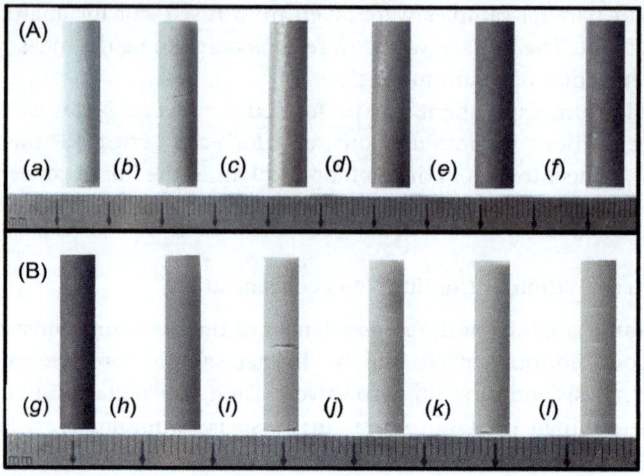

Fig. 1 Samples of titanium chloride in the NaCl–KCl melt with different concentrations of fluoride at 1023 K. The initial concentration of Ti(III) was 0.46×10^{-2} and the concentration of fluoride, $x_{F^-}^{total}/10^{-2}$, was (a) 0.0; (b) 2.6; (c) 5.3; (d) 6.4; (e) 7.7; (f) 9.0; (g) 10.0; (h) 10.5; (i) 11.4; (j) 12.1; (k) 13.0; (l) 13.8.

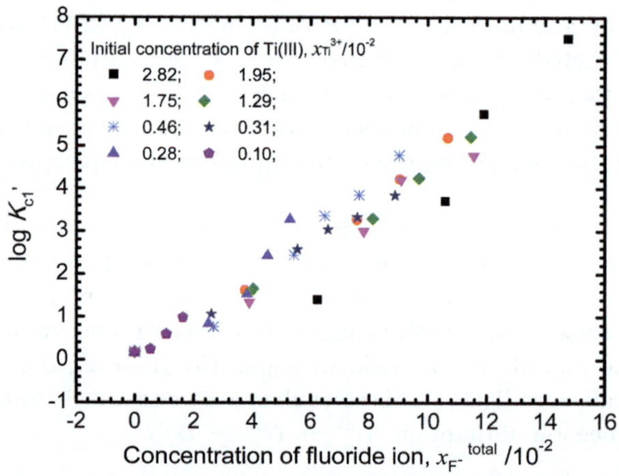

Fig. 2 The relationship between concentration of fluoride and apparent equilibrium constants (K'_{c1}). See key for different initial concentrations of Ti(III).

cation. Therefore, it is reasonable to assume that at low fluoride concentrations, most of the F^- ions are coordinated to Ti^{3+}, as in reaction (5)

$$Ti^{3+} + iF^- = TiF_i^{3-i},\ i = 1\text{–}6 \tag{5}$$

where i is coordination number (C. N.). Therefore, there will be two kinds of trivalent titanium ions, Ti(III) in the melt when fluoride is added into the chloride, TiF_i^{3-i} and Ti^{3+}. The former is the complex with F^-, the latter is "free" Ti^{3+} without coordination of F^-. It is worth noting that in the chloride melt there is not any real "free" Ti^{3+} ion, since each cation is always surrounded by some anions. It was reported that in an alkali chloride melt, the coordination number of Cl^- to Ti^{3+} is around 6.[14–17] Therefore the "free" Ti^{3+} here refers to the Ti(III) surrounded by Cl^- ($TiCl_6{}^{3-}$), and $TiF_i{}^{3-i}$ refers to complexes in which i F^- ions have replaced the position of Cl^- ($TiCl_{6-i}F_i{}^{3-}$). Therefore the coordination number, i, here is not a real coordination number of anions around Ti(III), but the number of F^- around Ti(III). The true coordination number of anions around Ti(III) is rather always 6.

With the presence of metallic titanium in the melt, the equilibrium between Ti^{2+}, Ti^{3+}, and Ti still exists as reaction (1). Therefore we can estimate the concentration of "free" Ti^{3+}, $x^{eq.}_{Ti^{3+}}$ according to eqn (3), by using the concentration of Ti^{2+} and the value of K_{c1} in the same chloride before the addition of KF ($K_{c1} = 1.55$). Then we can estimate the concentration of the complexes, $\sum x_{TiF_i{}^{3-i}}$, from the total concentration of Ti^{3+} obtained by the analysis, according to eqn (6):

$$\sum x_{TiF_i{}^{3-i}} = x^{total}_{Ti^{3+}} - x^{eq.}_{Ti^{3+}} \quad (6)$$

Fig. 3 shows the relationship between the concentration of fluoride ion added, $x^{total}_{F^-}$, and the total concentration of the complexes $TiF_i{}^{3-i}$, $\sum x_{TiF_i{}^{3-i}}$. It can be found that the concentration of the complexes increases with the addition of fluoride. However, there is not a simple relationship between them, and the concentration of the complex also depends on the initial concentration of titanium ions. Fig. 4 shows the relationship between the concentration of fluoride ion, total concentration of Ti^{3+} and concentration of complexes. It is clear from the figure that the concentration of the complexes, $\sum x_{TiF_i{}^{3-i}}$, increases with increasing total concentration of the Ti(III) as well. As can be found from Fig. 3, the ratio of the concentration of the total fluoride ion added to the concentration of complexes, $\alpha = x^{total}_{F^-}/\sum x_{TiF_i{}^{3-i}}$ ranged from 1 to 6. This indicates that the coordination number of the complexes is not only the same, but is between 1 and 6. Therefore, we assume that all 6 kinds of complexes with coordination numbers 1 to 6 exist in the melt. The equilibrium constants for complex formation can be expressed as:

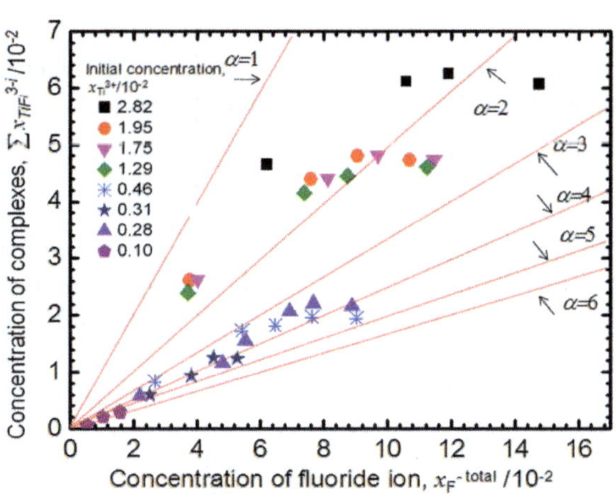

Fig. 3 The relationship between the concentrations of fluoride ion and complexes. See key for different initial concentrations of Ti(III).

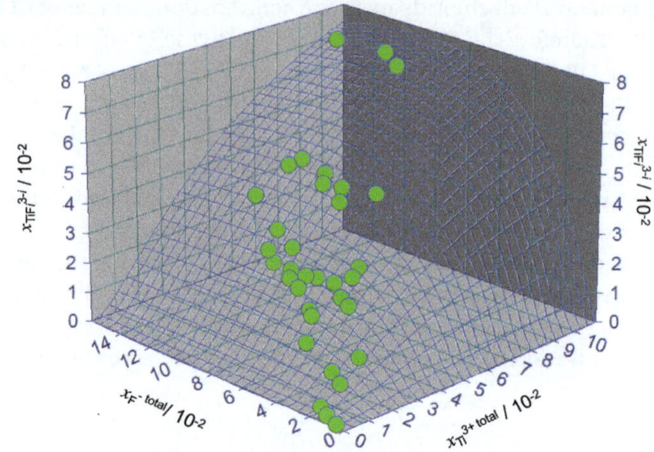

Fig. 4 The relationship between concentration of fluoride, concentrations of Ti(III) and concentration of complexes.

$$K_i = \frac{a_{TiF_i^{3-i}}}{(a_{F^-}^{eq.})^i a_{Ti^{3+}}^{eq.}} = \frac{x_{TiF_i^{3-i}}}{(x_{F^-}^{eq.})^i x_{Ti^{3+}}^{eq.}} \quad i = 1 - 6 \tag{7}$$

where a_i are activities of relevant species. They obey Henry's law under the condition of low concentration as the solutes. Therefore, the equilibrium constants for complex formation, K_i, were expressed in terms of concentration. $x_{F^-}^{eq.}$ refers to the concentration of "free" F^- ion, which means the F^- ion is not coordinated to Ti^{3+}. As described above, $x_{Ti^{3+}}^{eq.}$ can be calculated according eqn (3) using the concentration of Ti^{2+}, $x_{Ti^{2+}}^{eq.}$. The concentration of fluoride, $x_{F^-}^{total}$ can be expressed as eqn (8).

$$\begin{aligned} x_{F^-}^{total} &= 6K_6 x_{Ti^{3+}}^{eq.}(x_{F^-}^{eq.})^6 + 5K_5 x_{Ti^{3+}}^{eq.}(x_{F^-}^{eq.})^5 + 4K_4 x_{Ti^{3+}}^{eq.}(x_{F^-}^{eq.})^4 + 3K_3 x_{Ti^{3+}}^{eq.}(x_{F^-}^{eq.})^3 \\ &\quad + 2K_2 x_{Ti^{3+}}^{eq.}(x_{F^-}^{eq.})^2 + K_1 x_{Ti^{3+}}^{eq.}(x_{F^-}^{eq.}) + x_{F^-}^{eq.} \end{aligned} \tag{8}$$

Eqn (9) expresses the total concentration of Ti(III), $x_{Ti^{3+}}^{total}$, where K_i (i = 1–6) are equilibrium constants for formation of the complexes.

$$x_{Ti^{3+}}^{total} = \sum K_i x_{Ti^{3+}}^{eq.}(x_{F^-}^{eq.})^i + x_{Ti^{3+}}^{eq.} \quad i = 1 - 6 \tag{9}$$

The above equations apply for all points of experiment, and $x_{F^-}^{total}$, $x_{Ti^{3+}}^{eq.}$, and $x_{Ti^{3+}}^{total}$ were obtained for each experiment point. The equilibrium constants for the complexes, K_i, (i = 1–6), were obtained by least squares fitting using all the experiment points (34 points). Table 1 shows the results of the best fit of equilibrium constants for complex formation. Fig. 5 demonstrates the final result of the best fitting. A linear relationship between $\log\left[x_{TiF_i^{3-i}}/(x_{F^-}^{eq.})^i\right]$ and $\log x_{Ti^{3+}}^{eq.}$ (i = 1–6) was obtained for each complex with all the experiment points.

Thus, we can conclude that the formation equilibrium constants of the complexes, TiF_i^{3-i} (i = 1–6), can be revalued by the aforementioned best fitting

Table 1 The formation equilibrium constant and Gibbs free energies of complex formation

TiF_i^{3-i}	K_i	ΔG_i^0/kJ mol^{-1}	$\Delta G_i^0 i^{-1}$/kJ mol^{-1}
$i = 1$	3.9	−11.6	−11.6
$i = 2$	3.5×10^2	−49.8	−24.9
$i = 3$	4.5×10^{-5}	85.2	28.4
$i = 4$	4.1×10^4	−90.4	−22.6
$i = 5$	1.6×10^{-8}	133.1	26.6
$i = 6$	5.8×10^6	−132.4	−22.1

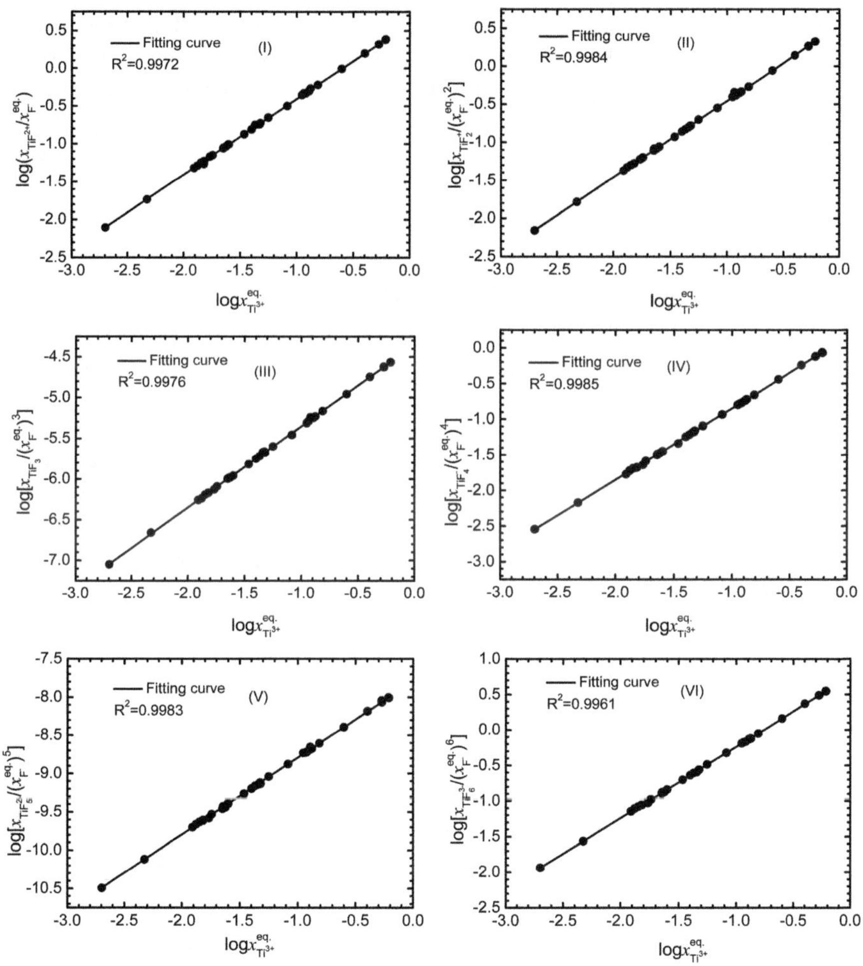

Fig. 5 The relationship between $\log\left[x_{TiF_i^{3-i}}/(x_{F^-}^{eq.})^i\right]$ and $\log x_{Ti^{3+}}^{eq.}$; (I) $i = 1$, (II) $i = 2$, (III) $i = 3$, (IV) $i = 4$, (V) $i = 5$, (VI) $i = 6$.

method. They are 3.9, 3.5×10^2, 4.5×10^{-5}, 4.1×10^4, 1.6×10^{-8} and 5.8×10^6, respectively. Using the equilibrium constants, the Gibbs free energies for complex formation can be obtained as

$$\Delta G_i^0 = -RT\ln(K_i), (i = 1\text{–}6) \tag{10}$$

where R is the gas constant and T is the absolute temperature. The results of the Gibbs free energies for complex formation reactions are also shown in Table 1. It can be found that the values of ΔG_1^0, ΔG_2^0, ΔG_4^0 and ΔG_6^0 are negative, while ΔG_3^0 and ΔG_5^0 are positive. This corresponds to the extremely small value of K_3 and K_5. It is well understandable according to coordination theory[15,16] that asymmetric coordination is uncommon. As mentioned above, the real coordination number of anions (Cl^-, F^-) around each Ti(III) is 6, i is the number of F^- ions which replace Cl^- positions. In the complex with 3 and 5 F^-, one of the F^- ions is coordinated opposite a Cl^- ion. This asymmetric structure will cause relatively higher chemical potential than a symmetric structure. Therefore, the mainly existential states of the complexes are TiF^{2+}, TiF_2^+, TiF_4^- and TiF_6^{3-} in the NaCl–KCl melt when fluoride ions were added. The value of $\Delta G_i^0/i$ is also shown in Table 1. This value corresponds to the change of the Gibbs free energies for the reactions of each F^- replacing the Cl^- ions. As can be found from the table, the values are very close to each other, for TiF_2^+, TiF_4^- and TiF_6^{3-}, which is reasonable because they are all symmetrically coordinated.

C. The equilibrium between Ti, Ti^{3+} and Ti^{4+}

In region (B), when the concentration of fluoride is high enough compared to the concentration of titanium ions, the concentration of Ti^{2+} was undetectable, while Ti^{4+} was detected.

Fig. 6 shows the relationship between concentrations of fluoride ion and the complexes of TiF_i^{3-i}. The symbols represent different initial concentrations of trivalent titanium ions, Ti(III), before the addition of fluoride. There are three series for initial concentrations of 2.8×10^{-3}, 3.1×10^{-3} and 4.6×10^{-3}. The solid symbols (■, ▲, ★) represent region (A), and the empty symbols (□, △, ☆) represent region (B).

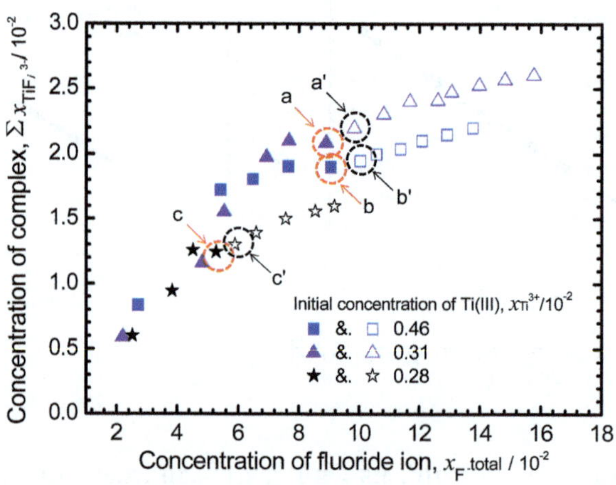

Fig. 6 The relationship between concentrations of fluoride ion and complexes. See key for different initial concentrations of Ti(III).

As described above, the addition of fluoride will cause complex formation, as in reaction (5). When a sufficient amount of fluoride was added into the melt, almost all of the Ti(III) ions will form the complexes, TiF_i^{3-i}. This will drive the concentration of "free" Ti^{3+} to a very low level, and then drive the concentration of Ti^{2+} to an extremely low and undetectable level. In this region the equilibrium between titanium ions and metallic titanium will follow reaction (2), and the equilibrium constant, K_{c2}, is defined by eqn (4). However, the apparent equilibrium constant, K'_{c2}, did not stay constant for different samples with different concentrations of titanium ions and fluoride when we accounted for the concentration of Ti^{4+} and Ti^{3+} obtained by analysis into $x_{Ti}{}^{4+}$ and $x_{Ti}{}^{3+}$ in eqn (4). This is because of the formation of complexes, as described above for TiF_i^{3-i} in the first part. K_{c2} will stay constant only when we account for the concentration of "free" Ti^{3+}, and Ti^{4+}, $x^{eq.}_{Ti^{4+}}$, $x^{eq.}_{Ti^{3+}}$, which is again the Ti(III) and Ti(IV) ions without coordination with F^-, but Cl^- in the first coordination. Therefore we need to consider the concentration of the complex for Ti^{4+} with F^- and TiF_i^{4-i} to estimate the concentration of "free" Ti^{4+}. The formation reaction of the complexes can be described by reaction (11).

$$Ti^{4+} + iF^- = TiF_i^{4-i} \tag{11}$$

Theoretically speaking, the coordination number, i, exists from 1 to 6 as in the case for Ti(III). However, as Ti(IV) is only detected in the melts with high fluoride concentration, we can presume that most of them possess a high coordination number. This was confirmed by calculating the concentration of Ti(III) in all forms of complexes, for the points with highest concentration of fluoride in the region (A) for each series, which is marked by "a", "b" and "c" in Fig. 6. The results are shown in Table 2.

As can be found from the table, the concentration of $TiF_6{}^{3-}$ is dominant, and the total concentration of the others is less than 4% for each series. Because in region (B), the concentration of fluoride is always higher than that in region (A), the ratio of $TiF_6{}^{3-}$ concentration to total Ti(III) concentration should be even higher than that at those three points. Therefore, it is reasonable to equate the concentration of Ti(III) to that of $TiF_6{}^{3-}$, and estimate the concentration of "free" Ti^{3+} in region (B). At the same time, it is also reasonable to assume that the complex of Ti(IV) with the coordination number of 6, $TiF_6{}^{2-}$, is dominating the concentration of Ti(IV). We can define the equilibrium constant of the complex as in eqn (12).

$$K'_6 = \frac{x_{TiF_6{}^{2-}}}{(x^{eq.}_{F^-})^6 x^{eq.}_{Ti^{4+}}} \tag{12}$$

Table 2 The concentration of complexes at different initial concentration of titanium

	Concentration, $x_i/10^{-2}$			
Point in Fig. 6	TiF^{2+}	$TiF_2{}^+$	$TiF_4{}^-$	$TiF_6{}^{3-}$
a	0.008	0.02	0.03	1.79
b	0.005	0.01	0.04	1.19
c	0.003	0.01	0.03	1.14

Therefore, the concentration of fluoride can be expressed as:

$$x_{F^-}^{total} = x_{F^-}^{eq.} + 6x_{TiF_6{}^{2-}} + 6x_{TiF_6{}^{3-}} \tag{13}$$

The concentration of $TiF_6{}^{3-}$ and $TiF_6{}^{2-}$ can be expressed as below:

$$x_{TiF_6{}^{3-}} = K_6(x_{F^-}^{eq.})^6 x_{Ti^{3+}}^{total} \Big/ \left[1 + (x_{F^-}^{eq.})^6 K_6\right] \tag{14}$$

$$x_{TiF_6{}^{2-}} = K'_6(x_{F^-}^{eq.})^6 x_{Ti^{4+}}^{total} \Big/ \left[1 + (x_{F^-}^{eq.})^6 K'_6\right] \tag{15}$$

Then, the eqn (16) can be obtained.

$$x_{F^-}^{total} = x_{F^-}^{eq.} + 6K'_6(x_{F^-}^{eq.})^6 x_{Ti^{4+}}^{total} \Big/ \left[1 + (x_{F^-}^{eq.})^6 K'_6\right] + 6K_6(x_{F^-}^{eq.})^6 x_{Ti^{3+}}^{total} \Big/ \left[1 + (x_{F^-}^{eq.})^6 K_6\right] \tag{16}$$

In eqn (16), the values of $x_{F^-}^{total}$, $x_{Ti^{3+}}^{total}$, $x_{Ti^{4+}}^{total}$ and K_6 were known for each experimental point, and $x_{F^-}^{eq.}$, which refers to the concentration of "free" F^- ion, can be estimated from K_6 and $x_{Ti^{3+}}^{total}$. By obtaining a best fit by the least squares method, the value of K'_6 was obtained as 1.51×10^6. Fig. 7 demonstrates the agreement of the experimental points with the best fit, in which $x_{Ti^{4+}}^{total}$, $x_{TiF_6{}^{2-}}$ were calculated for each point by using the fitting result of K'_6. As can be seen in the figure, excellent linear correlation between $\log\left[x_{TiF_6{}^{2-}}/(x_{F^-}^{eq.})^6\right]$ and $\log x_{Ti^{4+}}^{eq.}$ was obtained.

Using the value of K'_6, the Gibbs free energy for $TiF_6{}^{2-}$ formation can be calculated as -121.0 kJ mol^{-1}, as listed in Table 3.

By using the value of K_6 and K'_6 obtained above, we can now estimate the concentration of "free" Ti^{3+} and Ti^{4+} for each point of the experiment, and then calculate K_{c2} for reaction (2). This is the equilibrium between titanium metal and titanium ions of Ti(III) and Ti(IV) without coordination of F^-. The value of K_{c2} calculated in this way is plotted against the total concentration of Ti(III) in Fig. 8. As can be seen from the figure, the value of K_{c2} remains constant for all the

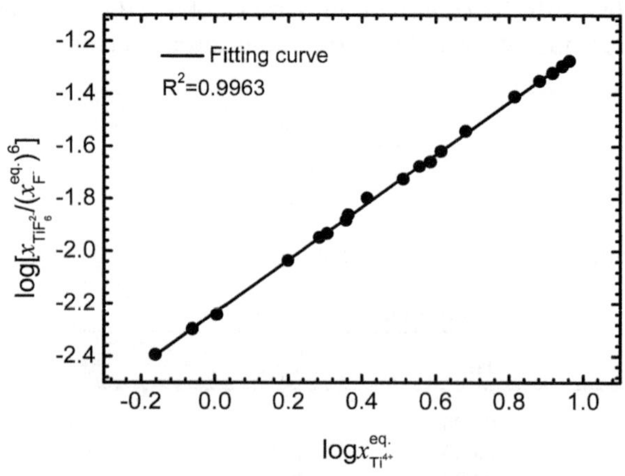

Fig. 7 The relationship between $\log\left[x_{TiF_6{}^{2-}}/(x_{F^-}^{eq.})^6\right]$ and $\log x_{Ti^{4+}}^{eq.}$.

Table 3 Complex and equilibrium constant for formation

Complex	Equilibrium constant, K'_6	ΔG_6^0/kJ mol^{-1}
TiF_6^{2-}	1.51×10^6	−121.0

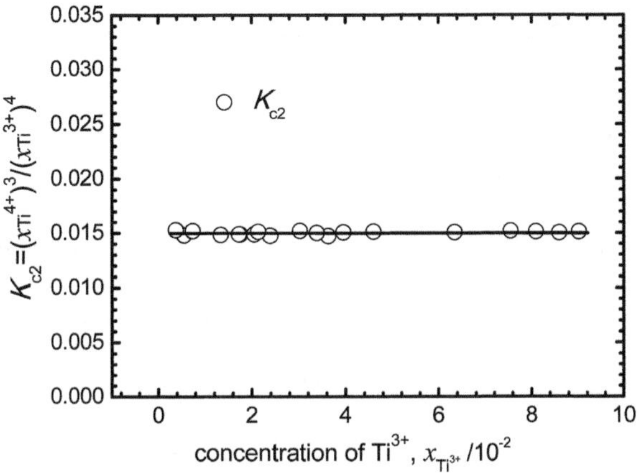

Fig. 8 The relationship between $x_{Ti}{}^{3+}$ and equilibrium constant for the reaction $4Ti^{3+} = 3Ti^{4+} + Ti$.

experimental points. The average value of K_{c2} is 0.015. Theoretically, this value is equivalent for the equilibrium reaction of reaction (2) in this chloride melt without the addition of fluoride. However, in the real pure chloride melt, the concentration of Ti(IV) will be undetectable because the value of K_{c2} is so low, and the concentration of Ti(III) is also limited because of the presence of titanium metal.

Conclusions

The detected titanium ions in equilibrium with metallic titanium are Ti^{2+} and Ti^{3+} in the chloride melt with low concentrations of fluoride. In the melts with high fluoride concentration, Ti^{2+} was undetectable and Ti^{4+} was detected. The influence of the fluoride ion on the equilibrium reaction between titanium metal and the ions was described in terms of the formation of complex ions of Ti^{3+} and Ti^{4+} with F^-. The equilibrium constants of the formation reaction of the complexes, TiF_i^{3-i}, were obtained. In the high fluoride concentration melts almost all Ti^{3+} and Ti^{4+} forms a complex with a coordination number of 6, TiF_6^{3-} and TiF_6^{2-}. The equilibrium constant for the formation of TiF_6^{2-} was also obtained. The equilibrium constant for the disproportionation reaction $4Ti^{3+} = 3Ti^{4+} + Ti$ in the chloride without the addition of fluoride was estimated as 0.015.

Acknowledgements

The authors gratefully acknowledge the support of the National Natural Science Foundation of China (No. 51322402) and the National High Technology Research and Development Program of China (863 Program, No. 2012AA062302).

Notes and references

1 G. M. Haarberg, W. Rolland and A. Sterten, *J. Appl. Electrochem.*, 1993, **23**, 217–224.

2 Q. Wang, Y. Li, S. Jiao and H. Zhu, *Electrochem. Commun.*, 2013, **35**, 135–138.

3 H. Sekimoto, Y. Nose, T. Uda and H. Sugimura, *Mater. Trans.*, 2010, **51**, 2121–2124.

4 H. Sekimoto, Y. Nose, T. Uda, A. Uehara, H. Yamana and H. Sugimura, *J. Alloys Compd.*, 2011, **509**, 5477–5482.

5 Q. Wang, J. Song, G. Hu, X. Zhu, J. G. Hou, S. Jiao and H. Zhu, *Metall. Mater. Trans. B*, 2013, **44**, 906–913.

6 X. Zhu, Q. Wang, J. Song, J. G. Hou, S. Jiao and H. Zhu, *J. Alloys Compd.*, 2014, **587**, 349–353.

7 J. Song, Q. Wang, X. Zhu, J. Hou, S. Jiao and H. Zhu, *Mater. Trans.*, 2014, **55**, 1299–1303.

8 G. Chen, M. Okido and T. Oki, *Electrochim. Acta*, 1987, **32**, 1637–1642.

9 F. Lantelme and A. Salmi, *J. Electrochem. Soc.*, 1993, **142**, 3451–3456.

10 F. R. Clayton, G. Mamantov and D. L. Manning, *J. Electrochem. Soc.*, 1973, **120**, 1193–1199.

11 A. Robin and R. B. Ribeiro, *J. Appl. Electrochem.*, 2000, **30**, 239–246.

12 L. P. Polyakova, P. T. Stangrit and E. G. Polyakov, *Electrochim. Acta*, 1986, **31**, 159–161.

13 Y. Castrillejo, A. M. Martinez, G. M. Haarberg, B. Borresen, K. S. Osen and R. Tunold, *Electrochim. Acta*, 1997, **42**, 1489–1494.

14 N. F. H. Bright and I. G. Wurm, *Can. J. Chem.*, 1958, **36**, 615–622.

15 J. G. Speight, *Lange's handbook of chemistry*, 2005, pp. 1357–1359.

16 D. M. Gruen and R. L. Mcbeth, *Pure Appl. Chem.*, 1963, **6**, 23–47.

17 G. Chen, M. Okido and T. Oki, *J. Appl. Electrochem.*, 1988, **18**, 80–85.

Faraday Discussions

PAPER

Formation of Si nanowires by the electrochemical reduction of SiO_2 with Ni or NiO additives

Sheng Fang, Han Wang, Juanyu Yang,* Bing Yu and Shigang Lu

Received 14th December 2015, Accepted 22nd January 2016
DOI: 10.1039/c5fd00222b

Various morphologies of silicon nanowires (SiNWs) were successfully prepared by the electrochemical reduction of silica mixed with different additives (Au, Ag, Fe, Co, Ni, and NiO, respectively). Straight SiNWs were extensively obtained by the electro-reduction of porous Ni/SiO_2 blocks in molten $CaCl_2$ at 900 °C. The SiNWs had a wide diameter distribution of 80 to 350 nm, and the Ni–Si droplets were found on the tips of the nanowires. The growth mechanism of SiNWs was investigated, which could reveal that the nano-sized Ni–Si droplets formed at the $Ni/SiO_2/CaCl_2$ three-phase interlines. Based on the mechanism proposed, NiO particles with sub-micrometer size were selected as the additive, and straight SiNWs with diameters of 60 to 150 nm were also prepared *via* the electrochemical process.

1. Introduction

Silicon nanowires (SiNWs) are of great interest due to their potential in advanced nano-electronics,[1] optoelectronics,[2,3] biological detection devices[4–6] and energy storage applications.[7–9] In the last few decades, reported methods involving the synthesis of SiNWs include chemical vapour deposition using catalyst metals,[10,11] electrodepositon,[12] laser ablation,[13,14] supercritical fluid solution-phase growth,[15–17] and metal-assisted wet chemical etching (MaCE).[18,19] However, some of these methods are not capable of producing SiNWs in bulk quantities, while others involving extreme conditions are not cost-effective.

Recently, it has been reported that SiNWs could be prepared *via* the electro-reduction of silica in molten $CaCl_2$ at medium temperatures.[20–24] During the electrolysis, silica was reduced to silicon at the cathode, and the electrochemical reaction was proposed to occur at and through the propagation of the conductor/$SiO_2/CaCl_2$ three-phase interlines.[25] As it has been suggested, this process provokes a novel industrial development of SiNWs production, since the increase of SiNWs yielded can be achieved by simply increasing the silica input used for

R&D Center for Vehicle Battery and Energy Storage, General Research Institute for Nonferrous Metals (GRINM), Beijing, 100088, P. R. China. E-mail: juanyuyang@163.com; Fax: +86-10-82241190

each run.[23,26] Both quartz plates and preformed porous pellets were used as cathodes, but it has been confirmed that the variety of raw materials has the influence on the aspect-ratio of SiNWs in the products.[20,22,24] For instance, SiNWs with a high-aspect-ratio of curvy morphology was synthesised by the electrolysis of porous pellets,[20] rather than by the electrolysis of quartz plates.[22] The major disadvantage of this approach is that the prepared SiNWs show a curved and largely crooked morphology, and also have many lattice defects. Due to the curved and irregular branches, it is difficult to achieve the controlled alignment and assembly of SiNWs into desired architectures.

In order to overcome this problem, Au nano-particles have been added to SiO_2 pellets.[21] It has been proven that more elongated SiNWs were prepared from the electrolysis of Au/SiO_2 pellets, compared to those from pure SiO_2 pellets. However, the diameter of the nanowires changes dramatically along the axial direction and increases significantly as the electrolysis proceeds. The promotion of longitudinal growth of SiNWs suggests that the nucleation and growth process of SiNWs could be changed by the addition of Au. A concept similar to the metal-catalyzed vapor–liquid–solid (VLS) nanowires processes was proposed by the authors.[21] Actually, the metal-catalyzed VLS growth is not a catalytic process. The seed metal will react with silicon and form a metal droplet which acts as a 'soft template' to collect the source material. When supersaturation of silicon in the droplet is reached, the excess silicon precipitates out of the droplet in the form of a nanowire. The critical process is the formation of nano-sized metal-silicon alloy droplets. The melting point of the metal-silicon alloy can be predicted by the phase diagram with a consideration of nano-sized effects. The nanostructure can be achieved by adding nano-sized seed metal particles[27] or gasification of the metal powders.[28] Thus, other nano-sized metal additives can be chosen on the basis of a phase diagram, if the growth mechanism of SiNWs after adding Au nanoparticles is similar to VLS mechanism. Indeed, it has been reported that Cu nanoparticles were added to the SiO_2 pellet to promote the growth of SiNWs with straight morphology since liquid Cu–Si alloy could be generated at the electrolysis temperature (900 °C).[29] Straight SiNWs became dominant in the electrolysed product obtained from the Cu/SiO_2 pellets. Thus, the concept of dissolution–supersaturation–precipitation can be used in this electroreduction process to lead the oriented growth of SiNWs.

However, it seems to be that using metal nanoparticles as additives is not suitable for the electrochemical reduction process, in terms of cost and safety issues during the mixing process. A recent report presented a successful generation of Ag–Si droplets on the surface of a silver substrate *via* the electroreduction of silica nanoparticles added into the molten $CaCl_2$.[30] This result may suggest that the nano-sized metal droplets can be produced by adding micrometer sized metal particles.

In this paper, micro-sized silver, gold, iron, cobalt, and both nickel powders and fibre were respectively added into the SiO_2 pellets to investigate the influence of metal additives on the morphology of the produced SiNWs. SiNWs with straight morphology were prepared from the electrolysis of Ni/SiO_2 blocks. The process of electrochemical reduction of SiO_2 and the growth mechanism of SiNWs after adding nickel powders were further investigated. Based on the mechanism proposed, nickel oxide was chosen as the additive. Results showed that nano-sized nickel–silicon droplets were formed at the $Ni/SiO_2/CaCl_2$ three-phase interlines during the electroreduction process, which suggested that this process could be a potential approach to synthesize nano-sized metal silicides.

2. Experimental

Silver powder (purity > 99.90%, size: 500 mesh), gold powder (purity > 99.90%, size: 1 μm), iron powder (purity > 99.0%, size: 45 μm), cobalt powder (purity > 99.50%, size: 300 mesh), nickel powder (purity > 99.7%, size: 2–10 μm), nickel fibre (purity > 99%, size: 20 μm) and nickel oxide (purity > 99.9%, size: 0.5–1.0 μm) were chosen as the additives.

SiO_2 powder (99.98%, particle size: 25–30 nm) was mixed with distilled water to form silica sol, and then metal additives or nickel oxide powder were respectively added into the sol at a desired $[SiO_2]/[Additives]$ ($[SiO_2]/[Additive] = 10$) mole ratio. The silica sol was dried at 80 °C for 2 days, followed by ball milling. For the metal additives, the preformed powders were hot-pressed (compacting pressure: 10 Mpa, hot-pressing temperature: 950 °C, dwell time: 75 min, purged with argon) into porous blocks with a size of 10 mm × 10 mm × 5 mm. For the nickel oxide additive, the preformed powders were pressed into porous pellets (40 mm in diameter, 3 mm in thickness) and sintered at 900 °C in the air for 2 hours. After that, pellets were cut into blocks with a size of 10 mm × 10 mm × 3 mm. All the blocks were drilled and threaded onto a molybdenum wire (99.95%, 2.5 mm in diameter) to form Mo-Additive/SiO_2 contacting electrodes. Calcium chloride was contained in a graphite crucible and then sealed in a stainless steel vessel which was continuously purged with argon gas (purity > 99.999%, O_2 < 2 ppm, H_2O < 1 ppm; 400 ml min^{-1}) during the electrolysis. The conducting experimental temperature was 900 °C. In order to remove impurities from the molten salt, pre-electrolysis was carried out between a graphite anode and a molybdenum rod cathode.

To prepare silicon nanowires, a constant voltage was applied for the electrochemical reduction of Additive/SiO_2 blocks, while the graphite crucible was used as the anode. After electrolysis, reduced blocks were raised from the melt and kept in the upper part of the reactor for cooling. The solidified salt adhered to the reduced blocks was removed by washing in distilled water in an ultrasonic bath. After washing, the products normally turned to powdery form, and then were collected and dried under vacuum for analysis.

Cyclic voltammograms (CVs) were recorded using a computer controlled PARSTAT 2273 Electrochemical System (Princeton Applied Research, USA). A graphite rod was used as the counter electrode, and a quartz tube Ag/AgCl electrode was fabricated and used as the reference electrode. A schematic diagram of this reference electrode is shown in Fig. 1. For the cyclic voltammetry experiments, modified molybdenum electrodes before and after loading nickel or Ni/SiO_2 powders were used as the working electrodes. Molybdenum wire with a diameter of 1.0 mm was used to prepare the modified electrode. The molybdenum wire was polished to remove the surface dirt. After that the wire was subjected to repeated bending until broken. Such broken wire had an end with a fresh and significantly rough surface. The rough end of the wire was then pressed manually into Ni–SiO_2 or Ni powders, after which the rough end was filled with particles of the powders. In the following text, this electrode is denoted as

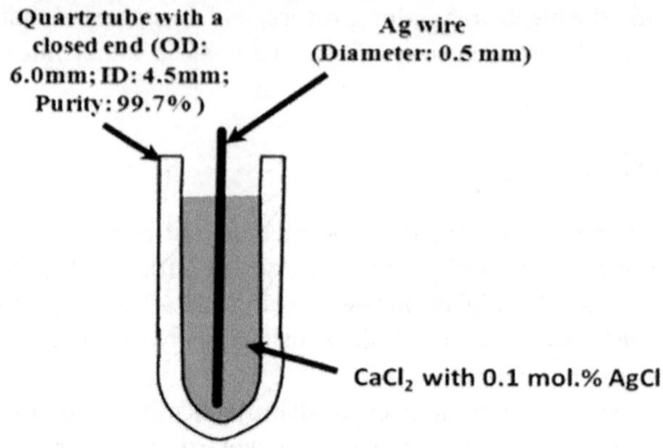

Fig. 1 A schematic diagram of the quartz tube Ag/AgCl reference electrode.

Mo–Ni or Mo–Ni/SiO_2. In the experiments, the immersion depth of the empty and powder loaded modified voltammetry was about 10 mm.

The synthesized powders were characterised by X-ray diffraction (XRD, X' Pert PRO MPD), field-emission scanning electron microscopy (FESEM, Hitachi S4800), transmission electron microscopy and high-magnification electron microscopy (HRTEM, FEI Tecnai F30), energy dispersive X-ray spectroscopy (EDS), and selected area electron diffraction (SAED).

3. Results and discussions

3.1 Influence of metal additives on the morphology of SiNWs

FESEM images of the products obtained *via* the electrochemical reduction of different metal/SiO_2 blocks are presented in Fig. 2. The results show that numerous curved and entangled SiNWs were prepared from the blocks with added Au, Ag, Fe and Co powder, respectively. These morphologies are similar to those observed from the products which were obtained from the molten salt electrolysis of pure SiO_2 pellets.[20] It seems to be that these metals cannot promote the oriented growth of SiNWs.

As shown in Fig. 2e and f, straight SiNWs can be found in the products prepared from Ni powder/SiO_2 and Ni fibre/SiO_2 blocks respectively. It should be noted that straight SiNWs are dominant in the products from the Ni powder/SiO_2 block, while only a few straight SiNWs are produced by the electrochemical reduction of the Ni fibre/SiO_2 block. Therefore, it indicates that nickel powder of micrometer size can be used to promote the growth of straight SiNWs. The differences in morphology of the products between adding nickel powders and adding nickel fibre into the raw materials, may be attributed to the different chemical properties of these two nickel particle additives. Based on the results above, the Ni/SiO_2 block was selected as the raw material to prepare SiNWs, and the properties of SiNWs were characterised in terms of morphology, crystal structure and composition. In order to optimise the preparation process, the growth mechanism was also further investigated.

Fig. 2 FESEM images of SiNWs produced by the electrolysis of metal/SiO_2 blocks at 1.5 V in molten $CaCl_2$ at 900 °C for 5 h. (a) Ag/SiO_2; (b) Au/SiO_2; (c) Fe/SiO_2; (d) Co/SiO_2; (e) Ni powder/SiO_2; (f) Ni fibre/SiO_2.

3.2 Characterisations on morphology, structure and composition

As shown in Fig. 3a, the SiNWs obtained from the Ni/SiO_2 blocks are long and straight with a wide diameter distribution ranging from 80 to 350 nm, and the diameters of most nanowires range from 80 to 250 nm. A bead can be observed on the tip of the nanowire as shown in the inset in Fig. 3a, which could be evidence for the tip-led growth mechanism. In addition, the XRD patterns displayed in Fig. 3b are indexed well to Si (JCPDS no. 27-1402) with a diamond cubic structure and $NiSi_2$ (JCPDS no. 43-0989) with a cubic fluorite structure. Since the starting mixtures contained excess silicon atoms compared to $NiSi_2$, it can be confirmed that the products have these two phases.

Except for SiNWs, byproducts with particle morphology are also found in the products and highlighted by an arrow as shown in Fig. 4a. The element

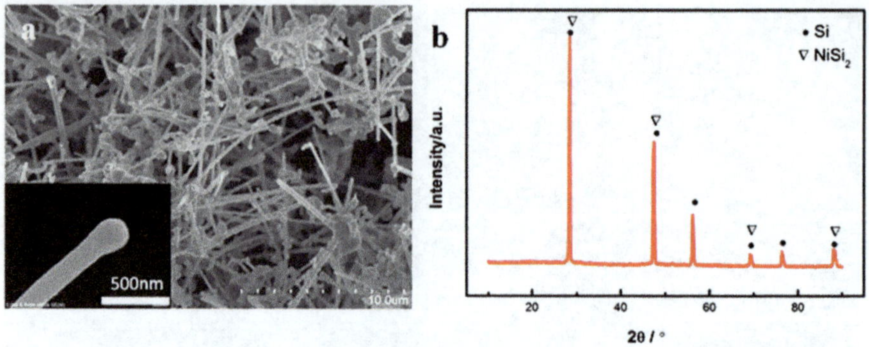

Fig. 3 (a) FESEM images of SiNWs produced by the electrolysis of Ni/SiO_2 blocks at 1.5 V in molten $CaCl_2$ at 900 °C for 5 h. Inset is a close-up view of a bead on the tip of a nanowire. (b) XRD patterns of the synthesised products.

compositions of the SiNW area and particle area were characterised respectively by EDS. Results of the EDX spectroscopy, as shown in Fig. 4b and c, indicate that the nanowire area is composed of Si and a small amount of O and Ni, while the particles contain more Ni. It can be revealed that concentrations of Ni, Si, and O are 1.67, 80.42, and 17.91 atom% in SiNW area, and 20.76, 60.07, and 19.71 atom % in particle area, respectively. Thus, the particle byproducts could be $NiSi_2$.

Fig. 5a shows a low-magnification TEM image revealing a straight silicon nanowire with a diameter of 155 nm. This nanowire is a single crystal with a cubic lattice structure indicated by SAED as shown in the inset in Fig. 5a. The high-resolution TEM image (Fig. 5b) reveals the well-defined lattice fringe of the (111) plane. The SiNWs obtained from the pure SiO_2 pellet are covered by a thin

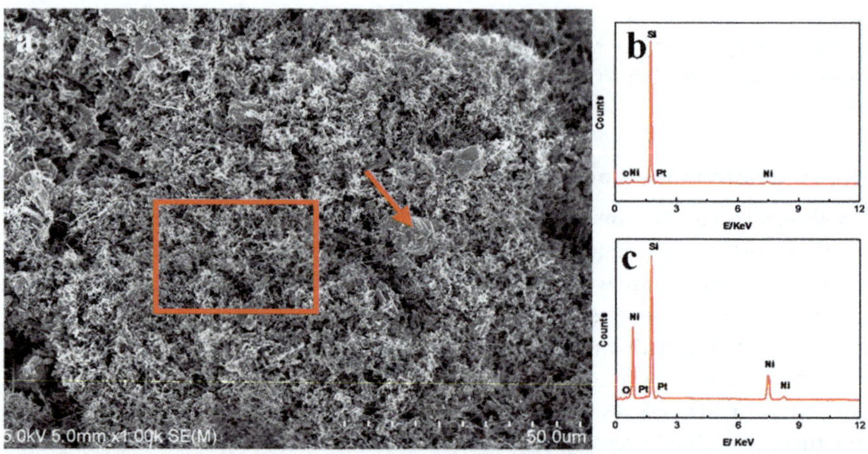

Fig. 4 (a) FESEM images of products produced by the electrolysis of Ni/SiO_2 blocks at 1.5 V in molten $CaCl_2$ at 900 °C for 5 h. The arrow indicates a particle byproduct. EDS results of the SiNW area (b) indicated by the square in (a) and the particle area (c) indicated by the arrow in (a).

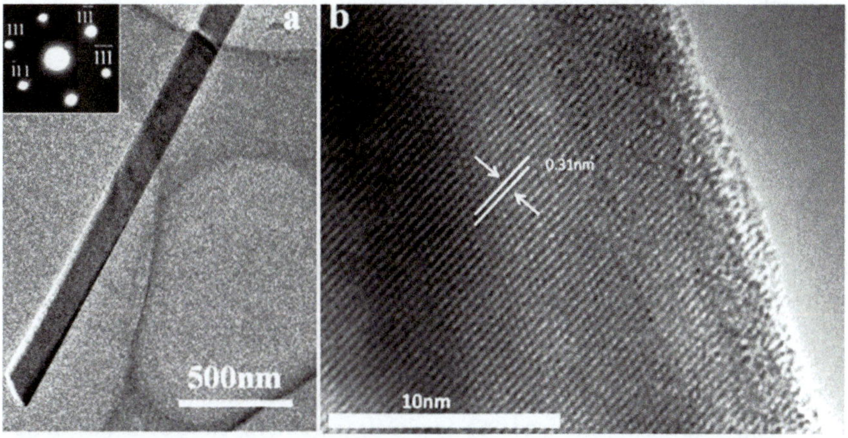

Fig. 5 (a) TEM image of a single silicon nanowire produced by the electroreduction of Ni/SiO_2 blocks at 1.5 V in molten $CaCl_2$ at 900 °C for 5 h. The inset shows the electron diffraction pattern of the silicon nanowire body. (b) High-resolution TEM taken from SiNW body.

amorphous SiO_2 outer sheath with a thickness of 5–6 nm,[20] while the silicon nanowire in this report shows a much thinner amorphous sheath (<1 nm).

Fig. 6a exhibits a low-magnification TEM image of a 330 nm diameter wide silicon nanowire with a sphere-like tip. According to the results of the TEM EDS line scan along the length of the nanowire, as shown in Fig. 6b, it should be silicon at the body, together with Ni at the tip. Thus, it can be suggested that the growth mechanism of the silicon nanowire is similar to the VLS (vapor–liquid–solid) mechanism, and nano-sized Ni–Si alloy should be generated. In that case, this electrolysis process could be a potential approach to synthesise nano-sized

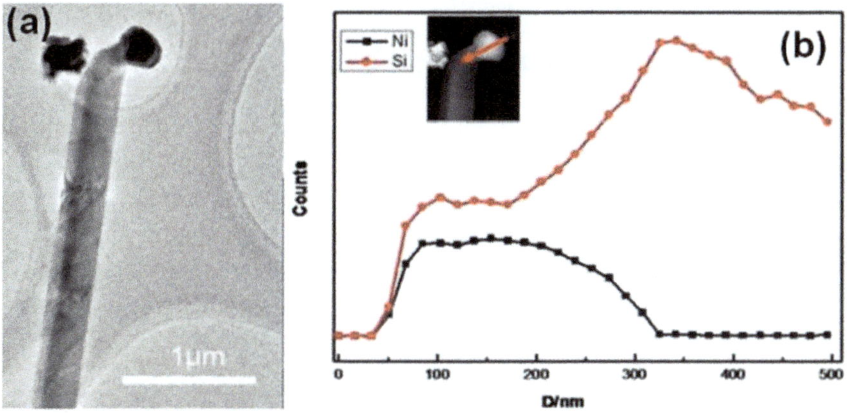

Fig. 6 (a) Low-magnification TEM image of a silicon nanowire with a diameter of 330 nm. (b) TEM EDS line scan (indicated by an arrow in the inset) of the silicon nanowire shown in (a).

metal-silicon alloy particles. Therefore, it is necessary to investigate the process of the formation of nano-sized nickel-silicon alloy and the growth of SiNWs.

3.3 Growth mechanism of SiNWs

The sintered blocks were immersed in molten $CaCl_2$ at 900 °C for 30 min, and then washed in distilled water, dried in a vacuum oven, and inspected under FESEM. The blocks before immersing have a blue colour, against the pale colour after immersing, as shown in Fig. 7a and c, respectively. The phase compositions of the blocks before and after immersing were then analyzed by XRD. It is confirmed that blocks both before and after immersing contain nickel and amorphous silica, as shown in Fig. 7b and d.

As shown in Fig. 8a, the silica in the block before immersion shows a particle morphology, while the immersed one demonstrates a network structure as shown in Fig. 8c. Our previous research has proven that this network structure was formed through a partial softening or melting process of nano-SiO_2 in molten $CaCl_2$ at 900 °C.[20] The nickel powders in the blocks before immersion have a smooth surface as presented in Fig. 8b, while melted silica can be found on the surface of the nickel particles as presented in Fig. 8d. It should be noted that once the nickel particle is linked to the conducting channel, SiO_2 at the Ni/SiO_2/$CaCl_2$ three-phase interline will be electrochemically reduced to silicon simultaneously.

The product obtained by the electrolysis of the Ni/SiO_2 blocks for 1 min was washed and then analysed by FESEM. Between the reacted and unreacted zone, an intermediate particle product is observed as shown in Fig. 9a. The high-

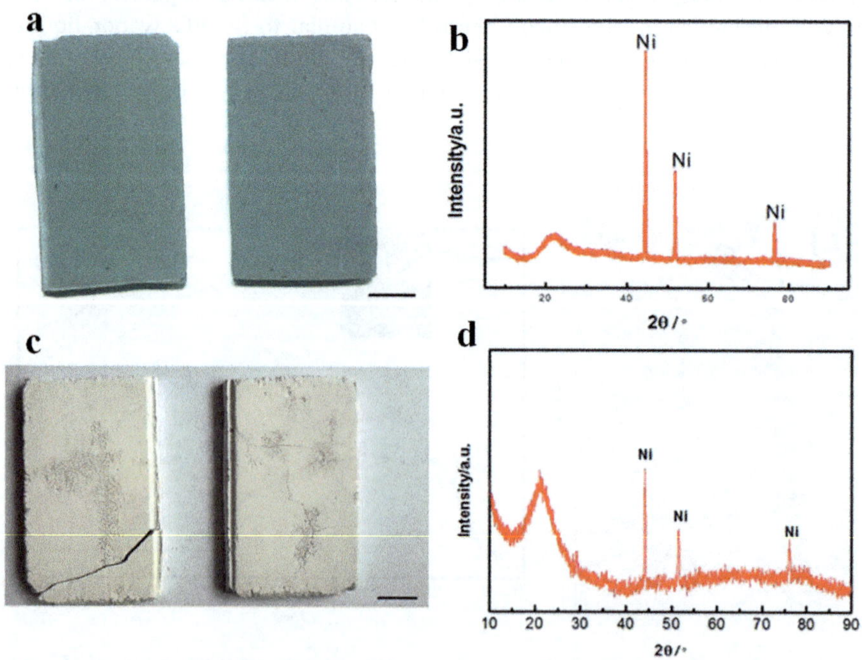

Fig. 7 Photographs of the block before (a) and after (c) immersion in molten $CaCl_2$ at 900 °C for 30 min. XRD patterns of the blocks before (b) and after (d) immersion.

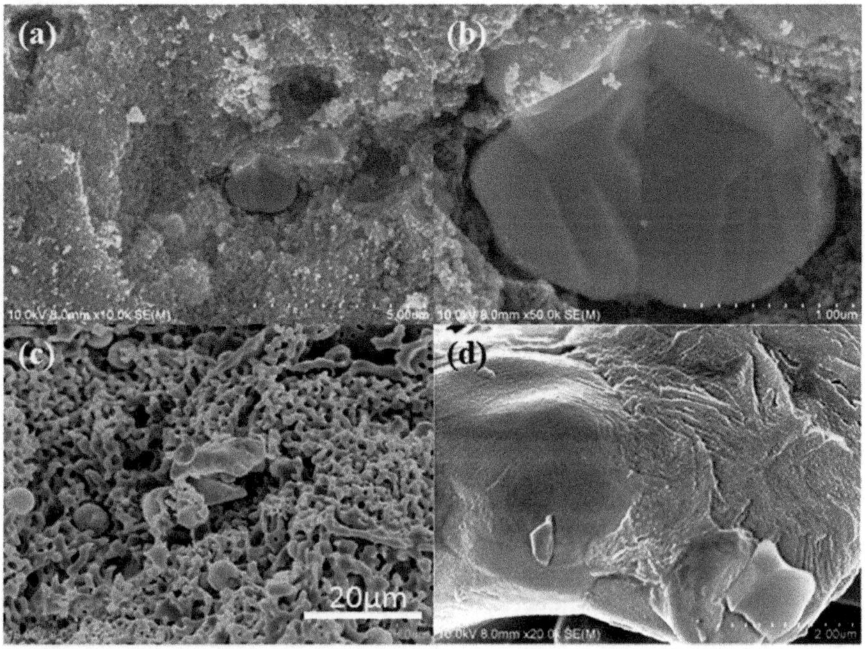

Fig. 8 (a) Cross-sectional FESEM image of Ni/SiO_2 blocks (a and b) before and (c and d) after immersing at 900 °C for 30 min.

magnification image (Fig. 9b) shows that beads with diameters ranging from 30 to 300 nm are found on the surface of the intermediate particle product. EDS results reveals that the concentrations of oxygen, nickel and silicon are 32.77, 19.38, and 47.85 atom% respectively in the particle area, which can indicate that this particle contains nickel silicide and silica. The beads on the surface may be the nickel–silicon alloy droplets, as their diameters are close to the diameter of the silicon nanowire.

During the electroreduction process, it is possible that the following reactions occur at the $Ni/SiO_2/CaCl_2$ three-phase interlines:

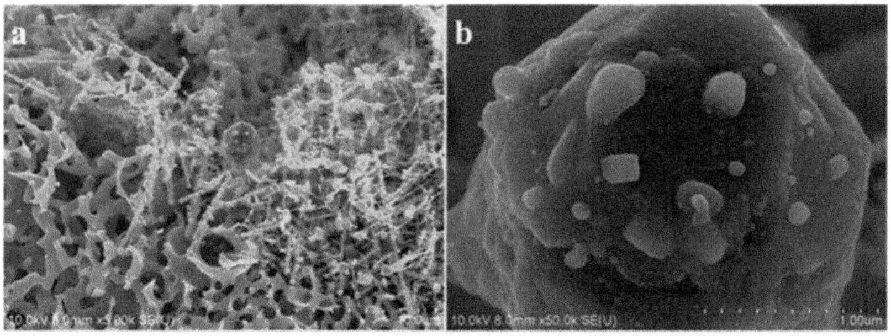

Fig. 9 (a) Cross-sectional FESEM image of a block after electrolysis at 1.5 V in molten $CaCl_2$ at 900 °C for 1 min. (b) High-magnification of the particle shown in (a).

$$SiO_2 + 4e^- \rightarrow Si + 2O^{2-}, 0.84\ V\ (vs.\ Ca^{2+}/Ca, 900\ °C) \quad (1)$$

$$SiO_2 + Ni + 4e^- \rightarrow NiSi + 2O^{2-}, 1.34\ V\ (vs.\ Ca^{2+}/Ca, 900\ °C) \quad (2)$$

$$2SiO_2 + Ni + 8e^- \rightarrow NiSi_2 + 4O^{2-}, 0.95\ V\ (vs.\ Ca^{2+}/Ca, 900\ °C) \quad (3)$$

Because the nickel silicides form spontaneously at elevated temperatures, the reduction of melted silica on the nickel surface to form nickel silicides should be thermodynamically much easier than the reduction of SiO_2 to Si as shown in the potential data (*vs.* Ca^{2+}/Ca; these data were calculated from HSC Chemistry version 6.1, Outotec). Thus, reaction (2) and reaction (3) are more likely to occur at the $Ni/CaCl_2/SiO_2$ three-phase interlines than reaction (1). The reaction here may be different from the reaction occurring at the Ag (or Au)/SiO_2/$CaCl_2$ three-phase interlines, since there is no Ag-silicide or Au-silicide phase.

The voltammetric analysis of molybdenum wire, Mo–Ni, and Mo–Ni/SiO_2 were conducted by cyclic voltammetry. Due to its satisfactory stability, Mo wire was employed to establish a background CV. Both Mo–Ni and Mo–Ni/SiO_2 were used to clarify the influence of Ni on the electrochemical behaviours of SiO_2. Typical CVs of Mo wire, Mo–Ni and Mo–Ni/SiO_2 electrode are presented in Fig. 10. All these experiments were conducted in molten $CaCl_2$ at 900 °C with the quartz tube Ag/AgCl reference electrode. In this paper, unless specified otherwise, all potential data are reported with reference to this Ag/AgCl reference electrode described in the experimental. It is clear that a pair of sharp cathodic and anodic waves (C1′ and A1′) was observed at −2.10 V (*vs. E* Ag^+/Ag) on the Mo wire, as shown in Fig. 10a. These waves could correspond to the deposition and dissolution of calcium. For the Mo–Ni electrode, the deposition of Ca almost started at a less positive potential (approximately −2.02 V). It has been reported that formation of Ca–Si intermetallic compounds *via* the reduction of Ca^{2+} over the reduction-generated silicon can occur at a more positive potential than the equilibrium potential of the Ca^{2+}/Ca couple.[31] Since the phase diagram of the Ni–Ca system shows the existence of several intermetallic compounds, the under potential reduction of Ca on the nickel may be possible. Thus, the deposition potential for Ca could be less positive for the Mo–Ni electrode. The waves C1 and A1 could

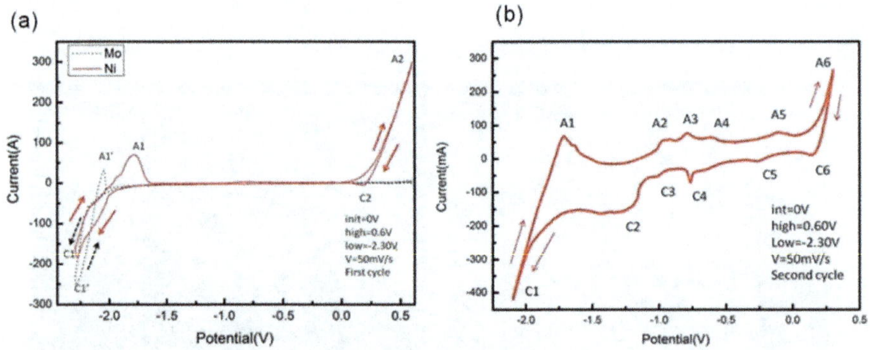

Fig. 10 Cyclic voltammograms of the (a) Mo, Mo–Ni, and (b) Mo–Ni/SiO_2 electrode in molten $CaCl_2$ at 900 °C (reference electrode: quartz tube Ag/AgCl reference electrode, scan rate: 50 mV s^{-1}).

correspond to the formation and deformation of a Ni–Ca alloy. Compared to the Mo electrode, wave A2 can be seen near the positive end of the potential scan for the Mo–Ni electrode. Waves C2 and A2 may correspond to the cathodic dissolution and anodic formation of chlorine. According to the thermodynamic data, the potential of SiO_2 to Si is 0.84 V (calculated from HSC Chemistry 6.1) which is more positive than that of Ca^{2+} to Ca. The reduction of SiO_2 could start at a potential of around −1.26 V. The CV of the Mo–Ni electrode shows no waves around this potential. Thus, Ni does not generate a new electrochemical reaction at the potential window for the electro-reduction of SiO_2.

Fig. 10b shows the CV of the Mo–Ni/SiO_2 electrode at a scan rate of 50 mV s^{-1} in molten $CaCl_2$ at 900 °C. Compared to the Mo electrode, the cathodic current peak occurs at a more positive potential (−1.62 V). This could be attributed to the under potential of Ca. Thus, waves C1 and A1 can correspond to the formation–deformation of Si–Ca alloy. Wave A6 and wave C6 were observed at almost the same potentials as wave A2 and wave C2 in Fig. 10a respectively. These waves could correspond to the dissolution and anodic formation of chlorine. In Fig. 10b, the potential for wave C2 is about −1.20 V which is close to the theoretical potential for reduction of SiO_2. It suggests that waves C2 and A2 could be attributed to the reduction of SiO_2 and oxidation of silicon. Two waves as indicated by C3 and C4 were observed at more positive potentials. The potentials for cathodic waves C3 and C4 are −1.02 V and −0.76 V, respectively. Both of the cathode reactions corresponding to waves C3 and C4 occur at more positive potentials. According to the thermodynamical datas shown in reaction (1) to (3), waves C3 and C5 are thought to correspond to reaction (3) and (2) respectively. In the anodic direction, current waves A3 and A4 were observed and could be attributed to the decomposition of Ni_xSi compounds. As such, it suggests that the nickel silicides could be formed *via* the direct electrochemical reduction of silica at the Ni/SiO_2/$CaCl_2$ three-phase interlines.

According to the binary Ni–Si phase diagram, it is obvious that the lowest temperature leading to the metal droplet tip-led growth of nanowire is 993 °C at which the liquid peritectic alloy (Ni 41 atom% and Si 59 atom%) can be formed. However, a thin layer of nickel deposited on the Si substrate can react with the substrate at temperatures above 930 °C, and then the nickel–silicon liquid alloy droplets can be formed due to the melting effect of small-size grains.[32] The softening or melting temperature of nano-sized alloy powders can be much lower than that of the corresponding bulk one. For instance, it has been reported that the melting point of nano-sized Ti–Fe alloy powders (particle size: 5–30 nm) decreased by 256.3–338.6 °C, compared to the melting point of the bulk Ti–Fe alloy.[33] Thus, due to the nanometer size of the reduction-generated nickel silicide, it is possible to form a Ni–Si droplet at the electrolytic temperature.

Fig. 11a shows that several silicon nanowires with their own beads at the tips are also connected to a particle with their roots. It can be explained that since the catalyst droplets are initially formed on the surface of nickel powders, the roots that are grown should be linked to the nickel particles. Further electrochemical reduction of silica will occur at the droplet/SiO_2/$CaCl_2$ three-phase interline. Continuous dissolution of the produced silicon will lead to the precipitation of silicon at the droplet–nickel interface, which can cause the separation of droplets from the nickel surface and then its extension to the SiO_2 network. As shown in Fig. 11b, a nanowire with a bead at the tip is found to grow straight along the

Fig. 11 Cross-sectional FESEM image of a block after the electrolysis at 1.5 V in molten $CaCl_2$ at 900 °C K for 1 min. (a) FESEM image of a silicon nanowire grew from the surface of a particle. The inset shows the magnification image of the nanowires. (b) Silicon nanowires grow along the path limited by the SiO_2 network.

network. Since SiO_2 is the only source of Si, the growth path of SiNWs is limited by the SiO_2 network. Thus, these results could be the evidence that straight SiNWs are formed through the dissolution–supersaturation–precipitation process.

Based on the results above, with consideration of the mechanism reported by Xiao *et al.*[25] and Nohira *et al.*,[21] the growth mechanism of the nanowire is proposed as shown in Fig. 12. The growth model can be explained as follows: (1) nano-sized Ni–Si droplets are formed on the surface of nickel *via* the direct electroreduction of silica at the Ni/SiO_2/$CaCl_2$ three-phase interlines; (2) the freshly formed droplets act as the new conducting sites, at which the silica is reduced to silicon. The generated silicon dissolves into the droplets immediately. Further reduction of silica and continuous dissolution of silicon in the droplets leads to the supersaturation of silicon, and eventually causes the precipitation of silicon at the droplet–nickel surface interface. Due to the high solubility of Si in

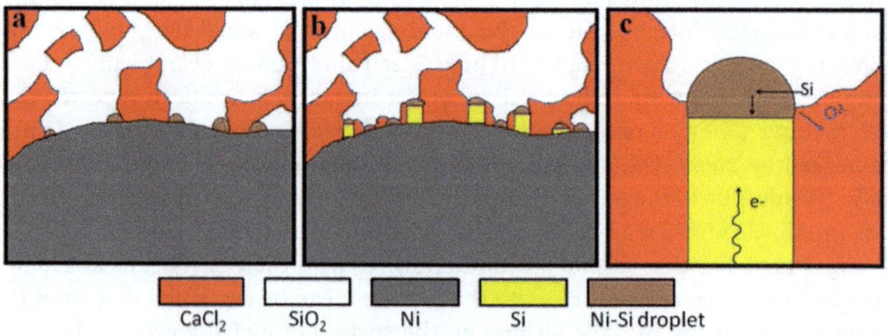

Fig. 12 Schematic drawings of the growth process of the SiNWs by the electrochemical reduction of the Ni/SiO_2 block. (a) Formation of nickel–silicon droplets at the Ni/SiO_2/$CaCl_2$ three-phase interline. (b) The droplet/SiO_2/$CaCl_2$ three-phase interlines moved to the SiO_2 network with growing SiNWs on the surface of the nickel particle. (c) Silica was reduced to silicon at the droplet/SiO_2/$CaCl_2$ three-phase interline and the silicon dissolved into the droplet. Tip-led growth was supported by the dissolution–supersaturation–precipitation process of silicon.

the Ni–Si droplet, the critical nuclei size of the precipitated silicon is comparable to the size of the droplet. This does not allow the spontaneous nucleation step to occur. Thus, the dissolution–supersaturation–precipitation process leads to the 1-D growth of silicon. Moreover, this process could separate the droplets from nickel, which causes the droplet/SiO_2/$CaCl_2$ three-phase interline to move to the SiO_2 network.

The critical process for the growth of straight SiNWs is the formation of nano-sized nickel–silicon droplets. The melted silica at the Ni/SiO_2/$CaCl_2$ three-phase interline could generate the nucleus with nanometer sizes, since silica has a character similar to the colloid or liquid phase. Most of the metal silicides can form spontaneously at elevated temperatures. In other words, it can be suggested that all nanometer sized metal silicides could be formed at the metal/SiO_2/$CaCl_2$ interline by using nano-sized silica as a raw material. However, formation of nano-sized metal silicides cannot seed the growth of straight SiNWs. In order to proceed the dissolution–supersaturation–precipitation process, liquid metal silicides should be formed, and the temperature for generating liquid alloy can be predicted based on the related phase diagram. It is worth mentioning that the value of the temperature can be lowered by using small sized raw materials. No straight SiNW was formed in the products from Fe/SiO_2 and Co/SiO_2 blocks, which may be attributed to the high eutectic point of the alloy. The theoretic temperatures required for Fe and Co additives are 307 °C and 359 °C higher than the electrolytic temperature, respectively. In addition, nano-sized droplets should be separated from metal particles by the precipitated silicon. Otherwise, droplets can not led the growth of SiNWs along the silica network. In other words, the initial alloy formed by the solid phase reaction of precipitated silicon and metal additives should be solid. Since the reaction occurs on the surface of the metal, the initial alloy could be metal rich phase. Taking nickel additive as an example, the temperature for liquid Ni rich alloy is high (>1143 °C), thus it is not likely to form a liquid alloy. Compared to nickel the meting point for Ag (962 °C) and Au (1064 °C) are lower. The temperature for the formation of liquid metal rich alloys are also low. The eutectic point for Ag–Si (Ag: 89.0; Si: 11.0 atom%) and Au–Si (Au: 81.4; Si: 18.6 atom%) eutectic alloy are 835 °C and 363 °C, respectively. Therefore, a liquid alloy could be formed on the surface of Ag and Au, which results in no straight SiNWs being formed.

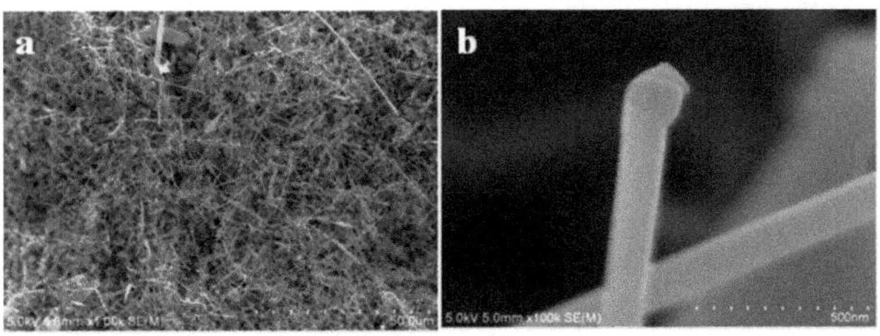

Fig. 13 (a) FE-SEM images of the prepared nanowires obtained by the electrochemical reduction of porous NiO/SiO_2 blocks at 900 °C in molten $CaCl_2$ at 1.5 V for 3 h. (b) Magnified image of the nanowires.

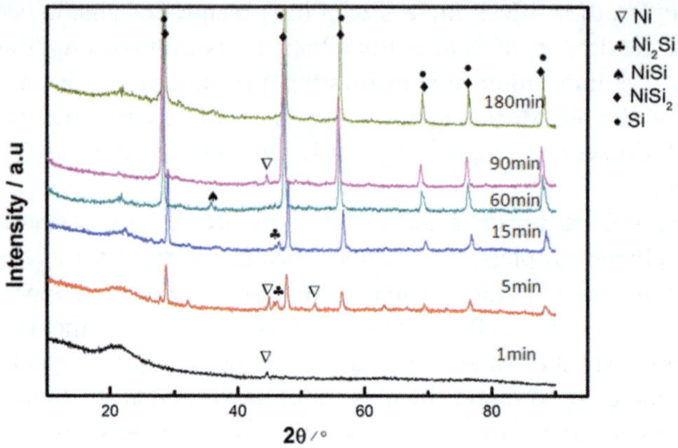

Fig. 14 XRD patterns of the products obtained by the electrolysis of NiO/SiO_2 blocks at 900 °C in molten $CaCl_2$ at 1.5 V with various electrolysis times.

3.4 SiNWs prepared *via* the electroreduction of NiO/SiO_2 blocks

The role of nickel powders in the Ni/SiO_2 block could be similar to the substrates used in the metal-catalyzed VLS growth process. In general, the agglomeration of metal droplets on the substrate results in the increase in diameter of the nanowire. Once connected to the conducting channels, a large number of Ni–Si droplets are formed on the surface of the nickel particles. The high electrolytic temperature may lead to the agglomeration of droplets, which causes the formation of SiNWs with larger diameters. Thus, in order to retard the agglomeration of the droplets, the propagation speed of $Ni/SiO_2/CaCl_2$ three-phase interlines has to be controllable. Specifically, the diameter of the nanowires is dependent on the quantity of nucleation. It should be mentioned that these can be achieved by using NiO particles as the additive. The decomposition voltage of NiO is 0.70 V at 900 °C, while the value of silica is 1.81 V. Thus, the reduction of NiO will occur at a more positive position than that of silica during the molten salt electrolysis. Moreover, NiO particles are believed to be more stable than Ni particles, in terms of their physicochemical properties. Therefore, nanosized or

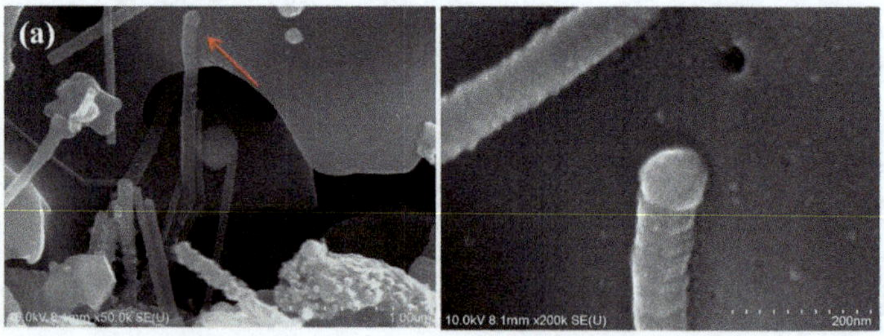

Fig. 15 Cross-sectional FESEM image of the products prepared from the electrolysis at 1.5 V in molten $CaCl_2$ at 900 °C for 5 min. (a) SiNWs grew along the silica network. (b) High-magnification image of the tip of the nanowire indicated by the arrow in (a).

sub-microsized NiO particles can be used to promote the formation of straight SiNWs in the electroreduction of silica.

Sub-micrometer sized NiO particles were selected as the additive to seed the growth of straight SiNWs in this work. As shown in Fig. 13a, nanowires with long and straight morphology were prepared by the electrochemical reduction of porous NiO/SiO_2 blocks at 1.5 V for 3 h in molten $CaCl_2$. The produced nanowires have a diameter distribution of 60 to 150 nm, and the length is upto 30 μm. It should be mentioned that the bead on the tip of nanowire is also found as shown in Fig. 13b.

The phase compositions of the products obtained by the electroreduction of NiO/SiO_2 blocks at 1.5 V with various electrolysis times were analysed by XRD. As displayed in Fig. 14, it is clear that the reduction process went through different stages from the initial formation of Ni to the formation of $NiSi_2$. This confirms that the decomposition value of NiO to Ni is more positive than that of SiO_2 to Si, and the products should contain Si and nickel silicides.

Fig. 15 shows the cross-sectional FESEM images of the products prepared from the electrolysis at 1.5 V in molten $CaCl_2$ at 900 °C for 5 min. As shown in Fig. 15a, SiNWs grew along the silica network. It is obvious that the nanowire also has a sphere-like tip as displayed in Fig. 15b. Thus, it can be suggested that the growth of the nanowire should be in agreement with the dissolution–supersaturation–precipitation process.

4. Conclusions

Various morphologies of SiNWs were successfully prepared by the electrochemical reduction of silica mixed with different additives (Au, Ag, Fe, Co, Ni, and NiO). Straight SiNWs were extensively obtained by the electro-reduction of porous Ni/SiO_2 blocks in molten $CaCl_2$ at 900 °C. These SiNWs had a wide diameter distribution of 80 to 350 nm, and Ni–Si droplets were found on the tips of the nanowires. The growth mechanism of SiNWs was investigated, which reveals that the nano-sized Ni–Si droplets formed at the $Ni/SiO_2/CaCl_2$ three-phase interline during the electroreduction process. Silica around the droplets was reduced to silicon atoms and immediately dissolved into the droplets. Continuous dissolution gave rise to the supersaturation of silicon in the droplets and then drove the precipitation of silicon. The dissolution and precipitation process led to the oriented growth of SiNWs. Based on the mechanism proposed, sub-micrometer sized NiO particles were selected as the additive, and also straight SiNWs with diameters of 60 to 150 nm were prepared *via* the electrochemical process. Thus, it may be suggested that this method can be an approach to produce nano-sized SiNWs and metal silicides in bulk quantities.

Acknowledgements

This study was financially supported by the National Natural Science Foundation of China (No. 51404032 and No. 51404030), the National High Technology Research and Development Program of China (863 Program) (No. 2013AA050904), and the Beijing Natural Science Foundation (No. 3154043).

Notes and references

1 Q. Li, X. Zhu, H. D. Xiong, *et al.* Silicon nanowire on oxide/nitride/oxide for memory application, *Nanotechnology*, 2007, **18**(23), 235204.

2 Y. Ahn, J. Dunning and J. Park, Scanning Photocurrent Imaging and Electronic Band Studies in Silicon Nanowire Field Effect Transistors, *Nano Lett.*, 2005, **5**(7), 1367–1370.

3 C. Yang, C. J. Barrelet, F. Capasso, *et al.* Single p-Type/Intrinsic/n-Type Silicon Nanowires as Nanoscale Avalanche Photodetectors, *Nano Lett.*, 2006, **6**(12), 2929–2934.

4 Z. Gao, A. Agarwal, A. D. Trigg, *et al.* Silicon Nanowire Arrays for Label-Free Detection of DNA, *Anal. Chem.*, 2007, **79**(9), 3291–3297.

5 W. Kim, J. K. Ng, M. E. Kunitake, *et al.* Interfacing Silicon Nanowires with Mammalian Cells, *J. Am. Chem. Soc.*, 2007, **129**(23), 7228–7229.

6 G. Zheng, F. Patolsky, Y. Cui, *et al.* Multiplexed electrical detection of cancer markers with nanowire sensor arrays, *Nature Biotechnology*, 2005, **23**(10), 1294–1301.

7 C. K. Chan, H. Peng, G. Liu, *et al.* High-performance lithium battery anodes using silicon nanowires, *Nat. Nanotechnol.*, 2008, **3**(1), 31–35.

8 H. Chen, Y. Xiao, L. Wang, *et al.* Silicon nanowires coated with copper layer as anode materials for lithium-ion batteries, *Journal of Power Sources 15th International Meeting on Lithium Batteries (IMLB)*, 2011, **196**(16), 6657–6662.

9 H. T. Nguyen, F. Yao, M. R. Zamfir, *et al.* Highly Interconnected Si Nanowires for Improved Stability Li-Ion Battery Anodes, *Adv. Energy Mater.*, 2011, **1**(6), 1154–1161.

10 M. K. Sunkara, S. Sharma, R. Miranda, *et al.* Bulk synthesis of silicon nanowires using a low-temperature vapor–liquid–solid method, *Appl. Phys. Lett.*, 2001, **79**(10), 1546–1548.

11 S. Christiansen, R. Schneider, R. Scholz, *et al.* Vapor–liquid–solid growth of silicon nanowires by chemical vapor deposition on implanted templates, *J. Appl. Phys.*, 2006, **100**(8), 084323.

12 J. Mallet, M. Molinari, F. Martineau, *et al.* Growth of Silicon Nanowires of Controlled Diameters by Electrodeposition in Ionic Liquid at Room Temperature, *Nano Lett.*, 2008, **8**(10), 3468–3474.

13 H. Z. Zhang, D. P. Yu, Y. Ding, *et al.* Dependence of the silicon nanowire diameter on ambient pressure, *Appl. Phys. Lett.*, 1998, **73**(23), 3396–3398.

14 Y. H. Tang, Y. F. Zhang, N. Wang, *et al.* Si nanowires synthesized from silicon monoxide by laser ablation, *J. Vac. Sci. Technol., B: Microelectron. Nanometer Struct.–Process., Meas., Phenom.*, 2001, **19**(1), 317–319.

15 T. Hanrath and B. A. Korgel, Supercritical Fluid–Liquid–Solid (SFLS) Synthesis of Si and Ge Nanowires Seeded by Colloidal Metal Nanocrystals, *Adv. Mater.*, 2003, **15**(5), 437–440.

16 D. C. Lee, T. Hanrath and B. A. Korgel, The Role of Precursor-Decomposition Kinetics in Silicon-Nanowire Synthesis in Organic Solvents, *Angew. Chem., Int. Ed.*, 2005, **44**(23), 3573–3577.

17 A. M. Chockla, K. C. Klavetter, C. B. Mullins, *et al.* Tin-Seeded Silicon Nanowires for High Capacity Li-Ion Batteries, *Chem. Mater.*, 2012, **24**(19), 3738–3745.

18 K. Q. Peng, J. J. Hu, Y. J. Yan, *et al.* Fabrication of Single-Crystalline Silicon Nanowires by Scratching a Silicon Surface with Catalytic Metal Particles, *Adv. Funct. Mater.*, 2006, **16**(3), 387–394.

19 Y. Hu, K. Peng, Z. Qiao, *et al.* Metal-Catalyzed Electroless Etching of Silicon in Aerated HF/H_2O Vapor for Facile Fabrication of Silicon Nanostructures, *Nano Lett.*, 2014, **14**(8), 4212–4219.

20 J. Yang, S. Lu, S. Kan, *et al.* Electrochemical preparation of silicon nanowires from nanometer silica in molten calcium chloride, *Chem. Commun.*, 2009, (22), 3273–3275.

21 Y. Nishimura, T. Nohira, K. Kobayashi, *et al.* Formation of Si Nanowires by Direct Electrolytic Reduction of Porous SiO_2 Pellets in Molten $CaCl_2$, *J. Electrochem. Soc.*, 2011, **158**(6), E55–E59.

22 W. Xiao, X. Wang, H. Yin, *et al.* Verification and implications of the dissolution–electrodeposition process during the electro-reduction of solid silica in molten $CaCl_2$, *RSC Adv.*, 2012, **2**(19), 7588–7593.

23 W. Xiao, X. Jin and G. Z. Chen, Up-scalable and controllable electrolytic production of photo-responsive nanostructured silicon, *J. Mater. Chem. A*, 2013, **1**(35), 10243–10250.

24 J. Zhao, J. Li, P. Ying, *et al.* Facile synthesis of freestanding Si nanowire arrays by one-step template-free electro-deoxidation of SiO_2 in a molten salt, *Chem. Commun.*, 2013, **49**(40), 4477–4479.

25 W. Xiao, X. Jin, Y. Deng, *et al.* Electrochemically driven three-phase interlines into insulator compounds: electroreduction of solid SiO_2 in molten $CaCl_2$, *ChemPhysChem*, 2006, **7**(8), 1750–1758.

26 T. Nohira, N. Kani, T. Tsuda, *et al.* Electrolytic Reduction of Solid SiO_2 in Molten $CaCl_2$ for the Production of Solar-grade Silicon, *ECS Trans.*, 2009, **16**(49), 239–245.

27 H. Tuan, D. C. Lee and B. A. Korgel, Nanocrystal-Mediated Crystallization of Silicon and Germanium Nanowires in Organic Solvents: The Role of Catalysis and Solid-Phase Seeding, *Angew. Chem., Int. Ed.*, 2006, **45**(31), 5184–5187.

28 S. Q. Feng, D. P. Yu, H. Z. Zhang, *et al.* The growth mechanism of silicon nanowires and their quantum confinement effect, *J. Cryst. Growth*, 2000, **209**(2–3), 513–517.

29 J. Yang, S. Lu, S. Kan, *et al.* Preparation of Silicon Nanowires by Electrochemical Method, *Chin. J. Inorg. Chem.*, 2009, **25**(4), 756–760.

30 S. K. Cho, F. F. Fan and A. J. Bard, Electrodeposition of Crystalline and Photoactive Silicon Directly from Silicon Dioxide Nanoparticles in Molten $CaCl_2$, *Angew. Chem., Int. Ed.*, 2012, **51**(51), 12740–12744.

31 W. Xiao, X. Jin, Y. Deng, *et al.* Rationalisation and optimisation of solid state electro-reduction of SiO_2 to Si in molten $CaCl_2$ in accordance with dynamic three-phase interlines based voltammetry, *J. Electroanal. Chem.*, 2010, **639**(1–2), 130–140.

32 D. P. Yu, Y. J. Xing, Q. L. Hang, *et al.* Controlled growth of oriented amorphous silicon nanowires *via* a solid–liquid–solid (SLS) mechanism, *Phys. E*, 2001, **9**(2), 305–309.

33 W. Gao, H. Yan, S. Sheng, *et al.* Phase Formation Regularities and Size Effect of Ti–Fe Alloy Nanometer Powders by Gas Evaporation, *Rare Met. Mater. Eng.*, 2007, **36**(2), 250–254.

Faraday Discussions

PAPER

Towards large scale preparation of graphene in molten salts and its use in the fabrication of highly toughened alumina ceramics

Ali Reza Kamali,* John Feighan and Derek J. Fray

Received 12th January 2016, Accepted 4th February 2016
DOI: 10.1039/c6fd00005c

Highly crystalline graphene nanosheets were reproducibly generated by the electrochemical exfoliation of graphite electrodes in molten LiCl containing protons. The graphene product has been successfully applied in several applications. This paper discusses the effect of molten salt produced graphene on the microstructures and mechanical properties of alumina articles produced by slip casting and pressureless sintering, which is one of the most convenient methods for the commercial production of alumina ceramics. In addition to graphene, graphite powder and multi-walled carbon nanotubes (CNTs) were also used to prepare alumina articles for comparative purposes. A graphene strengthening effect was realized through microstructural refinement and by influencing the formation of alumina nanorods during the sintering of α-Al_2O_3 articles. The fracture toughness of the sintered alumina articles increased to an impressive value of 6.98 MPa $m^{1/2}$ by adding 0.5 wt% graphene nanosheets. This was attributed to the unique microstructure obtained, comprised of micrometer sized alumina grains separated by alumina nanorods.

Introduction

During the last twenty years, molten alkali-metal halide salts have provided interesting opportunities for the production of high value materials including metals such as Ti,[1] W,[2] Zr[3] and U,[4] intermetallics such as NiTi,[5] ceramics such as $LaAlO_3$ (ref. 6) and $LiNbO_3$ (ref. 7) as well as new carbon materials such as carbon nanotubes,[8] nanodiamonds[9] and graphene.[10,11] Graphene has been the subject of much research, and a byword for interesting new materials and technologies since its first isolation in 2004. Most of this interest has been based on graphene's remarkable physical, chemical and mechanical properties which make graphene an ideal material for many applications including energy storage devices[12–15] as

Department of Materials Science and Metallurgy, University of Cambridge, 27 Charles Babbage Road, Cambridge CB3 0FS, UK. E-mail: ark42@cam.ac.uk; alirezakam@yahoo.com

well as high performance metal-, polymer- and ceramic-based structural composites.[16–18]

However, as it has also been pointed out by Raccichini *et al.*,[19] the most ambitious challenge facing the widespread application of graphene is the capability of large-scale production of high quality graphene. Primarily because the current high price of quality graphene makes it uneconomic for many applications. Furthermore, most of the potential applications cannot be appropriately evaluated due to the lack of availability of sufficient quantities of quality graphene, which is in turn due to the nonscalability of the current production methods.

In fact, most of the methods developed for the preparation of graphene, including a modified Hummers method followed by reduction processes,[20] liquid phase exfoliation of graphite[21] and chemical vapour deposition techniques[22] suffer from one or more disadvantages such as a low quality of graphene product, a low rate of production, or use of hazardous organic solvents.

It has recently been reported that graphite can be electrochemically exfoliated in molten salts to produce graphene nanosheets.[10,11] This process provides a unique opportunity for the large-scale fabrication of high quality graphene, and thus allows for many applications to be explored. This paper reports the molten salt generation of graphene and its effect on the microstructures and mechanical performances of alumina ceramics prepared by slip casting and pressureless sintering, which is a well known commercial method for manufacturing inexpensive ceramic articles.

Aluminium oxide (Al_2O_3) is a very important ceramic material due to its excellent combination of properties such as its high melting point, hardness, strength, stiffness, wear resistance and chemical inertness. These properties allow for alumina to be widely used in many applications, including in sealing rings, high temperature electrical insulators, sheaths for high voltage lamps, refractory materials and milling media.[23,24] Moreover, alumina ceramics possess high biocompatibility and hence have been used in the reconstruction of dental crowns,[25] jaws[26] and hip joints.[27] Although more than 3.5 million alumina implants have been implemented worldwide since 1990,[27] there have also been reports on the fracture of alumina implants, owing to the relatively low intrinsic fracture toughness of alumina (3–5 MPa $m^{1/2}$ (ref. 28 and 29)). Consequently, yttria-stabilised zirconia (YSZ) has been considered as an alternative to alumina for structural applications due to its higher fracture toughness brought about by a phase transformation toughening mechanism, enhancing its resistance to crack propagation.[27] YSZ ceramics, however, suffer from low temperature degradation which largely limits their long-term practical application.[27,30] As a result, the toughening of alumina ceramics has been the subject of a large number of studies, most of which have mainly been based on the incorporation of a toughening agent. In this regard, various agents and fabrication techniques have been used, leading to an increase in the fracture toughness for alumina composite products. For example, fracture toughnesses of 5.4, 4.8 and 5.0 MPa $m^{1/2}$ were achieved for alumina composites containing 5 wt% glass phase,[31] 13 vol% Zr + Ag[32] and 13 vol% Ni,[33] respectively, using cold isostatic pressing and pressureless sintering. Hot-pressing methods led to the fabrication of alumina composites containing 20 wt% YSZ[34] and 20 vol% Fe[35], with fracture toughnesses of 6.8 MPa $m^{1/2}$ and 6.6–10.2 MPa $m^{1/2}$, respectively.

Recently, graphene has been investigated as a toughening agent in alumina ceramics. The alumina–graphene ceramics were prepared by means of spark plasma sintering, and exhibited fracture toughness values in the range of 3.5–5.2 MPa $m^{1/2}$.[36–38] The graphene materials used in these investigations were prepared by a modified Hummers method followed by hydrazine reduction,[36] thermal exfoliation of intercalated graphite,[37] and liquid phase exfoliation of graphite.[38] Apart from the limitations associated with the preparation of the graphene materials used in these investigations, the fabrication methods employed to prepare the ceramic materials, *i.e.* hot pressing and spark plasma sintering processes, are costly and involve expensive equipment. Therefore, more economical methods for the preparation of both graphene and alumina–graphene ceramics are required in order to explore the future potential of graphene in the alumina ceramic industry. It is also worth noting that graphene materials produced by different techniques exhibit different properties. To address these issues, this paper investigates the effect of molten-salt produced graphene nanosheets on the microstructures and mechanical properties of alumina ceramics fabricated by means of slip casting and pressureless sintering.

Experimental

Materials

Three different carbon materials were selected for the various investigations, and these are: synthetic graphite powder (Sigma-Aldrich); commercial multiwalled carbon nanotubes (Thomas Swan, UK) produced by a gas phase process, and graphene nanosheets. The graphene nanosheets were prepared *via* an electrochemical method based on the intercalation of hydrogen into a graphite rod (Goodfellow 809-013-12, diameter 1.3 cm, length 10 cm, purity 99.997%) polarised in molten LiCl, allowing graphene layers to be exfoliated. The effect of these carbon materials on the microstructures and properties of alumina ceramics was investigated.

Preparation of alumina ceramics

In order to prepare the alumina ceramics, carbon materials were added to deionised water and dispersed using ultrasonication for 30 minutes. Then, sufficient amount of alumina powder (Showadenko chemicals, AL-160SG3, 99.9% purity) was added to create a slip of the desired amount of carbon, and the mixture was ultrasonicated for a further 30 minutes. The resulting slurry was ball milled at 90 rpm using a mixture of 13 mm and 20 mm diameter alumina balls for 12 hours, with a powder-to-ball weight ratio of 1 : 5. Then, small amounts of deionised water and dispersant (dispex, polyacrylate acid BASF Chemicals, AA4140, Germany) were added, in order to decrease the slurry's viscosity and enhance its cast ability, followed by a further 30 minutes of ball milling.

It should be noted that the amounts of alumina and water used depended on the amount of the carbon in the ceramic. For each sample alumina powder was added to give a slip of the desired amount of carbon. The quantity of water was chosen to give a roughly 2 : 1 alumina to water weight ratio. Pure alumina samples were also fabricated using the same processing route for comparative purposes.

The green cast bodies were then prepared using slip casting. For this, the slip was poured into plaster of Paris moulds, forming rectangular specimens upon loss of the water (see Fig. 8). The casts were sintered in a resistant furnace under a protective Ar atmosphere. For sintering, the temperature was raised from room temperature to 1650 °C with a heating rate of 5 °C min^{-1}, and kept at the maximum temperature for four hours.

Characterisation

Density measurements were performed *via* the Archimedes method. X-ray diffraction measurements were performed on a Bruker D8 diffractometer equipped with a position sensitive detector (LynxEye) and a graphite 2nd beam monochromator using Cu K_α radiation ($\lambda = 1.54$ Å). The flexural strength was examined by a three-point bending test using a Hounsfield 5 kN tester with a cross-head speed of 1 mm min^{-1}. Flexural strength σ was calculated using[39]

$$\sigma = \frac{3PL}{2bh^2} \tag{1}$$

where P is the load at the fracture point, L is the span length, b is the sample breadth and h is the sample thickness. Hardness and indentation fracture toughness were examined using Vickers indentation method. The values were measured from polished cross sections using the standard Vickers indentation procedure with a 50 N load and a 12 s dwell (Mitutoyo HM 100, Japan). The fracture toughness was determined by measuring the radial crack lengths produced by the Vickers indentations. Several measurements were conducted at different locations on the samples. Calculation of the fracture toughness was done using the equation proposed by Anstis[40]

$$K_{\mathrm{IC}} = 0.016\left(\frac{E}{H}\right)^{1/2}\frac{P}{c^{3/2}} \tag{2}$$

where E is the Young's modulus of the material, H is the hardness of the material, P is the indentation load and c is the indentation crack length. Radial crack lengths were measured using a standard optical microscope. The elastic modulus was determined by nanoindentation on polished test specimens (MML Cabinet, Berkovich tip).

Results and discussion

Characterization of the as received alumina powder

SEM micrographs of the alumina powder used for the slip casting process are shown in Fig. 1. The powder comprises irregular shaped particles of a mean size of about 500 nm. Moreover, it can be seen that individual particles within the powder tend to form micrometer sized agglomerates. The X-ray diffraction pattern of the alumina powder is shown in Fig. 2a. The XRD pattern indicates that the powder is pure α-Al_2O_3 (corundum), and that the (104) diffraction peak is the most intense in the XRD spectrum.

Fig. 1 Secondary electron SEM micrographs of the alumina powder used for the slip casting–pressureless sintering process.

Molten salt preparation of graphene

Molten salt assisted electrochemical exfoliation of graphite is based on the high temperature intercalation of hydrogen into the interlayer space of graphite crystallites. In order to prepare graphene, a graphite electrode was cathodically polarized in molten LiCl, containing hydrogen cations, to deliver a constant current of 32 A, corresponding to an initial cathode current density of about 1 A cm^{-2}. During the process, the space above the molten salt was purged with a stream of argon–5% hydrogen gas, with a flow rate of 200 mL min^{-1}, leading to the formation of hydrogen cations in molten LiCl, which is known to be able to dissolve hydrogen in the form of protons.[11,41,42]

Hydrogen cations formed are then discharged on the cathode to form hydrogen atoms, which subsequently intercalate into the interlayer space of the graphite electrode lattice. The combination of hydrogen atoms in the space between graphene layers of the graphite leads to the formation of hydrogen molecules with enough kinetic energy to overcome the van der Waals forces between the graphene layers. Fig. 3a shows a schematic representation of the electrochemical exfoliation of graphite in molten salts. The graphite cathode is

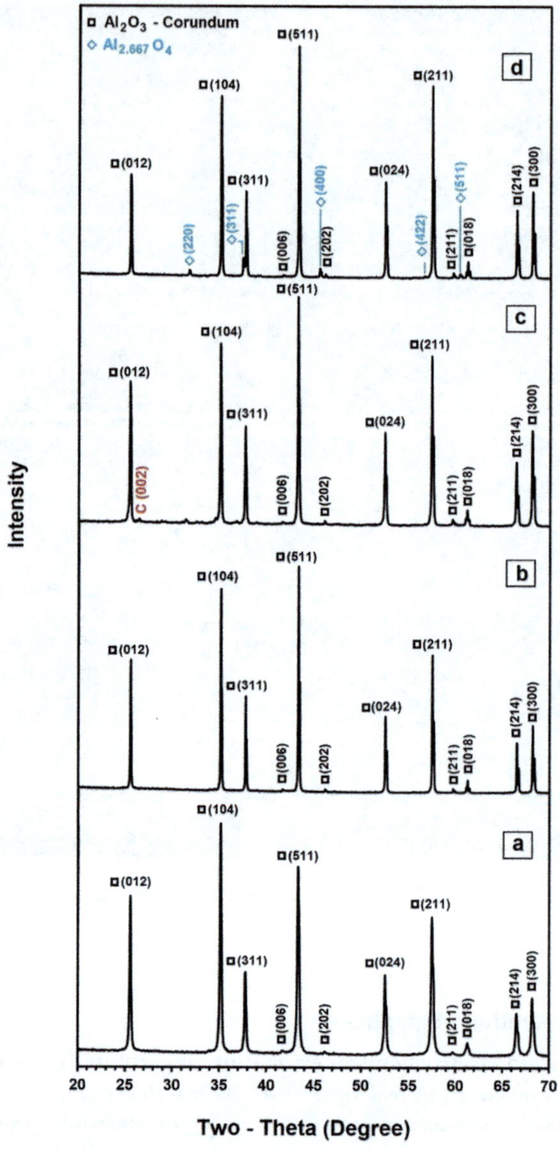

Fig. 2 X-ray diffraction patterns of (a) alumina powder used for the fabrication of ceramics, and those of ceramics prepared using (b) no additive, (c) 1 wt% graphite powder and (d) 1 wt% molten salt produced graphene. The lack of appearance of a (002) carbon peak in the alumina–graphene spectrum is due to the low dimensionality of the graphene material in the direction perpendicular to the *c*-axis, and also its low relative quantity.

gradually eroded because of its surface exfoliation, leading to the release of graphene nanosheets into the molten salt. Fig. 3b (upper panel) shows photographs of a graphite electrode at the different stages of the process. The graphene nanosheets produced can then be retrieved from the solidified salt by water washing and filtration. Fig. 3b (lower panel) shows the graphene nanosheets stored in glass jars while drying in an oven. It should be mentioned that the

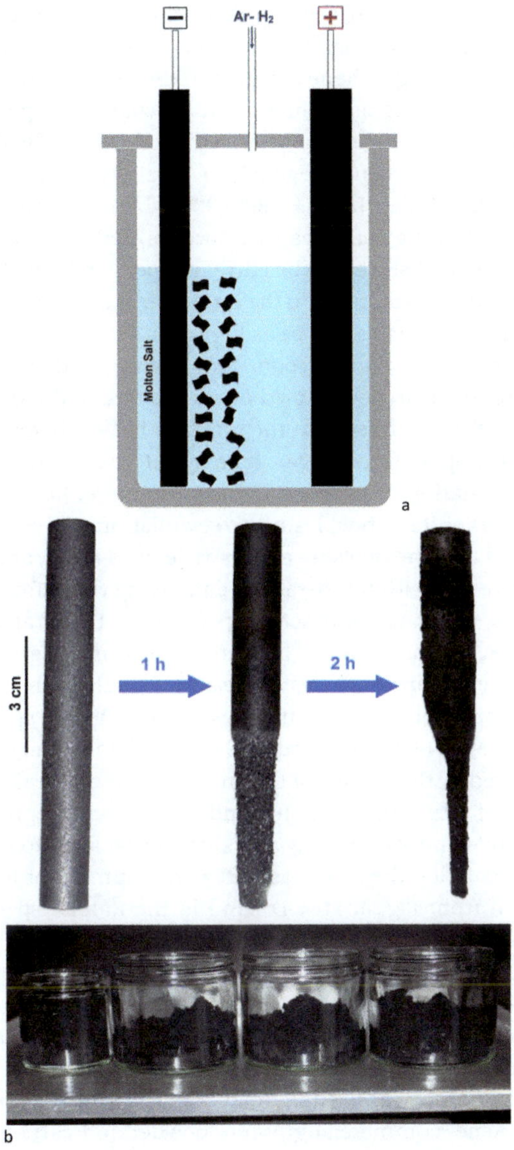

Fig. 3 (a) Schematic representation of the molten salt technique for the preparation of graphene. A cathodically polarised graphite electrode was eroded in molten LiCl containing hydrogen cations, brought about by the use of an Ar–5% H_2 flow. The eroded product, in the form of graphene nanosheets, can be retrieved from the solidified salt afterward. (b) (upper panel) Photographs of the graphite electrode at different stages of the process; (lower panel) the graphene nanosheets stored in glass jars during drying.

erosion of graphite cathodes in molten LiCl in the absence of hydrogen at both laboratory-[43] and bench-scales[44] has been reported in the literature, and is known to cause the formation of multi-walled carbon nanotubes and nanoparticles. In this case, it was proposed that lithium from molten salt deposits on and then intercalates into the graphite lattice, under the influence of a sufficiently negative electrode potential, leading to the expansion of the lattice. By increasing the amount of intercalated lithium, thereby enhancing the stress, the graphite lattice fragments, leading to the formation of various carbon nanoparticles and nanotubes.[45] The formation of graphene reported in this paper is, therefore, attributed to the presence of hydrogen according to the mechanism explained above.

It is also worth mentioning that although the reduction of hydrogen cations on graphite cathodes has been investigated for a variety of electrolytes, including molten potassium bisulphate (at 245–280 °C)[46] and water (at room temperature),[47] exfoliation of the graphite cathodes was not reported to occur. The exfoliating behaviour of hydrogen observed in the current paper can be explained by the high diffusion rate of hydrogen atoms into the interlayer space of the graphite lattice at 800 °C, which can be estimated[48] to be as high as 3.3×10^{-5} cm^2 s^{-1}; about 45 000 and 50 times more than that at room temperature and 280 °C, respectively. Therefore, the atomic hydrogen formed on the surface of the graphite cathode at 800 °C is more likely to diffuse into the graphite lattice, before combining with another hydrogen atom to form H_2 gas, thereby before escaping from the graphite surface. The combination of hydrogen atoms, therefore, has a greater chance to occur within the graphite lattice, leading to exfoliation of the graphite. It should also be considered that the pressure and particle movement, and thus the kinetic energy, of H_2 increases with temperature, enhancing the exfoliation process.

Fig. 4a shows the X-ray diffraction pattern of the graphene nanosheets produced. The low intensity of the (002) reflection in the graphene product is indicative of its low-dimensional crystalline structure along the *c*-axis. The Raman spectrum of the product is shown in Fig. 4b. The Raman spectrum is characterised by the presence of well defined D, G and 2D bands. In graphitic materials, the D peak is caused by the defects of the graphene layers, the G band is indicative of a high degree of structural ordering, and the 2D-band represents the second order of the D-mode and its intensity and shape are related to the stacking order of graphene sheets along the *c*-axis, as well as the number of layers.[49]

As can be seen from Fig. 4b, the D band in the Raman spectrum of the graphene product is very weak whilst the 2D band is a sharp and symmetric peak. These features are indicative of the high quality of the graphene product. This was further confirmed by direct SEM observations as shown in Fig. 5. According to these images, the product consists of exfoliated graphene sheets with an average sheet dimension of 2–3 μm.

The graphene product was used in the manufacturing of alumina ceramics. In addition to graphene, commercial graphite powder and carbon nanotubes were also used for comparison.

Commercial graphite powder and carbon nanotubes

The XRD and Raman analyses of the graphite powder and CNTs used in the manufacturing of alumina ceramics are shown in Fig. 4. The relatively sharp (002) peak in the XRD pattern of the graphite powder demonstrates the abundance of graphite flakes. All three graphite Raman bands can be identified in the Raman spectrum of the graphite powder. As can be seen, the 2D band in the spectrum is broad and asymmetric which further confirms the existence of many-layer graphite flakes. The SEM micrograph of the graphite powder is shown in Fig. 6. As can be depicted, the material contains randomly oriented graphite grains of several micrometers in size.

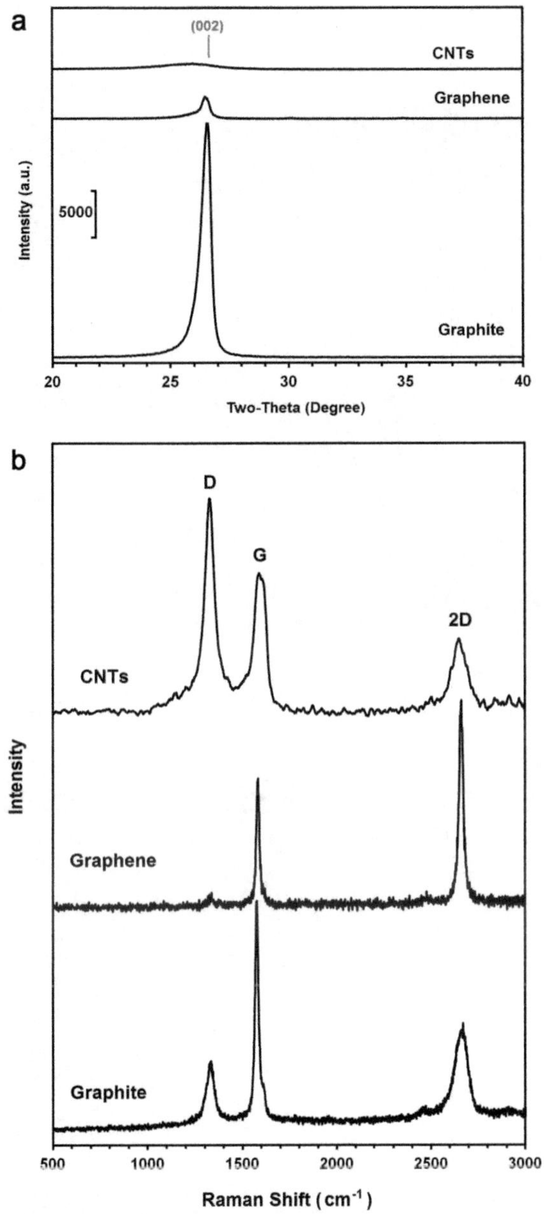

Fig. 4 (a) XRD patterns and (b) Raman spectra of the graphite powder, graphene product and CNTs. The low intensity of the C (002) XRD peak for the graphene material and CNTs is due to the low dimensionality and poor crystallinity of these carbon nanostructures in the direction perpendicular to the *c*-axis, respectively. The D Raman band is sharper and the 2D Raman band is shorter and asymmetric in graphite powder in comparison to those of graphene, indicating that the graphite powder comprises thick stacks of graphite flakes. The results also indicate that the graphene product is of a high quality.

According to Fig. 4a, the (002) reflection in the XRD pattern of the CNTs is weak and broad, and also shifted to lower angles compared to those recorded for the graphite powder and graphene product. This result, together with the presence of

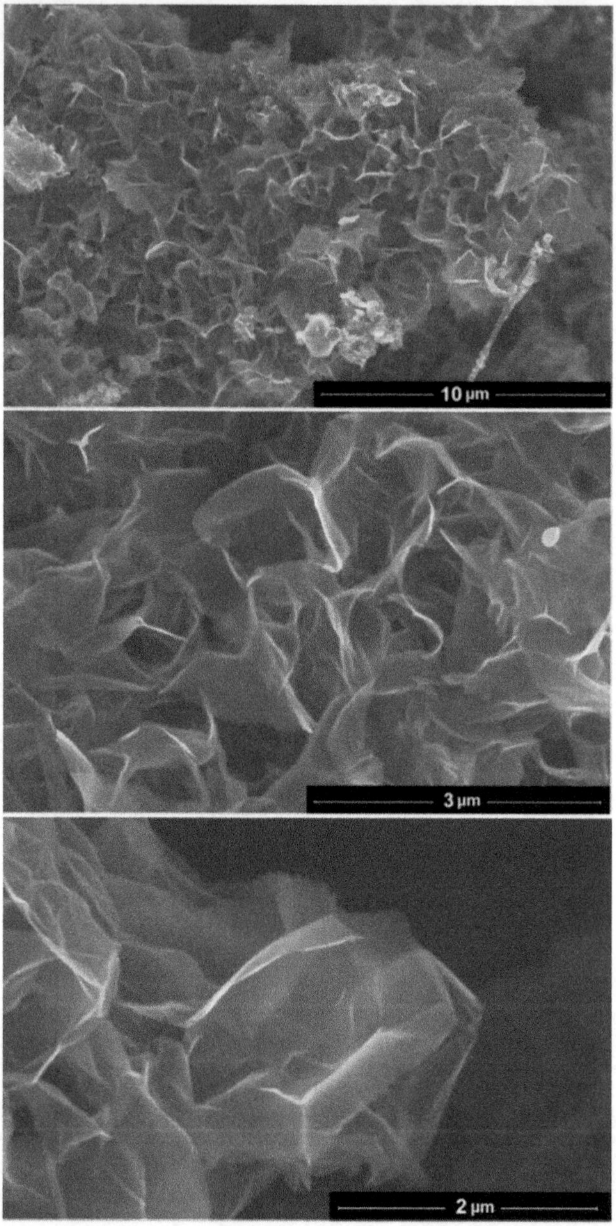

Fig. 5 Secondary electron SEM micrographs showing graphene layers of the product, produced by the exfoliation of graphite electrodes in molten salt. This material was used in the manufacturing of alumina ceramics.

an intense D band in the Raman spectrum of the CNTs, (displayed in Fig. 4b) is attributed to their low crystallinity. Secondary electron SEM micrographs of the CNTs, presented in Fig. 7, shows that the individual tubes tend to agglomerate and form large clusters up to several tens of micrometers in size.

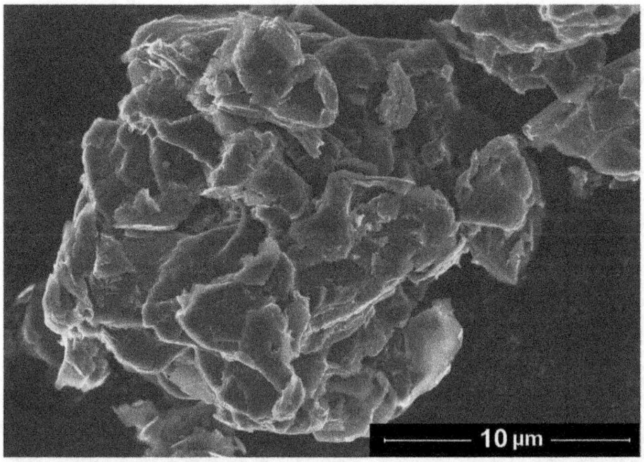

Fig. 6 Secondary electron SEM micrograph of the graphite powder used in the manufacturing of alumina–graphite ceramics.

Preparation of alumina ceramics

The conventional low-cost slip casting process followed by pressureless sintering was used to fabricate alumina matrix ceramics, employing the molten salt-produced graphene, graphite powder and CNTs in the fabrication process. Homogeneous slips could successfully be made for all three carbon materials. The slip-casted green bodies were then dried and sintered at 1650 °C.

It should be mentioned that the slip casting process is a unique and cost effective technique for producing green bodies, in which the suspension is easily added to a permeable mould that determines the shape of the object, as can be seen in Fig. 8. This manufacturing technique doesn't require expensive equipment and is capable of producing from one to hundreds of articles.

Despite the success of producing homogeneous suspensions using all of the carbon materials, upon sintering, however, the CNTs failed to remain dispersed, and instead concentrated in the centre of the sample, as shown in Fig. 9. This has been put down to the high affinity of the tubes for other tubes and weak alumina–nanotube bonding. Fig. 9 also shows the X-ray diffraction analyses of the two distinctive parts of the sample's cross section. The XRD pattern of the area closer to the edge of the cross section can be indexed to the diffraction of α-Al_2O_3. The XRD pattern obtained at the centre of the sample, however, reveals extra diffraction peaks, in addition to those of α-Al_2O_3. These diffraction peaks can be indexed to $Al_{2.667}O_4$ reflections. Considering the fact that this area contains CNTs, the formation of $Al_{2.667}O_4$, with a slightly lower value of O : Al atomic ratio in comparison with Al_2O_3, can be attributed to the surface reduction of Al_2O_3 by CNTs, which occurred during the sintering process. We will discuss this later in this paper.

Unlike CNTs, graphene and graphite remained mostly homogeneous on sintering, which can be attributed to the fact that graphene and graphite flakes have more difficulty diffusing through the alumina body during the sintering process in comparison with CNTs. This is in part owing to their wide and flat shapes, but also due to their larger contact areas generating stronger bonding to the alumina.

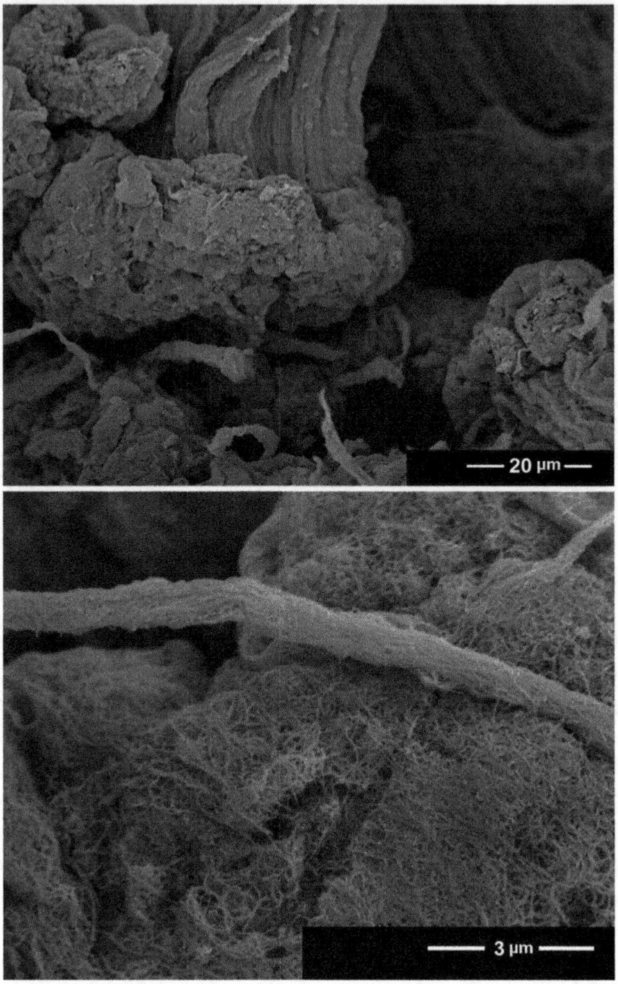

Fig. 7 Secondary electron SEM micrographs of the CNTs used in the manufacturing of alumina–graphite composites.

The fracture surfaces of the sintered samples are shown in Fig. 10–12, and their properties are shown in Table 1. As can be seen from Fig. 10a, sintering of an alumina sample led to the formation of a coarse-grained body with a grain size of 10–20 µm. The fracture toughness of the sample was measured to be 4.50 MPa $m^{1/2}$ which is in agreement with that of pressureless sintered alumina ceramics reported in the literature.[29] From Fig. 10b, no clear change in alumina grains size could be observed by adding 1 wt% graphite, although graphite flakes were well dispersed in between Al_2O_3 particles. The fracture toughness of this sample (4.31 MPa $m^{1/2}$) was slightly lower than that of alumina. This is because graphite flakes function as defects, hence lowering the toughness of the material. Fig. 10c shows the morphology of the sample fabricated by sintering the green body containing 1 wt% graphene. Surprisingly, although no visible sign of the presence of graphene could be observed, the grain sizes (2–3 µm) were found to be much smaller than

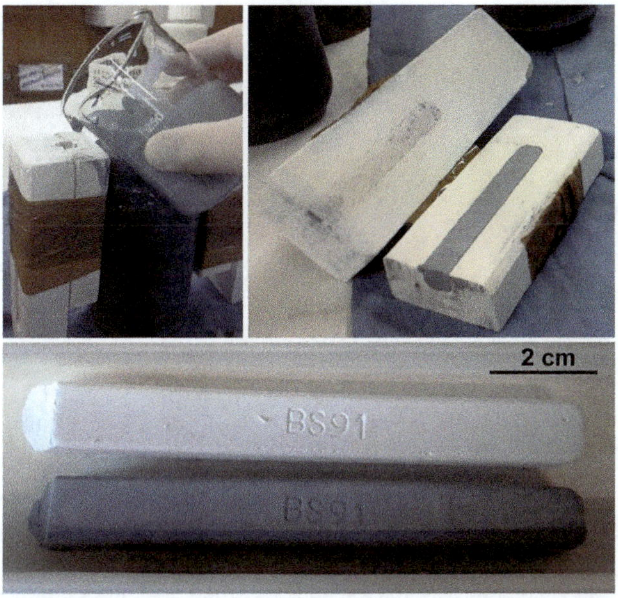

Fig. 8 The slip casting process was adopted to prepare alumina and alumina–carbon green bodies. The process includes the preparation of slips of the desired amount of alumina and carbon followed by casting of the slip into a two-piece plaster mould, forming rectangular specimens upon loss of the water.

those of alumina. Furthermore, the fracture toughness of the sample had an impressive value of 6.53 MPa $m^{1/2}$. Fig. 10c (lower panel) shows intergranular crack propagation observed in the sample. The reason why graphene nanosheets could not be observed by SEM is probably because the graphene sheets partially

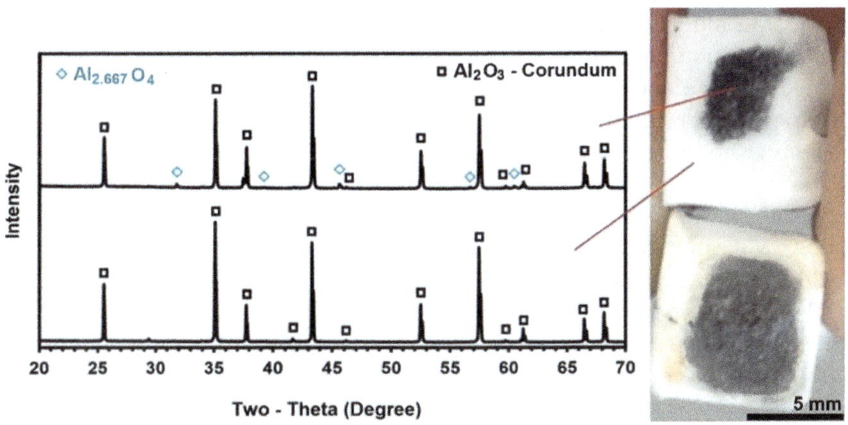

Fig. 9 Cross section of the sintered alumina–CNTs sample. The CNTs failed to remain dispersed during the sintering process and concentrated in the centre of the sample. The X-ray diffraction patterns of the two distinctive parts of the cross section are also presented.

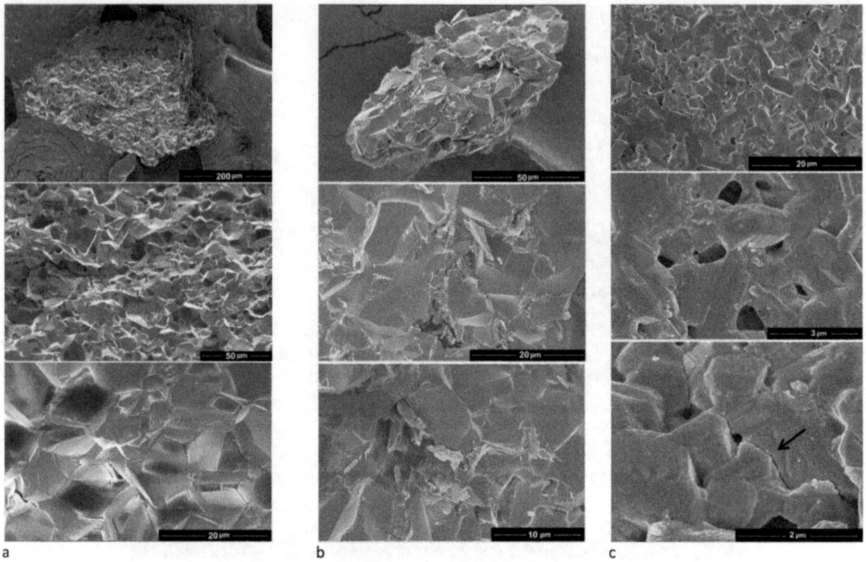

Fig. 10 Secondary electron SEM micrographs of sintered samples produced using (a) alumina, (b) alumina–1 wt% graphite, and (c) alumina–1 wt% graphene. The arrow shows intergranular crack propagation in the fine-grain microstructure.

cover the alumina particles. This could also explain the fine grain sizes observed in the graphene containing sample. The results obtained demonstrate the obvious microstructural refining effect of molten salt produced graphene nanosheets.

It is worth mentioning that the microstructural changes that occur during the heat treatment of alumina powders can be classified as coarsening and densification.[50] It is known that grain boundaries and volume diffusion contribute to the densification of compacts, whilst surface diffusion is the dominant coarsening mechanism during the sintering process.[51] The presence of graphene layers around the alumina particles during the sintering process could reduce the surface diffusion, and hence the microstructural coarsening is to some extent avoided. It should be noted that the effect of graphene on the toughening of the alumina matrix composites depends on the method used for the fabrication of the composite. It also strongly depends on the properties of the graphene used, which in turn depends on its fabrication process. For composites prepared by means of spark plasma sintering, different fracture toughness values of 3.5, 4.5 and 5.2 MPa $m^{1/2}$ have been achieved using graphene materials fabricated by liquid phase exfoliation,[37] thermal expansion/exfoliation of intercalated graphite[38] and a modified Hummers method followed by hydrazine reduction.[36] The high fracture toughness obtained in the molten salt produced graphene–alumina composite can be attributed to the high flexibility of the molten salt produced graphene nanosheets which enables them to be tightly folded on alumina particles, reducing the crack growth rate in the composite product.

In order to elucidate the effect of graphene content on the morphology and toughness of the alumina ceramics, two more composite green bodies with graphene contents of 2 and 0.5 wt% were also prepared by the same method of slip

Fig. 11 Secondary electron SEM micrographs of sintered samples produced using 2 wt% graphene. The arrows show intergranular crack propagation observed in the sample.

casting and pressureless sintering. Fig. 11 and 12 show SEM micrographs for the fracture surfaces of the 2 and 0.5 wt% samples, respectively. The morphology of the sample fabricated using 2 wt% graphene (Fig. 11) can be characterised by the presence of alumina grains having dimensions of less than 5 μm. The sample showed a fracture toughness of 6.40 MPa $m^{1/2}$. However, and interestingly, the morphology of the sample fabricated using 0.5 wt% graphene is distinct from all other samples and consists of alumina particles of about 6 μm that are connected to each other by alumina nanorods of 50–200 nm, according to Fig. 12. The fracture toughness of this sample was measured to be 6.98 MPa $m^{1/2}$. The high toughness of this sample can be attributed to its unique microstructure in which alumina grains are separated by alumina nanorods, reducing the likelihood of grain-boundary crack propagation. The evolution of this microstructure will be explained shortly in this paper.

X-ray diffraction patterns of the samples prepared by sintering of alumina, alumina–1 wt% graphite, as well as alumina–1 wt% graphene are shown in Fig. 2. From this figure, the pure alumina sample can be identified to be fully α-Al_2O_3 (corundum). The peaks in the XRD pattern of the alumina–1 wt% graphite sample can also be perfectly indexed to α-Al_2O_3. Furthermore, the (002) reflection of

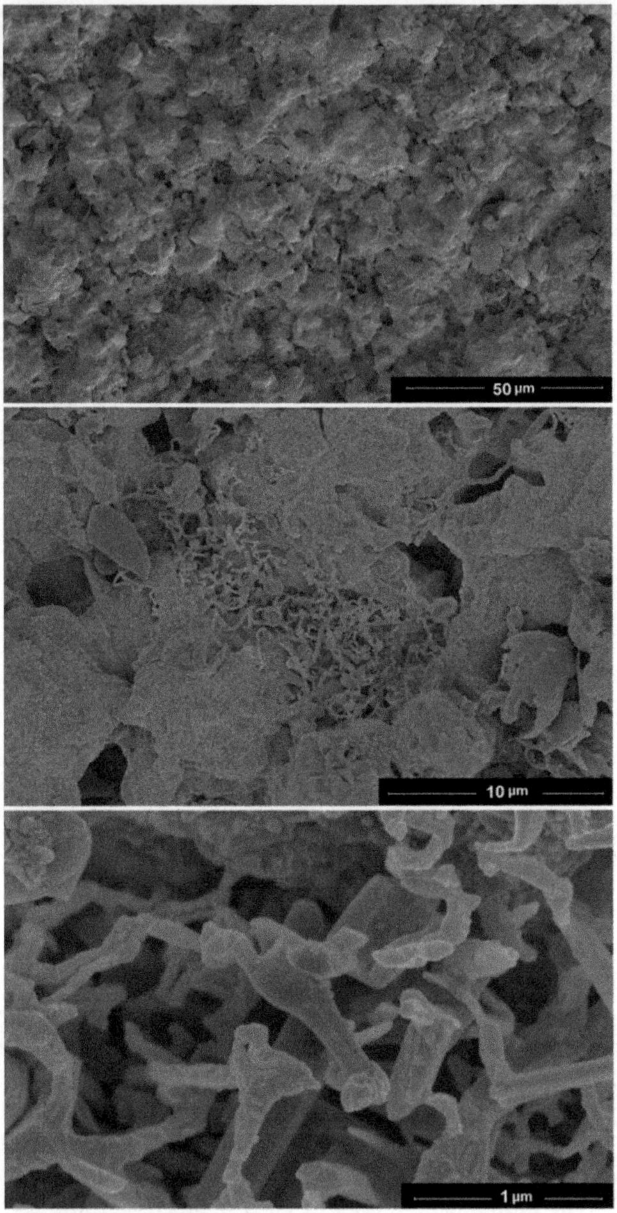

Fig. 12 Secondary electron SEM micrographs of sintered sample produced using 0.5 wt% graphene. The microstructure obtained is unique in the sense that micrometer-sized alumina particles are separated by alumina nanorods of 50–200 nm in diameter. This material exhibited an impressive fracture toughness of 6.98 MPa $m^{1/2}$, owing to its interesting morphology. Moreover, the alumina article produced is porous, which makes it attractive for biomedical applications.

graphite can be clearly seen at $2\theta = 26.44°$ and no other XRD peaks could be detected. The XRD result obtained is in agreement with the SEM micrographs shown in Fig. 10b, where the presence of graphite flakes between alumina grains

Table 1 Properties of the alumina bodies fabricated by the pressureless sintering method

Sample	Bulk density ($g\ cm^{-3}$)	K_{IC} ($MPa\ m^{1/2}$)	H_v (GPa)	Reference
Al_2O_3	3.79	4.50 ± 0.60	20.3 ± 3.8	Current study
Al_2O_3 + 1 wt% graphite	3.72	4.31 ± 0.24	9.8 ± 4.5	
Al_2O_3 + 0.5 wt% graphene	3.58	6.98 ± 0.54	19.3 ± 4.9	
Al_2O_3 + 1 wt% graphene	3.65	6.53 ± 0.65	13.6 ± 3.6	
Al_2O_3 + 2 wt% graphene	3.63	6.40 ± 0.01	10.6 ± 0.5	
Al_2O_3–0.1 wt% CNTs	—	5.3–5.7	15.1–16.3	55
Al_2O_3	—	2.8	19	56
Al_2O_3–(0.8–1.5) vol% graphene plates	—	3.6–4.5	18–18.5	56
Al_2O_3–1 vol% SiC	—	4–5	16.4	28
Al_2O_3	—	2.7–3.6	15.1	28

is visually confirmed. The XRD pattern of the alumina–1 wt% graphene sample is shown in Fig. 2d. The XRD reflections of α-Al_2O_3 can be easily identified in the pattern. No XRD peak corresponding to the (002) reflection of hexagonal graphite could be observed at $2\theta = 26$–$27°$. Moreover, the presence of weak reflections corresponding to $Al_{2.667}O_4$ is evident from the pattern. A similar XRD result was observed when CNTs were used in the manufacturing of alumina samples (Fig. 9). Likewise, the XRD patterns of the samples with initial concentrations of 0.5 and 2 wt% graphene could also be characterised by the presence of α-Al_2O_3 as the major phase, with minute amounts of $Al_{2.667}O_4$ (the diffraction patterns are not shown here). This result is interesting and demonstrates that, unlike graphite flakes, CNTs and graphene have a reducing effect on Al_2O_3, leading to the formation of oxygen vacancies on the Al_2O_3 surface. It should be mentioned that α-Al_2O_3 with oxygen vacancies has so far only been prepared using complicated techniques such as energetic heavy ion bombardment[52] and the aluminium vapour treatment[53] of α-Al_2O_3. Formation of the alumina nanorods, as observed in Fig. 12, can be attributed to the solid phase directional growth of α-Al_2O_3 with oxygen vacancies that occurred during sintering. The surface reduction of Al_2O_3 to $Al_{2.667}O_4$ with lower oxygen content provides a chemical driving force for surface diffusion and subsequent crystal growth of aluminium oxide into the one dimensional nanostructure. Similar solid state growth influenced by surface oxygen vacancies has also been observed for SnO_2.[54] Owing to their unique microstructures and mechanical properties, the resulting alumina articles fabricated using molten salt produced graphene can be considered for applications such as biomedical and dental implants, where the fracture toughness of the alumina article is critical.

The results obtained in this paper indicate that molten salt produced graphene can be employed to enhance the mechanical properties of alumina ceramics manufactured by the conventional slip casting and pressureless sintering method. These results are of specific importance for technical application, considering the scalability and economic feasibility of the molten salt process. This process consumes mainly graphite, H_2 and electricity. A part of the salt used evaporates from the crucible, due to the high operating temperature of about 800 °C, and deposits on the cooler inner surface of the inconel retort. Apart from this the consumption of LiCl is low, supporting the idea that the process mainly proceeds by the cathodic discharge of H^+ and not Li^+. At the end of the process a mixture of graphene and solidified salt is obtained. The salt is highly soluble in water, enabling one to separate LiCl in solution from graphene using vacuum filtration. LiCl can then be retrieved from the solution by a simple water evaporation method. Moreover, the Cl_2 gas evolved on the anode during the electrochemical process reacts with H_2 available in the atmosphere above the molten salt, and the HCl produced mostly dissolves into the melt providing hydrogen cations. The rate of production of graphene using this method is estimated to be 450 g of graphene per litre of molten salt per day. The process uses graphite electrodes which are readily available with diameters as large as 80 cm at a low cost of about US $4000 per metric ton. The specific energy consumption and the cost of producing high quality graphene can be estimated to be about 25 kW h kg^{-1} and US $10–20 kg^{-1}, respectively. The economic production of large quantities of high quality graphene in molten salts is the driving force for its evaluation in various applications.

Conclusions

High quality graphene nanosheets were prepared by the exfoliation of graphite in molten salt. The graphene product, graphite powder and CVD-produced CNTs were used as additives in the fabrication of alumina articles using the convenient method of slip casting and pressureless sintering. Graphene was shown to outperform other carbon additives with respect to enhancing the mechanical properties of pressureless sintered alumina. Addition of 0.5 wt% graphene led to the formation of a microstructure consisting of micrometer sized alumina particles separated by networks of alumina rods, exhibiting a superior toughness of 6.98 MPa $m^{1/2}$, a hardness of 19.3 GPa and a flexural strength of 368 MPa. The formation of alumina nanorods during the sintering process was ascribed to the directional growth of alumina crystallites influenced by the surface oxygen vacancies.

Acknowledgements

A. R. Kamali designed the present research, produced the graphene materials, contributed to the preparation of composite samples and characterisations, and prepared the manuscript. J. Feighan contributed to the preparation of composite samples, mechanical and physical measurements, and documentation. D. J. Fray commented on the manuscript. The Worshipful Company of Armourers and Brasiers is thanked for funding support for this research. Technical contribution from Mr M. Misson (Almath Crucibles Ltd) is greatly appreciated.

References

1 G. Z. Chen, D. J. Fray and T. W. Farthing, *Nature*, 2000, **407**, 361.
2 R. Abdulaziz, L. D. Brown, D. Inman, S. Simons, P. R. Shearing and D. J. L. Brett, *Electrochem. Commun.*, 2014, **41**, 44.
3 Y. Sakamura, M. Iizuka, S. Kitawaki, A. Nakayoshi and H. Kofuji, *J. Nucl. Mater.*, 2015, **466**, 269.
4 J. M. Hur, S. S. Hong and H. Lee, *J. Radioanal. Nucl. Chem.*, 2013, **295**, 851.
5 B. Jackson, M. Jackson, D. Dye, D. Inman and R. Dashwood, *J. Electrochem. Soc.*, 2008, **155**, E171–E177.

6 D. D. Jayaseelan, Z. LI, W. Edward Lee and S. Zhang, Molten Salt Synthesis of Ceramic Materials, *Molten Salts and Ionic Liquids: Never the Twain?*, ed. M. Gaune-Escard and K. R. Seddon, Wiley, 2012 .

7 A. R. Kamali and D. J. Fray, *Ceram. Int.*, 2014, **40**, 1835.

8 W. K. Hsu, J. P. Hare, M. Terrones, H. W. Kroto, D. R. M. Walton and P. J. F. Harris, *Nature*, 1995, **377**, 687.

9 A. R. Kamali and D. J. Fray, *Chem. Commun.*, 2015, **51**, 5594.

10 A. R. Kamali and D. J. Fray, *Carbon*, 2013, **56**, 121.

11 A. R. Kamali and D. J. Fray, *Nanoscale*, 2015, **7**, 11310–11320.

12 L. Hou, L. Lian, D. Li, G. Pang, J. Li, X. Zhang, S. Xiong and C. Yuan, *Carbon*, 2013, **64**, 149.

13 B. Jia and L. Zou, *Carbon*, 2012, **50**, 2315.

14 A. R. Kamali and D. J. Fray, *J. New Mater. Electrochem. Syst.*, 2010, **13**, 147.

15 G. Wang, X. Shen, J. Yao and J. Park, *Carbon*, 2009, **47**, 2049.

16 H. Wu, H. Li, G. Sun, S. Ma and X. Yang, *J. Mater. Chem. C*, 2015, **3**, 5457.

17 T. Kuilla, S. Bhadra, D. Yao, N. H. Kim, S. Bose and J. H. Lee, *Prog. Polym. Sci.*, 2010, **35**, 1350.

18 H. Porwal, S. Grasso and M. J. Reece, *Adv. Appl. Ceram.*, 2013, **112**, 443.

19 R. Raccichini, A. Varzi, S. Passerini and B. Scrosati, *Nat. Mater.*, 2015, **14**, 271.

20 V. Nicolosi, M. Chhowalla, M. G. Kanatzidis, M. S. Strano and J. N. Coleman, *Science*, 2013, **340**, 1226419.

21 J. S. Bunch, Y. Yaish, M. Brink, K. Bolotin and P. L. McEuen, *Nano Lett.*, 2005, **5**, 287.

22 K. S. Kim, Y. Zhao, H. Jang, S. Y. Lee, J. M. Kim, K. S. Kim, J. H. Ahn, P. Kim, I. Y. Choi and B. H. Hong, *Nature*, 2009, **457**, 706.

23 P. Boch and J. C. Niepce, *Ceramic Materials: Processses, Properties and Applications*, ISTE Ltd, UK, 2007.

24 R. B. Heimann, *Classic and Advanced Ceramics: From Fundamentals to Applications*, Wiley-VCH, 2010.

25 M. Øiloa and G. e. D. Quinn, *J. Mech. Behav. Biomed. Mater.*, 2016, **53**, 93.

26 T. Usami, M. Komatsu, H. Mizutani and T. Kaneda, *J. Oral Maxillofac. Surg.*, 1966, **46**, 507.

27 J. Chevalie and L. Gremillard, *J. Eur. Ceram. Soc.*, 2009, **29**, 1245–1255.

28 H. Reveron, O. Zaafrani and G. Fantozzi, *J. Eur. Ceram. Soc.*, 2010, **30**, 1351.

29 J. P. Auerkari, *Mechanical and physical properties of engineering alumina ceramics*, ulkaisija-Utgivare, Finland, 1996.

30 J. Chevalier, *Biomaterials*, 2006, **27**, 535.

31 Y. Q. Wu, Y. F. Zhang, G. Pezzotti and J. K. Guo, *J. Eur. Ceram. Soc.*, 2002, **22**, 159.

32 J. Lalande, S. Scheppokat, R. Janssen and N. Claussen, *J. Eur. Ceram. Soc.*, 2002, **22**, 2165.

33 W. H. Tuan and R. J. Brook, The Toughening of Alumina with Nickel Inclusions, *J. Eur. Ceram. Soc.*, 1990, **6**, 31–37.

34 S. W. Kim, S. L. Cockcroft, K. A. Khalil and K. Ogi, *Mater. Sci. Eng., A*, 2010, **527**, 4926.

35 P. A. Trusty and J. A. Yeomans, *J. Eur. Ceram. Soc.*, 1997, **17**, 495.

36 K. Wang, Y. Wang, Z. Fan, J. Yan and T. Wei, *Mater. Res. Bull.*, 2011, **46**, 315.

37 H. Porwal, P. Tatarko, S. Grasso, J. Khaliq, I. Dlouhy and M. J. Reece, *Carbon*, 2013, **64**, 359.

38 J. Liu, H. Yan and K. Jiang, *Ceram. Int.*, 2013, **39**, 6215.

39 *Standard Test Methods for Flexural Properties of Unreinforced and Reinforced Plastics and Electrical Insulating Materials*, ASTM Standard D790–10, ASTM International, West Conshohocken, 2010.

40 G. R. Anstis, P. Chantikul, B. R. Lawn and D. B. Marshall, *Am. Ceram. Soc. Bull.*, 1981, **64**, 533.

41 A. R. Kamali, D. J. Fray and C. Schwandt, *J. Therm. Anal. Calorim.*, 2011, **104**, 619.

42 A. R. Kamali, *Green Chem.*, 2016, **18**, 1952.

43 G. Z. Chen, X. D. Fan, A. Luget, M. S. P. Shaffer, D. J. Fray and A. H. Windle, *J. Electroanal. Chem.*, 1998, **446**, 6.

44 A. R. Kamali and D. J. Fray, *Carbon*, 2014, **77**, 835.

45 A. T. Dimitrov, G. Z. Chen, I. A. Kinloch and D. J. Fray, *Electrochim. Acta*, 2002, **48**, 91.

46 E. J. Balskus, W. E. Triaca and A. J. Aria, *Electrochim. Acta*, 1972, **17**, 45.

47 P. B. Vilanova, S. Puig, R. G. Olmos, A. V. Pons, L. Bañeras, M. D. Balaguer and J. Colprim, *Int. J. Hydrogen Energy*, 2014, **39**, 1297.

48 C. P. Herrero and R. Ramirez, *J. Phys. D: Appl. Phys.*, 2010, **43**, 255402.

49 M. S. Dresselhaus, A. Jorio and R. Saito, *Annu. Rev. Condens. Matter Phys.*, 2010, **89**, 108.

50 N. J. Shaw and R. J. Brook, *J. Am. Ceram. Soc.*, 1986, **69**, 107.

51 F. Raether and P. Schulze Horn, *J. Eur. Ceram. Soc.*, 2009, **29**, 2225.

52 F. Jollet, J. P. Duraud, C. Nogera, E. Dooryhee and Y. Langevin, *Nucl. Instrum. Methods Phys. Res.*, 1990, **B46**, 125.

53 A. I. Surdo, V. S. Kortov, V. A. Pustovarov and V. Y. Yakovlev, *Phys. Status Solidi*, 2005, **1**, 527.

54 A. R. Kamali and D. J. Fray, *Mater. Sci. Eng., B*, 2012, **177**, 819.

55 M. Michalek, K. Bodisova, M. Michalkova, J. Sedlaceka and D. Galusek, *Ceram. Int.*, 2013, **39**, 6543.

56 J. Liu, Y. Yang, H. Hasssinin, N. Jumbu, S. Deng, Q. Zuo and K. Jiang, *ACS Appl. Mater. Interfaces*, 2016, **8**, 2607.

Faraday Discussions

PAPER

Coordination numbers and physical properties in molten salts and their mixtures

Dario Corradini,[a] Paul A. Madden[b] and Mathieu Salanne*[a]

Received 14th December 2015, Accepted 28th January 2016
DOI: 10.1039/c5fd00223k

Mixtures of trivalent metal halides with alkali halides are involved in many technologies but, from a more fundamental and general perspective, are worthy of study as interesting systems in which to examine the relationship between atomic-scale structure and physical properties. Here we examine the relationship between the viscosity and local and longer range structural measures in such mixtures where the trivalent metal cations span a significant size range and exhibit different behaviours in the dependence of their viscosity on the mixture composition. We characterise the structure and dynamics of the first coordination shell and the relationship between its structural relaxation time and the shear relaxation time of the mixture (the Maxwell relaxation time). We are then led to an examination of the structure of the networks which progressively form between the trivalent metal cations as their concentration increases in the mixtures. Here we find significant differences between small and larger cations, sufficient to explain the different behaviour of their viscosities. We draw attention to the similarities and differences of these networks with those which form in highly viscous, glass-forming materials like BeF_2:LiF.

1 Introduction

Molten salt mixtures play a central role in several emerging technologies, such as molten salt nuclear reactors[1] and pyroprocessing of nuclear waste,[2–4] as well as in more established processes, like aluminium refinement.[5] Often the nature of the application involves extreme physical conditions, and the refinement of the selection of the salt requires knowledge and understanding of the way the physical properties respond to changes in composition, temperature *etc.* The accepted way to develop such understanding in modern liquid state science is *via* an appreciation of the link between atomic-scale structure and the properties of interest. Alongside the technological drive for designing suitable salts, "simple"

[a]*Sorbonne Universités, UPMC Univ Paris 06, CNRS, UMR 8234, PHENIX, F-75005, Paris, France. E-mail: mathieu.salanne@upmc.fr*
[b]*Department of Materials, University of Oxford, Parks Road, Oxford OX1 3PH, UK*

inorganic molten salt mixtures have many favourable characteristics for pursuing the structure/property paradigm at a fundamental level – and in detail. Regarded as mixtures of spherical ions, they are uncomplicated from the perspective of performing a range of directly interpretable experiments on structure. From a theoretical perspective, many important materials, such as the halide melts of a wide range of technologically relevant trivalent cations, can be accurately described on the basis of an ionic model[6,7] and, as we illustrate below, this can be used as the basis of simulation methods capable of accurately reproducing and predicting physical properties as well as structural information.[8–13] The structure–property relationship can then be explored *via* a detailed examination of the microscopic events in the simulation trajectory. In the ionic model, the most important parameter controlling the interionic interactions (and therefore, ultimately, the physical properties) is the ionic size. A further attraction of the trivalent metals is that within the set of more than 20 elements forming trivalent cations with ionic halides, there is a huge range of ionic sizes ranging from Al^{3+} to U^{3+}, and so it becomes possible to rationalise the evolution of the behaviour of the physical properties across the whole set. We will focus on chloride melts in what follows, which is what most experiments and simulation studies have focused on.

It has been known for many years[14–16] that when trivalent metal cations are dissolved in halide melts they are surrounded by a tightly coordinated shell of anions, as illustrated in Fig. 1. Although the experimental observables contain information about other structural features, a cornerstone of the traditional discussion of structure has been the coordination number, *i.e.* the number of anions in the coordination shell. In the crystalline state the coordination number is a well-defined quantity and in crystals of compounds of trivalent metal chlorides with alkali chlorides a coordination number of 6 around the trivalent cation (*i.e.* octahedral coordination) is found across a very wide range of cation sizes. When such compounds are melted, the Raman bands[17–19] associated with the vibrations of the coordination complex remain at close to the same frequency, even as the composition of the mixture is shifted towards the pure trivalent metal chloride, which suggests that octahedral coordination remains dominant in the

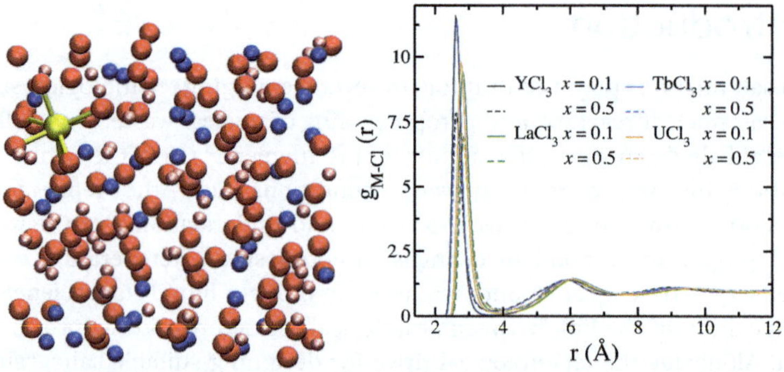

Fig. 1 Left: Snapshot of a system consisting of one U^{3+} ion dissolved in eutectic (LiCl–KCl)$_{eut}$ (blue: K^+, pink: Li^+, red: Cl^- and yellow: U^{3+}). The ionic bonds between the U atom and its first shell of Cl neighbors are shown with yellow sticks. Right: RDFs for the M^{3+}–Cl^- pairs, where M = Y, Tb, La and U at $T = 873$ K for two x_{MCl_3} mole fractions of MCl_3 in (LiCl–KCl)$_{eut}$.

melt. Early X-ray diffraction experiments on the melts of the pure materials were also interpreted on the basis of a well-defined coordination number of 6, expected to persist across the composition range of mixtures.[20] More recent experimental and simulation studies have challenged this viewpoint. Neutron scattering allowed structure factors to be measured to much larger values of the scattering vector than could be achieved in the early X-ray measurements and hence allowed for much better resolution of the full shape of the first peak of the M–Cl radial distribution function (RDF),[21–23] and thereby a more reliable value for the coordination number. The development of X-ray absorption fine structure measurements (EXAFS) on synchrotron sources allowed for the structure around a particular cation to be measured even in relatively dilute solutions[13] where, previously, only Raman observations had been possible. Finally, simulations have been refined to the point where, from a single simulation, the experimental data from all the experiments (X-ray,[12] neutron,[11] EXAFS,[13] Raman[24]) performed on a given melt can be accurately reproduced, giving the confidence that the uniquely detailed information available from an analysis of the ionic positions generated in the simulation trajectory is a faithful representation of the real material.

Such studies suggest a refinement of the traditional picture of a fixed coordination number. Firstly, in a melt the instantaneous coordination numbers around an ion take on a range of values around some preferred number.[13,24] For example, as discussed in ref. 13, the instantaneous coordination number around a La^{3+} ion in a $(LiCl–KCl)_{eut}$ chloride melt ranges from 6 to 8 for a low mole fraction $x_{LaCl_3} = 0.01$ and the experiment averages over their different contributions to the observable. Secondly, this distribution of instantaneous coordination numbers shifts with composition in mixtures, and in dilute solutions of a MCl_3 system in a weakly-coordinating alkali chloride, the M^{3+} coordination does tend towards a well-defined value, in the sense that the distribution narrows around this value.[13,24] Furthermore, this dilute solution value is overwhelmingly likely to be 6 in chloride melts even across the very wide range of cation sizes in the set of interest. These realisations indicate that if we are to fulfil the objective of understanding the structure–property relationship in mixtures, we will need to look well beyond the traditional mean coordination number measure of the former.

The picture which emerges of the ideal dilute solution then is of isolated octahedral coordination complexes around each M^{3+} cation floating in a sea of anions and alkali cations, as illustrated in Fig. 1. The cation could be regarded as having competed more strongly for the anions than the alkali cations and completed its ideal coordination structure. As the mixture becomes more concentrated, this picture has to evolve; for example, beyond a composition of MCl_3:3ACl, or $x_{MCl_3} = 0.25$ (A^+ represents an alkali cation), there are simply not enough chloride anions in the melt for isolated 6-coordinate clusters to form. Behaviour based on two extreme scenarios may follow, either the cation can reduce the number of anions in its coordination shell to remain isolated, or the individual cations might begin to share anions and have overlapping shells. There is good evidence to show that in the much studied case of cryolitic melts (AlF_3:NaF) the former happens,[25–30] but for the chloride melts of principal interest to us, there is an increasing degree of corner and edge-sharing between coordination polyhedra as the MCl_3 concentration increases, leading ultimately to a network structure.[13,24] Rather than decreasing the coordination number, this process leads to an increase in the mean coordination number for the larger cations, reaching 8 in La^{3+}, for example. The rate at which this increase happens as the concentration changes depends on the identity of the alkali cation[13] as well as the M^{3+} ionic size, indicating that the tendency to corner or edge-share reflects a competition for coordination to the anions by the M^{3+} and alkali cations.

In the remainder of this article we will illustrate the structure–property paradigm by showing how just one physical property, the viscosity, is influenced by these factors affecting the structure of the melt as we vary its chemical composition. At the pictorial level, one can illustrate the origins of this connection by imagining a shear applied to the structures pictured in Fig. 2 and trying to envisage how they will relax to a new configuration. This will involve some combination of a breaking or relative movement of the coordination complexes or, in the La picture, the stretching or breaking of the "bonds" which make up the network structure. Viscosity is an important physical property in some of the technologies mentioned at the outset. For example, in one design for a large scale molten salt reactor, the liquid core is pumped at a very high rate to ensure, amongst other things, sufficiently rapid heat transfer. Clearly, it is vital to understand how the viscosity of the core would change as its composition is altered by the nuclear reaction or the addition of fresh fuel or by a change in temperature.

2 The simulations

We have conducted molecular dynamics simulations on mixtures composed of MCl_3 (where M^{3+} is Y^{3+}, Tb^{3+}, La^{3+} and U^{3+}) dissolved in $(LiCl–KCl)_{eut}$, at three temperatures: $T = 873$ K, $T = 973$ K and $T = 1073$ K. These cations span a range of ionic sizes (Y^{3+} 90.0 pm, Tb^{3+} 92.3 pm, La^{3+} 103.2 pm and U^{3+} 102.5 pm (ref. 31)), which is of interest for the pyroprocessing studies, covering the rare earths as well as U^{3+}. The mole fraction of MCl_3 is varied from $x_{MCl_3} = 0$ to $x_{MCl_3} = 0.5$ in steps of $\delta x_{MCl_3} = 0.1$. The detailed compositions of the examined systems are summarized in Table 1. The eutectic composition $(LiCl–KCl)_{eut}$, $x_{LiCl} = 0.582$, is often chosen in

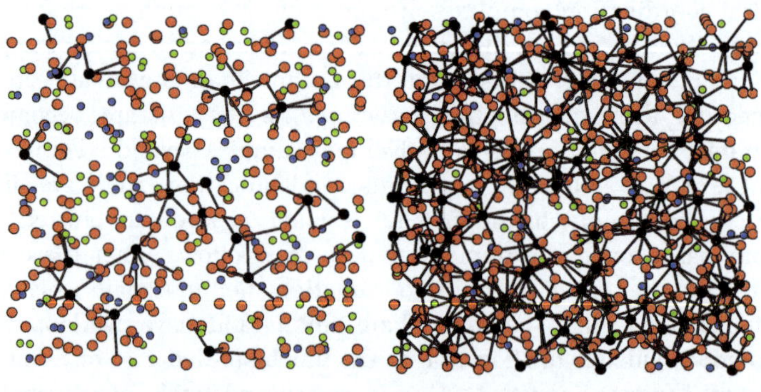

Fig. 2 Snapshots of a system consisting of $LaCl_3$ dissolved in eutectic $(LiCl–KCl)_{eut}$ at two different concentrations (blue: K^+, green: Li^+, red: Cl^- and black: La^{3+}). The ionic bonds between the La atoms and their first shells of Cl neighbors are shown with black sticks. Left: $x_{LaCl_3} = 0.1$, right: $x_{LaCl_3} = 0.5$.

Table 1 Molecular compositions of the considered systems for metal chloride mole fractions from $x_{MCl_3} = 0$ to $x_{MCl_3} = 0.5$. The KCl/LiCl ratio is approximately constant (0.712–0.720) and corresponds to the eutectic ratio (0.718).[32] The total number of ions in the simulation box is also reported

x_{MCl_3}	N_{MCl_3}	N_{LiCl}	N_{KCl}	N_{ions}
0.0	—	126	90	432
0.1	22	113	81	476
0.2	43	101	72	518
0.3	65	88	63	562
0.4	86	75	54	602
0.5	108	63	45	648

experimental and technological studies because of its low melting point, $T_m^{eut} = 625$ K.[32]

We use the polarizable ion model force field for performing our simulations.[6,7] In this model the ions carry their formal valence charges and the effect of the polarization of the anions by coulombic and short-range interactions with the cations is included. This has been shown to be essential to reproduce the experimental structural, thermodynamic and dynamical properties of the melts of interest. All of the parameters of the force field are provided in ref. 10. They have been validated in comparison with a wide range of experimentally obtained structural, thermodynamic and dynamical properties of these melts.[10–13,33] In particular, the activity coefficients of the trivalent ions of interest here were very successfully predicted by the simulations,[10] as has recently been confirmed in a series of electrochemical measurements.[34] Some of these liquids have also been studied using potentials generated on a purely first-principles basis with virtually identical results.[35] Other details of the simulation methods are very similar to those reported in previous works.[8–10,33]

3 Results

Viscosity

The shear viscosity is calculated from the integral over time of the autocorrelation function of the anisotropic elements of the stress tensor. This is known as the Green–Kubo relation and it reads:

$$\eta = \frac{V}{k_B T} \int_0^\infty \langle \sigma_{\alpha\beta}(0)\sigma_{\alpha\beta}(t) \rangle \mathrm{d}t \tag{1}$$

where k_B is the Boltzmann constant and $\sigma_{\alpha\beta}$ can be any of the five independent anisotropic components of the stress tensor (*e.g.* σ_{xy}). An average over these five components is performed to improve the statistics. In practice, the value of the integral in eqn (1) usually converges to a plateau in a few ps.[33] This allows for the accumulation of reliable statistics using simulation runs of reasonable length (of the order of ns). An example of this running integral is shown in Fig. 3. We also show a fit obtained by representing the stress-tensor correlation function as a decaying exponential

$$\langle \sigma_{\alpha\beta}(0)\sigma_{\alpha\beta}(t) \rangle \simeq A \exp(-t/\tau_M) \tag{2}$$

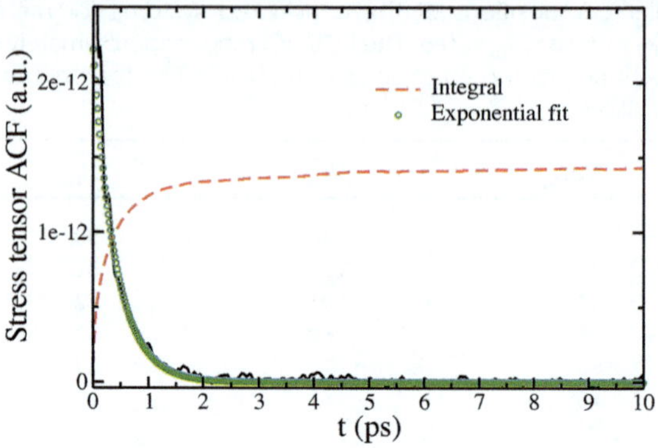

Fig. 3 Time evolution of the stress tensor correlation function in the system consisting of YCl_3–(LiCl–KCl)$_{eut}$ ($x_{YCl_3} = 0.1$) at 1073 K. The viscosity is calculated from the plateau value of the integral.

where τ_M is the Maxwell relaxation time, which we will discuss more fully below.

We show in Fig. 4 the shear viscosity as a function of the MCl_3 mole fraction, for the four cases considered, $M^{3+} = Y^{3+}$, Tb^{3+}, La^{3+} and U^{3+}. Although there is no experimental data for the compositions we simulate, we are confident on the predictive power of our polarizable force fields, which were validated against a wide range of experimentally available structural, thermodynamic and dynamical properties.[8–13] The temperature and concentration dependence of the Maxwell relaxation time τ_M follows that of the viscosity closely. We observe immediately the similarity in the behaviors of the viscosities for the pairs of cations with similar ionic radii. At all temperatures, the viscosities for all metal cations at mole fractions $x_{MCl_3} = 0.1$ and 0.2 take very similar values and start to

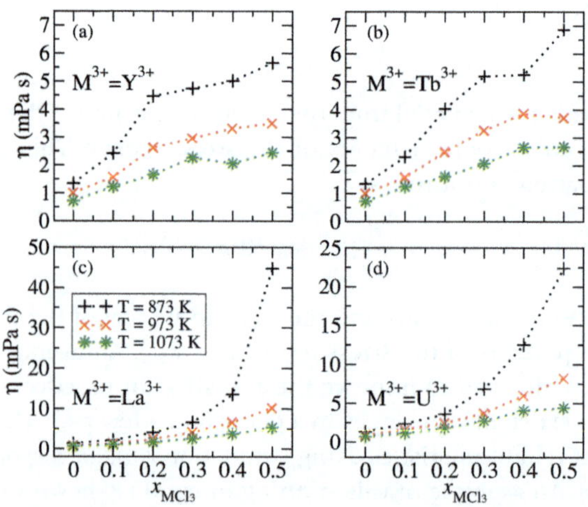

Fig. 4 Simulated shear viscosity η as a function of the MCl_3 mole fraction x_{MCl_3}. M^{3+} = (a) Y^{3+}; (b) Tb^{3+}; (c) La^{3+} and (d) U^{3+}. The curves η *vs.* x_{MCl_3} are shown for the three temperatures investigated: $T = 873$ K (plus signs), $T = 973$ K (crosses) and $T = 1073$ K (stars). The dotted lines joining the points are guides for the eye. Note that 1 mPa s = 1 cP.

become significantly larger for La^{3+}/U^{3+} compared to Y^{3+}/Tb^{3+} by $x_{MCl_3} = 0.4$. Thereafter the viscosity increases with MCl_3 concentration more steeply for the larger cations. Finally, we can also see that, despite the very similar ionic radii of La^{3+} and U^{3+} (with La^{3+} slightly bigger), a significantly larger value of the viscosity is reached in $LaCl_3$–(LiCl–KCl)$_{eut}$, for which η at $T = 873$ K and $x_{LaCl_3} = 0.5$ is about double that for the corresponding case in UCl_3–(LiCl–KCl)$_{eut}$.

Coordination numbers

We now move to the analysis of the structural properties of the systems in the hope of relating them to the behaviors of the viscosities. In Fig. 1 we have already shown the M^{3+}–Cl^- RDFs at $T = 873$ K for the lowest, $x_{MCl_3} = 0.1$, and for the highest, $x_{MCl_3} = 0.5$, metal cation salt content. The behavior of the structure at the other two temperatures investigated, $T = 973$ K and $T = 1073$ K, remains qualitatively similar. Furthermore, we wanted to show the differences in structure between the least and most viscous systems, thus selecting the two extremes of the mole fractions. Before describing the details of the RDFs, we immediately note that for the structural properties, the similarity of the results for the Y^{3+}/Tb^{3+} and La^{3+}/U^{3+} pairings is already evident.

The sharp first peak in the RDF indicates a tight coordination complex and its position increases with the ionic radius of the cation M^{3+}, as would be expected. In all cases, a decrease in the height of the first peak and of the depth of the first minimum is observed upon increasing the metal cation salt content. This indicates that when the mole fraction increases, the M^{3+}–Cl^- coordination shell become less well defined. Also note that the first minimum of the M^{3+}–Cl^- RDFs goes to zero only for Y^{3+}/Tb^{3+} at $x_{MCl_3} = 0.1$. In this case the M^{3+}–Cl^- first coordination shell is then expected to remain stable on relatively long timescales, while in the other cases frequent exchanges of chloride anions are expected between the first and the second coordination shells of the cations.

The running integral of the M–Cl RDF out to the first minimum gives the average coordination number of the cation in that mixture; as we commented in the introduction, this value was the focal point of many of the earlier structural studies. In the simulation we can examine the structure in more detail and, for each cation at each time step, we can determine the number of anions within a spherical shell whose radius corresponds to the position of the minimum in the RDF, and this constitutes an instantaneous coordination number for that cation. In Fig. 5 we plot the distributions of these coordination numbers for Y^{3+} and La^{3+}, as representatives of the small and large ions, at the two extreme concentrations and a temperature of 1073 K. We observe that, as anticipated in the introduction, the widths of these distributions narrow as the MCl_3 concentration is lowered into the dilute solution regime and that the means of the distributions approach 6. Elsewhere it has been shown how an experimental observable, such as the EXAFS spectrum, may be regarded as a superposition of spectra associated with these different coordination complexes.[13] At higher concentrations, the distributions broaden but the mean coordination number for Y^{3+} remains close to 6, whereas that for La^{3+} begins to increase. In other studies[13,24] the mean coordination number in pure $LaCl_3$ has been found to be close to 8.

The viscosity and mean coordination number both increase with increasing MCl_3 concentration; the results are summarised in Fig. 6. However, as we shall

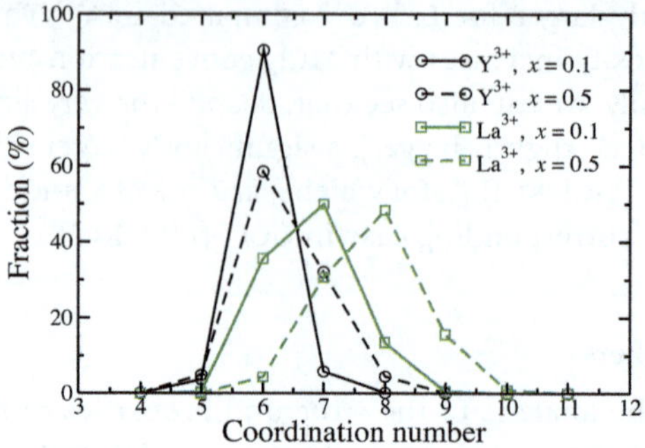

Fig. 5 The distributions of instantaneous coordination numbers in YCl_3 and $LaCl_3$ at the lowest and highest concentrations studied. The Y^{3+} distribution remains peaked at 6 across the concentration range, though it does broaden at higher concentrations. Even at the lowest concentration studied, the La^{3+} mean coordination number is larger than 6 and it shifts to still larger values at higher concentrations.

now demonstrate, this does not mean that the increase in coordination number should be seen as the *cause* of the rise in viscosity, rather that both are consequences of some other phenomenon affecting structure and dynamics.

Strength of the M–Cl ionic bonds

When the fluid is sheared it must relax to a new configuration and the timescale of this relaxation is the Maxwell relaxation time. Many types of microscopic events can contribute to this relaxation process and, in the present context, one might imagine that breaking up of the M^{3+} coordination shells could be involved. We have therefore examined the strength of the M^{3+}–Cl^- linkages.

Although there is no established criterion to define the "strength" of an ionic bond, their stability may be estimated[36,37] using the potential of the mean force that determines the exchange of Cl^- with the first solvation shell of M^{3+}, and can be evaluated by

$$\beta W^{\text{eff}}(r) = -\ln g_{\text{M-Cl}}(r) - 2\ln\left(\frac{r}{r_{\text{min}}}\right), \tag{3}$$

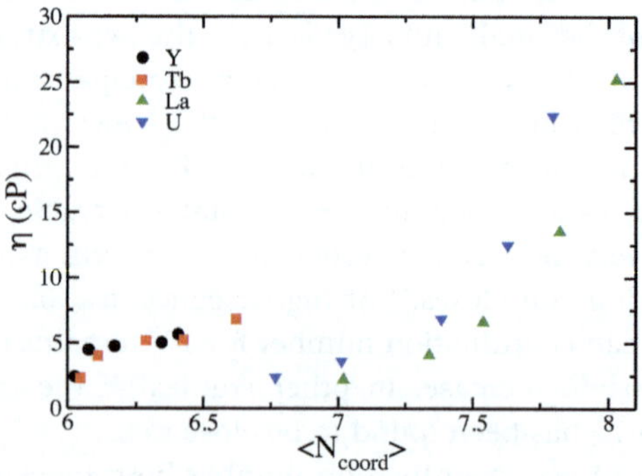

Fig. 6 Mean coordination number *vs.* viscosity for the different mixtures studied at 873 K.

where $\beta = 1/k_BT$, with k_B being the Boltzmann constant and T the temperature of the simulation, and r_{min} is a constant. These functions, shown in Fig. 7, display a deep first minimum at the position of the first peak of the RDF. To escape this minimum, a coordinated Cl^- anion must overcome an energy barrier ($E_{barrier}$).

Fig. 8 indicates that the height of the barrier decreases with increasing concentration and with increasing coordination number, *i.e.* the opposite trend to the viscosity. This result is rather counter-intuitive since one may expect that stronger bonds will lead to a larger resistance to shear. We will return later to an explanation of why this trend arises.

As an alternative measure of the stability of the first coordination shell, we analyze the rate of change of the instantaneous coordination number, which can be interpreted as the solvation shell lifetime. This is done by performing a cage correlation function analysis.[33,38] The first minimum of the M–Cl partial RDF is again used as a simple geometrical criterion to determine whether a chloride anion is located inside the first coordination shell of a cation at a given time. The number of counter-ions that have left the first coordination shell between time t to $t + \delta t$ is denoted $n_i^{out}(t, t + \delta t)$. The rate at which a single counterion (Cl^-) leaves the M^{3+} ion's solvation shell (the cage) is given by the cage-out correlation function,[38] which is defined as:

$$C^{out}(t) = \frac{1}{N_{M^{3+}}} \left\langle \sum_{i=1}^{N_{Cl^-}} \Theta\left(1 - n_i^{out}(0,t)\right) \right\rangle, \tag{4}$$

where Θ is the Heaviside step function. Examples of cage correlation functions for the cations of interest at a MCl_3 concentration of 0.3 are shown in Fig. 9. Note that the relaxation time is longer for the smaller cations, indicating a greater degree of stability for the first coordination shell in these cases, in agreement with the trend in the barrier height for breaking of the M^{3+}–Cl bond.

In Fig. 10 we have plotted the ratio of the Maxwell relaxation time to the cage correlation time against the barrier height for bond-breaking. The striking feature of the figure is that the cage relaxation time is much longer than the Maxwell time

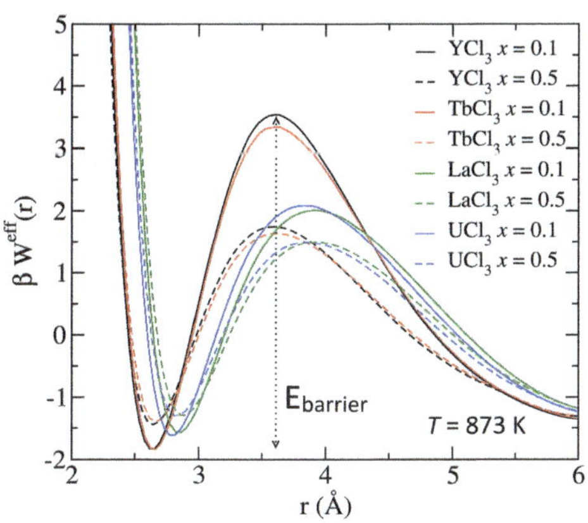

Fig. 7 The potential of the mean force for the breaking of an M^{3+}–Cl^- "bond" in the various mixtures studied, at 873 K. Notice that the barrier to dissociation is higher for the smaller cations and that its height falls at increasing concentrations.

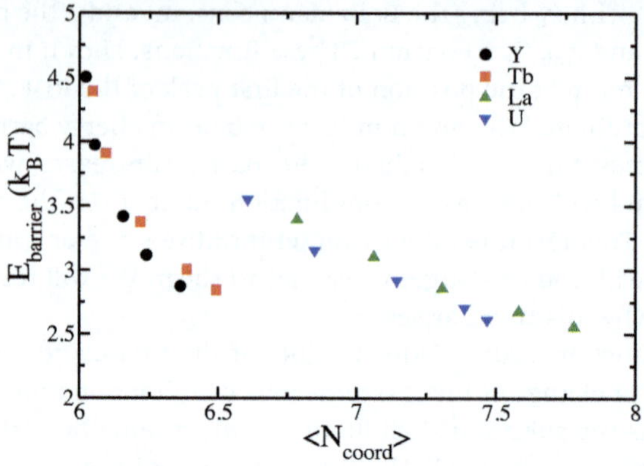

Fig. 8 The height of the barrier to dissociation *vs.* mean coordination number at 1073 K.

over most of the concentration range studied. In the relatively dilute solutions, then, the relaxation of the fluid to an applied shear is complete before the M^{3+} coordination shells have rearranged significantly – they behave as rigid units on the timescale of this relaxation. Only at the highest concentrations, where the fluid is becoming quite viscous due to the slow shear relaxation, do the timescales become comparable. This confirms that to find a link between structure and the behaviour of the viscosity we will need to look beyond the first coordination shell.

Network forming ability

In contrast to the MCl RDFs, the M^{3+}–M^{3+} RDFs for the larger and smaller cations differ markedly. In Fig. 11(a) we show the Y^{3+}–Y^{3+} RDFs and in Fig. 11(b) we show those for La^{3+}–La^{3+}. The temperature considered here is always $T = 873$ K and as previously the behavior for Tb^{3+}–Tb^{3+} is qualitatively analogous to the one for

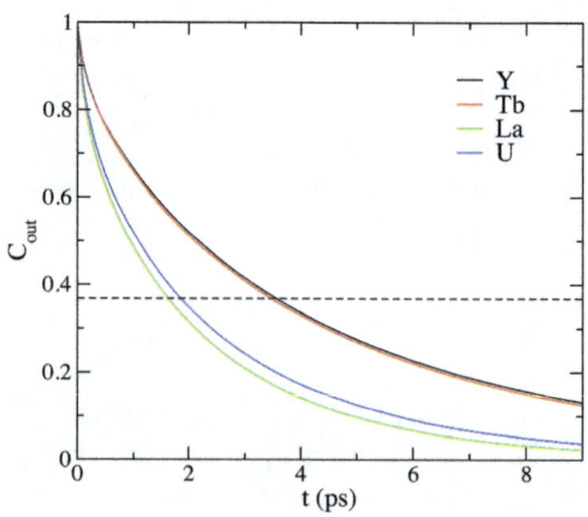

Fig. 9 Cage correlation functions for the cations of interest at a concentration of $x = 0.3$ and a temperature of 1073 K. Note that the relaxation times are longer for the smaller cations, in agreement with the expectation from the barrier heights in Fig. 7.

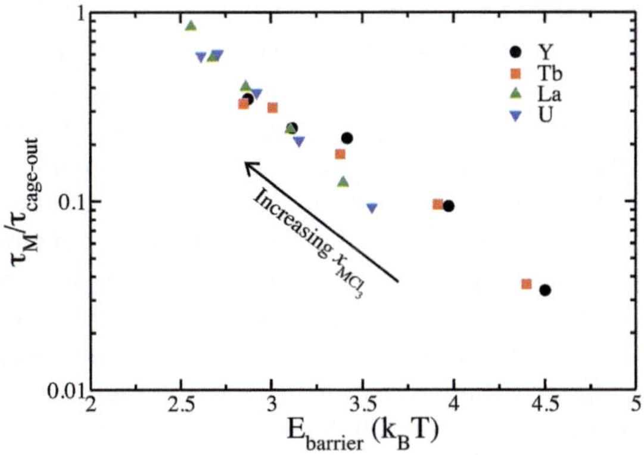

Fig. 10 The ratios of the cage and Maxwell relaxation times at 1073 K *vs.* the barrier heights determined above. Note that increasing barrier height is the direction of decreasing MCl_3 concentration.

Y^{3+}–Y^{3+}, while the behavior for U^{3+}–U^{3+} is similar to that for La^{3+}–La^{3+}, so these two cases are not shown for clarity.

If we focus initially on the Y–Y case for the most dilute $x = 0.1$ concentration, we see a small feature composed of two partially merged peaks at small Y–Y separations ($r \sim 5$ Å) and a much larger peak at $r \sim 8$ Å. Recalling that the most probable Y–Cl separation for the first coordination shell (from Fig. 1) is $r_{M\text{-}Cl} \sim 2.7$ Å, we can associate the first feature with pairs of cations with overlapping coordination shells. At this low concentration, the probability that a neighbouring pair of cations are involved in such an association is much smaller than the probability that they are unassociated (recall that the probability of being associated with a given RDF peak is proportional to the integral of r^2 times the RDF over the peak, so that one needs to notionally treble the area of the peak at larger r relative to the first to gain an estimate of the ratio of probabilities). Thus, to a first approximation, in this dilute case, the yttrium melt may be viewed as

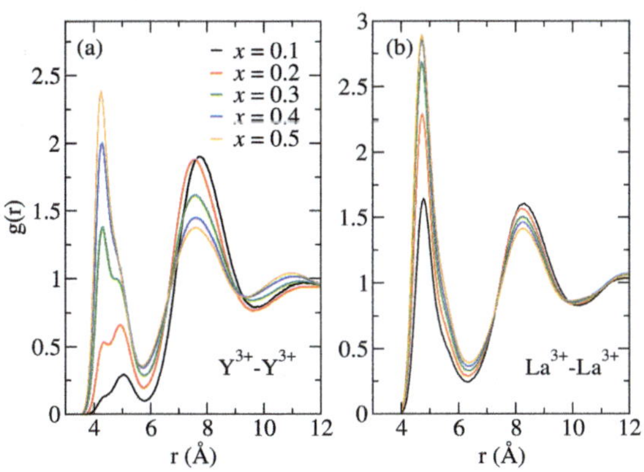

Fig. 11 RDFs for the pairs (a) Y^{3+}–Y^{3+} and (b) La^{3+}–La^{3+} at $T = 873$ K; the span of the MCl_3 mole fractions was investigated for all. The features appearing at $r < 6$ Å arise from pairs of cations with overlapping coordination shells.

a solution of separated YCl_6^{3-} ions dispersed in the solvent. We will see shortly that the double-peaked nature of the first feature arises from the presence of Y–Y pairs which share a single anion or two anions in their coordination shells, corresponding to the corner- and edge-sharing linkages illustrated in Fig. 2. The latter are associated with the peak at smaller r, $\sim$ 5 Å, in the RDF. As the concentration of YCl_3 in the mixture increases we see that the probability that a Y^{3+} ion is involved in these Cl-bridged linkages with neighbouring Y^{3+} ions increases dramatically. From a direct examination of the instantaneous configurations generated in the simulations, we may classify the M–M linkages depending on the number of anions that the two cations share and count the corner-sharing, edge-sharing and face-sharing (for three common anions) MCl_n polyhedra.[39] By performing this analysis, we obtain a majority of corner-sharing species and a minority of edge-sharing species at low YCl_3 concentrations, as shown in Fig. 12. For the higher concentrations of YCl_3, although we observe an increase in both peaks, the short-distance (edge-sharing) one grows more rapidly and becomes predominant by $x = 0.5$. We also note the formation of a few face-sharing species, although no characteristic peak is observed for them in the RDFs.

The La^{3+}–La^{3+} case behaves quite differently. Firstly, the intensity of the low r peak is considerably larger than for the YCl_3 mixtures, implying a significant degree of cross-linking even at the lowest concentrations studied. From the linkage analysis in Fig. 12 we observe that the three types (corner, edge and face) of sharing coexist, and that the variations of their relative extent is much smaller than in the case of Y^{3+}. Secondly, the first peak in the RDF is not split in the manner of the Y–Y case. However, by examining it more carefully, we can see that the first peak is quite asymmetric, and we can attribute the shoulder on its right side to the corner-sharing configurations. The implication is that the difference in M–M separations between the corner- and edge-sharing configurations is smaller for La–La than for Y–Y, and is simply not resolved.

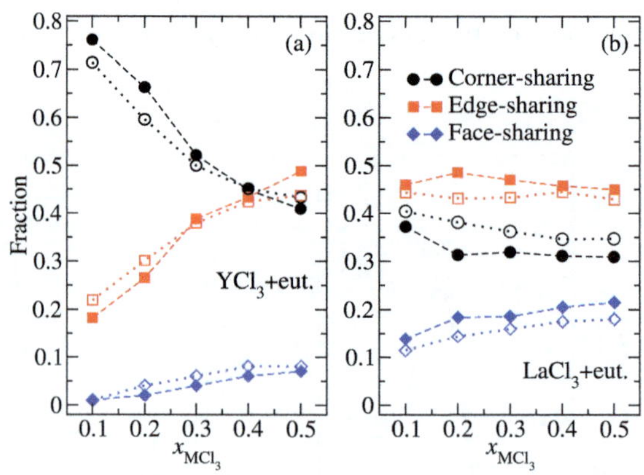

Fig. 12 Analysis of the number of anions shared between two M^{3+} cations within first shell distances. One anion indicates a corner-sharing configuration, two cations an edge-sharing configuration and three cations a face-sharing configuration. The analysis is shown for (a) YCl_3–(LiCl–KCl)$_{eut}$ and (b) $LaCl_3$–(LiCl–KCl)$_{eut}$. The values shown are for T = 873 K (solid symbols) and T = 1073 K (hollow symbols). At all temperatures and all MCl_3 mole fractions, the fraction of cases where the M^{3+}–Cl^- pair is not connected to another M^{3+} is about 0.05 for YCl_3 and 0.03 for $LaCl_3$ (not shown for clarity).

The proportion of M^{3+}–Cl^- bonds involved in these linkages becomes quite high, linking the cations into a network (as visualised in Fig. 2). We can illustrate this with a simple analysis for the Y^{3+} case, where the coordination number remains close to 6 across the concentration range. For the $x = 0.5$ concentration, the stoichiometry dictates that there are $4Cl^-$ ions per Y^{3+} and, counting a Cl^- anion involved in a linkage as contributing one half to each cation, we see that to complete the coordination shell of a cation, 4 out of the 6 anions must be involved in linkages to other cations. We can now see a reason why the barriers to dissociation and the lifetimes of the M^{3+}–Cl^- bonds decrease with increasing coordination. The measures we have used to characterise the strengths of these bonds average over all of the M^{3+}–Cl^- bonds in the system and therefore increasingly reflect the characteristics of the shared linkages at higher concentrations. We can also see that the tendency to increase coordination with increasing concentration must be linked to the propensity to form linkages of different types.

At $x = 0.5$, we see from Fig. 12 that the proportions of corner- and edge-shared linkages between Y^{3+} ions are roughly equal and, for ease of this simplified discussion, we will ignore the small number of face-sharing cases. Since the edge-sharing links account for two of the four anions in the coordination shell involved in linkages, we can expect that (on average) each Y^{3+} is participating in two corner-sharing linkages and one edge-sharing one, and therefore is connected to three other Y^{3+} cations. This degree of connectivity is sufficient to consider the cations to be linked into an extended network. Note too that, depending on the types of linkages formed (corners *vs.* edge *vs.* face), the number of cations linked to a given cation will vary and we should expect this to influence the character of the network and its strength, and we note the different proportions exhibited by the small and larger cations in our study (Fig. 12). Wilson[39,40] has discussed the relationship between fluidity and the concept of network connectivity in tetrahedrally-coordinated glass-formers.[41,42] In the field of amorphous solids a critical coordination number above 2.4 at the vertices of the network signals transition to a rigid state, and Wilson examined whether the same analysis can be applied to measures of fluidity in a liquid.

4 Discussion: cross-linking, network-structure and viscosity

The picture which has emerged from the above analyses is of an increasing degree of cross-linking between the cations as the concentration is increased, and the subsequent emergence of a network. The challenge now is to link the properties of the network to the rapid increase in viscosity at higher concentrations, with a view to pointing to reasons why, for example, the La/U-containing melts are more viscous than the Y/Tb ones at higher concentrations.

An analogous challenge was to account for the very strong concentration dependence of the viscosity in mixtures of BeF_2 with LiF – an important component of many proposed molten salt nuclear reactors.[1] In these systems the Be^{2+} ions are invariably tetrahedrally coordinated to four F^- species and, whilst present as isolated BeF_4^{2-} ions at low concentrations, they become increasingly linked by bridging F^- ions in exclusively corner-sharing arrangements as the concentration is increased.[8,9,43,44] As the concentration approaches the equimolar point, the viscosity increases sharply and becomes very high for pure BeF_2 – which is a good glass-forming material (unlike the pure MCl_3 systems). Although there are numerous differences in the structure compared to MCl_3 (well-defined 4-coordinate, only corner-sharing linkages), the relationship between the local dynamics of bond-breaking and the viscous relaxation is very similar to that exposed above for the MCl_3 : ACl mixtures.[45]

In particular, at low concentrations, where the highly charged cations are involved in simple $BeF_4{}^{2-}$ or $YCl_6{}^{3-}$ units or small oligomers of them, these units behave as rigid bodies on the timescale of viscous relaxation, which is dominated by solvent relaxation, and the cage relaxation time is much longer than the Maxwell time. As the concentration reaches the point where the degree of cross-linking between cations becomes large enough to support an extended network (the speciation of Be-containing ions was followed explicitly in the BeF_2 case), the viscosity increases markedly and the cage relaxation time becomes much more similar to the Maxwell time (quantitatively in the $LaCl_3$ case and exhibiting the same concentration dependence as for BeF_2). These similarities support the assertion that the viscous relaxation at higher concentrations is dominated by the reorganisation of the network in the MCl_3 mixtures. In the BeF_2 case, we showed that the Be–F cage time becomes a good measure of the timescale for breaking the Be–Be links in the network. We also showed that this latter timescale increased as the network became more complete at higher concentrations and suggested that this was because the constraints imposed by the surrounding network inhibited the breaking of the linkage – so that the individual links stiffen as the network becomes more extended.

Despite these similarities, there are significant differences in the behaviors of the viscosities between the BeF_2 and the MCl_3 cases, with the former becoming far more viscous. This may simply be a consequence of the greater strength of the Be–F bond, but may also reflect the differences in local and network structures. The MCl_3 systems exhibit a much higher mean coordination number and a more flexible coordination shell as well as several different (corner, edge and face sharing) types of interaction linkages – all of which suggest that other types of local reorganisation, more than the simple breaking of M–Cl bonds, may be possible and may allow the network configuration to relax. Wilson[39,40] has shown that the breaking of an edge-sharing linkage is substantially slower than for a corner-sharing one, which suggests that the presence of edge- and face-sharing linkages will stiffen the network; however, as noted above they also reduce the number of cations to which a given cation is linked, and hence also reduce the connectivity of the network. It is likely that a further examination of these factors will be necessary to fully explain the differences in behaviors of the viscosities of YCl_3 and $LaCl_3$.

Acknowledgements

This work was carried out with the European Commission financial support in the 7th framework program, under the collaborative project "SACSESS", Grant Agreement No. 323282.

References

1 C. Le Brun, *J. Nucl. Mater.*, 2007, **360**, 1–5.

2 T. Kato, T. Inoue, T. Iwai and Y. Arai, *J. Nucl. Mater.*, 2006, **357**, 105–114.

3 J. J. Katz, G. T. Seaborg and L. R. Morss, *The Chemistry of Actinide Elements*, Chapman and Hall, New York, 2nd edn, 1986, vol. 1.

4 H. On Nam and D. Morgan, *J. Nucl. Mater.*, 2015, **465**, 224–235.

5 K. Grjotheim, C. Krohn, M. Malinovsky, K. Matiasovsky and J. Thonstad, *Fundamentals of the Hall–Héroult Process*, Aluminium-Verlag, Dusseldorf, Germany, 1982.

6 P. A. Madden and M. Wilson, *Chem. Soc. Rev.*, 1996, **25**, 339.

7 M. Salanne and P. A. Madden, *Mol. Phys.*, 2011, **109**, 2299–2315.

8 R. J. Heaton, R. Brookes, P. A. Madden, M. Salanne, C. Simon and P. Turq, *J. Phys. Chem. B*, 2006, **110**, 11454–11460.

9 M. Salanne, C. Simon, P. Turq, R. J. Heaton and P. A. Madden, *J. Phys. Chem. B*, 2006, **110**, 11461–11467.

10 M. Salanne, C. Simon, P. Turq and P. A. Madden, *J. Phys. Chem. B*, 2008, **112**, 1177–1183.

11 F. Hutchinson, M. Wilson and P. A. Madden, *Mol. Phys.*, 2001, **99**, 811–824.

12 Y. Okamoto, P. Madden and K. Minato, *J. Nucl. Mater.*, 2005, **344**, 109–114.

13 Y. Okamoto, S. Suzuki, H. Shiwaku, A. Ikeda-Ohno, T. Yaita and P. A. Madden, *J. Phys. Chem. A*, 2010, **114**, 4664–4671.

14 M. Rovere and M. P. Tosi, *Rep. Prog. Phys.*, 1986, **49**, 1001–1081.

15 M. P. Tosi, D. L. Price and M.-L. Saboungi, *Annu. Rev. Phys. Chem.*, 1993, **44**, 173.

16 A.-L. Rollet and M. Salanne, *Annu. Rep. Prog. Chem., Sect. C: Phys. Chem.*, 2011, **107**, 88–123.

17 G. M. Photiadis, B. Borresen and G. N. Papatheodorou, *J. Chem. Soc., Faraday Trans.*, 1998, **94**, 2605–2613.

18 G. A. Voyiatzis, A. G. Kalampounias and G. N. Papatheodorou, *Phys. Chem. Chem. Phys.*, 1999, **1**, 4797–4803.

19 G. D. Zissi and G. N. Papatheodorou, *Phys. Chem. Chem. Phys.*, 2004, **6**, 4480–4489.

20 *X-Ray Diffraction Analysis of Ionic Liquids*, Molten Salt Forum, ed. H. Ohno, K. Igarashi, N. Umesaki and K. Furukawa, Trans Tech Publications, Switzerland, 1994, vol. 3.

21 J. C. Wasse and P. S. Salmon, *J. Phys.: Condens. Matter*, 1999, **11**, 1381–1396.

22 J. C. Wasse and P. S. Salmon, *J. Phys.: Condens. Matter*, 1999, **11**, 9293–9302.

23 H. E. Fischer, A. C. Barnes and P. S. Salmon, *Rep. Prog. Phys.*, 2006, **69**, 233–299.

24 W. J. Glover and P. A. Madden, *J. Chem. Phys.*, 2004, **121**, 7293–7303.

25 E. Robert, J. E. Olsen, V. Danek, E. Tixhon, T. Ostvold and B. Gilbert, *J. Phys. Chem. B*, 1997, **101**, 9447–9457.

26 B. Gilbert, E. Robert, E. Tixhon, J. Olsen and T. Ostvold, *Inorg. Chem.*, 1996, **35**, 4198–4210.

27 E. Robert, V. Lacassagne, C. Bessada, D. Massiot, B. Gilbert and J. Coutures, *Inorg. Chem.*, 1999, **38**, 214–217.

28 V. Lacassagne, C. Bessada, P. Florian, S. Bouvet, B. Ollivier, J.-P. Coutures and D. Massiot, *J. Phys. Chem. B*, 2002, **106**, 1862–1868.

29 Z. Akdeniz and P. A. Madden, *J. Phys. Chem. B*, 2006, **110**, 6683–6691.

30 S. Cikit, Z. Akdeniz and P. A. Madden, *J. Phys. Chem. B*, 2014, **118**, 1064–1070.

31 R. D. Shannon, *Acta Crystallogr., Sect. A: Cryst. Phys., Diffr., Theor. Gen. Crystallogr.*, 1976, **32**, 751–767.

32 A. S. Basin, A. B. Kaplun, A. B. Meshalkin and N. F. Uvarov, *Russ. J. Inorg. Chem.*, 2008, **53**, 1509–1511.

33 R. Brookes, A. Davies, G. Ketwaroo and P. A. Madden, *J. Phys. Chem. B*, 2005, **109**, 6485–6490.

34 K. Fukasawa, A. Uehara, T. Nagai, N. Sato, T. Fujii and H. Yamana, *J. Nucl. Mater.*, 2012, **424**, 17–22.

35 M. Salanne, B. Rotenberg, C. Simon, S. Jahn, R. Vuilleumier and P. A. Madden, *Theor. Chem. Acc.*, 2012, **131**, 1143.

36 O. Pauvert, M. Salanne, D. Zanghi, C. Simon, S. Reguer, D. Thiaudière, Y. Okamoto, H. Matsuura and C. Bessada, *J. Phys. Chem. B*, 2011, **115**, 9160–9167.

37 M. Levesque, V. Sarou-Kanian, M. Salanne, M. Gobet, H. Groult, C. Bessada and A.-L. Rollet, *J. Chem. Phys.*, 2013, **138**, 184503.

38 E. Rabani, J. D. Gezelter and B. J. Berne, *J. Chem. Phys.*, 1997, **107**, 6867–6876.

39 M. Wilson, *Phys. Chem. Chem. Phys.*, 2012, **14**, 12701–12714.

40 M. Wilson, *J. Phys.: Condens. Matter*, 2012, **24**, 284114.

41 M. Thorpe, *J. Non-Cryst. Solids*, 1983, **57**, 355–370.

42 M. Bauchy and M. Micoulaut, *EPL*, 2013, **104**, 56002.

43 J. Dai, D. Long, P. Huai and Q. Li, *J. Mol. Liq.*, 2015, **211**, 747–753.

44 J. Dai, H. Han, Q. Li and P. Huai, *J. Mol. Liq.*, 2016, **213**, 17–22.

45 M. Salanne, C. Simon, P. Turq and P. A. Madden, *J. Phys. Chem. B*, 2007, **111**, 4678–4684.

Faraday Discussions

PAPER

Effect of C_8mimPF_6 on miniemulsion polymerization for application in new latex coating products

Binjie Hu,*[a] Yiyang Kong,[a] Rongmeihui Zheng,[a] Jie Dong,[a] Kwang-Leong Choy[b] and Huanhuan Zhao[a]

Received 1st February 2016, Accepted 22nd February 2016
DOI: 10.1039/c6fd00009f

C_8mimPF_6, as a type of room temperature ionic liquid (RTIL) with non-volatility and a low melting point, may replace conventional coalescing agents in latex coatings, thus preventing volatile organic compound (VOC) emissions caused by coalescing agents. In this study, systematic investigations on the effect of various factors including initiator type, initiator concentration, temperature and C_8mimPF_6 concentration on the conversion of latex and droplet/particle size of a miniemulsion during polymerization have been conducted. The presence of C_8mimPF_6 has shown to have a marked effect on the reaction rate. Such an effect strongly depends on the type of initiator being used. For polymerization initiated by 2,2-azobis (isobutyronitrile) (AIBN), C_8mimPF_6 had a promoting effect on the reaction rate at low concentrations, but this effect might be reversed upon certain C_8mimPF_6 concentrations, *e.g.* 10 wt%. While initiated by H_2O_2/Vc, this promoting effect faded even at low C_8mimPF_6 concentrations. The different limiting factors, which determine the reaction rate with different types of initiator, may contribute to the results. For reactions initiated by hydrophobic AIBN, the reaction was dominated by kinetics. The presence of C_8mimPF_6 may cause an enhanced chain propagation rate and reduced chain termination rate, which may further contribute to the increase in reaction rate at lower concentrations of C_8mimPF_6. With hydrophilic H_2O_2/Vc, the resistance for the transfer of radicals into a droplet/particle might be increased significantly with increasing C_8mimPF_6 concentration due to a tighter interfacial structure at lower concentrations of C_8mimPF_6. Thus, such transfer of radicals may become a limiting step whilst the presence of C_8mimPF_6 increases the transfer resistance on radicals resulting in a decrease in reaction rate. The reaction temperature, which is related to the decomposition temperature of the initiator being used, was another factor affecting the conversion of latex and the size of latex particles. A higher temperature *e.g.* 50 °C promotes the coalescence of droplets/particles, and hence produces larger latex particles. In the presence of C_8mimPF_6, the reaction temperature could be significantly reduced to as low as 40 °C, which prevents phase separation. The final particle size depends on the nucleation mechanism as well as the coalescence of droplets/particles during polymerization.

[a]*Department of Chemical and Environmental Engineering, University of Nottingham Ningbo China, 199 Taikang East Road, Ningbo, China. E-mail: Binjie.HU@nottingham.edu.cn*
[b]*Centre for Materials Discovery, University College London, 20 Gower Street, London, WC1H 0AJ, UK*

1 Introduction

With increasing demand to reduce the use of volatile organic components (VOCs) worldwide, especially in coating products, more environmentally friendly coating products (*e.g.* latex coatings) have attracted increased attention. Latex coatings are produced by mixing latex, a dispersion of polymer particles in water, with pigments, fillers and additives. As water is used as the solvent in place of organic solvents, VOC emissions can be minimized. However, complete elimination of VOCs is currently impossible due to the addition of organic additives. Among these additives, coalescing agents are the main source of VOCs. They function as plasticizers that can effectively reduce the glass transition temperature (T_g) of a polymer, thus promoting the transformation from discrete particles to a continuous film. After film formation, the coalescing agents diffuse into the atmosphere as VOCs, and even release an unpleasant odour that lasts for days.[1]

Room temperature ionic liquids (RTILs) are a type of non-volatile organic salt, some of which have been proven to be efficient plasticizers that can reduce the T_g of polymers.[2,3] Therefore, RTILs could potentially be used as coalescing agents with low-VOC emissions, remaining in the film as a polymer/RTIL composite after film formation. RTILs were found to considerably enhance the thermal stability of composites, bringing about the possibility of applying flexible polymers at high temperatures.[4] In addition, the presence of a RTIL may improve the ionic conductivity of a film because of its excellent conductivity.[5–7] RTILs have also shown antimicrobial activities due to the electrostatic interactions between RTIL cations and bacterial cell walls,[8–10] which widened their application to marine antifouling coatings.[11] Based on the above studies, it can be postulated that certain RTILs might act as multi-functional coalescing agents with minimal VOC emissions, which would fit with the developmental trend for latex coating additives.

Hydrophobic RTILs may be used to improve the water resistance of films. However, hydrophobic RTILs cannot be directly dispersed in the water phase. It is possible to encapsulate RTILs inside particles *via* miniemulsion polymerization. The dominant nucleation mechanism for miniemulsion polymerization is droplet nucleation, where monomers are polymerized directly inside droplets.[12] Hydrophobic compounds present in the droplets gradually become covered by the polymer formed, and are eventually encapsulated inside particles. Hydrophobic compounds Miglyol 812, castor oil and *n*-heptane have been encapsulated successfully using this method.[13–15] Unlike ordinary hydrophobic compounds, RTILs may have an effect on the rate of polymerization. Miniemulsion polymerization is a kind of free radical polymerization, and the effect of RTILs on free radical polymerization has long been studied. Harrisson *et al.* first studied the radical polymerization of methyl methacrylate (MMA) in 1-butyl-3-methylimidazolium hexafluorophosphate (C_4mimPF_6) solution by pulse laser polymerization (PLP).[16,17] It was observed that the chain propagation rate of MMA increased with increasing C_4mimPF_6 concentration, while the chain termination rate decreased. The presence of C_4mimPF_6 increased the polarity of the medium which may have contributed to the charge-transfer structure of the transition states, thus lowering the activation energy of chain propagation. The decrease in the chain termination rate was owing to the increase in the viscosity of the medium, which in turn was due to the presence of viscous C_4mimPF_6. These two effects might explain why a high overall polymerization rate and a high molecular weight could be achieved when MMA was polymerized in a RTIL medium.[18] Alternative explanations for the RTIL effect were put forward, including a "protected" radical mechanism, where parts of the monomers are separated into extremely small domains by the barrier of the RTIL domain, and so the radical can be protected in the RTIL domain. There is then a partition between the radical protected in the RTIL domain and the radical formed in the monomer, which would thus prevent chain termination.[19]

Thus far, to the best of our knowledge, no report on applying RTILs as multifunctional additives for latex coatings is available in the literature. In this work, we systematically investigate the encapsulation of RTILs and their effect on miniemulsion polymerization. The effect of RTILs on the initial droplet size and stability of miniemulsions was reported in our earlier work.[20] In this study, 1-octyl-3-methylimidazolium hexafluorophosphate (C_8mimPF_6) was chosen as the target RTIL. Two kinds of initiators, hydrophobic 2,2'-azobis(isobutyronitrile) (AIBN) and hydrophilic hydrogen peroxide/ascorbic acid (H_2O_2/Vc) were used in the polymerization. The effects of various factors, including initiator type, initiator concentration, temperature and C_8mimPF_6 concentration, on product conversion and miniemulsion droplet/particle size at different reaction stages were studied.

2 Materials and methods

2.1 Chemicals

Methyl methacrylate (MMA, chemically pure (CP)) was purchased from Sinopharm Chemical Reagent Co., Ltd and purified by vacuum distillation to remove inhibitor. 2,2'-Azobis(isobutyronitrile) (AIBN, 97%) was purchased from Aladdin Industrial Inc and recrystallized using ethanol. L-Ascorbic acid (Vc, analytical reagent (AR)), hydrogen peroxide (H_2O_2, AR) and sodium dodecyl sulfonate (SDSO, CP) were purchased from Sinopharm Chemical Reagent Co., Ltd. Hexadecane (HD, 98%) was purchased from Aladdin Industrial Inc. 1-Octyl-3-methylimidazolium hexafluorophosphate (C_8mimPF_6, 99%) was supplied by Shanghai Cheng Jie Chemical Co., Ltd. Ultra-pure water with a resistivity of 18.2 MΩ cm^{-1} was used in all experiments.

2.2 Preparation of miniemulsion

An oil in water miniemulsion (O/W) was prepared with an oil to water phase volume ratio of 3 to 7. The oil phase contained 5 wt% HD, the remaining 95 wt% containing C_8mimPF_6 and MMA. Both C_8mimPF_6 and HD were infinitely miscible in MMA. After HD and C_8mimPF_6 were fully dissolved in MMA, the oil phase was added to the water phase which contained 30 mM SDSO in a 100 mL plastic container. The total volume of the mixture was 50 mL and it was kept at 40 °C. An ultrasound probe was then placed one centimeter below the surface of the mixture in the middle of the container. The output power of the sonicator (Xinzhi Scientz II) was set to 95 W. The miniemulsion was prepared by homogenizing the two phase mixture for six minutes in a state of one second on, one second off.

2.3 Miniemulsion polymerization

The miniemulsion polymerization was carried out in a four-necked flask equipped with a stirrer, a reflux condenser, a thermometer and a nitrogen inlet. The flask was immersed in a water bath in order to be able to control the reaction temperature. The miniemulsion was loaded into the flask and was gently stirred with bubbling nitrogen for about 1.5 h to remove oxygen in the flask. The temperature of the water bath was then set to reaction temperature. The method used to add the initiator depended on the type of initiator used. When H_2O_2/Vc was used, the H_2O_2/Vc solution with a H_2O_2 to Vc molar ratio of 1 to 1.3 was injected directly into the miniemulsion after the desired reaction temperature was reached, and the polymerization reaction then started. When AIBN was used, because it could not be added directly into the miniemulsion due to its hydrophobicity, it was dissolved in the oil phase before the preparation of the miniemulsion, and then the reaction temperature was adjusted to the desired temperature in order to start the polymerization. The reaction process was carried out under a nitrogen atmosphere.

2.4 Characterization

2.4.1 Determination of latex conversion. The conversion of latex was determined gravimetrically. The product was sampled at regular intervals during the reaction and stored in the refrigerator in order to terminate the reaction. After evaporating the monomer under ventilated conditions and storing the sample in an oven at 120 °C for 4 h, polymer, C_8mimPF_6 and SDSO remained in the solid product. As the content of C_8mimPF_6 and SDSO can be calculated according to raw material ratios, the yield of polymer could be obtained by subtracting the C_8mimPF_6 and SDSO contents from the solid product.

2.4.2 Droplet/particle size measurement. The droplet/particle sizes and size distributions within the miniemulsion were measured at different reaction times using the laser diffraction method (Mastersizer 3000, Malvern Inc). This technique has been widely accepted as a tool to evaluate droplet/particle size.[21–23] Samples were diluted with water in order to meet the requirement of obscuration. The diluted water solutions were saturated with SDSO and monomers in order to prevent coalescence during sample measurements.[24] All measurements were repeated three times.

2.4.3 Interfacial tension measurement. The interfacial tension between the oil and water phases was measured using the Du Noüy ring method (BZY-2 tensiometer, Hengping Instruments). The measurement was carried out at 40 °C, which was the same as the reaction temperature. Measurements were repeated three times.

3 Results and discussion

In this section, the effects of various factors, including the initiator type, initiator concentration, temperature and C_8mimPF_6 concentration on the conversion of latex as well as the droplet/particle size of the miniemulsion during polymerization are discussed. The reaction schemes for free radical polymerization of MMA initiated by AIBN and H_2O_2/Vc are illustrated below.

In free radical polymerization, the first step is the generation of a free radical *via* the decomposition of initiators, followed by chain propagation, chain termination or chain transfer.[25]

Radical generation: Generation of radicals from AIBN and H_2O_2/Vc is shown as follows.[25,26]

AIBN:

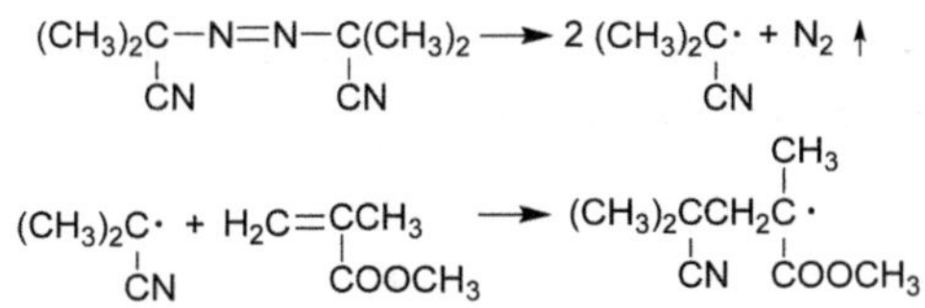

$(R_1\cdot)$

H_2O_2/Vc:

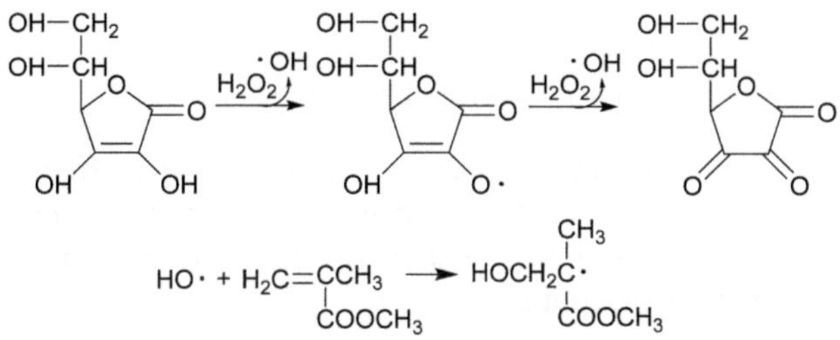

$(R_2\cdot)$

Chain propagation: Addition reactions between monomer radicals ($R_1^{\cdot}$ or $R_2^{\cdot}$) and MMA form new radicals. The newly formed radicals continue to react with MMA, forming long chain radicals. These are strong exothermal reactions with low activation energies.

AIBN:

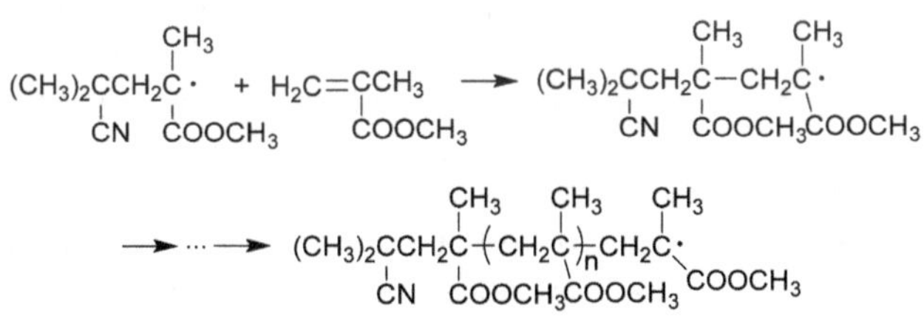

$(R_3\cdot)$

H_2O_2/Vc:

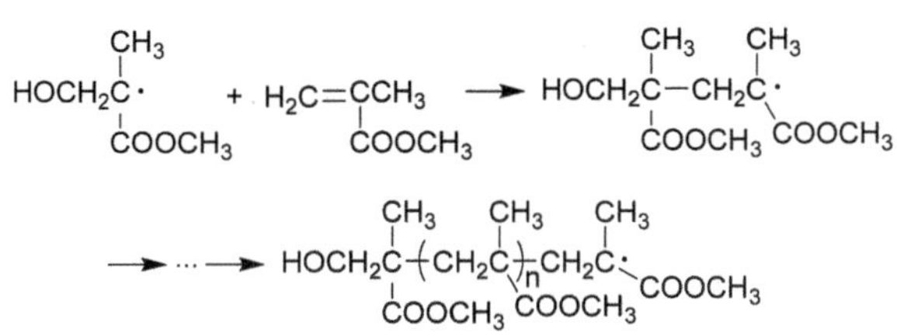

$(R_4\cdot)$

Chain termination: Highly active long chain radicals ($R_3^{\cdot}$ or $R_4^{\cdot}$) cannot exist stably, and are likely to terminate with other radicals. This bi-radical termination contains coupling termination and disproportionation termination. The long chain radical can be expressed in a more general form as given below.

Coupling termination means single electrons from two radicals bonding with each other through a covalent bond.

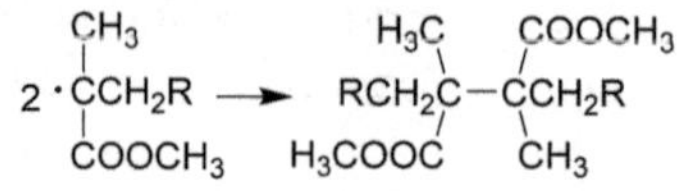

Disproportionation termination is a process in which one radical captures a hydrogen atom or another atom of another radical to terminate both radicals.

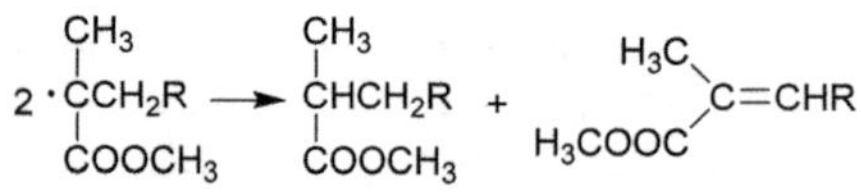

Chain transfer: Long chain radicals may capture an atom from a monomer, initiator, solvent or macromolecule, expressed as YK in the following scheme, and transfer an electron to the molecule that's losing an atom. A new radical, ˙K, forms and continues chain propagation.

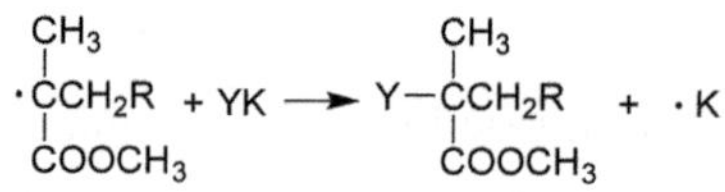

3.1 Factors affecting the conversion of latex in miniemulsion polymerization

3.1.1 Effect of initiator concentration on latex conversion. To study the effects of initiator concentration, miniemulsions containing 10 wt% C_8mimPF_6 (based on oil phase) were prepared and initiated using different concentrations of AIBN ranging from 0.1 mol% to 2 mol% (based on monomer) at 40 °C. The latex conversions after polymerization are shown in Table 1. We found that the conversion of latex increases whilst the reaction time decreases with increasing AIBN concentration from 0.1 mol% to 1.0 mol%. A further increase in AIBN concentration to 2.0 mol% had very little effect on the conversion of latex.

A hydrophilic initiator, H_2O_2/Vc, was also used to study such effects, using the same C_8mimPF_6 concentration and temperature as for the above. The latex conversions after polymerization are shown in Table 2. When the H_2O_2 concentration is 0.1 mol%, no reaction takes place. At such low concentrations, the hydroxyl radicals generated might be captured, owing to the presence of oxygen in the water phase, before they get the chance to react with the monomers. Generally, oxygen is an inhibitor that can capture radicals. In this study, it might react with Vc and produce superoxide. Such superoxide could effectively scavenge hydroxyl radicals.[26] At low H_2O_2 concentrations, the hydroxyl radicals generated from H_2O_2 might run out completely due to this effective scavenger, thus resulting in no reaction occurring. As the H_2O_2 concentration increases from 0.5 mol% to 2 mol%, there is a higher conversion of latex, *i.e.* 100% conversion is achieved. It was also observed that the reaction rate is sensitive to the concentration of initiator being used. An increase in the concentration of H_2O_2 from 0.5 mol% to 1 mol% results in a shorter reaction time for the same conversion

Table 1 Latex conversion in a miniemulsion containing 10 wt% C_8mimPF_6, initiated by different concentrations of AIBN at 40 °C

AIBN concentration (mol%)	Conversion (%)	Reaction time (min)
0.1	29.1	240
0.5	78.9	180
1.0	89.5	150
2.0	90.2	150

(100%). However, this enhancement in conversion rate and corresponding decrease in reaction time, owing to the increase in initiator concentration, diminishes once the initiator concentration reaches a certain value, *i.e.* 1 mol%. The results are consistent with the observations found for AIBN. This indicates that 1.0 mol% could be the optimal concentration to achieve both a high rate of reaction and a high conversion of latex.

3.1.2 Effect of temperature on latex conversion. In order to study the effects of temperature, a miniemulsion containing 10 wt% C_8mimPF_6 (based on oil phase) was prepared, and initiated by 1.0 mol% AIBN at temperatures ranging from 30 °C to 50 °C. Phase separation occurred above 55 °C.

Fig. 1 shows the yield of polymer as a function of reaction time for the miniemulsion containing 10 wt% C_8mimPF_6, after polymerization was initiated by AIBN at different temperatures. It was observed that carrying out the reaction at 50 °C resulted in the highest yield of polymer (95.1%) and the shortest reaction time (60 min), whilst at 40 °C a sharp increase in polymer yield was also observed within the first 60 min, followed by a slow increase in yield until 150 min, with a maximum latex conversion of 89.5%. From Fig. 1 it can also be noted that nearly no reaction occurs at 30 °C. The change in reaction rate observed in Fig. 1 was mainly caused by the difference in the decomposition rate of the initiator, which is strongly dependent on temperature. The half-life time of AIBN has been reported to be 80 min at 80 °C,[27] while this time was found to extend to 37 h when the temperature was reduced to 55 °C.[28] Clearly, higher temperatures promote a higher radical generation efficiency owing to the decomposition of AIBN, and effective temperatures range from 50 °C to 90 °C.[29–31] It is not surprising that no reaction takes place at 30 °C, as such a low temperature may cause inefficiency in radical generation, leading to a lack of radicals available for further polymerization. However, it is unexpected that such a high conversion of latex would be observed at 40 °C, which is below the efficient decomposition temperature of

Table 2 Latex conversion in a miniemulsion containing 10 wt% C_8mimPF_6, initiated by different concentrations of H_2O_2/Vc at 40 °C

H_2O_2 concentration (mol%)	Conversion (%)	Reaction time (min)
0.1	0.0	90
0.5	100.0	50
1.0	100.0	30
2.0	100.0	30

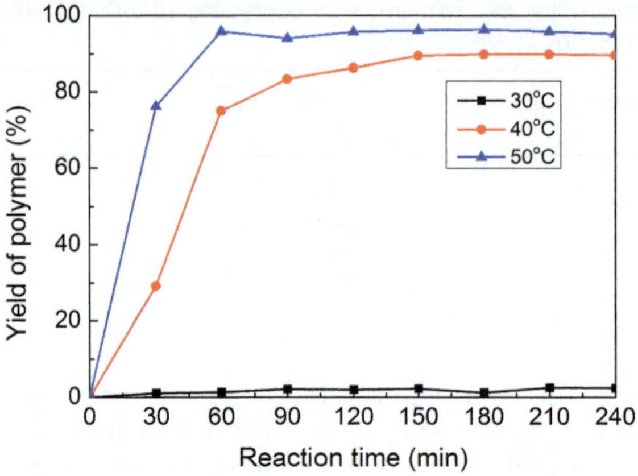

Fig. 1 Yield of polymer as a function of reaction time for miniemulsions containing 10 wt% C_8mimPF_6, after polymerization was initiated by AIBN at different temperatures.

AIBN. This could potentially be attributed to the presence of C_8mimPF_6, which will be discussed further in the following section.

A similar temperature effect was also observed for polymerization initiated by H_2O_2/Vc, as shown in Fig. 2. For temperatures ranging from 30 °C to 50 °C, higher reaction rates are observed for higher temperatures, but the final conversion of latex is approximately 100.0% for all conditions. It is interesting to note that by replacing AIBN with Vc/H_2O_2, polymerization could be completely carried out at 30 °C. As redox reactions between H_2O_2 and Vc have much lower activation energies than that of AIBN, radicals could be generated at much lower temperatures. Indeed, initiating polymerization by H_2O_2/Vc can be achieved at temperatures as low as 25 °C.[26] In addition, polymerization was found to occur even at 20 °C, which further indicates the potential effect that C_8mimPF_6 can have on the reaction process.

3.1.3 Effect of C_8mimPF_6 concentration on latex conversion. Following the observation of a temperature effect in the presence of C_8mimPF_6, it is necessary to investigate different concentrations of C_8mimPF_6, ranging from 0 wt% to 30 wt%, on miniemulsion polymerization initiated by 1.0 mol% AIBN at 40 °C in order to gain a better understanding of the role of C_8mimPF_6 in miniemulsion polymerization.

From Fig. 3, it can be observed that in the absence of C_8mimPF_6, no reaction occurs. However, adding as little as 1 wt% C_8mimPF_6 can trigger the reaction and give a final latex conversion of 83.1%. Different concentrations of C_8mimPF_6 show very different trends with respect to the conversion of latex, with 83.1% conversion achieved for 1 wt% C_8mimPF_6 in 240 min, as compared to 89.5% conversion with 10 wt% C_8mimPF_6 in 150 min. The reaction rate increases with increasing C_8mimPF_6 concentration up until 10 wt%, however, a further increase in C_8mimPF_6 concentration results in a decrease in reaction rate, although the final conversion, around 89.0%, remains similar.

The marked increase in reaction rate in the presence of 1 wt% C_8mimPF_6 might be attributed to two reasons: (1) it is possible that AIBN could not be decomposed at a temperature as low as 40 °C, leading to no reaction occurring

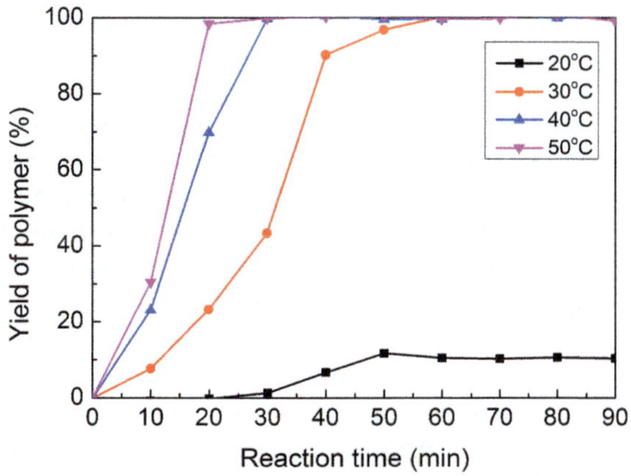

Fig. 2 Yield of polymer as a function of reaction time for miniemulsions containing 10 wt% C_8mimPF_6, after polymerization was initiated by H_2O_2/Vc at different temperatures.

when C_8mimPF_6 was absent; however, this decomposition may have been enhanced in the presence of C_8mimPF_6; or (2) C_8mimPF_6 may efficiently accelerate chain propagation rate, as well as reduce the termination rate of free radical polymerization, which has been observed for another RTIL with a similar structure, C_4mimPF_6.[16,17]

In order to explore whether the decomposition of AIBN would occur at 40 °C, polymerization in the absence of C_8mimPF_6 was extended to 72 h. A latex conversion value of more than 80% was achieved after 72 h. This implies that radicals can be generated by the decomposition of AIBN even at 40 °C, causing chain propagation. On the other hand, Thurecht *et al.* studied the decomposition of AIBN in C_4mimPF_6 and found that the decomposition rate constant was similar to that recorded in organic solvents such as toluene and xylene.[19] This indicates that the increase in reaction rate observed in Fig. 3 in the presence of C_8mimPF_6 is not likely to be attributable to an enhancement in the decomposition of AIBN at

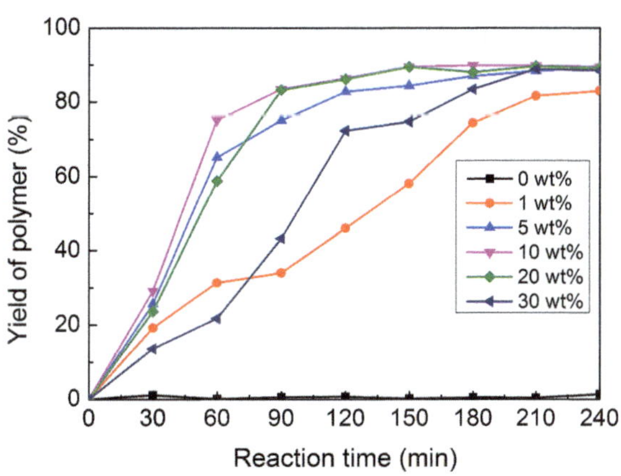

Fig. 3 Yield of polymer as a function of reaction time for miniemulsions containing different concentrations of C_8mimPF_6, after polymerization was initiated by AIBN at 40 °C.

the low temperature of 40 °C, but instead may be due to other free radical polymerization processes, *e.g.* chain propagation and chain termination. Exact details are yet to be uncovered. Thurecht *et al.* reported that when polymerization of MMA in 1-ethyl-3-methylimidazolium ethyl sulphate ($C_2mimEtSO_4$) was initiated by AIBN, some MMA molecules were separated into extremely small domains by the barrier of the $C_2mimEtSO_4$ domain, and that radicals could be protected in the $C_2mimEtSO_4$ domain. A partition was present between radicals protected in the $C_2mimEtSO_4$ domain and radicals formed in the monomer domains, and thus chain termination could be effectively prevented.[19] This could also be the case here, and when there are fewer radicals generated from AIBN at 40 °C, the protection of radicals from chain termination might be more important. Thus, more radicals may be able to coexist under the protection of C_8mimPF_6. In addition, RTILs can accelerate the rate of chain propagation.[16,17] Radical protection together with acceleration of chain propagation in the presence of C_8mimPF_6 might contribute to the large increase in reaction rate observed when the system contains only 1 wt% C_8mimPF_6.

In order to further determine the mechanism for the C_8mimPF_6 effect in the polymerization reaction, some basic kinetic theories for miniemulsion polymerization are worth considering. In miniemulsion polymerization, its process can be divided into two stages. At stage I, droplets transform into particles through droplet nucleation. The efficiency of droplet nucleation is normally very high. However, micellar nucleation and homogenization nucleation may still occur. At stage II, the effect of micellar nucleation and homogenization nucleation fades down, thus the number of particles in the aqueous phase would remain constant. Monomers stored in particles are continuously consumed with the progress of the polymerization process. Assuming the number of particles per unit volume of the aqueous phase remains constant during polymerization, the polymerization rate at stage II can be expressed as eqn (1):[32]

$$R_p = \frac{k_p[\mathrm{M}]\bar{n}N}{N_A}, \tag{1}$$

where R_p is the rate of polymerization; k_p is the chain propagation rate constant; [M] is the monomer concentration in the droplet/particle; $\bar{n}$ is the average effective radical number per droplet/particle; N is the number of droplets/particles per unit volume of aqueous phase in the miniemulsion; N_A is Avogadro's number.

From eqn (1), as N_A is a constant, the polymerization rate, R_p, is directly proportional to k_p, [M], $\bar{n}$ and N. The impact of C_8mimPF_6 on the individual terms could be further explicated.

For example, even by adding 1 wt% of C_8mimPF_6, the system may drastically increase the chain propagation rate (k_p). It may also cause a reduction in the rate of the chain termination resulting in more radicals coexisting in droplet/particle form, hence the number of radicals in droplets/particles, $\bar{n}$, would increase. The number of droplets/particles per unit volume of aqueous phase, N, which is inversely proportional to the droplet/particle size would also increase because of the much smaller droplet size produced by 1 wt% of C_8mimPF_6 shown in Fig. 12. With almost the same monomer concentration in droplets/particles [M], it is not surprising to observe a significant increase in the reaction rate in the presence of 1 wt% C_8mimPF_6.

As C_8mimPF_6 concentrations increase, other factors may also come into play. For example in the case of a miniemulsion containing above 20 wt% C_8mimPF_6, the volume of the oil phase is fixed so the concentration of monomer, [M] may be reduced. Likewise, the concentration of initiator (AIBN or H_2O_2/Vc), which had the same proportion as monomer being used, could be reduced significantly leading to a reduction in the rate of radical formation and hence the number of radicals in particles, $\bar{n}$ could be decreased. The impact of C_8mimPF_6 on the polymerization reaction rate may be a combined effect of k_p, [M], $\bar{n}$ and N. At the moment, we cannot fully quantify the term, k_p and $\bar{n}$, however, considering the change of k_p would not be significant when we increased the concentration of C_8mimPF_6 from 1 wt% to 30 wt%, we could assume they are constant. The change of $\bar{n}$ has a similar effect as the change of [M]. Thus the polymerization rate, R_p could be related to monomer concentration [M] and the number of particle, N, expressed as eqn (2):

$$R_p \propto [M]N, \tag{2}$$

where [M] can be calculated using the initial concentration of MMA added in the oil phase, and N can be calculated from eqn (3).

$$N = \frac{V_{oil}}{\frac{4}{3}\pi r^3 V_{water}}, \tag{3}$$

Since the oil phase volume, V_{oil} and the aqueous phase volume, V_{water} are fixed, the number of particles, N, is inversely proportional to the particle/droplet radius, r, (taken from 30 minutes, when the droplet/particle size become constant).

Based on eqn (2), the variation of the dimensional group of [M]N would reflect the change of R_p. [M]N as function of C_8mimPF_6 concentration is shown in Fig. 4.

It has been observed that the dimensional group [M]N at 1 wt% of C_8mimPF_6 is equal to 9×10^{-17} mol L^{-2} which is a much higher level and reaches the peak at 5 wt% of C_8mimPF_6 followed by a sharp decline as the concentration of C_8mimPF_6 increased from 10 wt% to 30 wt%. This trend implied that the polymerization rate, R_p, would change in a similar manner and this is consistent with the reaction rate reduction at a similar liquid condition as observed in Fig. 3.

To extend our understanding of the impact of C_8mimPF_6 on reaction rates, the hydrophilic type initiator, H_2O_2/Vc, (a mixture of 1.0 mol% H_2O_2 and 1.3 mol% Vc based on monomer), was used to replace AIBN with other conditions kept the same. The conversions of latex during miniemulsion polymerization at different concentrations of C_8mimPF_6 ranging from 0 wt% to 30 wt% were evaluated (Fig. 5). A 100% conversion of latex is achieved at all C_8mimPF_6 concentration ranges being studied. The time taken to complete the reaction is increased from 20 min to 50 min as the concentration of C_8mimPF_6 increases from 0 to 30 wt%. The overall reaction rate is ranked according to the C_8mimPF_6 concentrations as follows: 0 wt%, 1 wt% > 5 wt%, 10 wt% > 20 wt%, 30 wt%.

Fig. 5 exhibits very different effects of C_8mimPF_6 on the reaction rate of miniemulsion polymerization as compared to the one with hydrophobic AIBN as an initiator. By comparison, reaction rates between 0 wt% C_8mimPF_6 and 1 wt% C_8mimPF_6 (Fig. 5) showed no apparent increase. The different effects in the

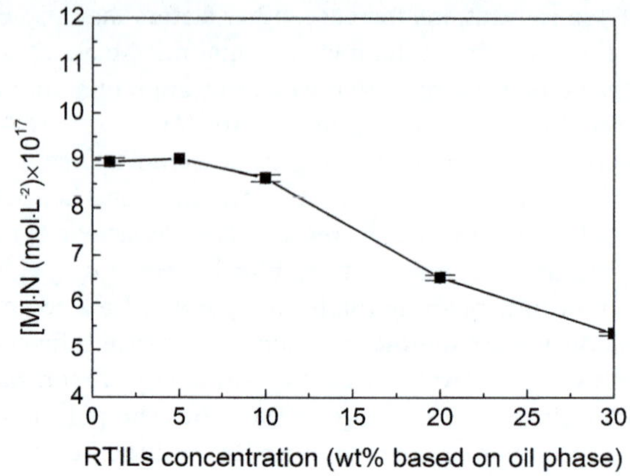

Fig. 4 Dimensional group [M]*N* as a function of C_8mimPF_6 concentration for miniemulsions initiated by AIBN.

presence of 1 wt% C_8mimPF_6 might be attributed to the fact that H_2O_2/Vc is hydrophilic. Miniemulsion polymerization can only occur when radicals generated in the aqueous phase transport to the reactant site, MMA, which is present in the oil phase. Hence, mass transfer of radicals between the dispersed phase and the continuous phase may play a more important role during the reaction process.[33] The number of radicals available for reaction in the droplets/particles for the H_2O_2/Vc system might be determined by the entry of radicals from the water phase. In other words, the interfacial barrier or resistance which is related to the interfacial structure between the water phase and droplet/particle could play a key role in the radical's transfer process. Our early research with similar liquid formula suggested that by adding 1 wt% of C_8mimPF_6 in the MMA oil phase caused a marked reduction in oil/water interfacial tension, from 3.94 mN m^{-1} to 0.31 mN m^{-1} (shown in Table 3) which was attributed to the synergistic effect between the C_8mim^+ cation and SDSO anion, and hence resulted in a tighter

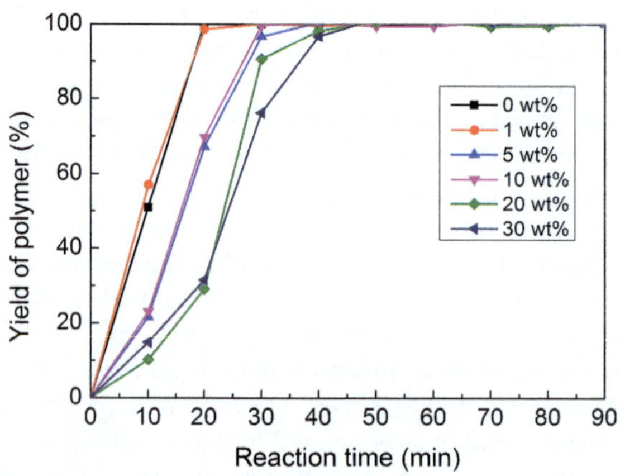

Fig. 5 Yield of polymer as a function of reaction time for miniemulsions containing different C_8mimPF_6 concentrations initiated by H_2O_2/Vc.

interfacial space alignment.[20] It has been reported in open literature that the tighter interfacial structure may reduce the entry rate of radicals into droplets.[34–36] Thus it would be expected that the number of radicals in droplets/particles, $\bar{n}$, would be reduced. Such negative effects could cancel the other positive kinetic effects, such as radical chain propagation and chain termination rate. This could be the main reason why the presence of 1 wt% of C_8mimPF_6 in miniemulsion polymerization initiated by AIBN exhibited a positive effect on the reaction rate whilst no apparent effect on reaction rate was detected for the reaction initiated by H_2O_2/Vc.

For polymerization with concentrations of C_8mimPF_6 higher than 1 wt%, the variation of interfacial tension is insignificant (Table 3), thus the difference of interfacial resistance for radical transportation may be treated as constant. With similar assumptions for k_p and $\bar{n}$, as mentioned in the discussion of Fig. 4, we may use dimensional group [M]N to evaluate the variation of reaction rate with different C_8mimPF_6 concentrations from 1 wt% to 30 wt%. Results shown in Fig. 6 exhibit a sharp decline on [M]N at C_8mimPF_6 concentration ranging from 1 wt% to 5 wt% and 10 wt% to 20 wt% whilst the changes in between the curves are rather flat. This trend appears to be in good agreement with the variation of reaction rate in the same liquid system. This confirms the mixed effect of C_8mimPF_6 on reaction rate with different types of initiator.

3.2 Factors affecting droplet/particle size of the miniemulsion during polymerization

In latex coating products, particle size is one of the important parameters to control product quality. To some extent, monitoring the droplet/particle size during polymerization could help to understand the mechanism of polymerization. In the following section, we will explicate a series of results for droplet/particle size with effects of temperature and C_8mimPF_6 concentration during polymerization.

3.2.1 Effect of temperature on droplet/particle size of miniemulsion during polymerization. To illustrate the mechanism for the temperature effect, the formula of miniemulsion and polymerization conditions were the same as the ones when we studied the effect of temperature on the conversion of latex.

Fig. 7 illustrates the variation of droplet/particle sizes for miniemulsion initiated by H_2O_2/Vc at different temperatures.

Table 3 Interfacial tensions between water phase and oil phase containing different concentrations of C_8mimPF_6 at 40 °C

C_8mimPF_6 (wt%)	Interfacial tension (mN m^{-1})
0	3.94
1	0.31
5	0.21
10	0.15
20	0.22
30	—

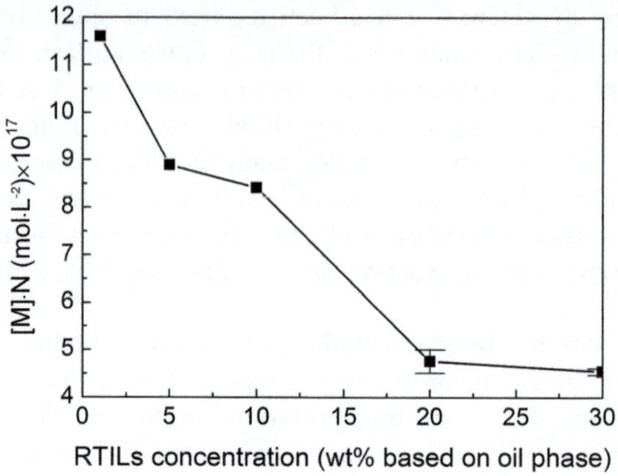

Fig. 6 Dimensional group [M]*N* as a function of C_8mimPF_6 concentration for miniemulsions initiated by H_2O_2/Vc.

A marked reduction of droplet/particle size is observed soon after the reaction took place, *e.g.* 10 min for 30 to 50 °C, 40 min for 20 °C. As the reaction progresses, the change in droplet/particle size behaves differently. *E.g.* at 20 °C, 30 °C and 40 °C, droplet/particle size remains unchanged. while at 50 °C, a sharp increase in droplet/particle size is found after 20 min, then the droplet/particle size remains unchanged until the reaction is completed. This suggests that in addition to the properties of C_8mimPF_6, which affected the initial droplet size of miniemulsion, the final particle size of the latex would depend on both process conditions such as temperature and the reaction mechanism.

A sharp increase in size at 50 °C was probably attributed to the high frequency of droplet/particle movement at 50 °C, hence the opportunity of droplet/particle collision would increase which induced droplet/particle coalescence. Similar observations have been also found for miniemulsion initiated by AIBN (Fig. 10), which shows the strong dependency of droplet size on temperature. This

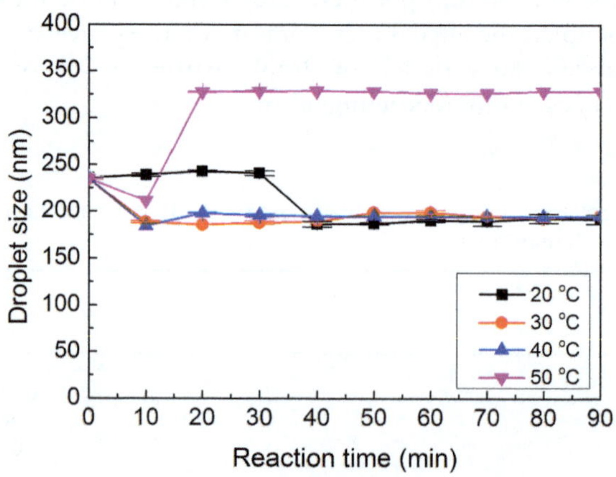

Fig. 7 Variation of droplet/particle sizes as a function of reaction time for miniemulsion containing 10 wt% C_8mimPF_6 at different temperatures initiated by H_2O_2/Vc.

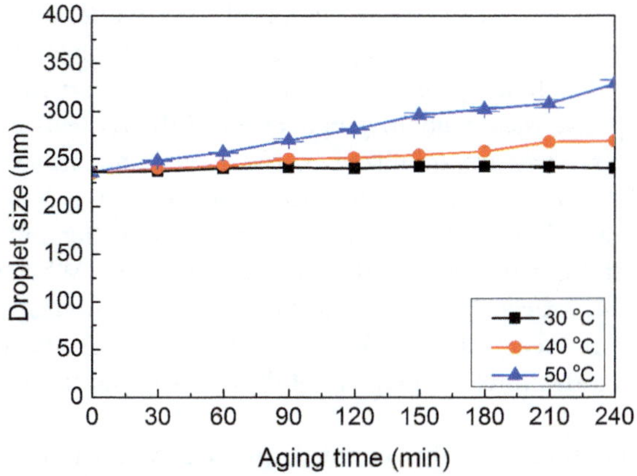

Fig. 8 Variation of droplet sizes as a function of storage time for miniemulsion containing 10 wt% C_8mimPF_6 at different temperatures in the absence of initiator.

indicates that the miniemulsion is more unstable at a higher temperature, *e.g.* 50 °C. Such instability caused by temperature can even be observed without polymerization. Fig. 8 shows the variation of droplet sizes as a function of storage time for miniemulsion containing 10 wt% C_8mimPF_6 at different temperatures in the absence of initiator. It is observed that the size of the droplet in the miniemulsion increased by 40% after being stored at 50 °C for 4 hours, whilst at 40 °C and 30 °C there was little change oin droplet size.

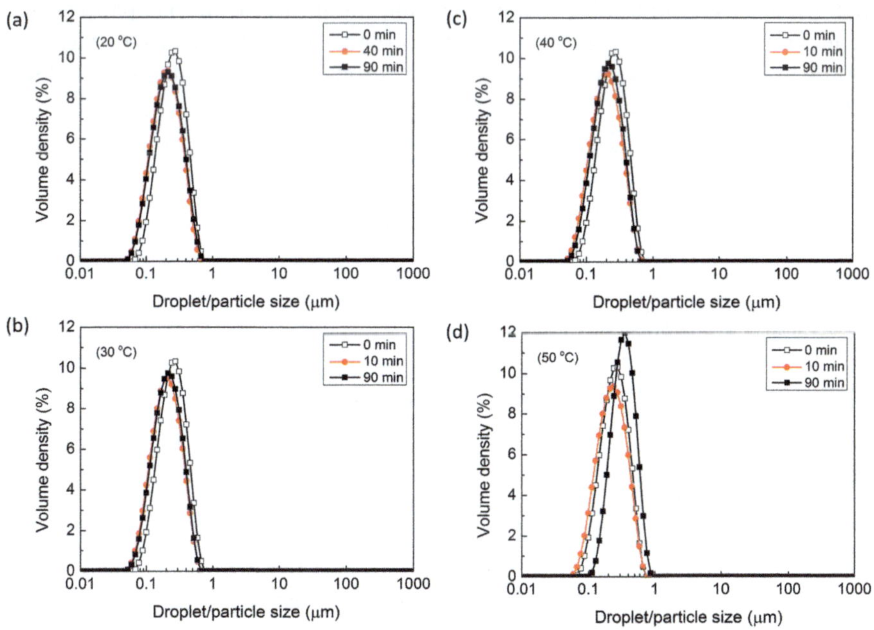

Fig. 9 Droplet/particle size distribution at different reaction time for miniemulsion containing 10 wt% C_8mimPF_6 at different temperatures initiated by H_2O_2/Vc. (a) 20 °C; (b) 30 °C; (c) 40 °C; (d) 50 °C.

The reduction in droplet size after the initial reaction period shown in Fig. 7 was probably attributed to the homogeneous nucleation or micellar nucleation mechanism.[12,37,38] For homogeneous nucleation, primary particles are generated in the aqueous phase *via* the chain propagation of the radicals with monomers dissolved in water. They are stabilized by the surfactant dissolved in water or released from the droplet surface. As to micellar nucleation, polymerization takes place in micelles and forms primary particles. In both micellar and homogeneous cases, a smaller size of particles would be quickly generated. When the number of such smaller latex particles increases significantly, the mean droplet/particle size is expected to be reduced. The droplet size distribution results shown in Fig. 9 also demonstrated the shift on the peak of droplet size distribution towards the smaller sizes.

For hydrophilic initiators, this is not surprising. As for micellar or homogeneous nucleation, both initiation and chain propagation requires radicals present in the water phase and this could be easily achieved with a hydrophilic initiator, such as H_2O_2/Vc. Similar inferences could also be made about miniemulsion polymerization of styrene initiated by the hydrophilic initiator sodium persulfate.[37]

In order to evaluate the effect of the nucleation mechanism on droplet/particle size, H_2O_2/Vc was replaced by the hydrophobic initiator, AIBN. The investigation of the droplet/particle size as a function of reaction time at similar temperature ranging from 30 °C to 50 °C under similar miniemulsion formula was conducted. The results are illustrated in Fig. 10. A small reduction in droplet/particle size was observed for both 50 °C and 40 °C in the initial 30 min. As the reaction progresses further, the droplet/particle size at 40 °C remains unchanged, while the droplet/particle size at 50 °C increases sharply, then remains constant after 60 min until the reaction is terminated. For miniemulsions at 30 °C, the droplet/particle size remains unchanged though the whole process may be attributed to the fact that nearly no reaction occurs. This indicates strong reaction influence on final particles size.

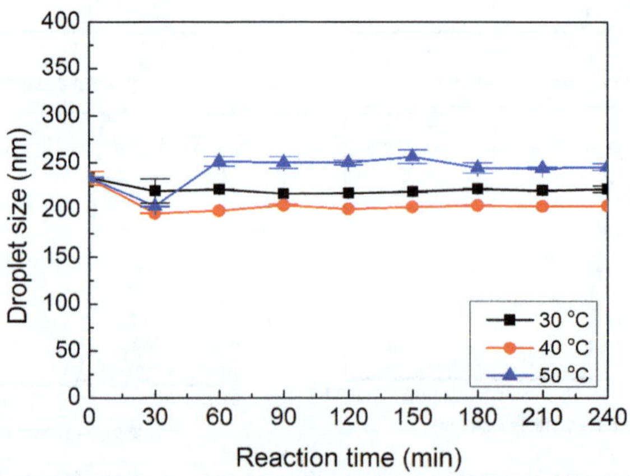

Fig. 10 Variation of droplet/particle sizes as a function of reaction time for miniemulsion containing 10 wt% C_8mimPF_6 at different temperatures initiated by AIBN.

Fig. 11 shows the droplet/particle size distribution for miniemulsions containing 10 wt% C_8mimPF_6 at different temperatures initiated by AIBN. It is found that for the reaction temperature at 50 °C, after 240 min reaction, a more multimodal distribution appears, whilst at 30 °C and 40 °C, only unimodal appears throughout the reaction. Some larger droplets/particles appear ranging from a few μm to 100 μm, result in an increase in mean droplet/particle size being observed in Fig. 10, though the peak distribution remained almost unchanged since the reaction started. This is a clear indication of an early stage of coalescence of droplets/particles. Compared with Fig. 10, it might be postulated that such multi-modal distribution due to the coalescence of droplets might occur at 60 min after the reaction and be kept constant until the end of reaction. This implies that, in addition to the influence of the reaction, the final latex particle size could be affected by the coalescence of droplet/particle induced by the temperature effect. The results appear to be consistent with ones observed in Fig. 7, where H_2O_2/Vc is used as the initiator.

By observing the main peaks at different reaction temperatures shown in Fig. 11, it can be seen that both size range and distribution peak are almost identical, which indicates different nucleation mechanisms compared to the miniemulsion initiated by H_2O_2/Vc. This suggests that droplet nucleation could play a key role for polymerization initiated by AIBN. This is not surprising since AIBN is hydrophobic. It would be expected that decomposition of AIBN to produce radicals should take place easily in the oil phase, where the monomer MMA was presented. Thus it would favour droplet nucleation, where initiation and chain propagation would occur in oil droplets. The final particles of latex would be copied from the oil droplets till termination of the reaction. Thus particle size remained constant.

On the other hand, the small reduction in mean droplet size for polymerization initiated by AIBN, suggests that it is possible that the micelle or homogeneous nucleation also occurred simultaneously. Possible explanations could be that the free radicals necessary for micellar or homogeneous nucleation were generated from the small fraction of hydrophobic initiator dissolved in the water phase.[39,40] Another source could be radicals desorbed from the droplet/particle and resided in the water phase, as the solubility of radical decomposed from AIBN in the water phase would increase significantly. It has been observed that isobutyronitrile which has the same structure as radicals generated from AIBN but in its molecular state is far more hydrophilic as compared with AIBN.[41] Thus it is possible that radicals generated from AIBN may diffuse quickly from droplets into the aqueous phase after the reaction started and they became the source of radicals for micellar or homogeneous nucleation.

3.2.2 Effect of C_8mimPF_6 concentration on droplet/particle size of miniemulsion during polymerization. In order to investigate the underlying mechanism for the C_8mimPF_6 effect on droplet/particle size of miniemulsion during polymerization, experiments were conducted using the same formula of miniemulsion and polymerization conditions as the experiments investigating the effect of C_8mimPF_6 on the conversion of latex.

Fig. 12 shows the variation of droplet/particle sizes as a function of reaction time for miniemulsions containing different C_8mimPF_6 concentrations initiated by AIBN. The reduction in droplet/particle size for miniemulsions containing 1, 5, 10, 20, 30 wt% of C_8mimPF_6 after the initial 30 min of starting the reaction is

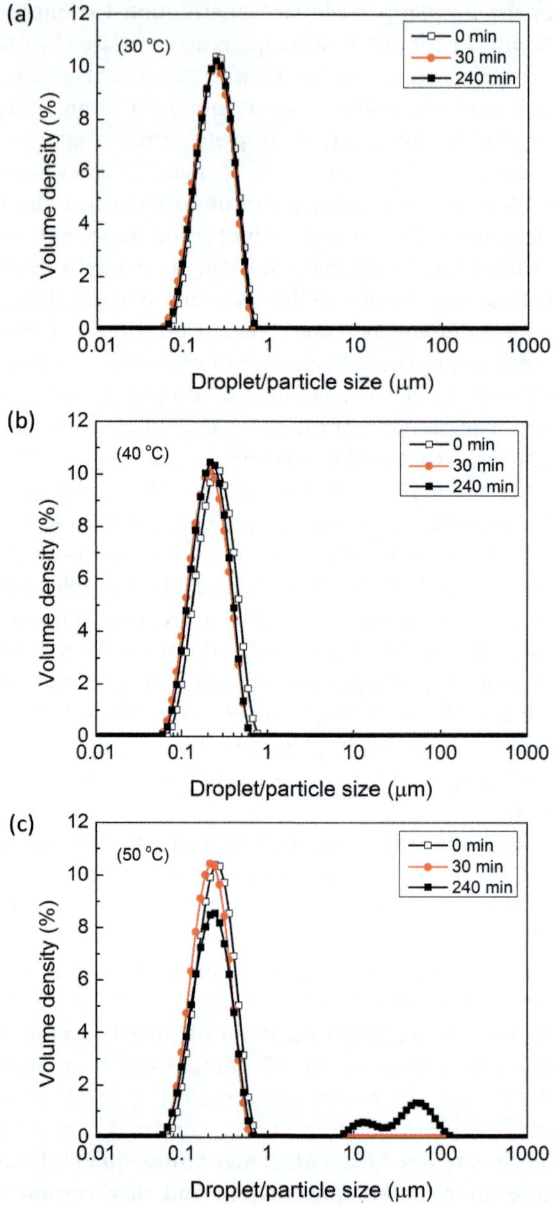

Fig. 11 Droplet/particle size distribution at different reaction times for miniemulsions containing 10 wt% C_8mimPF_6 at different temperatures initiated by AIBN. (a) 30 °C; (b) 40 °C; (c) 50 °C.

observed. This may be attributed to micellar or homogeneous nucleation as described in Section 3.2.1. With the progression of the reaction, the droplet/particle size for the miniemulsions containing 1 wt%, 5 wt%, and 10 wt% C_8mimPF_6 remains unchanged. This suggests that the droplet nucleation is a more dominating factor to influence the droplet/particle size, which is consistent with the one observed in Section 3.2.1. Such observations are very different to

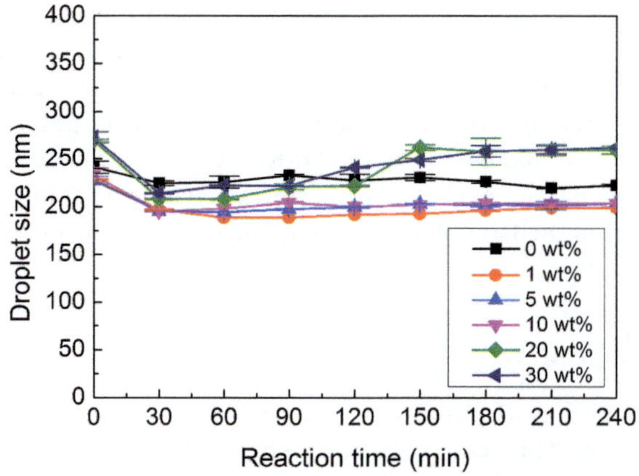

Fig. 12 Variation of droplet/particle sizes as a function of reaction time for miniemulsions containing different C_8mimPF_6 concentrations initiated by AIBN.

those without reactions, in which the change of C_8mimPF_6 concentration would cause remarkable differences in the rate of increase of droplet size during storage. Details about this work were reported in our early work.[20] Fig. 12 shows that the droplet/particle size increases significantly for miniemulsions with 20 wt% and 30 wt% C_8mimPF_6, as the reaction progresses. It is worth mentioning that with the increase in storage time (where no reaction took place), the droplet size of the miniemulsions containing higher concentrations of C_8mimPF_6 also increased to a similar magnitude.[20] This indicates that although the droplet nucleation appeared to be dominating, the final particle size of the latex was strongly influenced by the droplet/particle size at the early stage of the reaction and the storage stability.

Fig. 13 shows the variation of droplet/particle sizes as a function of reaction time for miniemulsions containing different C_8mimPF_6 concentrations initiated by H_2O_2/Vc. The decreasing trends are observed for all C_8mimPF_6 concentrations

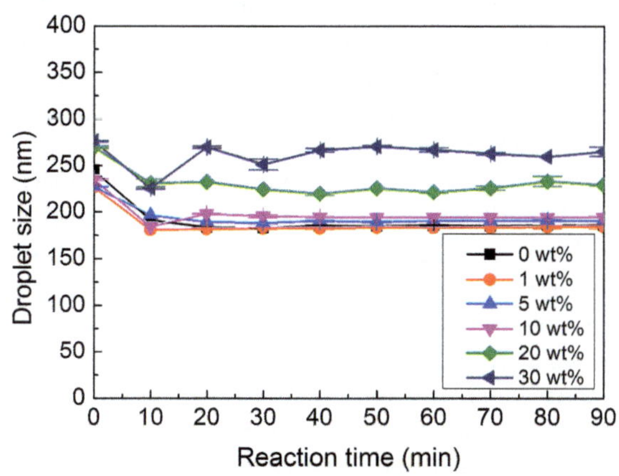

Fig. 13 Variation of droplet/particle sizes as a function of reaction time for miniemulsions containing different C_8mimPF_6 concentrations initiated by H_2O_2/Vc.

for the initial 10 min, which confirm the occurrence of micellar or homogeneous nucleation during reaction. As the reaction progresses, the droplet/particle size for miniemulsions containing different concentrations of C_8mimPF_6 ranging from 0 to 20 wt% remains constant, while at 30 wt% concentration of C_8mimPF_6, droplet/particle size increases significantly due to droplet coalescence. This confirms the dependency of the intrinsic storage stability of miniemulsion on the final particle size of the latex, which was supported by the system where AIBN was used as initiator.

4 Conclusions

A new latex containing C_8mimPF_6, a type of RTILs, has been developed for low VOC latex coatings. This was achieved by encapsulating C_8mimPF_6 inside particles *via* miniemulsion polymerization. Two types of initiators, hydrophobic AIBN and hydrophilic H_2O_2/Vc were chosen as target initiators. Systematic investigations on various factors including initiator type, initiator concentration, temperature, and C_8mimPF_6 concentration on conversion of latex and droplet/particle size of miniemulsion during polymerization were carried out.

It has been found that the conversion of latex and the reaction rate strongly depended on temperature, C_8mimPF_6 concentration and type of initiator being used. Higher temperatures resulted in higher reaction rates regardless of any initiator being used. For AIBN, the presence of C_8mimPF_6, reaction efficiency improved significantly at concentrations as low as 1 wt% C_8mimPF_6. A much higher conversion could also be found at 40 °C which was beyond the normal effective decomposition temperature range of AIBN (50 °C to 90 °C). The effect of RTILs can be attributed to the faster chain propagation rate, slower chain termination rate caused by radical protection function of RTILs. At a high concentration of C_8mimPF_6, *e.g.* 20 wt% and 30 wt%, combination effects such as monomer concentration in the droplet/particle, the average radical number per droplet/particle, and number of droplets/particles in miniemulsion all decreased with the increase of C_8mimPF_6 concentration which counter acted against the positive effect of C_8mimPF_6 on reaction rate, leading to the decrease in overall reaction rate.

For polymerization initiated by hydrophilic H_2O_2/Vc, a different behavior in terms of reaction rate was observed as compared to systems where AIBN was used. The presence of C_8mimPF_6 had a negative effect on reaction rates as compared to the systems without C_8mimPF_6 though the final conversions of latex were similar. It appears that the higher the concentration of C_8mimPF_6 in the droplet, the slower the reaction progress. We postulated that for polymerization of MMA, when the hydrophilic initiator was used, because particle formation was subject to generation of radicals, transportation of radicals from the water phase to droplet where monomer was present. Thus the overall reaction rate not only depended on the generation of radicals from decomposition of initiator, but also the transfer of radicals from the water phase to the reaction site of MMA, which was in droplet form. Such transfer could be the limiting step for the overall reaction. In this case, the presence of C_8mimPF_6, especially at lower concentrations, could result in a much tighter space alignment at the oil/water interface caused by the synergistic effect between C_8mim^+ cation and SDSO anion. The tighter interfacial structure would increase the transfer resistance for radicals entering the droplet, thus significantly slowing down the entry rate of radicals into the droplet. In this case, the reaction rate depended on effective transfer of radicals from the water phase to the droplet.

The simultaneous measurement on droplet/particle size during polymerization provided extra evidence on the effect of temperature and C_8mimPF_6 concentration and reaction mechanism, which associated with the change of droplet/particle size during reaction. It can be concluded that the final particle size of the latex strongly depended on the reaction mechanism, temperature and C_8mimPF_6 concentration. Higher temperatures (50 °C) and higher C_8mimPF_6 concentrations (20 wt% and 30 wt%) promoted droplet coalescence causing an increase in the particle size of the latex regardless of the type of initiator used. At lower temperatures (40 °C) and lower concentrations of C_8mimPF_6, where more stable droplets could be provided, the particle size of the latex tended to be influenced by the reaction mechanism, *e.g.* when homogeneous or micelle nucleation was important during polymerization initiated by a hydrophilic initiator, H_2O_2/Vc, a marked reduction in droplet/particle size compared to the initial droplet size could be obtained whilst for the reaction initiated by hydrophobic AIBN, the droplet/particle size reduction could be minimum due to different mechanism, *i.e.* droplet nucleation took place more frequently with a hydrophobic initiator.

Acknowledgements

One of the authors would like to acknowledge the Low Carbon Automation Manufacture Innovation Team 2011B81006 for PhD studentship and the Ningbo International Cooperation Program 2014D10021 for funding support.

Notes and references

1 Y.-M. Chang, W.-H. Hu, W.-B. Fang, S.-S. Chen, C.-T. Chang and H.-W. Ching, *J. Air Waste Manage. Assoc.*, 2011, **61**, 35–45.
2 M. P. Scott, C. S. Brazel, M. G. Benton, J. W. Mays, J. D. Holbrey and R. D. Rogers, *Chem. Commun.*, 2002, 1370–1371.
3 M. M. Mok, X. Liu, Z. Bai, Y. Lei and T. P. Lodge, *Macromolecules*, 2011, **44**, 1016–1025.
4 M. P. Scott, M. Rahman and C. S. Brazel, *Eur. Polym. J.*, 2003, **39**, 1947–1953.
5 A. Noda and M. Watanabe, *Electrochim. Acta*, 2000, **45**, 1265–1270.
6 K. Matsumoto and T. Endo, *Macromolecules*, 2008, **41**, 6981–6986.
7 M. A. B. H. Susan, T. Kaneko, A. Noda and M. Watanabe, *J. Am. Chem. Soc.*, 2005, **127**, 4976–4983.
8 L. Carson, P. K. W. Chau, M. J. Earle, M. A. Gilea, B. F. Gilmore, S. P. Gorman, M. T. McCann and K. R. Seddon, *Green Chem.*, 2009, **11**, 492–497.
9 K. M. Docherty and J. C. F. Kulpa, *Green Chem.*, 2005, **7**, 185–189.
10 S. N. Riduan and Y. Zhang, *Chem. Soc. Rev.*, 2013, **42**, 9055–9070.
11 Q. Ye, T. Gao, F. Wan, B. Yu, X. Pei, F. Zhou and Q. Xue, *J. Mater. Chem.*, 2012, **22**, 13123.
12 J. M. Asua, *Prog. Polym. Sci.*, 2002, **27**, 1283–1346.
13 A. P. Romio, C. Sayer, P. H. H. Araújo, M. Al-Haydari, L. Wu and S. R. P. da Rocha, *Macromol. Chem. Phys.*, 2009, **210**, 747–751.

14 F. R. Steinmacher, N. Bernardy, J. B. Moretto, E. I. Barcelos, P. H. H. Araújo and C. Sayer, *Chem. Eng. Technol.*, 2010, **33**, 1877–1887.

15 C. A. Capeletto, C. Sayer and P. H. H. de Araújo, *Macromol. Symp.*, 2012, **319**, 54–63.

16 S. Harrisson, S. R. Mackenzie and D. M. Haddleton, *Chem. Commun.*, 2002, 2850–2851.

17 S. Harrisson, S. R. Mackenzie and D. M. Haddleton, *Macromolecules*, 2003, **36**, 5072–5075.

18 K. Hong, H. Zhang, J. W. Mays, A. E. Visser, C. S. Brazel, J. D. Holbrey, W. M. Reichert and R. D. Rogers, *Chem. Commun.*, 2002, 1368–1369, DOI: 10.1039/b204319j.

19 K. J. Thurecht, P. N. Gooden, S. Goel, C. Tuck, P. Licence and D. J. Irvine, *Macromolecules*, 2008, **41**, 2814–2820.

20 Y. Kong, B. Hu, K.-L. Choy, X. Li and G. Chen, *Soft Matter*, 2015, **11**, 1293–1302.

21 Q. Zhang, Y. Shi, X. Zhan and F. Chen, *Colloids Surf., A*, 2012, **393**, 17–26.

22 M. Winkelmann and H. P. Schuchmann, *Particuology*, 2011, **9**, 502–505.

23 Y. Wang, Y. Zhang, T. Xia, W. Zhao and W. Yang, *Sol. Energy Mater. Sol. Cells*, 2014, **120**, 481–490.

24 Y. Meliana, L. Suprianti, Y. C. Huang, C. T. Lin and C.-S. Chern, *Colloids Surf., A*, 2011, **389**, 76–81.

25 P. J. Flory, *Principles of polymer chemistry*, Cornell University Press, Ithaca, New York, 1953.

26 M. Kitagawa and Y. Tokiwa, *Carbohydr. Polym.*, 2006, **64**, 218–223.

27 S. Perrier, P. Takolpuckdee and C. A. Mars, *Macromolecules*, 2005, **38**, 2033–2036.

28 F. Svec and J. M. J. Frechet, *Macromolecules*, 1995, **28**, 7580–7582.

29 S. Balke and A. Hamielec, *J. Appl. Polym. Sci.*, 1973, **17**, 905–949.

30 Z. Jia, Z. Wang, C. Xu, J. Liang, B. Wei, D. Wu and S. Zhu, *Mater. Sci. Eng., A*, 1999, **271**, 395–400.

31 T. Yamamoto, *Colloid Polym. Sci.*, 2013, **291**, 2741–2744.

32 N. Bechthold and K. Landfester, *Macromolecules*, 2000, **33**, 4682–4689.

33 C. S. Chern, *Prog. Polym. Sci.*, 2006, **31**, 443–486.

34 L. Gan, C. Chew and I. Lye, *Die Makromolekulare Chemie*, 1992, **193**, 1249–1260.

35 M. Mosca, F. Cuomo, F. Lopez and A. Ceglie, *Food Res. Int.*, 2013, **50**, 377–383.

36 Z. Cao, G. Shan, G. Fevotte, N. Sheibat-Othman and E. Bourgeat-Lami, *Macromolecules*, 2008, **41**, 5166–5173.

37 C.-S. Chern and Y.-C. Liou, *Polymer*, 1999, **40**, 3763–3772.

38 F. J. Schork, Y. Luo, W. Smulders, J. P. Russum, A. Butté and K. Fontenot, *Adv. Polym. Sci.*, 2005, **175**, 129–255.

39 C.-S. Chern and Y.-C. Liou, *J. Polym. Sci., Part A: Polym. Chem.*, 1999, **37**, 2537–2550.

40 P. C. Mørk and Y. Makame, *J. Polym. Sci., Part A: Polym. Chem.*, 1997, **35**, 2347–2354.

41 Y. Shang and G.-R. Shan, *AIChE J.*, 2012, **58**, 3135–3143.

Faraday Discussions

PAPER

Three dimensional ink-jet printing of biomaterials using ionic liquids and co-solvents†

Deshani H. A. T. Gunasekera, SzeLee Kuek, Denis Hasanaj, Yinfeng He, Christopher Tuck, Anna K. Croft and Ricky D. Wildman*

Received 11th December 2015, Accepted 14th January 2016
DOI: 10.1039/c5fd00219b

1-Ethyl-3-methylimidazolium acetate ($[C_2C_1Im][OAc]$) and 1-butyl-3-methylimidazolium acetate ($[C_4C_1Im][OAc]$) have been used as solvents for the dissolution and ink-jet printing of cellulose from 1.0 to 4.8 wt%, mixed with the co-solvents 1-butanol and DMSO. 1-Butanol and DMSO were used as rheological modifiers to ensure consistent printing, with DMSO in the range of 41–47 wt% producing samples within the printable range of a DIMATIX print-head used (printability parameter < 10) at 55 °C, whilst maintaining cellulose solubility. Regeneration of cellulose from printed samples using water was demonstrated, with the resulting structural changes to the cellulose sample assessed by scanning electron microscopy (SEM) and white light interferometry (WLI). These results indicate the potential of biorenewable materials to be used in the 3D additive manufacture process to generate single-component and composite materials.

Introduction

Additive manufacturing (3D printing)[1] is able to produce objects for a wide variety of applications in a material-efficient fashion with little or no waste, and has shown particular benefit in areas of high-value and bespoke manufacture, such as for medical devices.[2] Further enhancements to the manufacturing process can be achieved by sourcing the base materials from biorenewable sources. Biopolymers such as cellulose, lignin, wood *etc.* are particularly promising materials to be used in additive manufacturing, as they are biodegradable, biocompatible, can be sourced from waste streams, and offer interesting physical and chemical properties.[3,4] However, the low solubility of many biopolymers in water and other common solvents has limited their application to the additive manufacture

Faculty of Engineering, University of Nottingham, NG7 2RD, UK. E-mail: Ricky.Wildman@nottingham.ac.uk

† Electronic supplementary information (ESI) available: Sample calculations, apparatus used to fill the DIMATIX materials print cartridge, scanning electron microscopy images, white light interferometry images and data can be found in ESI. See DOI: 10.1039/c5fd00219b

process.[3,5] Ionic liquids have recently been introduced as an effective class of solvent for the dissolution of these biopolymers.[3–6]

Cellulose is the most abundant component of lignocellulose.[7,8] It is a linear syndiotactic polymer with D-glucopyranose (glucose) as the repeating unit attached by β-1,4-glycosidic bonds. Glucose, cellobiose and short oligomers of glucose (degree of polymerisation < 10 units) are soluble in water.[9] However, due to the presence of the strong glycosidic bonds and intra- and inter molecular hydrogen bonds, cellulose does not dissolve in water.[5]

Ionic liquids are defined as liquids that consist entirely of ions and have a melting/glass transition temperature less than 100 °C. An important characteristic of ionic liquids is the ability to vary the cation and anion to produce a wide variety of ionic liquids with different chemical and physical properties. Ionic liquids possess some unique combinations of properties such as negligible vapour pressure, high thermal stability, ionic conductivity, low surface tension and non-flammability, which can be exploited to generate enhanced manufacturing processes. These properties make them suitable to replace traditional solvents.[10] Several reviews have summarised the factors that affect the dissolution of cellulose in different ionic liquids.[3,4,6] These can be used as an initial guideline when selecting ionic liquids for cellulose dissolution.

Different additive manufacturing technologies have been introduced over the past few decades that use a variety of materials, *i.e.* polymers, metals, ceramics and composites to produce parts layer by layer.[2] Material jetting utilising piezo-based ink-jet technology is one of the most promising techniques that can be used to deposit drops continuously (continuous ink-jet printing) or when needed (drop-on-demand ink-jet printing).[11–13] Material jetting can use a variety of materials,[14] is versatile, and printed patterns can be easily changed.[15] These advantages make it a suitable technology to produce three-dimensional objects. Ink-jet printed biomaterials can be used in biomedical applications (tissues and organs)[16] and in smart materials (actuators and sensing system).[17]

Markstedt *et al.*,[18] demonstrated the first use of extrusion printing with cellulose and an ionic liquid ([C_2C_1Im][OAc]) to form three-dimensional objects. Cylinders reaching a maximum height of 25 mm were printed on a vertical agar support (made from agar and water).[18] Haverhals *et al.*,[17] reported simple ink-jet printing of a mixture of [C_2C_1Im][OAc] and acetonitrile onto cellulosic paper and exposed samples to varying conditions of heat and rinsing protocols. They concluded that it was possible to obtain printed materials with exceptional properties that could be used in a wide variety of applications.[17] Recently, similar work conducted by Sweely *et al.*,[19] showed the results of ink-jet printing a mixture of [C_2C_1Im][OAc], carbon black and microcrystalline cellulose on to a paper substrate (100% cotton). They suggested that adding a co-solvent (like acetonitrile, ethanol or DMSO) can both increase the amount of cellulose dissolved in an ionic liquid containing solution and reduce the overall viscosity.[19] This opens up the possibility of utilising such solutions in conventional 3D ink-jet printers, which generally require low (<20 mPa s) viscosities.

This paper describes the identification of mixtures of ionic liquids and organic co-solvents (DMSO and 1-butanol) for formulating materials for ink-jet based 3D printing of cellulose, with a view to identify solutions that could be exploited for other ionic liquid-soluble biopolymers. Rheological properties of the ionic liquid-co-solvent mixtures have been measured and assessed to inform the characteristics required for effective printing. Methods for subsequent removal of the ionic liquid from the final 3D-printed structure have also been evaluated.

Experimental

Chemicals

Ionic liquids: 1-ethyl-3-methylimidazolium acetate ($[C_2C_1Im][OAc]$) (purity ≥ 95%) and 1-butyl-3-methylimidazolium acetate ($[C_4C_1Im][OAc]$) (purity ≥ 95%) were purchased from Sigma Aldrich (United Kingdom) and used as received. Co-solvents: 1-butanol and dimethyl sulfoxide (DMSO) were purchased from Thermo Fisher Scientific (United Kingdom). Microgranular cellulose was purchased from Sigma Aldrich (United Kingdom) and dried at 60 °C in an oven (Lenton) for 3 hours before use. Deionised water was used in all the experiments.

Sample preparation

The mass of cellulose used was calculated by taking 2 and 5% of the ionic liquid mass (outlined in Table 1). The required quantity of cellulose was weighed using a four figure digital balance (Denver Instruments) and placed into a glass vial. The ionic liquid was measured out using a 1 mL ink-jet syringe and was added to the cellulose. To investigate the effect of adding a co-solvent on cellulose dissolution and the physical properties of the resultant solution, two different percentages of cellulose (calculated as above) and three different percentages of co-solvent (50, 70 and 90% of ionic liquid mass) were then added to samples containing cellulose and $[C_4C_1Im][OAc]$ (Table 2). The amount of cellulose, ionic liquid and co-solvent in all the samples were also calculated as a wt% (Tables 1 and 2), alongside mole fraction of ionic liquid relative to co-solvent (χ_{IL}) for the mixtures (Table 2). To create the final formulation, the mixtures were stirred using a magnetic stirrer at 800 rpm whilst heating in an oil bath at 100 °C (samples 1 to 4) and 80 °C (samples 5 to 10), and visually observed until no particulate matter remained (refer to ESI† for example calculations).

Viscosity and surface tension measurements

Viscosity measurements were carried out using a Malvern Kinexus rheometer, using a cone-plate geometry (angle 4°, diameter 40 mm, gap size 0.15 mm) and a cup and bob (cup size C14 (DIN), bob diameter 14 mm). Cone-plate geometry was used to measure the viscosities of $[C_2C_1Im][OAc]$, $[C_4C_1Im][OAc]$, deionised water and for all the samples in Tables 1 and 2, and cup and bob was used to measure the viscosities of DMSO and 1-butanol. Viscosities of the pure chemicals

Table 1 Formulations containing cellulose and ionic liquid

Sample	Cellulose (g)	$[C_4C_1Im][OAc]$ (g)	$[C_2C_1Im][OAc]$ (g)	Cellulose (%)	Cellulose (wt%)	Ionic liquid (wt%)
1	0.1688	8.4		2.0	2.0	98
2	0.4220	8.4		5.0	4.8	95
3	0.1643		8.2	2.0	2.0	98
4	0.4108		8.2	5.0	4.8	95

Table 2 Formulations containing cellulose, ionic liquid and co-solvent

Sample	Cellulose (g)	$[C_4C_1Im][OAc]$ (g)	DMSO (g)	1-Butanol (g)	Cellulose (%)	Co-solvent (%, χ_{IL})	Cellulose (wt%)	$[C_4C_1Im][OAc]$ (wt%)	Co-solvent (wt%)
5	0.1266	6.3	3.2		2.0	50, 0.44	1.3	66	33
6	0.1266	6.3	4.4		2.0	70, 0.36	1.2	58	41
7	0.1055	5.3	4.7		2.0	90, 0.30	1.0	52	47
8	0.1266	6.3		3.2	2.0	50, 0.43	1.3	66	33
9	0.1846	3.7	1.8		5.0	50, 0.44	3.2	65	32
10	0.3165	6.3		3.2	5.0	50, 0.43	3.2	65	32

were measured with shear rates of 10 to 1000 s^{-1} at 25 °C, and the viscosities of the mixtures (as identified in Tables 1 and 2) were measured with shear rates of 1 to 1000 s^{-1} at 60 °C. This was because the maximum working temperature for the DIMATIX print head was 70 °C. Viscosity was measured in triplicate and the average viscosity at shear rate 1000 s^{-1} was used, except for sample 2 where the average viscosity was calculated at a shear rate of 631 s^{-1}.

Surface tension of pure chemicals and mixtures were measured at room temperature using a KRUSS DSA100 drop shape analyser. Measurements were taken six times and the average value was recorded.

Ink-jet printing

Filling the DIMATIX materials print cartridge

Ink-jet printing was performed using a FUJIFILM DIMATIX DMP-2831 printer. The cartridge used (10 pL) was purchased from FUJIFILM DIMATIX. It consists of 16 nozzles, and has a nozzle diameter of 21 μm, and a built-in heater that can heat the cartridge up to 70 °C. Jetting is driven by a piezo element with amplitudes that can be varied between 0 and 40 V.

The apparatus used to fill the DIMATIX cartridge are shown in the ESI (Fig. S1†). The required volume of sample was initially measured using a 1 mL ink-jet syringe and transferred to the DIMATIX fluid module using a 5 μm nylon syringe filter (Cole–Parmer) and the fill needle (supplied with the cartridge). The nylon syringe filter was used to prevent any particles from agglomerating in the nozzle. When filling the DIMATIX fluid module, care was taken to reduce the formation of air bubbles by tapping into the fluid module when filling it. The fluid module was filled with 2 mL of sample. Finally, the DIMATIX jetting module was attached to the fluid module and installed into the DIMATIX printer. A cleaning pad (supplied with the cartridge) was also inserted into the DIMATIX printer.

Printing

DIMATIX Drop Manager Software was used to control the printing process. After installing the cartridge, a cleaning cycle was performed to prime the nozzles. This cycle ensures that additional air trapped inside the fluid module and cartridge is removed. Polyethylene terephthalate (PET) and glass were used as the substrates. Printing was carried out with the print head temperature set at 55 °C and droplet spacing of 150 μm (sabre angle 36.2°). Voltage, frequency and the waveform were adjusted to achieve stable drop formation. During printing, when the cleaning cycles were unable to unblock the nozzles, the print head was manually cleaned using a polyester wipe. Ink-jet printed samples were dried for approximately 18 hours under different conditions, one under room temperature and the other in the oven at 50 °C.

Cellulose regeneration

Cellulose was regenerated by adding 0.1 mL of deionised water at room temperature. Deionised water was left on ink-jet printed samples for different time intervals (30 minutes, 1 hour, 1 hour 30 minutes or 2 hours) before drying for approximately 18 hours at 50 °C.

Optical characterisation

Optical characterisation of the ink-jet printed samples was performed using a Nikon ECLIPSE LV100ND. Scanning electron microscope (SEM) images of untreated cellulose were obtained using a Hitachi TM3030 Table Top SEM and for ink-jet printed and regenerated samples a Philips XL30 SEM was used. Surface profile analysis of the ink-jet printed samples was performed using a white light interferometry (WLI) Bruker Contour GT. All ink-jet printed samples analysed using the SEM and WLI were coated with platinum to ensure reduced charging during SEM and provide reflectivity for white light interferometry.

Results and discussion

Rheological properties

Viscosities of $[C_2C_1Im][OAc]$, $[C_4C_1Im][OAc]$ and deionised water were measured using a cone-plate geometry and the viscosities of DMSO and 1-butanol were measured using cup and bob to avoid DMSO and 1-butanol evaporation.

The printability parameter, Z, which is a combination of physical properties of the fluid including density (ρ), viscosity (η) and surface tension (σ) and the characteristic length of the nozzle (d), (21 μm in this case) was calculated using eqn (1). To obtain stable drops during printing, it is recommended to have the printability parameter between 1 and 10.[13]

$$Z = \frac{\sqrt{\sigma \rho d}}{\eta} \tag{1}$$

Both $[C_2C_1Im][OAc]$ and $[C_4C_1Im][OAc]$ have very high viscosities that exceed the printing range for the DIMATIX ink-jet printer (12–18 mPa s). The surface tension of $[C_4C_1Im][OAc]$ is within the DIMATIX ink-jet printing range (28–32 mN m^{-1}). However, $[C_2C_1Im][OAc]$ has a higher value than the recommended range. Overall, these values result in the printability parameters for the two ionic liquids being much less than the acceptable range of 1 to 10 (Table 3).

Table 3 Rheological properties and printability parameter of pure chemicals

Chemical	Density (g cm^{-3})	Viscosity (mPa s)	Viscosity SD (±)	Surface tension (mN m^{-1})	Surface tension SD (±)	Printability parameter, Z
$[C_4C_1Im][OAc]$	1.1	278.00	15.74	30.30	0.78	0.093
$[C_2C_1Im][OAc]$	1.0	85.47	5.03	39.94	0.54	0.34
Deionised water	1.0	2.99	0.31	71.64	0.51	13
DMSO	1.1	4.11	1.03	29.99	1.01	6.4
1-Butanol	0.81	4.31	0.01	30.12	0.91	5.2

Dissolution and rheological properties of samples

The rheological properties of samples containing cellulose and ionic liquids ([C_4C_1Im][OAc] and [C_2C_1Im][OAc]) are shown in Table 4. The viscosity of sample 2 was measured at a lower shear rate of 631 s^{-1} due to a limitation in the rheometer when measuring samples with high viscosity. Sample density was calculated by dividing the total sample mass by the total sample volume (refer to ESI† for example calculations).

The viscosity of the ionic liquid and cellulose solutions were higher than the printing range (12–18 mPa s), even at 60 °C. Whilst increasing the temperature is known to decrease the viscosity,[20,21] temperature modulation in this printing system was limited by the maximum working temperature for the DIMATIX cartridge (70 °C). To decrease the viscosity, different concentrations of co-solvents were added, following the recommendations by Sweely *et al.*[19] As the amount of cellulose was increased, from 2.0 to 4.8 wt%, sample viscosity increased (Table 4). A similar trend was seen in a study conducted by Kosan *et al.*[22]

Cellulose dissolution in [C_4C_1Im][OAc] was further studied in the presence of different co-solvents (DMSO and 1-butanol). Samples containing DMSO dissolved easily compared to samples containing 1-butanol, where a cloudy solution formed, which resulted in a precipitate at room temperature. Therefore, the printability parameter for samples containing 1-butanol (samples 8 and 10) was not calculated.

Initially 33 wt% of DMSO was added and the viscosity was measured, then DMSO concentration was increased from 41 to 47 wt% to bring the sample viscosity within the ink-jet printing range (12–18 mPa s). It can be seen from the rheological properties summarised in Table 5, increasing the concentration of DMSO reduces the viscosity of the samples. For sample 9, viscosity was more than 90 mPa s at 60 °C, which was much higher than the printing range (12–18 mPa s) required.

Table 4 Rheological properties of samples containing ionic liquid and cellulose

Sample	Density (g cm^{-3})	Viscosity (mPa s)	Viscosity SD (±)
1	1.0	163.90	7.58
2	1.0	723.13	49.53
3	1.0	87.30	2.56
4	0.99	290.70	22.21

Table 5 Rheological properties and printability parameter of samples containing [C_4C_1Im][OAc], cellulose and DMSO

Sample	Density (g cm^{-3})	Viscosity (mPa s)	Viscosity SD (±)	Surface tension (mN m^{-1})	Surface tension SD (±)	Printability parameter, Z
5	1.1	28.49	2.12	36.71	0.18	1.0
6	1.1	19.64	1.13	35.58	0.64	1.4
7	1.1	15.66	1.32	36.06	0.41	1.8
9	1.0	91.27	2.21	37.15	0.43	0.31

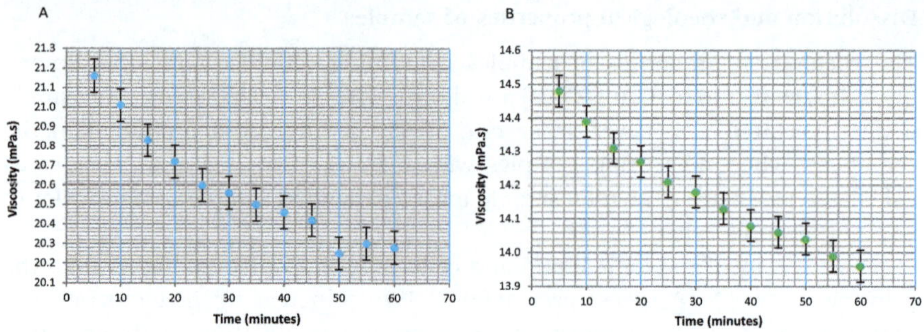

Fig. 1 Viscosity *vs.* time at fixed shear rate and temperature for (A) sample 6 (58 wt% [C_4C_1Im][OAc] + 1.2 wt% cellulose + 41 wt% DMSO) and (B) sample 7 (52 wt% [C_4C_1Im][OAc] + 1.0 wt% cellulose + 47 wt% DMSO).

Comparing the surface tension and viscosity of the as-received chemicals recorded in Table 3 with the samples in Table 5, it can be seen that when mixed to create the formulations given in Table 2, surface tension increases and the viscosity decreases. Surface tension and density recorded in Table 5 remain approximately the same and, the characteristic length of the nozzle being a constant, results in the printability parameter (Z) being mainly affected by the sample viscosity.

Based on Z, samples 6 and 7 (Table 5) were chosen for ink-jet printing. This was because their Z values were within the acceptable range (between 1 and 10) for stable drop formation.

Both the samples showed non-Newtonian behaviour. To further study the rheological behaviour exhibited by these fluids, viscosities of samples 6 and 7 were measured against time at a fixed shear rate (1000 s^{-1}) and temperature (60 °C). From the results (Fig. 1) it can be seen that samples 6 and 7 show thixotropic behaviour, exhibiting shear thinning.[23]

When the viscosity decreases under shearing it becomes easier for the ink to pass through the nozzle.[24] When shear is removed, viscosity increases[25] and the printed pattern is able to retain its shape.[24] This is a useful property of the formulated ink that can be used in ink-jet printing to eject droplets at a viscosity lower than that recorded in Table 5.

Ink-jet printing

When printing sample 6, the nozzles were blocked more easily than for sample 7, possibly due to sample 6 possessing a viscosity higher than the manufacturer's acceptable viscosity range of 12–18 mPa s. During printing, the drop leaves the nozzle as a jet with a long extended tail/ligament attached to the leading drop. The tail/ligament breaks after some time where part of the ink returns to the nozzle and the remainder joins the leading large drop, but on occasion it is possible that smaller satellite drops can form.[12,13] When observing the drops using the drop watcher in the DIMATIX Drop Manager Software, frequency, waveform and voltage were adjusted until acceptable drops that merged with the leading drop were formed. Jetting parameters for sample 7 are summarised in Table 6 and a schematic diagram of the waveform is shown in Fig. 2.

Table 6 Jetting parameters for sample 7 (52 wt% [C_4C_1Im][OAc] +1.0 wt% cellulose + 47 wt% DMSO)

Segment	Level (%)	Slew rate	Duration (μs)
1	0	0.65	3.584
2	100	0.93	3.712
3	27	0.60	3.392
4	40	0.80	0.832
Duration scalar	1		
Width	11.520 μs		
Maximum jetting frequency	1.5 kHz		
Jetting voltage	35 V		

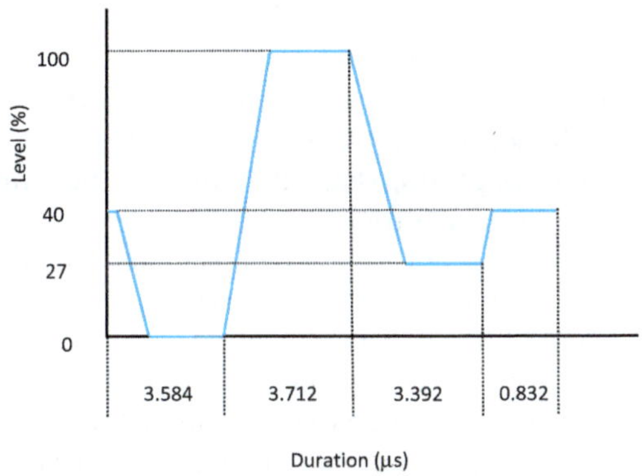

Fig. 2 Schematic diagram of the waveform described in Table 6.

The pattern printed on the PET and glass was a 3 mm × 3 mm square having an array of single drops with a droplet spacing of 150 μm (sabre angle 36.2°) and printed with the print head temperature at 55 °C. Printing was achieved using a single nozzle. Fig. 3 shows the array of drops printed using samples 6 and 7.

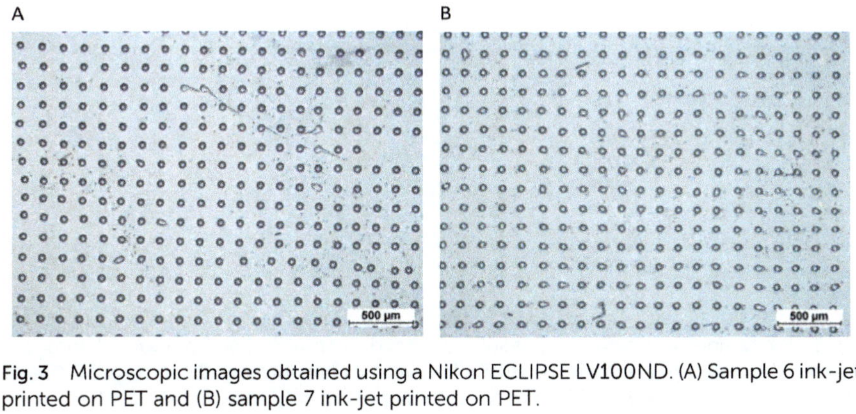

Fig. 3 Microscopic images obtained using a Nikon ECLIPSE LV100ND. (A) Sample 6 ink-jet printed on PET and (B) sample 7 ink-jet printed on PET.

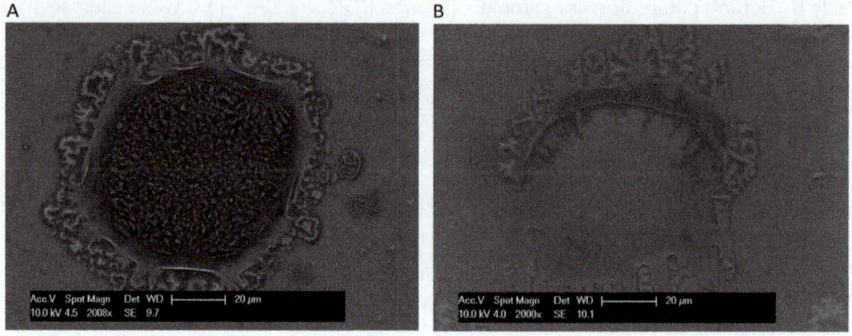

Fig. 4 SEM images obtained from a Philips XL30 scanning electron microscope. (A) Sample 6 was printed on PET and dried for approximately 18 hours at 50 °C and (B) sample 7 was printed on PET and dried for approximately 18 hours at 50 °C.

From the microscopic images obtained, it can be seen that some drops were missing (Fig. 3A), which may be due to a blockage in the nozzle. Small drops were also deposited next to the large drop (Fig. 3B) indicating the formation of smaller ligaments alongside the primary drop.

Scanning electron microscopy images of samples obtained after ink-jet printing samples 6 and 7 and drying for approximately 18 hours at 50 °C are shown in Fig. 4 (other SEM images can be found in ESI Fig. S2†). From the SEM images, it can be seen that there was more material deposited for ink-printed samples using formulation 6 compared to formulation 7. The principle difference between these two samples is the amount of DMSO. The DMSO composition in sample 7 was higher (47 wt%) than in sample 6 (41 wt%).

To further study the distribution of materials on the droplet, surface profile analysis was performed using white light interferometry. The surface profile images indicate that the height of the surface, corresponding to the amount of material deposited, differs (Fig. 5). This was further investigated by calculating the average height of the ink-jet printed droplet obtained from the surface profile data (Table 7). All WLI images and data can be found in ESI.†

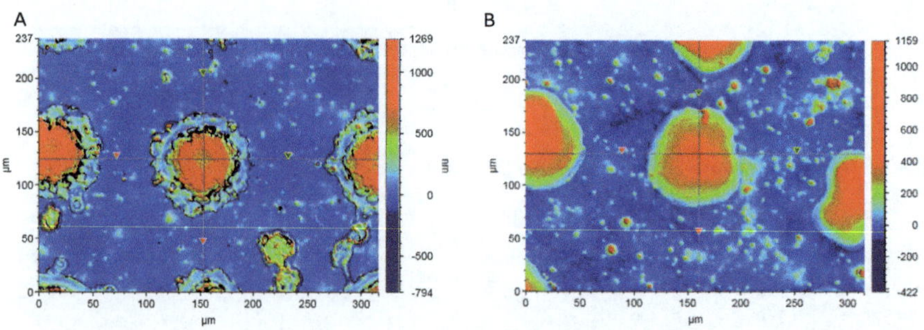

Fig. 5 White light interferometry results obtained from a Bruker Contour GT interferometer. (A) Sample 6 was printed on PET and dried for approximately 18 hours at 50 °C and (B) sample 7 was printed on PET and dried for approximately 18 hours at 50 °C.

Table 7 Average height of a droplet for ink-jet printed samples

Remarks	Average height in X direction (nm)	Average height in Y direction (nm)
Sample 6, dried for approximately 18 hours at room temperature	510	530
Sample 6, dried for approximately 18 hours at 50 °C	680	680
Sample 7, dried for approximately 18 hours at room temperature	180	84
Sample 7, dried for approximately 18 hours at 50 °C	340	380

From Table 7, it can be seen that samples printed using formulation 6 showed a higher height over both X and Y axes, compared to samples printed using formulation 7. This is consistent with the results obtained from the scanning electron microscope.

Cellulose regeneration

Cellulose has previously been regenerated by adding an anti-solvent such as either water,[26–28] ethanol, or acetone.[26] Zavrel *et al.*,[28] explained this phenomenon as solute-displacement, where the water forms a hydrodynamic shell around the ions present in the ionic liquid, preventing it from interacting with cellulose. This leads to the reformation of the hydrogen bonds in cellulose causing it to precipitate and forming a disordered structure.[28] The observed regeneration of cellulose points towards a mechanism to generate 3D structures through an ink-jet based deposition-regeneration process.[17,19]

The regeneration experiments were conducted by adding deionised water and leaving the inkjet printed samples for different time intervals. It was assumed that leaving deionised water for a longer time interval (2 hours) would remove more

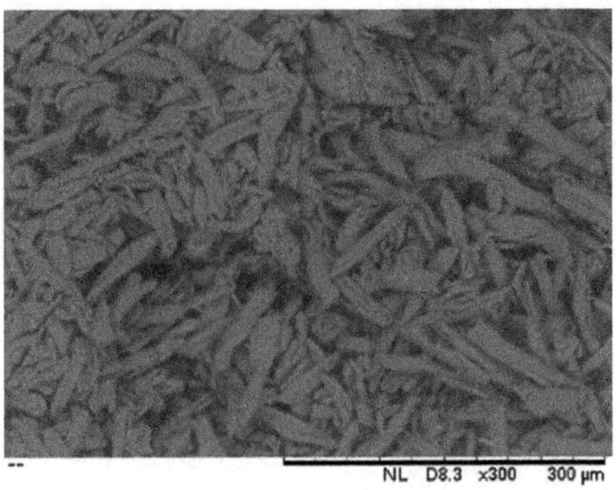

Fig. 6 SEM image obtained from a Hitachi TM3030 scanning electron microscope of untreated cellulose.

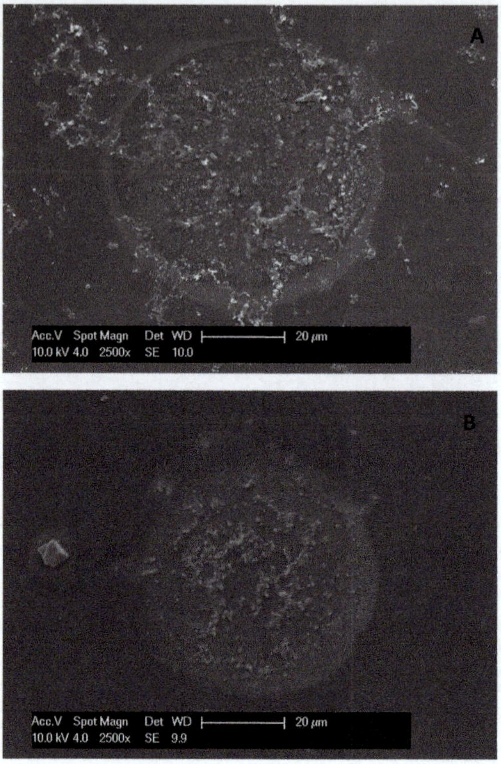

Fig. 7 SEM image obtained from a Philips XL30 scanning electron microscope. The samples were ink-jet printed on PET and regenerated by leaving deionised water for 1 hour and 30 minutes before drying for approximately 18 hours at 50 °C. (A) Sample 6 and (B) sample 7.

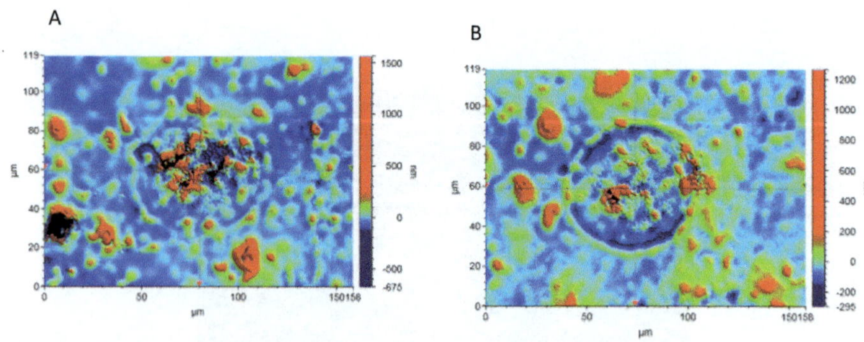

Fig. 8 White light interferometry results obtained from a Bruker Contour GT interferometer. The samples were ink-jet printed on PET. (A) Sample 6 was ink-jet printed and regenerated by leaving deionised water for 1 hour and 30 minutes before drying for approximately 18 hours at 50 °C and (B) sample 7 was ink-jet printed and regenerated by leaving in deionised water for 30 minutes before drying for approximately 18 hours at 50 °C.

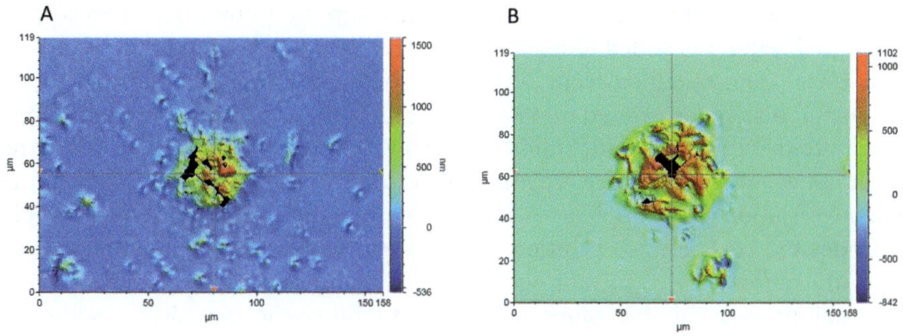

Fig. 9 White light interferometry results obtained from a Bruker Contour GT interferometer. The samples were ink-jet printed on glass. (A) Sample 6 was ink-jet printed and regenerated by leaving deionised water for 1 hour and 30 minutes before drying for approximately 18 hours at 50 °C and (B) sample 7 was ink-jet printed and regenerated by leaving deionised water for 30 minutes before drying for approximately 18 hours at 50 °C.

ionic liquid than leaving it for a short time interval (30 minutes). Untreated microgranular cellulose was used as a benchmark for a qualitative comparison of morphology (Fig. 6).

Fig. 7 shows the scanning electron microscopy image obtained from a regenerated ink-jet printed sample. As seen from the images, characterisation of the morphology of the regenerated cellulose was challenging, due to the presence of debris on and around the printed droplet. Further experiments must be carried out to distinguish the composition of the droplets.

Surface profile images obtained from the regenerated samples indicate that the central region of the drop has sunken, creating a concave residue (Fig. 8). A study conducted by Wang *et al.*,[29] found that $[C_4C_1Im][OAc]$ was responsible for PET degradation. In their work, 3.4 wt% of PET dissolved in $[C_4C_1Im][OAc]$ at 150 °C. Even though the ink-jet printed samples were treated at a maximum temperature of 55 °C in this study, the surface profile images in Fig. 8 suggest some degree of PET degradation.

Further ink-jet printing was carried out using glass as the substrate and the surface profile images, as shown in Fig. 9, reveal no evidence of a sunken central region in the drop. This shows the possibility of forming 3D structures on glass as an improved printing substrate, relative to PET.

Conclusions

In this study, cellulose dissolved in the ionic liquids $[C_4C_1Im][OAc]$ and $[C_2C_1Im][OAc]$ resulted in high viscosity solutions (>80 mPa s) that are not suitable for 3-D printing. Different concentrations of the co-solvent DMSO were able to bring these solutions into the DIMATIX ink-jet printing range, whilst maintaining cellulose solubility. The use of alternative co-solvents, such as 1-butanol, resulted instead in precipitation in the solutions studied here. The ink-jet printed samples indicated that some drops were missing and small ligaments were formed. This implies a blockage in the nozzle and formation of small ligaments that did not return to the nozzle or merge with the leading drop, but demonstrates that printing using such equipment is possible. Modifying the ink formulation and waveform would help to overcome this problem in future. The non-solvent deionised water was added to regenerate the ink-jet printed samples on both PET and glass. Data obtained from WLI studies exhibited structural distortion for regenerated samples printed on PET, potentially related to the dissolution of the substrate by either the ionic liquid or co-solvent. Glass substrates exhibited no such issues, and as such are recommended to be used instead of PET for future characterisation and applications. The preliminary work presented herein provides the basis for 3-D printing of biopolymeric materials, by demonstrating that utilisation of a co-solvent with an ionic liquid is able to modulate liquid printability to a suitable range, whilst maintaining the biopolymer solubility affected by the ionic liquid and affording materials that can be regenerated. Further characterisation must be conducted to distinguish the composition of the regenerated samples in more detail in order to apply this methodology to suitable end applications.

Acknowledgements

This work was made possible through funding from University of Nottingham and the Engineering and Physical Sciences Research Council, under grant EP/I033335/2 EPSRC Centre for Innovative Manufacturing in Additive Manufacturing. AKC wishes to acknowledge support from COST programme CM1206 EXIL.

References

1 M. Lukić, J. Clarke, C. Tuck, W. Whittow and G. Wells, *J. Appl. Polym. Sci.*, 2016, **133**, 42931.
2 N. Hopkinson, R. J. M. Hague and P. M. Dickens, *Rapid Manufacturing: An Industrial Revolution for the Digital Age*, Wiley, 2006.
3 H. Wang, G. Gurau and R. D. Rogers, *Chem. Soc. Rev.*, 2012, **41**, 1519–1537.
4 P. Mäki-Arvela, I. Anugwom, P. Virtanen, R. Sjöholm and J. P. Mikkola, *Ind. Crops Prod.*, 2010, **32**, 175–201.
5 A. Pinkert, K. N. Marsh, S. Pang and M. P. Staiger, *Chem. Rev.*, 2009, **109**, 6712–6728.
6 K. C. Badgujar and B. M. Bhanage, *Bioresour. Technol.*, 2015, **178**, 2–18.
7 H. Jørgensen, J. B. Kristensen and C. Felby, *Biofuels, Bioprod. Biorefin.*, 2007, **1**, 119–134.
8 A. Brandt, J. Gräsvik, J. P. Hallett and T. Welton, *Green Chem.*, 2013, **15**, 550–583.
9 T. v. d. Ven and L. Godbout, *Cellul.: Fundam. Aspects*, 2013, 143–178.
10 M. Freemantle, *An Introduction to Ionic Liquids*, Royal Society of Chemistry, 2010.
11 H. P. Le, *J. Imaging Sci. Technol.*, 1998, **42**, 49–62.
12 G. D. Martin, S. D. Hoath and I. M. Hutchings, *J. Phys.: Conf. Ser.*, 2008, 105.
13 B. Derby, *Annu. Rev. Mater. Res.*, 2010, **40**, 395–414.
14 C. K. Chua, K. F. Leong and C. S. Lim, *Rapid Prototyping: Principles and Applications*, World Scientific, 2010.
15 J. R. Castrejon-Pita, W. R. S. Baxter, J. Morgan, S. Temple, G. D. Martin and I. M. Hutchings, *Atomization Sprays*, 2013, **23**, 541–565.

16 S. V. Murphy and A. Atala, *Nat. Biotechnol.*, 2014, **32**, 773–785.
17 L. M. Haverhals, E. K. Brown, M. P. Foley, H. C. De Long and P. C. Trulove, *ECS Trans.*, 2012, **50**, 615–621.
18 K. Markstedt, J. Sundberg and P. Gatenholm, *3D Print. Add. Manufact.*, 2014, **1**, 115–121.
19 K. D. Sweely, E. T. Fox, E. K. Brown, L. M. Haverhals, H. C. De Long and P. C. Trulove, *ECS Trans.*, 2014, **64**, 575–582.
20 J. Vitz, T. Erdmenger, C. Haensch and U. S. Schubert, *Green Chem.*, 2009, **11**, 417–424.
21 O. A. El Seoud, V. C. da Silva, S. Possidonio, R. Casarano, E. P. G. Arêas and P. Gimenes, *Macromol. Chem. Phys.*, 2011, **212**, 2541–2550.
22 B. Kosan, C. Michels and F. Meister, *Cellulose*, 2008, **15**, 59–66.
23 H. A. Barnes, J. F. Hutton and K. Walters, *An Introduction to Rheology*, Elsevier, 1989.
24 R. Detsch, F. Uhl, U. Deisinger and G. Ziegler, *J. Mater. Sci.: Mater. Med.*, 2008, **19**, 1491–1496.
25 R. Faddoul, N. Reverdy-Bruas and J. Bourel, *J. Mater. Sci.: Mater. Electron.*, 2012, **23**, 1415–1426.
26 R. P. Swatloski, S. K. Spear, J. D. Holbrey and R. D. Rogers, *J. Am. Chem. Soc.*, 2002, **124**, 4974–4975.
27 T. Heinze, K. Schwikal and S. Barthel, *Macromol. Biosci.*, 2005, **5**, 520–525.
28 M. Zavrel, D. Bross, M. Funke, J. Büchs and A. C. Spiess, *Bioresour. Technol.*, 2009, **100**, 2580–2587.
29 H. Wang, Z. Li, Y. Liu, X. Zhang and S. Zhang, *Green Chem.*, 2009, **11**, 1568–1575.

Faraday Discussions

DISCUSSIONS

Advancement in knowledge of phenomena and processes: general discussion

Toru H. Okabe, Han Wang, Jianbang Ge, Yalan Liu, Jennifer Pringle, Wei Xiao, Babak Khalaghi, Andrew Doherty, Qian Xu, Dihua Wang, Paul Madden, Binjie Hu, Chaohui Wei, Kathie McGregor, John M. Slattery, Andrew Mount, Yiyang Kong, Ali Kamali, Linpo Yu, Xianbo Jin, Ye Liu, Bencan Tang, George Zheng Chen, Bin Dong, Ian Farnan, John Irvine, Hongmin Zhu, Yingjun Liu, Deshani Gunasekera and Juanyu Yang

DOI: 10.1039/C6FD90032A

Yalan Liu opened discussion of the paper by Hongmin Zhu: Why did you add F^- ions into the chloride molten salt?

Hongmin Zhu replied: We want to analyze the effect of F^- ion on the equilibrium of the titanium ions and metallic titanium. By adding F^- ion into a chloride molten salt, you can tune the coordination situation of anions around the cation. In molten salts each cation is always coordinated by some anions. When F^- ions are introduced into chloride salts, the F^- ions will be preferentially coordinated to the higher valance cations (in this case, Ti^{3+} and Ti^{4+}); this will drive the equilibrium of the disproportionation reaction to the higher valance cation side. Therefore, by analyzing carefully the effect of the F^- ion addition to the equilibrium, you will be able to know the relative stability (chemical potential) of the cations. This is also linked with the electrode reaction steps.

Yalan Liu enquired: If you have a Ti^{4+} ion interacting strongly with a F^- ion, is that good news for the reduction of Ti^{4+} back to Ti metal?

Hongmin Zhu responded: This will depend on the conditions at which you are going to reduce the titanium ion. If you start from Ti^{4+} as the feeding salt, you need to reduce it from its 4+ oxidation state to metallic Ti(0). In this case, the most important issue is that the electrode reaction needs to proceed directly from Ti^{4+} to Ti(0) without any intermediate valence ions (Ti^{3+}, Ti^{2+}). From this point of view, the strong interaction (band) between Ti^{4+} and F^- ions is good news, since this will drive the complex ion formation of Ti^{4+} with F^-, and suppress the formation of intermediate valence ions.

Toru H. Okabe said: I understand that your experimental system is very difficult to analyse and I very much appreciate your efforts for this work. I have some questions. Nowadays, Ti powder production is becoming more important for 3D printing and other near-net-shape applications. Can you apply your knowledge of disproportionation reactions of titanium ions to produce Ti powder? If applicable, are you going to create a temperature gradient in the reaction system? If so, what type of Ti powder can you produce using these disproportionation reactions? Does it have a spherical shape suitable for powder metallurgy?

Hongmin Zhu answered: This is a very good idea. Thanks for your great comment. Yes, this could be a good approach to making 3D powders. We can use the disproportionation reaction for the production of Ti powder. A simple approach is just to mix sponge Ti with some titanium chloride salt in molten salts. The disproportionation reaction will "transfer" the Ti from the sponge shape into a powder. That is, Ti reacts with high valence Ti ions to form lower valence Ti ions on the Ti sponge surface which are then transported to the bulk salt where the disproportionation reaction will happen, forming the Ti powder. In this way, we are able to form powders. We always see this during our experiments. As for the reaction, the temperature gradient may help the reaction, but since the driving force of the reaction is chemical potential, we do not even need any other design for the gradient. The powders that we always see during the experiments are very fine. We did not check the detailed morphologies of the particles.

George Zheng Chen remarked: I fully appreciate your fundamental studies which are rare in the current literature. When you propose different species resulting from coordination between titanium and fluoride ions, for example the dianion complex, do you have direct evidence for their formation? I guess the structures of such complex species might have been reported in the 1970s. You have described Ti^{4+} and Ti^{3+} containing species, how did you analyse and determine these valences?

Hongmin Zhu responded: In molten salts, every cation is always surrounded (coordinated) by some anions. It was reported that in the alkali chloride melt the first coordination number of the anion (Cl^-) to an alkali cation is 4–6. Therefore, each titanium ion, Ti(II), Ti(III), Ti(IV), is always coordinated to a certain number of anions (Cl^-, F^-). The overall first coordination number should stay the same. As a result, we concluded that the highest coordination number of F^- is 6 (please see our paper for further details). Therefore, we can assume that the total number of anions (Cl^- and F^-) around a titanium ion is 6. As we described in the paper, the "free" Ti^{3+} here refers to the Ti(III) surrounded only by Cl^- ($TiCl_6{}^{3-}$), and $TiF_i{}^{3-i}$ refers to complexes in which i F^- ions have replaced the positions of the Cl^- ions ($TiCl_{6\text{-}i}F_i{}^{3-}$). The coordination number, i, is not a real coordination number of anions around Ti(III), but rather the number of F^- ions around Ti(III). The true coordination number of anions around Ti(III) is always 6. This is also the same for Ti(II) and Ti(IV).

We found that the added fluoride ion is always preferentially coordinated to a higher valence titanium ion, Ti(III) or Ti(IV), due to the relatively lower chemical potential, which is fairly understandable.

George Zheng Chen asked: Looking at the reactions in your chemical analysis, what happens if you add HCl into the $TiCl_2$ solution? Can you achieve complete conversion of Ti^{2+} to Ti^{3+}? Why is this reaction not making Ti^{4+}? I have suspicions about the accuracy of the detection method which involves the measurement of the volume of a gas, *i.e.* the hydrogen gas. Regarding the oxidation of Ti^{2+} to Ti^{3+} by the Fe^{3+} ion, is this reaction irreversible? If not, what is the equilibrium constant for this reaction?

Hongmin Zhu replied: The reaction for the analysis goes completely and is not under equilibrium.

George Zheng Chen continued: I am impressed by your work. I think there is still the need for direct evidence of the different coordinated species that you have discussed. Going back to the analysis, was it carried out in an aqueous solution? Is $TiCl_2$ stable in water? How is the analysis of $TiCl_2$ in water linked to the actual concentration of $TiCl_2$ in the melt? Would Ti(II) change to other valences during the transfer from molten salt to water?

Hongmin Zhu answered: Thank you, Prof. Chen. Regarding the evidence for the different coordinated species, please refer to my above response to this question. Yes, the chemical analysis was carried out in an aqueous solution. We sampled the molten salt under equilibrium by extraction into a quartz tube. On top of the quartz tube was a quartz filter. The sample was soon quenched into a solid after passing through the filter (please see our previous report[1] for details of the sampling), and therefore, after sampling, the sampled salt is no longer under equilibrium. It is however reasonable to assume that the concentrations of the species remain unchanged because the sample is in the solid form. We analyzed the concentrations of Ti(II), Ti(III) and Ti(IV) in each sample by H_2 volumetric analysis, titration, and spectrophotometry, respectively. $TiCl_2$ is not stable in water, reacting with water to generate H_2. In a HCl solution, Ti(II) reacts as:

$$2Ti(II) + 2H^+ \rightarrow 2Ti(III) + H_2$$

This is exactly the principle for the Ti(II) analysis. The total concentration of Ti(III) in the solution was determined by titration using a $NH_4Fe(SO_4)_2$ aqueous solution:

$$Ti^{3+} + Fe^{3+} \rightarrow Ti^{4+} + Fe^{2+}$$

All of the reactions for the analysis are irreversible, such that they go completely, and this guarantees the accuracy of the analysis. The concentration of Ti^{4+} was determined by spectrophotometry. The total concentration of Ti ions was also double-checked by oxidizing all of the Ti(II) and Ti(III) to Ti(IV) with the addition of a HNO_3 solution. Through the above analysis methods, we determined the concentrations of Ti(II), Ti(III) and Ti(IV) in each sample. As with the analytical method for determining the titanium ions of different valences, the method described above is well established, and the accuracy for each analysis is also pretty good.

The most important point of this research is, however, the sampling. Since we are talking about the equilibrium of titanium ions with metallic titanium, the salt should be sampled with metallic titanium powder dispersed in the molten salt. However, the fine titanium powders in the sample will cause the generation of H_2 during the analysis of Ti(II). Therefore, we filtered the Ti powder during the sampling using the quartz filter, and kept the equilibrium with Ti metal by putting a Ti wire in the tube, and quickly quenching the sample. After the sample treatment we get reproducible results with good accuracy.

1 Q. Wang *et al.*, *Metall. Mater. Trans. B.*, 2013, **44B**, 906–913.

Toru H. Okabe commented: In aqueous solution, the Ti^{2+} ion is not thermodynamically stable because it is so electronegative compared to the H^+/H_2 equilibrium. When a Ti^{2+} containing salt is injected into the aqueous system, the Ti^{2+} ions are oxidized to Ti^{3+} by H^+. This means that Ti^{2+} reduces water to generate H_2 gas. The amount of H_2 gas evolved after the reduction of water should be equivalent to the amount of Ti^{2+} in the salt samples. Ti^{3+} ions can thermodynamically exist in aqueous solution, and the amount of Ti^{3+} can be determined by titration utilizing the Fe^{2+}/Fe^{3+} equilibrium. The amount of Ti^{4+} can be determined from the total Ti content in the sample using the determined amounts of Ti^{2+} and Ti^{3+}.

George Zheng Chen responded: Thank you for your explanation.

Prof. Zhu, your molten salt apparently contained a mixture of titanium complexes in different valences. When you dissolve this mixture in water, could the Ti(III) species be converted back to Ti(II) species? If so, the analysis of Ti(II) in the aqueous solution would not be the same as it was in the melt. There should be an equilibrium between the different valences of titanium in water which may not be the same as that in molten salt. Ti(II) can change concentration due to conversion from Ti(III), which means that the Ti(II) concentration determined in water may differ from that in molten salt. The total amount of titanium species does not change, but the equilibria in water and in molten salt will be different.

Hongmin Zhu answered: We analyzed the concentrations of Ti(II), Ti(III) and Ti(IV) for each sample. We usually see two cases, one is Ti(II) and Ti(III) (region A), and the other is Ti(III) and Ti(IV) (region B). Please see our paper for further detail.

Paul Madden addressed George Zheng Chen and Hongmin Zhu: George Chen has commented that the Ti^{2+} ions would be oxidised by water, and therefore queried as to how the titration with HCl solution to liberate hydrogen can be quantitative. Isn't the point that whether oxidised by reaction with water or by the HCl in solution, hydrogen will be liberated in each oxidation event so the amount of hydrogen collected is a measure of the amount of Ti^{2+} transferred into the solution. The role of the HCl is to drive the reaction to completion.

Hongmin Zhu replied: Yes, we are counting the total volume of H_2 generated during the oxidation of Ti^{2+} in the HCl solution. It does not make any difference whether the H_2 is generated from H_2O or HCl.

George Zheng Chen said: I thank Prof. Madden for this helpful comment and explanation.

George Zheng Chen opened discussion of Juanyu Yang's paper: I have some colleagues in China who are very skeptical about whether the FFC Cambridge

Process can be successful commercially. I think that your work demonstrates very well that this process can be successfully developed for the industrial production of silicon in controlled morphologies. From my heart, this is brilliant work, congratulations.

Han Wang replied: Thanks.

Wei Xiao remarked: You claimed the formation of single-crystal Si nanowires with the addition of Ni species. Are the formed Si nanowires single crystals or polycrystals with the addition of other metals?

Juanyu Yang responded: Single-crystal Si nanowires are formed with the addition of Ni, Cu, Fe and so on, but Si nanotubes are formed with the addition of Ag.

Xianbo Jin commented: I'm very impressed with your work and the preparation of silicon nanowires. As you said, your aim is to prepare silicon nanowires for Li batteries; have you carried out some battery tests? What were the results?

Juanyu Yang answered: Yes, we have carried out some battery tests. The Si/C nanowire electrodes exhibit an excellent cycling stability: after 160 cycles the electrode preserves 60.6% of the original capacity (~650 mAh g^{-1}).

Xianbo Jin queried: Did you add the carbon before or after electrolysis? Will you form silicon carbide?

Juanyu Yang replied: The carbon was added before electrolysis, and also it could be added after electrolysis. The formation of SiC is also related to the fabrication process of the cathode, not just the electrolytic conditions.

Wei Xiao added: The formation of Si and SiC will compete upon molten salt electrolysis. For solar cell applications, the purity of electrolytic silicon is important. In contrast, for lithium ion battery applications, the purity of the electrolytic silicon is adequate. Tesla has just issued a new model, using silicon as an additive to the negative electrode. I think that adding some carbon in the cathode is a good idea to prepare Si–C composites for lithium ion batteries.

Xianbo Jin responded: Yes, I agree, for the case where Si, as opposed to SiC, is still the main electrolysis product, which seems to be the case according to Dr Yang's description.

Ali Kamali commented: I'm impressed with the quality of the silicon that you produce. What is the main challenge towards scaling up the process? Do you have any limitations towards size-up in terms of quantity of materials?

Han Wang said: Technically, there is no problem with the scaling up of this process. The problems that we have met so far are mainly with respect to the engineering, and fortunately, we have not reached the ceiling of limitation for size-up.

Dihua Wang asked: You used Ni and NiO powders as the additives in the preform; did you observe any differences in the structures and morphologies of the products when using Ni powder and NiO powder?

Juanyu Yang answered: The mean diameter of the SiNWs produced using NiO powder is smaller than that for those produced using Ni powder.

George Zheng Chen enquired: At 800 °C, I think that nickel oxide nanoparticles, and even microparticles, would sinter into larger particles. Do you agree? If so, how would this affect the electrode process?

Han Wang replied: Yes, you are absolutely right. However, during our preparation of the cathode pellet, the amount of nickel oxide added is much lower than the amount of silicon dioxide. Each nickel oxide particle can be completely surrounded by silicon dioxide. Thus, during the sintering, the initial particle size of the nickel oxide will not change.

Dihua Wang questioned: According to your Si nanowire growth mechanism, will the initial particle size of the Ni affect the morphology of the product? How does it affect the diameter of the Si nanowires?

Juanyu Yang responded: Yes, the initial Ni particle size will affect the morphology of the product. Ni additives with larger particle sizes could lead to SiNWs with larger diameters.

Chaohui Wei communicated: What are the advantages of using Ni/NiO in the reduction process compared to other additives?

Juanyu Yang communicated in reply: The melting point of Ni (1453 °C) is higher than the electrolysis temperature (900 °C). Ni additives can be solid during the reduction process, which gives rise to reaction points (Ni/SiO_2/$CaCl_2$ three-phase interlines) for generating nano-sized nickel silicide. The nano-sized nickel silicide can form a nano-sized peritectic alloy. The lowest temperature that leads to the metal droplet tip-led growth of nanowires is 993 °C, at which the liquid peritectic alloy (Ni 41 at% and Si 59 at%) can be formed. This temperature is higher than the electrolysis temperature but is close to it. Thus, a nano-sized peritectic alloy can be liquid while alloys with large particle sizes are solid, resulting in the formation of a solid–liquid interface on the surface of the Ni particles. The nano-sized peritectic alloy is separated from the solid alloy by the precipitation of silicon. After that, nano-sized droplets can be formed.

Chaohui Wei communicated: I notice that in Fig. 4a of your paper you claim that there is some $NiSi_2$ by-product; what percentage of by-product is formed? Will it have some further effect on the functionality and application of the Si nanowires?

Juanyu Yang communicated in response: The content of $NiSi_2$ is 33.92 wt% in theory. For Li-ion battery applications, the $NiSi_2$ by-product may improve the cycling performance, since as an inert phase it can buffer the expansion of the silicon during the lithiation process.

Chaohui Wei communicated: In your conclusions, it is said that these SiNWs have a wide diameter distribution of 80–350 nm. Does this mean that your product has mixed lengths of 80–350 nm or that you can control the specific length range? If the latter, how can you control it?

Juanyu Yang communicated in reply: It does not mean that the product has mixed lengths of 80–350nm. So far, we can't control the lengths of the SiNWs. It means that the product has mixed diameters of 80–350 nm.

Chaohui Wei communicated: Have you tried to amplify the experiment for trial tests? If so, how did it go and what are the main problems?

Juanyu Yang communicated in response: No, we haven't.

Qian Xu opened a general discussion of Ali Kamali's paper: You used graphite as the raw material for preparing graphene — is there anything special about the graphite, such as its purity, density or structure?

Ali Kamali answered: Important properties of the graphite feed material which can affect the morphology and quality of the graphene product include the degree of crystallinity, purity, grain size, density, conductivity and physical strength.

Jianbang Ge remarked: This is very meaningful work! Due to the specific properties of graphene, it may be easy to absorb other ions and aggregation may occur easily; can you comment on this?

Ali Kamali responded: Thank you. This method is based on a cathodic exfoliation mechanism that avoids excessive oxidation of graphite during the process, meaning that the graphene product is of a high structural quality. However, in general, the edges of graphene sheets are capable of absorbing ions from molten salts or aqueous solutions.

Jennifer Pringle asked: What are the voltages at the negative and positive electrodes? What is the process occurring at the positive electrode? Can you alter the voltage parameters to control the size of your graphene flakes and the conductivity?

Ali Kamali replied: The potential differences between the pseudo-reference electrode and the graphite electrodes during the exfoliation process were -1 to -2V for the cathode and about 3V for the anode. The cathodic voltage can affect the exfoliation rate.

Jennifer Pringle enquired: Can you change the Ar/H_2 ratio to give more control over the graphene synthesis process? It seems that you are physically breaking up the starting material, which is a hard thing to control. Possibly the quality of the graphite starting material is most important in determining the properties of the final product?

Ali Kamali answered: The Ar/H_2 ratio can affect the rate of exfoliation.

George Zheng Chen remarked: This is an interesting paper. When you were making graphene by electrolysis in molten LiCl, how was the water vapour introduced, and how do you control the amount of vapour introduced?

Ali Kamali replied: The water vapour can be introduced in the retort by passing the inlet argon gas through a U-shaped glass tube partially filled with water. The amount of vapour introduced can be controlled by changing the flow rate of argon gas.

George Zheng Chen asked: Would you please explain the mechanism of your proposed proton intercalation? What is the driving force apart from electrostatic interaction? Lithium metal has a high tendency to react with carbon, which is the fundamental driving force for lithium (ion) intercalation into graphite. Is there any evidence that hydrogen gas will react with graphite? Further, if proton intercalation indeed occurred, how would it help with the destruction of the structure of the graphite cathode, particularly considering that the size of a proton (and also a hydrogen atom) is so small?

Ali Kamali answered: Please consider that hydrogen cations are discharged on the graphite surface under the influence of the cathodic potential. The intercalation mechanism proposed is, therefore, based on the high-temperature intercalation of atomic hydrogen into the interlayer space of graphite, followed by the combination of hydrogen atoms in the space between graphene layers of the graphite leading to the formation of hydrogen molecules, and subsequently, the exfoliation of the graphite into graphene nanosheets. Hydrogen molecules formed in the interlayer space of graphite possess enough kinetic energy to overcome the van der Waals forces between the graphene layers.

George Zheng Chen commented: We did similar work in Cambridge prior to 1998 and obtained TEM and HRTEM images showing graphenes in the electrolysis products, although our purpose then was to make carbon nanotubes.[1] In that case, we did not use any H_2 or water. I also observed large carbon debris, resulting from using high cell voltages or currents that led to intercalation that was too fast, and electro-reduction of Li^+ or Na^+ ions in the graphite structure. Under proper control of electrolysis, uniform erosion of the graphite cathode could be achieved, producing very high yields of carbon nanotubes, particularly in molten NaCl. Did you observe similar changes on the graphite cathode in your experiments? Because we obtained graphenes in molten salt in the absence of water or hydrogen gas, I am suspicious of your claim that protons play such a large role in your process. Speed of intercalation is more important. Would you please comment on that?

1 G. Z. Chen, X. Fan, A. Luget, M. S. P. Shaffer, D. J. Fray and A. H. Windle, *J. Electroanal. Chem.*, 1998, **446**, 1–6.

Ali Kamali responded: Thank you for your comment and your question. According to my observations, high-quality graphene nanosheets, such as those shown in the SEM images in Fig. 5 of our paper or the TEM images in Fig. 6 of **ref. 1** below, can be produced in the presence of a considerable amount of hydrogen or water in the system. As also evident from the micrographs shown in your 1998 paper and other earlier studies concerning the cathodic polarisation of graphite in molten LiCl, much less to no well-defined graphene can be produced when nominally dry argon is used, and therefore the production of graphene had not been reported in these studies. It should also be taken into account that usually it is very difficult to completely remove moisture from the reactor as LiCl is very hygroscopic and easily absorbs moisture from the atmosphere, especially in the humid weather we have in the UK! Therefore, it is likely that some level of moisture gets involved during the molten salt process, even if it is not intended.

1 A. R. Kamali and D. J. Fray, *Nanoscale*, 2015, 7, 11310.

George Zheng Chen said: I have a different opinion. You claimed that hydrogen or protons were doing the job in the intercalation process. To break up the graphite structure, a large amount of charge is needed. If you convert the charge to the number of protons, you will find that it may not be possible to provide enough protons in the molten salt. The total concentration of protons in the molten salt is too low. At a cell voltage of 5 V, I believe that your current density could reach beyond 1 A cm^{-2}. In the industry process for aluminium extraction, the current densities on the anode and cathode are both lower than 1 A cm^{-2}. How can the small amount of protons in the molten salt maintain such high current densities? The mechanism must involve the intercalation of Li^+ ions. Would you please comment on this consideration?

Ali Kamali answered: Many thanks for this comment. During the molten salt process, hydrogen cations are continuously fed into the molten salt *via* water or hydrogen gas. Considering the high solubility of HCl in molten LiCl,[1,2] and the volume of the molten salt used, hydrogen cations can be sufficiently available to support high currents leading to the exfoliation of graphite into graphene. If dry argon were used during the molten salt process, such graphene nanosheets might not be produced, showing the effect of hydrogen cations on the process. I also observed that a negligible amount of carbon product can be produced when LiF is used instead of LiCl, whilst the cathodic reaction in both cases should be the reduction of lithium ions. In my opinion, hydrogen molecules formed by the combination of intercalated hydrogen atoms are responsible for the peeling off of graphite flakes into few layer graphene nanosheets. Please consider that this proposed mechanism does not reject the possibility of the deposition or intercalation of Li during the electrolysis process, but rather introduces the role of hydrogen.

1 A. R. Kamali and D. J. Fray, *Nanoscale*, 2015, 7, 11310.
2 Unpublished work.

George Zheng Chen commented: The melting temperature of LiF (*ca.* 845 °C) is much higher than that of LiCl (*ca.* 610 °C). Intercalation would not be the same at 845 °C as at 610 °C. Temperature plays an important role, and the LiCl *vs.* LiF comparison may not be appropriate here.

Ali Kamali replied: I have tested both molten LiCl and LiF at the same temperature under dry argon. LiCl was able to produce a large quantity of carbon nanomaterials, whilst LiF wasn't.

George Zheng Chen asked: Could the concentration of protons in the molten salt sustain the current density that you observed? Would you be able to give us the concentration of protons in molarity, for example?

Ali Kamali responded: LiCl containing molten salts have a high solubility for HCl.[1] My observations confirm that mono-component LiCl shows a high solubility for HCl at high temperatures, so that hydrogen cations formed can deliver the high currents required for the efficient exfoliation of graphite.

1 N. Q. Minh and B. J. Welch, *J. Electroanal. Chem.*, 1978, **92**, 179–189.

George Zheng Chen said: Considering thermodynamics, I doubt that HCl could dissolve in the melt in a large amount. It is common knowledge that hydrolysis occurs due to LiCl reacting with water at elevated temperature to produce Li_2O and HCl. The produced HCl then evaporates as a gas. You have observed significant erosion on the graphite cathode, which corresponds to a large current that has to be maintained by a high concentration of protons in the melt.

Ali Kamali replied: LiCl is a unique salt in terms of its reactivity with water at high temperatures. Dehydration of lithium hydrates occurs at temperatures below 200 °C to form anhydrous lithium chloride. However, the tendency of LiCl to react with water (through hydrolysis) becomes more pronounced once the melting point has been exceeded, since both LiOH and HCl formed can easily dissolve in the molten LiCl. It is known that HCl is soluble in LiCl-based molten salts and that the dissolved HCl is dissociated into protons and chloride ions. Furthermore, the diffusion coefficient of the protons arising from the dissolved HCl is an order of magnitude higher than for most other ions in molten salts. Please see **ref. 1** below and references therein, especially 106–112.

1 A. R. Kamali and D. J. Fray, *Nanoscale*, 2015, 7, 11310.

Andrew Mount queried: If the issue is whether hydrogen ions can provide the current required to produce hydrogen for the proposed mechanism, could the mass transfer limited current for hydrogen reduction not be measured and this then used to determine whether the current can be maintained solely by hydrogen reaction, or whether lithium ion reduction/intercalation may also be involved?

Ali Kamali answered: According to experimental results, a mechanism involving the reduction of hydrogen cations on the graphite surface and diffusion of hydrogen atoms in the graphite lattice followed by their combination is thought to be responsible for the exfoliation of graphite into graphene nanosheets. Lithium intercalation by itself may not lead to the formation of graphene sheets such as is observed when hydrogen is involved. More detailed theoretical and experimental investigations on the mechanism should be conducted.

Chaohui Wei communicated: Your paper compares graphene made in molten salts with graphite powder and multi-walled carbon nanotubes (MWCNTs); what characterization results and analyses show that the graphene is better than graphite and CNTs? Have you compared the graphene that you have made with graphene made by other methods?

Ali Kamali communicated in reply: In the paper presented here, graphene produced in molten salt, graphite powder and CVD-produced CNTs were used as additives in the fabrication of alumina articles using slip casting — a pressureless sintering technique. Molten salt-produced graphene was shown to outperform the other carbon additives by enhancing the mechanical properties of pressureless sintered alumina. Addition of a small amount of graphene led to the formation of a nanostructured alumina article with a superior toughness. CNTs failed to remain dispersed during the sintering process and concentrated in the centre of the sample. Graphene materials produced by other methods were not used in this study. However, the results obtained have been compared with the data available in the literature where other forms of graphene have been used.

Chaohui Wei communicated: When you wash the graphene nanosheets from the solidified salts, what percentage of product do you obtain from the salts? Is the recyclability of the salts affected?

Ali Kamali communicated in response: LiCl is highly soluble in water and, therefore, can be readily washed off. The carbon product can be almost entirely retrieved from the salt. The salt can then be recycled and reused.

Chaohui Wei communicated: Why do you use an Ar–5% H_2 flow during the process?

Ali Kamali communicated in reply: Hydrogen in Ar–H_2 gas is used as the source of hydrogen cations in the molten salt, which are then discharged on the graphite cathode and subsequently lead to the exfoliation of graphite. Please see the exfoliation mechanism explained in the paper.

Andrew Mount communicated: Can this process be tuned to functionalise the edge of the graphene sheets? Can such functionalisation be used to control aggregation and tune the structure and resulting properties of the graphene product?

Ali Kamali communicated in response: This is an interesting question. Post-functionalisation of graphene nanosheets produced is of course possible. *In situ* functionalisation of graphene in molten salt is an interesting issue to be investigated.

Andrew Mount opened discussion of the paper by Paul Madden: From Fig. 4, have you calculated activation energies for viscosity? Concentrating on the low values of x appropriate to our measurements, the value seems to be of the order of 40 kJ mol^{-1}, comparable with those observed for the diffusion of redox active species in our paper (DOI: 10.1039/c6fd00002a). Is this the case, as it would support the primary effect on diffusion being the viscosity?

Paul Madden responded: In dilute solutions, the trivalent cations tend to be surrounded by a coordination shell to form an isolated complex (not linked to other complexes in the way involved in network formation) that remains intact on the timescales of the local viscous relaxation processes around a moving ion. As such, you would expect the diffusion coefficient to be quite well represented by Stokes' law with a "hydrodynamic radius" given by that of the whole coordination complex. Hence, you would expect the activation energy for diffusion and the viscosity to be very similar. See **ref. 1**, for example.

1 R. Brookes, A. Davies, G. Ketwaroo and P. Madden, *J. Phys. Chem. B*, 2005, **109**, 6485.

Ian Farnan said: I was interested in the mechanism of viscous flow. You observe various time-dependent processes such as cage lifetimes/Maxwell lifetimes; does the cage remain intact? In silicate liquids we see anions move away from one cation, making other cations more highly coordinated — do you see anything similar to that in your simulations?

Paul Madden replied: Yes, the anions are indeed swapping between the coordination shells of different cations, and the timescale of this is the cage lifetime (this is the timescale on which the identities of the ions in the coordination shell of a cation change). In the dilute solutions, this timescale is considerably longer than the Maxwell relaxation time, meaning that these coordination shell exchange processes cannot be contributing on the timescale of the processes which determine the viscosity. At higher concentrations, when the coordination shells are linked to form a network, the network relaxation dominates the viscous relaxation process and, of course, breaking and forming the cation and anion linkages is integral to this. However, because the network is so entangled, breaking a single "bond" is not enough. So, although the two processes will track each other (in *e.g.* their temperature dependencies), a single bond-breaking picture of the relaxation is too simplistic an interpretation. We studied the relationship between these times in the paper on BeF_2 referred to in our paper.

Ian Farnan enquired: Could I ask about how the Maxwell relaxation time is derived? Do you assume a shear modulus at infinite frequency for the liquid, or do you derive a time scale from the integral of the velocity auto-correlation function (diffusive) time scale?

Paul Madden answered: The viscosity is the time integral of the stress tensor autocorrelation function (as in the equation in the paper) –not the velocity autocorrelation function. The relaxation of this function is therefore that of the shear relaxation processes which determine the viscosity and the characteristic relaxation time is the Maxwell time. The viscosity could be regarded as the product of the zero time value of the correlation function (which is the high frequency shear modulus) and the relaxation time (strictly, the integrated area under the normalised correlation function which is the same as the relaxation time for an exponentially decaying function). Usually, it is the temperature and concentration dependence of the relaxation time which dominates the behaviour of the viscosity, with the dependence of the shear modulus on these variables being less significant.

Babak Khalaghi asked: In your paper you have mentioned that when the mixtures become concentrated, two different scenarios may occur with respect to the coordination number; either the cation can reduce the number of anions in its coordination shell to remain isolated, or the individual cations might begin to share anions and have overlapping shells. In the case of cryolitic melts (AlF_3–NaF) the former scenario occurs, whereas for chloride melts the second one occurs. Could you please explain why the cryolitic melts behave like this?

Paul Madden responded: I think there are two significant physical factors involved in answering your interesting question. Firstly, the fluoride ion seems to be less inclined to become involved in the kind of bridging linkages between cations which result in the formation of oligomers and networks than the chloride (as illustrated in the paper) — this may be related to the larger polarizability of the chloride anion. Secondly, the Al^{3+} cation is considerably smaller than the cations considered here, so that, although in the crystalline cryolite (elpasolite structure) it is six coordinate (just like LaK_3Cl_6), it is probably over-coordinated compared to its preferred coordination number with F (whereas I would maintain that La^{3+} is under-coordinated with Cl). Hence, as the concentration of a solution is increased from very dilute and the cation faces the dilemma of how to adapt to the shortage of free anions with which to form its ideal coordination structure, the two systems respond in different ways — the Al^{3+} reduces the number of anions in its coordination shell, whereas the La forms linkages which seem to have the consequence that the coordination number increases, as illustrated in the paper. We have studied the cryolitic melts using just the same methodologies.[1]

1 S. Cikit, Z. Akdeniz and P. A. Madden, *J. Phys. Chem. B*, 2014, **118**, 1064–1070.

John Slattery queried: Some people have discussed the viscosity of molten salts using hole theory. Based on your simulations, do you feel that would be a valid way to look at the system that you have studied?

Paul Madden answered: I think that your question raises an important point of principle about what is required to establish an actual mechanism for a process. Using hole theory (or indeed any other idealised model of the local structure and dynamics) one can derive relationships between observable quantities (*e.g.* viscosity/conductivity) which might help to correlate the behaviour of the quantities. However, this does not prove that the underlying picture itself is correct. To really establish a mechanism, one needs to follow the microscopic events which contribute to the phenomenon of interest. Viscous relaxation processes have been examined in inordinate detail in fluids close to the glass transition — the role of many types of microscopic events have been examined in great detail (including the role of holes and bond-breaking in network materials) and suffice it to say that nothing straightforward emerges. Those interested might start with a very nice paper by Reichman and co-workers[1] and use that as the starting point for a search back through the literature.

1 A. Widmer-Cooper, H. Perry, P. Harrowell and D. R. Reichman, *J. Chem. Phys.*, 2009, **131**, 194508.

John Slattery commented: It would seem to be inconsistent with the idea that you're breaking the network structure under sheer stress.

Qian Xu questioned: How many ions in the simulation cell did you take into account in your calculation? Does the number of ions in the cell have any effect on the assessment of the properties of molten salts?

Paul Madden replied: You have raised an important point. The numbers of ions in the simulations conducted for this work are given in the paper. In general, we know from theory (and practice) that there is a correction to the value of the diffusion coefficient which is proportional to $1/L$, where L is the length of the simulation cell. Hence, for a definitive value, you do simulations at several cell sizes and extrapolate to $L =$ infinity. One would expect the same to be true for other transport coefficients. Based on these considerations, we would expect the transport coefficients quoted in the paper to be within 10% of the asymptotic value. However, there is another wrinkle for the type of liquids under discussion here which form networks which (in the most concentrated solutions) extend throughout the cell. If the whole extended network were to remain intact on the timescale of the viscous relaxation processes then the system would be behaving like a polymeric system and one would have to start worrying about another set of considerations to decide if the length and timescales of the simulation were appropriate. However, this does not apply here, but could become problematic when simulating a strong network system like SiO_2.

Qian Xu asked: How long does the structure that you calculated for the molten chloride last?

Paul Madden responded: It depends on which structure you mean and the concentration of the melt. The lifetime of the coordination complex (meaning the time on which the coordinating anions exchange with the bulk) around a cation is given by the cage correlation time which we discuss in the paper (see Fig. 9) and is on the order of 4–5 picoseconds in these systems — this is the relevant structure in the more dilute solutions. We have not studied the lifetimes of the more complex oligomeric species and the network in detail in these systems (in the manner done for $LiF{:}BeF_2$ in **ref. 1**). However, the case that we make in the paper is that in the more concentrated solutions the structural relaxation reflected in the value of the viscosity is the relaxation of the network, and hence you would expect the Maxwell relaxation time to mirror the timescale of the network relaxation (see Fig. 10).

1 M. Salanne, C. Simon, P. Turq and P. A. Madden, *J. Phys. Chem. B*, 2007, **111**, 4678–4684.

Andrew Mount enquired: Given previous discussions on metal solubility in liquid salts, can this approach be used to simulate the addition of individual metal atoms derived from molten salt cations to determine the effects of metal solubility on the thermodynamic and physical properties?

Paul Madden said: The answer is yes, in principle! When the metal atom is dissolved in the molten salt, we might, at first sight, think that this would simply generate a cation and a free(ish) electron. In the case of alkali metals dissolving in alkali halides this is what happens. In this case, the problem for simulation would be to treat the ion and its interaction with other cations and anions in the system (which we have seen can be done in the paper I presented) and the electron and its interaction with the molten salt, which requires a quantum mechanical description. At low concentrations, where the electrons are isolated from each other, they are localised in f-centre-like solvation complexes, and this can be captured at relatively low levels of quantum mechanical theory in computer simulations (see *e.g.* early work by Parrinello and co-workers). At higher concentrations the electrons begin to interact, still as localised entities, forming bipolarons *etc.*, where capturing the quantum mechanical "correlation effects" required for a good description necessitates a higher level of theory which is difficult to incorporate in finite temperature simulations. Ultimately, at high concentration, the electrons delocalise to form a metallic fluid, which again can be described in a fully quantum mechanical simulation using density functional theory.

In the more relevant (to our discussion) case of alkaline earth dissolution, I believe that something different happens. Rather than forming isolated Ca^{2+} *etc.* cations, I believe that more complex species like $(Ca_2)^{2+}$ (which is isoelectronic with the stable K_2 molecule) and larger charged clusters are formed. I would suggest that the ability to form such species, in the molten salt, is the reason for the subtle chemistry for the dissolution of the alkaline earth metals (*e.g.* why Ca dissolves in $CaCl_2$, but not Mg in $MgCl_2$). That is, there are subtle interactions between the cations and the electrons which need a high level of quantum mechanics (correlation effects) to capture accurately as one is looking at the relationship between different possible localised states of the electron in these systems and all at finite temperatures. I do not think that adequate calculations have yet been done.

Toru H. Okabe queried: Is it possible to observe metal solubility in molten salts by spectroscopic methods? It seems very difficult experimentally. If possible, what is the container? When analyzing, for example, the Ca–$CaCl_2$ system, materials for containers are limited because the metal dissolving salt is chemically very active, and everything has to be done in a highly reducing atmosphere.

Paul Madden answered: Given the importance of the phenomenon of metal dissolution, it seems to me that it would be important to undertake some systematic studies of the phenomenon. I am not aware of any work of the type you mention on the alkaline earth systems, but there have been some very beautiful and extensive examinations of systems like K–KCl by Werner Freyland and co-workers. See, for example, **ref. 1** and the references therein. I commented previously, in an answer to a question by Prof. Mount, on the issues to be faced by studying these systems by computer simulation.

1 S. Dogel, W. Freyland, H. Hippler, D. Nattland, C. Nese and A.-N. Unterreiner, *Phys. Chem. Chem. Phys.*, 2003, **5**, 2934–2937.

George Zheng Chen opened a general discussion of Binjie Hu's paper: Your work is very novel and practically important. Emulsion formation depends on the stability of the small spheres and requires the environment to be right, charge balance in particular. When you introduce an ionic liquid, does it help or not help with the stability of the emulsion?

Binjie Hu responded: Our previous work showed that adding small amounts of ionic liquid (*e.g.* 10 wt%) can enhance the stability of miniemulsions due to the increased adsorption of surfactants on the interface;[1] such an effect can be extended to emulsions. However, adding too much ionic liquid (*e.g.* 30 wt%) may destabilize the emulsion due to the unbalance of charge on the interface caused by excessive ionic liquid. Such charge balance may depend on the type of surfactant, pH of the liquid phase and type of ionic liquid being used. The zeta potential value could be a good indicator for surface charge balance.

1 Y. Kong, B. Hu, K.-L. Choy, X. Li and G. Chen, *Soft matter*, 2015, **11**, 1293–1302.

George Zheng Chen said: The matching properties between the solvent and polymers used and the added ionic liquid are important to make a good paint. The addition of ionic liquid is not always helpful, is it?

Binjie Hu answered: That's correct. One of the results in our paper showed that adding too much RTIL in the oil phase caused an increase in droplet size and hence destabilized the miniemulsion, resulting in uncontrollable droplet sizes.

George Zheng Chen enquired: There is the opportunity to combine ionic liquid and surfactant *via* synthesis. If you have the surfactant moiety made into the structure of either or both the anion and cation of the ionic liquid, do you think that it could be a good way to develop the technology?

Binjie Hu replied: Yes, it is possible. In fact, in our paper, the measurements of interfacial tension at different concentrations of C_8mimPF_6 (ionic liquid) in the oil phase showed that adding a small amount of C_8mimPF_6 caused a reduction in interfacial tension between the O/L interface, which confirm the function of interfacial activity for C_8mimPF_6.

George Zheng Chen remarked: It is important that a paint dries after it has been applied to a substrate. Will the fact that ionic liquids do not evaporate easily affect the drying of paint and thus the paint quality?

Binjie Hu responded: No. We dried samples for yield estimations after polymerisation. The drying temperature was 60 °C under vacuum and the sample was dried out after a few hours. Thus, we believe that the ionic liquid is present as an additive in the formula and that after polymerization it no longer freely exists in the liquid phase, *i.e.* it changes its original state and becomes part of the polymer produced. In this case, only water is the solvent, which needs to be evaporated during film formation.

Andrew Doherty commented: In the old days people used linseed oil to make paint and oil cloth (linoleum). In sunlight and/or air, the oil polymerises so the paint dries or, more correctly, cures due to a polymerisation reaction. This is still common practice in the coatings industry. Ionic liquids could be synthesised with a secondary functional group which would allow them to cure *via in situ* polymerisation upon photo-initiation.

Binjie Hu said: This is an interesting point. Linseed oil would act as a drying oil in paint to allow solidification to take place during film formation. This is mainly due to the unsaturated bond which forms a network during film formation. To have a similar function, it might be helpful to synthesise ILs with a secondary group containing an unsaturated bond.

Andrew Doherty queried: In your paper you use the term miniemulsion. That's a new term as I've never heard it before. Normally they are described as simply emulsions or microemulsions, the latter being used even for nano-scale systems. Recently, the term "nanosomes" has been introduced to describe nano-scale emulsions. What is the distinction between miniemulsions, microemulsions and nanosomes, *i.e.* what is the droplet size difference?

Binjie Hu answered: Based on the range of droplet sizes, emulsions can be classified into emulsion (>1 μm), miniemulsion (50–500 nm) and microemulsion (10–100 nm).[1] Nanoemulsion is a colloidal dispersion system composed of small droplets in the range 50–200 nm.[2] The term "nanosomes" has been introduced to describe nano-scale emulsions. It would be useful to know where this statement comes from before I can provide further explanations.

1 C. D. Anderson and E. S. Daniels, *Emulsion Polymerisation and Latex Applications*, Rapra Review Reports, Shrewsbury, UK, 2003.
2 T. Tadros, R. Izquierdo, J. Esquena and C. Solans, *Adv. Colloid Interface Sci.*, 2004, **108**, 303–318.

Andrew Doherty asked: How did you measure the size of the emulsion?

Binjie Hu replied: Do you mean the particle size of the emulsion? It was measured using a Mastersizer 3000 instrument. Please refer to section 2.4.2 of our paper.

Ali Kamali enquired: You have used ILs to fabricate coatings in an environmentally friendly way. What is the environmental impact of the IL itself, especially when you imagine that you want to use it on a larger scale? What are the environmental concerns in terms of using ILs?

Binjie Hu responded: The environmental impact of using ILs, especially on a larger scale, is in the reduction of emissions generated by additives such as conventional coalescence agents, making the coating product emission-free. The concerns in terms of using ILs may be 1) the higher overall cost because of the higher price of RTILs, if the type of RTIL selected is only available in small scale production, 2) the method of fabrication of the RTIL in this formula limits its broad application to all types of coating products because the type of RTIL used can stabilise the miniemulsion in this formula, but may not be suitable for other monomers or additives being used, and 3) potential toxicity on the ecosystem after being discharged if the toxicity of the product has not been examined carefully.

John Slattery commented: Our understanding of the biodegradation of ILs is still at a relatively early stage. The ones used here are likely to be poorly biodegradable. In addition, there is some suggestion that they could be toxic in aquatic environments. The high stabilities of these ILs are probably good for their application in paints, but may not be so good for the environment.

Binjie Hu said: This is true if the environmental issue we are referring to is biodegradability. The advantage of using ILs as additives in coating formulae is that they may contribute less emissions compared to traditional additives, *e.g.* plasticizer, because of their lower vapour pressures. If biodegradability of the paint is the key requirement, then other natural material resources, such as biomass to form resin, could be used to replace the current additives used in paint formulae. In addition, for certain applications such as antibacterial marine coatings, biodegradability is not desirable as it would undermine the overall performance of the coating.

Ye Liu remarked: It's a good idea to use ILs to modulate the properties of the coating. But you use PF_6^- as the counter-anion in your work, which may be toxic in the open air since it isn't stable enough and will result in HF release. Have you thought of replacing PF_6^- with other counter-anions such as acetate?

Binjie Hu answered: In fact, in our work, the original state of the RTILs might have changed after polymerisation. So even if the RTILs being used in the raw material have PF_6^- counter-anions, which may result in HF release, the toxicity of the polymerised RTIL which is associated with polymer could be reduced. Further examination of the final product, rather than the original RTILs being used, is required.

John Slattery said: I believe that there is also some evidence of cation toxicity with imidazolium-based ionic liquids.

Bencan Tang commented: I think there is concern in the audience regarding toxicity — in the old days they used plant oil in paint, maybe this could be one of the solutions to address the problem — modifying the structure of the ionic liquid and polymerising it could possibly be a way to solve the toxicity problem

Binjie Hu replied: Yes, I agree. This is something which was discussed in response to Ye Liu's question.

John Irvine commented: Toxicity isn't all bad — another word for this is antibacterial. Have you thought about that with respect to ionic liquids in paints?

Binjie Hu responded: Indeed. One of the potential functions of ionic liquids being developed in our work is antibacterial, if a suitable type of RTIL can be selected.

Andrew Doherty noted: Another way of looking at Professor Irvine's point regarding the antimicrobial activity of ionic liquids is their biocompatibility. If you can develop ionic liquids that are biocompatible you'll be able to manufacture many unique high-value consumer products from biomaterial feedstocks such as cellulose. One can easily imagine tissue engineering scaffolds, wound management/healing products and personal hygiene products, *etc.*

Binjie Hu said: This might be a good area in which to extend further research that focuses on the biocompatibility of ionic liquids, comparing against substances used in conventional products.

Chaohui Wei communicated: Is it easy to remove the initiators after the process, and if so, how?

Binjie Hu communicated in reply: The initiator should decompose after reaction and thus there is no need for it to be removed.

Chaohui Wei communicated: How do you detect the yield of polymer produced with different reaction times?

Binjie Hu communicated in response: Please refer to section 2.4.1 "Determination of conversion of latex" of our paper.

Ye Liu opened the discussion of the paper by Deshani Gunasekera†: After you used the IL for 3D printing, did you have to recycle the IL?

Deshani Gunasekera responded: In this study the ionic liquid was not recycled after regeneration since the amount of anti-solvent (*i.e.* water) added to each printed pattern was only 0.1 mL. However, when printing higher magnitude structures the possibility of recycling the ionic liquid will need to be considered.

Ye Liu asked: What kind of organic solvent do you use together with the IL for recycling and why?

Deshani Gunasekera replied: Swatloski *et al.* previously showed that it was possible to use acetone and ethanol to precipitate cellulose.[1] Other organic solvents like acetonitrile, dichloromethane[2] and methanol[3] can also be used. These organic solvents are polar in nature and their interactions with ions are similar to those of water. Cellulose can precipitate in the presence of these solvents, whilst they can also be used to dissolve ionic liquid;[1–3] the organic solvent can subsequently be removed from the ionic liquid by evaporation.[4]

1 R. P. Swatloski, S. K. Spear, J. D. Holbrey and R. D. Rogers, *J. Am. Chem. Soc.*, 2002, **124**, 4974–4975.
2 D. A. Fort, R. C. Remsing, R. P. Swatloski, P. Moyna, G. Moyna and R. D. Rogers, *Green Chem.*, 2007, **9**, 63–69.
3 A. P. Dadi, S. Varanasi and C. A. Schall, *Biotechnol. Bioeng.*, 2006, **95**, 904–910.

† Deshani Gunasekera's paper was presented by Majd Eshtaya, *University of Nottingham, Nottingham, UK.* All responses were communicated in reply.

4 N. Sun, M. Rahman, Y. Qin, M. L. Maxim, H. Rodríguez and R. D. Rogers, *Green Chem.*, 2009, **11**, 646–655.

Ye Liu enquired: Can the recovered IL be reused successfully?

Deshani Gunasekera answered: Recycled ionic liquid has previously been shown to be reusable.[1] However, when using recycled ionic liquid it is necessary to check for the presence of impurities such as water.

1 S. Varanasi, C. A. Schall, A. P. Dadi, J. Anderson, K. Rao, G. Kumar and P. Paripati, US8030030 B2, 2011.

Ye Liu commented: In China, 3D printing is very popular. It is a hot topic and an important research area.

John Slattery queried: What is the viscosity limit for these kinds of printers?

Deshani Gunasekera responded: The viscosity range for a Dimatix DMP-2831 is between 12 to 18 mPa s. In addition, ink-jet printability is usually defined in terms of the Ohnesorge number, or more commonly its inverse, the "printability" index, which as a guide needs to be within 1 and 10.

Kathie McGregor asked: Solvents such as DMSO are highly coordinating; is there a particular reason you have used DMSO? Is it possible to use ethanol?

Deshani Gunasekera said: DMSO was selected due to previous evidence for its dissolution of cellulose.[1] Ethanol can be used as an anti-solvent to precipitate out cellulose rather than as a solvent for cellulose dissolution. As future work we will investigate the possibility of printing these samples without adding a co-solvent, which will avoid the use of toxic co-solvents like DMSO.

1 T. Heinze and T. Liebert, *Prog. Polym. Sci.*, 2001, **26**, 1689–1762.

Bin Dong enquired: My research is on the synthesis of energy storage materials with nanoscale primary structures and micro-scale secondary structures. The morphology of the material influences its performance significantly. From these SEM images, the particle size is quite big, about 20 microns. Is it possible to use 3D printing to synthesise nanoscale materials so that we can control the morphology and improve the energy storage properties easily?

Deshani Gunasekera replied: We believe that it may be possible to produce nanoscale materials (using different technologies such as multiphoton lithography or aerosol jetting), but it would require significant investigation and control over the precipitation of the cellulose and the removal of the ionic liquid.

Yiyang Kong asked: Does the ionic liquid have an effect on the contact angle between the cellulose solution and the polyethylene terephthalate (PET) or glass substrates? I am wondering whether the change in contact angle may influence the difference in missing drops shown in Fig. 3 of your paper.

Deshani Gunasekera responded: The missing drops were due to nozzle blockages rather than due to an issue with the contact angle — on occasion nozzles fail and a drop is not deposited.

Yiyang Kong remarked: After the evaporation of co-solvent, the sample contained only ionic liquid and cellulose. What was the distribution of cellulose in the ionic liquid? Did phase separation occur between them, meaning that one phase was rich in ionic liquid and another phase was rich in cellulose?

Deshani Gunasekera answered: This was not investigated in this study as the focus was on the morphology of cellulose after regeneration compared to untreated cellulose (see Fig. 6).

Ali Kamali questioned: The purity of the IL you used is mentioned to be more than 95%. What would the remaining ~5% be? Is this impurity organic or inorganic? Does this have an impact on the process?

Deshani Gunasekera replied: Water was the major impurity that was present in the ionic liquids studied. This was approximately less than 1.8% (determined by Karl Fischer titration). The presence of water may reduce cellulose dissolution and sample viscosity (which may require reformulating the sample to bring it within the ink-jet printing range) and affect the morphology of the regenerated cellulose. As we move forward to printing films and other structures, determining the effects of impurities on the microstructure and physical properties will become necessary.

Ali Kamali added: Can you purify your IL?

Deshani Gunasekera replied: The ionic liquid can be purified by vacuum evaporation to remove any moisture.[1]

1 E. Quijada-Maldonado, S. van der Boogaart, J. H. Lijbers, G. W. Meindersma and A. B. de Haan, *J. Chem. Thermodyn.*, 2012, **51**, 51–58.

Binjie Hu said: The raw ionic liquid materials are only 95% pure. The influence of impurities on the sample structure have not been properly addressed. From the open literature,[1] the presence of 5 wt% of water could affect the crystallinity of cellulose in 1-ethyl-3-methylimidazolium acetate, and thus physical properties such as viscosity and even sample structure may be changed. The co-solvent effect described in this paper may not be only attributable to co-solvent. The effect of water can not be ignored.

1 C. Olsson, A. Idström, L. Nordstiern, G. Westman, *Carbohydr. Polym.*, 2014, **99**, 438–446.

Deshani Gunasekera responded: Yes, we agree that the presence of water cannot be ignored. We performed Karl Fischer titration on the ionic liquids, which showed the water content of 1-ethyl-3-methylimidazolium acetate to be

1.67% (± 0.138) and 1-butyl-3-methylimidazolium acetate to be 1.54% (± 0.00153). Successful dissolution of cellulose suggests that this amount of water had a minimal/negligible effect on this system.

John Slattery commented: Adding cellulose to the ionic liquid is very likely to introduce a significant amount of water anyway. The water content in the ionic liquid may not be such a huge issue for this application.

Deshani Gunasekera said: It is possible that adding cellulose has introduced a small amount of water which could have a minor effect on the physical properties, but it did not appear to affect the dissolution of the cellulose in the ionic liquid.

George Zheng Chen noted: For 3D printing, the ink needs good mechanical, chemical and thermal stabilities. Would cellulose based ink be sufficiently stable for 3D printing?

Deshani Gunasekera answered: The ink was observed to be stable for long periods (~1 h), suggesting that it may be sufficiently stable for 3D printing.

George Zheng Chen queried: Do you think that lignin might be better than cellulose for 3D printing? I think that cellulose is bigger and hence may not be ideal, whilst lignin may be better but it is a polymer. What behaviour would you predict if lignin is mixed into the printing ink?

Deshani Gunasekera replied: All things being the same, lignin may result in a lower viscosity ink, requiring less co-solvent to bring the viscosity into range.

George Zheng Chen remarked: According to some conference presentations I have heard, it seems that the solubility of biomass in ionic liquids might not be a big issue for making the printing ink, which needs to be quite viscous so that the printed object can be mechanically stable to maintain the printed structure and shape. What is the range of viscosity of the ink to enable a stable printed body?

Deshani Gunasekera responded: Our constraints for viscosity are determined by ink-jet printing and fall in the range 12–18 mPa s.

Binjie Hu communicated: On page 3 of your paper you state that the mixtures were stirred using 800 rpm. Why is such a high speed used? This agitation speed could be high enough to generate air bubbles and hence may affect the liquid properties such as viscosity, as well as the structure of the sample. Thus, the homogeneous liquid phase may become a two phase mixture.

Deshani Gunasekera replied: For the dissolution experiments that were carried out at 800 rpm there was no formation of air bubbles or separation of the mixture. The exception was for 1-butanol, where a precipitate formed and as a result it was deemed unsuitable as a co-solvent.

George Zheng Chen commented: Using ILs has several benefits for materials and energy applications, such as good solubility of biomass and low volatility. The issue may be viscosity for 3D printing. The low volatility is linked to the large molecular size of cations and anions which may also contribute to increasing the viscosity. High viscosity means low ionic conductivity which is unfavourable for electrochemical applications.

John Slattery said: Isn't it possible to just wash the salt off with an appropriate solvent?

George Zheng Chen responded: Obviously this is possible in a laboratory, but not an easy job for paint on a substrate with a large area.

Linpo Yu remarked: When we want to wash out ionic liquids from a product, we will choose a proper solvent, like acetonitrile, and then wash the raw product with the solvent several times to dissolve the ionic liquids.

John Slattery asked: Which solvent systems do you use?

Linpo Yu responded: A mixture of 1-butyl-methylpyrrolidinium tri(pentafluoroethyl)trifluorophosphate and gamma-butyrolactone containing lithium perchlorate.

George Zheng Chen opened a general discussion: Professor Madden's work is brilliant, covering both ionic liquids and molten salts. For research, we would like to better understand the similarities and differences between these two types of liquid salt. Professor Madden, would you please offer a comment based on your modelling research?

Paul Madden replied: Thank you for the compliment! I think that "ionic liquids" and "molten salts" are both molten salts. In answer to another question I pointed to work on the similarities between them at the level of the most general thermodynamic behaviour (*e.g.* the work by Woodcock and co-workers) which demonstrates the central role of the coulombic interactions between ions. However, this similarity at the fundamental level may not be very helpful when considering what we need to know under the labels of the two subfields. The molecular complexity of the anions and cations in the "ionic liquids" introduces new interactions, like hydrogen bonding, not present in the "molten salts" and these are responsible for the subtle features which affect the properties of the ionic liquids as solvents or in affecting their physical properties and the process of crystallisation. These subtleties and the effects they give rise to become the objectives of the study of ionic liquids. The underlying coulombic interactions set the stage on which these subtleties are played out but, because the range of thermodynamic states of interest in ionic liquid studies is limited to a fairly narrow temperature range close to freezing (often limited by the thermal stability of the ions) their effects are masked. In molten salts, on the other hand, the ions are spherical - so that coulombic interactions are dominant. Ion size becomes the important parameter which can be used to relate the properties of different materials. The relevant crystalline phases are often easy to predict.

Paul Madden commented: It has been argued that the overall thermodynamic behaviour of "ionic liquids" is the same as that of "molten salts", see **ref. 1 and 2**, for example. The difficulty of sustaining this perspective is that, because ionic liquids tend to thermally decompose at high temperatures, the range of thermodynamic states on which data can be compared is quite limited. Close to freezing, the general thermodynamic picture becomes affected by such details as the relative energies of different crystalline polymorphs.

1 R. Babarao, J. Jiang and L. V. Woodcock, *Ind. Eng. Chem. Res.*, 2011, **50**, 234–238.
2 R. Babarao, J. Jiang and L. V. Woodcock, *Ind. Eng. Chem. Res.*, 2011, **50**, 227–233.

George Zheng Chen queried: At conferences, I often hear reports on building phase diagrams of molten salts from computer modelling, and the results agree well with experimental findings. However, such studies seem rare for ionic liquids. Is it because the ionic liquid research is still progressing, or are there particular difficulties in modelling ionic liquids in terms of their thermal and equilibrium behaviour?

Paul Madden responded: My answer here follows on from that to the previous question. In the molten salts, the significant interactions are (relatively) straightforwardly represented and the complexities of molecular shape avoided. Accurate prediction of material properties on the basis of a systematic approach to modelling is possible. Furthermore the properties of different materials can readily be related through a consideration of the ionic size and the number of possible crystalline phases is limited, so that generic phase diagrams for a whole class of materials involving different cations and anions can be presented. Doing this across different classes of ionic liquids would be difficult because of the subtle interactions mentioned above, though relating the properties of liquids which differ through minor changes in *e.g.* the aliphatic tail of the cation should be possible.

George Zheng Chen said: I would like to address this to people who are experts in the experimental study of ionic liquids. In high temperature molten salts, viscosity seems not to be a great topic of research, and the liquid usually behaves like water in terms of flowability when the working temperature is about 50 °C higher than the melting temperature of the salt. A change of 50 °C is not a big issue when working with molten salts whose liquid temperature range is often as wide as several hundred degrees Celsius. However, for ionic liquids which are supposed to work at room temperature, a 50 °C temperature variation may be too large. For example, in our lab, heating and cooling an ionic liquid between room temperature and 80 °C for several hundred cycles could turn the colourless liquid into a yellowish liquid. Thus, without increasing the temperature, the viscosity of ionic liquids could be an issue for practical uses, *e.g.* as the electrolyte for supercapacitors. Is there any common approach that would be effective without losing the merits of ionic liquids? Also, can the viscosity of ionic liquids be predicted from computer modelling?

John Slattery answered: You're right, sometimes viscosity increases when you change the composition with IL mixtures. We are working on a system like this at the moment and I think the explanation may be related to what Paul was saying about structure in his molten-salt mixtures. The effects that we are seeing in ILs are based on different molecular interactions, but the concept of breaking a network structure under sheer stress may apply here too.

Andrew Mount commented: Following on from Paul Madden's observations on clustering, in contrast to earlier statements that ionic liquids have negligible vapour pressures, I observe that Earle *et al.* demonstrated the distillation of ionic liquids.[1] This clearly shows the importance of neutral aggregates (clusters), the likely distilling species in these systems.

1 M. J. Earle, J. M. S. S. Esperança, M. A. Gilea, J. N. Canongia Lopes, L. P. N. Rebelo, J. W. Magee, K. R. Seddon and J. A. Widegren, *Nature*, 2006, **439**, 831-834.

Yingjun Liu said: Thanks for Paul's talk and Andrew's comments. I am led to think about the case of $AlCl_3$ and Al_2Cl_6. How can the cluster and complex be differentiated in molten salts? In other words, is there a clear boundary between Al_2Cl_6 and $Al_2Cl_7^-$ in molten salts?

Paul Madden commented: An atomic scale snapshot of a molten salt mixture, such as $AlCl_3$–KCl, will reveal quite a number of ionic species, $AlCl^{4-}$, Al_2Cl^{7-} *etc.*, which interconvert rapidly and with average concentrations which are determined by the chemical composition of the melt. In pure liquid $AlCl_3$ you would find a lot of Al_2Cl_6, and this results in a high vapour pressure and low boiling point because Al_2Cl_6 is a neutral molecule and therefore not held into the liquid by the coulombic interactions (*i.e.* by the influence of the Madelung potential). In a KCl-rich mixture, on the other hand, the equilibrium will be pushed towards the ionic species, like $AlCl^{4-}$ and Al_2Cl^{7-}, and consequently this behaves more like a normal molten salt with a relatively low vapour pressure (see *e.g.* The 'ionic' to 'molecular' transitions in $AlCl_3$ and $FeCl_3$ as predicted by an ionic interaction model.[1]

In a simulation, you can characterise which ions are associated with each other at a given instant by looking at their separations. As you can see from the cation–anion radial distribution functions in Fig. 1 of the paper, the first coordination shell is well defined and well separated from the second shell of non-coordinated ions. In similar rdfs for $AlCl_3$-containing mixtures, these peaks are even more sharply defined. So that, by examining the instantaneous separations between a cation and neighbouring anions in the simulation, you can count the number of anions coordinated to a given cation using a distance criterion, and distinguish between bridging anions, as in Al–Cl–Al and "dangling" ones, Al–Cl. You can also see which cations are linked to each other by bridging anions, as in Fig. 11 of the paper. By examining all the associations of all the cations you can characterise the speciation.[2] These species are all interconverting and the timescale of this can be quite rapid – tens of picoseconds. The interesting challenge is to relate the nature of these species and the way they interconvert to the physical properties of the melt (as we attempt to do in the paper).

1 F. Hutchinson, M. K. Walters, A. J. Rowley and P. A. Madden, *J. Chem. Phys.*, 1999, **110**, 5821.
2 M. Salanne, C. Simon, P. Turq, and P. A. Madden, *J. Phys. Chem. B*, 2007, **111**, 4678–4684.

Faraday Discussions

PAPER

A perspective on liquid salts for energy and materials

J. T. S. Irvine

Received 24th July 2016, Accepted 26th July 2016
DOI: 10.1039/c6fd00173d

Liquid salts comprising molten salts and ionic liquids offer important media to address both energy and materials challenges. Here we review topics presented in this Faraday Discussion volume related to improved electrowinning of metals, optimisation of processes, new electrochemical device concepts, chemistry in ionic liquids, conversion of biomass, carbon chemistry and nuclear applications. The underlying phenomenology is then reviewed and commentary given. Some future applications are then discussed, further exemplifying the high potential rewards achievable from these chemistries.

Introduction

Molten salt chemistry is an area of considerable technical importance; however, it has somewhat slipped from the mainstream of chemical focus over the decades. With the advent of ionic liquid chemistry and rediscovery of the power of molten salt chemical processes, attention is increasingly turning back to molten salt chemistry, especially in areas such as energy materials.

Molten salts consist entirely of ions and are characterised by high melting points (*e.g.* >200 °C) due to strong bonding between anions and cations. In the solid state they exhibit very low ionic conductivity and negligible electronic conductivity. Typically there is a very large volume change on melting (~25%) creating significant free volume in the molten state. They exhibit high ionic conductivity and negligible electronic conductivity with modest viscosity. Application of a voltage above decomposition of melt (~2–3 V) results in decomposition, *e.g.* for LiCl this causes the production of lithium on the cathode and evolution of chlorine on the anode.

Ionic liquids are closely related to molten salts, although they tend to be liquid at or near to ambient temperature. They offer good solvation characteristics and high ionic conductivity but are not necessarily highly polar.

Molten salts are widely utilised in many important applications. Examples include the production of aluminium through cryolite electrolysis in the Hall–Heroult process, the production of lithium through the electrolysis of lithium

School of Chemistry, University of St. Andrews, St Andrews, Fife KY16 9 ST, UK. E-mail: jtsi@st-andrews.ac.uk

chloride[1] and the production and purification of titanium as in the FFC Cambridge process.[2]

Molten salt chemistry research can contribute to a reduction in energy costs and use in a number of ways. Firstly by improving efficacy of existing processes for producing high energy metals (*e.g.* Al, Li, Si and Ti), secondly by developing new metallurgical processes, thirdly by bringing forward new energy concepts that offer improved storage and conversion of energy and fourthly through creating new energy materials that can enable this. Here we consider various possible approaches to achieve these objectives.

Improved electro-winning

Whilst the activity available in molten salt chemistry greatly enables the technology, this also creates major issues in achieving stable operation, especially relating to corrosion. For example, providing an inert anode for electro-winning a metal from an oxide is a fairly stringent test. To achieve this, the following material properties are required:[3]

1. Insolubility in a molten salt electrolyte up to 1000 °C
2. Thermal stability up to 1000 °C with adequate resistance to thermal shocks
3. Resistance to the oxygen produced at the anode
4. Low electrical resistivity
5. Low oxygen over-potential
6. Resistance toward fluoridation/chlorination
7. Does not contaminate the electro-won metal
8. Adequate mechanical strength and stable electrical connections
9. Low cost and easy fabrication into large shapes

This often involves formation of a stable corrosion layer at the surface that blocks further degradation. There is also interest in novel compositions such as Nb doped TiO_2 (ref. 3) to improve electrode stability.

Efficient electro-reduction of Ta_2O_5 to Ta has been achieved at a graphite electrode in a $MgCl_2$–NaCl–KCl melt at 700 °C *via* a TaO_2 intermediate. This new composition has been tuned to utilise the $MgCl_2$ to support the metal production and increase the longevity of the anode, with MgO and Cl_2 being the anodic products, avoiding decomposition of the carbon anode.[4] Similarly Mg produced *via* Mg reduction/deoxidation in electrolysis of $MgCl_2$ has been utilised to remove residual oxygen from a Ti cathode yielding less than 100 ppm(O) Ti.[5]

Optimisation of processes

The production of aluminium from its ore is energy intensive and has a large carbon footprint. One approach to reduce CO_2 output is to utilise natural gas instead of carbon at the anode in the electro-winning of aluminium from alumina in molten fluorides. This produces water as well as carbon dioxide and effectively reduces the CO_2 output by a factor of 2.[6]

A solid oxide ion conducting membrane has been applied to a number of metal carbide syntheses in $CaCl_2$ at 900 °C.[7] The membrane separates the anodic and cathodic processes and affords more precise control of the carbon stoichiometry. The process can also yield intricate nanostructured carbides as shown schematically in Fig. 1.

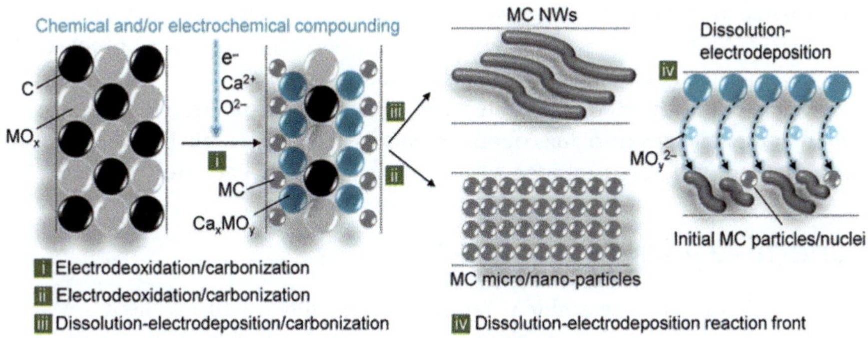

Fig. 1 Schematic representation of the route from metal oxide to metal carbide *via* electro-reduction of MO_x/C MCs in molten $CaCl_2$, reproduced by permission from ref. 7.

In another approach, electrochemical reduction of SiO_2 in the presence of NiO has been shown to yield Si nanowires with diameters 60–150 nm.[8] Here, silica was reduced to silicon at the Ni–Si alloy droplet/SiO_2/$CaCl_2$ three-phase interphase and dissolved into the alloy droplets, in a tip growth process of Si nanowires, Fig. 2.

Devices

Molten salt chemistry is proving particularly powerful for developing nanomaterials for energy applications. This is highly important in energy storage devices such as batteries and supercapacitors. Supercapacitors act to store charge at high surface area interfaces and tend to operate over a potential range. Their best mode of operation is in storing large amounts of energy to be delivered in a short timescale, *i.e.* in power driven applications. Batteries tend to store energy within the materials and, at least for traditional insertion/intercalation chemistry, operate over well-defined potential ranges. Batteries are best suited to storing energy over longer timescales than supercapacitors. With the development of conversion chemistries,[9] this distinction has become blurred with materials offering characteristics intermediate between batteries and supercapacitors. Chen has formalised this concept in terms of a supercapattery[10] and has reported this concept using an ionic liquid solution of $LiClO_4$.[11] An interesting example of

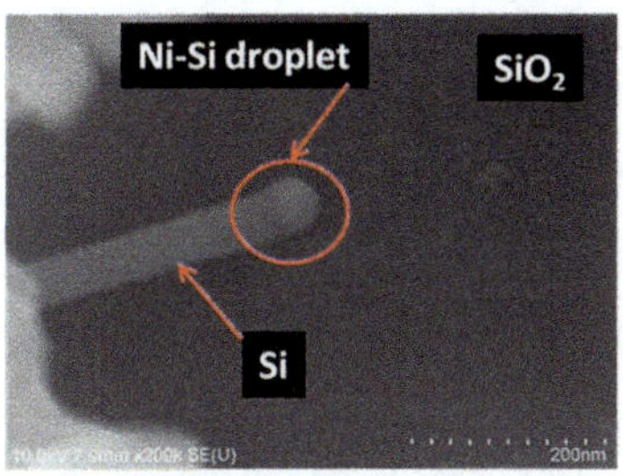

Fig. 2 Tip growth of Si nanowires on SiO_2 electroreduction in molten $CaCl_2$ facilitated by the presence of NiO, reproduced by permission from ref. 8.

such behaviour with considerable charge storage spread over 3 V is in AlFe nanowires.[12] Here, the electrode was based on very fine nanowires prepared by electrodeposition on copper in an $AlCl_3$ based ionic liquid.

Pringle *et al.* reported an exciting new concept, thermoelectrochemical cells wherein the changes in redox couple potentials of a redox ionic liquid electrolyte with temperature were exploited to generate electricity in an effective device.[13]

Chemistry in ionic liquids

Ionic liquids offer near ambient temperature systems with excellent solvating ability to perform chemistry in the absence of water. A highly efficient and stable system for hydro-formylation has been developed,[14] this involves a biphasic system with a rhodium catalyst dissolved in one ionic liquid phase and a diphosphine functionalised ionic liquid in the other. This chemistry is now being applied to a range of processes including processing of biomass.

Biomass

Lignin is the most abundant biopolymer after cellulose and is particularly difficult to convert into useful chemicals in an economically sustainable manner. Finding alternative technologies to process this would have great benefit to concepts such as biorefineries. In an innovative approach reported by the Nottingham team, the mediator 2,2′-azino-bis(3-ethylbenthiazoline-6-sulfonic acid) diammonium salt (ABTS) has been evaluated by means of cyclic voltammetry for enhancing the oxidation of the three types of lignin – organosolv, Kraft and lignosulfonate, in the ionic liquid 1-ethyl-3-methylimidazolium ethyl sulfate, $[C_2mim][C_2SO_4]$,[15] Fig. 3.

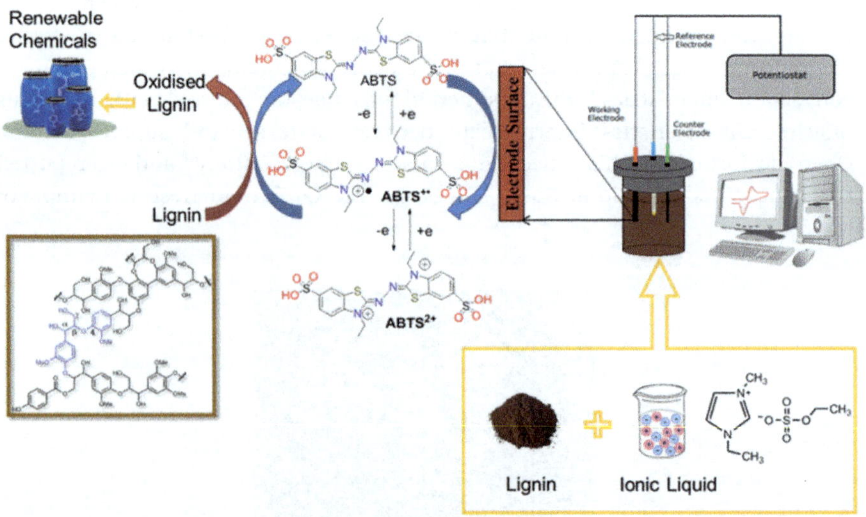

Fig. 3 Scheme showing electro-oxidation of lignins in ionic liquid, reproduced by permission from the Liquid Salts for Energy and Materials Faraday Discussion presentation.[17]

In a different approach, carbon nano-sheets were obtained from a typical waste biomass, used coffee beans, through heating in carbonate and chloride molten salts.[16] These carbons showed high capacitance and demonstrated good oxygen reduction kinetics when tested in a symmetric capacitor and alkaline cell modes, respectively. Carbonate derived carbons had a smaller surface area, they exhibited less surface functional groups and increased order in structure and performance was improved in both modes.

Carbons

An alternative approach to CO_2 capture *via* gas sequestration is to convert CO_2 to carbon. This can be achieved *via* the electroreduction of CO_2 in a molten lithium carbonate/chloride mixture.[18] Morphologies can be controlled by modifying electrochemical conditions, including substrates, and a range of quasi spherical and nanofibrous particles have been achieved in short term electrolysis studies.

In a similar approach, nanoflakes were produced on electrolysis of CO_2 in LiCl/KCl molten electrolytes that also contained carbonates.[19]

An interesting route to high quality graphene sheets has been found to involve electrochemical exfoliation of graphite in molten LiCl. This process was found to be greatly enhanced by the presence of a little water in the molten chloride.[20]

An important process that operates in reverse is the Direct Carbon Fuel Cell (DCFC). This concept attempts to convert carbon sources directly to CO_2 electrochemically and so to obtain maximal conversion efficiency.[21] This process can in one configuration utilise a molten carbonate mediator at a solid oxide electrode to achieve high performance. Commonly, CO is a major product, *via* Boudouard processes, reducing potential efficiency. This contrasts with the reverse systems mentioned in preceding paragraphs that convert CO_2 to C with minimal CO. It is important to note, however, that the carbon capture experiments tend to be at lower temperature which would reduce the influence of the CO producing reactions.

A remarkable observation is the spontaneous combustion of a carbonate/xerogel spent mixture after DCFC testing under ambient conditions.[22] This is a strong indication of the presence of exotic, highly reactive species such as peroxide in molten salt chemistry.

Nuclear applications

Pyrochemical reprocessing of spent nuclear fuels is an important application involving molten salts. In a typical process LiCl–KCl molten salt is used as the medium for electrorefining of spent fuels. Liquid metal cathodes, such as Cd, Bi and Al, are utilised to extract and separate actinides and lanthanides and this can be suitably monitored utilising cyclic voltammetry identifying a series of intermetallic compositions in model Ce and La systems.[23] Molten salt stable microelectrodes have been developed *via* microfabrication techniques to perform *in situ* electroanalysis and utilised to characterise intermetallic formation in model molten Bi systems.[24] The thermal diffusivity of various molten carbonate electrolytes has also been determined as a function of temperature using a novel laser flash system.[25]

High temperature ^{35}Cl NMR has been utilised to probe changes in local structure on dissolving the model compound $CeCl_3$ in LiCl.[26] Coordination can also be probed by techniques such as EXAFS, and molecular dynamic simulations provide highly complementary information to such experimental measurements on coordination chemistry of dissolved lanthanides or actinides.[27]

Underlying phenomena/chemistry

^{35}Cl NMR has also been utilised to investigate the changes in liquid dynamics through probing spin lattice relaxation time across the molten LiCl–KCl system. Surprisingly the variation in T1 activation energy was not linear with composition and instead a maximum was observed near the eutectic composition.[26] As motion might be expected to be easier in the lower melting eutectic region, this suggests that location of the maximum E_a at the eutectic is coincidental. Perhaps this may relate to the mixed alkali effect,[28] supposed to be related to entropic effects, which shows a hindrance to ionic motion in liquids or glasses on introduction of secondary ions, *i.e.* K in molten LiCl.

In the chloride melts used for electro-refining of titanium, titanium is usually present in 2+ or 3+ states. Addition of fluoride has been shown to facilitate Ti formation through facilitating disproportionation reactions such as $3Ti^{2+} \rightarrow 2Ti^{3} + Ti$.[29]

Comments

There is a rich and diverse chemistry that can occur in molten salt electrolytes, that gives rise to their many important potential applications. Species that are almost unthinkable in more familiar systems such as conventional aqueous and non-aqueous electrolytes, or indeed in solid state reactions, are encountered in the molten state. This can include small amounts of dissolved metals in melts, as well as peroxide and superoxide species. The Lux–Flood definition of acid–base theory is helpful in understanding some of these chemistries.[30] This definition describes an acid as an oxide ion (O^{2-}) acceptor and a base as an oxide ion donor.

$$MgO\ (base) + CO_2\ (acid) \rightarrow MgCO_3$$

This can help to explain the suitability of molten carbonates and hydroxides to grow high oxidation state crystals *e.g.* peroxide, Cu^{3+} and Bi^{5+}.[31] Another example is the attack of zirconia electrolytes by peroxidic species in molten carbonate assisted direct carbon fuel cells. In fact, this seems only to be a problem in oxidising conditions and is not a problem under the reducing conditions encountered under operating conditions.[21] In molten hydroxides and carbonates we can anticipate the presence of species such as superoxide, O_2^{-}; peroxide, O_2^{2-}; and percarbonate, CO_4^{2-} especially on the anodic side.[32] The presence of small concentrations of dissolved alkali metals is also possible in cathodic environments, giving further opportunities for "exotic" chemistry.

A useful concept to further explore the nature of such processes in molten salts is optical basicity. This concept was first developed by John Duffy for solid oxides and glasses parametrising the basicity through spectroscopic probe ions.[33] This has recently been transferred to low melting salts such as ionic liquids (ILs).[34] These were doped with $Mn(NTf_2)_2$ (NTf_2 = bis(trifluoromethanesulfonyl)imide) as a probe of the IL anion basicity by means of UV–Vis spectroscopy. The $^6A_1 \rightarrow {}^4A, {}^4E(G)$ transition band of Mn^{2+} is influenced by the electron donor properties, *i.e.*, Lewis basicity, of the coordinating ligands but remains—to the first order—independent from the ligand field. IL anions showed an order of $BF_4^- < ClO_4^- < NTf_2^- \sim MeSO_3^- < NO_3^- <$ $DCA^- < TFA^- < EtSO_4^- \sim OTos^- < Cl^- < Br^-$ with respect to rising optical basicity.

Immediate applications

There are many applications of ionic liquids and molten salts that may come to fruition in the near future. For ionic liquids, examples include use in paints to reduce volatile organic compounds[35] and as a solvent in 3D ink jet printing.[36]

Molten salt electrochemistry also finds many important applications such as electrorefining of Neodymium, which is so important in wind turbine magnets, and in molten LiF electrolytes.[37] A particularly exciting application is the production of ammonia from nitrogen using renewable electricity. A new approach involving reduction of nitrogen in a cathode compartment with water oxidation in the anode compartment, with H-transfer through a separating membrane, has recently been described, Fig. 4.[38]

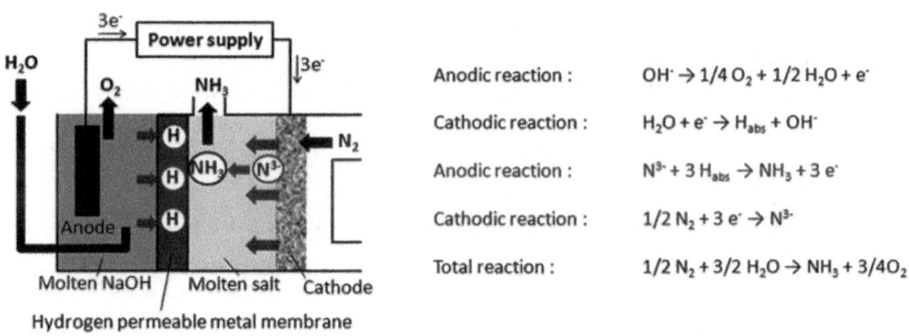

Fig. 4 Schematic of ammonia production in separated cells using molten hydroxide and chloride electrolytes, reproduced by permission from ref. 38.

Conclusions

Liquid salts offer a very exciting environment for developing new chemical and electrochemical approaches. Environments are often aggressive and hard to manage; however, this offers new routes and functionalities to find solutions that may not otherwise be accessible. In short this is an important area that is re-emerging with great potential for new advances.

Acknowledgements

The author thanks EPSRC for platform support, EP/K015540/1 and the Royal Society for a Wolfson Merit Award WRMA 2012/R2.

References

1 A. Cox and D. J. Fray, *Trans. Inst. Min. Metall., Sect. C*, 1997, **106**, C123.

2 D. J. Fray, G. Z. Chen and T. W. Farthing, *Nature*, 2000, **407**, 361.

3 G. Snook, K. McGregor, A. J. Urban, M. R. Lanyon, R. Donelson and M. I. Powinceby, *Faraday Discuss.*, 2016, DOI: 10.1039/c5fd00235d.

4 Y. Yuan, W. Li, H. Chen, Z. Wang, X. Jin and G. Z. Chen, *Faraday Discuss.*, 2016, DOI: 10.1039/c5fd00231a.

5 T. Okabe, Y. Hamanaka and Y. Taninouchi, *Faraday Discuss.*, 2016, DOI: 10.1039/c5fd00229j.

6 G. M. Haarberg, B. Khalaghi and T. Mokkelbost, *Faraday Discuss.*, 2016, DOI: 10.1039/c5fd00239g.

7 X. Zou, K. Zheng, X. Lu, Q. Xu and Z. Zhou, *Faraday Discuss.*, 2016, DOI: 10.1039/c5fd00221d.

8 S. Fang, H. Wang, J. Yang, B. Yu and S. Lu, *Faraday Discuss.*, 2016, DOI: 10.1039/c5fd00222b.

9 P. Poizot, S. Laruelle, S. Grugeon, L. Dupont and J.-M. Tarascon, *Nature*, 2000, **407**, 496.

10 B. Akinwolemiwa, C. Peng and G. Z. Chen, *J. Electrochem. Soc.*, 2015, **162**, A5054.

11 L. Yu and G. Z. Chen, *Faraday Discuss.*, 2016, DOI: 10.1039/c5fd00232j.

12 G. Chen, Y. Chen, H. Wang, Q. Guo and B. Li, *Faraday Discuss.*, 2016, DOI: 10.1039/c5fd00211g.

13 J. He, D. Al-Masri, D. R. MacFarlane and J. M. Pringle, *Faraday Discuss.*, 2016, DOI: 10.1039/c5fd00238a.

14 Y.-Q. Li, H. Liu, P. Wang, D. Yang, X. Zhao and Y. Liu, *Faraday Discuss.*, 2016, DOI: 10.1039/c5fd00195a.

15 M. Eshtaya, A. Ejigu, G. Stephens, D. A. Walsh, G. Z. Chen and A. K. Croft, *Faraday Discuss.*, 2016, DOI: 10.1039/c5fd00226e.

16 B. Lu, J. Zhou, Y. Song, H. Wang, W. Xiao and D. Wang, *Faraday Discuss.*, 2016, DOI: 10.1039/c5fd00215j.

17 M. Eshtaya, A. Ejigu, G. Stephens, D. A. Walsh, G. Z. Chen and A. K. Croft, private communication.

18 J. Ge, L. Hu, Y. Song and S. Jiao, *Faraday Discuss.*, 2016, DOI: 10.1039/c5fd00217f.

19 B. Deng, Z. Chen, M. Gao, Y. Song, K. Zheng, J. Tang, W. Xiao, X. Mao and D. Wang, *Faraday Discuss.*, 2016, DOI: 10.1039/c5fd00234f.

20 A. R. Kamali and D. J. Fray, *Nanoscale*, 2015, **7**, 11310.

21 C. Jiang, J. Ma, A. D. Bonaccorso and J. T. S. Irvine, *Energy Environ. Sci.*, 2012, **5**, 6973.

22 X. Yue, A. Arenillas and J. T. S. Irvine, *Faraday Discuss.*, 2016, DOI: 10.1039/c6fd00001k.

23 Y. L. Liu, K. Liu, L. Y. Yuan, Z. F. Chai and W. Q. Shi, *Faraday Discuss.*, 2016, DOI: 10.1039/c5fd00220f.

24 D. Corrigan, J. Elliott, E. Blair, S. Reeves, I. Schmueser, A. J. Walton and A. R. Mount, *Faraday Discuss.*, 2016, DOI: 10.1039/c6fd00002a.

25 X. An, J. Cheng, P. Zhang, Z. Tang and J.-Q. Wang, *Faraday Discuss.*, 2016, DOI: 10.1039/c5fd00236b.

26 H. Zhang and I. Farnan, *Faraday Discuss.*, 2016, DOI: 10.1039/c6fd00003g.
27 D. Corradini, P. Madden and M. Salanne, *Faraday Discuss.*, 2016, DOI: 10.1039/c5fd00223k.
28 J. O. Isard, *J. Non-Cryst. Solids*, 1968, **1**, 235.
29 J. Song, Q. Wang, J. Wu, S. Jiao and H. Zhu, *Faraday Discuss.*, 2016, DOI: 10.1039/c6fd00007j.
30 H. Flood and T. Forland, *Acta Chem. Scand.*, 1947, **1**, 592.
31 T. T. Nguyen and H. C. zur Loye, *J. Cryst. Growth*, 1997, **172**, 182.
32 L.-J. Chen, C.-J. Lin, J. Zuo, L.-C. Song and C.-M. Huang, *J. Phys. Chem. B*, 2004, **108**, 7553.
33 J. A. Duffy and M. D. Ingram, *J. Non-Cryst. Solids*, 1976, **21**, 373.
34 S. Pitula and A. Mudring, *Phys. Chem. Chem. Phys.*, 2010, **12**, 7056.
35 B. Hu, Y. Kong, R. Zheng, J. Dong, K.-L. Choy and H. Zhao, *Faraday Discuss.*, 2016, DOI: 10.1039/c6fd00009f.
36 D. H. A. T. Gunasekera, S. Kuek, D. Hasanaj, Y. He, C. Tuck, R. D. Wildman and A. K. Croft, *Faraday Discuss.*, 2016, DOI: 10.1039/c5fd00219b.
37 C. Huang, X. Liu, Y. Gao, S. Liu and B. Li, *Faraday Discuss.*, 2016, DOI: 10.1039/c6fd00014b.
38 Y. Ito, T. Nishikiori and H. Tsujimura, *Faraday Discuss.*, 2016, DOI: 10.1039/c5fd00237k.

Poster titles

Development of $M_{n+1}AX_n$ phase materials for use in molten salt reactors, **D. Cooper, K. Abrams, H. Lambert, C. Jantzen, M. Rainforth, M. Ogden, T. Abram and K. R. Whittle**, *University of Sheffield, United Kingdom*

Bio-electrocatalytic ionic liquids, **A. P. Doherty and L. Graham**, *Queen's University of Belfast, United Kingdom*

Energy harvesting and storage application of ambient-temperature liquid salts, **A. P. Doherty, L. Graham, E. Marley, K. K. Wagner, D. L. Officer, J. Chen and G. G. Wallace**, *Queen's University of Belfast, United Kingdom*

Electrochemical properties of Si–Mn alloy prepared by molten salt electrolysis from SiO_2–MnO_2, **Y. Gao, Y. Sun, Z. Qin and X. Wang**, *North China University of Science and Technology*, *China*

Production of Ti-6Al-4V *via* molten salt electrolysis: roles of metal Co-oxides in the process, **D. Hu, A. J. Stevenson and G. Z. Chen**, *University of Nottingham Ningbo, China*

Investigation of the performance of potential primary/secondary circuit materials for molten salt reactors, **C. Jantzen, T. Abram, D. Engleberg and R. Harrison**, *University of Manchester, United Kingdom*

Methane-supplied gas anodes for electrowinning of aluminium, **B. Khalaghi, H. Gudbrandsen, O. S. Kjos, K. S. Osen, T. Mokkelbost and G. M. Haarberg**, *Norwegian University of Science and Technology*, *Norway*

Effect of alkyl spacer on the ionic conductivity of polyanion-type ionic liquids, **O. Kuzmina and H. Ohno**, *Imperial College London*, *United Kingdom*

Electrochemical synthesis of NbC–Sn composite powder in molten chloride, **H. Li, Q. Song, Q. Xu, Y. Chen and J. Meng**, *Northeastern University*, *China*

Bifunctional ionic liquid containing phosphine and Lewis acidic phosphonium for tandem hydroformylation-acetalization, **Y.-Q. Li, H. Liu, P. Wang and Y. Liu**, *East China Normal University*, *China*

Separation of uranium and lanthanum on a liquid Ga electrode in LiCl–KCl eutectic, **K. Liu, J. Pang, Y. Liu, L. Luo, L. Yuan, Z.-F. Chai and W.-Q. Shi**, *Institute of High Energy Physics*, *China*

Electrochemistry of Tb(III)/Tb(0) couple at liquid Zn electrode and electrochemical properties of Tb–Zn alloys in molten salt, **L.-X. Luo, Y.-L. Liu, N. Liu,**

L.-Y. Yuan, Z.-F. Chai and W.-Q. Shi, *Institute of High Energy Physics, Chinese Academy of Sciences, China*

Electrochemical deposition of TaC film on Ta substrate in molten LiCl–KCl–K_2CO_3, **Q. Song, J. Meng, H. Li, Q. Xu and Y. Chen**, *Northeastern University*, *China*

Metal nitrides for electrochemical ammonia synthesis in molten salt systems, **T. Sudmeier, I. McPherson, S. C. E. Tsang and I. Wilkinson**, *University of Oxford*, *United Kingdom*

Electrochemical extraction of copper and iron from sulfide or minerals in molten NaCl–KCl salts, **M. Tan, Z. Wang and X. Jin**, *Wuhan University*, *China*

Direct electrochemical reduction of chalcopyrite in molten KCl–NaCl, **M. Tan, Z. Wang and X. Jin**, *Wuhan University*, *China*

Ru-complex ligated by phosphine-functionalized ionic liquids for tandem hydroformylation-hydrogenation of 1-octene, **P. Wang, H. Liu, Y.-Q. Li and Y. Liu**, *East China Normal University*, *China*

Redox behavior of zirconium in molten LiF–KF–ZrF_4, **L. Xu, Q. Xu, Y. Xiao, A. van Sanwijk, Q. Song and Y. Yang**, *Northeastern University*, *China*

Electrochemistry of acetylide anion and anodic formation of carbon film in a LiCl–KCl–$CaCl_2$ melt, **Y. Chen, Q. Xu and Q. Song**, *Shanghai University*, *China*

NiS nanobelts as positive materials for Al-ion batteries, **Z. Yu and S. Jiao**, *University of Science and Technology Beijing*, *China*

Environment friendly Sb-based positive electrodes for high performance liquid metal batteries, **H. Li, X. Guo, K. Wang, Z. Li and K. Jiang**, *Huazhong University of Science and Technology*, *China*

Electrochemical synthesis of advanced energy materials in molten salts, **K. Wang, S. Huang, W. Li, H. Li and K. Jiang**, *Huazhong University of Science and Technology*, *China*

Effect of the electric field on electrolysis efficiency of TiO_2 electrolysis in molten $CaCl_2$, **M. Hu, Z. Qu, C. Bai, X. Lv and G. Qiu**, *Chongqing University*, *China*

Molten fluorides purification for FHRs, **Y. Zuo, S. Zhao, R. Tang and Y. Wang**, *Shanghai Institute of Applied Physics*, *China*

Synchrotron-radiation technology studies on the molten salt structure, **L. Zhang and J. Wang**, *Shanghai Institute of Applied Physics*, *China*

Thermal physical properties of molten salts in MSR, **J.-H. Cheng, X.-H. An, K. Wang, H.-Q. Yin and P. Zhang**, *Shanghai Institute of Applied Physics*, *China*

Preparation of nanostructured lamellar materials in molten salts, **J. Zhou, W. Xiao and D. Wang**, *Wuhan University*, *China*

Molten salt structure analysis by absorption spectroscopy, **T. Su, P. Zhang, H.-T. Liu and J. Wang**, *Shanghai Institute of Applied Physics*, *China*

Corrosion behaviour of different construction materials in flinak with addition of fluorides salts based on elements of the corroded materials, **M. Boča, V. Pavlík and M. Kontrík**, *Slovak Academy of Sciences*, *Slovakia*

CO_2 enhanced processing of oil shale, **L. Tang, T. Wu and C. Peng**, *University of Nottingham Ningbo*, *China*

The Skinner Prize for the best poster was jointly awarded to Tim Sudmeier of University of Oxford, United Kingdom, for his poster on metal nitrides for electrochemical ammonia synthesis in molten salt systems, and Qiushi Song of Northeastern University, China, for her poster on electrochemical deposition of TaC film on Ta substrate in molten $LiCl–KCl–K_2CO_3$.

List of participants

Mr Bamidele Akinwolemiwa, *University of Nottingham Ningbo China*, *China*
Mrs Xuehui An, *Shanghai Institute of Applied Physics, Chinese Academy of Sciences*, *China*
Dr Chunhao Bao, *AMETEK Measurement Technology*, *China*
Miss Biyash Bhattacharya, *University of Nottingham Ningbo China*, *China*
Dr Miroslav Boca, *Institute of Inorganic Chemistry, Slovak Academy of Sciences*, *Slovakia*
Mr Gang Chen, *East China University of Science and Technology*, *China*
Professor George Chen, *University of Nottingham*, *United Kingdom*
Ms Ying Chen, *Northeastern University*, *China*
Mr Jinhui Cheng, *Shanghai Institute of Applied Physics, Chinese Academy of Sciences*, *China*
Mr Daniel Cooper, *University of Sheffield*, *United Kingdom*
Dr Andrew Doherty, *Queens University Belfast*, *United Kingdom*
Mr Bin Dong, *University of Nottingham Ningbo China*, *China*
Miss Majd Eshtaya, *University of Nottingham*, *United Kingdom*
Dr Ian Farnan, *University of Cambridge*, *United Kingdom*
Professor Yun Gao, *North China University of Science and Technology*, *China*
Mr Jianbang Ge, *University of Science and Technology Beijing*, *China*
Mr Xiaolin Guo, *Huazhong University of Science and Technology*, *China*
Professor Geir Martin Haarberg, *Norwegian University of Science and Technology*, *Norway*
Dr Binjie Hu, *University of Nottingham Ningbo China*, *China*
Dr Di Hu, *University of Nottingham Ningbo China*, *China*
Professor Meilong Hu, *Chongqing University*, *China*
Ms Tiki Hu, *Royal Society of Chemistry*, *China*
Mr Shu Huang, *The Huazhong University of Science and Technology*, *China*
Mrs Xiani Huang, *University of Nottingham Ningbo China*, *China*
Mr Nasiru Ibrahim, *University of Nottingham Ningbo China*, *China*
Professor John Irvine, *University of St Andrews*, *United Kingdom*
Professor Yasuhiko Ito, *IMSEP Company, Ltd.*, *Japan*
Mr Craig Jantzen, *University of Manchester*, *United Kingdom*
Dr Cairong Jiang, *University of St Andrews*, *United Kingdom*
Professor Shuqiang Jiao, *University of Science and Technology Beijing*, *China*
Dr Xianbo Jin, *Wuhan University*, *China*
Dr Ali Kamali, *University of Cambridge*, *United Kingdom*
Mr Babak Khalaghi, *Norwegian University of Science & Technology*, *Norway*
Mr Yiyang Kong, *University of Nottingham Ningbo China*, *China*
Dr Olga Kuzmina, *Imperial College London*, *United Kingdom*
Ms Amy Lam, *Royal Society of Chemistry*, *China*
Miss Hongmei Li, *Northeastern University*, *China*
Dr Wei Li, *Guangdong Academy of Sciences*, *China*
Miss Yong-Qi Li, *East China Normal University*, *China*

Mr Huan Liu, *East China Normal University*, *China*
Dr Huijun Liu, *Institute of Metal Research, Chinese Academy of Sciences*, *China*
Mr Kui Liu, *Institute of High Energy Physics, Chinese Academy of Sciences*, *China*
Mr Xiaolong Liu, *East China University of Science and Technology*, *China*
Dr Yalan Liu, *Institute of High Energy Physics, Chinese Academy of Sciences*, *China*
Professor Ye Liu, *East China Normal University*, *China*
Dr Yingjun Liu, *University of Cambridge*, *United Kingdom*
Mr Liangping Long, *University of Nottingham Ningbo China*, *China*
Miss Sarah Lu, *Royal Society of Chemistry*, *China*
Mrs Lixia Luo, *Institute of High Energy Physics, Chinese Academy of Sciences*, *China*
Professor Paul Madden, *The Queen's College, Oxford*, *United Kingdom*
Dr Richard Massey, *Royal Society of Chemistry*, *United Kingdom*
Dr Kathie McGregor, *Commonwealth Scientific and Industrial Research Organisation*, *Australia*
Dr Heather Montgomery, *Royal Society of Chemistry*, *United Kingdom*
Professor Andrew Mount, *The University of Edinburgh*, *United Kingdom*
Professor Toru Okabe, *University of Tokyo*, *Japan*
Ms Hongmei Peng, *Royal Society of Chemistry*, *China*
Professor Zhangquan Peng, *Changchun Institute of Applied Chemistry, Chinese Academy of Sciences*, *China*
Dr Jennifer Pringle, *Deakin University*, *Australia*
Mr Zhengfeng Qu, *Chongqing University*, *China*
Ms Wendy Ren, *University of Nottingham Ningbo China*, *China*
Dr Victoria Richards, *Royal Society of Chemistry*, *United Kingdom*
Ms Mia Shen, *University of Nottingham Ningbo China*, *China*
Professor Weiqun Shi, *Institute of High Energy Physics, Chinese Academy of Sciences*, *China*
Dr John Slattery, *The University of York*, *United Kingdom*
Dr Qiushi Song, *Northeastern University*, *China*
Mr Tim Sudmeier, *University of Oxford*, *United Kingdom*
Mr Mingsheng Tan, *Wuhan University*, *China*
Dr Bencan Tang, *University of Nottingham Ningbo China*, *China*
Dr Hiromitsu Urakami, *Royal Society of Chemistry*, *Japan*
Professor Dihua Wang, *Wuhan University*, *China*
Professor Han Wang, *General Research Institute for Non-ferrous Metals*, *China*
Professor Jian-Qiang Wang, *Shanghai Institute of Applied Physics, Chinese Academy of Sciences*, *China*
Dr Jinxia Wang, *Northeastern University*, *China*
Mr Peng Wang, *East China Normal University*, *China*
Miss Qian Wang, *Institute of Metal Research, Chinese Academy of Sciences*, *China*
Zhaowen Wang, *Dongbei University*, *China*
Miss Chaohui Wei, *University of Nottingham Ningbo China*, *China*
Yanke Wu, *General Research Institute for Nonferrous Metals*, *China*
Dr Wei Xiao, *Wuhan University*, *China*
Dr Hongwei Xie, *Northeastern University*, *China*
Dr Junli Xu, *Northeastern University*, *China*
Mr Liang Xu, *Northeastern University*, *China*
Professor Qian Xu, *Shanghai University*, *China*
Professor Juanyu Yang, *General Research Institute for non Ferrous Metals*, *China*
Dr Linpo Yu, *University of Nottingham Ningbo China*, *China*
Mr Zhanglong Yu, *China Automotive Battery Research Institute Co., Ltd.*, *China*
Miss Zhijing Yu, *University of Science and Technology Beijing*, *China*
Mr Yating Yuan, *Wuhan University*, *China*

Dr Xiangling Yue, *University of St Andrews*, *United Kingdom*
Dr Peng Zhang, *Shanghai Institute of Applied Physics, Chinese Academy of Sciences*, *China*
Professor Hongmin Zhu, *University of Science and Technology Beijing*, *China*
Dr Xingli Zou, *Shanghai University*, *China*